Lecture Notes in Computer Science 7975

Commenced Publication in 1973
Founding and Former Series Editors:
Gerhard Goos, Juris Hartmanis, and Jan van Leeuwen

T0223961

Beniamino Murgante   Sanjay Misra
Maurizio Carlini   Carmelo M. Torre
Hong-Quang Nguyen   David Taniar
Bernady O. Apduhan   Osvaldo Gervasi (Eds.)

Computational Science and Its Applications – ICCSA 2013

13th International Conference
Ho Chi Minh City, Vietnam, June 24-27, 2013
Proceedings, Part V

 Springer

Volume Editors

Beniamino Murgante, Università degli Studi della Basilicata, Potenza, Italy
E-mail: beniamino.murgante@unibas.it

Sanjay Misra, Covenant University, Canaanland OTA, Nigeria
E-mail: sanjay.misra@covenantuniversity.edu.ng

Maurizio Carlini, Università degli Studi della Tuscia, Viterbo, Italy
E-mail: maurizio.carlini@unitus.it

Carmelo M. Torre, Politecnico di Bari, Italy
E-mail: torre@poliba.it

Hong-Quang Nguyen, Int. University VNU-HCM, Ho Chi Minh City, Vietnam
E-mail: htphong@hcmiu.edu.vn

David Taniar, Monash University, Clayton, VIC, Australia
E-mail: david.taniar@infotech.monash.edu.au

Bernady O. Apduhan, Kyushu Sangyo University, Fukuoka, Japan
E-mail: bob@is.kyusan-u.ac.jp

Osvaldo Gervasi, University of Perugia, Italy
E-mail: osvaldo@unipg.it

ISSN 0302-9743   e-ISSN 1611-3349
ISBN 978-3-642-39639-7   e-ISBN 978-3-642-39640-3
DOI 10.1007/978-3-642-39640-3
Springer Heidelberg Dordrecht London New York

Library of Congress Control Number: 2013942720

CR Subject Classification (1998): C.2.4, C.2, H.4, F.2, H.3, D.2, F.1, H.5, H.2.8, K.6.5, I.3

LNCS Sublibrary: SL 1 – Theoretical Computer Science and General Issues

Typesetting: Camera-ready by author, data conversion by Scientific Publishing Services, Chennai, India

Printed on acid-free paper

Springer is part of Springer Science+Business Media (www.springer.com)

Preface

These multiple volumes (LNCS volumes 7971, 7972, 7973, 7974, and 7975) consist of the peer-reviewed papers from the 2013 International Conference on Computational Science and Its Applications (ICCSA2013) held in Ho Chi Minh City, Vietnam, during June 24–27, 2013.

ICCSA 2013 was a successful event in the International Conferences on Computational Science and Its Applications (ICCSA) conference series, previously held in Salvador, Brazil (2012), Santander, Spain (2011), Fukuoka, Japan (2010), Suwon, South Korea (2009), Perugia, Italy (2008), Kuala Lumpur, Malaysia (2007), Glasgow, UK (2006), Singapore (2005), Assisi, Italy (2004), Montreal, Canada (2003), (as ICCS) Amsterdam, The Netherlands (2002), and San Francisco, USA (2001).

Computational science is a main pillar of most of the present research, industrial, and commercial activities and plays a unique role in exploiting ICT innovative technologies; the ICCSA conference series have been providing a venue to researchers and industry practitioners to discuss new ideas, to share complex problems and their solutions, and to shape new trends in computational science.

Apart from the general track, ICCSA 2013 also included 33 special sessions and workshops, in various areas of computational sciences, ranging from computational science technologies, to specific areas of computational sciences, such as computer graphics and virtual reality. We accepted 46 papers for the general track, and 202 in special sessions and workshops, with an acceptance rate of 29.8%. We would like to express our appreciation to the Workshops and Special Sessions Chairs and Co-chairs.

The success of the ICCSA conference series, in general, and ICCSA 2013, in particular, is due to the support of many people: authors, presenters, participants, keynote speakers, Workshop Chairs, Organizing Committee members, student volunteers, Program Committee members, International Liaison Chairs, and people in other various roles. We would like to thank them all. We would also like to thank Springer for their continuous support in publishing ICCSA conference proceedings.

May 2013

David Taniar
Beniamino Murgante
Hong-Quang Nguyen

Message from the General Chairs

On behalf of the ICCSA Organizing Committee it is our great pleasure to welcome you to the proceedings of the 13th International Conference on Computational Science and Its Applications (ICCSA 2013), held June 24–27, 2013, in Ho Chi Minh City, Vietnam.

ICCSA is one of the most successful international conferences in the field of computational sciences, and ICCSA 2013 was the 13th conference of this series previously held in Salvador da Bahia, Brazil (2012), in Santander, Spain (2011), Fukuoka, Japan (2010), Suwon, Korea (2009), Perugia, Italy (2008), Kuala Lumpur, Malaysia (2007), Glasgow, UK (2006), Singapore (2005), Assisi, Italy (2004), Montreal, Canada (2003), (as ICCS) Amsterdam, The Netherlands (2002), and San Francisco, USA (2001).

The computational science community has enthusiastically embraced the successive editions of ICCSA, thus contributing to making ICCSA a focal meeting point for those interested in innovative, cutting-edge research about the latest and most exciting developments in the field. It provides a major forum for researchers and scientists from academia, industry and government to share their views on many challenging research problems, and to present and discuss their novel ideas, research results, new applications and experience on all aspects of computational science and its applications. We are grateful to all those who have contributed to the ICCSA conference series.

For the successful organization of ICCSA 2013, an international conference of this size and diversity, we counted on the great support of many people and organizations.

We would like to thank all the workshop organizers for their diligent work, which further enhanced the conference level and all reviewers for their expertise and generous effort, which led to a very high quality event with excellent papers and presentations.

We especially recognize the contribution of the Program Committee and local Organizing Committee members for their tremendous support, the faculty members of the School of Computer Science and Engineering and authorities of the International University (HCM-VNU), Vietnam, for allowing us to use the venue and facilities to realize this highly successful event. Further, we would like to express our gratitude to the Office of the Naval Research, US Navy, and other institutions/organizations that supported our efforts to bring the conference to fruition.

We would like to sincerely thank our keynote speakers who willingly accepted our invitation and shared their expertise.

We also thank our publisher, Springer-Verlag, for accepting to publish the proceedings and for their kind assistance and cooperation during the editing process.

Finally, we thank all authors for their submissions and all conference attendees for making ICCSA 2013 truly an excellent forum on computational science, facilitating an exchange of ideas, fostering new collaborations and shaping the future of this exciting field.

We thank you all for participating in ICCSA 2013, and hope that you find the proceedings stimulating and interesting for your research and professional activities.

Osvaldo Gervasi
Bernady O. Apduhan
Duc Cuong Nguyen

Organization

ICCSA 2013 was organized by The Ho Chi Minh City International University (Vietnam), University of Perugia (Italy), University of Basilicata (Italy), Monash University (Australia), and Kyushu Sangyo University (Japan).

Honorary General Chairs

Phong Thanh Ho   International University (VNU-HCM),
                          Vietnam
Antonio Laganà   University of Perugia, Italy
Norio Shiratori   Tohoku University, Japan
Kenneth C.J. Tan   Qontix, UK

General Chairs

Osvaldo Gervasi   University of Perugia, Italy
Bernady O. Apduhan   Kyushu Sangyo University, Japan
Duc Cuong Nguyen   International University (VNU-HCM),
                         Vietnam

Program Committee Chairs

David Taniar   Monash University, Australia
Beniamino Murgante   University of Basilicata, Italy
Hong-Quang Nguyen   International University (VNU-HCM),
                         Vietnam

Workshop and Session Organizing Chair

Beniamino Murgante   University of Basilicata, Italy

Local Organizing Committee

Hong Quang Nguyen   International University (VNU-HCM),
                         Vietnam (Chair)
Bao Ngoc Phan   International University (VNU-HCM),
                         Vietnam

Van Hoang   International University (VNU-HCM),
                               Vietnam
Ly Le   International University (VNU-HCM),
                               Vietnam

International Liaison Chairs

Jemal Abawajy   Deakin University, Australia
Ana Carla P. Bitencourt   Universidade Federal do Reconcavo da Bahia,
                               Brazil
Claudia Bauzer Medeiros   University of Campinas, Brazil
Alfredo Cuzzocrea   ICAR-CNR and University of Calabria, Italy
Marina L. Gavrilova   University of Calgary, Canada
Robert C.H. Hsu   Chung Hua University, Taiwan
Andrés Iglesias   University of Cantabria, Spain
Tai-Hoon Kim   Hannam University, Korea
Sanjay Misra   University of Minna, Nigeria
Takashi Naka   Kyushu Sangyo University, Japan
Ana Maria A.C. Rocha   University of Minho, Portugal
Rafael D.C. Santos   National Institute for Space Research, Brazil

Workshop Organizers

Advances in Web-Based Learning (AWBL 2013)

Mustafa Murat Inceoglu   Ege University, Turkey

Big Data: Management, Analysis, and Applications (Big-Data 2013)

Wenny Rahayu   La Trobe University, Australia

Bio-inspired Computing and Applications (BIOCA 2013)

Nadia Nedjah   State University of Rio de Janeiro, Brazil
Luiza de Macedo Mourell   State University of Rio de Janeiro, Brazil

Computational and Applied Mathematics (CAM 2013)

Ana Maria Rocha   University of Minho, Portugal
Maria Irene Falcao   University of Minho, Portugal

Computer-Aided Modeling, Simulation, and Analysis (CAMSA 2013)

Jie Shen   University of Michigan, USA
Yanhui Wang   Beijing Jiaotong University, China
Hao Chen   Shanghai University of Engineering Science,
                               China

Computer Algebra Systems and Their Applications (CASA 2013)

Andres Iglesias   University of Cantabria, Spain
Akemi Galvez   University of Cantabria, Spain

Computational Geometry and Applications (CGA 2013)

Marina L. Gavrilova   University of Calgary, Canada
Han Ming Huang   Guangxi Normal University, China

Chemistry and Materials Sciences and Technologies (CMST 2013)

Antonio Laganà   University of Perugia, Italy

Cities, Technologies and Planning (CTP 2013)

Giuseppe Borruso   University of Trieste, Italy
Beniamino Murgante   University of Basilicata, Italy

Computational Tools and Techniques for Citizen Science and Scientific Outreach (CTTCS 2013)

Rafael Santos   National Institute for Space Research, Brazil
Jordan Raddickand   Johns Hopkins University, USA
Ani Thakar   Johns Hopkins University, USA

Econometrics and Multidimensional Evaluation in the Urban Environment (EMEUE 2013)

Carmelo M. Torre   Polytechnic of Bari, Italy
Maria Cerreta   Università Federico II of Naples, Italy
Paola Perchinunno   University of Bari, Italy

Energy and Environment - Scientific, Engineering and Computational Aspects of Renewable Energy Sources, Energy Saving and Recycling of Waste Materials (ENEENV 2013)

Maurizio Carlini   University of Viterbo, Italy
Carlo Cattani   University of Salerno, Italy

Future Computing Systems, Technologies, and Applications (FISTA 2013)

Bernady O. Apduhan   Kyushu Sangyo University, Japan
Rafael Santos   National Institute for Space Research, Brazil
Jianhua Ma   Hosei University, Japan
Qun Jin   Waseda University, Japan

Geographical Analysis, Urban Modeling, Spatial Statistics (GEOG-AN-MOD 2013)

Giuseppe Borruso   University of Trieste, Italy
Beniamino Murgante   University of Basilicata, Italy
Hartmut Asche   University of Potsdam, Germany

International Workshop on Biomathematics, Bioinformatics and Biostatistics (IBBB 2013)

Unal Ufuktepe   Izmir University of Economics, Turkey
Andres Iglesias   University of Cantabria, Spain

International Workshop on Agricultural and Environmental Information and Decision Support Systems (IAEIDSS 2013)

Sandro Bimonte   IRSTEA, France
Andr Miralles   IRSTEA, France
Franois Pinet   IRSTEA, France
Frederic Flouvat   University of New Caledonia, New Caledonia

International Workshop on Collective Evolutionary Systems (IWCES 2013)

Alfredo Milani   University of Perugia, Italy
Clement Leung   Hong Kong Baptist University, Hong Kong

Mobile Communications (MC 2013)

Hyunseung Choo   Sungkyunkwan University, Korea

Mobile Computing, Sensing, and Actuation for Cyber Physical Systems (MSA4CPS 2013)

Moonseong Kim   Korean Intellectual Property Office, Korea
Saad Qaisar   NUST School of Electrical Engineering and
                              Computer Science, Pakistan

Mining Social Media (MSM 2013)

Robert M. Patton   Oak Ridge National Laboratory, USA
Chad A. Steed   Oak Ridge National Laboratory, USA
David R. Resseguie   Oak Ridge National Laboratory, USA
Robert M. Patton   Oak Ridge National Laboratory, USA

Parallel and Mobile Computing in Future Networks (PMCFUN 2013)

Al-Sakib Khan Pathan International Islamic University Malaysia, Malaysia

Quantum Mechanics: Computational Strategies and Applications (QMCSA 2013)

Mirco Ragni Universidad Federal de Bahia, Brazil
Vincenzo Aquilanti University of Perugia, Italy
Ana Carla Peixoto Bitencourt Universidade Federal do Reconcavo da Bahia, Brazil
Roger Anderson University of California, USA
Frederico Vasconcellos
 Prudente Universidad Federal de Bahia, Brazil

Remote Sensing Data Analysis, Modeling, Interpretation and Applications: From a Global View to a Local Analysis (RS 2013)

Rosa Lasaponara Institute of Methodologies for Environmental Analysis - National Research Council, Italy
Nicola Masini Archaeological and Monumental Heritage Institute - National Research Council, Italy

Soft Computing for Knowledge Discovery in Databases (SCKDD 2013)

Tutut Herawan Universitas Ahmad Dahlan, Indonesia

Software Engineering Processes and Applications (SEPA 2013)

Sanjay Misra Covenant University, Nigeria

Spatial Data Structures and Algorithms for Geoinformatics (SDSAG 2013)

Farid Karimipour University of Tehran, Iran and Vienna University of Technology, Austria

Software Quality (SQ 2013)

Sanjay Misra Covenant University, Nigeria

Security and Privacy in Computational Sciences (SPCS 2013)

Arijit Ukil Tata Consultancy Services, India

Technical Session on Computer Graphics and Geometric Modeling (TSCG 2013)

Andres Iglesias   University of Cantabria, Spain

Tools and Techniques in Software Development Processes (TTSDP 2013)

Sanjay Misra   Covenant University, Nigeria

Virtual Reality and Its Applications (VRA 2013)

Osvaldo Gervasi   University of Perugia, Italy
Lucio Depaolis   University of Salento, Italy

Wireless and Ad-Hoc Networking (WADNet 2013)

Jongchan Lee   Kunsan National University, Korea
Sangjoon Park   Kunsan National University, Korea

Warehousing and OLAPing Complex, Spatial and Spatio-Temporal Data (WOCD 2013)

Alfredo Cuzzocrea   Istituto di Calcolo e Reti ad Alte Prestazioni -
                                      National Research Council, Italy and
                                      University of Calabria, Italy

Program Committee

Jemal Abawajy	Deakin University, Australia
Kenny Adamson	University of Ulster, UK
Filipe Alvelos	University of Minho, Portugal
Hartmut Asche	University of Potsdam, Germany
Md. Abul Kalam Azad	University of Minho, Portugal
Assis Azevedo	University of Minho, Portugal
Michela Bertolotto	University College Dublin, Ireland
Sandro Bimonte	CEMAGREF, TSCF, France
Rod Blais	University of Calgary, Canada
Ivan Blecic	University of Sassari, Italy
Giuseppe Borruso	University of Trieste, Italy
Yves Caniou	Lyon University, France
José A. Cardoso e Cunha	Universidade Nova de Lisboa, Portugal
Carlo Cattani	University of Salerno, Italy
Mete Celik	Erciyes University, Turkey
Alexander Chemeris	National Technical University of Ukraine "KPI", Ukraine
Min Young Chung	Sungkyunkwan University, Korea
Gilberto Corso Pereira	Federal University of Bahia, Brazil
M. Fernanda Costa	University of Minho, Portugal

Wenny Rahayu	La Trobe University, Australia
Jerzy Respondek	Silesian University of Technology, Poland
Ana Maria A.C. Rocha	University of Minho, Portugal
Humberto Rocha	INESC-Coimbra, Portugal
Alexey Rodionov	Institute of Computational Mathematics and Mathematical Geophysics, Russia
Cristina S. Rodrigues	University of Minho, Portugal
Haiduke Sarafian	The Pennsylvania State University, USA
Ricardo Severino	University of Minho, Portugal
Jie Shen	University of Michigan, USA
Qi Shi	Liverpool John Moores University, UK
Dale Shires	U.S. Army Research Laboratory, USA
Ana Paula Teixeira	University of Tras-os-Montes and Alto Douro, Portugal
Senhorinha Teixeira	University of Minho, Portugal
Graça Tomaz	University of Aveiro, Portugal
Carmelo Torre	Polytechnic of Bari, Italy
Javier Martinez Torres	Centro Universitario de la Defensa Zaragoza, Spain
Giuseppe A. Trunfio	University of Sassari, Italy
Unal Ufuktepe	Izmir University of Economics, Turkey
Mario Valle	Swiss National Supercomputing Centre, Switzerland
Pablo Vanegas	University of Cuenca, Equador
Paulo Vasconcelos	University of Porto, Portugal
Piero Giorgio Verdini	INFN Pisa and CERN, Italy
Marco Vizzari	University of Perugia, Italy
Krzysztof Walkowiak	Wroclaw University of Technology, Poland
Robert Weibel	University of Zurich, Switzerland
Roland Wismüller	Universität Siegen, Germany
Xin-She Yang	National Physical Laboratory, UK
Haifeng Zhao	University of California, Davis, USA
Kewen Zhao	University of Qiongzhou, China

Additional Reviewers

Antonio Aguilar	Universitat de Barcelona, Spain
José Alfonso Aguilar Caldern	Universidad Autnoma de Sinaloa, Mexico
Vladimir Alarcon	Geosystems Research Institute, Mississippi State University, USA
Margarita Alberti	Universitat de Barcelona, Spain
Vincenzo Aquilanti	University of Perugia, Italy
Takefusa Atsuko	National Institute of Advanced Industrial Science and Technology, Japan
Raffaele Attardi	University of Napoli Federico II, Italy

Sansanee Auephanwiriyakul	Chiang Mai University, Thailand
Assis Azevedo	University of Minho, Portugal
Thierry Badard	Université Laval, Canada
Marco Baioletti	University of Perugia, Italy
Daniele Bartoli	University of Perugia, Italy
Paola Belanzoni	University of Perugia, Italy
Massimiliano Bencardino	University of Salerno, Italy
Priyadarshi Bhattacharya	University of Calgari, Canada
Massimo Bilancia	University of Bari, Italy
Gabriele Bitelli	University of Bologna, Italy
Letizia Bollini	University of Milano Bicocca, Italy
Alessandro Bonifazi	University of Bari, Italy
Atila Bostam	Atilim University, Turkey
Maria Bostenaru Dan	University of Bucharest, Romania
Thang H. Bui	Ho Chi Minh City University of Technology, Vietnam
Michele Campagna	University of Cagliari, Italy
Francesco Campobasso	University of Bari, Italy
Maurizio Carlini	University of Tuscia, Italy
Simone Caschili	University College of London, UK
Sonia Castellucci	University of Tuscia, Italy
Filippo Celata	University of Rome La Sapienza, Italy
Claudia Ceppi	Polytechnic of Bari, Italy
Ivan Cernusak	Comenius University of Bratislava, Slovakia
Maria Cerreta	University of Naples Federico II, Italy
Aline Chiabai	Basque Centre for Climate Change, Spain
Andrea Chiancone	University of Perugia, Italy
Eliseo Clementini	University of L'Aquila, Italy
Anibal Zaldivar Colado	Universidad Autonoma de Sinaloa, Mexico
Marco Crasso	Universidad Nacional del Centro de la provincia de Buenos Aires, Argentina
Ezio Crestaz	Saipem, Italy
Maria Danese	IBAM National Research Council, Italy
Olawande Daramola	Covenant University, Nigeria
Marcelo de Alemida Maia	Universidade Federal de Uberlândia, Brazil
Roberto De Lotto	University of Pavia, Italy
Lucio T. De Paolis	University of Salento, Italy
Pasquale De Toro	University of Naples Federico II, Italy
Hendrik Decker	Universidad Politécnica de Valencia, Spain
Margherita Di Leo	Joint Research Centre, Belgium
Andrea Di Carlo	University of Rome La Sapienza, Italy
Arta Dilo	University of Twente, The Netherlands
Alberto Dimeglio	CERN, Switzerland
Young Ik Eom	Sungkyunkwan University, South Korea
Rogelio Estrada	Universidad Autonoma de Sinaloa, Mexico
Stavros C. Farantos	University of Crete, Greece

Cristian Mateos	National University of the Center of the Buenos Aires Province, Argentina
Giovanni Mauro	University of Trieste, Italy
Giovanni Millo	Generali Group, Italy
Fernando Miranda	University of Minho, Portugal
Nazri MohdNawi	Universiti Tun Hussein Onn Malaysia, Malaysia
Danilo Monarca	University of Tuscia, Italy
Antonio Monari	University of Bologna, Italy
Rogerio Moraes	Department of Communication and Information Technology of Brazilian Navy, Brazil
Luiza Mourelle	Universidade do Estado do Rio de Janeiro, Brazil
Andrew Nash	Vienna Transport Strategies, Austria
Ignacio Nebot	University of Valencia, Spain
Nadia Nedjah	University of Rio de Janeiro, Brazil
Alexandre Nery	State University of Rio de Janeiro, Brazil
Van Duc Nguyen	Hanoi University of Science and Technology, Vietnam
José Luis Ordiales Coscia	Universidad Nacional del Centro de la Provincia de Buenos Aires, Argentina
Michele Ottomanelli	Polytechnic of Bari, Italy
Padma Polash Paul	University of Calgary, Canada
Francesca Pagliara	University of Naples Federico II, Italy
Marco Painho	Universidade Nova de Lisboa, Portugal
Dimos Pantazis	Technological Educational Institution of Athens, Greece
Enrica Papa	University of Naples Federico II, Italy
Jason Papathanasiou	University of Macedonia, Greece
Maria Paradiso	University of Sannio, Italy
Sooyeon Park	Korea Polytechnic University, South Korea
Juan Francisco Peraza	Universidad Autonoma de Sinaloa, Mexico
Massimiliano Petri	University of Pisa, Italy
Cassio Pigozzo	Universidade Federal da Bahia, Brazil
François Pinet	National Research Institute of Science and Technology for Environment and Agriculture, France
Stefan Porschen	University of Cologne, Germany
Tolga Pusatli	Cankaya University, Turkey
Md. Obaidur Rahman	Dhaka University of Engineering and Technology (DUET), Bangladesh
Syed Muhammad Raza	COMSATS University, Pakistan
Isabel Ribeiro	University of Porto, Portugal
Eduard Roccatello	3DGIS srl, Italy
Cristina Rodrigues	University of Minho, Portugal
Daniel Rodriguez	University of Alcalá, Spain

Yong-Wan Roh	Korean Intellectual Property Office, South Korea
Luiz Roncaratti	Universidade de Brasilia, Brazil
Marzio Rosi	University of Perugia, Italy
Francesco Rotondo	Polytechnic of Bari, Italy
Catherine Roussey	National Research Institute of Science and Technology for Environment and Agriculture, France
Rafael Oliva Santos	Universidad de La Habana, Cuba
Valentino Santucci	University of Perugia, Italy
Dario Schirone	University of Bari, Italy
Michel Schneider	Institut Supérieur d'Informatique de Modélisation et de leurs Applications, France
Gabriella Schoier	University of Trieste, Italy
Francesco Scorza	University of Basilicata, Italy
Nazha Selmaoui	Université de la Nouvelle-Calédonie, New Caledonia
Ricardo Severino	University of Minho, Portugal
Vladimir V. Shakhov	Institute of Computational Mathematics and Mathematical Geophysics SB RAS, Russia
Sungyun Shin	National University Kunsan, South Korea
Minhan Shon	Sungkyunkwan University, South Korea
Ruchi Shukla	University of Johannesburg, South Africa
Luneque Silva Jr.	State University of Rio de Janeiro, Brazil
V.B. Singh	University of Delhi, India
Michel Soares	Federal University of Uberlândia, Brazil
Changhwan Son	Sungkyunkwan University, South Korea
Henning Sten Hansen	Aalborg University, Denmark
Emanuele Strano	University of the West of England, UK
Madeena Sultana	Jahangirnagar University, Bangladesh
Setsuo Takato	Toho University, Japan
Kazuaki Tanaka	Kyushu Institute of Technology, Japan
Xueyan Tang	Nanyang Technological University, Singapore
Sergio Tasso	University of Perugia, Italy
Luciano Telesca	IMAA National Research Council, Italy
Lucia Tilio	University of Basilicata, Italy
Graça Tomaz	Instituto Politécnico da Guarda, Portugal
Melanie Tomintz	Carinthia University of Applied Sciences, Austria
Javier Torres	Universidad de Zaragoza, Spain
Csaba Toth	University of Calgari, Canada
Hai Tran	U.S. Government Accountability Office, USA
Jim Treadwell	Oak Ridge National Laboratory, USA

Chih-Hsiao Tsai	Takming University of Science and Technology, Taiwan
Devis Tuia	Laboratory of Geographic Information Systems, Switzerland
Arijit Ukil	Tata Consultancy Services, India
Paulo Vasconcelos	University of Porto, Portugal
Flavio Vella	University of Perugia, Italy
Mauro Villarini	University of Tuscia, Italy
Christine Voiron-Canicio	Université Nice Sophia Antipolis, France
Kira Vyatkina	Saint Petersburg State University, Russia
Jian-Da Wu	National Changhua University of Education, Taiwan
Toshihiro Yamauchi	Okayama University, Japan
Iwan Tri Riyadi Yanto	Universitas Ahmad Dahlan, Indonesia
Syed Shan-e-Hyder Zaidi	Sungkyunkwan University, South Korea
Vyacheslav Zalyubouskiy	Sungkyunkwan University, South Korea
Alejandro Zunino	National University of the Center of the Buenos Aires Province, Argentina

Sponsoring Organizations

ICCSA 2013 would not have been possible without tremendous support of many organizations and institutions, for which all organizers and participants of ICCSA 2013 express their sincere gratitude:

Ho CHi Minh City International University, Vietnam
(http://www.hcmiu.edu.vn/HomePage.aspx)

University of Perugia, Italy
(http://www.unipg.it)

 MONASH University   Monash University, Australia
(http://monash.edu)

 KIU
九州産業大学
KYUSHU SANGYO UNIVERSITY   Kyushu Sangyo University, Japan
(www.kyusan-u.ac.jp)

University of Basilicata, Italy (http://www.unibas.it)

The Office of Naval Research, USA
(http://www.onr.navy.mil/Science-technology/onr-global.aspx)

ICCSA 2013 Invited Speakers

Dharma Agrawal
University of Cincinnati, USA

Manfred M. Fisher
Vienna University of Economics and Business, Austria

Wenny Rahayu
La Trobe University, Australia

Selecting LTE and Wireless Mesh Networks for Indoor/Outdoor Applications

Dharma Agrawal*

School of Computing Sciences and Informatics, University of Cincinnati, USA
dharmaagrawal@gmail.com

Abstract. The smart phone usage and multimedia devices have been increasing yearly and predictions indicate drastic increase in the upcoming years. Recently, various wireless technologies have been introduced to add flexibility to these gadgets. As data plans offered by the network service providers are expensive, users are inclined to utilize freely accessible and commonly available Wi-Fi networks indoors.

LTE (Long Term Evolution) has been a topic of discussion in providing high data rates outdoors and various service providers are planning to roll out LTE networks all over the world. The objective of this presentation is to compare usefulness of these two leading wireless schemes based on LTE and Wireless Mesh Networks (WMN) and bring forward their advantages for indoor and outdoor environments. We also investigate to see if a hybrid LTE-WMN network may be feasible. Both these networks are heterogeneous in nature, employ cognitive approach and support multi hop communication. The main motivation behind this work is to utilize similarities in these networks, explore their capability of offering high data rates and generally have large coverage areas.

In this work, we compare both these networks in terms of their data rates, range, cost, throughput, and power consumption. We also compare 802.11n based WMN with Femto cell in an indoor coverage scenario, while for outdoors; 802.16 based WMN is compared with LTE. The main objective is to help users select a network that could provide enhanced performance in a cost effective manner.

---

* More information can be found at http://www.iccsa.org/invited-speakers

Neoclassical Growth Theory, Regions and Spatial Externalities

Manfred M. Fisher*

Vienna University of Economics and Business, Austria
manfred.fischer@wu.ac.at

Abstract. The presentation considers the standard neoclassical growth model in a Mankiw-Romer-Weil world with externalities across regions.

The reduced form of this theoretical model and its associated empirical model lead to a spatial Durbin model, and this model provides very rich own- and cross-partial derivatives that quantify the magnitude of direct and indirect (spillover or externalities) effects that arise from changes in regions characteristics (human and physical capital investment or population growth rates) at the outset in the theoretical model.

A logical consequence of the simple dependence on a small number of nearby regions in the initial theoretical specification leads to a final-form model outcome where changes in a single region can potentially impact all other regions. This is perhaps surprising, but of course we must temper this result by noting that there is a decay of influence as we move to more distant or less connected regions.

Using the scalar summary impact measures introduced by LeSage and Pace (2009) we can quantify and summarize the complicated set of non-linear impacts that fall on all regions as a result of changes in the physical and human capital in any region. We can decompose these impacts into direct and indirect (or externality) effects. Data for a system of 198 regions across 22 European countries over the period 1995 to 2004 are used to test the predictions of the model and to draw inferences regarding the magnitude of regional output responses to changes in physical and human capital endowments.

The results reveal that technological interdependence among regions works through physical capital externalities crossing regional borders.

---

* More information can be found at http://www.iccsa.org/invited-speakers

Global Spatial-Temporal Data Integration to Support Collaborative Decision Making

Wenny Rahayu*

La Trobe University, Australia
W.Rahayu@latrobe.edu.au

Abstract. There has been a huge effort in the recent years to establish a standard vocabulary and data representation for the areas where a collaborative decision support is required. The development of global standards for data interchange in time critical domains such as air traffic control, transportation systems, and medical informatics, have enabled the general industry in these areas to move into a more data-centric operations and services. The main aim of the standards is to support integration and collaborative decision support systems that are operationally driven by the underlying data.

The problem that impedes rapid and correct decision-making is that information is often segregated in many different formats and domains, and integrating them has been recognised as one of the major problems. For example, in the aviation industry, weather data given to flight en-route has different formats and standards from those of the airport notification messages. The fact that messages are exchanged using different standards has been an inherent problem in data integration in many spatial-temporal domains. The solution is to provide seamless data integration so that a sequence of information can be analysed on the fly.

Our aim is to develop an integration method for data that comes from different domains that operationally need to interact together. We especially focus on those domains that have temporal and spatial characteristics as their main properties. For example, in a flight plan from Melbourne to Ho Chi Minh City which comprises of multiple international airspace segments, a pilot can get an integrated view of the flight route with the weather forecast and airport notifications at each segment. This is only achievable if flight route, airport notifications, and weather forecast at each segment are integrated in a spatial temporal system.

In this talk, our recent efforts in large data integration, filtering, and visualisation will be presented. These integration efforts are often required to support real-time decision making processes in emergency situations, flight delays, and severe weather conditions.

---

* More information can be found at http://www.iccsa.org/invited-speakers

Table of Contents – Part V

Computational Methods, Algorithms, and Scientific Applications

High Performance Computing and Networks

Geometric Modeling, Graphics, and Visualization

Advanced and Emerging Applications

Information Systems and Technologies

Improvements of the Methods of Radiation Fields Numerical Modeling on the Basis of Mirror Reflection Principle

Oleg I. Smokty

St. Petersburg Institute of Informatics and Automation,
Russian Academy of Sciences, St. Petersburg, Russia
soi@iias.spb.su

Abstract. Fundamental properties of the angular-spatial symmetry of radiation fields in the uniform slab of a finite optical thickness are used for improvement of the numerical methods and algorithms of the classical radiative transfer theory. A new notion of so called photometrical invariants is introduced. The basic boundary-value problem of the radiative transfer theory is reformulated in new terms for the subsequent simplification of algorithms of numerical modeling methods such as spherical harmonics, discrete ordinates, Gauss-Seidel and Case methods. This simplification leads to two-fold decrease of the ranks of linear algebraic equations with simultaneous reduction of numerical modeling intervals connected with angular and spatial variables.

Keywords: mirror refelection principle, photometrical invariants, boundary-value problem, numerical algorithms, rank of linear equations system, iteration, spherical harmonics, discrete ordinates, Gauss-Seidel and Case methods.

1   Introduction

Until recently theoretical investigations and numerical modeling of the radiation fields angular spatial distribution in a uniform slab of a finite optical thickness have been carried out in the framework of traditional radiative transfer theory notions which cover the basic functions incending structure and properties of external and internal radiation fields symmetry [1 – 3]. Firstly, for the analytical description of radiation fields in the upward and downward directions the two classical basic functions are generally used: $\varphi(\eta, \tau_0)$ and $\psi(\eta, \tau_0)$ – Ambarzumian's functions, or $X(\eta, \tau_0)$ and $Y(\eta, \tau_0)$ – Chandrasekhar's functions. Secondly, the rigorous separation of angular variables in the brightness coefficients and investigation of their angular-spatial symmetry is carried out only for a unique fixed optical level in the slab. In this case, the symmetrical properties are usually considered in the sense of the optical reciprocity theorem, i.e. radiation fields do not change with the mutual translation of the detector's position and that of the radiation source. However, the problem of radiation fields angular and spatial symmetry is not limited to that. The radiation field in a

B. Murgante et al. (Eds.): ICCSA 2013, Part V, LNCS 7975, pp. 1–16, 2013.
© Springer-Verlag Berlin Heidelberg 2013

uniform medium possesses the inner fundamental property of the angular spatial symmetry relative to the medium's natural optical axis of symmetry [4]. In the case of a uniform slab, the symmetry axis coincides with the geometrical middle of the slab. From the mathematical point of view, the above mentioned properties of symmetry give rise to new notions in the radiative transfer theory, namely new elements and constructions of the radiation field – so called photometrical invariants [5 – 7].

It should be noted such situation is determined only by the optical uniformity of the slab and by the mathematical properties of the adequate structure expressed by photometrical invariants. From the physical point of view the advantage of the introduced photometrical invariants becomes evident through symmetrization of independent radiation sources (e.g. the Sun), i.e. through their mirror images relative to the middle of the uniform slab. In a particular case of the uniform slab's external boundaries, the significance of the symmetrization of the Sun's radiation has been demonstrated in [8].

The present paper is devoted to the use of the main mathematical properties of the photometrical invariants for improving and simplification of the radiation modeling and numerical algorithms of the classical radiative transfer theory [9].

2   The Mirror Reflection Principle in the Radiative Transfer Theory

Let us consider a uniform slab of the total optical thickness τ_0 in which the processes of absorption and multiple light scattering with albedo of single scattering Λ and phase function $P(\cos\gamma)$ take place. The distribution and capacity of primary energy sources are arbitrary. Let the function $I(\tau,\eta,\zeta,\varphi,\tau_0)$ be the intensity of radiation at the optical depth τ at the vision angle $\theta = \arccos\eta$ and solar angle $\theta_0 = \arccos\zeta$ to the outward vector normal to the external boundary (τ_0), and the azimuth angle φ.

From the traditional consideration of the multiple light scattering processes in a uniform slab (excluding its boundaries) on the basis of the classical radiative transfer theory treatments and methods used, as a rule, only one arbitrary optical level τ is fixed for radiation field medium. With such approach, the angular and spatial structure of any linear combination of upward and downward radiation exhibits not its inherent but latent properties of the mirror symmetry of the total radiation field of the uniform slab [3].

To display them it is necessary in the first place to "break" the arbitrary optical fixed level τ for the initial slab's radiation field into two levels: τ and $(\tau_0 - \tau)$ symmetrical about the middle($\frac{1}{2}\tau_0$) of the slab (see Fig. 1, A). Then one needs to consider the upward and downward radiations in the mirror directions (η and $-\eta$) at each fixed symmetrical levels within the chosen system of coordinates [7].

Within the context of the above statements, let us consider the mirror radiation field relative to the slab's midplane ($\frac{1}{2}\tau_0$) at the mirror optical levels τ_* and $(\tau_0 - \tau)_*$ in the mirror vision directions η_* and $-\eta_*$ (see Fig. 1, B).

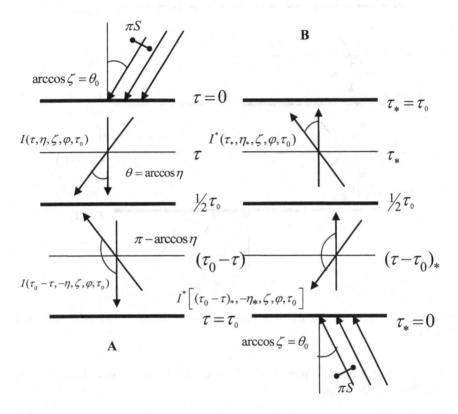

Fig. 1. Initial radiation (A) and mirror reflection (B) (relative to the slab's middle level $\frac{1}{2}\tau$) radiation fields

By uniting the initial and mirror radiation fields into one field (see Fig. 2), we come to mirror symmetry treatment of the classical radiative transfer problem for the slab of a finite optical thickness.

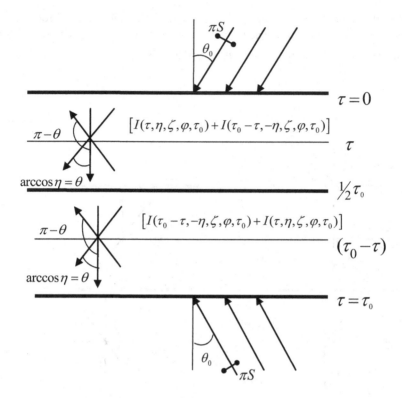

Fig. 2. The united initial (A) and mirror reflection (B) radiation fields in the coordinates system of Fig.1 (A)

In such a way, new basic photometrical values $I^{\pm}$ can be introduced according to the following relationships:

$$\left| I(\tau,\eta,\zeta,\varphi,\tau_0) \pm I(\tau_0-\tau,-\eta,\zeta,\varphi,\tau_0) \right| = I^{\pm}(\tau,\eta,\zeta,\varphi,\tau_0) \qquad (1)$$

Formed in such manner, the new values (photometrical invariants), unlike the traditional photometrical values of the classical radiative transfer theory [1 – 3], represent the optical uniformity of the radiation field in an arbitrary uniform slab in the evident mathematical form. Only they possess the main property of angular-spatial symmetry which is characteristic for any mirror image: photometrical invariant do not change if two arbitrary symmetrical levels τ and $(\tau_0-\tau)$ in the medium are rearranged and vision directions of radiation η are substituted by mirror directions $(-\eta)$ simultaneously:

$$I^{\pm}(\tau,\eta,\zeta,\varphi,\tau_0) = I^{\pm}(\tau_0-\tau,-\eta,\zeta,\varphi,\tau_0) \qquad (2)$$

To put it another way, in mirror vision directions (η) and $(-\eta)$ they are equal at the same distances from the middle of the slab $(\frac{1}{2}\tau_0)$. This basic property of photometrical invariants is a consequence of the mirror reflection principle formulated in [5 – 7]. According to the principle, the photometric invariants $I^{\pm}(\tau,\eta,\zeta,\varphi,\tau_0)$ in the uniform slab are pseudo intensities in the photometrically symmetrized radiative transfer theory. In this case, at each fixed optical level of the medium, the values $I^{\pm}$ are equal correspondingly to the sums or differences of the initial radiation field I and its mirror reflection I^*:

$$\left|I(\tau,\eta,\zeta,\varphi,\tau_0)\pm I^*(\tau_*,\eta_*,\zeta,\varphi,\tau_0)\right| = I^{\pm}(\tau,\eta,\zeta,\varphi,\tau_0), \tag{3}$$

where $\tau_* = \tau_0 - \tau$ and $\eta_* = -\eta$. In the case of optical uniformity of the given slab, the values I^* can be expressed as appropriate values of the initial radiation field:

$$I^* = I(\tau_0 - \tau, -\eta, \zeta, \varphi, \tau_0) \tag{4}$$

As shown below, the use of photometrical invariants (3) and linear transformations of the initial radiation field intensities $I(\tau,\eta,\zeta,\varphi,\tau_0)$ connected with the variables spatial shift $\tau \Rightarrow \tau_0 - \tau$ and angular rotation $\eta \Rightarrow -\eta$ allow to represent more rationally the analytical structure of the solutions to the basic boundary-value problem of the radiative transfer theory and at the same time to simplify the appropriate algorithms of numerical modeling of radiation fields.

3   Mirror Symmetrization of the Basic Boundary-Value Problem

The use of angular-spatial mirror symmetry of radiation fields in uniform slabs with arbitrary distributions of primary energy sources allows to modify the basic boundary-value problem of the classical radiative transfer theory in terms of photometrical invariants I^+ and I^- (2 – 3). Indeed, from the traditional form using the basic boundary-value problem in the radiative transfer theory (without a reflecting bottom) we have [3]:

$$\eta \frac{dI(\tau,\eta,\zeta,\varphi,\tau_0)}{\tau} = -I(\tau,\eta,\zeta,\varphi,\tau_0) + \tag{5}$$

$$+\frac{\Lambda}{4\pi}\int_0^{2\pi}d\varphi'\int_{-1}^{1}I(\tau,\eta',\zeta,\varphi-\varphi'\tau_0)P(\eta,\eta',\varphi')d\eta' + g(\tau,\eta,\zeta,\varphi),$$

$$I(0,\eta,\zeta,\varphi,\tau_0) = I(\tau_0,-\eta,\zeta,\varphi,\tau_0) = 0, \quad \eta > 0. \tag{6}$$

Let us now consider the symmetric levels τ and τ_* and mirror symmetric directions η and η_* in accordance with the above mentioned mirror reflection principle. Transforming τ to $\tau_* = (\tau_0 - \tau)$ and η to $\eta_* = (-\eta)$ in (5), we get

$$\eta_* \frac{dI^*(\tau_*,\eta_*,\zeta,\varphi,\tau_0)}{d\tau_*} = -I^*(\tau_*,\eta_*,\zeta,\varphi,\tau_0) +$$

$$+ \frac{\Lambda}{4\pi} \int_0^{2\pi} d\varphi' \int_{-1}^1 I^*(\tau_*,\eta_*',\zeta,\varphi-\varphi',\tau_0) P(\eta_*,\eta_*',\varphi') d\eta_*' + g^*(\tau_*,\eta_*,\zeta,\varphi). \tag{7}$$

$$I^*(0,\eta^*,\zeta,\tau_0) = I^*(\tau_0,-\eta^*,\zeta,\tau_0) = 0 \tag{8}$$

Here we have taken into account the known mirror symmetry property of the phase function $P(\cos\gamma)$. This fundamental property, as a result of the reciprocity theorem, is the mathematical expression of local mirror symmetry of single light scattering processes in the elementary of a uniform slab [4]:

$$P(\eta,\eta',\varphi-\varphi') = P(-\eta,-\eta',\varphi-\varphi'). \tag{9}$$

Taking into account (3) and (9), after appropriate linear transformations of the boundary conditions for (6), from (5 – 8) we obtain:

$$\eta \frac{dI^\pm(\tau,\eta,\zeta,\varphi,\tau_0)}{d\tau} = I^\pm(\tau,\eta,\zeta,\varphi,\tau_0) +$$

$$+ \frac{\Lambda}{4\pi} \int_0^{2\pi} d\varphi' \int_{-1}^1 I^\pm(\tau,\eta',\zeta,\varphi-\varphi',\tau_0) P(\eta,\eta',\varphi') d\eta' + g^\pm(\tau,\eta,\zeta,\varphi), \tag{10}$$

$$I^\pm(0,\eta,\zeta,\varphi,\tau_0) = I^\pm(\tau_0,\eta,\zeta,\varphi,\tau_0) = 0, \tag{11}$$

where

$$g^\pm(\tau,\eta,\zeta,\varphi) = |g^*(\tau_*,\eta_*,\zeta,\varphi) \pm g^*(\tau,\eta,\zeta,\varphi)|. \tag{12}$$

Thus we get two separate but equivalent boundary-value problems for photometric invariants (3) given in two different forms, namely classical initial (5 – 6) and symmetrized modified (10 – 12). From (10 – 12) we can see that the current variables τ and η for the modified problems vary in more narrow intervals $D_1 = \{\tau \in [0,\tau_0] \cap \eta \in [0,1]\}$ or alternatively $D_2 = \{\tau \in [0,\frac{1}{2}\tau_0] \cap \eta \in [-1,1]\}$, whereas in the classical case (5 – 6) we have $D = \{\tau \in [0,\tau_0] \cap \eta \in [-1,1]\}$. As a result it is possible to make the well-known numerical methods and algorithms of the modern radiative transfer theory more efficient from the point of view of applied discretization and approximation procedures. Furthermore, as shown below, the ranks of appropriate algebraic systems used in modified treatment (10 – 12) will be half as low as in the classical theory [9].

4   Simplification of Numerical Methods of the Radiative Transfer Theory

As follows from the above considerations, the applied importance of the mirror reflection principle is that it allows to substantially simplify the use of the well-known and methods and algorithms of numerical modeling in the classical radiative transfer theory [9]. Below this statement will be considered in detail in view of the spherical harmonics method, discrete ordinate method, Gauss-Seidel method, and Case method.

Spherical Harmonics Method. Let us apply the classical spherical harmonics method [10] to azimuthal harmonics $I_{\pm}^{m}(\tau,\eta,\zeta,\tau_{0})$ of radiation intensity invariants (3) for solving the symmetrized boundary-value problem (10 – 12):

$$I_{\pm}(\tau,\eta,\zeta,\varphi,\tau_{0}) = I_{\pm}^{0}(\tau,\eta,\zeta,\tau_{0}) + 2\sum_{m=1}^{N} I_{\pm}^{m}(\tau,\eta,\zeta,\tau_{0})\cos m\varphi, \tag{13}$$

$$P(\cos\gamma) = P^{0}(\cos\eta,\zeta) + 2\sum_{m=1}^{N} P^{m}(\eta,\zeta)\cos m\varphi \tag{14}$$

at the external primary energy sources in the form direct solar rays [3]:

$$g_{\pm}^{m}(\tau,\eta,\xi) = \Lambda S/4\{P_{\bullet}^{m}(\eta,\xi)\exp(-\tau/\xi) + P^{m}(-\eta,\xi)\exp[-(\tau_{0}-\tau)/\xi]\}. \tag{15}$$

Then we have

$$I_{\pm}^{m}(\tau,\eta,\zeta,\tau_{0}) = \sum_{n=m}^{N}(2n+1)A_{n,\pm}^{m}(\tau)P_{n}^{m}(\eta), \tag{16}$$

$$A_{n,\pm}^{m}(\tau) = A_{n}^{m}(\tau)\pm(-1)^{n+m}A_{n}^{m}(\tau_{0}-\tau), \tag{17}$$

where $N = 2p+m-1 > M$, p is a whole number, M is the total number of azimuth harmonics, $P_{n}^{m}(\tau)$ - associated Legendre polynomials. In contrast to the classical scheme of this method, the unknown function coefficients $A_{n,\pm}^{m}(\tau)$ are obtained not from the $2p$ boundary conditions (11) but already from the p boundary conditions in the following form:

$$\int_{0}^{1} I_{\pm}^{m}(0,\eta)P_{m+2j-1}^{m}(\eta)d\eta = 0, \tag{18}$$

$$\int_{0}^{1} I_{\pm}^{m}(\tau_{0},-\eta)P_{m+2j-1}^{m}(\eta)d\eta = 0, \tag{19}$$

$$j = 1,...,p, \quad m = 0,1,...N$$

According to the standard general algorithm, we get the following expression for functions $A_{n,\pm}^m(\tau)$:

$$A_{n,\pm}^m(\tau) = \sum_{i=1}^p \left[k_{i,\pm}^m g_n^m(v_i)e^{v_i\tau} \pm k_{-i,\pm}^m g_n^m(-v_i)e^{-v_i\tau} \right] + h_n^m e^{-\tau/\zeta} \pm f_n^m e^{\frac{\tau_0-\tau}{\zeta}}, \quad (20)$$

where v_n represents characteristic roots, and the values $k_{n,\pm}^m$ are defined from the boundary conditions (18 – 19). In that case, according to the classical scheme, for each of 2p values of the index i the following system of linear algebraic equations of 2p rank can be formulated for finding the unknown values $g_n^m, n = m, m+1,...N$:

$$(n-m)g_{n-1}^m(v_i)v_i + z_n g_n^m(v_i) + (n+m+1)g_{n+1}^m(v_i)v_i = 0,$$
$$i = \pm 1, ... \pm p, n = m, ..., N, \quad (21)$$

where $g_{m-1}^m(v_i^m) = g_{N+1}^m(v_i^m) = 0$ and the symbol z_n is used as follows:

$$z_n = 2n+1-\Lambda x_n, n \le M \; ; \; z_n = 2n+1, n > M, \quad (22)$$

From the the condition $\Delta(v_i) = 0$ of solvability of system (21) , we can find the values of characteristic roots v_i :

$$\Delta(v_i) = \begin{vmatrix} z_n & (2n+1)v_i & 0 & 0 & ...0 \\ v_i & z_{n+1} & (2n+2)v_i & 0 & ...0 \\ 0 & 2v_i & z_{n+2} & (2n+3)v_i & ...0 \\ ... & ... & ... & ... & ... \\ 0 & 0 & 0 & (N-n)v_i & z_N \end{vmatrix} = 0 \quad (23)$$

$$\Delta(-v_i) = \Delta(v_i), \quad v_{-i} = -v_i, \quad i = 1,...,p. \quad (24)$$

After finding g_n^m , we can find 2p unknown values k_i^m in (20) from the following system of 2p rank algebraic equations:

$$\sum_{n=m}^N (2n+1)\left[h_n^m + \sum_{i=-p}^p k_{i,\pm}^m g_n^m(v_i) \right] \int_0^1 P_n^m(\eta)P_{m+2j-1}^m(\eta)d\eta = 0 , \quad (25)$$

$$\sum_{n=m}^N (2n+1)\left[h_n^m + \sum_{i=-p}^p k_{i,\pm}^m g_n^m(v_i) \right](-1)^{n+m-1} \int_0^1 P_n^m(\eta)P_{m+2j-1}^m(\eta)d\eta = 0 . \quad (26)$$

Now let us use the main invariance property (2). As a basic result of its application in the considered method, we find the following important relations:

$$A_n^m(\tau) = (-1)^{n+m} A_n^m(\tau_0 - \tau), \quad (27)$$

$$k_{i,\pm}^m g_n^m(v_i) = (-1)^{n+m} k_{-i,\pm}^m g_n^m(-v_i^m) e^{-v_i \tau_0},$$ (28)

$$h_n^m = (-1)^{n+m} f_n^m.$$ (29)

By applying the relations (27 - 29) to (20), we finally obtain:

$$A_{n,\pm}^m(\tau) = \sum_{i=1}^{p} k_{i,\pm}^m g_n^m(v_i) \left[e^{v_i \tau} \pm (-1)^{n+m} e^{-v_i(\tau_0 - \tau)} \right] + h_n^m \left[e^{-\tau/\zeta} \pm (-1)^{n+m} e^{-\frac{\tau_0 - \tau}{\zeta}} \right],$$ (30)

$$n = m, m+1,..., N.$$

Thus, the representation (30) contains not 2p but only p unknown values $k_{i,\pm}^m$ and $g_n^m(v_i)$. Furthermore, if according to the classical scheme the coefficients $g_n^m(v_i)$ can be found from a 2p system for each of the 2p values of v_i, then for the symmetrized boundary-value problem (10 – 12) this is done only p times for positive values of $v_i > 0, i = 1, 2,..., p$. The resulting total number of coefficients $A_{n,\pm}^m(\tau)$ in representation (17) remains the same as in the classical interpretation of the spherical harmonics method (2p), whereas the use of the mirror reflection principle allows to halve the number of the unknown coefficients k_i^m and g_n^m in this representation.

Discrete Ordinate Method. Without sacrificing the generality of further considerations, for the sake of simplicity let us assume that $P(\cos \gamma) = 1$ and $S = 1$. In accordance with the given method [2], let us represent the integral element in equation (5) in the following form:

$$\int_{-1}^{+1} I^\pm(\tau,\eta) d\eta \approx \sum_{j=-n}^{n} a_j^\pm I^\pm(\tau,\eta_j),$$ (31)

where the discrete values $\eta_j, j = \pm1,..., \pm n$ are the roots of Legendre polynomials $P_{2n}(\eta_j) = 0$, and moreover the following relations are fulfilled: $a_{-j} = a_j, \eta_{-j} = -\eta_j$, $j = 1,..., n$.

Applying the classical algorithms of the discrete ordinate method [2] to solution of the boundary-value problem (10 – 12), we get

$$I^\pm(\tau,\eta_i,\zeta,\tau_0) = \sum_{s=1}^{n} \left[\frac{\beta_s^\pm}{1 + k_{\pm,s}\eta_i} e^{-k_{\pm,s}\tau} \pm \frac{\beta_{-s}^\pm}{1 - k_{\pm,s}\eta_i} e^{k_{\pm,s}\tau} \right] + \alpha_i(\eta_i) e^{-\tau/\zeta} \pm \gamma_i(\eta_i) e^{-\frac{\tau_0 - \tau}{\zeta}},$$ (32)

$$i = \pm1,..., \pm n,$$

where the number of $\pm k_{\pm,s}$ is 2n, and they are different, not equal to 0, characteristic roots $(s = 1, 2,..., n)$; the 2n unknown coefficients $\beta_s^\pm, s = \pm1, \pm2,..., \pm n$ are found from the boundary conditions of the boundary-value problem (10 – 12) described for n angular vision directions $0 < \eta_j \in [0,1]$. The use of the invariance property (2) allows to halve the rank of the above mentioned algebraic systems and the total number of the desired coefficients $\beta_s^\pm, s = \pm 1, \pm 2, ..., \pm n$. Indeed, applying it, we obtain the following equivalence relations from (32):

$$\alpha_i(-\eta) = \gamma_{-i}, \tag{33}$$

$$\beta_s^\pm = \beta_{-s}^\pm e^{k_s \tau_0}. \tag{34}$$

Therefore, in representation (32) for each of the 2n functions $I^\pm(\tau, \eta, \zeta, \tau_0)$ there are not 2n unknown coefficients, namely $\beta_s^\pm, s = \pm 1, \pm 2, ..., \pm n$, as in the classical scheme of the discrete ordinate method, but only n coefficients:

$$I^\pm(\tau, \eta_i, \tau_0) = \sum_{s=1}^{n} \beta_s^\pm \left[\frac{e^{-k_{\pm,s}\tau}}{1 + k_{\pm,s}\eta_i} + \frac{e^{-k_{\pm,s}(\tau_0 - \tau)}}{1 - k_{\pm,s}\eta_i} \right] + \alpha_i e^{-\tau/\zeta} \pm \gamma_i e^{\frac{\tau_0 - \tau}{\zeta}}. \tag{35}$$

In other words, the rank of the linear equation system (35) required for finding the coefficients $\beta_s^\pm$ by using the mirror reflection principle is half as high as that in the classical treatment [2]. In this case, if α_i and γ_i are known, the use the unique boundary condition ($I^\pm(0, \eta_j, \zeta, \tau_0) = 0$, $\eta_j \in [0,1]$) of symmetrized boundary-value problem (10 – 12), which is described for n angular directions $\eta_i \in [0,1], i = 1, 2, ..., n$, gives the following algebraic equation system required for finding n unknown coefficients $\beta_s^\pm$:

$$\sum_{s=1}^{n} \beta_s^\pm \left[\frac{1}{1 + k_s \eta_j} + \frac{e^{-k_{\pm,s}\tau_0)}}{1 - k_{\pm,s}\eta_j} \right] + \alpha_i(\eta_j) + \gamma_i(\eta_j) e^{\frac{-\tau_0}{\zeta}} = 0. \tag{36}$$

In this case the coefficients α_i and γ_i are considered to be known and can be found in the same way as in the classical scheme of the discrete ordinate method by using relation (33). It should be emphasized that the application of the mirror reflection principle and the angular discretization method in solving linear integral equations [11] will similarly lead to algebraic systems with a rank two times lower than in their classical treatment [3].

Gauss-Seidel Method. In this method [9], using the above mentioned invariance treatment of basic boundary-value problem (10 – 12) and invariance property (2) for azimuthal harmonics $I_\pm^m$, we have

$$I_\pm^m(\tau, \eta, \zeta, \tau_0) = \int_0^\tau B_\pm^m(\tau', \eta, \zeta, \tau_0) e^{-\frac{\tau - \tau'}{\eta}} \frac{d\tau'}{\eta}, \tag{37}$$

$$m = 0, 1, ..., N, \quad \tau \in [0, \tau_0], \quad \eta \in [0,1].$$

where the values $B_{\pm}^{m}(\tau,\eta,\zeta,\tau_0)$ are the azimuthal harmonics of the source function invariants $B_{\pm}(\tau,\eta,\zeta,\varphi,\tau_0)$. It should be emphasized that according to the classical treatment of the basic boundary-value problem [9], the current variables τ and η are within the following intervals: $\tau \in [0,\tau_0]$ and $\eta \in [-1,1]$ unlike in (37).

Using the solution of boundary-value problem (5 – 6) as in (37), for azimuthal harmonics of the source function $B_{\pm}^{m}$ we have the following representation:

$$B_{\pm}^{m}(\tau,\eta,\zeta,\tau_0) = \frac{\Lambda}{2}\left[\int_0^1 P^m(\eta,\eta')I_{\pm}^{m}(\tau,\eta',\zeta,\tau_0)d\eta' \pm \int_0^1 P^m(-\eta,\eta')I_{\pm}^{m}(\tau_0-\tau,\eta',\zeta,\tau_0)d\eta'\right] +$$

$$+\frac{\Lambda}{4}\left[P^m(\eta,\zeta)e^{-\tau/\zeta} \pm P^m(-\eta,\zeta)e^{-\frac{\tau_0-\tau}{\zeta}}\right], \quad m = 0,1,...,N. \tag{38}$$

Applying simultaneously the classical scheme of the Gauss-Seidel method [9] and invariance treatment of the basic boundary-value problem in the radiative transfer theory as in (37 – 38), we have the following representation for arbitrarily chosen but fixed optical level τ_1:

$$I_{\pm}^{m}(\tau,\eta,\zeta,\tau_o) = I_{\pm}^{m}(\tau_1,\eta,\zeta,\tau_o)e^{-\frac{(\tau-\tau_1)}{\eta}} + \int_{\tau_1}^{\tau} B_{\pm}^{m}(\tau',\eta,\zeta,\tau_0)e^{-\frac{(\tau-\tau')}{\eta}}\frac{d\tau'}{\eta} \tag{39}$$

$$\tau > \tau_1 \geq 0, \quad m = 0,1,...,N.$$

The iteration with integral approximation according to the known Gauss fourmula by three-point levels of discretization at τ, similarly to the classical scheme, take the following form:

$$I_{\pm,(n+1)}^{m}(\tau_{i+2},\eta,\zeta,\tau_0) = I_{\pm,(n)}^{m}(\tau,\eta,\zeta,\tau_0)e^{-\frac{2\Delta\tau}{\eta}} + B_{\pm,(n)}^{m}(\tau_{i+1},\eta,\zeta,\tau_0)(1-e^{-\frac{2\Delta\tau}{\eta}}), \tag{40}$$

$$i = 1,...,N-2.$$

In that case the values of τ_{i+1} are calculated from $\tau = 0$ to $\tau = \tau_0$ and we have

$$I_{\pm,(n)}^{m}(\tau_1,\eta,\zeta,\tau_0) = 0, \quad \forall n \geq 0, \tau \in [0,\tau_0] \cap \eta \in [0,1]$$

$$B_{\pm,(n)}^{m}(\tau_{i+1},\eta,\zeta,\tau_0) = \frac{\Lambda}{2}\left[\begin{array}{c}\int_0^1 P^m(\eta,\eta')I_{\pm,(n)}^{m}(\tau,\eta',\zeta,\tau_o)\pm \\ \pm\int_0^1 P^m(-\eta,\eta')I_{\pm,(n)}^{m}(\tau_0-\tau,\eta',\zeta,\tau_0)d\eta'\end{array}\right] + \tag{41}$$

$$+\frac{\Lambda}{4}\left[P^m(\eta,\zeta)e^{-\tau/\zeta} \pm P^m(-\eta,\zeta)e^{-\frac{\tau_0-\tau}{\eta}}\right], \quad m = 0,1,...,N.$$

The spatial and angular discretizaton of integrals in (40 – 41) is performed according to Gauss' formula in a standard way with the use of the weight function w_k :

$$B_{\pm,(n)}^m(\tau,\eta_j,\zeta,\tau_0) = \frac{\Lambda}{2}\sum_{k=1}^{L} w_k [P^m(\eta_j,\eta_k)I_{\pm,(n)}^m(\tau,\eta_k,\zeta,\tau_0)\pm$$

$$\pm P^m(-\eta_j,\eta_k)I_{\pm,(n)}^m(\tau_0-\tau,\eta_k,\zeta,\tau_0)]+\frac{\Lambda}{4}\left[P^m(\eta_j,\zeta)e^{-\tau/\zeta}\pm P^m(-\eta_j,\zeta)e^{-\frac{\tau_0-\tau}{\zeta}}\right]$$

(42)

Finally, in case of numerical modeling, the values of upward and downward radiation intensity $I^m(\tau,\eta,\zeta,\tau_0)$ and $I^m(\tau_0-\tau,-\eta,\zeta,\tau_0)$ can be calculated by the following formulae:

$$I^m(\tau,\eta,\zeta,\tau_0) = \frac{1}{2}\left[I_+^m(\tau,\eta,\zeta,\tau_0)+I_-^m(\tau,\eta,\zeta,\tau_0)\right], \eta > 0 \tag{43}$$

$$I^m(\tau_0-\tau,-\eta,\zeta,\tau_0) = \frac{1}{2}\left[I_+^m(\tau,\eta,\zeta,\tau_0)-I_-^m(\tau,\eta,\zeta,\tau_0)\right], \eta > 0. \tag{44}$$

Let us introduce now the following notations:

$$I_i^{\pm,j} = I^{\pm}(\tau_i,\eta_j),$$
$$i = 0,1,...,N; j = \pm 1,...,\pm M, \tag{45}$$

assuming $\Delta\eta_k = \Delta\eta_{-k} = \Delta\eta$. Then having performed discretization of the current variables $\Delta\tau$ and $\Delta\eta$ in accordance with the Gauss-Seidel method algorithm [9] and approximization of integrals in (40 – 41), for upward $(\eta < 0)$ and down-ward $(\eta > 0)$ radiation, we have:

$$I_{i+2}^{\pm,j} = I_i^{\pm,j}e^{-\frac{2\Delta\tau}{\eta_j}}+\frac{\Lambda}{2}\sum_{k=1}^{M}\left(P_{k,j}I_{i+1}^{\pm,k}+P_{k,-j}I_{i+1}^{\pm,-k}\right)\Delta\eta\left(1-e^{-\frac{2\Delta\tau}{\eta_j}}\right)+f_{i+1}^{\pm,j}, \tag{46}$$

$$I_i^{\pm,-j} = I_{i+2}^{\pm,-j}e^{-\frac{2\Delta\tau}{\eta_j}}+\frac{\Lambda}{2}\sum_{k=1}^{M}\left(P_{k,-j}I_{i+1}^{\pm,k}+P_{k,j}I_{i+1}^{\pm,-k}\right)\Delta\eta\left(1-e^{-\frac{2\Delta\tau}{\eta_j}}\right)+f_{i+1}^{\pm,-j}, \tag{47}$$

$i = 0,1,...,N; j = 1,2,...,M.$

Here the levels i and $N-i$ are symmetrical. Assuming $i = N-i-2$ in (47), we have

$$N_{N-i-2}^{\pm,-j} = I_{N-i}^{\pm,-j}e^{-\frac{2\Delta\tau}{nj}}+\frac{\Lambda}{2}\sum_{k=1}^{M}\left(P_{k,-j}I_{N-i-1}^{\pm,k}P_{k,j}I_{N-i-1}^{\pm,-k}\right)\Delta\eta\left(1-e^{-\frac{2\Delta\tau}{\eta_j}}\right)+f_{N-i-1}^{\pm,-j}, \tag{48}$$

where

$$P_{-k,-j} = P_{k,j} = \frac{1}{\Delta\eta_k} \int_{\Delta\eta_k} P(\eta_j,\eta) d\eta', \tag{49}$$

$$f_{i+1}^{\pm,j} = \frac{\Lambda}{4} P(\eta_j,\zeta) e^{-\frac{\tau_{i+1}}{\zeta}}, \tag{50}$$

$$f_{i+1}^{\pm,-j} = \frac{\Lambda}{4} P(-\eta_j,\zeta) e^{-\frac{\tau_i+1}{\zeta}}. \tag{51}$$

Now let us introduce the following relations:

$$V_{i,j}^{\pm} = I_i^{\pm,j} \pm I_{N-i}^{\pm,-j}, \tag{52}$$

$$W_{i,j}^{\pm} = I_i^{\pm,-j} \pm I_{N-i}^{\pm,j}. \tag{53}$$

Then from (46 – 47) we obtain

$$V_{i+2,j}^{\pm} = V_{i,j}^{\pm} e^{-\frac{2\Delta\tau}{n_j}} + \frac{\Lambda}{2} \sum_{k=1}^{M} \left(P_{k,j} V_{i+1,k}^{\pm} \pm P_{k,-j} W_{i+1,k}^{\pm} \right) \Delta\eta \left(1 - e^{-\frac{2\Delta\tau}{n_j}} \right) + f_{i+1}^{\pm,j} \pm f_{N-i-1}^{\pm,-j}, \tag{54}$$

$$W_{i,j}^{\pm} = W_{i+2,j}^{\pm} e^{-\frac{2\Delta\tau}{n_j}} + \frac{\Lambda}{2} \sum_{k=1}^{M} \left(P_{k,j} W_{i+1,k}^{\pm} \pm P_{k,-j} V_{i+1,k}^{\pm} \right) \Delta\eta \left(1 - e^{-\frac{2\Delta\tau}{n_j}} \right) + f_{i+1}^{\pm,-j} \pm f_{N-i-1}^{\pm,j}. \tag{55}$$

Afterwards, we can then use the invariance property in the form of relation

$$V_{i,j}^{\pm} = \pm W_{N-i,j}^{\pm} \tag{56}$$

and the following boundary conditions

$$V_{0,j}^{\pm} = 0, \tag{57}$$

$$W_{N,j}^{\pm} = 0. \tag{58}$$

In (57) $V_{N,j}^{\pm}$ is known, and in (58) $W_{0j}^{\pm} = V_{Nj}^{\pm}$.

By substituting (56) in (55), we get

$$V_{i+2,j}^{\pm} = V_{i,j}^{\pm} e^{-\frac{2\Delta\tau}{n_j}} + \frac{\Lambda}{2} \sum_{k=1}^{M} \left(P_{k,j} V_{i+1,k}^{\pm} + P_{k,-j} V_{N-i-1,k}^{\pm} \right) \Delta\eta \left(1 - e^{-\frac{2\Delta\tau}{n_j}} \right) + f_{i+1}^{\pm,j} \pm f_{N-i-1}^{\pm,-j} \tag{59}$$

under boundary conditions (57). The Gauss-Seidel iteration scheme is realized according to the following algorithm: at each nth step the value $I_{i+2}^{\pm,j}$ is found through $I_i^{\pm,j}$ and $I_{i+1}^{\pm,j}$ for the same n^{th} iteration, and the value $I_{i+1}^{\pm,k}$ is found from the $(n-1)^{th}$ iteration. Thus, for finding of values $V_j^{\pm}$ and $W_j^{\pm}$ we obtain the linear system of equations (59) of the rank $(N \times M)$ which is half as small as that in the classical version of the method under consideration. In this case we can also use a more narrow interval of modeling for the current angular variables $\eta_j \in [0,1]$ at $\tau_i \in [0, \tau_0]$ or alternatively $\tau_i \in [0, \frac{1}{2}\tau_0]$ at $\eta_j \in [1, -1]$.

Case Method. According to this method [12], the solutions of the basic non-symmetrized boundary-value problem (5 – 6) for the azimuthal harmonics of radiation intensities $I^m(\tau, \eta, \zeta, \tau_0)$ can be represented in the form of expansion in the full and orthogonal system of eigenfunctions φ_ν^m of uniform radiative transfer equation (5):

$$I^m(\tau, \eta, \zeta, \tau_0) = \frac{\Lambda}{2} \left(\alpha_m \frac{e^{-k_m \tau}}{1-k_m \eta} + \beta_m \frac{e^{k_m \tau}}{1+k_m \eta} \right) + \\ + \int_0^1 A^m(\nu) e^{-\tau/\nu} \, \varphi_\nu^m(\eta) d\eta + \int_0^1 A^m(-\nu) e^{\tau/\nu} \, \varphi_{-\nu}^m(\eta) d\eta \tag{60}$$

under boundary conditions:

$$I^m(0,\eta) = \delta(\eta - \zeta), \quad \eta > 0, \tag{61}$$

$$I^m(\tau_0, \eta) = 0, \quad \eta > 0. \tag{62}$$

The values $\pm k^m \in [0,1[$ are the roots of a corresponding characteristic equation [11], the coefficients α_m, β_m and function $A^m(\pm\nu)$ are the sought-for quantities. In the classical scheme of the considered method, the unknown coefficients α^m, β^m and functions $A^m(\nu), A^m(-\nu)$ are found from boundary conditions and from the orthogonality relations of eigenvalues φ_ν^m, with the following conditions being hold: $\varphi_{-\nu}^m(\eta) = \varphi_\nu^m(-\eta)$ and $\varphi_{-\nu}^m(\eta) = \varphi_\nu^m(\eta)$.

Using the above formulated mirror reflection principle and appropriate invariants (2), in place of (60) we find

$$I_{\pm}^m(\tau, \eta, \zeta, \tau_0) = \frac{\Lambda}{2} \left[\left(\alpha_m \pm \beta_m e^{k_m \tau_0} \right) \frac{e^{-k_m \tau}}{1-k_m \tau} + \left(\alpha_m e^{-k_m \tau_0} \pm \beta_m \right) \frac{e^{k_m \tau_0}}{1+k_m \eta} \right] + \\ + \int_0^1 \left[A^m(\nu) + A^m(-\nu) e^{\tau_0/\nu} \right] \varphi_\nu(\eta) d\eta \pm \int_0^1 \left[A^m(\nu) e^{-\frac{\tau_0}{\nu}} + A^m(-\nu) \right] \varphi_{-\nu}(\eta) e^{\tau/\nu} d\nu \tag{63}$$

Let us now introduce new values according to the relations:

$$\gamma_{\pm}^m = \alpha_m \pm \beta_m e^{k\tau_0}, \tag{64}$$

$$A_{\pm}^{m}(v) = A^{m}(v) \pm A^{m}(-v)e^{\frac{\tau_{0}}{v}} \tag{65}$$

Then using (64 – 65), instead of (63) we have finally

$$I_{\pm}^{m}(\tau,\eta,\zeta,\tau_{0}) = \gamma_{\pm}^{m}\frac{\Lambda}{2}\left(\frac{e^{-k_{m}\tau_{0}}}{1-k_{m}\eta} \pm \frac{e^{-k_{m}(\tau_{0}-\tau)}}{1+k_{m}\eta}\right) +$$
$$+ \int_{0}^{1}A_{\pm}^{m}(v)\left[\varphi_{v}^{m}(\eta)e^{-\frac{\tau}{v}} \pm \varphi_{-v}^{m}(\eta)e^{-\frac{\tau_{0}-\tau}{v}}\right]dv \tag{66}$$

under the boundary condition $I_{\pm}^{m}(0,\eta,\zeta,\tau_{0}) = \delta(\eta-\zeta), \eta > 0$. It should be noted that invariance property (64 – 65) can be independently deduced from (60) and (66).

Thus, the use of the symmetrized version of boundary-value problem (10 – 12) in place of (5 – 6) makes it necessary to find only two unknown values $\gamma_{\pm}^{m}$ and $A_{\pm}^{m}(v)$ instead of four unknown values $\alpha^{m}, \beta^{m}, A^{m}(v)$ and $A^{m}(-v)$ as in the classical scheme of Case's method [12].

5   Conclusion

The above considerations have clearly demonstrated that the introduction of mirror reflection principle into the scalar radiation transfer theory allows to substantially simplify the main methods and the numerical algorithms of this theory. For example, the rank of algebraic equation systems occurring in their application is halved as compared with the traditional classical schemes. Besides, basic variable intervals of numerical modeling for radiation intensities can be substantially narrowed for current optical depth τ and angular values η, namely $\tau_{i} \in [0,\tau_{0}]$ and $\eta_{j} \in [0,1]$, or alternatively $\tau_{i} \in [0,\frac{1}{2}\tau_{0}]$ and $\eta_{j} \in [-1,1]$ In case of modeling the radiation fields in natural media of a large optical thickness $\tau_{0} \gg 1$ (aerosol and cloud atmospheres, oceans) and strongly elongated phase functions $P(\cos\gamma)$ at numbers $M \gg 1$ and $N \gg 1$, this provides great advantages in carrying out of appropriate numerical calculations for applied investigations connected with radiative correction, calibration and comprehensive quantative analysis of the Earth and planets remote sensing data from outer space [13].

It should be noted that the above analysis can also be applied in the case of vector (polarized) radiation fields and to other numerical methods used in the radiative transfer theory (e.g. the Hunt-Grant matrix method [14]).

References

1. Ambarzumian, V.A.: Scientific Works, vol. 1, Erevan (1960). Амбарцумян В.А. Научные труды, том 1. Ереван (1960)
2. Shandrasekhar, S.: Radiative Transfer. Clarendon Press, Oxford (1950)
3. Sobolev, V.V.: Light Scattering in Planetary Atmospheres. Nauka, Moscow (1972); Соболев В.В. Рассеяние света в атмосферах планет. Наука. Москва (1972)
4. van de Hulst, H.C.: Light Scattering by Small Particles. Dover Publ. Inc., New York (1981)
5. Smokty, O.I.: Modeling Radiation Fields in Problems of Space Spectrophotometry. Nauka, Leningrad (1986); Смоктий О.И. Моделирование полей излучения в задачах космической спектрофотометрии. Наука, Ленинград (1986)
6. Smokty, O.I.: Photometrical Invariants in the Radiative Transfer Theory. In: IGARSS 1993, pp. 1960–1961. Kogakuin University, Tokyo (1993)
7. Smokty, O.I.: Development of Radiative Transfer Theory on the Basis of Mirror Symmetry Principle. In: Current Problems in Atmospheric Radiation, IRS 2001, pp. 341–342. A. Deepak Publ. Co., Hampton (2001)
8. Hovenier, J.W.: A Unified Treatment of the Reflected and Transmitted Intensities of Homogeneous Plane-Parallel Atmosphere. Astron. and Astrophys. 68, 230–250 (1978)
9. Lenoble, J. (ed.): Radiative Transfer in Scattering and Absorbing Atmospheres: Standard Computational Procedures. A. Deepak Publ. Co., Hampton (1985)
10. Davison, B.: Neutron Transport Theory. Clarendon Press, Oxford (1958)
11. Smokty, O.I.: Multiple Polarized Light Scattering in a Uniform Slab: New Invariant Constructions and Smmetry Relations. In: Current Problems in Atmospheric Radiation, IRS 2008, pp. 97–100. Amer. Inst. of Physics, New York (2009)
12. Case, K.M., Zweifel, P.F.: Linear Transport Theory. Addison-Wesley Publ. Co., Reading (1967)
13. Kondratyev, K.Y., Kozoderov, V.V., Smokty, O.I.: Remote Sensing of the Earth from Space: Atmospheric Correction. Springer, New York (1992)
14. Grant, I.P., Hunt, G.E.: Discrete Space Theory of Radiative Transfer Theory and its Application to Problems in Planetrary Atmospheres. J. Atmos. Sci. 26, 963–972 (1969)

A Multiphase Convection-Diffusion Model for the Simulation of Interacting Pedestrian Flows

Hartmut Schwandt[1], Frank Huth[1], Günter Bärwolff[1], and Stefan Berres[2]

[1] Technische Universität Berlin, Fakultät II, Institut für Mathematik,
Sekr. MA 6-4, Strasse des 17. Juni 135, 10623 Berlin, Germany
{schwandt,huth,baerwolf}@math.tu-berlin.de
[2] Universidad Catolica de Temuco, Faculdad de Ingeniera,
Departamento de Cs. Matematicas y Fsicas,
Rudecindo Ortega 02950, Temuco, Chile
sberres@uct.cl

Abstract. The simulation of pedestrian flow has become an important tool in the fields of planning and operation of public spaces like airports or shopping malls. While evacuation situations have been deeply investigated in numerous publications, the interaction of distinct pedestrian flows still needs more consideration. In this paper, we develop a macroscopic model for the simulation of interacting, more precisely intersecting pedestrian flows by a multiphase convection-diffusion approach. The convection corresponds to a movement towards a strategic direction whereas the diffusion corresponds to a tactical movement that avoids jams. Different populations moving in different directions are represented by different phases. Numerical experiments demonstrate the qualitative behaviour of the simulation model.

Keywords: multiphase model, convection-diffusion equation, pedestrian flow, simulation, macroscopic model.

1 Introduction

The modelling and the simulation of pedestrian traffic has become an important tool not only for scientific purposes. Its importance has increased continuously also as a tool for management decisions in the fields of planning and operation of airports, railway stations, sport stadiums, public events and manifestations, shopping malls etc. The main general goals are trouble-free operation and security aspects. While evacuation scenarios, which can be interpreted as single destination problems, have been studied quite intensively, multi destination problems where distinct streams of pedestrians move from one or more starting points to multiple destinations still need more investigation. In particular, the crossing of pedestrian streams has not yet been deeply investigated.

In the description and modelling of pedestrian behaviour we distinguish two basic model classes: microscopic and macroscopic models [6]. In microscopic

B. Murgante et al. (Eds.): ICCSA 2013, Part V, LNCS 7975, pp. 17–32, 2013.

models, pedestrians are considered as individual objects interacting with each other, while in macroscopic models, pedestrian behaviour is analyzed in terms of more global properties of a continuous stream. Macroscopic models focus on the balancing relationships of particle density interpreting pedestrians as particles, flow intensity and flow speed etc. As a third class mesoscopic models combine the main properties of the two former. For a detailed, comprehensive overview of both vehicular and pedestrian traffic and the main modelling and simulation approaches, in particular also macroscopic models, we refer to [10].

The modelling of pedestrian flows by macroscopic approaches has received various contributions in the last decade, see [2,11,13], e.g. In dense pedestrian crowds, pedestrians behave quite similarly to gas particles. This was one important reason to model pedestrian flows by fluid physical models from gas or fluid dynamics. Most of the research in the macroscopic area of pedestrian simulation is, therefore, focussed on the discussion and development of general partial differential equations, one- and two-dimensional in space, based on physical principles like mass, momentum and energy balance. Social and physical force models resulting from microscopic modelling are included as outer force terms in only a few papers. In [9], e.g., two-dimensional density and pressure fields were illustrated, which were mostly obtained by collecting results of microscopic social force models.

With some heuristic assumptions, models have been derived based on the Boltzmann equation and conservation laws ([14], [15], e.g.). They can be considered as special cases of the continuity or density equation and the momentum/velocity equation. The classical Euler equation (without viscous terms) can be derived assuming the velocity distribution to be approximately equal to the equilibrium/Gaussian distribution. Models using the Navier-Stokes equation can be derived without the equilibrium assumption. They differ from classical fluid or gas dynamic equations mostly by the source term in the momentum equation to describe social and physical forces acting on and due to pedestrians.

On the other hand, there exists a large number of microscopic approaches based on ordinary differential equations, cellular automata or graph theory. See again [10]. Among them, cellular automata are prominent since they use intuitive physical assumptions and are quickly implemented. They have been adapted to both vehicular traffic ([7], e.g.) and pedestrian dynamics ([3], e.g.). An approach described in [4] will be used in future work of the present context. The advantage of cellular automata and similar discrete models is that they are very flexible in engineering ad-hoc assumptions. In addition, they allow agent (person or car) tracking rather than only counting densities.

A new hybrid approach [5,16] describes a combined microscopic and macroscopic modelling in a multiscale framework using a measure-theoretic approach. The crowd velocity is composed as a sum of a macro-scale desired velocity and a micro-scale interaction velocity that accounts for an explicit local control in a small neighborhood. A control mechanism considers a local movement that is based on the situation in a small neighborhood and is directed away from a center of mass or, similarly, towards a so-called inverse center of mass.

All approaches have their advantages and disadvantages as they consider different aspects. In particular, microscopic approaches are useful due to the treatment of individual behaviour and local effects which can not be covered by macroscopic models, macroscopic approaches due to the simulation of global and mass effects which are not reflected by microscopic models.

Simulation models for *vehicular traffic* have a more or less one-dimensional character as cars move in lanes on streets allowing cross-directional flow only at distinct crossing points. In contrast, pedestrian flow allows a genuine spatial structure: pedestrian movement can be directed principally to any direction and it is strongly influenced by human behaviour. Therefore, simulation models for pedestrian traffic are two-(or even three-)dimensional.

A well known problem which has been thoroughly studied in particular by microscopical models is the *escape* or *evacuation problem*, where the task is to model escape panic ([8], e.g.). In an escape situation, all individuals de facto try to reach the same destination. Therefore, escape problems can be formally considered as single destination problems which can be modelled by a unique pedestrian stream. The modelling of multi-destination problems characterized by at least two distinct pedestrian streams remains a specifically difficult and not yet deeply investigated challenge, in particular if one assumes that each stream has its own, unique target direction. Both microscopic and macroscopic approaches are suitable.

We develop a macroscopic multiphase model using two-dimensional convection-diffusion systems. The intention is to show that this approach is well suited for the description and simulation of various aspects of pedestrian traffic, especially in the case of multiple intersecting streams. The main goal of the present work in this context is to find an adequate and robust simulation model for the detection of severe jam and congestion situations. This paper has to be considered as a further step to a more general simulation model which is under development. We start with a purely macroscopic approach and investigate up to which extent aspects of pedestrian flow can be adequately modelled in the above context. The limitations of purely macroscopic models we can expect are due to the fact that pedestrians are not "infinitely small' gas molecules and are not compressible. They require a minimal physical space, the range of possible values for the walking speed is restricted and, most important, their movements are also strongly influenced by individual behaviour and decisions. As a consequence, macroscopic models have to be complemented by additional elements (see [5,16] or [17], e.g.).

2   Modelling

The idea of considering traffic flow in the context of continuum mechanics was first used in the Lighthill-Whitham-Richards model [15], where vehicular traffic, which has an essentially one-dimensional nature, is modelled by the scalar one-dimensional conservation law

$$\rho_t + f(\rho)_x = 0, \tag{1}$$

where $\rho = \rho(x, t)$ denotes a locally averaged car density at time t and position x. The flux function

$$f(\rho) = a\,\rho\,V(\rho) \tag{2}$$

encapsulates the model assumptions concerning the average velocity of cars. The constant a corresponds to the maximum velocity and $V = V(\rho)$ is a normalized density-dependent velocity function. Greenshield's model

$$V(\rho) = 1 - (\rho/\rho_{\max})^n, \ n > 0, \ \rho_{\max} \in (0, 1) \tag{3}$$

and

$$V(\rho) = 1 - \rho \tag{4}$$

are two simple examples of this function. A first generalization of (1) to two-dimensional multiple-species pedestrian flow was proposed in [12]. It is based on mass equations having the form

$$\varrho_t + f(\varrho)_x + g(\varrho)_y = 0, \tag{5}$$

where f and g are the fluxes in x and y directions, resp., and where $\rho = (\rho_1, \rho_2)^{\mathrm{T}}$ with the local densities $\rho_i \equiv \rho_i(x, y, t), i = 1, 2$, of the respective flows. The reference [12] can be considered as a first modelling of pedestrian flow by continuum theory. It treated a specific situation with a preliminary use of application specific numerical methods, but did not elaborate the dynamics of phase separation which are essential for the modelling of crossing pedestrian flows.

In [1], we have started to consider a further step towards a multi-phase model based on convection-diffusion systems in which each phase models a distinct stream of pedestrians. The main goal consists in a better comprehension of the effects when two or more streams intersect. We further assume that the streams start at different locations and that the members of each stream move to the target of that stream, i.e. particles do not change their strategical direction by moving to the target of another stream: in this context streams do not merge.

2.1 Basic Approach

Pedestrian flow can be considered as a transport problem which is principally governed by a mass balance. Assume $n, n \in \mathbb{N}$, distinct pedestrian flows/species. Let Ω be an open sufficiently smooth bounded domain in $\mathbb{R}^2$ and $(0, T)$ an open interval. For $(x, y) \in \Omega$, $t \in (0, T)$ the equation

$$\frac{\partial \varrho_i}{\partial t} + \nabla \cdot (\varrho_i\, v_i) = 0, \qquad i = 1, \ldots, n, \tag{6}$$

describes the mass flow where t denotes time, $\varrho_i \in [0, 1]$, $v_i = v_i(\varrho_1, \ldots, \varrho_n), 1 \leq i \leq n$, the local densities and speeds, respectively, of the i-th pedestrian flow/species. Obviously, (1) and (5) formally represent special cases of this approach.

Following the idea of (5) we can introduce flux functions $\boldsymbol{f}_i : \mathbb{R}^{n+2} \to \mathbb{R}^2$,

$$\boldsymbol{f}_i(\varrho) = a_i \varrho_i V(\varrho) \boldsymbol{d}_i, \quad \varrho = \sum_{i=1}^{n} \varrho_i, \quad i = 1, \ldots, n, \tag{7}$$

which depend on the local pedestrian densities $\varrho_i \in [0,1]$, the total pedestrian density $\varrho \in [0,1]$, an appropriate function $V(\varrho)$ like in (3) or (4), constants a_i corresponding to actual maximum velocities (the usual pedestrian walking speed does normally not exceed $a \leq 1.4\,m/s$, see [9] or [10], e.g.), the velocity magnitude $V \in [0,1]$, and species specific unit vector fields $\boldsymbol{d}_i$ that indicate the directions. The model needs to be completed by appropriate initial conditions $\varrho_i(t = 0) = \varrho_i^{(0)}$. The equation (6) can now be replaced by

$$\frac{\partial \varrho_i}{\partial t} + \nabla \cdot \boldsymbol{f}_i(\varrho_1, \ldots, \varrho_n; x, y) = 0, \; i = 1, \ldots, n. \tag{8}$$

The introduction of the fluxes $\boldsymbol{f}_i$ enables us to distinguish the different species (flows) explicitly by properties like preferred walking direction or speed. However, (8) has a drawback in the present context of modelling intersecting pedestrian flows. When two (or more) flows cross each other, it is inevitable that, in particular in the presence of higher densities, pedestrians (particles) frequently compete for the same physical position. The model does not provide a means for permitting evasion movements or withdrawal actions. The situation can be compared to that of blind persons which encounter an obstacle with their canes and stop without taking any further action. This model results in early and frequent jams which are not necessarily realistic. A first remedy consists in the introduction of a Laplace term

$$\frac{\partial \varrho_i}{\partial t} + \nabla \cdot \boldsymbol{f}_i(\varrho_1, \ldots, \varrho_n; x, y) = \varepsilon \Delta \varrho_i, i = 1, \ldots, n, \tag{9}$$

to allow for some diffusion. The practical result will be that the pedestrians will try to move undirectedly as long as they find a minimal space to move to in situations like a jam, i.e. any space left will be occupied. In our metaphor of blind persons, they will take the first space they find with their canes.

In the sequel we will generalize (9) to

$$\frac{\partial \varrho_i}{\partial t} + \nabla \cdot \boldsymbol{f}_i(\varrho_1, \ldots, \varrho_n; x, y) = \sum_{j=1}^{n} \nabla \cdot (b_{ij}(\varrho_1, \ldots, \varrho_n) \nabla \varrho_j), \quad i = 1, \ldots, n, \tag{10}$$

where the $b_{ij} \equiv b_{ij}(\varrho_1, \ldots, \varrho_n)$, $1 \leq i, j \leq n$, denote the components of a diffusion matrix $\boldsymbol{B}$. The model (9) is included with $b_{ij} = \varepsilon \, \delta_{i,j}$, $i, j = 1, \ldots, n$. The approach (9) results in a constant diffusion covering the available space rather equally using any free space. This is not yet an adequate model for distinct crossing streams of pedestrians whose members all try to continue to their target as part of their stream after having passed through a potential intersection area. Diffusion does not adequately model the necessary separation of the phases (pedestrian streams) after the crossing.

2.2   Modelling the Flux

The key for a more realistic modelling of intersectio situations is yeld by a better approach for the modelling of pedestrian behaviour in the presence of crowded situations. We assume that pedestrians, instead of creating an uncontrolled and rather uniform diffusion, will try to evade from crowded spaces. This effect will be modelled as a function of the local density, i.e. our model will be based on the assumption that pedestrians avoid densely populated areas by modifying their motion into the direction of the negative gradient of the total local density ϱ. This local orientation reflects the behaviour of blind persons with white canes, who generally stick to their planned direction, but modify it by taking a direction away from a congestion, i.e. they "detect" the gradient.

When modelling the above transport problem, we have to reflect two aspects of pedestrian behaviour that correspond to strategic and tactic decision making, resp.: The pedestrian has a target, which he tries to reach, but he might also be forced to deal with local problems like high densities. To take account of these two aspects, we restart vom (8) and replace the flux functions $\boldsymbol{f}_i$ in the (convective) term $\nabla \cdot \boldsymbol{f}_i(\dots)$ by introducing the total flux

$$\phi_i(\varrho_1,\dots,\varrho_n) = \boldsymbol{f}_i^s(\varrho_1,\dots,\varrho_n) + \boldsymbol{f}_i^t(\varrho_1,\dots,\varrho_n) \qquad (11)$$

defined by two components:

- a strategic component $\boldsymbol{f}_i^s$, which reflects the disposition to follow a strategic goal of reaching a certain destination on a desired path,
- and a tactical component $\boldsymbol{f}_i^t$ of the flux, which locally avoids densely populated areas.

Both components are modelled as a product of density, velocity and direction,

$$\boldsymbol{f}_i^s = \varrho_i V^{(S)} \boldsymbol{d}_i^s, \qquad \boldsymbol{f}_i^t = \varrho_i V^{(T)} \boldsymbol{d}_i^t, \qquad (12)$$

where $\boldsymbol{d}_i^s, \boldsymbol{d}_i^t$ denote the direction unit-vector fields defining a desired (strategic) and an adapted tactical walking direction, respectively. Standard strategic directions are either opposite or perpendicular. For a two-species model ($n = 2$) the most simple choices are given by $\boldsymbol{d}_1^s = (1,0)^T$, $\boldsymbol{d}_2^s = (-1,0)^T$ and $\boldsymbol{d}_1^s = (1,0)^T$, $\boldsymbol{d}_2^s = (0,1)^T$, respectively. More sophisticated strategic directions are aligned to a potential field $\boldsymbol{P}$ with $\boldsymbol{d}_1^s = \boldsymbol{P}_x$, $\boldsymbol{d}_2^s = \boldsymbol{P}_y$.

The tactical direction is modelled as

$$\boldsymbol{d}_i^t = \begin{cases} -\nabla \varrho / |\nabla \varrho| & \text{for} \quad |\nabla \varrho| \geq 1, \\ -\nabla \varrho & \text{for} \quad |\nabla \varrho| < 1. \end{cases} \qquad (13)$$

using a partial normalization by which unrealistic "escape" velocities can be avoided in the model.

The magnitudes of the speeds in these directions are denoted by $V^{(S)} \equiv V^{(S)}(\varrho), V^{(T)} \equiv V^{(T)}(\varrho)$. Whereas $V^{(S)}$ is weighting the strategic part of the pedestrian way, $V^{(T)}$ is weighting the tactical part. A generic assumption for $V^{(S)}$ is to be decreasing

$$V^{(S)}(0) = 1, \quad (V^{(S)})' \leq 0, \quad V^{(S)}(1) = 0, \qquad (14)$$

describing a throttling effect at higher concentrations: the more persons are in a given region, the more they get stuck on their way towards the strategic direction. The tactical velocity $V^{(T)}$ is assumed to increase

$$V^{(T)}(0) = 0, \quad (V^{(T)})' \geq 0, \quad V^{(T)}(1) = 1. \qquad (15)$$

At higher concentrations the tendency to evade increases: The more persons are blocking the way, the stronger is the tendency to move on along an alternative trajectory. In view of the the symmetry of (14), (15) we can impose the constraint

$$V^{(S)} + V^{(T)} = 1 \qquad (16)$$

to model the realistic assumption that the partitioning of the flux into a strategically caused and a tactically caused directed part also results in a partitioning of a total velocity (normalized to 1). The equality (16) expresses that a pedestrian has an individual level of moving activity which he partitions on the two alternatives of moving to the desired target or evading from jams.

Given the velocity throttling (14) for $V^{(S)}$, the additional condition (15) is weaker than taking (16). Condition (16) does not constrain the qualitative behaviour but slightly reduces the degree of freedom in (14), which is helpful in absence of other modelling guidelines.

We now replace the fluxes $\boldsymbol{f}_i$ in (8) by total fluxes $\boldsymbol{\phi}_i$ from (11), and we get

$$\frac{\partial \varrho_i}{\partial t} + \nabla \cdot \left\{ \varrho_i \left[a_i V^{(S)} \boldsymbol{d}_i^s + b_i V^{(T)} \boldsymbol{d}_i^t \right] \right\} = 0 \qquad (17)$$

with constants $a_i, b_i, i = 1, \ldots, n$, reflecting maximal velocities. The actual speeds can be modelled by the parameters a_i and b_i. Defining

$$\chi(\varrho) = \begin{cases} 1/|\nabla \varrho| & \text{if} \quad |\nabla \varrho| \geq 1, \\ 1 & \text{if} \quad |\nabla \varrho| < 1, \end{cases} \qquad (18)$$

we can interpret (17) as special case of (10) with $\boldsymbol{f}_i \equiv \boldsymbol{f}_i^s$, $i = 1, \ldots, n$, and $b_{ij}(\varrho_1, \ldots, \varrho_n) = b_i \varrho_i V^{(T)} \chi(\varrho)$, $i, j = 1, \ldots, n$. The undirected diffusion in (8) has now been replaced by a flux opposite to the gradient of the total density, i.e. away from the latter. In this paper, we choose $V^{(S)} = 1 - \varrho$. Assuming (16), this yields $V^{(T)} = \varrho$, hence

$$b_{ij}(\varrho_1, \ldots, \varrho_n) = b_i \varrho_i \varrho \chi(\varrho), \quad i, j = 1, \ldots, n. \qquad (19)$$

Note that according to the above assumptions (14), (15) and (16), other choices for $V^{(S)}$ are possible. Using $\varrho = \sum_{i=1}^{n} \varrho_i$ we can finally write (17) in this case by adapting (10) as a convection-diffusion system in vector form

$$\frac{\partial \varrho}{\partial t} + \nabla \cdot \boldsymbol{f}(\varrho) = \nabla \cdot \left(B(\varrho) \nabla \varrho \right), \qquad (20)$$

where the diffusion matrix has the form

$$B(\varrho) = \varrho \begin{pmatrix} b_1 \varrho_1 & \cdots & b_1 \varrho_1 \\ \vdots & & \vdots \\ b_n \varrho_n & \cdots & b_n \varrho_n \end{pmatrix} \chi(\varrho) \tag{21}$$

and where $\varrho = (\varrho_1, \ldots, \varrho_n)^T$, $\nabla \varrho = (\nabla \varrho_1, \ldots, \nabla \varrho_n)^T$. For $n = 2$, (20) reduces with $\varrho = \varrho_1 + \varrho_2$ to

$$\frac{\partial \varrho_1}{\partial t} + \nabla \cdot \boldsymbol{f}_1(\varrho_1, \varrho_2) = \nabla \cdot \left(b_1 \, \varrho_1 \, \varrho \, \chi(\varrho) \left(\nabla \varrho_1 + \nabla \varrho_2 \right) \right)$$

$$= \nabla \cdot \left(b_1 \, \varrho_1 \, \varrho \, \chi(\varrho) \, \nabla \varrho \right),$$

$$\frac{\partial \varrho_2}{\partial t} + \nabla \cdot \boldsymbol{f}_2(\varrho_1, \varrho_2) = \nabla \cdot \left(b_2 \, \varrho_2 \, \varrho \, \chi(\varrho) \, \nabla \varrho \right).$$

2.3 Modelling Remarks

We note that in (9) the flux is not only produced by the transport terms $\nabla \cdot \boldsymbol{f}_i$, but also by the divergence term: We recall that for (9) we have a diffusion matrix $B(\varrho) \equiv B = \varepsilon \, I$, I identity matrix, i.e. additional flux components

$$\hat{\boldsymbol{f}}_i^{\,div} \equiv \hat{\boldsymbol{f}}_i^{\,div}(\varrho_i) = \varepsilon \cdot \nabla \varrho_i, i = 1, \ldots, n, \tag{22}$$

formally analogous to (12) and total fluxes

$$\phi_i = \boldsymbol{f}_i + \hat{\boldsymbol{f}}_i^{\,div}, \quad i = 1, \ldots, n. \tag{23}$$

Note that the fluxes $\hat{\boldsymbol{f}}_i^{\,div}$ do not depend on the total density, but only on the density of the respective stream. This models an uncoupled diffusion which appears not to be realistic for a model for pedestrian behaviour.

Similarly, we can discuss the "gradient" model (20), (22). The equation (17) contains a second transport term by the definition:

$$\hat{\boldsymbol{f}}_i^{\,grad} \equiv \hat{\boldsymbol{f}}_i^{\,grad}(\varrho_1, \varrho_2) = b_i \, \varrho_i \, \varrho \, \chi(\varrho) \, \nabla \varrho, \tag{24}$$

with $V^{(T)} = b_i \, \varrho$, $\boldsymbol{d}_i^{(t)} = \chi(\varrho) \, \nabla \varrho$ and total fluxes

$$\phi_i = \boldsymbol{f}_i + \hat{\boldsymbol{f}}_i^{\,grad}, \quad i = 1, \ldots, n. \tag{25}$$

These fluxes depends on the total density ϱ.

3 Numerical Examples

By the following examples we illustrate the fundamental difference in the behaviour of the "divergence" or Δ-model (9) and the "gradient" or ∇-model (20) for a simple, but instructive 2D situation of two pedestrian streams crossing each other in an angle of 180°. Due to the lack of comparable work in that area and in view of the fact that the simple transport equation (6) is not a reasonable approach in this context, we have to restrict our comparison to these two models For $n = 2$ these models reduce to

$$\frac{\partial \varrho_i}{\partial t} + \nabla \cdot \boldsymbol{f}_i(\varrho_1, \varrho_2; x, y) = \varepsilon \Delta \varrho_i, \ i = 1, 2, \tag{26}$$

for the Δ-model with $a_i = 1, b_i = \epsilon = 0.01$, and

$$\frac{\partial \varrho_i}{\partial t} + \nabla \cdot \boldsymbol{f}_i(\varrho_1, \varrho_2; x, y) = \nabla \cdot \left(\varrho_i \, \varrho \, \chi(\varrho) \nabla \varrho \right), \ i = 1, 2, \tag{27}$$

for the ∇-model with $a_i = b_i = 1$. We use "normalized" parameters a, b to illustrate the different qualitative behaviour. Comparing the parameters b in the models, we note that $b = \epsilon = 0.01$ in the Δ-model is chosen to be much smaller than $b = 1$ in the ∇-model. This choice results from the experimental observation that a larger ε reflects a larger diffusion spreading pedestrians rather quickly over the available space. This effect contradicts the assumption of pedestrian streams oriented towards a specific target. It can be weakened by a smaller ε, which, however, sometimes dramatically reduces the overall movement. On the other hand, the gradient term directly depends on the density which is a desired effect. We choose the domain $\Omega = (-1, 1)^2$ and define the subboundaries:

$$\Gamma_w = \{(x, y) : x = -1, y \in [-.3, .3]\}$$

$$\Gamma_e = \{(x, y) : x = 1, y \in [-.3, .3]\}$$

$$\Gamma_c = \partial \Omega \setminus (\Gamma_w \cup \Gamma_e).$$

Γ_w and Γ_e define combined entries and exits, while Γ_c denotes walls. Stream 1 and 2, resp., enter the domain by the entry Γ_w and Γ_e, resp., and leave it by exit Γ_e and Γ_w, resp. We start with an empty domain, i.e.

$$\varrho_i(0, x, y) = 0 \quad \forall (x, y) \in \Omega, \quad i = 1, 2,$$

where we recall that $\varrho_i \equiv \varrho_i(x, y, t)$. As boundary conditions, we use for both models (26) and (27) for entries and exits

$$\boldsymbol{n} \cdot \nabla \rho_i = 0 \quad \text{on} \ = \begin{cases} \Gamma_e \ \text{for} \ i = 1 \\ \Gamma_w \ \text{for} \ i = 2 \end{cases} \text{(outflux boundary)} , \tag{28}$$

$$\rho_i = \text{constant on} \ = \begin{cases} \Gamma_w \ \text{for} \ i = 1 \\ \Gamma_e \ \text{for} \ i = 2 \end{cases} \text{(influx boundary)}$$

and for walls Γ_c

$$\begin{aligned} \rho_i &= 0 & \text{for the } \nabla\text{-model (27)}, \\ \boldsymbol{n} \cdot \nabla \rho_i &= 0 & \text{for the } \Delta\text{-model (26)}, \end{aligned} \tag{29}$$

where n denotes the outwards directed normal. The boundary condition $n \cdot \nabla \rho_i = 0$ signifies "no diffusion", but the flux which is constant at the outflux boundaries and directed by d_i away from the walls. The constant flow at the influx boundary is not mandatory, but sufficient for the present context to illustrate the differences between the models. We assume that the pedestrians try to reach their respective exit on the shortest path from every point in the domain. This leads to the side effect, that the corners of the exits are especially sensible for congestion. Fig. 1 illustrates the resulting direction fields of the two streams. The (strategic) direction is perpendicular to the walls, oriented inside on the entries and outside at the exits, i.e.

$$
\boldsymbol{n} \cdot \boldsymbol{d}_i = \begin{cases} -1 & \text{for influx boundaries,} \\ 1 & \text{for outflux boundaries,} \\ 0 & \text{for walls.} \end{cases} \tag{30}
$$

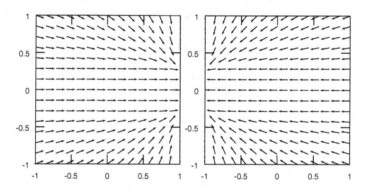

Fig. 1. Direction fields of stream 1 (left), 2 (right)

For (27) we have $\boldsymbol{d}_i^{(s)} \equiv \boldsymbol{d}_i$ and, in addition, $\boldsymbol{n} \cdot \boldsymbol{d}_i^{(t)} = 0$. In principle, other choices of boundary values are possible. The choice of the boundary values, however, does not have the highest priority for the qualitative discussion of intersection and congestion situations. In our context, the total transport capacity $\varrho V^{(S)}(\varrho) = \varrho (1 - \varrho)$ reaches its maximum of 0.25 at $\varrho = 0.5$. Therefore, assuming a symmetric flow, a larger flow for higher densities $\varrho_1, \varrho_2 > 0.25$ is impossible. As the considered operators counter the given influx, either by diffusion or by a movement opposite to the density gradient, only limited fluxes are permitted by the two models. Therefore, beyond some model specific threshold higher densities at the entry do not produce higher fluxes anymore. Beyond a theoretical maximum flux of 0.25, which would imply a symmetric flux of 0.125 per species, if the flux is balanced between both species, the models should produce a transition into congestion state.

The numerical results are based on a finite volume discretization using the OpenFOAM package (see www.openfoam.com). We refer the discussion of details of discretization and implementation to a subsequent paper.

In the following tables, we denote entries/exits and the two models by

$$\Gamma_{io}, io \in \{e, w\}, \quad mod \in \{div, grad\}.$$

The flow at the entries *in* and exits *out* can be obtained with (22), (24) from

$$\phi_i^{mod} \equiv \phi_i^{mod}(t) \overset{\cdot}{=} \boldsymbol{n} \cdot (\boldsymbol{f}_i(t) + \boldsymbol{f}_i^{mod}(t)), \, i = 1, 2.$$

In analogy to (11) the term $\boldsymbol{f}_i(t) + \boldsymbol{f}_i^{mod}(t)$ defines a total flux with a strategic and a tactical component. The average flow at the entries/exits is given by

$$\Phi_i^{mod,io}(t) = \frac{\int_{\Gamma_{io}} \phi_i^{mod}(t) \, d\Gamma_{io}}{|\Gamma_{io}|}, \, i = 1, 2, \, io = e, w,$$

at time t, where $|\Gamma_{io}|$, $i = 1, 2$, $io = e, w$, denotes the length of Γ_{io}. The following tables illustrate the flow at the entries/exits by some numerical experiments. The indicated numerical values for Φ_i^{mod} have been computed from the stiffness matrix of the finite volume discretization. We consider densities ϱ_i to be critical if beyond the respective values we do not anymore observe steady-state solutions. Tab. 1 and Fig. 2 illustrate a configuration with initial densities for which the ∇-model yields a situation close to becoming critical. The densities near the exits become significant, however, flow is still possible. We still get a steady-state solution with symmetric flow of both streams.

Table 1. ∇-model. $t_1 = 1, t_e = 115.047, \rho_{1,2}|_{\Gamma_{in}} = 0.159375.$

i	io	$\max_t \left|\Phi_i^{mod}(t)\right|$	$\Phi_i^{mod,io}(t_1)$	$\Phi_i^{mod,io}(t_e)$
1	$in = w$	0.079688	−0.079688	−0.079688
1	$out = e$	0.079688	0.000001	0.079688
2	$out = w$	0.079688	0.000001	0.079688
2	$in = e$	0.079688	−0.079688	−0.079688

Similar results can be observed for the Δ-model. Note that in subcritical configurations both operators reproduce similar fluxes of 0.79688 ((∇-model) and 0.077610 (Δ-model, test not shown), respectively, at the transition density to critical situations. These fluxes are clearly smaller than 0.125, the value which could be expected theoretically. Most likely this is due to the uneven distribution of the density at the combined entries/exits (see Fig. 2, Tab. 1). Note that we chose model specific different entry densities $\rho_i|_{\Gamma_{in}} = 0.159375$ for the ∇-model and $\rho_i|_{\Gamma_{in}} = 0.203125$ for the Δ-model (test not shown here) such that entry densities close to the critical transition point are used. In this respect the Δ-operator estimates the maximum permissible density for a stable flux significantly higher.

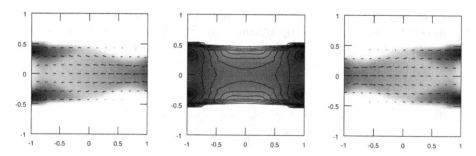

Fig. 2. ∇-model. Densities at entries $\rho_{1,2}|_{in} = 0.159375$. Densities $\varrho_1, \varrho, \varrho_2$ (left to right) and flux vectors for steady-state solution on Ω.

Table 2. ∇-model. $t_1 = 1, t_e = 5000.001$, $\rho_{1,2}|_{\Gamma_{in}} = 0.1625$.

i	io	$\max_t\left|\Phi_i^{mod}(t)\right|$	$\Phi_i^{mod,io}(t_1)$	$\Phi_i^{mod,io}(t_e)$
1	$in = w$	0.081250	−0.081250	−0.000000
1	$out = e$	0.079373	0.000001	−0.000000
2	$out = w$	0.079373	0.000001	−0.000000
2	$in = e$	0.081250	−0.081250	−0.000000

Tab. 2 and Fig. 3 show some time steps of a supercritical situation for the ∇-model. The space near the entries/exits is filled more and more until the entries are nearly blocked by the pedestrians arriving from the respective other stream. The streams become asymmetric as one stream begins to dominate and to block the other entry. Finally, one stream completely fills the available space. The situation is sensible even to minor asymmetries. The most important result is that congestion is predicted by the ∇-model in contrast to the Δ-model: in the former, all fluxes completely cease when total congestion occurs. Tab. 3 and Fig. 4 show some time steps of a comparable supercritical situation for the Δ-model. Similarly to the ∇-model, for supercritical densities one species blocks the entry of the other after some time and pushes it out. In contrast to the latter, even in the case of complete congestion at the entries/exits, there subsist fluxes in Ω. Diffusion effects subsist where they should not occur in reality. In the present examples there can be observed flows mainly near to the upper and lower walls.

Table 3. Δ-model. $t_1 = 1, t_e = 480.085$, $\rho_{1,2}|_{\Gamma_{in}} = 0.225$.

i	io	$\max_t\left|\Phi_i^{mod}(t)\right|$	$\Phi_i^{mod,io}(t_1)$	$\Phi_i^{mod,io}(t_e)$
1	$in = w$	0.103629	−0.103629	−0.016800
1	$out = e$	0.076538	0.000008	0.016800
2	$out = w$	0.077008	0.000008	0.011489
2	$in = e$	0.103629	−0.103629	−0.011489

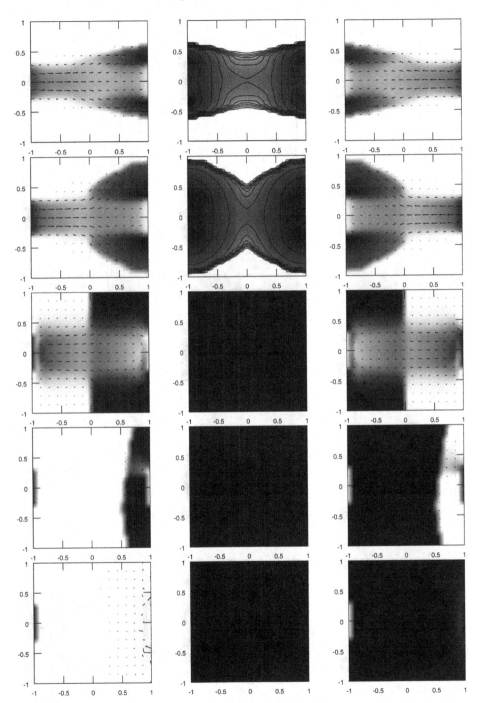

Fig. 3. ∇-model. Densities at entries $\rho_{1,2}|_{in} = 0.1625$. Densities $\varrho_1, \varrho, \varrho_2$ (left to right) and the flux vectors of streams 1, 2. Time steps $t = 50, 75, 100, 1000, 5000$ (top-down).

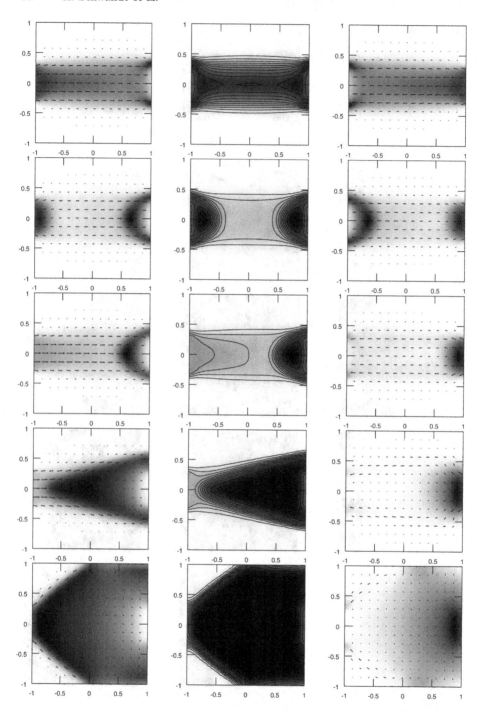

Fig. 4. Δ-model. Densities at entries $\rho_{1,2}|_{in} = 0.225$. Densities $\varrho_1, \varrho, \varrho_2$ (left to right) and the flux vectors of streams 1, 2. Time steps: 5, 10, 30, 40, 481 (top-down).

Mathematically it is interesting that these diffusion effects can have a periodical behaviour in specific situations. Near to the upper and lower walls, we can, for example observe empty "bubbles" moving periodically in the x-direction. For the modelling of pedestrian behaviour these diffusion effects resulting in subsisting fluxes even in the case of complete congestion illustrate that the Δ-model is not suitable for the modelling of pedestrian flow.

4   Conclusion

Pedestrian flow can be modelled by macroscopic models up to a certain extent. In contrast to "usual" flow problems, pedestrians are "particles" which have a significant physical extension, limitations in speed and other parameters and an individual behaviour. Pedestrian movement is influenced by discrete effects which cannot be covered completely by macroscopic models. In view of this, the modelling of jam and congestion situations is particularly important. The transport equation (6) leads to a complete stop of any movement whenever particles encounter an obstacle or other particles. The Δ-model (9) models jam or congestion situations by introducing diffusion in order to create evasion movements. The result, however, is a padding of empty spaces. This contradicts the goal of simulation models for crossing pedestrian streams, i.e. distinct particle streams each of which tries to reach an own target without merging with other streams. This requires a clear separation of the phases (streams) in particular after crossings. The Δ-model is not useful to achieve this goal. The ∇-model is based on the assumption that pedestrians try to reach a specific target, but are flexible enough to change their direction in the case of jam or congestion. The movement towards the target can be temporarily replaced by a withdrawal away from higher densities, i.e. in a direction opposite to the gradient of total density. This model respects phase separation and signals, in particular, congestion situations while the Δ-model still predicts flow where it should not occur anymore. In summary, higher input densities lead to deadlock and strong phase separation in case of the ∇-operator and to many effects with little phase separation, but not to a complete deadlock (as should be expected) in case of the Δ-operator. In the next step we will develop the ∇-model with the goal of forcing more the redirection towards the original targets after evasion movements.

5   Outlook

In several tests, the ∇-model has proved to be robust and to be rather reliable in the prediction of congestion. The withdrawal effect away from higher densities replaces undirected diffusion and reflects a behaviour which can be observed in reality. The next step will be the development of a model component redirecting the pedestrians after a withdrawal towards their originally intended target while remaining in their stream, i.e. without moving to another stream. This will be achieved by the introduction of potential lines which have to be dynamically updated, most probably also by integrating current local information from a microscopical model.

References

1. Berres, S., Ruiz-Baier, R., Schwandt, H., Tory, E.M.: An adaptive finite-volume method for a model of two-phase pedestrian flow. Networks and Heterogeneous Media 6, 401–423 (2011)
2. Bruno, L., Tosin, A., Tricerri, P., Venuti, F.: Non-local first-order modelling of crowd dynamics: A multidimensional framework with applications. Appl. Math. Model. 35, 426–445 (2011)
3. Burstedde, C., Klauck, K., Schadschneider, A., Zittartz, J.: Simulation of pedestrian dynamics using a two-dimensional cellular automaton. Physica A: Stat. Mech. Appl. 295, 507–525 (2001)
4. Chen, M., Bärwolff, G., Schwandt, H.: A derived grid-based model for simulation of pedestrian flow. J. Zhejiang Univ.: Science A10, 209–220 (2009)
5. Cristiani, E., Piccoli, B., Tosin, A.: Multiscale modelling of granular flows with application to crowd dynamics. Multiscale Model. Simul. 9, 155–182 (2011)
6. Daamen, W., Bovy, P.H.L., Hoogendoorn, S.P.: Modelling pedestrians in transfer stations. In: Schreckenberg, M., Sharma, S.D. (eds.) Pedestrian and Evacuation Dynamics, pp. 59–73. Springer, Heidelberg (2002)
7. Esser, J., Schreckenberg, M.: Microscopic simulation of urban traffic based on cellular automata. Int. J. Mod. Phys. C 8, 1025–1036 (1997)
8. Helbing, D., Farkas, I., Vicsek, T.: Simulating dynamical features of escape panic. Nature 407, 487–490 (2000)
9. Helbing, D., Buzna, L., Johansson, A., Werner, T.: Self-organized pedestrian crowd dynamics: experiments, simulations, and design solutions. Transportation Science 39, 1–24 (2005)
10. Helbing, D., Farkas, I., Vicsek, T.: Traffic and related self-driven many-particle systems. Reviews of Modern Physics 73, 1067–1141 (2001)
11. Hoogendoorn, S.P., Daamen, W.: Self-organization in pedestrian flow. Traff. Granul. Flow 3, 373–382 (2005)
12. Hughes, R.L.: A continuum theory for the flow of pedestrians. Transp. Res. B 36, 507–535 (2002)
13. Jiang, Y., Zhang, P., Wong, S.C., Liu, R.: A higher-order macroscopic model for pedestrian flows. Physica A 389, 4623–4635 (2010)
14. Kerner, B.S., Konhäuser, P.: Structure and parameters of clusters in traffic flow. Physical Review E 50, 54–83 (1994)
15. Lighthill, M.J., Whitham, G.B.: On kinematic waves. II. A theory of traffic flow on long crowded roads. Proc. Roy. Soc. London Ser. A 229, 317–345 (1955)
16. Piccoli, B., Tosin, A.: Time-evolving measures and macroscopic modelling of pedestrian flow. Arch. Ration. Mech. Anal. 199, 707–738 (2011)
17. Slawig, T., Bärwolff, G., Schwandt, H.: Simulation of pedestrian flows for traffic control systems. In: Li, E.Y. (ed.) Proc. of the 7th International Conference on Information and Management Sciences (IMS 2008) (Urumtschi, 12. 8. - 19. 8. 08). Series on Information and Management Sciences, vol. 7, pp. 360–374. California Polytechnic State University, California (2008)

Aspects of Linearity in Cryptographic Sequence Generators*

Amparo Fúster-Sabater

Information Security Institute, C.S.I.C.
Serrano 144, 28006 Madrid, Spain
amparo@iec.csic.es

Abstract. In the present work, it is shown that the sequences obtained from a cryptographic sequence generator, the so-called shrinking generator, are just particular solutions of a kind of linear difference equations. Moreover, all these sequences are simple linear combinations of m-sequences weighted by other primary sequences. This fact establishes a subtle link between irregular decimation and linearity that can be conveniently exploited in the cryptanalysis of such sequence generators. These ideas can be easily extended to other decimation-based cryptographic generators as well as to interleaved sequence generators.

Keywords: decimated sequence, linear difference equation, stream cipher, cryptography.

1 Introduction

Symmetric key ciphers are usually divided into two large classes: stream ciphers and block-ciphers depending on whether the encryption function is applied either to each individual bit or to a block of bits, respectively.

At the present moment, stream ciphers are the fastest among the encryption procedures so they are implemented in many technological applications e.g. the encryption algorithm RC4 [17] used in Wired Equivalent Privacy (WEP) as a part of the IEEE 802.11 standards, the encryption function E0 in Bluetooth specifications [1] or the recent proposals HC-128 or Rabbit from the eSTREAM Project [20] that are included in the latest release versions of CyaSSL (lightweight open source embedded implementation of the SSL/TLS protocol) [21].

From a short secret key (known only by the two interested parties) and a public algorithm (the sequence generator), stream cipher procedures consist in generating a long sequence of seemingly random bits. Such a sequence is called the keystream sequence. For encryption, the sender performs the bit-wise (Exclusive-OR) XOR operation among the bits of the original message or plaintext and the

---

* This work was supported by CDTI (Spain) under Project Cenit-HESPERIA as well as by Ministry of Science and Innovation and European FEDER Fund under Project TIN2011-25452/TSI.

B. Murgante et al. (Eds.): ICCSA 2013, Part V, LNCS 7975, pp. 33–47, 2013.

keystream sequence. The result is the ciphertext to be sent. For decryption, the receiver generates the same keystream, performs the same bit-wise XOR operation between the received ciphertext and the keystream sequence and recovers the original message.

Most keystream generators are based on Linear Feedback Shift Registers (LFSRs) [8]. They are linear structures characterized by their length (the number of memory cells), their characteristic polynomial (the feedback function) and their initial state (the seed or key of the cryptosystem). Their output sequences, the so-called m-sequences, are combined by means of nonlinear functions to produce pseudorandom sequences of cryptographic application [4]. Combinational generators, nonlinear filters, clock-controlled generators or irregularly decimated generators are just some of the most popular keystream sequence generators [5], [9], [16].

Inside the family of irregularly decimated generators, the most representative example is the *shrinking generator* proposed by Coppersmith, Krawczyk and Mansour [2] that includes two different LFSRs. Extensions of this sequence generator are the self-shrinking generator [15] involving only one LFSR and the generalized self-shrinking generator described in [11]. Irregularly decimated generators produce good cryptographic sequences characterized by long periods, accurate correlation features, excellent run distribution, balancedness [6], simplicity of implementation, etc. The underlying idea of this kind of generators is the irregular decimation of a m-sequence according to the bits of another one. The result of this decimation process is a binary sequence that will be used as keystream sequence in the cryptographic procedure.

In this work, it is shown that the sequences generated by shrinking generators are particular solutions of homogeneous linear difference equations. In fact, all those sequences are just linear combinations of m-sequences weighted by binary binomial coefficients, thus cryptographic parameters of such sequences (e.g. period, linear complexity or balancedness) can be analyzed in terms of linear equation solutions. Other methods of studying such parameters are based on new techniques to handle efficiently Vandermonde matrices, see for instance [10], [18], [19].

It must be noticed that, although these sequences have been irregularly decimated, in practice they are simple solutions of linear equations. This fact establishes a subtle link between irregular decimation and linearity that could be conveniently exploited in cryptanalytic terms.

The paper is organized as follows. Notation and basic concepts are introduced in Section 2. The description of the shrinking generator is given in Sections 3. The kind of linear difference equations considered is introduced in section 4. In section 5, the output sequences of shrinking generators are presented as solutions of linear difference equations while a method of reconstruction of such sequences is developed in Section 6. Finally, conclusions in Section 7 end the paper.

2   Fundamentals and Basic Notation

In this section, we provide some basic notation and definitions that will be used throughout the paper.

Let p be a prime, $q = p^m$ where m is a positive integer, and let $GF(q)$ denote a finite field (Galois Field) with q elements. The order of an element $\alpha \in GF(q)$ denoted by $\mathrm{ord}(\alpha)$ is the smallest positive integer k such that $\alpha^k = 1$. An element α with order $q - 1$ is called a primitive element in $GF(q)$. The primitive elements are exactly the generators of $GF^*(q)$, the multiplicative group consisting of the nonzero elements of $GF(q)$. Thus, a finite field $GF(q)$ consists of 0 and appropriate powers of a primitive element.

Let $\{s_n\} = (s_0, s_1, s_2, \ldots)$ $n \geq 0$ be a sequence over $GF(p)$ if $s_n \in GF(p)$, $\forall n$. The sequence $\{s_n\}$ is periodic if and only if there exists an integer $T > 0$ such that $s_{n+T} = s_n$ holds for all $n \geq 0$.

Let r be a positive integer, and let $c_0, c_1, \ldots, c_{r-1}$ be given elements of the finite field $GF(p)$. A sequence $\{s_n\}$ of elements of $GF(p)$ satisfying the relation

$$s_{n+r} = c_1 s_{n+r-1} + c_2 s_{n+r-2} + \ldots + c_{r-1} s_{n+1} + c_r s_n, \qquad n \geq 0 \qquad (1)$$

is called a rth-order linear recurring sequence in $GF(p)$. The terms $s_0, s_1, \ldots, s_{r-1}$, which determine uniquely the rest of the sequence, are referred to as the initial values. A relation of the form given in (1) is called a rth-order homogeneous linear recurrence relation. The monic polynomial of degree r

$$f(x) = x^r + c_1 x^{r-1} + c_2 x^{r-2} + \ldots + c_{r-1} x + c_r \in GF(p)[x] \qquad (2)$$

is called the characteristic polynomial of the linear recurring sequence and the sequence $\{s_n\}$ is said to be generated by $f(x)$. The minimal polynomial of $\{s_n\}$ is the polynomial of least degree whose linear recurrence relation is satisfied by such a sequence. For a survey of linear recurring sequences over finite fields, the reader is referred to [13].

The generation of linear recurring sequences can be implemented on Linear Feedback Shift Registers (LFSRs). These devices with r memory cells handle information in the form of elements of $GF(p)$ and they are based on shifts and linear feedback. The output of the LFSR is the string of elements $(s_0, s_1, s_2, \ldots)$ received in intervals of one time unit. If the characteristic polynomial of the linear recurring sequence is primitive, then the LFSR is called maximal-length LFSR and its output sequence has period $2^r - 1$, see [8]. This output sequence is called m-sequence or maximal sequence.

The linear complexity (LC) of a sequence $\{s_n\}$ is defined as the length of the shortest LFSR that can generate such a sequence or equivalently the order of the shortest linear recurrence relation satisfied by such a sequence. In a general sense, linear complexity is related with the amount of sequence that is needed to determine the whole sequence. In cryptographic terms, linear complexity must be as large as possible [7]. The recommended value is approximately half the period $LC \simeq T/2$.

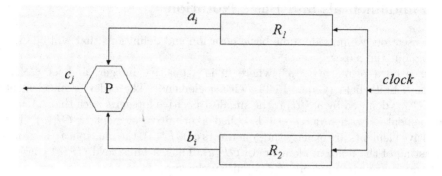

Fig. 1. General scheme of the shrinking generator

In the remaining of this paper, we will consider sequences defined exclusively over the binary field ($p = 2$ and $q = 2^m$) denoted by $GF(2)$ where the extension field will be denoted by $GF(2^m)$. It should be noticed that the analysis provided here can be extended to sequences over any prime extension.

3   A Cryptographic Generators Based on Decimation: The Shrinking Generator

The most important example of irregularly decimated sequence generator is next introduced.

The shrinking generator is a binary sequence generator [2] composed of two maximal-length LFSRs: a control register, called R_1, that decimates the sequence produced by the other register, called R_2, see Fig. 1. We denote by L_j ($j = 1, 2$) their corresponding lengths and by $P_j(x)$ ($j = 1, 2$) their corresponding characteristic polynomials [8].

The sequence produced by the LFSR R_1, that is $\{a_i\}$, controls the bits of the sequence produced by R_2, that is $\{b_i\}$, which are included in the output sequence $\{c_j\}$ (*shrunken sequence*), according to the following decimation rule denoted by P:

1. If $a_i = 1 \implies c_j = b_i$
2. If $a_i = 0 \implies b_i$ is discarded.

A simple example illustrates the behavior of this structure.

Example 1: Let us consider the following LFSRs:

1. R_1 of length $L_1 = 3$, characteristic polynomial $P_1(x) = 1 + x^2 + x^3$ and initial state $IS_1 = (1, 0, 0)$. The sequence generated by R_1 is $\{1, 0, 0, 1, 1, 1, 0\}$ with period $T_1 = 2^{L_1} - 1 = 7$.
2. R_2 of length $L_2 = 4$, characteristic polynomial $P_2(x) = 1 + x + x^4$ and initial state $IS_2 = (1, 0, 0, 0)$. The sequence generated by R_2 is $\{1, 0, 0, 0, 1, 0, 0, 1, 1, 0, 1, 0, 1, 1, 1\}$ with period $T_2 = 2^{L_2} - 1 = 15$.

The output sequence $\{c_j\}$ is given by:

- $\{a_i\} \rightarrow 1\,0\,0\,1\,1\,1\,0\,1\,0\,0\,1\,1\,1\,0\,1\,0\,0\,1\,1\,1\,0\,1$ .....
- $\{b_i\} \rightarrow 1\,\underline{0}\,\underline{0}\,0\,1\,0\,0\,1\,1\,\underline{0}\,1\,0\,1\,1\,1\,\underline{1}\,0\,0\,0\,1\,\underline{0}\,\underline{0}$ .....
- $\{c_j\} \rightarrow 1\,0\,1\,0\,1\,1\,0\,1\,1\,0\,0\,1\,0$ .....

According to rule P, the underlined bits $\underline{0}$ or $\underline{1}$ in $\{b_i\}$ are discarded. In brief, the shrunken sequence produced by the shrinking generator is an irregular decimation of the sequence generated by R_2 controlled by the bits of R_1.

It can be proved [2] that the period of the shrunken sequence is

$$T = (2^{L_2} - 1)2^{(L_1-1)}, \tag{3}$$

and its linear complexity LC satisfies the following inequality

$$L_2\,2^{(L_1-2)} < LC \leq L_2\,2^{(L_1-1)}. \tag{4}$$

A simple calculation, based on the fact that every state of R_2 coincides once with every state of R_1, allows one to compute the number of 1's in the shrunken sequence. Such a number is constant and equal to:

$$No.\ 1's = 2^{(L_2-1)}2^{(L_1-1)}. \tag{5}$$

Comparing (3) with (5), it can be deduced that the shrunken sequence is a quasi-balanced sequence.In addition, the output sequence has some nice distributional statistics too [2]. Therefore, this scheme is suitable for practical implementation of stream cipher cryptosystems and pattern generators.

4   Linear Difference Equations

In this section, the kind of linear difference equations we are dealing with will be introduced.

The linear recurrence relation given in (1) can be expressed as a linear difference equation

$$\left(E^r + \sum_{j=1}^{r} c_j\,E^{r-j}\right) z_n = 0, \qquad n \geq 0 \tag{6}$$

where $z_n \in GF(2)$ is the n-th term of a binary sequence $\{z_n\}$ that satisfies the previous equation, E is the shifting operator which operates on the terms z_n of a solution sequence, i.e. $E^j z_n = z_{n+j}$, the coefficients c_j are binary coefficients $c_j \in GF(2)$, r is an integer and the operations of (6) are the defined operations over $GF(2)$; namely, addition and multiplication modulo 2 (XOR and AND logic operations, respectively). The r-degree characteristic polynomial of (6) is

$$f(x) = x^r + \sum_{j=1}^{r} c_j\,x^{r-j}. \tag{7}$$

If $f(x)$ is an irreducible polynomial in $GF(2)[x]$ of degree r, then $f(x)$ has a root α in $GF(2^r)$. Furthermore, all the roots of $f(x)$ are simple and are given by the r distinct elements in the extension field ([13], Th. 2.14, pp. 49-50):

$$\alpha, \; \alpha^2, \; \alpha^{2^2}, \ldots, \; \alpha^{2^{(r-1)}} \in GF(2^r). \tag{8}$$

In this case, $A_0 \alpha^n$ is a solution of (6) where $A_0 \in GF(2^r)$ is an arbitrary constant. Since the polynomial (7) has r roots, there are r linearly independent solutions to (6), and the general solution is a linear combination of these solutions with r arbitrary constants $A_0, A_1, \ldots, A_{r-1} \in GF(2^r)$ determined by the initial values $z_0, z_1, \ldots, z_{r-1}$. Thus, the general solution can be written as

$$z_n = \sum_{j=0}^{r-1} A_j \, \alpha^{2^j n}, \qquad n \geq 0 \tag{9}$$

where $A_j = (A_{j-1})^2$ $(j = 1, 2, \ldots, r-1)$ for z_n to be in $GF(2)$, see [3]. Therefore, the equation (9) is simplified to

$$z_n = \sum_{j=0}^{r-1} A^{2^j} \, \alpha^{2^j n}, \qquad n \geq 0 \tag{10}$$

with $A \in GF(2^r)$.

Let us generalize the previous linear difference equation to a more complex kind of linear difference equation whose roots have a multiplicity greater than 1. In fact, we are going to consider difference equations of the form

$$(E^r + \sum_{j=1}^{r} c_j \, E^{r-j})^p \, z_n = 0, \qquad n \geq 0 \tag{11}$$

p being an integer $p > 1$. The characteristic polynomial $f_c(x)$ of this kind of equation is

$$f_c(x) = f(x)^p = (x^r + \sum_{j=1}^{r} c_j \, x^{r-j})^p. \tag{12}$$

In this case, the roots of $f_c(x)$ are the same as those of the polynomial $f(x)$, that is $\alpha, \alpha^2, \alpha^{2^2}, \ldots, \alpha^{2^{(r-1)}}$, but with multiplicity p. If α is a root of multiplicity p, then the expression $\binom{n}{i} A_i \, \alpha^n$ with $i = 0, 1, 2, \ldots, p-1$ provides the p linearly independent solutions of (11) associated with the root α, where $\binom{n}{i}$ is a binomial coefficient reduced modulo 2 and the arbitrary constants $A_i \in GF(2^r)$, see [12]. Therefore, the general solution of the equation (11) is a linear combination of the $p \cdot r$ independent solutions that can be written as follows

$$z_n = \sum_{i=0}^{p-1} \binom{n}{i} A_i \, \alpha^n + \sum_{i=0}^{p-1} \binom{n}{i} (A_i)^2 \, \alpha^{2n} + \ldots + \sum_{i=0}^{p-1} \binom{n}{i} (A_i)^{2^{r-1}} \, \alpha^{2^{r-1} n}, \tag{13}$$

where each term corresponds to the p independent solutions associated with the root α^{2^j} $(j = 0, 1, \ldots, r - 1)$, respectively. As before the same relation among arbitrary constants applies here for z_n to be in $GF(2)$. In a more compact way, the equation (13) can be written as

$$z_n = \sum_{i=0}^{p-1} \left(\binom{n}{i} \sum_{j=0}^{r-1} A_i^{2^j} \alpha^{2^j n} \right), \qquad n \geq 0 \tag{14}$$

where $A_0, A_1, \ldots, A_{p-1} \in GF(2^r)$.

In brief, the n-th term of a solution sequence $\{z_n\}$ of (11) is the addition of the n-th term of each one of the p sequences $\{\sum_{j=0}^{r-1} A_i^{2^j} \alpha^{2^j n}\}$ $(0 \leq i < p)$ weighted by binomial coefficients.

Table 1. Binomial coefficients, primary sequences and periods T_i

Binomial coeff.	Primary sequences	T_i
$\binom{n}{0}$	$S_0 = \{1, 1, 1, 1, 1, 1, 1, 1 \sim\}$	$T_0 = 1$
$\binom{n}{1}$	$S_1 = \{0, 1, 0, 1, 0, 1, 0, 1 \sim\}$	$T_1 = 2$
$\binom{n}{2}$	$S_2 = \{0, 0, 1, 1, 0, 0, 1, 1 \sim\}$	$T_2 = 4$
$\binom{n}{3}$	$S_3 = \{0, 0, 0, 1, 0, 0, 0, 1 \sim\}$	$T_3 = 4$
$\binom{n}{4}$	$S_4 = \{0, 0, 0, 0, 1, 1, 1, 1 \sim\}$	$T_4 = 8$
$\binom{n}{5}$	$S_5 = \{0, 0, 0, 0, 0, 1, 0, 1 \sim\}$	$T_5 = 8$
$\binom{n}{6}$	$S_6 = \{0, 0, 0, 0, 0, 0, 1, 1 \sim\}$	$T_6 = 8$
$\binom{n}{7}$	$S_7 = \{0, 0, 0, 0, 0, 0, 0, 1 \sim\}$	$T_7 = 8$
$\cdots$	$\cdots$	$\cdots$

Let us now analyze the binomial coefficients modulo 2. Indeed, when n takes successive values $n \geq 0$ each binomial coefficient $\binom{n}{i}$ $(n \geq i \geq 0)$ defines a *primary sequence* S_i with constant period T_i. In Table 1, the first binomial coefficients with their corresponding primary sequences and periods are depicted. From Table 1, it is immediate to notice that the generation of such sequences follows a simple formation rule. Indeed, the S_i primary sequence associated with $\binom{n}{i}$ for $(2^k \leq i < 2^{k+1})$ (k being an integer) has a period $T_i = 2^{k+1}$ and its digits are:

1. The first 2^k bits are 0's.
2. The other 2^k bits are the first 2^k bits of the primary sequence S_{i-2^k}.

According to this rule, primary sequences can be easily generated.

5 Shrunken Sequences as Solutions of Linear Difference Equations

Now the main results concerning shrunken sequences and linear difference equations are introduced.

Theorem 1. *The sequence generated by a shrinking generator, the so-called shrunken sequence, is a particular solution of the homogeneous linear difference equation:*

$$(E^{L_2} + \sum_{j=1}^{L_2} c_j \, E^{L_2-j})^p \, z_n = 0, \qquad p = 2^{L_1-1}, \tag{15}$$

whose characteristic polynomial is $f_c(x) = f(x)^p$ where $f(x)$ is the L_2-degree polynomial $(x+\alpha^N)(x+\alpha^{2N})\ldots(x+\alpha^{2^{(L_2-1)}N})$, L_2 is the length of the register R_2 and N is an integer given by $N = 2^0 + 2^1 + \ldots + 2^{L_1-1}$, L_1 being the length of the control register R_1.

Sketch of Proof: The shrunken sequence can be written as an interleaved sequence [9] made out of an unique m-sequence starting at different points and repeated $2^{(L_1-1)}$ times. Such a sequence is obtained from the sequence $\{b_i\}$ in register R_2 taking digits separated a distance $2^{L_1} - 1$, that is the period of the sequence $\{a_i\}$ in register R_1. See the general scheme of the shrinking generator in section 3. On the other hand, if $f_c(x)$ is the characteristic polynomial of the shortest linear recurrence relation satisfied by the shrunken sequence, then $f_c(x)$ is of the form:

$$f_c(x) = f(x)^{LC} \tag{16}$$

where LC is its linear complexity (the order of the shortest linear recurrence relation satisfied by such a sequence). At the same time, it is a well known fact that the linear complexity of a periodic sequence is lesser or equal than its period [8], [12]. Thus, for any shrunken sequence $LC \le 2^{L_1-1}$ and the polynomial of the shortest linear recurrence relation $f_c(x)$ divides the characteristic polynomial of (15). Therefore, the shrunken sequence satisfies the equation (15) and is a particular solution of this homogeneous linear difference equation.   □

Thus, the shrunken sequence is a linear combination of a unique m-sequence, which depends on the registers R_2 and R_1, weighted by binomial coefficients $\binom{n}{i}$. Let us see an illustrative example of this main result.

5.1 An Illustrative Example

Example 2: Let us consider the following shrinking generator characterized by:

1. R_1 of length $L_1 = 3$, characteristic polynomial $P_1(x) = 1 + x^2 + x^3$ and initial state $IS_1 = (1, 1, 1)$. The sequence generated by R_1 is $\{1, 1, 1, 0, 1, 0, 0\}$ with period $T_1 = 2^{L_1} - 1 = 7$.

2. R_2 of length $L_2 = 4$, characteristic polynomial $P_2(x) = 1 + x + x^4$ and initial state $IS_2 = (1, 1, 1, 1)$. The sequence generated by R_2 is $\{1, 1, 1, 1, 0, 1, 0, 1, 1\ 0, 0, 1, 0, 0, 0\}$ with period $T_2 = 2^{L_2} - 1 = 15$.

The shrunken sequence $\{c_n\}$ is given by:

- $\{a_i\} \rightarrow$ 1 1 1 0 1 0 0 1 1 1 0 1 0 0 1 1 1 0 1 0 0 1 1 1 0 1 0 0 1 1 .....
- $\{b_i\} \rightarrow$ 1 1 1 1 0 1 0 1 1 0 0 1 0 0 0 1 1 1 1 0 1 0 1 1 0 0 1 0 0 0 .....
- $\{c_j\} \rightarrow$ 1 1 1 0 1 1 0 1 0 1 1 1 0 1 1 0 0 0 .....

According to Theorem 1, $N = 2^0 + 2^1 + 2^2 = 7$, $p = 4$, $f(x) = 1 + x + x^4$ and $f_c(x) = (1 + x + x^4)^4$. Thus, the shrunken sequence satisfies the equation

$$c_n = \sum_{i=0}^{3} \left(\binom{n}{i} \sum_{j=0}^{3} A_i^{2^j}\, \alpha^{2^j n} \right), \qquad n \geq 0 \tag{17}$$

for particular values of the parameters A_i. In brief, for this example the shrunken sequence is the linear combination of 4 times the m-sequence whose characteristic polynomial is $f(x) = 1 + x + x^4$ starting at different points (determined by the corresponding coefficients A_i) and weighted by the binomial coefficients.

Table 2. Shrunken sequence as linear combination of m-sequences weighted by primary sequences

Shrunken seq. $\{c_n\}$	$\{S_0\}$	$\{S_1\}$	$\{S_2\}$	$\{S_3\}$	$\{M_0\}$	$\{M_1\}$	$\{M_2\}$	$\{M_3\}$
1	○	■	■	■	1	0	1	0
1	○	○	■	■	0	1	0	0
1	○	■	○	■	1	1	0	0
0	○	○	○	○	0	0	1	1
1	○	■	■	■	1	1	1	0
1	○	○	■	■	1	0	0	0
0	○	■	○	■	1	1	1	1
1	○	○	○	○	1	1	0	1
0	○	■	■	■	0	1	1	0
1	○	○	■	■	0	1	1	1
1	○	■	○	■	0	0	1	0
1	○	○	○	○	1	0	1	1
0	○	■	■	■	0	0	0	1
1	○	○	■	■	0	1	0	1
1	○	■	○	■	1	0	0	1
0	○	○	○	○	1	0	1	0

Table 2 synthesizes the decomposition of the shrunken sequence into the four shifted versions $\{M_i\}$ ($0 \leq i < 3$) of the same m-sequence as well as depicts the primary sequences $\{S_i\}$ ($0 \leq i < 3$) that bit-wise weight the corresponding

m-sequences. The symbol $\bigcirc$ represents the bit 1 while the symbol $\blacksquare$ represents the bit 0. In this way,

1. The first term of the shrunken sequence c_0 equals the first term of the m-sequence $\{M_0\}$.
2. The second term of the shrunken sequence c_1 corresponds to the XOR of the second terms of the m-sequences $\{M_0\}$ and $\{M_1\}$.
3. The third term of the shrunken sequence c_2 equals the XOR of the third terms of the m-sequences $\{M_0\}$ and $\{M_2\}$.
4. The fourth term of the shrunken sequence c_3 equals the XOR of the fourth terms of the m-sequences $\{M_0\}$, $\{M_1\}$, $\{M_2\}$ and $\{M_3\}$.
5. And so on for the successive terms of the shrunken sequence y their corresponding primary sequences $\{S_i\}$.

6 Reconstruction of Shrunken Sequences in Terms of Linear Difference Equations

Making use of the fact that the shrunken sequence is a particular solution of a linear difference equation and, consequently, a linear combination of p m-sequences, the shrunken sequence $\{c_n\}$ can be univocally reconstructed from the knowledge of a certain number of its bits. In fact, each m-sequence satisfies the same linear recurrence relation given by the characteristic polynomial $f(x)$. From the knowledge of the first 2^L bits $(c_0, c_1, \ldots, c_{2^L-1})$, the p m-sequences can be determined. Later, the bit-wise XOR of all the weighted m-sequences gives rise to the output sequence of the shrinking generator.

- From $c_0, c_p, c_{2p}, \ldots, c_{(p-1)p}$ and the linear recurrence relation, the sequence $\{M_0\}$ is determined.
- From $c_1, c_{p+1}, c_{2p+1}, \ldots, c_{(p-1)p+1}$ and $\{M_0\}$, the sequence $\{M_1\}$ is determined.
- From $c_2, c_{p+2}, c_{2p+2}, \ldots, c_{(p-1)p+2}$ and $\{M_0\}$, the sequence $\{M_2\}$ is determined.
- From $c_3, c_{p+3}, c_{2p+3}, \ldots, c_{(p-1)p+3}$, $\{M_0\}$, $\{M_1\}$ and $\{M_2\}$, the sequence $\{M_3\}$ is determined.
- $\ldots$ $\ldots$
- From $c_{p-1}, c_{2p-1}, c_{3p-1}, \ldots, c_{p^2-1}$, $\{M_0\}$, $\{M_1\}$, $\{M_2\}$, $\ldots$ and $\{M_{p-2}\}$, the sequence $\{M_{p-1}\}$ is determined.

The reconstruction method can be applied to *Example 2*.

In this case, $p = 4$, $(c_0, c_1, \ldots, c_{15}) = (1, 1, 1, 0, 1, 1, 0, 1, 0, 1, 1, 1, 0, 1, 1, 0)$ (see Table 2) and the *jth*-term of the sequence $\{M_i\}$ is denoted by m_j^i. Now we proceed as follows:

− *Computation of $\{M_0\}$:*
From $(c_0, c_4, c_8, c_{12}) = (1,1,0,0)$, we set $c_0 = m_0^0$, $c_4 = m_4^0$, $c_8 = m_8^0$ and $c_{12} = m_{12}^0$. Then, from these 4 terms of $\{M_0\}$ and the linear recurrence relation given by $f(x)$, that is $m_{n+4}^0 = m_{n+1}^0 + m_n^0$, $n \geq 0$, the sequence $\{M_0\}$ is determined. See Table 3.

− *Computation of $\{M_1\}$:*
From $(c_1, c_5, c_9, c_{13}) = (1,1,1,1)$, we set:

$$
\begin{array}{lll}
c_1 = m_1^0 + m_1^1 & 1 = 0 + m_1^1 & m_1^1 = 1 \\
c_5 = m_5^0 + m_5^1 & 1 = 1 + m_5^1 & m_5^1 = 0 \\
c_9 = m_9^0 + m_9^1 & 1 = 0 + m_9^1 & m_9^1 = 1 \\
c_{13} = m_{13}^0 + m_{13}^1 & 1 = 0 + m_{13}^1 & m_{13}^1 = 1
\end{array}
$$

Then, from these 4 terms of $\{M_1\}$ $(m_1^1, m_5^1, m_9^1, m_{13}^1) = (1,0,1,1)$ and the linear recurrence relation given by $f(x)$, the sequence $\{M_1\}$ is determined. See Table 4.

− *Computation of $\{M_2\}$:*
From $(c_2, c_6, c_{10}, c_{14}) = (1,0,1,1)$, we set:

$$
\begin{array}{lll}
c_2 = m_2^0 + m_2^2 & 1 = 1 + m_2^2 & m_2^2 = 0 \\
c_6 = m_6^0 + m_6^2 & 0 = 1 + m_6^2 & m_6^2 = 1 \\
c_{10} = m_{10}^0 + m_{10}^2 & 1 = 0 + m_{10}^2 & m_{10}^2 = 1 \\
c_{14} = m_{14}^0 + m_{14}^2 & 1 = 1 + m_{14}^2 & m_{14}^2 = 0
\end{array}
$$

Then, from these 4 terms of $\{M_2\}$ $(m_2^2, m_6^2, m_{10}^2, m_{14}^2) = (0,1,1,0)$ and the linear recurrence relation given by $f(x)$, the sequence $\{M_2\}$ is determined. See Table 5.

− *Computation of $\{M_3\}$:*
From $(c_3, c_7, c_{11}, c_{15}) = (0,1,1,0)$, we set:

$$
\begin{array}{lll}
c_3 = m_3^0 + m_3^1 + m_3^2 + m_3^3 & 0 = 0 + 0 + 1 + m_3^3 & m_3^3 = 1 \\
c_7 = m_7^0 + m_7^1 + m_7^2 + m_7^3 & 1 = 1 + 1 + 0 + m_6^3 & m_6^3 = 1 \\
c_{11} = m_{11}^0 + m_{11}^1 + m_{11}^2 + m_{11}^3 & 1 = 1 + 0 + 1 + m_{11}^3 & m_{11}^3 = 1 \\
c_{15} = m_{15}^0 + m_{15}^1 + m_{15}^2 + m_{15}^3 & 0 = 1 + 0 + 1 + m_{13}^3 & m_{15}^3 = 0
\end{array}
$$

Then, from these 4 terms of $\{M_3\}$ $(m_3^3, m_7^3, m_{11}^3, m_{15}^3) = (1,1,1,0)$ and the linear recurrence relation given by $f(x)$, the sequence $\{M_3\}$ is determined. See Table 6.

Once the m-sequences $\{M_i\}$ have been determined, the repetition of such sequence jointly with the primary sequence $\{S_i\}$ allows one to obtain the successive bits of the shrunken sequence.

Table 3. Determination of the m-sequence $\{M_0\}$

Shrunken seq. $\{c_n\}$	$\{S_0\}$	$\{S_1\}$	$\{S_2\}$	$\{S_3\}$	$\{M_0\}$	$\{M_1\}$	$\{M_2\}$	$\{M_3\}$
1	○	■	■	■	1	-	-	-
-	○	○	■	■	0	-	-	-
-	○	■	○	■	1	-	-	-
-	○	○	○	○	0	-	-	-
1	○	■	■	■	1	-	-	-
-	○	○	■	■	1	-	-	-
-	○	■	○	■	1	-	-	-
-	○	○	○	○	1	-	-	-
0	○	■	■	■	0	-	-	-
-	○	○	■	■	0	-	-	-
-	○	■	○	■	0	-	-	-
-	○	○	○	○	1	-	-	-
0	○	■	■	■	0	-	-	-
-	○	○	■	■	0	-	-	-
-	○	■	○	■	1	-	-	-
-	○	○	○	○	1	-	-	-

Table 4. Determination of the m-sequence $\{M_1\}$

Shrunken seq. $\{c_n\}$	$\{S_0\}$	$\{S_1\}$	$\{S_2\}$	$\{S_3\}$	$\{M_0\}$	$\{M_1\}$	$\{M_2\}$	$\{M_3\}$
-	○	■	■	■	-	0	-	-
1	○	○	■	■	0	1	-	-
-	○	■	○	■	-	1	-	-
-	○	○	○	○	-	0	-	-
-	○	■	■	■	-	1	-	-
1	○	○	■	■	1	0	-	-
-	○	■	○	■	-	1	-	-
-	○	○	○	○	-	1	-	-
-	○	■	■	■	-	1	-	-
1	○	○	■	■	0	1	-	-
-	○	■	○	■	-	0	-	-
-	○	○	○	○	-	0	-	-
-	○	■	■	■	-	0	-	-
1	○	○	■	■	0	1	-	-
-	○	■	○	■	-	0	-	-
-	○	○	○	○	-	0	-	-

Table 5. Determination of the m-sequence $\{M_2\}$

Shrunken seq. $\{c_n\}$	$\{S_0\}$	$\{S_1\}$	$\{S_2\}$	$\{S_3\}$	$\{M_0\}$	$\{M_1\}$	$\{M_2\}$	$\{M_3\}$
-	○	■	■	■	-	-	1	-
-	○	○	■	■	-	-	0	-
1	○	■	○	■	1	-	0	-
-	○	○	○	○	-	-	1	-
-	○	■	■	■	-	-	1	-
-	○	○	■	■	-	-	0	-
0	○	■	○	■	1	-	1	-
-	○	○	○	○	-	-	0	-
-	○	■	■	■	-	-	1	-
-	○	○	■	■	-	-	1	-
1	○	■	○	■	0	-	1	-
-	○	○	○	○	-	-	1	-
-	○	■	■	■	-	-	0	-
-	○	○	■	■	-	-	0	-
1	○	■	○	■	1	-	0	-
-	○	○	○	○	-	-	1	-

Table 6. Determination of the m-sequence $\{M_3\}$

Shrunken seq. $\{c_n\}$	$\{S_0\}$	$\{S_1\}$	$\{S_2\}$	$\{S_3\}$	$\{M_0\}$	$\{M_1\}$	$\{M_2\}$	$\{M_3\}$
-	○	■	■	■	-	-	-	0
-	○	○	■	■	-	-	-	0
-	○	■	○	■	-	-	-	0
0	○	○	○	○	0	0	1	1
-	○	■	■	■	-	-	-	0
-	○	○	■	■	-	-	-	0
-	○	■	○	■	-	-	-	1
1	○	○	○	○	1	1	0	1
-	○	■	■	■	-	-	-	0
-	○	○	■	■	-	-	-	1
-	○	■	○	■	-	-	-	0
1	○	○	○	○	1	0	1	1
-	○	■	■	■	-	-	-	1
-	○	○	■	■	-	-	-	1
-	○	■	○	■	-	-	-	1
0	○	○	○	○	1	0	1	0

7   Conclusions

In this work, it has been shown that the sequence obtained from shrinking generators are particular solutions of homogeneous linear difference equations with binary coefficients. This fact allows one:

1. To analyze the structural properties of the shrunken sequence from the point of view of the linear difference equations.
2. To reconstruct the whole shrunken sequence.

It must be noticed that, although shrunken sequences are generated from LF-SRs by means of irregular decimation, in practice they are simple solutions of linear equations. This fact establishes a subtle link between irregular decimation and linearity that can be conveniently exploited in the cryptanalysis of such keystream generators. A natural extension of this work is the generalization of this procedure to many other cryptographic sequences:

1. The sequences generated by the self-shrinking generator and the generalized self-shrinking generator as examples of decimation-based keystream generators.
2. The so-called interleaved sequences, as they present very similar structural properties to those of the sequences obtained from irregular decimation generators.

In brief, linear difference equations can contribute very efficiently to the cryptanalysis of keystream sequences in stream ciphers.

References

1. Bluetooth, Specifications of the Bluetooth system, Version 1.1, http://www.bluetooth.com/
2. Coppersmith, D., Krawczyk, H., Mansour, Y.: The Shrinking Generator. In: Stinson, D.R. (ed.) CRYPTO 1993. LNCS, vol. 773, pp. 22–39. Springer, Heidelberg (1994)
3. Dickson, L.E.: Linear Groups with an Exposition of the Galois Field Theory, pp. 3–71. Dover, New York (1958), An updated reprint can be found at http://www-math.cudenver.edu/~wcherowi/courses/finflds.html
4. eSTREAM, the ECRYPT Stream Cipher Project, Call for Primitives, http://www.ecrypt.eu.org/stream/
5. Fúster-Sabater, A., Caballero-Gil, P.: Strategic Attack on the Shrinking Generator. Theoretical Computer Science 409(3), 530–536 (2008)
6. Fúster-Sabater, A., Caballero-Gil, P., Delgado-Mohatar, O.: Deterministic Computation of Pseudorandomness in Sequences of Cryptographic Application. In: Allen, G., Nabrzyski, J., Seidel, E., van Albada, G.D., Dongarra, J., Sloot, P.M.A. (eds.) ICCS 2009, Part I. LNCS, vol. 5544, pp. 621–630. Springer, Heidelberg (2009)
7. Fúster-Sabater, A.: Generation of Pseudorandom Binary Sequences with Controllable Cryptographic Parameters. In: Murgante, B., Gervasi, O., Iglesias, A., Taniar, D., Apduhan, B.O. (eds.) ICCSA 2011, Part I. LNCS, vol. 6782, pp. 563–572. Springer, Heidelberg (2011)

8. Golomb, S.W.: Shift Register-Sequences. Aegean Park Press, Laguna Hill (1982)
9. Gong, G.: Theory and Applications of q-ary Interleaved Sequences. IEEE Trans. Information Theory 41(2), 400–411 (1995)
10. Lee, K., O'Sullivan, M.E.: List decoding of Hermitian codes using Gröbner bases. Journal of Symbolic Computation 44(12), 1662–1675 (2009)
11. Hu, Y., Xiao, G.: Generalized Self-Shrinking Generator. IEEE Trans. Inform. Theory 50, 714–719 (2004)
12. Key, E.L.: An Analysis of the Structure and Complexity of Nonlinear Binary Sequence Generators. IEEE Trans. Informat. Theory 22(6), 732–736 (1976)
13. Lidl, R., Niederreiter, H.: Introduction to Finite Fields and Their Applications. Cambridge University Press, Cambridge (1986)
14. Massey, J.L.: Shift-Register Synthesis and BCH Decoding. IEEE Trans. Informat. Theory 15(1), 122–127 (1969)
15. Meier, W., Staffelbach, O.: The Self-Shrinking Generator. In: De Santis, A. (ed.) EUROCRYPT 1994. LNCS, vol. 950, pp. 205–214. Springer, Heidelberg (1995)
16. Menezes, A.J., et al.: Handbook of Applied Cryptography. CRC Press, New York (1997)
17. Rivest, R.L.: The RC4 Encryption Algorithm. RSA Data Sec., Inc. (March 1998)
18. Respondek, J.S.: On the confluent Vandermonde matrix calculation algorithm. Applied Mathematics Letters 24(2), 103–106 (2011)
19. Respondek, J.S.: Numerical recipes for the high efficient inverse of the confluent Vandermonde matrices. Applied Mathematics and Computation 218(5), 2044–2054 (2011)
20. Robshaw, M., Billet, O. (eds.): New Stream Cipher Designs. LNCS, vol. 4986. Springer, Heidelberg (2008)
21. Yet Another SSL (YASSL), http://www.yassl.com

An Incremental Linear Programming Based Tool for Analyzing Gene Expression Data

Satish Chandra Panigrahi, Md. Shafiul Alam, and Asish Mukhopadhyay*

School of Computer Science, University of Windsor
401 Sunset Avenue, Windsor, ON, Canada
{panigra,alam9}@uwindsor.ca,
asishm@cs.uwindsor.ca

Abstract. The availability of large volumes of gene expression data from microarray analysis (cDNA and oligonucleotide) has opened a new door to the diagnoses and treatments of various diseases based on gene expression profiling. In this paper, we discuss a new profiling tool based on linear programming. Given gene expression data from two subclasses of the same disease (e.g. leukemia), we are able to determine efficiently if the samples are linearly separable with respect to triplets of genes. This was left as an open problem in an earlier study that considered only pairs of genes as linear separators. Our tool comes in two versions - offline and incremental. Tests show that the incremental version is markedly more efficient than the offline one. This paper also introduces a gene selection strategy that exploits the class distinction property of a gene by separability test by pairs and triplets. We applied our gene selection strategy to 4 publicly available gene-expression data sets. Our experiments show that gene spaces generated by our method achieves similar or even better classification accuracy than the gene spaces generated by t-values, FCS(Fisher Criterion Score) and SAM(Significance Analysis of Microarrays).

Keywords: Gene expression analysis, DNA microarrays, linear separation, tissue classification.

1 Introduction

The availability of large volumes of gene expression data from microarray analysis (cDNA and oligonucleotide) has opened a new door to the diagnoses and treatments of various diseases based on gene expression profiling.

In a pioneering study, Golub et al [1] identified a set of 50 genes that can distinguish an unknown sample with respect to 2 kinds of leukemia with a low classification error rate. Following this work, other researchers attempted to replicate this effort in the diagnoses of other diseases. There were several notable successes. van't Veer *et al.* [2] found that 231 genes are significantly related to breast cancer. Their FDA approved MammaPrint uses 70 genes as biomarkers

* This research is supported by an NSERC Discovery Grant.

B. Murgante et al. (Eds.): ICCSA 2013, Part V, LNCS 7975, pp. 48–64, 2013.
© Springer-Verlag Berlin Heidelberg 2013

to predict the relapse of breast-cancer in patients whose condition has been detected early [2]. Khan *et al.* [3] found 96 genes to classify small, round, blue-cell cancers. Ben-Dor *et al.* [4] used 173-4,375 genes to classify various cancers. Alon *et al.* [5] used 2,000 genes to classify colon cancers.

A major bottleneck with any classification scheme based on gene expression data is that while the sample size is small, numbering in hundreds, the feature space is much larger, running into tens of thousands of genes. Using too many genes as classifiers results in over-fitting, while using too few leads to under-fitting. Thus the main difficulty of this effort is one of scale: the number of genes is much larger than the number of samples. The consensus is that genes numbering between 10 and 50 may be sufficient for good classification [1, 6].

In [7], Computational Geometry tools were used for testing the linear separability of gene expression data by pairs of genes. Applying their tool to 10 different publicly available gene-expression data-sets, they determined that 7 of these are highly separable. From this they inferred that there might be a functional relationship "between separating genes and the underlying phenotypic classes". Their method of linear separability, applicable to pairs of genes only, checks for separability incrementally. For separable datasets, the running time is quadratic in the sample size m.

In [8], we proposed a different geometric tool for testing the separability of gene expression data sets. This is based on a linear programming algorithm of Megiddo [9–11] that can test linear separability with respect to a fixed set of genes in time proportional to the size of the sample set. In this paper, we extend this work to testing separability with respect to triplets of genes. Since most gene sets do not separate the sample expression data, we have proposed and implemented an incremental version of Megiddo's scheme that terminates as soon as linear inseparability is detected. The usefulness of such incremental algorithm to detect inseparability in gene expression dataset is also observed by [7]. The performance of the incremental version turned out to be better than the offline version when we tested the separability of 5 different data sets by pairs/triplets of genes. In the paper we have also conclusively demonstrated that linear separability can be put to good use as feature for classification.

The major contributions of this paper can be summarized as follows:

1. An offline adaptation of Megiddo's algorithm to test separability by gene pairs/triplets, fully implemented and tested.
2. An incremental version of Megiddo's algorithm that is particularly useful for gene expression datasets, fully implemented and tested.
3. Demonstration of the usefulness of linear separability as a tool to build a good classifier with application to concrete examples.

For the completeness of the paper, in the following section we briefly discuss about LP formulation of separability [8].

2   LP Formulation of Separability

We have m samples, m_1 from a cancer type C_1 and $m_2(= m - m_1)$ from a cancer type C_2 (for example, m_1 from ALL and m_2 from AML [1]). Each sample is a point in a d-dimensional Euclidean space, whose coordinates are the expression values of the samples with respect to the d selected genes. This d-dimensional space is called the *primal space*. If a hyperplane in this primal space separates the sample-points of C_1 from those of C_2, then the test group of genes is a linear separator and the resulting linear program in dual space has a feasible solution.

Suppose there is a separating hyperplane in primal space and, say, the sample points of C_1 are above this plane, while the sample points of C_2 are below (Fig. 1(a) is a 2-dimensional illustration of this). Fig. 1(b) shows that the separating line maps to a point inside a convex region. The set of all points inside this convex region make up the feasible region of a linear program in dual space and correspond to all possible separating lines in primal space.

Thus there is a separating hyperplane in primal space if the resulting linear program in dual space has a feasible solution. Note, however, that we will have to solve 2 linear programs since it is not known a priori if the m_1 samples of C_1 lie above or below the separating hyperplane H.

Formally, one of these linear programs in d-dimensional dual space $(u_1, u_2, \ldots, u_d)$ is shown below:

$$\text{minimize } u_d$$

$$
\begin{aligned}
p_1^i u_1 + \ldots + p_{d-1}^i u_{d-1} - u_d - p_d^i < 0, \ i = 1, ..., m_1 \\
p_1'^i u_1 + \ldots + p_{d-1}'^i u_{d-1} - u_d - p_d'^i > 0, \ i = 1, ..., m_2
\end{aligned}
\tag{1}
$$

where $(p_1^i, p_2^i, \ldots, p_d^i)$ is the i-th sample point from C_1, and the first set of m_1 linear inequalities express the conditions that these sample points are above the separating plane, while the second set of m_2 linear inequalities express the conditions that the sample points $(p_1'^i, p_2'^i, \ldots, p_d'^i)$ from C_2 are below this plane. The linear inequalities above that describe the linear program are called constraints, a term that we shall also use from now on.

Megiddo in [9, 10] and Dyer in [11] both proposed an ingenious prune-and-search technique for solving the above linear program that, for fixed d (dimension of the linear program), takes time linear in (proportional to) the number of constraints. Over the next two sections, we discuss how the above LP-framework achieve the more limited goal of testing separability of the samples from the two input classes by an offline algorithm and an incremental one, both of which are based on an adaptation Megiddo's (and Dyer's) technique.

This approach is of interest for two reasons: (a) in contrast to the algorithm of Unger and Chor [7], the *worst case* running time of this algorithm is linear in the sample size; and (b) in principle it can be extended to study the separability of the sample classes with respect to *any number of genes*.

In what follows, we adopt a coloring scheme to refer to the points that represent the samples: those in the class C_1 are colored blue and make up the set S_B, while those in C_2 are colored red and make up the set S_R.

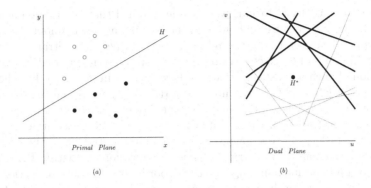

Fig. 1. A separating line H in primal space is a feasible solution H^* in dual space

3   Offline Approach

As Megiddo's algorithm is central to our discussion, we briefly review this algorithm for $d = 2$ and refer the reader to [10] for the cases $d \geq 3$. A bird's eye view is this: in each of $\log m$ iterations it prunes away at least a quarter of the constraints that do not determine the optimum (minimum in our formulation), at the same time reducing the search space (an interval on the u-axis) in which the optimal solution lies.

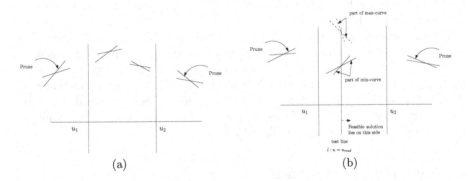

Fig. 2. (a) Pruning Constraints (b) Testing Feasibility

Definition: If $f_1, f_2, ..., f_n$ is a set of real single-valued functions defined in an interval $[a, b]$ on the real line, their point wise minimum (maximum) is another function f such that for every $x \in [a, b]$
$$f(x) = min(f_1(x), f_2(x), ..., f_n(x))\ (f(x) = max(f_1(x), f_2(x), ..., f_n(x)))$$
Let us call the point wise minimum (maximum) of the heavy (light) lines in Fig.1(b) the *min-curve (max-curve)*. These are also called the upper and lower envelope respectively.

Assume that after i iterations we have determined that the minimum lies in the interval $[u_1, u_2]$. Let us see how to prune redundant constraints from the set of constraints that determine the min-curve. We make an arbitrary pairing of the bounding lines of these constraints. With respect to the interval $[u_1, u_2]$, the intersection of such a constraint pair can lie as shown in Fig. 2(a). For the ones that lie to the left (right) of the line $u = u_1 (u = u_2)$, we prune the constraint whose bounding line has larger (smaller) slope. We can likewise prune redundant constraints from the set of constraints that determine the max-curve.

In order to further narrow down the interval on the u-axis where the minimum lies, for all other pairs of constraints whose intersections lie within the interval $[u_1, u_2]$, we find the median u_{med} of the u-coordinates of the intersections and let $l : u = u_{med}$ be the line with respect to which we test for the location of the minimum. We do this test by examining the intersections of the min-curve and max-curve with l. This is accomplished by using the residual constraint sets that implicitly define the min-curve and max-curve. From the relative positions of these intersections and the slopes of the bounding lines of the constraints that determine these intersections, we can determine on which side of l, the minimum lies (see Fig. 2(b)). Next we prune a constraint from each pair whose intersections lie within $[u_1, u_2]$ but on the side opposite to which the minimum lies. Because of our choice of the test-line, we are guaranteed to throw a quarter of the constraints from those that determine these intersections. We now reset the interval that contains the minimum to $[u_{med}, u_2]$ or $[u_1, u_{med}]$.

The above algorithm allows us to determine feasibility as soon as we have found a test line such that the intersection of the min-curve with l lies above its intersection with the max-curve.

We have implemented the above offline algorithm both in 2 and 3 dimensions from scratch. Ours is probably the first such implementation in 3 dimensions.

Offline algorithms are effective for determining linear separability. However, as most of the gene-pairs and gene-triplets are not linearly separating, incremental algorithms would be more efficient than offline ones. This was also observed by Unger and Chor [7]. In view of this, in the next section we discuss in details an incremental version of the above algorithm.

4   Incremental Approach

The following obvious but useful theorem (true in any dimension $d \geq 1$) underlies our algorithm in dual space.

Theorem 1. *Let S'_B and S'_R be arbitrary subsets of S_B and S_R respectively. If S'_B and S'_R are linearly inseparable, then so are S_B and S_R.*

Proof. Straightforward, since if S_B and S_R are linearly separable, then so are S'_B and S'_R.

First, we choose a small constant number of lines from each of the duals of S_R and S_B, and use the offline approach of the previous section to determine if there is a feasible solution to this constant-size problem. If not, we declare infeasibility (Theorem 1) and terminate. Otherwise, we have an initial feasible region and a test-line $l : u = \overline{u}$.

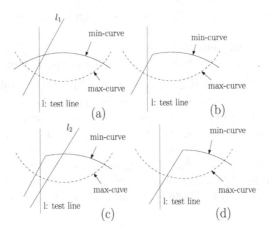

Fig. 3. *Updation of min-curve (a) addition of line l_1 (b) test line continues to pass through feasible region on updating of min-curve (c) addition of line l_2 (d) test line goes out of feasible region on updating of min-curve*

We continue, adding a line from one of the residual sets S_A^* or S_B^*, also chosen randomly. Several cases arise. This line (a) either becomes a part of the boundary of the feasible region, or (b) leaves it unchanged or (c) establishes infeasibility, in which case the algorithm terminates. Case(a) spawns two sub-cases as shown in Fig. 3. (a.1) the test line l still intersects the feasible region. (a.2) the test line l goes outside the feasible region. The lines belonging to the case (b) always leads to a condition mentioned in (a.1).

If the test line l goes outside the feasible region, constraint-pruning is triggered. This consists of examining pairs of constraints whose intersections lie on the side of l that does not include the feasible region. One of the constraints of each such pair does not intersect the feasible region and is therefore eliminated from further consideration. However, if l lies inside the new feasible interval, we continue to add new lines.

If we are able to add all lines without hitting case (c), then we have a feasible region and hence a separating line in the primal space.

A formal description of the iterative algorithm is as below.

Algorithm. IncrementallySeparatingGenepairs
Input: Line duals S_R^* and S_B^* of the point sets S_R and S_B.
Output: LP feasible or infeasible.

1: Choose $S_R^{*'} \subset S_R^*$ and $S_B^{'*} \subset S_B^*$ so that $\left| S_R^{*'} \right| = \left| S_B^{*'} \right| = 2$.

2: Apply the offline approach to $S_R^{*'}$ and $S_B^{*'}$, distinguishing between the following cases:

Case 1: If infeasible then report this and halt.

Case 2: If feasible then return the vertical test line l and continue with Step 3.

3: Repeatedly add a line from $\left| S_R^* - S_R^{*'} \right|$ or $\left| S_B^* - S_B^{*'} \right|$ until no more lines remain to be added or there exists no feasible point on the test line l.

4: If there is a feasible point on the test line l we report separability and halt.

5: If there is no feasible point on the test line l, determine on which side of l the feasible solution lies.

6: Update $S_R^{*'}$ and $S_B^{*'}$ by eliminating a line from each pair whose intersection does not lie in the feasible region and was earlier used to determine l.

7: Update $S_R^{*'}$ and $S_B^{*'}$ by including all those lines added in step 3 and go to Step 2.

When S_R and S_B are linearly separable the running time of the incremental algorithm is linear in the total number of inputs. Otherwise, as the algorithm terminates when a line added that reveals inseparability, the time complexity for this case is linear in the number of lines added so far.

In an Appendix we provide the pseudo code for the extension of this incremental algorithm to 3 dimensions.

Theorem 2. *If m is the total number of samples then time complexity of the incremental algorithm is $O(m)$.*

Proof. In each iteration, the algorithm prunes one quarter of the constraints (i.e. samples) from the current set $S_R^{*'} \cup S_B^{*'}$. The time complexity of each iteration is $O(m)$. The run-time $T(m)$ satisfies the recurrence $T(m) = O(m) + T(\frac{3m}{4})$, whose solution is $T(m) = O(m)$.

5   Gene Selection

Gene selection is an important preprocessing step for the classification of gene expression dataset. This helps (a) to reduce the size of the gene expression dataset and improve classification accuracy; (b) to cut down the presence of noise in the gene expression dataset by identifying informative genes; and (c) to improve the computation by removing irrelevant genes that not only add to the computation time but also make classification harder.

5.1   Background

In this subsection we briefly discuss some popular score functions used for gene selection. We compare these with our gene selection method, proposed in the next section. A simple approach to feature selection is to use the correlation between gene expression values and class labels. This method was first proposed by Golub et al [1]. The correlation metric defined by Nguyen and Rocke [12] and by Golub et al [1] reflects the difference between the class mean relative to standard deviation within the class. High absolute value of this correlation metric favors those genes that are highly expressed in one class as compared to the other class, while their sign indicates the class in which the gene is highly expressed. We have chosen to select genes based on a t-statistic defined by Nguyen and Rocke [12].

For ith gene, a t-value is computed using the formula

$$t_i = \frac{\mu_1^i - \mu_2^i}{\sqrt{\frac{\sigma_1^{i\,2}}{n_1} + \frac{\sigma_2^{i\,2}}{n_2}}} \tag{2}$$

where n_k, μ_k^i and $\sigma_k^{i\,2}$ are the sample size, mean and variance of ith gene respectively of class $k = 1, 2$.

Another important feature selection method is based on the Fisher Score [13] [14]. The Fisher Score Criterion(FCS) for ith gene can be defined as

$$F_i = \frac{n_1(\mu_1^i - \mu^i)^2 + n_2(\mu_2^i - \mu^i)^2}{n_1(\sigma_1^i)^2 + n_2(\sigma_2^i)^2} \tag{3}$$

where n_k, μ_k^i and $\sigma_k^{i\,2}$ are the sample size, mean and variance of ith gene respectively of class $k = 1, 2$. μ^i represents mean of the ith gene.

Significance Analysis of Microarrays (SAM) proposed by Tusher et al [15] is another important gene filter technique for finding significant genes in a set of microarray experiments. The SAM score for each gene can be defined as

$$M_i = \frac{\mu_2^i - \mu_1^2}{s_i + s_0} \tag{4}$$

For simplicity the correcting constant s_0 is set to 1 and s_i is computed as follows

$$s_i = \left[\left(\frac{1}{n_1} + \frac{1}{n_2} \right) \frac{\left\{ \sum_{j \in C_1}(x_j^i - \mu_1^i)^2 + \sum_{j \in C_2}(x_j^i - \mu_2^i)^2 \sum_{j \in C_2} \right\}}{(n_1 + n_2 - 2)} \right]^{\frac{1}{2}} \tag{5}$$

where x_j^i is jth sample of ith gene. The classes 1 and 2 are represented by C_1 and C_2. Similarly n_k, μ_k^i and $\sigma_k^{i\,2}$ are the sample size, mean and variance of ith gene respectively of class $k = 1, 2$. For the purpose of generating significant genes by SAM we have used the software written by Chu et al [16] which is publicly available at http://www-stat.stanford.edu/ tibs/clickwrap/sam/academic.

6   A New Methodology for Gene Selection

To find a set of genes of suitable size that is large enough to be robust against noise and small enough to be applied to the clinical setting, we propose a simple gene selection strategy based on an individual gene ranking approach. This consists of two steps: *coarse filtration*, followed by *fine filtration*.

6.1   Coarse Filtration

The purpose of coarse filtration is to remove most of the attributes that contribute to noise in the gene expression dataset. This noise can be categorized into *(i) biological noise* and *(ii) technical noise* [17]. Biological noise refers to the genes in gene expression dataset that are irrelevant for classification. Technical noise refers to errors incurred at various stages during data preparation.

For coarse filtration we follow an established approach based upon t-metric discussed in the previous section. Following a general consensus [1, 6], we chose to select a sufficient number genes that can be further considered for fine filtration. This is a set of 100 genes obtained by taking 50 genes with the largest positive t-values and another 50 genes with the smallest negative t-values.

6.2   Fine Filtration

One of the problems with the above correlation metric is that the t-value is calculated from the expression values of a single gene, ignoring the information available from the other genes. To rectify this, we propose the following scheme.

Let $\Delta = \{g_1, g_2, ..., g_n\}$ be a gene expression dataset with n genes. For a gene $g_i \in \Delta$, let $S_i = \{g_j | (g_i, g_j) \text{ is an LS(Linearly Separable) pair}, g_j \in \Delta \text{ and } i \neq j\}$. In words, S_i consists of all genes that form linearly separable pairs with g_i. For each gene $g_i \in \Delta$, its P_i-value is set to be $P_i = |S_i|$.

The intuition underlying the above definition is that the informative genes have quite different expression values in the two classes. If such genes exist in the gene expression data set then the above ranking strategy will assign the highest rank to those genes.

A drawback of this gene selection method is that it is applicable only to those gene expression datasets that have linearly separable pairs. For those datasets that have few linearly separable pairs, such as Lung Cancer [18] and Breast Cancer [2], we can extend the definition, using linearly separable gene triplets.

For a gene $g_i \in \Delta$, set $Q_i = \{(g_j, g_k) | (g_i, g_j, g_k) \text{ is an LS(Linearly Separable)}$ triplet, $g_j, g_k \in \Delta$, and $i \neq j \neq k\}$, In words, Q_i consists of all gene-pairs (g_j, g_k) that make up a linearly separable triplet with the gene g_i. For each gene $g_i \in \Delta$, define $T_i = |Q_i|$. Clearly, T_i lies between 0 and $^{n-1}C_2$.

7   Results and Discussions

In this paper we have developed an offline as well as an incremental version of a geometric tool to test linear separability of pairs and triplets of genes, followed

Table 1. Five Gene Expression Datasets

	Dataset	No. of Genes	Total Samples
1.	Lung Cancer [18]	12533	181(31+150)
2.	Leukemia [19]	12582	52(24+28)
3.	SRBCT [3]	2308	43(23+20)
4.	Colon [5]	2000	62(40+22)
5.	Breast Cancer [2]	21682	77(44+33)

by a simple gene selection strategy that uses this tool to rank the genes within a gene expression dataset. Based upon this ranking, we choose a suitable number of top-scoring genes to construct a good classifier.

We demonstrate the usefulness of the proposed methodology by testing with five publicly available gene expression datasets: (a) Lung Cancer [18] (b) Leukemia Data [19] (c) SRBCT [3] (d) Colon Data [5] (e) Breast Cancer [2]. Table 1 shows number of samples belong to different datasets and the number of samples from each class appear in the parenthesis.

The 100 genes that we select from each of these datasets in the *Coarse Filtration* step effectively prunes away most of the attributes(genes) that are irrelevant for classification. On the other hand, this number is large enough to provide us with a number attributes(genes) that may be over fitting for classifier construction.

To get the best subset of genes for good classification we chose to populate the attribute space with 5, 10, 15, 25 and 30 genes from each dataset by applying *Fine Filtration*. The choices of these attribute/feature-space sizes are somewhat arbitrary but the chosen attribute/feature-spaces are sufficiently large in comparison to the size of the sample spaces as Table 1 shows.

The computational time of the *Fine Filtration* step depends upon the geometric tool that we use to check the separability of gene expression data. In this paper, we have presented linear time incremental algorithms for both gene pairs and gene triplets. In order to illustrate the effectiveness of this approach we ran both versions (offline and incremental) on each of the five datasets obtained by *Coarse Filtration*. The computing platform was a Dell inspiron 1545 model-Intel Core2 Duo CPU, 2.00 GHz and 2 GB RAM, running under Windows Vista. The run-time efficiency of the incremental version over the offline one is evident from Table 2.

A group of genes that is being tested for linear separability may include a gene that is a perfect 1-D separator with TNoM score zero, using the terminology of [4]. In this case, such a group will provide a positive separability test. In order to exclude such groups, we checked for the existence of such 1-D separators, and found that no such genes exists in the above datasets. Likewise, if a gene pair shows linear separability then all gene triplets that include these gene pairs will also be linearly separating. In order to count gene triplets that exhibit pure 3-D linear separability, we avoid testing gene triplets that include a linearly separable pair. Thus our 3-D test results shown here include only such gene triplets. We call such gene triplets as Perfect Linearly Separable

Table 2. 2-D and 3-D Separability Test with Runtime

| Dataset | 2-D Separability Test with RT | | | | 3-D Separability Test with RT | | | |
	% LSP	RT of offline in msec	RT of incr. in msec	Impr. of incr. over offline	% PLST	RT of offline in msec	RT of incr. in msec	Impr. of incr. over offline
Lung Cancer	0.72%	4617	1537	66.71%	0.946%	1114467	166662	85.04%
Leukemia	11.92%	1138	418	63.27%	3.72%	115791	49263	57.45%
SRBCT	8.86%	987	356	63.93%	4.11%	92825	52080	43.89%
Colon	0%	1328	275	79.29%	0%	170143	39955	76.516
Breast Cancer	0.93%	1606	440	72.6%	0.137%	274141	83913	69.39%

Abbreviations RT: Run Time, Incr.: Incremental, Impr.: Improvement

Triplet(*PLST*). The percentage of Linearly Separable Pairs(*LSP*), Linearly Separable Triplets(*LST*) and Perfect Linearly Separable Triplets(*PLST*) are calculated using the formulas below.

$$\% \ of \ LSP = \frac{\# \ of \ LSP}{Total \ possible \ LSP} \times 100 = \frac{\# \ of \ LSP}{^2C_n} \times 100 \qquad (6)$$

$$\% \ of \ LST = \frac{\# \ of \ LST}{Total \ possible \ LST} \times 100 = \frac{\# \ of \ LST}{^3C_n} \times 100 \qquad (7)$$

$$\% \ of \ PLST = \frac{\# \ of \ PLST}{Total \ possible \ LST - ((\# \ of \ LSP) \times (n-2))} \times 100$$
$$= \frac{\# \ of \ PLST}{^3C_n - ((\# \ of \ LSP) \times (n-2))} \times 100 \qquad (8)$$

where n is total number genes in the gene expression data set.

The above formulas show that the total number of triplets relative to the *PLST*s is much higher than the total number of pairs relative to the *LSP*s. Thus the increase in the actual number of *PLST* over the number of *LSP* is suppressed by the high value of the denominator in the former case. The separability test shows that Colon Data [5] has neither any *LSP* nor any *PLST*. The Lung Cancer [18] and Breast Cancer [2] datasets have a few *LSP*, whereas the number of *PLST* is respectively 41 and 5 times (approximately) the number of *LSP*. The Leukemia Data [19] and SRBCT [3] show a good number of *LSP*, while the number of *PLST* is respectively 6 and 11 times (approximately) the number of *LSP*.

The motive underlying our gene selection strategy is to identify if a gene, jointly with some other genes, has the class distinction property or not. In the current study, we identify the class distinction property by separability tests where we restricted the group size to pairs and triplets. As the Colon Data [5] did not show any positive separability result we continued our study with the remaining four gene expression datasets. This result in Colon Data [5] is not surprising at all since according to Alon et al. [5] some samples such as T2, T30, T33, T36, T37, N8, N12, N34 in Colon Data have been identified as outliers and presented with anomalous muscle-index. This confirms the uncertainty of these samples.

To continue, in the *Fine Filtration* stage we use the incremental version of our algorithm to test separability by gene pairs and assign a P_i value to a gene $g_i \in \Delta$. Based on the ranking, we choose a set of top-scoring genes to populate five different feature spaces of size 5, 10, 15, 20, 25 and 30. If more than one gene have same rank then we choose an arbitrary gene from that peer group. To compare our method with other selection methods such as t-metric, FCS and SAM, we populate similar feature spaces respectively.

For classification we used machine learning tools supported by WEKA version 3.6.3 [20]. We used the following two classifiers :(a) *Support Vector Classifier*: WEKA SMO class implements John C. Platt's [21] sequential minimal optimization algorithm for training a support vector classifier. We used a linear kernel. (b) *Bayes Network Classifier*: Weka BayesNet class implements Bayes Network learning using various search algorithms and quality measures [22]. We have chosen Bayes Network classifier based on $K2$ for learning structure [23]. Both of the above classifiers normalized the attributes by default to provide a better classification result. We used a 10-fold cross-validation[24] for prediction. As suggested by Kohavi [24] we have used ten-fold stratified cross-validation. In stratified cross-validation the folds are stratified so that they contain approximately same proportions of labels as original datasets.

A comparative classification accuracy of the feature spaces generated from P-values, t-values, FCS and SAM is shown in Fig. 4. The results clearly show that the gene spaces generated by P-values yields a good classifier. Specifically, the feature spaces of sizes 10, 15, 20, 25 and 30 generated by the P-values perform mostly better than or as good compared to the feature spaces generated by the t-values, FCS and SAM.

To illustrate the performance of the classifiers with respect to the feature spaces generated by the T-values we considered two datasets with few *LSP*, such as Lung Cancer [18] and Breast Cancer [2]. To make sure that the dataset has no *LSP* we removed all genes that are responsible for pair separability in feature the selection process. Then feature spaces of size 5, 10, 15, 20, 25 and 30 are populated based upon the T-values. The classification results are shown in Fig. 4. It is interesting to note that the feature space generate from lung Cancer [18] dataset by T-values achieves similar or even better classification accuracy as compared to t-values, FCS and SAM. In Fig. 5 we have shown the classification accuracy of feature space confined to 25 and 30.

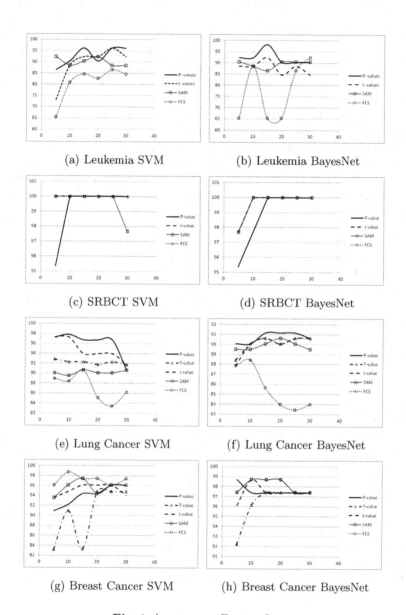

(a) Leukemia SVM   (b) Leukemia BayesNet

(c) SRBCT SVM   (d) SRBCT BayesNet

(e) Lung Cancer SVM   (f) Lung Cancer BayesNet

(g) Breast Cancer SVM   (h) Breast Cancer BayesNet

Fig. 4. Accuracy vs Feature Space

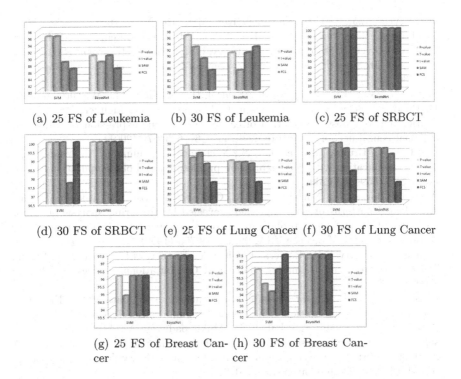

(a) 25 FS of Leukemia   (b) 30 FS of Leukemia   (c) 25 FS of SRBCT

(d) 30 FS of SRBCT   (e) 25 FS of Lung Cancer (f) 30 FS of Lung Cancer

(g) 25 FS of Breast Can- (h) 30 FS of Breast Cancer   cer

Fig. 5. Classifier Accuracy of gene expression dataset on 25 and 30 Feature Space(FS)

8   Conclusions

We presented a gene selection strategy to achieve a high classification accuracy. The gene selection strategy exploits the class distinguishing property of genes by testing separability by pairs and triplets. To test for separability we have provided two versions of a linear time algorithm, and demonstrated that the run-time of the incremental version is markedly better than that of the offline version. The importance of the given method lies in the fact that it can be easily extended to higher dimensions, allowing us to test if groups of genes of size greater than 3 can separate the datasets. In the current study, we have limited the separability tests to gene pairs and triplets and used this criterion to rank the genes.

References

1. Golub, T.R., Slonim, D.K., Tamayo, P., Huard, C., Gaasenbeek, M., Mesirov, J.P., Coller, H., Loh, M.L., Downing, J.R., Caligiuri, M.A., Bloomfield, C.D., Lander, E.S.: Molecular classification of cancer: Class discovery and class prediction by gene expression monitoring. Science 286(5439), 531–537 (1999)

2. van 't Veer, L.J., Dai, H., van de Vijver, M.J., He, Y.D., Hart, A.A.M., Mao, M., Peterse, H.L., van der Kooy, K., Marton, M.J., Witteveen, A.T., Schreiber, G.J., Kerkhoven, R.M., Roberts, C., Linsley, P.S., Bernards, R., Friend, S.H.: Gene expression profiling predicts clinical outcome of breast cancer. Nature 415(6871), 530–536 (2002)
3. Khan, J., Wei, J.S., Ringnér, M., Saal, L.H., Ladanyi, M., Westermann, F., Berthold, F., Schwab, M., Antonescu, C.R., Peterson, C., Meltzer, P.S.: Classification and diagnostic prediction of cancers using gene expression profiling and artificial neural networks. Nature medicine 7(6), 673–679 (2001)
4. Ben-Dor, A., Bruhn, L., Friedman, N., Nachman, I., Schummer, M., Yakhini, Z.: Tissue classification with gene expression profiles. Journal of Computational Biology 7(3-4), 559–583 (2000)
5. Alon, U., Barkai, N., Notterman, D.A., Gish, K., Ybarra, S., Mack, D., Levine, A.J.: Broad patterns of gene expression revealed by clustering analysis of tumor and normal colon tissues probed by oligonucleotide arrays. Proceedings of the National Academy of Science USA, 96, 6745–6750 (1999)
6. Kim, S., Dougherty, E.R., Barrera, J., Chen, Y., Bittner, M.L., Trent, J.M.: Strong feature sets from small samples. Journal of Computational Biology 9, 127–146 (2002)
7. Unger, G., Chor, B.: Linear separability of gene expression data sets. IEEE/ACM Transactions on Computational Biology and Bioinformatics 7, 375–381 (2010)
8. Alam, M.S., Panigrahi, S., Bhabak, P., Mukhopadhyay, A.: A multi-gene linear separability of gene expression data in linear time. In: Short Abstracts in ISBRA 2010: 6th International Symposium on Bioinformatics Research and Applications, pp. 51–54, May 23-26, Connecticut (2010)
9. Megiddo, N.: Linear-time algorithms for linear programming in R^3 and related problems. SIAM J. Comput. 12(4), 759–776 (1983)
10. Megiddo, N.: Linear programming in linear time when the dimension is fixed. J. ACM 31, 114–127 (1984)
11. Dyer, M.E.: Linear time algorithms for two- and three-variable linear programs. SIAM J. Comput. 13(1), 31–45 (1984)
12. Nguyen, D.V., Rocke, D.M.: Tumor classification by partial least squares using microarray gene expression data. Bioinformatics 18(1), 39–50 (2002)
13. Bishop, C.M.: Neural Networks for Pattern Recognition. Oxford University Press, New York (1995)
14. Zhang, D., Chen, S., Zhou, Z.H.: Constraint score: A new filter method for feature selection with pairwise constraints. Pattern Recogn. 41(5), 1440–1451 (2008)
15. Tusher, V.G., Tibshirani, R., Chu, G.: Significance analysis of microarrays applied to the ionizing radiation response. Proceedings of the National Academy of Sciences of the United States of America 98(9), 5116–5121 (2001)
16. Gil, C., Jun, L., Balasubramanian, N., Robert, T., Virginia, T.: In: "Significance Analysis of Microarrays" Users guide and technical document. Stanford University, Stanford CA 94305
17. Lu, Y., Han, J.: Cancer classification using gene expression data. Inf. Syst. 28(4), 243–268 (2003)
18. Gordon, G.J., Jensen, R.V., Hsiao, L.L., Gullans, S.R., Blumenstock, J.E., Ramaswamy, S., Richards, W.G., Sugarbaker, D.J., Bueno, R.: Translation of Microarray Data into Clinically Relevant Cancer Diagnostic Tests Using Gene Expression Ratios in Lung Cancer and Mesothelioma. Cancer Research 62(17), 4963–4967 (2002)

19. Armstrong, S.A., Staunton, J.E., Silverman, L.B., Pieters, R., den Boer, M.L., Minden, M.D., Sallan, S.E., Lander, E.S., Golub, T.R., Korsmeyer, S.J.: Mll translocations specify a distinct gene expression profile that distinguishes a unique leukemia. Nature Genetics 30, 41–47 (2002)
20. Hall, M., Frank, E., Holmes, G., Pfahringer, B., Reutemann, P., Witten, I.H.: The WEKA data mining software: An update. SIGKDD Explorations 11, 11–18 (2009)
21. Platt, J.C.: In: Fast training of support vector machines using sequential minimal optimization, pp. 185–208. MIT Press, Cambridge (1999)
22. Bayesian network classifiers in WEKA. Internal Notes 11(3), 1–23 (2004)
23. Cooper, G.F., Herskovits, E.: A bayesian method for the induction of probabilistic networks from data. Machine Learning 9, 309–347 (1992)
24. Kohavi, R.: A study of cross-validation and bootstrap for accuracy estimation and model selection. In: Proceedings of the 14th International Joint Conference on Artificial Intelligence, vol. 2, pp. 1137–1143. Morgan Kaufmann Publishers Inc., San Francisco (1995)

Appendix: Incremental Code for Linearly Separating Gene Triplets

Algorithm. IncrementalySeparatingGeneTriplets
Input: Plane duals S_R^* and S_B^* of the point sets S_R and S_B.
Output: LP feasible or infeasible.
1: Initialize $S_R'^* \subset S_R^*$ and $S_B'^* \subset S_B^*$. We can choose $|S_R'^*| = |S_B'^*| = 4$.
2: Apply Megiddo's approach to $S_R'^*$ and $S_B'^*$. We distinguish with following cases
Case 1: If infeasible then report the inseparability and halt.
Case 2: If feasible then return a vertical test plane $\overline{U}$ (or $\overline{V}$) and continue with step 3.
3: Repeatedly add a constraint from $|S_R^* - S_R'^*|$ or $|S_B^* - S_B'^*|$ and initiate an incremental 2-D approach on vertical test plane $\overline{U}$ (or $\overline{V}$). We distinguish with following cases
Case 1: If the test plane $\overline{U}$ (or $\overline{V}$) is feasible after all the constraints being considered then report separability and halt.
Case 2: If the test plane $\overline{U}$ (or $\overline{V}$) is infeasible then solve two 2D linear program to determine which side of test plane the feasible region lies.
Case 2.1: If any one of 2D linear program is feasible then identify the side of feasible solution and continue with step 4.
Case 2.2: If both 2D linear program are not feasible or both are feasible then report the inseparability and halt.
4: Identify a second vertical test plane $\overline{V}$ (or $\overline{U}$) and initiate an incremental 2-D approach by resuming the addition of constraints from $|S_R^* - S_R'^*|$ or $|S_B^* - S_B'^*|$ excluding those which are already being considered with $\overline{U}$ (or $\overline{V}$). In case if we do not have any second vertical test plane then continue with step 5. We distinguish with following cases

Case 1: If the test plane $\overline{V}$ (or $\overline{U}$) is feasible after all the constraints being considered then report separability and halt.

Case 2: If the test plane $\overline{V}$ (or $\overline{U}$) is infeasible then solve two 2D linear program to determine which side of test plane the feasible region lies.

Case 2.1: If any one of 2D linear program is feasible then identify the side of feasible solution and continue with step 5.

Case 2.2: If both 2D linear program are not feasible or both are feasible then report the inseparability and halt.

5: Update $S'_R{}^*$ and $S'_B{}^*$ by eliminating a constraint from each coupled line which does not pass through the feasible quadrant $\overline{U}$ and $\overline{V}$.

6: Update $S'_R{}^*$ and $S'_B{}^*$ by including all those constraints considered in step 3 and step 4.

7: Repeat the algorithm for updated set of $S'_R{}^*$ and $S'_B{}^*$.

An Accurate Numerical Method
for Systems of Differentio-Integral Equations

Jian-Jun Shu

School of Mechanical & Aerospace Engineering, Nanyang Technological University,
50 Nanyang Avenue, Singapore 639798
mjjshu@ntu.edu.sg
http://www.ntu.edu.sg/home/mjjshu

Abstract. A very simple and accurate numerical method which is applicable to systems of differentio-integral equations with quite general boundary conditions has been devised. Although the basic idea of this method stems from the Keller Box method, it solves the problem of systems of differential equations involving integral operators not previously considered by the Keller Box method. Two main preparatory stages are required: (i) a merging procedure for differential equations and conditions without integral operators and; (ii) a reduction procedure for differential equations and conditions with integral operators. The differencing processes are effectively simplified by means of the unit-step function. The nonlinear difference equations are solved by Newton's method using an efficient block arrow-like matrix factorization technique. As an example of the application of this method, the systems of equations for combined gravity body force and forced convection in laminar film condensation can be solved for prescribed values of physical constants.

Keywords: numerical method, differentio-integral equations.

1 Introduction

There are a large variety of numerical methods which are used to solve mathematical physical problems. Two particular methods, the Box scheme and the Crank-Nicolson scheme, seem to dominate in most practical applications. Keller [1] himself preferred and stressed the Box scheme. This scheme was devised in Keller [2] for solving diffusion problems, but it has subsequently been applied to a broad class of problems. It has been tested extensively on laminar flows, turbulent flows, separating flows and many other such problems [3]. These have almost invariably been 'one-boundary-layer flow' problem which are described only by differential equations and differential boundary conditions. The numerical method has not previously been adapted to multilayer problem which are more naturally described as differentio-integral equations and conditions.

In engineering, there are many multilayer problems involving the presence of distinct phases or immiscible fluids. The 'guess' method has usually been used to solve the problem [4]. The general procedure by which solutions are obtained

B. Murgante et al. (Eds.): ICCSA 2013, Part V, LNCS 7975, pp. 65–76, 2013.

may be outlined as follows: The values of some parameters are guessed so that the solutions can be carried out from one layer to another using a standard numerical method for one-layer flows. A check is then made as to whether the guessed values are correct. If not, a new guess is made and the entire calculation is repeated until the acceptable, correct values are obtained. The chief drawback of this method is its basis in guesswork. A particular problem may be sensitive to the correct guess and, moreover, it is difficult to adopt this method to non-similar flows.

For non-similar multilayer flows, systems of differentio-integral equations often need to be solved. Here a very simple and accurate numerical scheme which is applicable to quite general single or multiple boundary layer flow problems has been devised. The method shall be illustrated by showing its application in some detail to nonsimilar plane laminar incompressible double boundary layers and in particular to the equations for combined gravity body force and forced convection in laminar film condensation [5]. There is no difficulty in adopting the methods to laminar flows involving more than two layer flows and to turbulent flows (using the eddy viscosity and eddy conductivity formulations), but such calculations are not included here.

For incompressible laminar double boundary layer flow over a plane surface, the double-boundary-layer equations can be reduced to the dimensionless forms [5]:

$$\frac{\partial^3 f}{\partial \eta^3} + \alpha(\xi)f\frac{\partial^2 f}{\partial \eta^2} + \beta(\xi)\left[\gamma(\xi) - \left(\frac{\partial f}{\partial \eta}\right)^2\right] = 2\xi p(\xi)\left[\frac{\partial f}{\partial \eta}\frac{\partial^2 f}{\partial \xi \partial \eta} - \frac{\partial^2 f}{\partial \eta^2}\frac{\partial f}{\partial \xi}\right]$$

$$\frac{1}{P_r}\frac{\partial^2 \theta}{\partial \eta^2} + \alpha(\xi)f\frac{\partial \theta}{\partial \eta} = 2\xi p(\xi)\left[\frac{\partial f}{\partial \eta}\frac{\partial \theta}{\partial \xi} - \frac{\partial \theta}{\partial \eta}\frac{\partial f}{\partial \xi}\right] \qquad 0 < \eta < \eta_\delta(\xi)$$

$$\frac{\partial^3 f^*}{\partial \eta^{*3}} + \alpha(\xi)f^*\frac{\partial^2 f^*}{\partial \eta^{*2}} - \beta(\xi)\left(\frac{\partial f^*}{\partial \eta^*}\right)^2 = 2\xi p(\xi)\left[\frac{\partial f^*}{\partial \eta^*}\frac{\partial^2 f^*}{\partial \xi \partial \eta^*} - \frac{\partial^2 f^*}{\partial \eta^{*2}}\frac{\partial f^*}{\partial \xi}\right] \qquad \eta^* > 0$$

$$H_0\left[\left(\frac{\partial \theta}{\partial \eta}\right)_{\eta=0} + P_r\alpha(\xi)\int_0^{\eta_\delta(\xi)}\frac{\partial f}{\partial \eta}\theta d\eta + 2\xi P_r p(\xi)\int_0^{\eta_\delta(\xi)}\left(\frac{\partial^2 f}{\partial \eta \partial \xi}\theta + \frac{\partial f}{\partial \eta}\frac{\partial \theta}{\partial \xi}\right)d\eta\right]$$

$$+\alpha(\xi)\left(f\right)_{\eta=\eta_\delta(\xi)} + 2\xi p(\xi)\left(\frac{\partial f}{\partial \xi}\right)_{\eta=\eta_\delta(\xi)} + 2\xi p(\xi)\left(\frac{\partial f}{\partial \eta}\right)_{\eta=\eta_\delta(\xi)}\frac{d\eta_\delta(\xi)}{d\xi} = 0$$

where $\xi \geq 0$ is a transformed streamwise variable. η, $0 < \eta < \eta_\delta(\xi)$, measures distance in the inner boundary layer whose thickness is $\eta_\delta(\xi)$, an unknown function of ξ. $\eta^* \geq 0$ measures distance in the outer boundary layer. $f(\xi, \eta)$ and $f^*(\xi, \eta^*)$ are proportional to the inner and the outer stream functions respectively. $\theta(\xi, \eta)$ is dimensionless temperature in the inner boundary layer. $\alpha(\xi)$, $\beta(\xi)$, $\gamma(\xi)$ and $p(\xi)$ are the physical parameters, which depend on ξ. H_0 is a physical constant and P_r denotes the Prandtl number.

The most general boundary conditions are of the forms

$$f(\xi, 0) = 0, \quad \frac{\partial f(\xi, 0)}{\partial \eta} = 0, \quad \theta(\xi, 0) = 1, \quad \theta(\xi, \eta_\delta(\xi)) = 0,$$

$$C_0 f^*(\xi, 0) = f(\xi, \eta_\delta(\xi)), \quad C_1 \frac{\partial f^*(\xi, 0)}{\partial \eta^*} = \frac{\partial f(\xi, \eta_\delta(\xi))}{\partial \eta},$$

$$C_2 \frac{\partial^2 f^*(\xi, 0)}{\partial \eta^{*2}} = \frac{\partial^2 f(\xi, \eta_\delta(\xi))}{\partial \eta^2}, \quad \frac{\partial f^*(\xi, +\infty)}{\partial \eta^*} = U_e(\xi)$$

where C_0, C_1 and C_2 are the physical constants and $U_e(\xi)$ is the external velocity field. The last boundary conditions, which provide the prerequisites for the solvability of the equations, are

$$\frac{\partial^2 f^*(\xi, +\infty)}{\partial \eta^{*2}} = 0 \quad \text{and} \quad \frac{\partial^3 f^*(\xi, +\infty)}{\partial \eta^{*3}} = 0.$$

Using these, it is easily shown from the equations that

$$-\beta(\xi) U_e(\xi) = 2\xi p(\xi) \frac{dU_e(\xi)}{d\xi}. \tag{1}$$

Therefore $U_e(\xi)$ is a solution of the above first order equation with at most an arbitrary constant of integration. This constant may be determined by the value of $U_e(0)$. When $\xi > 0$, $U_e(\xi)$ is a function determined by both the coefficients of the equations and the value of $U_e(0)$ rather than an arbitrary function. For this reason, the last boundary condition may be rewritten as

$$\frac{\partial f^*(0, +\infty)}{\partial \eta^*} = U_e(0) \quad \text{and} \quad \frac{\partial^2 f^*(\xi, +\infty)}{\partial \eta^{*2}} = 0 \quad \xi > 0.$$

Notice that the first, second and third of the governing equations are differential equations, whilst the fourth equation is a differentio-integral equation. The differencing scheme shall be separately examined for the purely differential systems before examining a scheme for the differentio-integral system.

2 Differential Merging Scheme

In order to make all unknown functions appear as explicit forms in both equations and conditions, the new variables are defined

$$F(\xi, \phi) = f(\xi, \eta), \quad \Theta(\xi, \phi) = \theta(\xi, \eta), \quad \phi = \eta/\eta_\delta(\xi),$$

$$F^*(\xi, \phi^*) = f^*(\xi, \eta^*), \quad \phi^* = 1 + \eta^*/\eta_\delta(\xi).$$

The differential equations and all boundary conditions transform to the following systems of equations

$$\frac{\partial^3 F}{\partial \phi^3} + \alpha(\xi)\eta_\delta(\xi) F \frac{\partial^2 F}{\partial \phi^2} + \beta(\xi)\eta_\delta(\xi) \left[\gamma(\xi)\eta_\delta^2(\xi) - \left(\frac{\partial F}{\partial \phi}\right)^2 \right]$$

$$= 2\xi p(\xi) \left[\eta_\delta(\xi) \frac{\partial F}{\partial \phi} \frac{\partial^2 F}{\partial \xi \partial \phi} - \frac{d\eta_\delta(\xi)}{d\xi} \left(\frac{\partial F}{\partial \phi}\right)^2 - \eta_\delta(\xi) \frac{\partial^2 F}{\partial \phi^2} \frac{\partial F}{\partial \xi} \right]$$

$$\frac{1}{P_r}\frac{\partial^2 \Theta}{\partial \phi^2} + \alpha(\xi)\eta_\delta(\xi)F\frac{\partial \Theta}{\partial \phi} = 2\xi p(\xi)\eta_\delta(\xi)\left[\frac{\partial F}{\partial \phi}\frac{\partial \Theta}{\partial \xi} - \frac{\partial \Theta}{\partial \phi}\frac{\partial F}{\partial \xi}\right] \qquad 0 < \phi < 1$$

$$\frac{\partial^3 F^*}{\partial \phi^{*3}} + \alpha(\xi)\eta_\delta(\xi)F^*\frac{\partial^2 F^*}{\partial \phi^{*2}} - \beta(\xi)\eta_\delta(\xi)\left(\frac{\partial F^*}{\partial \phi^*}\right)^2$$

$$= 2\xi p(\xi)\left[\eta_\delta(\xi)\frac{\partial F^*}{\partial \phi^*}\frac{\partial^2 F^*}{\partial \xi \partial \phi^*} - \frac{d\eta_\delta(\xi)}{d\xi}\left(\frac{\partial F^*}{\partial \phi^*}\right)^2 - \eta_\delta(\xi)\frac{\partial^2 F^*}{\partial \phi^{*2}}\frac{\partial F^*}{\partial \xi}\right] \qquad \phi^* > 1$$

with boundary conditions

$$F(\xi,0) = 0, \qquad \frac{\partial F(\xi,0)}{\partial \phi} = 0, \qquad \Theta(\xi,0) = 1, \qquad \Theta(\xi,1) = 0,$$

$$C_0 F^*(\xi,1) = F(\xi,1), \qquad C_1\frac{\partial F^*(\xi,1)}{\partial \phi^*} = \frac{\partial F(\xi,1)}{\partial \phi},$$

$$C_2\frac{\partial^2 F^*(\xi,1)}{\partial \phi^{*2}} = \frac{\partial^2 F(\xi,1)}{\partial \phi^2},$$

$$\frac{\partial F^*(0,+\infty)}{\partial \phi^*} = U_e(0)\eta_\delta(0) \quad \text{and} \quad \frac{\partial^2 F^*(\xi,+\infty)}{\partial \phi^{*2}} = 0 \qquad \xi > 0.$$

We merge $F(\xi,\phi)$ and $F^*(\xi,\phi^*)$ into a unitary function and introduce a continuation of the definition domain of $\Theta(\xi,\phi)$ to the infinite region as follows

$$g(\xi,\phi) = \begin{cases} F(\xi,\phi) & 0 \le \phi < 1 \\ F^*(\xi,\phi^*) & \phi = \phi^* > 1 \end{cases}$$

$$\varphi(\xi,\phi) = \begin{cases} \Theta(\xi,\phi) & 0 \le \phi < 1 \\ 0 & \phi > 1 \end{cases}.$$

It is obvious that $g(\xi,\phi)$ and $\varphi(\xi,\phi)$ are not defined at $\phi = 1$, but there are two limits, left limit and right limit, at $\phi = 1$ for them. In terms of the new variables, the transformed equations in the domain $\xi \ge 0$, $\phi \ge 0$ are

$$\frac{\partial^3 g}{\partial \phi^3} + \alpha(\xi)\eta_\delta(\xi)g\frac{\partial^2 g}{\partial \phi^2} + \beta(\xi)\eta_\delta(\xi)\left[\gamma(\xi)\eta_\delta^2(\xi)H(1-\phi) - \left(\frac{\partial g}{\partial \phi}\right)^2\right]$$

$$= 2\xi p(\xi)\left[\eta_\delta(\xi)\frac{\partial g}{\partial \phi}\frac{\partial^2 g}{\partial \xi \partial \phi} - \frac{d\eta_\delta(\xi)}{d\xi}\left(\frac{\partial g}{\partial \phi}\right)^2 - \eta_\delta(\xi)\frac{\partial^2 g}{\partial \phi^2}\frac{\partial g}{\partial \xi}\right]$$

$$\frac{1}{P_r}\frac{\partial^2 \varphi}{\partial \phi^2} + \alpha(\xi)\eta_\delta(\xi)g\frac{\partial \varphi}{\partial \phi} = 2\xi p(\xi)\eta_\delta(\xi)\left[\frac{\partial g}{\partial \phi}\frac{\partial \varphi}{\partial \xi} - \frac{\partial \varphi}{\partial \phi}\frac{\partial g}{\partial \xi}\right]$$

with boundary conditions

$$g(\xi,0) = 0, \qquad \frac{\partial g(\xi,0)}{\partial \phi} = 0, \qquad \varphi(\xi,0) = 1, \qquad \varphi(\xi,1^-) = 0, \qquad \varphi(\xi,1^+) = 0,$$

$$C_0 g(\xi, 1^+) = g(\xi, 1^-), \quad C_1 \frac{\partial g(\xi, 1^+)}{\partial \phi} = \frac{\partial g(\xi, 1^-)}{\partial \phi},$$

$$C_2 \frac{\partial^2 g(\xi, 1^+)}{\partial \phi^2} = \frac{\partial^2 g(\xi, 1^-)}{\partial \phi^2},$$

$$\frac{\partial g(0, +\infty)}{\partial \phi} = U_e(0) \eta_\delta(0) \quad \text{and} \quad \frac{\partial^2 g(\xi, +\infty)}{\partial \phi^2} = 0 \qquad \xi > 0,$$

$$\varphi(\xi, +\infty) = 0$$

where the superscripts $-$ and $+$ mean the left limit and the right limit respectively. Two new conditions for $\varphi(\xi, \phi)$ are added to maintain consistency between the unknown functions and conditions. $H(x)$ is the unit-step function, also known as the Heaviside function or Heaviside's step function, which is usually defined by

$$H(x) = \begin{cases} 0 & x < 0 \\ \frac{1}{2} & x = 0 \\ 1 & x > 0 \end{cases}.$$

Using the same idea, one may also merge the three functions, $F(\xi, \phi)$, $F^*(\xi, \phi^*)$ and $\Theta(\xi, \phi)$, into a unitary function, but it is not helpful in solving the resultant difference equations hereafter because the differential equation for $\Theta(\xi, \phi)$ involves the unknown function $F(\xi, \phi)$.

The above is the main preparatory stage for introducing a merging procedure for the differential equations and conditions. Its principal requirements are based on the five following points:

(i) All unknown functions appear explicitly in both the equations and the conditions.

(ii) Unknown functions, which possess the same physical property and are independent of each other in their differential equations, are merged. These may only include one which is defined over an infinite domain.

(iii) All unknown functions with finite domain - those newly merged and the unmerged remainder - are continued into new unknown functions with infinite domain. In a broad sense, that a function with finite domain is continued into another new function with infinite domain is the same as having the function with finite domain merge with a known solution of an auxiliary equation over an infinite domain.

(iv) The necessary and appropriate conditions to maintain consistency are added.

(v) Boundary conditions at $+\infty$ may be replaced by asymptotic representations or alternatively a sufficiently large finite domain may be chosen, at the outer edge of which, the boundary conditions are understood to be satisfied. For simplicity, the latter approach has been adopted in the present work.

Now the equations are written as a first order system by introducing the new dependent variables $u(\xi, \phi)$, $v(\xi, \phi)$ and $w(\xi, \phi)$ as follows:

$$\frac{\partial g}{\partial \phi} = u$$

$$\frac{\partial u}{\partial \phi} = v$$

$$\frac{\partial v}{\partial \phi} + \alpha(\xi)\eta_\delta(\xi)gv + \beta(\xi)\eta_\delta(\xi)\left[\gamma(\xi)\eta_\delta^2(\xi)H(1-\phi) - u^2\right]$$

$$= 2\xi p(\xi)\left[\eta_\delta(\xi)u\frac{\partial u}{\partial \xi} - \frac{d\eta_\delta(\xi)}{d\xi}u^2 - \eta_\delta(\xi)v\frac{\partial g}{\partial \xi}\right]$$

$$\frac{\partial \varphi}{\partial \phi} = w$$

$$\frac{1}{P_r}\frac{\partial w}{\partial \phi} + \alpha(\xi)\eta_\delta(\xi)gw = 2\xi p(\xi)\eta_\delta(\xi)\left[u\frac{\partial \varphi}{\partial \xi} - w\frac{\partial g}{\partial \xi}\right].$$

Boundary conditions now become

$$g(\xi,0) = 0, \quad u(\xi,0) = 0, \quad \varphi(\xi,0) = 1, \quad \varphi(\xi,1^-) = 0, \quad \varphi(\xi,1^+) = 0,$$

$$C_0 g(\xi,1^+) = g(\xi,1^-), \quad C_1 u(\xi,1^+) = u(\xi,1^-), \quad C_2 v(\xi,1^+) = v(\xi,1^-),$$

$$u(0,\phi_\infty) = U_e(0)\eta_\delta(0) \quad \text{and} \quad v(\xi,\phi_\infty) = 0 \quad \xi > 0, \quad \varphi(\xi,\phi_\infty) = 0.$$

We place an arbitrary rectangular net of points (ξ_n, ϕ_j) on $\xi \geq 0$, $0 \leq \phi \leq \phi_\infty$ and use the notation:

$$\xi_0 = 0, \quad \xi_n = \xi_{n-1} + k_n, \quad n = 1, 2, \cdots\cdots;$$
$$\phi_0 = 0, \quad \phi_j = \phi_{j-1} + h_j, \quad j = 1, 2, \cdots, J_1, \cdots, J_2; \tag{2}$$

where $\phi_{J_1} = 1$, $\phi_{J_2} = \phi_\infty$. As a result of the discontinuity for $g(\xi,\phi)$ and non-differentiality of $\varphi(\xi,\phi)$ at $\phi = 1$, the point $\phi = 1$ must be included as a mesh-point. No additional restrictions need be placed on the meshwidths h_j and k_n except this requirement. Because $g(\xi,\phi)$ and $\varphi(\xi,\phi)$ have two values at $\phi = 1$, there are two ways to construct the scheme. One is that the function has two values at the same point $\phi = 1$, that is, the left limit and the right limit. The other is that the point $\phi = 1$ is thought to be two points with zero distance between them, that is, $h_{J_1+1} = 0$, and the function values at the left point and the right point are the left limit and the right limit respectively. Here the former one is preferred and implemented. Note that, for any function $z(\xi,\phi)$, $z_{J_1}^n$ represents the left limit or the right limit hereafter.

If $(g_j^n, u_j^n, v_j^n, \varphi_j^n, w_j^n)$ are to approximate (g, u, v, φ, w) at (ξ_n, φ_j), the difference approximations are defined, for $1 \leq j \leq J_2$, by

$$\frac{g_j^n - g_{j-1}^n}{h_j} = u_{j-1/2}^n \tag{3}$$

$$\frac{u_j^n - u_{j-1}^n}{h_j} = v_{j-1/2}^n \tag{4}$$

$$\frac{v_j^{n-1/2} - v_{j-1}^{n-1/2}}{h_j} + \alpha_{n-1/2}\left(\eta_\delta(\xi)gv\right)_{j-1/2}^{n-1/2}$$

$$+\beta_{n-1/2}\gamma_{n-1/2}H(J_1 - j + 1/2)\left(\eta_\delta^3(\xi)\right)_{j-1/2}^{n-1/2} - \beta_{n-1/2}\left(\eta_\delta(\xi)u^2\right)_{j-1/2}^{n-1/2}$$

$$= 2\xi_{n-1/2}p_{n-1/2}\left[\left(\eta_\delta(\xi)u\right)_{j-1/2}^{n-1/2}\frac{u_{j-1/2}^n - u_{j-1/2}^{n-1}}{k_n}\right.$$

$$\left. - \left(u^2\right)_{j-1/2}^{n-1/2}\frac{\eta_\delta(\xi_n) - \eta_\delta(\xi_{n-1})}{k_n} - \left(\eta_\delta(\xi)v\right)_{j-1/2}^{n-1/2}\frac{g_{j-1/2}^n - g_{j-1/2}^{n-1}}{k_n}\right] \qquad (5)$$

$$\frac{\varphi_j^n - \varphi_{j-1}^n}{h_j} = w_{j-1/2}^n \qquad (6)$$

$$\frac{1}{P_r}\frac{w_j^{n-1/2} - w_{j-1}^{n-1/2}}{h_j} + \alpha_{n-1/2}\left(\eta_\delta(\xi)gw\right)_{j-1/2}^{n-1/2}$$

$$= 2\xi_{n-1/2}p_{n-1/2}\left[\left(\eta_\delta(\xi)u\right)_{j-1/2}^{n-1/2}\frac{\varphi_{j-1/2}^n - \varphi_{j-1/2}^{n-1}}{k_n}\right.$$

$$\left. - \left(\eta_\delta(\xi)w\right)_{j-1/2}^{n-1/2}\frac{g_{j-1/2}^n - g_{j-1/2}^{n-1}}{k_n}\right] \qquad (7)$$

where $\xi_{n-1/2} = \frac{\xi_n + \xi_{n-1}}{2}$, $\alpha_{n-1/2}$, $\beta_{n-1/2}$, $\gamma_{n-1/2}$ and $p_{n-1/2}$ are the values of $\alpha(\xi)$, $\beta(\xi)$, $\gamma(\xi)$ and $p(\xi)$ at $\xi_{n-1/2}$ respectively. For any function $z(\xi,\phi)$, a notation has been introduced for averages and intermediate values as

$$z_{j-1/2}^n = \frac{z_j^n + z_{j-1}^n}{2}$$

$$z_j^{n-1/2} = \frac{z_j^n + z_j^{n-1}}{2}$$

$$z_{j-1/2}^{n-1/2} = \frac{z_j^n + z_{j-1}^n + z_j^{n-1} + z_{j-1}^{n-1}}{4}.$$

Note that the first, second and fourth equations are centered at $(\xi_n, \phi_{j-1/2})$ while the third and fifth equations are centered at $(\xi_{n-1/2}, \phi_{j-1/2})$. Since the first, second and fourth equations do not have a ξ derivative, they can be differenced about the point $(\xi_n, \phi_{j-1/2})$. It is found in practice that this damps high frequency Fourier error components better than differencing about $(\xi_{n-1/2}, \phi_{j-1/2})$.

The boundary conditions become simply:

$$g_0^n = 0, \quad u_0^n = 0, \quad \varphi_0^n = 1, \quad \varphi_{J_1^-}^n = 0, \quad \varphi_{J_1^+}^n = 0,$$

$$C_0 g_{J_1^+}^n = g_{J_1^-}^n, \quad C_1 u_{J_1^+}^n = u_{J_1^-}^n, \quad C_2 v_{J_1^+}^n = v_{J_1^-}^n, \qquad (8)$$

$$u_{J_2}^0 = U_e(0)\eta_\delta(0) \quad \text{and} \quad v_{J_2}^n = 0 \quad n = 1, 2, \cdots, \quad \varphi_{J_2}^n = 0.$$

3   Differentio-Integral Reduction Scheme

The same transformation is made as in the merging procedure for differential equations and conditions. The differentio-integral equation is transformed to the following equation

$$
H_0 \left[\left(\frac{\partial \Theta}{\partial \phi} \right)_{\phi=0} + P_r \alpha(\xi)\eta_\delta(\xi) \int_0^1 \frac{\partial F}{\partial \phi} \Theta d\phi \right.
$$

$$
\left. + 2\xi P_r p(\xi)\eta_\delta(\xi) \int_0^1 \left(\frac{\partial^2 F}{\partial\phi\partial\xi}\Theta + \frac{\partial F}{\partial \phi} \frac{\partial \Theta}{\partial \xi} \right) d\phi \right] + \alpha(\xi)\eta_\delta(\xi) \left(F \right)_{\phi=1}
$$

$$
+ 2\xi p(\xi)\eta_\delta(\xi) \left(\frac{\partial F}{\partial \xi} \right)_{\phi=1} = 0
$$

or

$$
H_0 \left[(w)_{\phi=0} + P_r \alpha(\xi)\eta_\delta(\xi) \int_0^1 u\varphi d\phi + 2\xi P_r p(\xi)\eta_\delta(\xi) \int_0^1 \left(\frac{\partial u}{\partial \xi}\varphi + u\frac{\partial \varphi}{\partial \xi} \right) d\phi \right]
$$

$$
+ \alpha(\xi)\eta_\delta(\xi) (g)_{\phi=1} + 2\xi p(\xi)\eta_\delta(\xi) \left(\frac{\partial g}{\partial \xi} \right)_{\phi=1} = 0.
$$

The main preparatory stage for developing a difference for the differentio-integral equation is a reduction procedure. Its principal requirements are based on the two following points:

(i) The transformations used for the differentio-integral equations are the same as those for the differential equations and conditions.

(ii) The formulation should always ensure that all limits of integration are constants.

The difference approximations are defined by

$$
H_0 \left\{ w_0^{n-1/2} + P_r \alpha_{n-1/2} \sum_{j=1}^{J_1} (\eta_\delta(\xi)u\phi)_{j-1/2}^{n-1/2} h_j \right.
$$

$$
+ 2\xi_{n-1/2} P_r p_{n-1/2} \sum_{j=1}^{J_1} \left[(\eta_\delta(\xi)\varphi)_{j-1/2}^{n-1/2} \frac{u_{j-1/2}^n - u_{j-1/2}^{n-1}}{k_n} \right.
$$

$$
\left. + (\eta_\delta(\xi)u)_{j-1/2}^{n-1/2} \frac{\varphi_{j-1/2}^n - \varphi_{j-1/2}^{n-1}}{k_n} \right] h_j \left. \right\} + \alpha_{n-1/2} (\eta_\delta(\xi)g)_{J_1}^{n-1/2}
$$

$$
+ \xi_{n-1/2} p_{n-1/2} \left[\eta_\delta(\xi_n) + \eta_\delta(\xi_{n-1}) \right] \frac{g_{J_1}^n - g_{J_1}^{n-1}}{k_n} = 0. \tag{9}
$$

Note that the equation is centered at the line $\xi = \xi_{n-1/2}$.

4   Solution of the Difference Equations

The nonlinear difference equations are solved recursively starting with $n = 0$ (on $\xi = \xi_0 = 0$). In the case of $n = 0$, we retain (3), (4) and (6) with $n = 0$ and simply alter (5), (7) and (9) by setting $\xi_{n-1/2} = 0$ and using superscripts $n = 0$ rather than $n - 1/2$ in the remaining terms. The resultant difference equations, are then easily solved by the scheme described below and accurate approximations to the solution of the systems of the ordinary differential equations are obtained.

The resultant linear systems of equations can be written in the block arrow-like matrix form as:

$$A^{(i)} \Delta^{(i)} = q^{(i)} \qquad i = 0, 1, 2, \cdots \cdots \tag{10}$$

where

$$\Delta^{(i)} \equiv \left(\delta_0^{(i)^T}, \cdots, \delta_{J_1^-}^{(i)^T}, \delta_{J_1^+}^{(i)^T}, \cdots, \delta_{J_2}^{(i)^T}, \delta \eta_\delta^{(i)}(\xi_n), \delta S_1, \delta S_2, \delta S_3, \delta S_4 \right)^T$$

$$q^{(i)} \equiv \left(0, 0, 0, r_{1/2}^{(i)^T}, \cdots, r_{J_1-1/2}^{(i)^T}, 0, 0, 0, 0, 0, r_{J_1+1/2}^{(i)^T}, \cdots, r_{J_2-1/2}^{(i)^T}, 0, 0, \gamma_\delta^{(i)}, 0, 0, 0, 0 \right)^T.$$

The matrix $A^{(i)}$ can be expanded by four extra rows and columns to form a block arrow-like matrix with 5×5 blocks if we include four extra unknowns S_1, S_2, S_3 and S_4 and four extra equations $S_1 = 0$, $S_2 = 0$, $S_3 = 0$ and $S_4 = 0$. The ordering of equations is (i) the three boundary conditions (8) at $\phi = 0$, (ii) the equations (3)-(7) at the centered location $j = 1/2, \cdots, J_1 - 1/2$, (iii) the five boundary conditions (8) at $\phi = 1$, (iv) the equations (3)-(7) at the centered location $j = J_1 + 1/2, \cdots, J_2 - 1/2$, (v) the two equations (8) at $\phi = \phi_\infty$, (vi) the equation (9) and (vii) the four dummy variable equations. We now have $5J_2 + 15$ equations and unknowns and the matrix $A^{(i)}$ has the form

$$A^{(i)} = \begin{pmatrix} A_0 & C_0 & & & & & & & D_0 \\ B_1 & A_1 & C_1 & & & & & & D_1 \\ & B_2 & A_2 & C_2 & & & & & D_2 \\ & & \cdot & \cdot & \cdot & & & & \cdot \\ & & & \cdot & \cdot & \cdot & & & \cdot \\ & & & & \cdot & \cdot & \cdot & & \cdot \\ & & & & & B_{N-1} & A_{N-1} & C_{N-1} & D_{N-1} \\ & & & & & & B_N & A_N & D_N \\ E_0 & E_1 & E_2 & \cdot & \cdot & \cdot & E_{N-1} & E_N & A_{N+1} \end{pmatrix}$$

where $N = J_2 + 1$. The matrix $A^{(i)}$ is called a block arrow-like matrix here due to its shape. A suitable algorithm for solving the matrix equation in (10) is the direct factorization method of block arrow-like matrix [6-12]. Algorithmically this is simply a modification of the usual solution of a block tridiagonal system to include the block arrow-like matrix and can be made efficient by taking account of the zeros appearing in matrices, comparing with the classical methods [13-16].

5   An Example of Application

We choose

$$\alpha(\xi) = \frac{1 + 24\xi}{(1 + \xi)^{1/4}(1 + 16\xi)^{3/4}}, \quad \beta(\xi) = \frac{\xi(17 + 32\xi)}{2(1 + \xi)^{5/4}(1 + 16\xi)^{3/4}},$$

$$\gamma(\xi) = \frac{4\,(1+\xi)^{1/2}\,(1+16\xi)^{1/2}}{17+32\xi}, \quad p(\xi) = \frac{(1+16\xi)^{1/4}}{(1+\xi)^{1/4}},$$

$U_e(0) = 1$ and the constants $C_0 = \frac{1}{\lambda\omega}$, $C_1 = \frac{1}{\lambda^2}$, $C_2 = \frac{1}{\lambda^3\omega}$ where λ and ω are two physical constants. Integrating (1) gives

$$U_e(\xi) = \frac{1}{(1+\xi)^{1/4}\,(1+16\xi)^{1/4}}.$$

The choice yields the equations for combined gravity body force and forced convection in laminar film condensation [5]. Consequently, the present problem in fact involves four physical constants H_0, P_r, λ and ω. Before the numerical solutions of the foregoing equations are discussed, it is worthwhile to consider the extreme case when the liquid film is very thin. In this case, perturbation techniques may be used and the following approximate formulas are obtained in the form

$$H_0 = \frac{0.2348}{\lambda^3\omega}\eta_\delta^3(0) + O\left(\eta_\delta^5(0) + \xi\right) \qquad \text{as} \quad \xi \longrightarrow 0$$
$$H_0 = \eta_\delta^4(+\infty) + O\left(\eta_\delta^8(+\infty) + \xi^{-1/2}\right) \qquad \text{as} \quad \xi \longrightarrow +\infty. \qquad (11)$$

The detailed mathematical analysis and physical background in this problem have been explained elsewhere [5]. When $H_0 = 0.008191$, $P_r = 10$, $\lambda = 1$ and $\omega = 10$ are chosen, it is found that the liquid film is very thin. The results can be compared with the analytical approximate formulas (11), so it is employed as a test case. A coarse grid of dimensions 42×80 is taken on the $(\xi,\ \phi)$ domain with each cell being divided into 1, 2, 3, 4 sub-cells for the two directions simultaneously and the iteration error is assigned as 0.5×10^{-13}. The values of ξ_n used are shown in Table 1. Note that the point $\phi = 1$, which is equivalent to the interface between liquid phase and vapor phase, must be included as a coarse mesh-point.

The most sensitive and the most important quantity in such calculations is $\eta_\delta(\xi)$ from which the liquid film thickness of condensate layer is determined. Thus only this quantity shall be examined in the discussion of results. All other computed quantities behaved similarly. Table 1 depicts the flavor of these results. From the convergence of the extrapolation process described in the last section, the absolute error is 2.8×10^{-8}. From this data, the results seem accurate to 7 decimal places which is more than sufficient for most purposes.

The approximate formulas (11) predict that $\eta_\delta(0) \approx 0.704$ and $\eta_\delta(+\infty) \approx 0.301$ for $H_0 = 0.008191$, $P_r = 10$, $\lambda = 1$ and $\omega = 10$, whereas the data from Table 1 display that $\eta_\delta(0) = 0.661$ and $\eta_\delta(+\infty) = 0.301$. Despite the fact which higher-order terms have been omitted when η_δ in (11) is calculated, a good agreement is obtained between the purely numerical computations and the analytical approximate formulas for thin film theory, even though $\eta_\delta(0) = 0.661$ is a moderate, rather than a thin value.

Table 1.

ξ	$\eta_\delta(\xi)$	ξ	$\eta_\delta(\xi)$	ξ	$\eta_\delta(\xi)$
0.0	0.661	0.3	0.414	15	0.305
10^{-6}	0.661	0.4	0.396	20	0.303
10^{-5}	0.661	0.6	0.374	30	0.302
10^{-4}	0.661	0.8	0.360	60	0.301
5×10^{-4}	0.659	1.0	0.351	100	0.301
10^{-3}	0.657	1.3	0.341	250	0.301
4×10^{-3}	0.645	1.6	0.335	10^3	0.301
0.01	0.624	2.0	0.329	10^4	0.301
0.025	0.585	2.5	0.324	10^5	0.301
0.05	0.542	3.2	0.319	10^6	0.301
0.075	0.513	4.0	0.316	10^8	0.301
0.1	0.492	5.0	0.312	10^{10}	0.301
0.15	0.462	7.0	0.309	10^{16}	0.301
0.2	0.441	10	0.306	10^{24}	0.301

6 Conclusion

We have demonstrated how to obtain the results of high accuracy for systems
of differentio-integral equations having multilayer flows by modifying the Keller
Box method and incorporating extrapolation. The technique can be used as the
basis for more complex problems.

References

1. Keller, H.B.: Numerical Methods in Boundary-Layer Theory. Annual Review of
   Fluid Mechanics 10, 417–433 (1978)
2. Keller, H.B.: A New Difference Scheme for Parabolic Problems. In: Numerical Solution of Partial Differential Equations — II, Synspade 1970, pp. 327–350. Academic
   Press, Inc. (1971)
3. Keller, H.B., Cebeci, T.: Accurate Numerical Methods for Boundary-Layer Flows.
   II: Two-Dimensional Turbulent Flows. AIAA Journal 10(9), 1193–1199 (1972)
4. Koh, J.C.Y., Sparrow, E.M., Hartnett, J.P.: The Two Phase Boundary Layer in
   Laminar Film Condensation. International Journal of Heat and Mass Transfer 2,
   69–82 (1961)
5. Shu, J.-J., Wilks, G.: Mixed-Convection Laminar Film Condensation on a Semi-
   Infinite Vertical Plate. Journal of Fluid Mechanics 300, 207–229 (1995)
6. Shu, J.-J., Wilks, G.: Heat Transfer in the Flow of a Cold, Two-Dimensional Vertical Liquid Jet against a Hot, Horizontal Plate. International Journal of Heat and
   Mass Transfer 39(16), 3367–3379 (1996)
7. Shu, J.-J., Pop, I.: Inclined Wall Plumes in Porous Media. Fluid Dynamics Research 21(4), 303–317 (1997)
8. Shu, J.-J., Pop, I.: On Thermal Boundary Layers on a Flat Plate Subjected to a
   Variable Heat Flux. International Journal of Heat and Fluid Flow 19(1), 79–84
   (1998)

9. Shu, J.-J., Pop, I.: Thermal Interaction between Free Convection and Forced Convection along a Vertical Conducting Wall. Heat and Mass Transfer 35(1), 33–38 (1999)
10. Shu, J.-J., Wilks, G.: Heat Transfer in the Flow of a Cold, Axisymmetric Vertical Liquid Jet against a Hot, Horizontal Plate. Journal of Heat Transfer-Transactions of the ASME 130(1), 012202 (2008)
11. Shu, J.-J., Wilks, G.: Heat Transfer in the Flow of a Cold, Two-Dimensional Draining Sheet over a Hot, Horizontal Cylinder. European Journal of Mechanics B-Fluids 28(1), 185–190 (2009)
12. Shu, J.-J., Wilks, G.: Heat Transfer in the Flow of a Cold, Axisymmetric Jet over a Hot Sphere. Journal of Heat Transfer-Transactions of the ASME 135(3), 032201 (2013)
13. Gantmacher, F.R.: The Theory of Matrices, 3rd edn. Chelsea Publishing Company (1984)
14. Respondek, J.S.: On the confluent Vandermonde matrix calculation algorithm. Applied Mathematics Letters 24(2), 103–106 (2011)
15. Respondek, J.S.: Numerical recipes for the high efficient inverse of the confluent Vandermonde matrices. Applied Mathematics and Computation 218(5), 2044–2054 (2011)
16. Bellman, R.: Introduction to Matrix Analysis. Literary Licensing, LLC (2012)

Option Pricing under an Exponential Lévy Model Using Mathematica

Aslam Aly El-Faïdal Saib[1], Arshad Ahmud Iqbal Peer[1], and Muddun Bhuruth[2]

[1] Department of Applied Mathematical Sciences, University of Technology, Mauritius
{asaib,apeer}@umail.utm.ac.mu
[2] Department of Mathematics, University of Mauritius, Reduit, Mauritius
mbhuruth@uom.ac.mu

Abstract. Exponential Lévy option pricing models are capable of explaining the jump patterns and the leptokurtic features of market option prices and the Merton model with normally distributed jump sizes is one such model which has been much considered in the literature. For this model, we develop a numerical method based on Mathematica's NDSolve function. Functional programming capabilities in Mathematica allows the development of an efficient code and numerical evidence of the accuracy and efficiency are given. Our code forms part of a computational environment that regroups several algorithms for pricing European and American options. Examples illustrating some capabilities of the software are given.

Keywords: European options, American options, Merton model, Partial integro-differential equations, Mathematica.

1 Introduction

The Black-Scholes model [3] cannot represent the risk associated with abrupt market movements and Black-Scholes implied volatilities vary with the strike and maturity which is inconsistent with the constant volatility assumption of the model. Among various extensions to the Black-Scholes model, one approach that leads to a better fit to market option prices is the inclusion of discontinuous Poisson jumps in the diffusion process. Appropriate parameter choices in such jump-diffusion models can generate implied volatility surfaces generally observed in market option prices and the Merton model [9] in which the distribution of jump sizes in the compound Poisson process has a normal distribution has attracted considerable interest in the literature. This model is a finite activity exponential Lévy model and the partial differential equation approach for contingent claim valuation requires the solution of a partial integro-differential equation (pide).

There has been considerable interest in developing fast and accurate numerical option pricing algorithms under the Merton model. Some new methods that have been proposed recently include the radial basis function method based on differential quadrature by Saib, Tangman and Bhuruth [13] and two and three time-level discretisations by Salmi and Toivanen [14].

B. Murgante et al. (Eds.): ICCSA 2013, Part V, LNCS 7975, pp. 77–90, 2013.

Multinomial trees for solving the pricing problem under the Merton model are generally only first-order accurate and stability requirements limit the time step. Second-order convergence rates can be achieved by using finite difference approximations. The numerical scheme proposed by Andersen and Andreasen [2] uses an alternating direction implicit (ADI) scheme in which the first step consists of computing the non-local integral term. Another second-order numerical scheme was developed by Almendral and Oosterlee [1]. This scheme uses the second-order backward differentiation formula for the time discretization and the fully discrete system of equations is solved using an iterative scheme. A different approach which leads to an unconditionally stable scheme was proposed by d'Halluin, Forsyth and Vetzal [8]. This scheme uses a fixed point iteration for solving the discretized system obtained by a fully implicit discretization of the pde part of the pricing equation. A more efficient approach proposed by Briani, Natalini and Russo [4] uses an implicit-explicit (IMEX) scheme where the non-local term is treated in an explicit manner and this circumvents the problem of dense matrix inversion associated with the non-local integral term.

Finite difference approximations yielding higher convergence rates have been proposed. A spectral discretisation in connection with an exponential time discretization was employed by Tangman, Gopaul and Bhuruth [15] and a high-order finite element scheme using quadratic elements was developed in [10]. This scheme yields fourth-order accurate numerical solutions. All the above numerical procedures use the Fast Fourier Transform (FFT) to compute the non-local integral term.

Carr and Mayo [5] outline various problems associated with the Fast Fourier Transform and they develop a more efficient technique for evaluating the correlation integral. They note that for the Merton model, the integral in the pide at a particular time point can be computed as a translation of a solution of a heat equation with initial values equal to the solution of the pide at that instant.

The semi-discretisation of the spatial derivatives in the pricing equation gives rise to tridiagonal matrices. For matrices with special structures, Respondek [11,12] has proposed efficient numerical algorithms for solving such matrix problems which arise in various applications.

This work proposes a numerical method based on Mathematica's NDSolve function for solving the Merton pide. The numerical method described forms part of a computational environment developed under Mathematica that gives fast approximations to option prices under the Merton model. Efficiency of the algorithms is achieved through the use of functional programming for coding the numerical schemes and the software allows comparison of the performances of some of the best known techniques. Our implementations for the jump-diffusion model forms part of a Mathematica package for option pricing and additional capabilities of the software are illustrated.

An outline of this paper is as follows. In §2, we describe the pricing equation under the Merton model. In §3, we present our solution strategy using the NDSolve function and we describe some existing numerical methods against which comparisons are drawn. Numerical results for the Merton model are presented in §4 and further illustrations of the capabilities of the Mathematica-based software are given in §5. Our conclusions are presented in §6.

2   The Merton Model

We assume an arbitrage-free market model with a single risky asset with price process $\{S_t\}_{t\in[0,T]}$ following an exponential Lévy model of the form $S_t = S_0 e^{X_t}$ on the filtered probability space $\left(\Omega, \mathcal{F}, \{\mathcal{F}_t\}_{t\in[0,T]}, \mathbb{P}\right)$ where the Lévy process $\{X_t\}_{t\in[0,T]}$ has dynamics given by

$$X_t = \left(\mu - \frac{\sigma^2}{2}\right) t + \sigma W_t^{\mathbb{P}} + J_t. \tag{1}$$

In (1), μ is a drift term, $\{J_t\}_{t\in[0,T]}$ is a jump process of the form $J_t = \sum_{k=1}^{N_t} Y_k$ with $\{N_t\}_{t\in[0,T]}$ denoting a Poisson process with intensity $\lambda > 0$ and $\{Y_k\}_{k\geq 1}$ are independent observations from a jump size variable Y. The $\mathbb{P}$-Brownian motion W_t, the Poisson process N_t and the jump magnitudes $\{Y_k\}_{k\geq 1}$ with common cumulative distribution function F are assumed to be independent. The Merton model assumes that the jump size Y has a normal distribution with mean μ_J and standard deviation σ_J.

The existence of an equivalent martingale measure $\mathbb{Q}$ imposes the restriction $\mu = r - \lambda\eta$ where r is the risk-free interest rate and

$$\eta = \int_{\mathbb{R}} (e^y - 1)\, dF(y).$$

Under $\mathbb{Q}$, the Lévy process (1) has the dynamics

$$X_t = \left(r - \frac{\sigma^2}{2} - \lambda\eta\right) t + \sigma W_t^{\mathbb{Q}} + \sum_{k=1}^{N_t} Y_k, \tag{2}$$

and the discounted price process $\widehat{S}_t = e^{-rt} S_t$ is a $\mathbb{Q}$-martingale.

Now consider a European call option with strike price K and payoff $g(S_T) = (S_T - K)^+$ where $c^+ = \max(c, 0)$. The Black-Scholes price $V_{\mathrm{BS}}(S_t, t, \sigma, r)$ for the European call is given by

$$V_{\mathrm{BS}}(S_t, t, \sigma, r) = S_t \Phi(d_1) - K e^{-r(T-t)} \Phi(d_2),$$

where Φ is the cumulative distribution function of the standard normal distribution with mean zero and unit standard deviation and

$$d_1 = \frac{\ln \frac{S_t}{K} + \left(r - \frac{1}{2}\sigma^2\right)(T-t)}{\sigma\sqrt{T-t}}, \quad d_1 = d_2 + \sigma\sqrt{T-t}.$$

2.1   Analytical Solution for a European Call

A closed form expression for the European call price under the Merton model can be derived using the Markov property of the Lévy process X_t. The time t price $V(S_t, t)$ of this call option is given by

$$V(S_t, t) = e^{-r(T-t)} \mathbb{E}^{\mathbb{Q}} \left[g \left(S_t e^{X_{T-t}} \right) \right]. \tag{3}$$

Evaluating the expectation in (3) and letting $\tau = T - t$ gives

$$V(S_t, t) = \sum_{n=0}^{\infty} \frac{1}{n!} e^{-\bar{\lambda}\tau} (\bar{\lambda}\tau)^n) V_{\text{BS}}(S_t, t, \sigma_n, r_n), \tag{4}$$

where $r_n = r - \lambda\eta + n\left(\mu_J + \sigma_J^2/2\right)/\tau$ and $\bar{\lambda} = (1+\eta)\lambda$. This closed-form expression is used to assess the numerical accuracy of a particular scheme.

2.2   The Pricing Equation

Now let $S_t = e^x$, $\tau = T - t$, and let $v(x, \tau) = V(S_t, t)$. Then, it can be shown that v solves the pide [6]

$$\frac{\partial v}{\partial \tau} = \frac{1}{2}\sigma^2 \frac{\partial^2 v}{\partial x^2} + \left(r - \lambda\eta + \lambda\mu_J - \frac{1}{2}\sigma_J^2 \right) \frac{\partial v}{\partial x} - rv$$
$$+ \int_{\mathbb{R}} \left[v(x+y, \tau) - v(x, \tau) - y\frac{\partial v}{\partial x} \right] \nu(dy),$$

where ν is the Lévy measure given by $\nu(dx) = \lambda F(dx)$. Since for Merton's model, $\int_{\mathbb{R}} y\nu(dy) = \lambda\mu_J$, we find that the pricing pide in log price is given by

$$\frac{\partial v}{\partial \tau} = \frac{\sigma^2}{2} \frac{\partial^2 v}{\partial x^2} + \left(r - \frac{\sigma^2}{2} - \lambda\eta \right) \frac{\partial v}{\partial x} - (r + \lambda) v + \lambda \int_{\mathbb{R}} v(x+y, \tau) f(y) dy. \tag{5}$$

For a European call option, we need to solve (5) using the initial condition $v(x, 0) = g(e^x)$ and the boundary conditions are given by

$$v(x, \tau) \to e^x - Ke^{-r\tau}, \ x \to \infty, \quad v(x, \tau) \to 0, \ x \to -\infty.$$

3   Numerical Algorithms

The technique proposed in this paper differs from the iterative scheme of Almendral and Oosterlee in the way the semi-discrete system of equations is solved. The use of Mathematica for implementation allows the use of the `NDSolve` function and this leads to a more efficient technique than the iterative approach. Our approach also offers possibilities of using high-order finite element or finite difference discretisations of the local part of the pricing problem.

3.1   The NDSolve Based Algorithm

In the following, we let $\tilde{g}(x) = g(e^x)$, $\kappa = (r - \sigma^2/2 - \lambda\eta)$. We then rearrange (5) in the form

$$\frac{\partial v}{\partial \tau} + \mathcal{D}v = \mathcal{I}v, \quad (x, \tau) \in \mathbb{R} \times (0, T],$$
$$v(x, 0) = \tilde{g}(x), \quad x \in \mathbb{R}, \tag{6}$$

where

$$\mathcal{D}v = -\frac{\sigma^2}{2}v_{xx} + (r + \lambda)\,v,$$

$$\mathcal{I}v = \mathcal{I}_1 v + \mathcal{I}_2 v; \quad \mathcal{I}_1 v = \kappa v_x \quad \text{and} \quad \mathcal{I}_2 v = \lambda \int_{\mathbb{R}} v(x + y, \tau) f(y) dy.$$

To solve the pricing problem numerically, we need to localise the problem (6) to a finite domain $\Omega_a = [-a, a]$. We then use the boundary conditions $v(-a, \tau) = 0$ and $v(a, \tau) = e^a - Ke^{-r\tau}$. However the non-local nature of the integral term requires the behavior of option prices outside the computational domain Ω_a. Letting $z = x + y$, we have

$$\int_{\mathbb{R}} v(x + y, \tau) f(y) dy = \int_{\Omega_a} v(z, \tau) f(z - x)\, dz + \int_{\mathbb{R} \backslash \Omega_a} v(z, \tau) f(z - x)\, dz.$$

Assume that $\mu_J = 0$ and let

$$\varepsilon(x, a, \tau) = \int_a^\infty (e^z - Ke^{-r\tau}) f(z - x)\, dx.$$

It is then seen that

$$\varepsilon(x, a, \tau) = e^{x + \frac{\sigma_J^2}{2}} \Phi\left(\frac{x - a + \sigma_J^2}{\sigma_j}\right) - Ke^{-r\tau}\Phi\left(\frac{x - a}{\sigma_J}\right).$$

The semi-discrete equations are then obtained as follows. We consider a uniform discretization of Ω_a with mesh size $h = 2a/n$ and we let $\Omega_h = \{x_i\}_{i=0}^n$ where $x_i = -a + ih$.

We first consider the approximation of the integral term in the pide (6). Using the trapezoidal rule on Ω_a and taking as ordinates the grid points $\{x_j\}_0^n$ of Ω_h, we find that

$$\int_{\mathbb{R}} v(z, \tau) f(z - x_i) dz \approx h \sum_{j=1}^{n-1} v(x_j, \tau) f_{ij} + \frac{h}{2} v(a, \tau) f_{in} + \varepsilon(x_i, a, \tau), \tag{7}$$

where $f_{ij} = f(x_j - x_i)$.

Using central difference approximations, the discretisation matrices $\mathcal{D}^h$ and $\mathcal{I}_1^h$ corresponding to the derivative terms $\mathcal{D}$ and $\mathcal{I}_1$, respectively are tridiagonal and we write them as

$$\mathcal{D}^h = \text{tridiag}\left[-\frac{\sigma^2}{2h^2}, \, r + \lambda + \frac{\sigma^2}{h^2}, \, -\frac{\sigma^2}{2h^2}\right],$$

$$\mathcal{I}_1^h = \text{tridiag}\left[-\frac{\kappa}{2h}, \, 0, \, \frac{\kappa}{2h}\right],$$

Now let

$$\mathbf{v}(\tau) = [v\,(x_1, \tau), \, \ldots, v\,(x_{n-1}, \tau)]^T, \quad \mathbf{v}(0) = [\tilde{g}(x_1), \ldots, \tilde{g}(x_{n-1})]^T.$$

We then use `NDSolve` to solve the semi-discrete equation

$$\mathbf{v}'(\tau) = A\mathbf{v}(\tau) + B\mathbf{v}(\tau) + b(\tau), \quad 0 \le \tau \le T, \tag{8}$$

where the matrix A is given by

$$A = \text{tridiagonal}\left[\frac{\sigma^2}{2h^2} - \frac{\kappa}{2h}, \, -\frac{\sigma^2}{h^2} - (r + \lambda), \, \frac{\sigma^2}{2h^2} + \frac{\kappa}{2h}\right],$$

and the matrix B has entries $b_{ij} = \lambda h f_{ij}$ for $1 \le i, \, j \le n-1$. The right-hand-side vector $b(\tau)$ is such that $b(\tau) = [b_1(\tau), \, b_2(\tau), \, \ldots, b_{n-1}(\tau)]^T$ with

$$b_i(\tau) = \lambda \varepsilon\,(x_i, \, a, \, \tau) + \frac{1}{2}\lambda h v\,(a, \, \tau)\, f_{in}, \, 1 \le i \le n - 2,$$

and $b_{n-1}(\tau) = \lambda \varepsilon\,(x_{n-1}, \, a, \, \tau) + \frac{1}{2}\lambda h v\,(a, \, \tau)\, f_{n-1, n} + \left(\frac{\sigma^2}{2h^2} + \frac{\kappa}{2h}\right) v\,(a, \, \tau).$

3.2   Existing Algorithms

Almendral and Oosterlee [1] use the second-order backward differentiation formula to solve (8). To describe this fully discrete scheme, we let $k = T/m$ denote the uniform time step and denote $\tau_l = lk$. Also let $v(x_i, \tau_l) = v_i^l$, $b_i^l = b_i\,(\tau_l)$ and denote the vector $\mathbf{v}^l = [v_1^l, \, v_2^l, \, \ldots, v_{n-1}^l]^T$. Then using the approximations

$$v_i'(\tau) = \begin{cases} \frac{1}{2k}\left(3v_i^l - 4v_i^{l-1} + v_i^{l-2}\right), & 2 \le l \le m; \\ \frac{1}{k}\left(v_i^l - v_i^{l-1}\right), & l = 1. \end{cases},$$

we need to solve the linear system given by

$$C\mathbf{v}^l = \tilde{b}^l, \tag{9}$$

where $C = \alpha I - kA - kB$, $\tilde{b}^l = \left[\tilde{b}_1, \, \tilde{b}_2, \ldots, \tilde{b}_{n-1}\right]$ with

$$\alpha = \begin{cases} 1, \, l = 1; \\ \frac{3}{2}, \, l \ge 2. \end{cases}, \quad \tilde{b}_i = k b_i^l + 2v_i^l - \frac{1}{2}v_i^{l-1}, \, 2 \le i \le n - 1,$$

and $\tilde{b}_1 = k b_1^l + v_i^{l-1}$.

The matrix C in (9) being dense, Almendral and Oosterlee use a regular splitting of the form $C = Q - R$ to solve the linear system iteratively. The splitting considered takes the matrix Q to be tridiagonal and this is done by extracting the three main diagonals of C. The matrix R is then approximated to a Toeplitz matrix and embedded in a circulant matrix to enable the use of the FFT algorithm.

Cont and Voltchkova [7] proposed an implicit-explicit approach in which the local term in the pide is treated using an implicit technique. Let $v^l = v(x, \tau_l)$ so that the resulting scheme is given by

$$\frac{v^{l+1} - v^l}{k} = \frac{1}{2} \left[\mathcal{D}^h(v^l) + \mathcal{D}^h(v^{l+1}) \right] + I^h(v^l).$$

This scheme is monotone, consistent and unconditionally stable, when the integral is evaluated using the trapezoidal rule. This method which we denote IMEX-Euler converges to the viscosity solution of the pide but it is not very efficient.

A higher-order implicit-explicit scheme using Runge-Kutta time stepping (mid-point IMEX-RK) was developed by Briani, Natalini and Russo [4]. The rationale behind the method is the splitting of the pide into its stiff and non-stiff parts. The Midpoint IMEX-RK method is stable under the time step restriction $k = \mathcal{O}(k^{4/3})$.

The alternating direction implicit scheme of Andersen and Andreasen uses two-fractional steps to solve the pricing equation (5). To briefly describe this method, let the derivative operator $\bar{D}$ be given by

$$\bar{D} = \frac{\sigma^2}{2} \frac{\partial^2}{\partial x^2} + (r - \lambda\eta) \frac{\partial}{\partial x}.$$

Then writing the problem (5) in the form

$$\frac{\partial v}{\partial \tau} - \bar{D}v = -rv + \lambda f \circledast v,$$

where $f \circledast v$ denotes a convolution between f and v, we consider the operator splitting procedure

$$\left(\frac{2}{k} - \bar{D}^h \right) v^{j+\frac{1}{2}} = \left(\frac{2}{k} - r \right) v^j + \lambda f \circledast v^j,$$

$$\left(\frac{2}{k} + r \right) v^j - \lambda f \circledast v^j = \left(\frac{2}{\Delta\tau} + \bar{D}^{\Delta x} \right) v^{j+\frac{1}{2}},$$

where $\bar{D}^h$ is the centered difference approximation to $\bar{D}$. Since the integral term represents a convolution integral, it can be evaluated using FFT. To minimize the wrap around effect that crops in with the use of FFT because of the non-periodicity of the functions, it is assumed that the solution beyond the computational domain behaves as a linear function in x that can be evaluated analytically.

Tangman, Gopaul and Bhuruth [15] use an exponential time integration technique to obtain an exact in time approximation to the pide (5). Being exact in time, the exponential time integration scheme is unconditionally stable and a one time step computation gives unconditional second order accuracy with central difference approximations. A fourth order accurate solution can be obtained if a spectral method based on a clustered Chebychev grid is employed.

Carr and Mayo [5] observed that the solution of the integral part is equal to a translation of the fundamental solution of the heat equation

$$\frac{\partial v}{\partial \tau'} = \frac{\partial^2 v}{\partial x^2}, \tag{10}$$

over $-\infty \leq x \leq \infty$ and $0 \leq \tau' \leq \sigma_j^2/2$. That is, to obtain an approximation to the non-local integral, it simply suffices to solve the heat equation numerically with initial values $u(\tau)$ that represents the solution of the pide at time τ. In order to retain second order accuracy in time, The pide is evaluated using the Crank Nicolson scheme. Setting $Dv = (\mathcal{D} + \mathcal{I}_1)v$, the scheme

$$\frac{v_i^{j+1} - v_i^j}{k} = \frac{1}{2}\left(D\left(v_i^{j+1}\right) + \left(v_i^j\right)\right) + \frac{1}{2}\left(\mathcal{I}_2\left(v_i^{j+1}\right) + \mathcal{I}_2\left(v_i^j\right)\right),$$

is solved via the Picard iteration:

$$\left[I - \frac{k}{2}D\right]v_i^{j+1,\,l+1} = \left[I + \frac{k}{2}D\right]v_i^j + \frac{k}{2}\left[\mathcal{I}_2\left(v_i^{j+1,\,l}\right) + \mathcal{I}_2\left(v_i^j\right)\right],$$

with respect to l, where $l = 0,\,1,\,\ldots$ and $v_i^{j+1,\,0} = v_i^j$. This method does not require the extension of the computational grid as it is the case with most methods involving FFT. This is because imposing boundary conditions in (10) avoids the truncation error that exists when we approximate the integral with infinite limits over our finite localized domain.

4   Numerical Results

We present the results of some numerical tests performed using Mathematica version 7 running on a computer with 256 MB of RAM and 2.8 GHZ processor. Corresponding to a solution $v(x, \tau)$ of (5) for $r = 0$, the solution $\hat{v}(x, \tau) = e^{-r\tau}v(x + r\tau, \tau)$ solves the same pide for $r \neq 0$. We have therefore assumed that $r = 0$ in the numerical computations.

Our first example illustrates the performance of our technique based on NDSolve for pricing European call options under the Merton model. The parameters chosen in the model are $\sigma = 0.2$, $\mu_J = 0$, $\sigma_J = 0.5$, $\lambda = 0.1$. We price two European calls with maturities $T = 1$ and $T = 2$ respectively and the strike price for both options is $K = 1$. We have chosen $a = 4$ for implementing the numerical method and we report on the computed price at the strike K.

The results are given in Table 1 where we have also shown the convergence rate C_r computed by taking the ratio of the differences between the solutions computed at successive grid refinements. The results indicate that errors decrease by an approximate factor of four which indicate second order convergence of the numerical solutions.

The second numerical test compares the performances of the schemes of Almendral and Oosterlee (AA), Carr and Mayo (CM) and NDSolve. We price a European call option with a maturity of $T = 0.5$ and strike $K = 1$. The parameters for the Merton model are $\lambda = 0.1$, $\sigma = 0.2$, $\sigma_J = 0.5$ and the exact price is 0.0645374. Our implementation of the methods allow the output of the price computed to a user-specified accuracy and we report on the cpu timings required by each method to yield numerical solutions with a certain accuracy. We have run all the schemes with $m = 500$ time steps and the results are shown in Table 2.

Table 1. Results for European call options

n	$T = 1$				$T = 2$			
	Price	Error	Rate C_r	cpu(s)	Price	Error	Rate C	CPU(s)
33	0.075878	1.826×10^{-2}	–	0.125	0.124202	1.276×10^{-2}	–	0.141
65	0.0899775	4.158×10^{-3}	4.411	0.5	0.134197	2.766×10^{-3}	4.77	0.531
129	0.0931739	9.616×10^{-4}	4.411	3.325	0.136291	6.721×10^{-4}	4.14	3.343
257	0.0938985	2.370×10^{-4}	4.073	13.328	0.136796	1.671×10^{-4}	4.04	17.063
Exact	0.0941355				0.1369631			

The results in Table 2 indicate that the numerical scheme using NDSolve is faster than the Almendral-Oosterlee and Carr-Mayo schemes. For example, NDSolve computes a numerical solution with error 7.6×10^{-6} in less than one-fifth of a second. In comparison, the Carr-Mayo numerical solution has an error of 9.4×10^{-6} and requires a computational time of approximately 2.8 seconds. Using NDSolve we also have the option of tuning the AccuracyGoal and the PrecisionGoal to reach the desired ℓ^{∞} error faster.

Table 2. Comparison of numerical schemes

Accuracy	AA		CM		NDSolve	
	Price	cpu(s)	Price	cpu(s)	Price	cpu(s)
10^{-2}	0.043097	0.453	0.052707	0.578	0.092497	0.001
10^{-3}	0.063135	0.609	0.055768	0.563	0.071575	0.016
10^{-4}	0.064198	1.001	0.063650	0.766	0.065149	0.047
10^{-5}	0.064453	1.343	0.064441	1.203	0.064613	0.078
10^{-6}	0.064532	3.375	0.064528	2.797	0.064545	0.156

Among these methods, we observe that the exponential time integration scheme is much faster than the other three methods. The reason for this is that for European options, the computation can be carried out in a single time step.

Table 3. Performances of numerical schemes

	IMEX-Euler		IMEX-RK		ADI		ETI	
Accuracy	Price	cpu(s)	Price	cpu(s)	Price	cpu(s)	Price	cpu(s)
10^{-2}	0.078236	0.156	0.082427	0.297	0.027523	0.109	0.033263	0.001
10^{-3}	0.073641	0.172	0.071750	0.359	0.057997	0.530	0.059668	0.016
10^{-4}	0.065183	0.140	0.065532	0.532	0.064199	2.792	0.063639	0.047
10^{-5}	0.064621	0.188	0.064568	1.203	0.064453	11.78	0.064443	0.050
10^{-6}	0.064542	0.250	0.064546	1.313	0.064532	559.5	0.064528	0.062

Table 3 shows the numerical results when we price the same option as in the second example using the implicit-explicit, alternating direction implicit and exponential time integration schemes.

We next give the root mean square error in the evaluation of a European call option price using n spatial steps with parameters $K = 10$, $T = 1.0$, $\sigma = 0.2$, $\sigma_J = 0.5$, $\lambda = 0.1$. The option prices are evaluated at 1950 equally spaced stock prices, with $S \in [8, 12]$ and a fixed number of temporal steps $m = 1024$. The root mean squared error is evaluated using the in-built Mathematica function RootMeanSquare.

Table 4. Root mean square errors

	Root Mean Square Errors			
n	ADI	CM	AA	ETI
2^5	1.26(-1)	1.27(-1)	1.26(-1)	3.59(-4)
2^6	3.08(-2)	3.09(-2)	3.08(-2)	1.95(-5)
2^7	7.63(-3)	7.63(-3)	7.64(-3)	1.25(-6)
2^8	1.90(-3)	1.88(-3)	1.94(-3)	1.56(-7)
2^9	4.76(-4)	4.44(-4)	5.41(-4)	7.53(-8)

Table 5 shows computed prices obtained for an at-the-money American put option under Merton's model. The strike price is $K = 1.0$ and the maturity is $T = 1.0$. Results are for the case when the parameters in the Merton model are $\sigma = 0.2$, $r = 0.1$, $\sigma_J = 0.5$ and $\lambda = 0.1$. The results indicate that all the schemes are computationally expensive when a solution requiring a higher accuracy is required.

Table 5. American put prices

	AA		ADI		CM		IMEX-Euler	
Accuracy	Price	cpu(s)	Price	cpu(s)	Price	cpu(s)	Price	cpu(s)
10^{-2}	0.039757	0.390	0.039761	0.624	0.039613	1.919	0.039583	0.983
10^{-3}	0.054408	0.686	0.054411	1.217	0.054386	3.151	0.054285	0.999
10^{-4}	0.058171	2.433	0.058176	6.037	0.058177	13.43	0.058051	1.092
10^{-5}	0.058352	4.711	0.058359	20.22	0.058361	34.60	0.05834	83.01
10^{-6}	0.058419	53.431	0.058415	602.5	-	-	-	-

5   The Mathematica Option Pricing Package

Our implementation of numerical algorithms for option pricing forms part of a general package developed under the Mathematica environment.

We briefly describe some aspects of this package. The `OptionPricing.m` file needs to be put in a directory accessible to Mathematica. We set the default to be a folder named "OptionPricing", which needs to be created in the "ExtraPackages" sub-directory of the "AddOns" directory within the Mathematica software installation domain of the system. Once the package has been stored, it may be loaded in the Mathematica front-end using the command:

```
<< OptionPricing`OptionPricing`
```

When the package is loaded, Mathematica outputs the following information:

```
The Package OptionPricing has been loaded. Sat 15 Dec 2012 18:41:49
This package has been created using Mathematica v. 8.
You are currently running Mathematica v. 8.

Please execute ?OptionPricing`OptionPricing`* for a list of the functions loaded
```

The package has a built-in documentation and usage information on a function is available from the command '?'. For instance `?MACALL` outputs:

```
MACALL[T, K, S, σ, r, λ, σJ]  prices a European call option with maturity
   T and strike K using the estimated volatility σ, riskfree interest rate r, jump intensity
   λ and variance of jumps σJ, at the stock price S, under Merton's analytical formula.
```

5.1   European Options

As described earlier, the package implements several numerical algorithms for European and American options under Merton's jump diffusion model. The `NDSCALL` and `NDSPUT` functions have been implemented using the Mathematica in-built `NDSolve` function. To evaluate the price of a European put option with maturity $T = 2.0$ and $S = K = 100$, $\sigma = 0.1$, $r = 0.2$, $\lambda = 0.1$, $\sigma_J = 0.5$ using, for instance, the `NDSPUT` function, one needs to evaluate the command:

```
In[2]:= NDSPUT[2.0, 100.0, 100.0, 0.1, 0.2, 0.1, 0.5]

Out[2]= 0.782138
```

European options with specified accuracy levels can be priced using the `EUCA` function. Alternatively, one can find prices at different orders of accuracy by passing a list of error tolerances to `EUCA`, using the `Map` function.

5.2   Graphical Manipulations

Several functions have been designed to provide user friendly graphical overviews. The function MEP is designed to provide a flexible manipulation of the parameters for pricing European options. The MEP function employs the Mathematica built-in Manipulate function to produce the interface of Figure 1. To price European options using the ADI method with the MEP function, we have:

In[3]:= **MEP["ADI", 1.0, 0.1, 0.1, 0.1, 0.1, 1.0]**

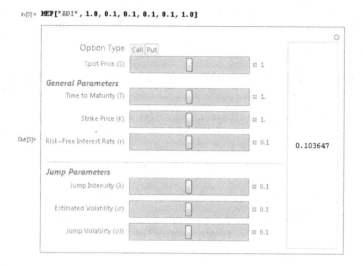

Fig. 1. The MEP function

In[5]:= **? MEG**

MEG["x", K, r, λ, σ, σJ, T] *outputs an interface to Manipulate European option prices Plot using method x. For instance MEG["ADI", 1.0, 0.1, 0.1, 0.3, 0.5, 1.0] outputs an interface to manipulate call and put option price plots using the ADI method with the specified parameter values.*

The function MEG, gives the user a graphical vision of the effect of the parameters change on the option price. As compared to the MEP function, MEG takes values of S in the range $[0.5\,K, 1.5\,K]$, where K is the strike. In the output, the user has in particular, the leisure of manipulating the option values for a range of the jump parameters. In particular, this helps when assessing the effect of abrupt changes in the parameters of the underlying. Figure 2 displays the graphical output for the ADI scheme:

In[2]:= **MEG["ADI", 1.0, 0.1, 0.1, 0.3, 0.5, 1.0]**

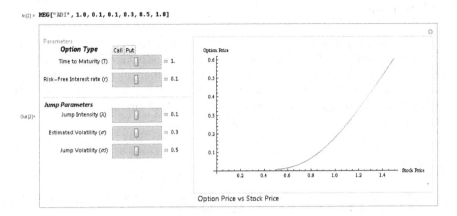

Fig. 2. The MEG interface

In[7]:= **CWA["ETD", 1.0, 1.0, 0.1, 0.1, 0.1, 0.1]**

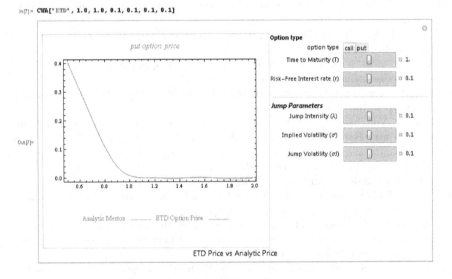

Fig. 3. The CWA function

Under some circumstances, one may further wish to see how changes in parameter values affect option prices and how the approximation prices vary in comparison to the benchmark prices. The CWA function outputs a graphical interface using the built-in Manipulate and ListlinePlot functions. The output consists of the plot of the European option prices computed using a specified method **x**, superimposed on the benchmark price plot following the analytical solution of Merton [9]. For instance, for a European put option, we have:

The function CWA gives an overview over the global convergence of the method. From the above plot, we can obtain information on the general accuracy of the ETD method and we can see the spot prices at which the errors are more pronounced.

6 Conclusion

We considered the pricing of European and American options under the Merton jump-diffusion model. We developed a numerical scheme using the in-built Mathematica function NDSolve and we provided numerical evidence that the methods performs well for pricing European options. We described some aspects of a Mathematica package which we have developed for pricing options and we illustrated some capabilities of the developed package.

References

1. Almendral, A., Oosterlee, C.W.: Numerical Valuation of Options with Jumps in the Underlying. Appl. Numer. Math. 53, 1–18 (2005)
2. Andersen, L., Andreasen, J.: Jump Diffusion Processes: Volatility Smile Fitting and Numerical Methods for Option Pricing. Rev. Derivatives Research 4, 231–262 (2000)
3. Black, F., Scholes, M.S.: The Pricing of Options and Corporate Liabilities. J. Polit. Econ. 81, 637–659 (1973)
4. Briani, M., Natalini, R., Russo, G.: Implicit-Explicit Numerical Schemes for Jump-Diffusion Processes. Calcolo 44, 33–57 (2007)
5. Carr, P., Mayo, A.: On the Numerical Evaluation of Option Prices in Jump Diffusion Processes. Eur. J. Finance 13, 353–372 (2007)
6. Cont, R., Tankov, P.: Financial Modelling with Jump Processes. Chapman and Hall, Boca Raton (2004)
7. Cont, R., Voltchkova, E.: A Finite Difference Scheme for Option Pricing in Jump Diffusion and Exponential Lévy Models. SIAM J. Numer. Anal. 43, 1596–1626 (2005)
8. d'Halluin, Y., Forsyth, P.A., Vetzal, K.R.: Robust Numerical Methods for Contingent Claims under Jump Diffusions. IMA J. Numer. Anal. 25, 87–112 (2005)
9. Merton, R.C.: Option Pricing when Underlying Stock Returns are Discontinuous. J. Financ. Econ. 3, 25–144 (1976)
10. Rambeerich, N., Tangman, D.Y., Gopaul, A., Bhuruth, M.: Exponential Time Integration for Fast Finite Element Solutions of some Financial Engineering Problems. J. Comput. Appl. Math. 224, 668–678 (2009)
11. Respondek, J.S.: Numerical Recipes for the High Efficient Inverse of the Confluent Vandermonde Matrices. Appl. Math. Comp. 218, 2044–2054 (2011)
12. Respondek, J.S.: On the Confluent Vandermonde Matrix Calculation Algorithm. Appl. Math. Lett. 24, 103–106 (2011)
13. Saib, A.A.E.F., Tangman, Y., Bhuruth, M.: A New Radial Basis Functions Method for Pricing American Options under Merton's Jump-Diffusion Model. Int. J. Computer Math. 89, 1164–1185 (2012)
14. Salmi, S., Toivanen, J.: Comparison and Survey of Finite Difference Methods for Pricing American Options under Finite Activity Jump-Diffusion models. Int. J. Computer Math. 89, 1112–1134 (2012)
15. Tangman, D.Y., Gopaul, A., Bhuruth, M.: Exponential Time Integration and Chebychev Discretisations Schemes for Fast Pricing of Options. Appl. Numer. Math. 58, 1309–1319 (2008)

Modeling and Numerical Simulation of Multi-destination Pedestrian Crowds

Günter Bärwolff, Tobias Ahnert, Minjie Chen, Frank Huth, Matthias Plaue, and Hartmut Schwandt

Technische Universität Berlin, Institut für Mathematik
Straße des 17. Juni 136, 10623 Berlin, Germany
{baerwolff,ahnert,chenmin,huth,plaue,schwandt}@math.tu-berlin.de
http://www.math.tu-berlin.de/projekte/smdpc

Abstract. In this paper we collect two parts of a research project on the pedestrian flow modeling. Rapid growth in the volume of public transport and the need for its reasonable, efficient planning has made the description and modeling of transport and pedestrian behaviors an important research topic in the last twenty years. Comparatively little attention has been paid to the problem of pedestrian crowd behaviors in geometries with multiple destinations: each of the possibly many pedestrians moves to one out of a number of destinations. The objective of the present study is to investigate pedestrian behaviors in such a context. The central problem is the modeling of crossing pedestrian streams. In view of a desirable practical relevance, realistic, i.e. rather complex geometries are studied in this context.

Keywords: macroscopic models, pedestrian density and flow measurement, human crowd experiments, intersecting pedestrian flows.

1 Introduction

In a close cooperation with engineers we investigate the mathematical description of pedestrian movement by

1. data generation with real world experiments for model validation and parameter calibration,
2. development of
   (a) microscopic mathematical models,
   (b) macroscopic mathematical models.

We managed the experiments with about 300 students of the TU Berlin and recorded the different constellations of pedestrian streams by a multi-trace recorder as a base for generation of individual tracks and density estimation.

Respecting microscopic models we discuss grid-based approaches rooted in cellular automata and a second ansatz with a combination of a force-based and a graph-based approach. The microscopic models are discussed in detail for example in [1] and [2].

B. Murgante et al. (Eds.): ICCSA 2013, Part V, LNCS 7975, pp. 91–106, 2013.

In this paper we focus on macroscopic models based on a set of pedestrian-specific coupled partial differential equations (pde's). The first discussed model is based on mass balance quantities in consideration include pedestrian density and fluxes which give information about the velocities of pedestrian groups.

In a second model we consider multispecies pedestrian model based on a 3d multiphase incompressible fluid flow model

2   A Compressible Macroscopic Model

This macroscopic approach is based on a set of pedestrian-specific coupled partial differential equations. The equations are not derived from the Euler-/Navier-Stokes equations known from fluid and gas dynamics. The specific situation of multi-destination pedestrian crowds with crossing streams requires the development of appropriately adapted methods. This has been targeted by the use of simple heuristics.

Typical applications of these approaches include real-world scenarios like airports, shopping malls, buildings of middle- to large size etc., where the participants (i.e. the pedestrians) do not exhibit an overall unanimity and (may) have different and multiple destinations.

Beyond the modeling of the above-mentioned problems, a particular aim of this project will be the development, implementation and test of appropriate computer-based simulation models.

2.1   The Transport Equation

Perceiving pedestrian flows as a transport problem, we start with the governing equation that describes the mass flow:

$$\frac{\partial \rho_i}{\partial \vartheta} + \nabla \cdot (\rho_i v_i) = 0. \tag{1}$$

In this document, ϑ denotes the time, and $i \in \{1, \ldots, n\}$ where n is the number of pedestrian "types" or "species" distinguished by certain properties. The desired velocity would be a frequent example of such a property. Furthermore, ρ_i is the current density and v_i the current velocity of a species in a given computational domain.

Since pedestrian dynamics cannot be described as a pure physical phenomenon, the parts of the equation can do well with some discussion.

2.2   Measuring Pedestrian Density

In physics, mass density ρ has units $[\rho] = \frac{\text{mass}}{\text{volume}}$. However, this does not seem to be the best fit for the problem considered here. The definition $[\rho] = \frac{\text{mass}}{\text{area}}$ used, for example, in [3] similarly includes "mass" which the authors need to model inertial effects, the latter are not included in our model.

Concerning pedestrian inertial mass, we require the following assumption:

Due to the smoothness of the controlling fields we assume that it is not necessary to describe mass (inducing inertial behavior in the model). In this way we assume that the pedestrians may follow to (adapt speed and heading) the controlling fields without significant lag by means of internal impetus, decision and physical strength.

Therefore, one natural way to measure the pedestrian density in this setting is to use $[\rho] = \frac{\text{pedestrians}}{\text{area}}$, which implicitly relies on a certain amount of homogeneity of the pedestrian crowd in the considered sample. An even smarter approach is presented in [4] by defining the area that a specific pedestrian occupies. This area depends on, e.g., whether the pedestrian is a child, an adolescent or an adult, what kind of the pedestrian wears (summer or winter clothes), how much luggage the pedestrian carries and so on. This yields an appropriate dimensionless measure $[\rho] = \frac{\text{pedestrians' needed area}}{\text{available area}}$. Since the area occupied by a pedestrian is not always readily available for input, we choose the former definition of density as pedestrians per area.

In our model we use normalized densities: $\rho_i, \rho \in [0,1]$ with $\rho = \sum_{i=1}^{n} \rho_i$. A value of $\rho = 1$ would for instance mean $5.4 \frac{\text{pedestrians}}{\text{area}}$ according to [5] and up to $10 \frac{\text{pedestrians}}{\text{area}}$ according to other sources (see, e.g., [6] for a discussion).

2.3   Transport Velocity

The primary goal is to find a sensible functional relationship $v_i = v_i(\rho_1, \ldots, \rho_n)$ that yields a nonlinear system for realistic cases.

In the literature, one frequently discriminates between a planned (e.g., "external" in [7] or "tactical" in [6]) and an instantaneous (e.g., "intelligent" in [7] or "operational" in [6]) velocity. In our opinion, this differentiation makes sense in the context of categorizing the cause of an action taken by a pedestrian.

Our approach is slightly more pragmatic and accounts for three different types of decisions. Pedestrians:

1. choose a direction they wish to go,
2. choose a speed for walking in the chosen direction based on local conditions,
3. alter speed and walking direction in order to locally avoid densely populated areas (prefer the direction of $-\nabla\rho$).

Therefore, we decompose the velocity as follows:

$$v_i = a_i V d_i^i - b_i W d^l, \tag{2}$$

where:

$V \in [0,1]$ is a normalized speed determined by a fundamental diagram (see below).

d_i^i is a unit vector field pointing into the direction of the desired heading.

d^l is a directional vector field for local correction (not necessarily of unit length, see below). Since it depends on the total density ρ, it is common to all pedestrian species.

a_i and b_i are constants: a_i stands for the absolute value of the wished velocity,
b_i is a measure for avoiding regions of high density.
$W = 1 - V$ reflects the operational shift from wanting to reach the desired target
to reacting to local encounters with other pedestrians at high densities.

In summary this the term $a_i V d_i^l$ stands for the gradient driven part of the
velocity and $b_i W d^l$ describes the influence of high density regions on the velocity
of pedestrians.

A model for two pedestrian species with just the $a_i V d_i^l$ term present has
been investigated in [8] with a focus on discussing the mathematical foundation.
This investigation highlights some shortcomings of the model with respect to
the simulation of real-world scenarios. The authors suggest to introduce (cross)
diffusion terms in order to solve these problems. Since the meaning of these terms
in the context of real-life applications seems to be obscure, here we venture to
introduce the $b_i W d^l$ term.

2.4   Desired Heading

The term d_i^l describes the pedestrian's choice of a walking direction, and is
based on spatial information in the vicinity and the global environment of the
pedestrian:

$$\Delta \phi_j^{(i)}(\vartheta) = r_j^{(i)}(\vartheta), \tag{3}$$

$$\phi_i(\vartheta) = \sum_j \phi_j^{(i)}(\vartheta), \tag{4}$$

$$d_i^l(\vartheta) = \begin{cases} \frac{\nabla \phi_i(\vartheta)}{|\nabla \phi_i(\vartheta)|} & \text{if } |\nabla \phi_i(\vartheta)| \neq 0, \\ \text{random unit vector} & \text{if } |\nabla \phi_i(\vartheta)| = 0. \end{cases}$$

With the formula (4) we add the different potentials coming from global influences like the shape of the considered domain or local influences like the local
density.

The subscripted character j of $\phi_j^{(i)}(\vartheta)$ denotes the influence type (for example global, local etc.), and the superscripted character i denotes the considered
species.

Therefore, according to the assumption of continuous influences, d_i^l is based
on source and boundary terms of j (partially) solved Poisson equations for each
pedestrian species i (reflecting j different influencing factors). The $\alpha_j^{(i)}$ are constant weights, and the $f_j^{(i)}(\vartheta)$ are source terms derived from spatially distributed
information—for example, the density $\rho_k(\vartheta)$ of some pedestrian species. This
kind of flow, the direction of which is derived from a potential, has been investigated in [9].

The parameters in the above equations have to be chosen very specifically,
and finding appropriate settings is one of the open tasks for the application of
this model.

With a special choice of the right hand side of the equation (3) we can model global or local influences by potentials. A detailed discussion of this point is given in [10].

2.5 Introducing the Gradient Term

There are a number of possible approaches to introduce a gradient term. Notable are the following two:

$$d^l(\vartheta, x) = \begin{cases} \frac{\nabla \rho(\vartheta, x)}{|\nabla \rho(\vartheta, x)|} & \text{if } |\nabla \rho(\vartheta, x)| > 0, \\ 0 & \text{if } |\nabla \rho(\vartheta, x)| = 0, \end{cases} \tag{5}$$

$$d^l(\vartheta, x) = \begin{cases} \frac{\nabla \rho(\vartheta, x)}{|\nabla \rho(\vartheta, x)|} & \text{if } |\nabla \rho(\vartheta, x)| > 1, \\ \nabla \rho(\vartheta, x) & \text{else.} \end{cases} \tag{6}$$

Concerning (5) it has to be noted that $d^l(\vartheta, x)$ is not necessarily continuous with respect to x at points where $|\nabla \rho(\vartheta, x)| = 0$ holds. Another obvious problem with this term is that it may show large scatter where the density is high. This can lead to the violation of the condition $\rho_i, \rho \in [0, 1]$ because of numerical overshooting. Another risk is the numerical oscillation of the solution (which might be interpreted as remaining erratic pedestrian activity at high densities in certain situations).

This term might be viewed as carrying some random disturbances as discussed in [11,12], which produces less effective motion (stronger clogging tendency due to the "freezing by heating effect" discussed there). The measurement of the mobility there did not show an increase of flux with more "thermal" motion at all. This is due to the inhibition of lane formation because the effect of lane formation yields an enhancement of flux. The observed "freezing by heating effect" has been considered for modeling panic situations, where it might well make sense. However, such scenarios are beyond the scope of this paper.

The gradient term defined by (6) is more likely to be the rule applied by pedestrians under normal conditions, because it is more efficient than the term given by Eq. (5).

Another argument in favor of (6) is given by the fact that the key idea of the macroscopic approach is to average the behavior of several pedestrians and smooth out random disturbances caused by single pedestrians at sufficiently large scales.

2.6 Walking Speed and Fundamental Diagram

A uni-directional flux can be defined by $J = \rho_i V(\rho_1, \ldots, \rho_n) d_i$. For the case of $J = \rho V(\rho) d$ the three quantities are related by a fundamental diagram. Fundamental diagrams have been determined by a number of authors, with a relatively wide range of different results that cannot be used to deduce a general law. According to [6], the values found in the literature for the maximum pedestrian density, where movement is possible at all, vary from $3.8/m^2 - 10/m^2$. Another controversially discussed issue is how V depends on whether movement is uni-directional or multi-directional.

For more details, see [13] and [14].

Despite the issues described above, we have evaluated the impact of different fundamental diagrams on quantitative and qualitative properties of the solutions. The differences are large enough to indicate the need for a better approximation in this respect. The fundamental diagrams that we have tested are:

$$V(\rho) = 1 - \rho, \; V(\rho) = (1 - \rho)^2, \; V(\rho) = 1 - \rho^2,$$
$$V(\rho) = 1 - \exp(-1.913/5.4(1/\rho - 1)) \, . \tag{7}$$

Note that, compared to [5, p. 65], Eq. (7) employs the normalization conditions $V \in [0, 1]$ and $\rho \in [0, 1]$.

2.7   Simulation Example - 90° Encounter

The above discussed boundary value problem completed by appropriate boundary conditions (see [10]) is solved using the Finite Volume package OpenFOAM. In Fig. 1 shows the results of two pedestrian streams crossing with an angle of 90°. A formation of dynamically reconfiguring clusters can be observed.

3   A Multispecies Pedestrian Model Based on a 3d Multiphase Incompressible Fluid Flow Model

The idea to simulate pedestrian flow by the application of fluid dynamics equations has a certain history in that field. This approach is based on the application of partial differential equations, which makes it a macroscopic method. The need to simulate several different species of pedestrians self-evident, which has not been met very well by numerical simulations in the macroscopic category. The basis of the description of non dense pedestrian movement by incompressible fluid flow models consists in the introduction of an empty phase as a species of a multiphase system of distinct phases. In the following we describe the mathematical model and modifications of the multiphase InterFoam-solver of the OpenFOAM library, which makes it applicable in this field and present results that show capabilities and limitations of the modified solver.

We introduce now a new technique for the simulation of several species in macroscopic simulation of pedestrian crowds. The focus is on the modeling of several species with different destination and the ability to intersect each other rather than on a precise reconstruction of known pedestrian phenomena for prediction purposes. We proceed by first presenting the mathematical model followed by a concrete implementation and some results. Based on the discussions in [15], [16] and [17] we choose the incompressible Navier-Stokes equations as a starting point of our model and add boundary conditions and transport equations to allow an intermixing and separation of different species.

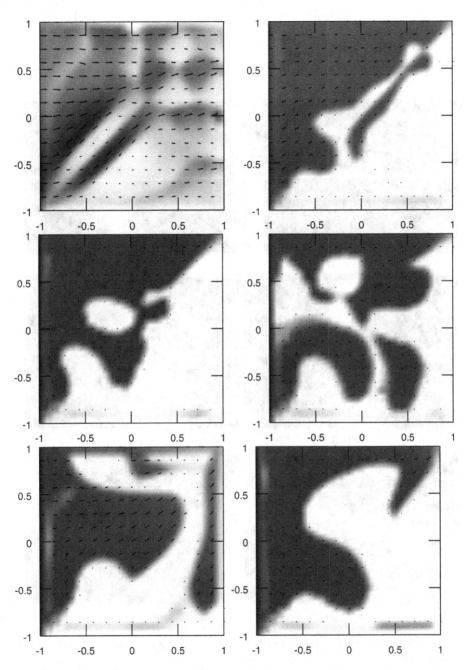

Fig. 1. Time steps 5, 10, 20, 40, 60 and 80 of the simulation of a 90° encounter of two species. Shown are density and flux of one species coming from left. The length of the arrow indicates flux strength, the grade of darkness indicates the density of the species. The crossing species coming from bottom is located in the light gray or white regions of the area

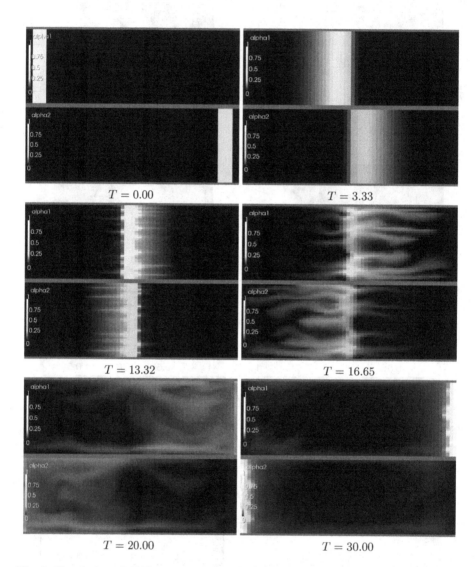

Fig. 2. Simulation of 180° crossing with $\max(\mathbf{u}) = 10.0$, $\mathbf{v}_{ps} = 0.04$, $\max(|\mathbf{f}|) = 1000$ as parameters

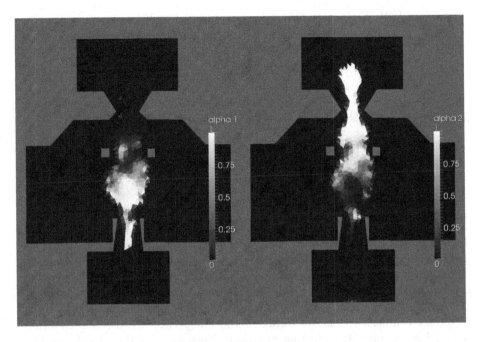

Fig. 3. Simulation done with a complex geometry inspired by real world experiments

3.1 The Mathematical Model

We use the non-stationary, incompressible Navier-Stokes equations

$$\rho \frac{\partial \mathbf{v}}{\partial t} + \rho \mathbf{v} \cdot \nabla \otimes \mathbf{v} + \nabla p - \tag{8}$$

$$\nabla \cdot (\mu(\nabla \otimes \mathbf{v}) + \mu(\nabla \otimes \mathbf{v})^T) = \mathbf{f} \tag{9}$$

$$\nabla \cdot \mathbf{v} = \mathbf{0} \tag{10}$$

combined with a volume of fluid (VOF) method as a starting point to simulate $N_p \in \mathbb{N}$ different pedestrian species. Let $\mathbb{P} = \{1, \ldots, N_p\}$ be the set of indices of pedestrian groups, then the VOF method keeps track of the species' positions by introducing one fraction function per species

$$\alpha_i(\mathbf{x}) \in [0, 1], \tag{11}$$

that describes the fill level at position $\mathbf{x} \in \Omega$ of species $i \in \mathbb{P}$. The fraction function can be discontinuous, especially when discretized for implementation purposes. We demand the sum of all fraction functions to be one, i.e.

$$\sum_{i \in \mathbb{P}} \alpha_i = 1. \tag{12}$$

A standard VOF method uses the velocity computed by (9) with $\rho = \sum_{i \in \mathbb{P}} \rho_i \alpha_i$, $\mu = \sum_{i \in \mathbb{P}} \mu_i \alpha_i$ and changes every α_i by solving the transport equation

$$\frac{\partial \alpha_i}{\partial t} + \mathbf{v} \cdot \nabla \alpha_i = 0 \text{ for all } i \in \mathbb{P}. \tag{13}$$

In the course of pedestrian simulation we tried to simulate group crossing. Therefore, it was necessary to solve three modeling problems:

1. simulation of spaces without a pedestrian species
2. distinct species forces
3. separation of species

3.2 Empty Spaces

An empty space is simulated by using a pedestrian group $f \in \mathbb{P}$, $\mathbb{P}_{\mathrm{wf}} = \mathbb{P} \backslash \{f\}$. This so-called fill-species is able to leave Ω by flowing through an additional dimension, i.e. for a two-dimensional Ω the third dimension or z-axis. It is therefore necessary to simulate a three dimensional domain for a two dimensional problem.

The inflow and outflow over the third dimension is implemented using special boundary conditions that are aware of the fill-species. We use a solver that is based on an operator splitting approach. Therefore, we have to choose two boundary conditions; one for the velocity and one for the pressure variable.

The boundary condition for the velocity is defined as

$$\mathbf{n} \cdot \mathbf{v} = 0, \text{ for } \mathbf{n} \cdot \mathbf{\Phi} \geq 0, \alpha_f = 0 \tag{14}$$
$$\mathbf{n} \cdot \nabla(\mathbf{n} \cdot \mathbf{v}) = 0 \text{ otherwise}, \tag{15}$$

where $\mathbf{\Phi}$ is the velocity value adjacent to the boundary condition face from the last pressure correction step.

The pressure boundary condition is defined as

$$p = \begin{cases} p_0 - \frac{1}{2}\rho \|\mathbf{v}\|^2, & \text{for } \mathbf{n} \cdot \mathbf{\Phi} < 0 \\ p_0, & \text{for } \mathbf{n} \cdot \mathbf{\Phi} \geq 0, \ \alpha_f > 0 \end{cases} \tag{16}$$
$$\mathbf{n} \cdot \nabla p = 0, \text{ for } \mathbf{n} \cdot \mathbf{\Phi} \geq 0, \ \alpha_f = 0 \tag{17}$$
$$\tag{18}$$

on the z-axis. The other sides of the domain can be chosen as slip boundary conditions.

3.3 Species Forces

Each species of the intersection of pedestrians needs to have a distinct destination. Therefore the need to implement species specific forces and velocities arises. Each pedestrian species $i \in \mathbb{P}_{wf}$ has a desired velocity $\mathbf{v}_i^d$, that is the velocity a pedestrian species has without the influences of other pedestrian species.

The desired velocity gets transformed into a resulting force for the right hand side in the Navier-Stokes equation (9). Following the nomenclature by Helbing et al. for microscopic models (cf. [18], [19]), we introduce a so-called behavioral force

$$\mathbf{f} := C_2(\alpha^{bil})\Big(C_1(\alpha^{bil}) \sum_{i\in\mathbb{P}_{wf}} \alpha_i \mathbf{v}_i^d - \mathbf{v}\Big), \tag{19}$$

with

$$\alpha^{bil} := \sum_{i\in\mathbb{P}_{wf}} \alpha_i \tag{20}$$

and add it to the right hand side of the Navier-Stokes equations (9). The functions C_1 and C_2 control the pedestrian behaviour, e.g. a choice of

$$C_1(\alpha^{bil}) := (1 - \alpha^{bil}) \tag{21}$$
$$C_2(\alpha^{bil}) := \alpha^{bil} \tag{22}$$

approximates the pedestrian fundamental diagram.

3.4 Separation of Species

The seperation of species is not naturally given by the discretized VOF method. Equation (13) does not provide a mean of separation of once mixed cells due to the fact we compute until now only a global velocity $\mathbf{v}$ in the Navier-Stokes equations (9). Thus, we introduce an additional transport equation

$$\frac{\partial \alpha_i}{\partial t} - \nabla \cdot \Big(C_3(\alpha_f)\frac{\mathbf{v}_i^d}{\|\mathbf{v}_i^d\|}\alpha_i\Big) = 0 \tag{23}$$

for all $i \in \mathbb{P}_{wb}$ followed by

$$\alpha_f = 1 - \sum_{i\in\mathbb{P}_{wf}} \alpha_i \tag{24}$$

with C_3 defining the magnitude of the separation velocity with a typical value of

$$C_3(\alpha_f) = \begin{cases} 0.01, & \text{for } \alpha_f > 0 \\ 0, & \text{for } \alpha_f = 0. \end{cases} \tag{25}$$

3.5 Implementation

The Navier-Stokes equation is solved using the so-called Pressure Implicit with Splitting Operators (PISO) algorithm [20] (use of the multiphaseInterFoam solver in OpenFOAM [21]). The solver consists mainly of three distinct steps. The velocity predictor step, the pressure correction loop and the fraction function adjustments. It further implements a surface tension force, which has been disabled for our experiments, but might be used in combination with our model, too.

We need to introduce some notation to proceed. We will call $\mathcal{E}$ the set of all velocity nodes and $\mathcal{N}(i), i \in \mathcal{E}$ the set of all neighbor nodes of i, that is the nodes whose cell share a face with the cell of i. Let us denote by V_i the volume of a cell for node $i \in \mathcal{E}$.

3.6   The Velocity Predictor Step

Let ρ_g and μ_g be defined as

$$\rho_g = \sum_{i \in \mathbb{P}} \rho_i \alpha_i, \quad \mu_g = \sum_{i \in \mathbb{P}} \mu_i \alpha_i, \tag{26}$$

where μ_i and ρ_i are species dependent and $\mathbf{f}$ be computed by (19). For the most simple case we use the explicit Euler method, so equation (9) becomes

$$\int_{V_i} \rho_g \frac{\mathbf{v}^{n+1} - \mathbf{v}^n}{\Delta t} \, d\mathbf{x} + \int_{\partial V_i} (\mathbf{n} \cdot \rho_g \mathbf{\Phi}^n) \mathbf{v}^n \, ds + \tag{27}$$

$$\int_{V_i} \nabla p^n \, d\mathbf{x} - \int_{V_i} \nabla \cdot (\mu_g (\nabla \otimes \mathbf{v}^n) + \mu_g (\nabla \otimes \mathbf{v}^n)^T) \, d\mathbf{x} \tag{28}$$

$$= \int_{V_i} \mathbf{f}^n \, d\mathbf{x} \tag{29}$$

in a finite volume context, where $\mathbf{\Phi}$ is the velocity interpolated to the faces using the values from neighbor cells and n symbolizes the current time step. OpenFOAM is using a kind of Rhie-Chow interpolation for flux fields, which we will symbolize by Π.

Then the algebraic equation for a single cell $i \in \mathcal{E}$ of (29) becomes

$$a_i \mathbf{v}_i + \sum_{n \in \mathcal{N}(i)} a_n \mathbf{v}_n = \mathbf{b}_i - \nabla p_i \tag{30}$$

in discretized form, where $a_i \in \mathbb{R}$ are the coefficients for $\mathbf{v}_i$ and $\mathbf{b}$ represents the right hand side of the algebraic equation without the pressure.

3.7   The Pressure Correction Loop

Let A be the diagonal matrix containing all a_i from equation (30), that is for $k = \lfloor j/3 \rfloor$ let $(A)_{jj} = a_k$ and $(A)_{ij} = 0$ for $i \neq j$ and further let H be the vector containing all $a_n \mathbf{v}_n$ and the right hand side $\mathbf{b}_i$, that is

$$H_{3i+j} = (- \sum_{n \in \mathcal{N}(i)} a_n \mathbf{v}_n + \mathbf{b}_i)_j \text{ for } j \in \{1, 2, 3\}. \tag{31}$$

This H-operator is common for OpenFOAM-based implementations. We then compute a Jacobi step for $\mathbf{v}$ with

$$\mathbf{v}_{\text{jac}} = A^{-1} H \tag{32}$$

Next, we compute $\mathbf{\Phi} = \Pi(\mathbf{v}_{\text{jac}} + A^{-1}\mathbf{f})$ followed by

$$\nabla \cdot (A^{-1} \nabla p^{n+1}) = \nabla \cdot \mathbf{\Phi} \tag{33}$$

to compute the new p^{n+1}.

The face flux Φ is then corrected by

$$\Phi^{n+1} = \Pi(A^{-1}H - A^{-1}\nabla p) \tag{34}$$

followed by the correction of the velocity

$$\mathbf{v}^{n+1} = \mathbf{v}_{\text{jac}} - \nabla p. \tag{35}$$

The boundary conditions (14) and (16) are used for the z-axis in equations (9) and (33), respectively. The velocity's boundary conditions have been set to slip at non-penetrable walls and boundary conditions for the pressure have been chosen as zero gradient. The pressure correction loop is repeated until the pressure converges or a maximum number of rounds is reached.

3.8   Adjustment of the Fraction Function

The computation of $\mathbf{v}^{n+1}$ allows the adjustment of the fraction function via (13). It follows the separation of the species by solving (23) and (24). Usually a downwind scheme should be used for the evaluation of C_3, so it is set depending on the α_f value in the target cell.

When the fraction function has been adjusted, the velocity predictor step continues with the next time step.

3.9   Numerical Results

We produced simulation results for quadratic geometries with an orthogonal mesh and on a more complex geometry inspired by real world experiments in the TU Berlin [22].

Fig. 2 shows the results for a quadratic area with two species crossing in 180 degrees.

As can be seen from Fig. 2 the species cross each other, show stripe formation, create lanes and reach their destination on opposite walls. At the end of the simulation the species are completely separated. It should, however, be noted, there are several effects originating in the impulse conservation, which are rather unnatural for crowd simulation. For example the the occurrence of a splash at the moment the species hit a wall with larger values of $\mathbf{v}$, which is due to the impulse conservation and can be seen at time $T = 20.0$ in Fig. 2. There, one is able to see species one splashing back at the bottom wall. Further, the masses have a non-neglectable acceleration time, which is in contrast to pedestrians behavior.

We made real world experiments which can be used to test parameters and validate the numerical results. In 2010 and 2011 we performed several experiments with up to four crowd groups that were crossing in a predefined area. The experiments have been recorded on video and we were able to observe common crowd phenomena like lane formation and isolated groups (c.f. [22]). It also allowed us to get quantitative results for evaluation purposes by video analysis [22].

Therefore, we made numerical simulation on a mesh with a geometry similar to the control area in the real world experiments. The simulation in the control area shows lane formation and congestion before an entrance, see Fig. 3. The origin of the congestion lays in the very static desired velocities we are currently using. A more dynamical desired velocity that better models pedestrian long and short sight behavior will be the subject of future work.

Experiments showed the fill-species and the pedestrian species should have the same density ρ. Otherwise, we may create artificial impulses through the separation step that could move heavier species to a place with higher velocity. Although different ρ values for different species will work, the impulse balance should be kept in mind.

We were also able to implement very basic in- and outlet boundary conditions for multiple species, i.e. the fill-species and a pedestrian species. For inlet boundaries we used a fixed value condition for the velocity together with the pressure boundary condition (16). For outlet boundary conditions we used (16) and (14) for the pressure and velocity, respectively. Further research should be put in in- and outlet boundary conditions for more complex in- and outflow scenarios of pedestrianis, e.g. the rate of flow should be conifigurable depending on the fill rate of cells next to the inlet boundary.

4   Discussion

In Section 2 we presented a macroscopic model based on the mass balance with velocity ansatzes respecting local and global influences. The numerical simulation results of the densities and the fluxes (pedestrian velocities) matched the experimental data with a good quality. Especially typical patterns of crossing pedestrian groups and time scales of egress-like processes were very good reproduced. The simulation results are promising and in our further research we shall investigate the detailed description of model parameters with the aim of a reduction of the present heuristic parameters. For special constellations like the genesis of small groups moving close together the fundamental diagrams must be appropriately modified.

In Section 3 we introduced a new ansatz for the simulation of pedestrian crossing and multispecies simulation. The implementation is based on the incompressible Navier-Stokes equations with a volume of fluid ansatz that has been altered by special boundary conditions for the pressure and the velocity as well as an added transport equation for the separation of intermixed species. The proposed model allowed us to reproduce common pedestrian crossing effects like stripe and lane formation. It also made possible to simulate higher numbers (more than two) of pedestrian species.

The model showed impulse effects originating from the Navier-Stokes equations, which are unnatural for pedestrian behavior. Therefore, it is the subject of future work to use a different set of equations and to study the stability and conservation properties of the solver in more detail. Another topic is the implementation of open boundaries for the in- and outflow of pedestrians in the simulation.

References

1. Chen, M.J., Bärwolff, G., Schwandt, H.: A study of step calculations in traffic cellular automaton models. In: 13th International IEEE Conference on Intelligent Transportation Systems, pp. 747–752 (2010), http://page.math.tu-berlin.de/~chenmin/pub/cbs100709.pdf (an electronic version) (accessed May 16, 2013)
2. Lämmel, G., Plaue, M.: Getting out of the way: Collision avoiding pedestrian models. In: PED 2012 Conference Proceedings (2012)
3. Moussaïd, M., Helbing, D., Theraulaz, G.: How simple rules determine pedestrian behavior and crowd disasters. PNAS 108(17), 6884–6888 (2011)
4. Predtechenskii, V.M., Milinskii, A.I.: Planning for Foot Traffic Flow in Buildings. Amerind Publishing, New Delhi (1978); Translation of Proekttirovanie Zhdanii s. Uchetom Organizatsii Dvizheniya Lyuddskikh Potokov. Stroiizdat, Moscow (1969)
5. Weidmann, U.: Transporttechnik der Fußgänger – transporttechnische Eigenschaften des Fußgängerverkehrs (Literaturstudie). Schriftenreihe der IVT 90 (March 1993) (in German)
6. Schadschneider, A., Klingsch, W., Kluepfel, H., Kretz, T., Rogsch, C., Seyfried, A.: Evacuation dynamics: Empirical results, modeling and applications. Encyclopedia of Complexity and Systems Science, 3142–3176 (2009)
7. Cristiani, E., Piccoli, B., Tosin, A.: Modeling self-organization in pedestrians and animal groups from macroscopic and microscopic viewpoints. In: Bellomo, N., Naldi, G., Pareschi, L., Toscani, G. (eds.) Mathematical Modeling of Collective Behavior in Socio-Economic and Life Sciences. Modeling and Simulation in Science, Engineering and Technology, pp. 337–364. Birkhäuser, Boston (2010)
8. Berres, S., Ruiz-Baier, R., Schwandt, H., Tory, E.M.: An adaptive finite-volume method for a model of two-phase pedestrian flow. Networks and Heterogeneous Media (NHM) 6 (2011)
9. Hughes, R.L.: A continuum theory for the flow of pedestrians. Transportation Research Part B 36, 507–535 (2002)
10. Huth, F., Bärwolff, G., Schwandt, H.: A macroscopic multiple species pedestrian flow model based on heuristics implemented with finite volumes. In: PED 2012 Conference Proceedings (2012)
11. Helbing, D., Farkas, I.J., Vicsek, T.: Freezing by heating in a driven mesoscopic system. Phys. Rev. Lett. 84(6), 1240–1243 (2000)
12. Radzihovsky, L., Clark, N.A.: Comment on "Freezing by heating in a driven mesoscopic system". Phys. Rev. Lett. 90(18), 189603 (2003)
13. Huth, F., Bärwolff, G., Schwandt, H.: Some fundamental considerations for the application of macroscopic models in the field of pedestrian crowd simulation. Preprint ID 2012/16 (2012), http://www.math.tu-berlin.de/menue/forschung/veroeffentlichungen/preprints_2012
14. Huth, F., Bärwolff, G., Schwandt, H.: Fundamental diagrams and multiple pedestrian streams. Preprint ID 2012/17 (2012), http://www.math.tu-berlin.de/menue/forschung/veroeffentlichungen/preprints_2012/
15. Bärwolff, G., Slawig, T., Schwandt, H.: Modeling of pedestrian flows using hybrid models of euler equations and dynamical systems. In: AIP Conference Proceedings, vol. 936(1), pp. 70–73 (2007)
16. Henderson, L.F.: The Statistics of Crowd Fluids. Nature 229(5284), 381–383 (1971)

17. Ahnert, T., Bärwolff, G., Schwandt, H.: A Multispecies Macroscopic Pedestrian Model approximated by a 3d incompressible Flow. In: Proceedings of the 7th International Conference on Information and Management Sciences 2012, Dunhuang/China. Series of Information and Management Sciences, vol. 7, pp. 475–480. California Polytechnic State University Press, Pomona (2012)
18. Helbing, D., Molnár, P.: Social force model for pedestrian dynamics. Phys. Rev. E 51, 4282–4286 (1995)
19. Johansson, A., Helbing, D., Shukla, P.: Specification of the social force pedestrian model by evolutionary adjustment to video tracking data. Advances in Complex Systems (10), 271–288 (2007)
20. Issa, R.I.: Solution of the implicitly discretised fluid flow equations by operator-splitting. Journal of Computational Physics (62) (1986)
21. OpenCFD Ltd: Open∇FOAM: The open source CFD toolbox (2010), http://www.openfoam.com (accessed today)
22. Plaue, M., Chen, M., Bärwolff, G., Schwandt, H.: Trajectory extraction and density analysis of intersecting pedestrian flows from video recordings. In: Stilla, U., Rottensteiner, F., Mayer, H., Jutzi, B., Butenuth, M. (eds.) PIA 2011. LNCS, vol. 6952, pp. 285–296. Springer, Heidelberg (2011)

A Highly Efficient Implementation on GPU Clusters of PDE-Based Pricing Methods for Path-Dependent Foreign Exchange Interest Rate Derivatives

Duy-Minh Dang[1], Christina C. Christara[2], and Kenneth R. Jackson[2]

[1] David R. Cheriton School of Computer Science,
University of Waterloo, Waterloo, ON, N2L 3G1, Canada
dm2dang@uwaterloo.ca
[2] Department of Computer Science,
University of Toronto, Toronto, ON, M5S 3G4, Canada
{ccc,krj}@cs.toronto.edu

Abstract. We present a highly efficient parallelization of the computation of the price of exotic cross-currency interest rate derivatives with path-dependent features via a Partial Differential Equation (PDE) approach. In particular, we focus on the parallel pricing on Graphics Processing Unit (GPU) clusters of long-dated foreign exchange (FX) interest rate derivatives, namely Power-Reverse Dual-Currency (PRDC) swaps with FX Target Redemption (FX-TARN) features under a three-factor model. Challenges in pricing these derivatives via a PDE approach arise from the high-dimensionality of the model PDE, as well as from the path-dependency of the FX-TARN feature. The PDE pricing framework for FX-TARN PRDC swaps is based on partitioning the pricing problem into several independent pricing sub-problems over each time period of the swap's tenor structure, with possible communication at the end of the time period. Finite difference methods on non-uniform grids are used for the spatial discretization of the PDE, and the Alternating Direction Implicit (ADI) technique is employed for the time discretization. Our implementation of the pricing procedure on a GPU cluster involves (i) efficiently solving each independent sub-problem on a GPU via a parallelization of the ADI timestepping technique, and (ii) utilizing MPI for the communication between pricing processes at the end of the time period of the swap's tenor structure. Numerical results showing the efficiency of the parallel methods are provided.

1 Introduction

In the current era of wildly fluctuating exchange rates, cross-currency interest rate derivatives, especially FX interest rate hybrid derivatives, referred to as hybrids, are of enormous practical importance. In particular, long-dated (maturities of 30 years or more) FX interest rate hybrids, such as Power-Reverse Dual-Currency (PRDC) swaps, are among the most liquid cross-currency interest rate derivatives [1]. The pricing of PRDC swaps, especially those with FX Target Redemption (TARN), is a subject of great interest in practice, especially among financial institutions. In a PRDC swap

B. Murgante et al. (Eds.): ICCSA 2013, Part V, LNCS 7975, pp. 107–126, 2013.

with a TARN feature, the sum of all FX-linked PRDC coupon amounts paid to date is recorded, and the underlying swap is terminated pre-maturely on the first date of the tenor structure when the accumulated PRDC coupon amount, including the coupon amount scheduled on that date, has reached or exceeded a pre-determined target cap. Hence, this exotic feature is usually referred to as a FX-TARN.

As FX interest rate derivatives, such as PRDC swaps, are exposed to movements in both the spot FX rate and the interest rates in both currencies, multi-factor pricing models having at least three factors, namely the domestic and foreign interest rates and the spot FX rate, must be used for the valuation of such derivatives. A popular choice for pricing PRDC swaps is Monte-Carlo (MC) simulation. However, this approach has several major disadvantages, such as slow convergence for problems in low-dimensions, i.e. fewer than five dimensions, and the limitation that the price is obtained at a single point only in the domain, as opposed to the global character of the Partial Differential Equation (PDE) approach. In addition, MC methods usually suffer from difficulty in computing accurate hedging parameters, such as delta and gamma, especially when dealing with the FX-TARN feature [2]. On the other hand, the pricing of these derivatives via the PDE approach is not only mathematically challenging but also very computationally intensive, due to (i) the "curse of dimensionality" associated with high-dimensional PDEs, and (ii) the complexities in handling path-dependent exotic features.

Over the last few years, the rapid evolution of Graphics Processing Units (GPUs) into powerful, cost-efficient, programmable computing architectures for general purpose computations has provided application potential beyond the primary purpose of graphics processing. In computational finance, although there has been great interest in utilizing GPUs in developing efficient pricing architectures for computationally intensive problems, the applications mostly focus on MC simulations applied to option pricing (e.g. [3, 4, 5]). The literature on utilizing GPUs in pricing financial derivatives via a PDE approach is rather sparse, with scattered work, such as [6, 7, 8, 9, 10]. The literature on GPU-based PDE methods for pricing cross-currency interest rate derivatives is even less developed.

In our paper [11], an efficient PDE pricing framework for pricing FX-TARN PRDC swaps is introduced in the public domain. The approach is to use an auxiliary path-dependent state variable to keep track of the accumulated PRDC coupon amount. This allows us to partition the pricing problem of these derivatives into several independent pricing sub-problems over each period of the swap's tenor structure, each of which corresponds to a discretized value of the auxiliary variable, with possible communication at the end of each time period.

In this paper, we describe a highly efficient parallelization of the PDE-based computation developed in [11] for the price of FX interest rate swaps with the FX-TARN feature. We adopt the three-factor pricing model proposed in [12]. Our implementation involves two levels of parallelism. The first is to use a cluster of GPUs together with the Compute Unified Device Architecture (CUDA) Application Programming Interface (API) to solve the afore-mentioned independent sub-problems simultaneously, each on a separate GPU. Since the main computational task associated with each sub-problem is the solution of the model three-dimensional PDE, the second level of parallelism is exploited via a highly efficient GPU-based parallelization of the ADI timestepping technique developed in our paper [7] for the solution of the model PDE. In addition, we utilize the Message Passing Interface (MPI) [13], a widely used message passing library standard, for efficient communication between the pricing processes at the end of each time period. The results of this paper show that GPU clusters can provide a significant increase in performance over GPUs when pricing exotic cross-currency interest rate derivatives with path-dependence features. Although we primarily focus on a three-factor model, many of the ideas and results in this paper can be naturally extended to higher-dimensional applications with constraints.

The remainder of this paper is organized as follows. In Section 2, we briefly describe PRDC swaps with FX-TARN features, then introduce a three-factor pricing model and the associated PDE. Discretization methods and a PDE-based pricing algorithm for FX-TARN PRDC swaps are discussed in Section 3. A parallelization of the pricing algorithm on GPU clusters for FX-TARN PRDC swaps is described in detail in Section 4. Numerical results are presented and discussed in Section 5. Section 6 concludes the paper and outlines possible future work.

2   Power-Reverse Dual-Currency Swaps

2.1   Introduction

Essentially, PRDC swaps are long-dated swaps (maturities of 30 years or more) which pay FX-linked coupons, i.e. PRDC coupons, referred to as the *coupon leg*, in exchange for London Interbank Offered Rate (LIBOR) floating-rate payments, referred to as the *funding leg*. Both the PRDC coupon and the floating rates are applied on the domestic currency principal N_d. There are two parties involved in the swap: the *issuer* of PRDC coupons (the receiver of the floating-rate payments – usually a bank) and the *investor* (the receiver of the PRDC coupons). We investigate PRDC swaps from the perspective of the issuer of PRDC coupons. Since a large variety of PRDC swaps are traded, for the sake of simplicity, only the basic structure is presented here.

To be more specific, we consider the tenor structure

$$T_0 = 0 < T_1 < \cdots < T_\beta < T_{\beta+1} = T, \nu_\alpha = T_\alpha - T_{\alpha-1}, \alpha = 1, 2, \ldots, \beta + 1, \quad (2.1)$$

where ν_α represents the year fraction between $T_{\alpha-1}$ and T_α using a certain day counting convention, such as the Actual/365 day counting one [14]. Unless otherwise stated, in this paper, the sub-scripts "d" and "f" are used to indicate domestic and foreign, respectively. Let $P_d(t, \overline{T})$ be the price at time $t \le \overline{T}$ in domestic currency of a domestic zero-coupon discount bond with maturity $\overline{T}$, and face value one unit of domestic currency. Note that, $P_d(t, \overline{T}) \le 1$ and $P_d(\overline{T}, \overline{T}) = 1$. For use later in the paper, define

$$T_{\alpha+} = T_\alpha + \delta \text{ where } \delta \to 0^+, \quad T_{\alpha-} = T_\alpha - \delta \text{ where } \delta \to 0^+, \quad (2.2)$$

i.e. $T_{\alpha-}$ and $T_{\alpha+}$ are instants of time just before and just after the date T_α, respectively.

Given the tenor structure (2.1), for a "vanilla" PRDC swap, at each time $\{T_\alpha\}_{\alpha=1}^\beta$, there is an exchange of a PRDC coupon amount for a domestic LIBOR floating-rate payment. More specifically, the funding leg pays the amount $\nu_\alpha L_d(T_{\alpha-1}, T_\alpha) N_d$ at

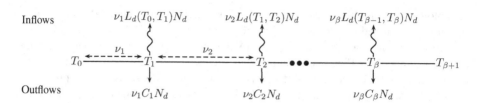

Fig. 1. Fund flows in a "vanilla" PRDC swap. Inflows and outflows are from the perspective of the PRDC coupon issuer, usually a bank.

time T_α for the period $[T_{\alpha-1}, T_\alpha]$. Here, $L_d(T_{\alpha-1}, T_\alpha)$ denotes the domestic LIBOR rate for the period $[T_{\alpha-1}, T_\alpha]$, as observed at time $T_{\alpha-1}$. This rate is simply-compounded and is defined by [14]

$$L_d(T_{\alpha-1}, T_\alpha) = \frac{1 - P_d(T_{\alpha-1}, T_\alpha)}{\nu_\alpha P_d(T_{\alpha-1}, T_\alpha)}. \tag{2.3}$$

Note that $L_d(T_{\alpha-1}, T_\alpha)$ is set at time $T_{\alpha-1}$, but the actual floating leg payment for the period $[T_{\alpha-1}, T_\alpha]$ does not occur until time T_α.

Throughout the paper, we denote by $s(t)$ the spot FX rate prevailing at time t. The PRDC coupon rate C_α, $\alpha = 1, 2, \ldots, \beta$, of the coupon amount $\nu_\alpha C_\alpha N_d$ issued at time T_α for the period $[T_\alpha, T_{\alpha+1}]$, $\alpha = 1, 2, \ldots, \beta$, has the structure

$$C_\alpha = \max\left(c_f \frac{s(T_\alpha)}{f_\alpha} - c_d, 0\right), \tag{2.4}$$

where c_d and c_f respectively are constant domestic and foreign coupon rates. The scaling factor f_α is usually set to the forward FX rate $F(0, T_\alpha)$ defined by [14]

$$F(0, T_\alpha) = \frac{P_f(0, T_\alpha)}{P_d(0, T_\alpha)} s(0), \tag{2.5}$$

which follows from no-arbitrage arguments. A diagram of fund flows in a "vanilla" PRDC swap is presented in Figure 1.[1]

By letting $h_\alpha = \dfrac{c_f}{f_\alpha}$, and $k_\alpha = \dfrac{c_d}{c_f} f_\alpha$, the PRDC coupon rate C_α can be viewed as a call option on FX rates, since, in this case, C_α reduces to

$$C_\alpha = h_\alpha \max(s(T_\alpha) - k_\alpha, 0). \tag{2.6}$$

As a result, the PRDC coupon leg in a "vanilla" PRDC swap can be viewed as a portfolio of long-dated options on the spot FX rate, i.e. long-dated FX options.

In a FX-TARN PRDC swap, the PRDC coupon amount, $\nu_\alpha C_\alpha N_d$, $\alpha = 1, 2, \ldots$, is recorded. The PRDC swap is pre-maturely terminated on the first date $T_{\alpha_e} \in \{T_\alpha\}_{\alpha=1}^\beta$ when the accumulated PRDC coupon amount, including the coupon amount scheduled on that date, reaches or exceeds a pre-determined target cap, hereinafter denoted by

---

[1] Note that in the above setting, the last period $[T_\beta, T_{\beta+1}]$ of the swap's tenor structure is redundant, since there is no exchange of fund flows at time $T_{\beta+1}$. However, to be consistent with [12], we follow the same notation used in [12].

a_c. That is, the associated underlying PRDC swap terminates immediately on the first date $T_{\alpha_e} \in \{T_\alpha\}_{\alpha=1}^\beta$ when $\sum_{\alpha=1}^{\alpha_e} \nu_\alpha C_\alpha N_d \geq a_c$. In this paper, we discuss the case when the early termination is determined by the equality, i.e. $\sum_{\alpha=1}^{\alpha_e} \nu_\alpha C_\alpha N_d = a_c$. Note that, in this case, the last PRDC coupon amount could possible get truncated, due to the cap a_c. A description of other variations of FX-TARN PRDC swaps, as well as the financial motivation for these derivatives can be found in [11].

We conclude this subsection by noting that, usually, there is a settlement in the form of an initial fixed-rate coupon between the issuer and the investor at time T_0 that is not included in the description above. This signed coupon is typically the value at time T_0 of the swap to the issuer, i.e. the value at time T_0 of all net fund flows in the swap, with a positive value of the fixed-rate coupon indicating a fund outflow for the issuer or a fund inflow for the investor, i.e. the issuer pays the investor. Conversely, a negative value of this coupon indicates a fund inflow for the issuer.

2.2 The Model and the Associated PDE

We consider the multi-currency model proposed in [12]. We denote by $s(t)$ the spot FX rate, and by $r_i(t), i = d, f$, the domestic and foreign short rates, respectively. Under the domestic risk-neutral measure, the dynamics of $s(t), r_d(t), r_f(t)$ can be described by [15]

$$\frac{ds(t)}{s(t)} = (r_d(t) - r_f(t))dt + \gamma(t, s(t))dW_s(t),$$
$$dr_d(t) = (\theta_d(t) - \kappa_d(t)r_d(t))dt + \sigma_d(t)dW_d(t), \tag{2.7}$$
$$dr_f(t) = (\theta_f(t) - \kappa_f(t)r_f(t) - \rho_{fs}(t)\sigma_f(t)\gamma(t, s(t)))dt + \sigma_f(t)dW_f(t),$$

where $W_d(t), W_f(t)$, and $W_s(t)$ are correlated Brownian motions with $dW_d(t)dW_s(t) = \rho_{ds}dt$, $dW_f(t)dW_s(t) = \rho_{fs}dt$, $dW_d(t)dW_f(t) = \rho_{df}dt$. The short rates follow the mean-reverting Hull-White model [16] with deterministic mean reversion rates and volatility functions, respectively, denoted by $\kappa_i(t)$ and $\sigma_i(t)$, for $i = d, f$, while $\theta_i(t)$, $i = d, f$, also deterministic, capture the current term structures. The local volatility function $\gamma(t, s(t))$ for the spot FX rate has the functional form [12]

$$\gamma(t, s(t)) = \xi(t)\left(\frac{s(t)}{\ell(t)}\right)^{\varsigma(t)-1}, \tag{2.8}$$

where $\xi(t)$ is the relative volatility function, $\varsigma(t)$ is the time-dependent constant elasticity of variance (CEV) parameter and $\ell(t)$ is a time-dependent scaling constant which is usually set to the forward FX rate $F(0, t)$, for convenience in calibration [12]. Let $u \equiv u(s, r_d, r_f, t)$ denote the domestic value function of a PRDC swap at time t, $T_{\alpha-1} \leq t < T_\alpha, \alpha = \beta, \ldots, 1$. Given a terminal payoff at maturity time T_α, then on $\mathbb{R}_+ \times \mathbb{R} \times \mathbb{R} \times [T_{\alpha-1}, T_\alpha)$, u satisfies the PDE [15][2]

---

[2] Here, we assume that u is sufficiently smooth on the domain $\mathbb{R}_+ \times \mathbb{R} \times \mathbb{R} \times [T_{\alpha-1}, T_\alpha)$.

$$\frac{\partial u}{\partial t} + \mathcal{L}u \equiv \frac{\partial u}{\partial t} + \frac{1}{2}\gamma^2(t,s(t))s^2\frac{\partial^2 u}{\partial s^2} + \frac{1}{2}\sigma_d^2(t)\frac{\partial^2 u}{\partial r_d^2} + \frac{1}{2}\sigma_f^2(t)\frac{\partial^2 u}{\partial r_f^2}$$

$$+ \rho_{ds}\sigma_d(t)\gamma(t,s(t))s\frac{\partial^2 u}{\partial s \partial r_d} + \rho_{fs}\sigma_f(t)\gamma(t,s(t))s\frac{\partial^2 u}{\partial s \partial r_f} + \rho_{df}\sigma_d(t)\sigma_f(t)\frac{\partial^2 u}{\partial r_d \partial r_f}$$

$$+ (r_d - r_f)s\frac{\partial u}{\partial s} + \Big(\theta_d(t) - \kappa_d(t)r_d\Big)\frac{\partial u}{\partial r_d} + \Big(\theta_f(t) - \kappa_f(t)r_f - \rho_{fs}\sigma_f(t)\gamma(t,s(t))\Big)\frac{\partial u}{\partial r_f}$$

$$- r_d u = 0.$$

(2.9)

Since we solve the PDE backward in time, the change of variable $\tau = T_\alpha - t$ is used. Under this change of variable, the PDE (2.9) becomes

$$\frac{\partial u}{\partial \tau} = \mathcal{L}u \qquad (2.10)$$

and is solved forward in τ. The pricing of cross-currency interest rate derivatives in general, and PRDC swaps in particular, is defined in an unbounded domain

$$\{(s, r_d, r_f, \tau) | s \geq 0, -\infty < r_d < \infty, -\infty < r_f < \infty, \tau \in [0, T]\}, \qquad (2.11)$$

where $T = T_\alpha - T_{\alpha-1}$. Here, $-\infty < r_d < \infty$ and $-\infty < r_f < \infty$, since the Hull-White model can yield any positive or negative value for the interest rate. To solve the PDE (2.10) numerically by FD methods, we truncate the unbounded domain into a finite-sized computational one

$$\{(s, r_d, r_f, \tau) \in [0, s_\infty] \times [-r_{d,\infty}, r_{d,\infty}] \times [-r_{f,\infty}, r_{f,\infty}] \times [0, T]\} \equiv \Omega \times [0, T], \qquad (2.12)$$

where s_∞, $r_{d,\infty}$ and $r_{f,\infty}$ are sufficiently large [17].

Since payoffs and fund flows are deal-specific, we defer specifying the terminal conditions until Section 3. The difficulty with choosing boundary conditions is that, for an arbitrary payoff, they are not known. A detailed analysis of the boundary conditions is not the focus of this paper; we leave it as a topic for future research. For this paper, we impose Dirichlet-type "stopped process" boundary conditions where we stop the processes $s(t), r_f(t), r_d(t)$ when any of the three hits the boundary of the finite-sized computational domain. Thus, the value on the boundary is simply the discounted payoff for the current values of the state variables [11]

3 Numerical Methods

In this section, we briefly discuss a PDE-based pricing method for FX-TARN PRDC swaps. The reader is referred to our paper [11] for more details.

3.1 Discretization of the Model PDE

Let the number of sub-intervals be $n + 1$, $p + 1$, $q + 1$, and l in the s-, r_d-, r_f-, and τ-directions, respectively. We use a fixed, but not necessarily uniform, spatial grid together with dynamically chosen timestep sizes. For the discretization of the space variables in the differential operator $\mathcal{L}$ of (2.10), we employ FD *central* schemes on non-uniform grids in the interior of the rectangular domain Ω. More specifically, the first and second partial derivatives of the space variables in (2.10) are approximated by the standard three-point stencils *central* FD schemes, while the cross-derivatives in (2.10) are approximated by a nine-point (3×3) FD stencil.[3]

For the time discretization of the PDE (2.10), we employ the ADI timestepping technique based on the Hundsdorfer and Verwer (HV) splitting approach [18], henceforth referred to as the *HV scheme*. Note that the study of the HV scheme for mixed derivatives high-dimensional PDEs is found in [19]. Let $\mathbf{u}^m$ denote the vector of values of the unknown prices at time τ_m on the mesh Ω that approximates the exact solution $u^m = u(s, r_d, r_f, \tau_m)$. We denote by $\mathbf{A}^m$ the matrix of size $npq \times npq$ arising from the FD discretization of the differential operator $\mathcal{L}$ at τ_m.

Following the HV approach, we decompose the matrix $\mathbf{A}^m$ into four sub-matrices: $\mathbf{A}^m = \mathbf{A}_0^m + \mathbf{A}_1^m + \mathbf{A}_2^m + \mathbf{A}_3^m$. The matrix $\mathbf{A}_0^m$ is the part of $\mathbf{A}^m$ that comes from the FD discretization of the cross-derivative terms in (2.10), while the matrices $\mathbf{A}_1^m$, $\mathbf{A}_2^m$ and $\mathbf{A}_3^m$ are the three parts of $\mathbf{A}^m$ that correspond to the spatial derivatives in the s-, r_d-, and r_f-directions, respectively. The term $r_d u$ in $\mathcal{L}u$ is distributed evenly over $\mathbf{A}_1^m$, $\mathbf{A}_2^m$ and $\mathbf{A}_3^m$. Starting from $\mathbf{u}^{m-1}$, the HV scheme generates an approximation $\mathbf{u}^m$ to the exact solution u^m, $m = 1, \ldots, l$, by[4]

$$\begin{cases}
\mathbf{v}_0 = \mathbf{u}^{m-1} + \Delta\tau_m(\mathbf{A}^{m-1}\mathbf{u}^{m-1} + \mathbf{g}^{m-1}), & \text{(3.1a)} \\[2mm]
(\mathbf{I} - \theta\Delta\tau_m\mathbf{A}_i^m)\mathbf{v}_i = \mathbf{v}_{i-1} - \theta\Delta\tau_m\mathbf{A}_i^{m-1}\mathbf{u}^{m-1} \\
\qquad\qquad + \theta\Delta\tau_m(\mathbf{g}_i^m - \mathbf{g}_i^{m-1}), \quad i = 1, 2, 3, & \text{(3.1b)} \\[2mm]
\tilde{\mathbf{v}}_0 = \mathbf{v}_0 + \dfrac{1}{2}\Delta\tau_m(\mathbf{A}^m\mathbf{v}_3 - \mathbf{A}^{m-1}\mathbf{u}^{m-1}) \\
\qquad\qquad + \dfrac{1}{2}\Delta\tau_m(\mathbf{g}^m - \mathbf{g}^{m-1}), & \text{(3.1c)} \\[2mm]
(\mathbf{I} - \theta\Delta\tau_m\mathbf{A}_i^m)\tilde{\mathbf{v}}_i = \tilde{\mathbf{v}}_{i-1} - \theta\Delta\tau_m\mathbf{A}_i^m\mathbf{v}_3, \quad i = 1, 2, 3, & \text{(3.1d)} \\[2mm]
\mathbf{u}^m = \tilde{\mathbf{v}}_3. & \text{(3.1e)}
\end{cases}$$

In (3.1), the vector $\mathbf{g}^m$ is given by $\mathbf{g}^m = \sum_{i=0}^3 \mathbf{g}_i^m$, where $\mathbf{g}_i^m$ are obtained from the boundary conditions corresponding to the respective spatial derivative terms.

When solving the PDE (2.10) backward in time over each time period of the swap's tenor structure, for damping purposes, we first apply the HV scheme with $\theta = 1$ for the first few (usually two) initial timesteps, and then switch to $\theta = \frac{1}{2} + \frac{1}{6}\sqrt{3}$ for the remaining timesteps.

3.2 Timestep Size Selector

We use a simple, but effective, timestep size selector, where, given the current stepsize $\Delta\tau_m$, $m \geq 1$, the new stepsize $\Delta\tau_{m+1}$ is given by [11]

$$\begin{cases}
\Delta\tau_{m+1} = \left(\min_{1 \leq \iota \leq npq} \left[\dfrac{\texttt{dnorm}}{\frac{|\mathbf{u}_\iota^m - \mathbf{u}_\iota^{m-1}|}{\max(N, |\mathbf{u}_\iota^m|, |\mathbf{u}_\iota^{m-1}|)}} \right] \right) \Delta\tau_m, \\[4mm]
\Delta\tau_{m+1} = \min\{\Delta\tau_{m+1}, T - \tau_m\}.
\end{cases} \tag{3.2}$$

---

[3] On uniform grids, the nine-point FD stencil reduces to a four-point one.

[4] This is the scheme (1.4) in [19] with $\mu = \frac{1}{2}$.

Here, dnorm is a user-defined target relative change, and the scale N is chosen so that the method does not take an excessively small stepsize where the value of the option is small. Normally, for option values in dollars, $N = 1$ is used. We use $N = 1$ for PRDC swap pricing too. In all our experiments, we used $\Delta \tau_1 = 10^{-2}$ and dnorm $= 0.3$ on the coarsest grids. The value of dnorm is reduced by two at each refinement, while $\Delta \tau_1$ is reduced by four.

3.3   A PDE Pricing Algorithm

Denote by $a(t)$, $0 \leq a(t) < a_c$, the auxiliary path-dependent state variable which represents the accumulated PRDC coupon amount. The value of a FX-TARN PRDC swap depends on four stochastic state variables, namely $s(t)$, $r_d(t)$, $r_f(t)$ and the path-dependent variable $a(t)$. It is important to note that, since $a(t)$ changes only on the dates $\{T_\alpha\}_{\alpha=1}^\beta$, the pricing PDE does not depend on $a(t)$ (see (2.9)). For presentation purposes, we further adopt the following notation: $a_{\alpha+} \equiv a(T_{\alpha+})$, $a_{\alpha-} \equiv a(T_{\alpha-})$.

Pricing FX-TARN PRDC swaps via a PDE approach is highly challenging due to the path-dependency of the TARN feature and the backward nature of a PDE approach. We observe that, over each period $[T_{(\alpha-1)+}, T_{\alpha-}]$ of the swap's tenor structure, the backward procedure, which computes the solution backward in time from $T_{\alpha-}$ to $T_{(\alpha-1)+}$, needs to be invoked only if the swap is still alive at time $T_{(\alpha-1)+}$, i.e. if $a_{(\alpha-1)+}$ satisfies $0 \leq a_{(\alpha-1)+} < a_c$. Since we progress backward in time and the variable $a(t)$ is path-dependent, we do not know the exact value of $a_{(\alpha-1)+}$. However, since $0 \leq a_{(\alpha-1)+} < a_c$, we can discretize the variable a, as we do with other spatial variables. To this end, we partition the interval $[0, a_c]$ into $w + 1$ sub-intervals having non-uniform gridpoints,

$$0 = a_0 < a_1 < \ldots < a_w < a_{w+1} = a_c, \tag{3.3}$$

where the gridpoints are denser toward a_c. The PDE pricing framework for a FX-TARN PRDC swap involves

(a) across each date $\{T_\alpha\}_{\alpha=\beta}^1$ and for each discretized value a_y of the variable a, applying certain updating rules to (i) take into account the fund flows scheduled on that date; (ii) reflect changes in the accumulated PRDC coupon amount, and the possibility of early termination; and (iii) obtain terminal conditions for the solution of the PDE from time $T_{\alpha-}$ to $T_{(\alpha-1)+}$.

(b) over each period $[T_{(\alpha-1)+}, T_{\alpha-}]$, $\alpha = \beta, \ldots, 1$, of the swap's tenor structure, for each discretized value a_y of the variable a, solving the model PDE (2.9) backward in time from $T_{\alpha-}$ to $T_{(\alpha-1)+}$, with the corresponding terminal condition obtained from the above step.

Remark 1. To improve the efficiency of the numerical methods, for the solution of the model PDE, we use non-uniform grids. We denote by Δ_α^y, $y = 0, \ldots, w$, the non-uniform three-dimensional grids used for the solution of the PDE corresponding to a_y over the time period $[T_{(\alpha-1)+}, T_{\alpha-}]$ in (b) above. The non-uniform grids Δ_α^y are more refined around $r_d(0)$ and $r_f(0)$ in the r_d- and the r_f-directions, respectively. In the s-direction, the grids Δ_α^y, are more refined around the strike k_α and around the value of s at which the early termination occurs, hereinafter denoted by b_α^y. Note that, within $[T_{(\alpha-1)+}, T_{\alpha-}]$, k_α is the same for all sub-problems, but b_α^y, $y = 0, \ldots, w$, are not. Both k_α and b_α^y, $y = 0, \ldots, w$, change from one time period to the next. In our implementation, we apply linear interpolation along the s- and a-directions to switch between spatial grids (see Lines 5 and 10 of Algorithm 3.1).

Let $u_\alpha(t; a)$ represent the value at time t of a FX-TARN PRDC swap that has (i) $\{T_{\alpha+1}, \ldots, T_\beta\}$ as pre-mature termination opportunities, i.e. the swap is still alive at time T_α; and (ii) the total accumulated PRDC coupon amount, including the coupon amount scheduled on T_α, is equal to $a < a_c$. In particular, the quantity $u_0(T_0; 0)$ is the value of the FX-TARN PRDC swap we are interested in at time T_0. Also let $u_\alpha^{y, \widetilde{\alpha}}(t; a)$, $y = 0, \ldots, w$, $\widetilde{\alpha} = \beta, \ldots, 1$, represent an approximation to $u_\alpha(t; a)$ at gridpoints of the computational grid $\Delta_{\widetilde{\alpha}}^y$. In general, the indices $(y, \widetilde{\alpha})$ denote the associated computational grid $\Delta_{\widetilde{\alpha}}^y$, $y = 0, \ldots, w$, $\widetilde{\alpha} = \beta, \ldots, 1$. A backward pricing algorithm for FX-TARN PRDC swaps is presented in Algorithm 3.1.

4   Efficient Implementation on Clusters of GPUs

4.1   GPU Device Architecture

A GPU is a hierarchically arranged multiprocessor unit, in which several scalar processors are grouped into a smaller number of streaming multiprocessors (SMs). Each SM has shared memory accessed by all its scalar processors. In addition, the GPU has global (device) memory (slower than shared memory) accessed by all scalar processors on the chip, as well as a small amount of cache for storing constants. According to the programming model of CUDA, which we adopt, the host (CPU/master) uploads the intensive work to the GPU as a single program, called the *kernel*. Multiple copies of the kernel, referred to as *threads*, are then distributed to the available processors, where they are executed in parallel. Within the CUDA framework, threads are grouped into *threadblocks*, which are in turn arranged on a *grid*. Threads in a threadblock run on at most one multiprocessor, and can communicate with each other efficiently via the shared memory, as well as synchronize their executions. For a more detailed description of the GPU, interested readers are referred to [20].

4.2   GPU Cluster

All of the experiments in this paper were carried out on a GPU cluster with the following specifications:
- The cluster has 22 (server) nodes, each of which consists of two quad-core Intel "Harpertown" host systems with Intel Xeon E5430 CPUs running at 2.66GHz, with a total of 8GB of memory shared between the two quad-core Xeon processors. Thus, there are 44 hosts available. All the nodes are interconnected via 4x DDR Infiniband (16 Gigabytes/s).
- The GPU portion of the cluster is composed of 11 NVIDIA S1070 GPU servers, each of which contains two pairs of Tesla 10-series (T10) GPUs. Thus, there are 44 GPUs available. Each pair of the T10 GPUs is attached to a node via a PCI Express 2.0x16

Algorithm 3.1 Backward algorithm for computing FX-TARN PRDC swaps.

1: construct Δ_β^y; set $u_\beta(T_{\beta+}; a_y) = 0$, $y = 0, \ldots, w$;
2: **for** $\alpha = \beta, \ldots, 1$ **do**
3:     **for** each a_y, $y = 0, \ldots, w$, **do**
4:         set $\qquad\qquad\qquad\qquad \bar{a}_y = a_y + \min(a_c - a_y, \nu_\alpha C_\alpha N_d);$ $\qquad\qquad$ (3.4)

5:         set
$$u_{\alpha-1}^{y,\alpha}(T_{\alpha+}; \bar{a}_y) = \begin{cases} 0 & \text{if } \bar{a}_y \geq a_c, \\ \dfrac{\bar{a}_y - a_{\bar{y}}}{a_{\bar{y}+1} - a_{\bar{y}}} u_\alpha^{y,\alpha}(T_{\alpha+}; a_{\bar{y}+1}) + \dfrac{a_{\bar{y}+1} - \bar{a}_y}{a_{\bar{y}+1} - a_{\bar{y}}} u_\alpha^{y,\alpha}(T_{\alpha+}; a_{\bar{y}}) \\ \qquad\qquad \text{if } a_{\bar{y}} \leq \bar{a}_y \leq a_{\bar{y}+1}, \bar{y} \in \{0, \ldots, w\}, \end{cases}$$
(3.5)

where $u_\alpha^{y,\alpha}(T_{\alpha+}; a_{\bar{y}})$ and $u_\alpha^{y,\alpha}(T_{\alpha+}; a_{\bar{y}+1})$ are obtained by linear interpolation along the s-direction on $u_\alpha^{\bar{y},\alpha}(T_{\alpha+}; a_{\bar{y}})$ and $u_\alpha^{\bar{y}+1,\alpha}(T_{\alpha+}; a_{\bar{y}+1})$, respectively;
6:         set
$$\hat{u}_{\alpha-1}^{y,\alpha}(T_{\alpha-}; a_y) = u_{\alpha-1}^{y,\alpha}(T_{\alpha+}; \bar{a}_y) - \min(a_c - a_y, \nu_\alpha C_\alpha N_d);$$
(3.6)

7:         solve the PDE (2.9) with the terminal condition (3.6) from $T_{\alpha-}$ to $T_{(\alpha-1)+}$ using the ADI scheme (3.1) for each time τ_m, $m = 1, \ldots, l$, with the timestep size $\Delta\tau_m$ selected by (3.2), to obtain $\hat{u}_{\alpha-1}^{y,\alpha}(T_{(\alpha-1)+}; a_y)$;
8:         **if** $\alpha \geq 2$ **then**
9:             construct $\Delta_{\alpha-1}^y$
10:             linearly interpolate $\hat{u}_{\alpha-1}^{y,\alpha}(T_{(\alpha-1)+}; a_y)$ along the s-direction to obtain $\hat{\hat{u}}_{\alpha-1}^{y,\alpha-1}(T_{(\alpha-1)+}; a_y)$;
11:             set $\quad u_{\alpha-1}^{y,\alpha-1} T_{(\alpha-1)+}; a_y) = \hat{\hat{u}}_{\alpha-1}^{y,\alpha-1}(T_{(\alpha-1)+}; a_y) + (1 - P_d(T_\alpha))N_d;$ $\quad$ (3.7)
12:         **else**
13:             set $\quad u_{\alpha-1}^{y,\alpha} T_{(\alpha-1)+}; a_y) = \hat{u}_\alpha^{y,\alpha-1}(T_{(\alpha-1)+}; a_y) + (1 - P_d(T_\alpha))N_d;$ $\quad$ (3.8)
14:         **end if**
15:     **end for**
16: **end for**
17: set $u_0(T_0; 0) = u_0(T_{0+}; 0);$

---

link. As such, there is a T10 GPU per quad-core Xeon processor, and thus each host has a GPU associated with it, and vice-versa.

Each NVIDIA Tesla T10 GPU consists of 4GB of global memory, 30 independent SMs, each containing 8 processors running at 1.44GHz, a total of 16384 registers, and 16 KB of shared memory per SM.

4.3   GPU-Based Parallel Pricing Framework

The key point in Algorithm 3.1 is that, over each time period $[T_{(\alpha-1)+}, T_{\alpha-}]$ of the tenor structure, we have multiple, entirely independent, pricing sub-problems (processes) to solve, each of which corresponds to a discrete value a_y, $y = 0, \ldots, w$. Hence, within each time period of the tenor structure, it is natural to assign each of the $w + 1$ pricing processes to a separate host/GPU. However, communication between these pricing processes is required across each date of the tenor structure, due to the interpolation (3.5), along the a-direction.

In the following presentation, we assume that the total number of available hosts of the cluster is at least $w + 1$, each host having a respective GPU associated with it. Under the MPI framework, assume that a group of $w + 1$ parallel pricing processes has been created, with the y-th process being associated with the discrete value a_y, $y = 0, \ldots, w$. Here, the quantities y, $y = 0, \ldots, w$, are referred to as *ranks* of the processes in the group. For each instance of $\alpha, \alpha = \beta, \ldots, 1$, to proceed from T_α to $T_{\alpha-1}$, assume that the values $u_\alpha^{y,\alpha}(T_{\alpha+}; a_y)$, $y = 0, \ldots, w$, have been computed at the previous period of the tenor structure, and are available in the yth host/GPU. Also assume that the appropriate kernels have been launched by the hosts on the respective GPUs. Then, the parallel implementation of Algorithm 3.1 for one instance of α can be described by the following stages:

Stage 1: each thread in each GPU updates its quantity $\bar{a}_y$ via (3.4), then determines the ranks of those processes from which it will require to receive data in order to apply the interpolation (3.5); each GPU appropriately collects the ranks' data from all its threads, so that each process knows collectively the ranks of those processes from which it will require to receive data to apply (3.5);

Stage 2: each host copies the ranks' data from its GPU global memory to the host memory.

Stage 3: the hosts perform communication amongst each other via MPI, so that each host receives the data needed for the interpolation (3.5) associated with the host's process.

Stage 4: each host copies the relevant data form its host memory to its GPU global memory.

Stage 5: each thread in each GPU carries out the interpolation (3.5).

Stage 6: each thread in each GPU computes the PRDC coupons via (3.6).

Stage 7: each GPU solves its associated PDE (2.9) from $T_{\alpha-}$ to $T_{(\alpha-1)+}$ with the terminal condition obtained from Stage 6.

Stage 8: each thread in each GPU (possibly) applies linear interpolation along the s-direction as given on Line 10 of Algorithm 3.1.

Stage 9: each thread in each GPU computes the funding payments via (3.7) or (3.8).

Note that, Stage 3 involves communication among hosts using MPI, while all other stages take place in each host/GPU, in parallel with and independently from other hosts/GPUs.

We now give more details of the implementation of the above stages. For presentation purposes, we denote by $\mathbf{u}_{\alpha+}^y$ the vector of data corresponding to a_y, $y = 0, \ldots, w$, i.e. the vector of data of the process y, available at time $T_{\alpha+}$ as it results from the computations during the last time period $[T_{\alpha+}, T_{(\alpha+1)-}]$.

4.4   Stages 1 and 2

For each process $y, y = 0, \ldots, w$, i.e. for each host/GPU, assume that we have an array of size $w + 1$ in the host memory, referred to as the array $RECV\_FROM$. The $\bar{y}$th entry of the array $RECV\_FROM$ corresponds to the discrete value $a_{\bar{y}}$, $\bar{y} = 0, \ldots, w$, i.e. it corresponds to the process with rank $\tilde{y}$ of the group. The entries of the array are of binary type, and are pre-set to a certain value, e.g. 0. The array is copied from the host memory to the device memory before the kernel of Stage 1 is launched.

We partition the computational grid of size $n \times p \times q$ into 2-D blocks of size $n_b \times p_b$. We let the kernel generate a $\texttt{ceil}\left(\frac{n}{n_b}\right) \times \texttt{ceil}\left(\frac{pq}{p_b}\right)$ grid of threadblocks, where $\texttt{ceil}$ denotes the ceiling function. All gridpoints of a $n_b \times p_b$ 2-D block are assigned to one threadblock only, with one thread for each gridpoint.

Each thread of a threadblock of the kernel launched in this stage computes the quantity $\bar{a}_y$ associated with it via (3.4). If the quantity $\bar{a}_y$ satisfies $a_{\tilde{y}} \leq \bar{a}_y \leq a_{\tilde{y}+1}$ for some $\tilde{y} \in \{y, \ldots, w\}$, the thread then changes the pre-set values of the $\tilde{y}$ and $(\tilde{y}+1)$st entries in the array $RECV\_FROM$ to 1. This procedure essentially marks the ranks of the processes from which some data are required by process y. Note that no data loadings from the global memory are required for this procedure.

The approach adopted here suggests a $(w+1-y)$-iteration loop in the kernel. During each iteration, each threadblock works with a pair of $a_{\tilde{y}}$ and $a_{\tilde{y}+1}$. Note that, although it may happen that multiple threads try to write to the same memory location of an entry of the array at the same time, it is guaranteed that one of the writes will succeed. Although we do not know which one, it does not matter for our purposes. Consequently, this approach suffices and works well.

After the kernel of Stage 1 has ended, Stage 2 takes place, in which the array $RECV\_FROM$ is copied back to the host memory for use in Stage 3.

4.5   Stages 3 and 4

At this point, each host has the array $RECV\_FROM$ corresponding to its process. Next, each process is to determine the ranks of those processes which need its data.

To handle this issue, consider a fictitious $(w+1) \times (w+1)$ matrix, for which the $\tilde{y}$th row, $\tilde{y} = 0, \ldots, w$, is the array $RECV\_FROM$ of the process of rank $\tilde{y}$. We observe that the yth column of this matrix, referred to as the array $SEND\_TO$, marks the ranks of processes which need the yth process data.

To form the array $SEND\_TO$ in each host, all hosts perform collective communication via MPI, essentially a parallel matrix transposition using the function $\texttt{MPI\_Alltoall}(\cdots)$.

Now, each process has in its host memory the arrays $RECV\_FROM$ and $SEND\_TO$, in addition to the vector $\mathbf{u}_{\alpha+}^y$. Thus, each process can easily perform data exchange with the appropriate processes, by looping through all the "marked" entries of the arrays $RECV\_FROM$ and $SEND\_TO$. In our implementation, we use $\texttt{MPI\_Send}(\cdots)$ and $\texttt{MPI\_Recv}(\cdots)$.

At this point, process y has in its host memory all the vectors of data it needs to carry out the interpolation scheme (3.5). By the data exchange procedure described above, these vectors are stored in a buffer in increasing order with respect to their associated ranks (or discrete values of a). For presentation purposes, we assume that a total of $k - 1$, $k \geq 1$, vectors of data were fetched by process y from other processes during Stage 3. We denote the sorted by index list of k vectors, including the vector $\mathbf{u}_{\alpha+}^y$, by $\{\mathbf{u}_{\alpha+}^{y_1}, \ldots, \mathbf{u}_{\alpha+}^{y_k}\}$, where y_j, $j = 1, \ldots, k$, are in $\{y, \ldots, w\}$, with $y_1 = y$, and $y_1 < y_2 < \cdots < y_k$. This concludes Stage 3.

In Stage 4, these vectors are then copied from the process' host memory to the global memory of the respective GPU, before the kernel for Stage 5 is launched.

4.6   Stages 5 and 6

In Stage 5, for a GPU-based implementation of the interpolation procedure, we adopt the same partitioning approach and assignment of gridpoints to threads as in Stage 1 described earlier. Recall that, in Stage 1, each thread has already computed the quantity $\bar{a}_y$ associated with it using (3.4). The interpolation (3.5) can be achieved by a k-iteration loop in the kernel. During the jth iteration of the k-iteration loop in the kernel, each thread in a threadblock performs linear interpolations, first along the s-direction, then along the a-direction, using the corresponding values in $\mathbf{u}_{\alpha+}^{y_j}$ and $\mathbf{u}_{\alpha+}^{y_{j+1}}$. Note that full memory coalescence is achieved for the data loading of this stage [21].

In Stage 6, using the same partitioning, each thread then computes the PRDC coupons via (3.6), independently from the others.

4.7   Stage 7

We now discuss a GPU-based parallel algorithm for the solution of the model PDE problem. The parallelism in a GPU for this stage is based on an efficient paralleliza-tion of the computation of each timestep of the ADI scheme (3.1a)–(3.1d) developed in our paper [7]. Below, we summarize our implementation. For details and discus-sions of related issues, such as memory coalescing and possible improvements, of our implementation, we refer the reader to [7].

4.7.1   ADI timestepping on GPUs

The HV scheme (3.1a)–(3.1d) can be divided into two phases. The first phase consists of a forward Euler step (predictor step (3.1a)), followed by three implicit, but unidirec-tional, corrector steps (3.1b), the purpose of which is to stabilize the predictor step. The second phase (i.e. (3.1c)-(3.1d)) restores second-order convergence of the discretization method if the model PDE contains mixed derivatives. Step (3.1e) is trivial. With respect to the CUDA implementation, the two phases are essentially the same; they can both be decomposed into matrix-vector multiplications and solving independent tridiagonal systems. Hence, for brevity, we only summarize our GPU parallelization of the first phase. For presentation purposes, let

$$\mathbf{w}_i = \Delta\tau_m \mathbf{A}_i^{m-1}\mathbf{u}^{m-1} + \Delta\tau_m(\mathbf{g}_i^{m-1} - \mathbf{g}_i^m), \quad i = 0, 1, 2, 3,$$

$$\widehat{\mathbf{A}}_i^m = \mathbf{I} - \theta\Delta\tau_m\mathbf{A}_i^m, \quad \widehat{\mathbf{v}}_i = \mathbf{v}_{i-1} - \theta\mathbf{w}_i, \quad i = 1, 2, 3,$$

and notice that $\mathbf{v}_0 = \mathbf{u}^{m-1} + \sum_{i=0}^{3} \mathbf{w}_i + \Delta\tau_m\mathbf{g}^m$. It is worth noting that the vectors $\mathbf{w}_i, \mathbf{v}_i, i = 0, 1, 2, 3$, and $\widehat{\mathbf{v}}_i, i = 1, 2, 3$, depend on τ, but, to simplify the notation, we do not indicate the superscript for the timestep index. Our CUDA implementation of the first phase consists of the following steps:

1. Step a.1: Compute the matrices $\mathbf{A}_i^m, i = 0, 1, 2, 3$, and $\widehat{\mathbf{A}}_i^m, i = 1, 2, 3$, and the

vectors $\mathbf{w}_i$, $i = 0, 1, 2, 3$, and $\mathbf{v}_0$.

2. Step a.2: Set $\widehat{\mathbf{v}}_1 = \mathbf{v}_0 - \theta\mathbf{w}_1$ and solve $\widehat{\mathbf{A}}_1^m\mathbf{v}_1 = \widehat{\mathbf{v}}_1$;

3. Step a.3: Set $\widehat{\mathbf{v}}_2 = \mathbf{v}_1 - \theta\mathbf{w}_2$ and solve $\widehat{\mathbf{A}}_2^m\mathbf{v}_2 = \widehat{\mathbf{v}}_2$;

4. Step a.4: Set $\widehat{\mathbf{v}}_3 = \mathbf{v}_2 - \theta\mathbf{w}_3$ and solve $\widehat{\mathbf{A}}_3^m\mathbf{v}_3 = \widehat{\mathbf{v}}_3$;

First phase - Step a.1

We partition the computational grid of size $n \times p \times q$ into three-dimensional (3-D) blocks of size $n_b \times p_b \times q$, each of which can be viewed as consisting of q two-dimensional (2-D) blocks, referred to as *tiles*, of size $n_b \times p_b$. For Step a.1, we let the kernel generate a $\texttt{ceil}\left(\frac{n}{n_b}\right) \times \texttt{ceil}\left(\frac{p}{p_b}\right)$ grid of threadblocks. Each of the threadblocks, in turn, consists of a total of $n_b p_p$ threads arranged in 2-D arrays, each of size $n_b \times p_b$. All gridpoints of a $n_b \times p_b \times q$ 3-D block are assigned to one threadblock only, with one thread for each stack of q gridpoints. Note that, since each 3-D block has a total of q $n_b \times p_b$ tiles and each threadblock is of size $n_b \times p_b$, the approach that we use here suggests a q-iteration loop in the kernel. During each iteration of this loop, each thread of a threadblock carries out all the computations/work associated with one gridpoint, and each threadblock processes one $n_b \times p_b$ tile.

Regarding the construction of the matrices $\mathbf{A}_i^m$, $i = 0, 1, 2, 3$, and $\widehat{\mathbf{A}}_i^m$, $i = 1, 2, 3$, note that each of these matrices has a total of npq rows, with each row corresponding to a gridpoint of the computational domain. Our approach is to assign each of the threads to assemble q rows of each of the matrices (a total of three entries per row of each matrix, since all matrices are tridiagonal). More specifically, during each iteration of the q-iteration loop in the kernel, each group of $n_b p_b$ rows corresponding to a tile is assembled in parallel by a $n_b \times p_b$ threadblock, with one thread for each row. That is, a total of np consecutive rows are constructed in parallel by the threadblocks during each iteration.

Regarding the parallel computation of the vectors $\mathbf{w}_i$, $i = 0, 1, 2, 3$, it is important to emphasize that, to calculate the values corresponding to gridpoints of the kth tile (i.e. the tile on the kth s-r_d plane), the data of the two adjacent tiles in the r_f-direction (i.e. the $(k-1)$st and the $(k+1)$st tiles) are needed as well. Since 16KB of shared memory available per multiprocessor are not sufficient to store many data tiles, each threadblock works with three data tiles of size $n_b \times p_b$ at a time and proceeds in the r_f-direction. As a result, we utilize a three-plane loading strategy. More specifically, during the kth iteration of the q-iteration loop in the kernel, assuming the data corresponding to the kth and $(k-1)$st tiles in the shared memory from the previous iteration, each threadblock

1. loads from the global memory into its shared memory the old data (vector $\mathbf{u}^{m-1}$) corresponding to the $(k+1)$st tile,

2. computes and stores new values (vectors $\mathbf{w}_i$, $i = 0, 1, 2, 3$ and $\mathbf{v}_0$) for the kth tile using data of the $(k-1)$st, kth and $(k+1)$st tiles,

3. copies the newly computed data of the kth tile (vectors $\mathbf{w}_i$, $i = 1, 2, 3$ and $\mathbf{v}_0$) from the shared memory to the global memory, and frees the shared memory locations taken by the data of the $(k-1)$st tile, so that they can be used in the next iteration.

Note that the data loading approach for Step a.1 is not fully coalesced, although it is highly effective. (We believe it is impossible to attain full memory coalescing for the data-loading part of this phase.)

First phase - Steps a.2, a.3, a.4

The data partitioning for each of Steps a.2, a.3 and a.4 is different from that for Step a.1 and is motivated by the block structure of the tridiagonal matrices $\widehat{\mathbf{A}}_i^m$, $i = 1, 2, 3$, respectively. For example, $\widehat{\mathbf{A}}_1^m$ has pq diagonal blocks, each block being $n \times n$ tridiagonal, thus the solution of $\widehat{\mathbf{A}}_1^m \mathbf{v}_1 = \widehat{\mathbf{v}}_1$, i.e. Step a.2, is computed by first partitioning $\widehat{\mathbf{A}}_1^m$ and $\widehat{\mathbf{v}}_1$ into pq independent $n \times n$ tridiagonal systems, and then assigning each tridiagonal system to one of the pq threads generated, i.e. each thread is assigned n gridpoints along the s-direction.

Regarding the memory coalescing for Steps a.2, a.3 and a.4, note that, in the current implementation, the data between Steps a.1, a.2, a.3 and a.4 are ordered in the s-, then the r_d-, then the r_f-direction. As a result, the data partitionings for the tridiagonal solves in the r_d- and r_f-direction, i.e. for solving $\widehat{\mathbf{A}}_i^m \mathbf{v}_i = \widehat{\mathbf{v}}_i$, $i = 2, 3$, allow full memory coalescence, while the data partitioning for solving $\widehat{\mathbf{A}}_1^m \mathbf{v}_1 = \widehat{\mathbf{v}}_1$ does not.

4.7.2 Timestep Selector on GPUs

As for the timestep selector (3.2), the key part in implementing it on the GPU involves finding the minimum element of an array of real numbers. In this regard, we adapt the parallel reduction technique discussed in [22]. The idea is to partition the array into multiple sub-arrays of size s_t, each of which is assigned to a 1-D threadblock of the same size. During the first kernel launch, each threadblock carries out the reduction operation via a tree-based approach to find the minimum of the corresponding sub-array and writes the intermediate result to a location in an array in the global memory. This array of intermediate minimum elements is then processed in the same manner by passing it on to a kernel again. This process is repeated until the array of partial minimums can be handled by a kernel launch with only one threadblock of size s_t, after which the minimum element of the initial array is found. More details about the implementation of the timestep selector can be found in our paper [8].

4.8 Stages 8 and 9

The GPU-based implementation for these stages is straightforward, since each thread of a threadblock can work independently from the others, i.e. neither communication between threads nor between processes is required. We use the same partitioning approach and assignment of gridpoints to threads employed in Stage 1. This approach allows for full memory coalescence of the loading of data from the global memory.

5 Numerical Results

As parameters to the model, we consider the same interest rates, correlation parameters, and the local volatility function as given in [12]. The domestic (JPY) and foreign (USD) interest rate curves are given by $P_d(0, T) = \exp(-0.02 \times T)$ and $P_f(0, T) = \exp(-0.05 \times T)$. The volatility parameters for the short rates and correlations are given by $\sigma_d(t) = 0.7\%$, $\kappa_d(t) = 0.0\%$, $\sigma_f(t) = 1.2\%$, $\kappa_f(t) = 5.0\%$, $\rho_{df} = 25\%$, $\rho_{ds} = -15\%$, $\rho_{fs} = -15\%$. The initial spot FX rate is set to $s(0) = 105.00$, and

Table 1. The parameters $\xi(t)$ and $\varsigma(t)$ for the local volatility function (2.8). (Table C in [12].)

	period (years)									
	(0, 0.5]	(0.5, 1]	(1, 3]	(3, 5]	(5, 7]	(7, 10]	(10, 15]	(15, 20]	(20, 25]	(25, 30]
$\xi(t)$	9.03%	8.87%	8.42%	8.99%	10.18%	13.30%	18.18%	16.73%	13.51%	13.51%
$\varsigma(t)$	-200%	-172%	-115%	-65%	-50%	-24%	10%	38%	38%	38%

the initial domestic and foreign short rate are 0.02 (2%) and 0.05 (5%), respectively, which follows from the respective interest rate curve. The parameters $\xi(t)$ and $\varsigma(t)$ for the local volatility function are assumed to be piecewise constant and given in Table 1. Note that the forward FX rate $F(0, t)$ defined by (2.5) and $\theta_i(t)$, $i = d, f$, in (2.7), and the domestic LIBOR rate (2.3) are fully determined by the above information [14].

We consider the tenor structure (2.1) that has the following properties: (i) $\nu_\alpha = 1$ (year), $\alpha = 1, \ldots, \beta + 1$ and (ii) $\beta = 29$ (years). Features of the PRDC swap are: the domestic and foreign coupons are $c_d = 2.25\%, c_f = 4.50\%$ and $c_d = 8.1, c_f = 9.00\%$, with the cap a_c being set to 50% and 10%, respectively, of the notional.

The truncated computational domain Ω is defined by setting $s_\infty = 5s(0) = 525.0$, $r_{d,\infty} = 10r_d(0) = 0.2$, and $r_{f,\infty} = 10r_f(0) = 0.5$. The grid sizes and the number of timesteps reported in the tables in this section are for each time period of the Table 1. Note that, since the timestep size selector (3.2) is used, the number of timesteps reported is the average number of timesteps for all sub-problems over all time periods of the swap's tenor structure.

We report the quantity "value", which is the value of the financial instrument. In pricing PRDC swaps, this quantity is expressed as a percentage of the notional N_d. Since in our case, an accurate reference solution is not available, to provide an estimate of the convergence rate of the algorithm, we also compute the quantity "$\log_\eta$ ratio" which provides an estimate of the convergence rate of the algorithm by measuring the differences in prices on successively finer grids, referred to as "change". More specifically, this quantity is defined by

$$\log_\eta \text{ratio} = \log_\eta \left(\frac{u_{approx}(\Delta x) - u_{approx}(\frac{\Delta x}{\eta})}{u_{approx}(\frac{\Delta x}{\eta}) - u_{approx}(\frac{\Delta x}{\eta^2})} \right),$$

where $u_{approx}(\Delta x)$ is the approximate solution computed with discretization stepsize Δx. For second-order methods, such as those considered in this paper, the quantity $\log_\eta$ ratio is expected to be about 2.

5.1 Convergence of Computed Prices

In this subsection, we demonstrate the correctness of our implementation. In Table 2, we present pricing results for FX-TARN PRDC swaps for two different combinations of c_d, c_f and a_c. In both cases, the number of sub-intervals in the a-direction is 30, i.e. $w = 29$ in (3.3). We note, for both cases, the computed prices exhibit second-order convergence, as expected from the ADI timestepping methods and the interpolation scheme.

Table 2. Values of the FX-TARN PRDC swap. The total of GPUs used is $w + 1 = 30$.

l	$n+1$	$p+1$	$q+1$	$c_d = 8.1, c_f = 9.00\%, a_c = 10\%$			$c_d = 2.25\%, c_f = 4.50\%, a_c = 50\%$		
				value	change	$\log_2$	value	change	$\log_2$
(τ)	(s)	(r_d)	(r_f)	(%)		ratio	(%)		ratio
6	30	15	15	18.521			-4.487		
12	60	30	30	18.609	8.8e-04		-4.409	7.8e-04	
23	120	60	60	18.631	2.2e-04	1.9	-4.389	2.0e-04	1.9
47	240	120	120	18.637	5.9e-05	1.9	-4.384	5.4e-05	1.9

The central question, of course, is whether the approximations of prices of FX-TARN PRDC swaps computed by the PDE method converge to the exact prices. To verify this, we compare our PDE-computed prices with prices obtained using MC simulations. More specifically, using MC simulations, with 10^6 simulation paths for the spot FX rate, the timestep size being $1/512$, and using antithetic variates as the variance reduction technique, the benchmark prices for the FX-TARN PRDC swaps are 18.638% (std. dev. = 0.021), and −4.383% (std. dev. = 0.020), respectively for the case $c_d = 8.1, c_f = 9.00\%$ and $c_d = 2.25\%, c_f = 4.50\%$[5]. The 95% confidence intervals for the two cases are $[18.635, 18.641]$ and $[-4.386, -4.379]$, respectively, which contain our PDE-computed prices.

For the case $c_d = 2.25\%, c_f = 4.50\%$, the investor should pay a net coupon of about 4.384% of the notional to the issuer. (Note the negative values in this case.) However, for the case $c_d = 8.1, c_f = 9.00\%$, the issuer should pay the investor a net coupon of about 18.631% of the notional.

5.2 Performance Results

For FX - TARN PRDC swaps, due to the high computational requirements of the pricing algorithm, which make sequentially CPU-based computation practically infeasible, we do not develop CPU-based numerical methods in this case. Instead, we focus on numerical methods on a GPU cluster and on a single GPU. In this section, we provide details of the GPU versus GPU cluster performance comparison in pricing FX-TARN PRDC swaps.

Additional statistics collected in this subsection include the following. The quantities "GPU time" and "MPI-GPU time" respectively denote the total computation times, in seconds (s.), on a single GPU and on the GPU cluster with specifications as in Subsection 4.2 using MPI. The quantity "MPI-GPU speed up" is defined as the ratio of the "GPU time" over the respective "MPI-GPU time". The quantity "MPI-GPU efficiency" is defined as

$$\text{MPI-GPU efficiency} = \frac{1}{w+1} \frac{\text{GPU time}}{\text{MPI-GPU time}},$$

which represents the standard (fixed) efficiency of the parallel algorithm using $w + 1$ GPUs of the cluster.

---

[5] Our sequential code written in MATLAB for MC simulations took about 2 days to finish.

Table 3 presents some selected timing results for FX-TARN PRDC swaps for the case $c_d = 2.25\%$, $c_f = 4.50\%$ and $a_c = 50\%$. The timing results for the other case are approximately the same, and hence omitted. Note that the times in the brackets are the total times required for data exchange between processes using MPI functions.

It is evident that the MPI-GPU implementation on the cluster are significantly more efficient than the single-GPU implementation, with the asymptotic speedups being about 25 when using 30 GPUs (15 nodes) of the cluster. Note that, our single-GPU implementation typically attains a speed up of about 30-31 times over a CPU implementation for the largest grid considered here [6, 7]. This means that a sequentially CPU-based solver for the FX-TARN PRDC swap would take approximately 170000 (s.) ($\approx 5421.1 \times 32$), or about 2 days to finish. In practical situations, such time requirements are prohibitive.

It is important to emphasize that the GPU-MPI efficiency increases with finer grid sizes (Table 3, from 60% to 87%). This is to be expected, since a fixed number of GPUs, i.e. 30 GPUs, is used for all the experiments, whereas the problem size is increasing, allowing the GPUs to be used more efficiently.

Table 3. Timing results for the FX-TARN PRDC swaps for the case $c_d = 2.25\%$, $c_f = 4.50\%$ and $a_c = 50\%$. The times in the brackets are those required for data exchange between processes using MPI functions.

l	n	p	q	GPU	MPI-GPU		
					time	speed-	effi-
(τ)	(s)	(r_d)	(r_f)	(s.)	(s.)	up	ciency
12	60	30	30	114.5	6.1 (0.3)	19.3	60%
23	120	60	60	520.7	21.3 (1.8)	24.1	81%
47	240	120	120	5421.1	206.8 (8.2)	26.3	87%

6   Conclusions and Future Work

This paper presents a parallelization on clusters of GPUs of the PDE-based computation of the price of FX interest rate swaps with the FX-TARN feature under a three-factor model. Our PDE approach is to partition the pricing problem into several independent pricing sub-problems over each time period of the swap's tenor structure, with possible communication at the end of the time period. Our implementation of the pricing procedure on clusters of GPU involves (i) efficiently solving each independent sub-problems on a GPU via a parallelization of the ADI timestepping technique, and (ii) utilizing MPI for the communication between pricing processes at the end of each time period of the swap's tenor structure. The results of this paper show that GPU clusters can provide a significant increase in performance over GPUs, when pricing exotic cross-currency interest rate derivatives with path-dependence features.

From a modeling perspective, it is desirable to impose stochastic volatility on the FX rate so that the market-observed FX volatility smiles are more accurately approximated [6]. This enrichment to the current model leads to a time-dependent PDE in four state variables – the spot FX rate, domestic and foreign interest rates, and volatility. In such an application, a our proposed parallel pricing method is expected to deliver even larger speedups and better performance when pricing path-dependent foreign exchange interest rate derivatives.

References

[1] Sippel, J., Ohkoshi, S.: All power to PRDC notes. Risk Magazine 15(11), 1–3 (2002)

[2] Piterbarg, V.V.: TARNs: Models, Valuation, Risk Sensitivities. Wilmott Magazine 14, 62–71 (2004)

[3] Abbas-Turki, L.A., Vialle, S., Lapeyre, B., Mercier, P.: High dimensional pricing of exotic European contracts on a GPU cluster, and comparison to a CPU cluster. In: Proceedings of the 2nd International Workshop on Parallel and Distributed Computing in Finance, pp. 1–8. IEEE Computer Society (2009)

[4] Murakowski, D., Brouwer, W., Natoli, V.: CUDA implementation of barrier option valuation with jump-diffusion process and Brownian bridge. In: Proceedings of the ACM/IEEE International Conference for High Performance Computing, Networking, Storage, and Analysis, pp. 1–4. IEEE Computer Society (2010)

[5] Tian, Y., Zhu, Z., Klebaner, F.C., Hamza, K.: Pricing barrier and American options under the SABR model on the graphics processing units. Concurrency and Computation: Practice and Experience, 867–879 (2012)

[6] Dang, D.M., Christara, C., Jackson, K.: Graphics processing unit pricing of exotic cross-currency interest rate derivatives with a foreign exchange volatility skew model. Journal of Concurrency and Computation: Practice and Experience (to appear, 2013), http://onlinelibrary.wiley.com/doi/10.1002/cpe.2824/abstract

[7] Dang, D.M., Christara, C., Jackson, K.: A parallel implementation on GPUs of ADI finite difference methods for parabolic PDEs with applications in finance. Canadian Applied Mathematics Quarterly (CAMQ) 17(4), 627–660 (2009)

[8] Dang, D.M., Christara, C., Jackson, K.: An efficient graphics processing unit-based parallel algorithm for pricing multi-asset American options. Journal of Concurrency and Computation: Practice and Experience 24(8), 849–866 (2012)

[9] Egloff, D.: GPUs in financial computing part III: ADI solvers on GPUs with application to stochastic volatility. Wilmott, 50–53 (March 2011)

[10] Egloff, D.: Pricing financial derivatives with high performance finite difference solvers on GPUs. In: Hwu, W.-M.W. (ed.) GPU Computing Gems Jade Edition. Applications of GPU Computing Series, pp. 309–322 (2012)

[11] Dang, D.M., Christara, C., Jackson, K., Lakhany, A.: An efficient numerical PDE approach for pricing foreign exchange interest rate hybrid derivatives. To appear in the Journal of Computational Finance (2012), http://ssrn.com/abstract=2028519

[12] Piterbarg, V.: Smiling hybrids. Risk Magazine 19(5), 66–70 (2006)

[13] Gropp, W., Lusk, E., Skjellum, A.: Using MPI-2: Advanced Features of the Message Passing Interface, 1st edn. MIT Press (1999)

[14] Andersen, L.B., Piterbarg, V.V.: Interest Rate Modeling, 1st edn. Atlantic Financial Press (2010)

[15] Dang, D.M., Christara, C.C., Jackson, K., Lakhany, A.: A PDE pricing framework for cross-currency interest rate derivatives. In: Proceedings of the 10th International Conference in Computational Science (ICCS). Procedia Computer Sciences, vol. 1, pp. 2371–2380. Elsevier (2010)

[16] Hull, J., White, A.: One factor interest rate models and the valuation of interest rate derivative securities. Journal of Financial and Quantitative Analysis 28(2), 235–254 (1993)

[17] Haentjens, T., In 't Hout, K.J.: Alternating direction implicit finite difference schemes for the Heston-Hull-White partial differential equation. Journal of Computational Finance 16(1), 83–110 (2012)

[18] Hundsdorfer, W.: Accuracy and stability of splitting with stabilizing corrections. Appl. Numer. Math. 42, 213–233 (2002)

[19] In 't Hout, K.J., Welfert, B.D.: Unconditional stability of second-order ADI schemes applied to multi-dimensional diffusion equations with mixed derivative terms. Appl. Numer. Math. 59, 677–692 (2009)

[20] NVIDIA: NVIDIA Compute Unified Device Architecture: Programming Guide Version 3.2. NVIDIA Developer Web Site (2010),
http://developer.nvidia.com/object/gpucomputing.html

[21] Dang, D.M.: Modeling multi-factor financial derivatives by a Partial Differential Equation approach with efficient implementation on Graphics Processing Units. PhD thesis, Department of Computer Science, University of Toronto, Toronto, Ontario, Canada (2011)

[22] Harris, M., Sengupta, S., Owens, J.D.: Parallel prefix sum (scan) with CUDA. In: GPU Gems 3, pp. 851–877. NVIDIA (2007)

Extreme Load Estimation for a Large Wind Turbine Using CFD and Unsteady BEM

Thanh-Toan Tran, Dong-Hyun Kim[*], and Kang-Sik Bae

Dept. of Aerospace and System Engineering, Gyeongsang National University, Korea
(trantoan2008,dhk)@gnu.ac.kr

Abstract. The numerical approach of an aerodynamic and structural load acting on NREL 5MW baseline wind turbine blade have been performed by using computational fluid dynamics (CFD) and unsteady blade element momentum theory (UBEM). Also, the optimal parameters required to design for wind turbine including a nacelle and a tower are investigated in this study. The computational model applied here was based on the Reynolds Average Navier-Stokes (RANS) equation using sliding mesh technique that is the most accurate method for simulating flows in multiple moving reference frames, but also the most computationally demanding. The k-ω shear-stress transport (k-ω SST) turbulent model is used for the unsteady state computations. An aerodynamic thrust and power have been computed along with the azimuth angle variation of the blade for an extreme coherent gust (ECG), an extreme operation gust (EOG) condition, based on a predefined function. In application of UBEM, three-dimensional stall model has been adopted to calculate load components on the wind turbine rotor. Structural loads considering gravity and rotational inertia are calculated from both UBEM method and CFD. The aerodynamic power and thrust of HAWT along with the wind speed variation ranging from 8 to 25m/sec are compared to the published data for the validation with a good agreement.

Keywords: NREL 5MW Wind Turbines, Computation Fluid Dynamics (CFD), Blade loading, Structure calculation.

1 Introduction

The purpose of this work is to build accurate aerodynamic analysis method for the calculation of various design loads using advanced computational fluid dynamics (CFD). In this paper, the results for an extreme coherent gust (ECG), and the extreme operating gust (EOG) within recurrence periods of 50 years have been successfully applied for using CFD method. Also, an aerodynamic analysis tool based on the unsteady blade element momentum theory (UBEM) is developed based on the previous works by Hansen et. al. [1-2] and applied to calculation of critical design loads in current study. Three-dimensional stall model obtained by Lindenburg [3] is adopted instead of using Beddoes-Leishman dynamic stall model [4]. There are essential design load conditions defend by international designguide lines such as IEC61400-1, GL, and DNV need to be accurately determined using numerical analysis tools. The

---

[*] Corresponding author.

B. Murgante et al. (Eds.): ICCSA 2013, Part V, LNCS 7975, pp. 127–142, 2013.
© Springer-Verlag Berlin Heidelberg 2013

traditional blade element momentum theory (BEM) [5] based on the application of two-dimensional aerodynamic properties has still some limitations to accurately consider various design load conditions. There were previous studies about the hybrid BEM-CFD method normally called as actuator disk method which roughly represents an extension of conventional BEM, integrating with Euler or Navier Stokes equations [6-7]. There were other previous good works for the aerodynamic analysis using BEM and CFD methods [8-15]. However, it is very hard to find previous works using unsteady CFD technology in oder to acculately calculate critical design loads considering extreme direction change of wind and operating gust conditions. In current study, in order to evaluate accuracy of simulation, the aerodynamic power and thrust of the NREL 5-MW baseline rotor blade has been simulated for velocity ranging from 8m/s to 25m/s. The results were compared with previous studies that also conducted by CFD solver. The validation results show a good agreement with published data by Sørensen et al [12]. It is importantly shown in this study that the aerodynamic loads of the blade can be enormously changed under extreme wind conditions. Aerodynamic power and thrust tend to vast oscillate due to extreme wind conditions. Additionally, integrated loads on blade coordinate system that includes aerodynamic load, structure load have also conducted. In order to obtain those force components, the vectors of blade coordinate system on global coordinate system were identified using Euler angles rules. Total six integrated force components of blade on its coordinate systems that almost engineer may be really interested, have also carried out in current study.

2   Numerical Backgound

2.1   CFD Theory

With sliding mesh method, computational domain is divided in two regions that are treated separately: the turbine blade region and outside of its region. The grid in the turbine blade region rotates with turbine blade. In the other hand, ouside of turbine blade region remains stationary. The two grid slide past each other at a cylindrical interface.

With respect to dynamic meshes, the standard conservation equation for continuity and momentum [16] for incompressible flow can be written as a tensor form

$$\frac{\partial}{\partial x_j}(u_j - v_j) = 0 \tag{1}$$

$$\frac{\partial}{\partial t}(\rho u_i) + \frac{\partial}{\partial x_j}\rho(u_j - v_j)\tilde{u} = -\frac{\partial p}{\partial x_i} + \frac{\partial \rho}{\partial x_i}[\tau_{ij} + R_{ij}] \tag{2}$$

where the viscous stress tensor and deformation tensor are defined as follows

$$\tau_{ij} = 2\mu\left[S_{ij} - \frac{1}{3}\delta_{ij}\frac{\partial u_k}{\partial x_k}\right] \tag{3}$$

$$S_{ij} = \frac{1}{2}\left[\frac{\partial u_i}{\partial x_j} + \frac{\partial u_j}{\partial x_i}\right] \tag{4}$$

Also,

$$\tilde{u} = u_j - u_{g,j} \tag{5}$$

Here, u_j is the fluid velocity in a stationary reference frame and v_j is the velocity component arising from mesh motion. And, $u_{g,j}$ is the grid velocity. And turbulence Reynolds stress tensor R_{ij} must be modeled in order to close Eq. (2). This tensor may be approximated by Following Boussinesq hypothesis:

$$R_{ij} \cong \mu_T \left[S_{ij} - \frac{2}{3} \frac{\partial u_k}{\partial x_k} \delta_{ij} \right] - \frac{2}{3} (\rho k) \delta_{ij} \tag{6}$$

In this study, Eqs.(1~6) are solved by the finite-volume method. For the discretization of RANS equations, second-order upwind scheme is applied. Additionally, the shear-stress transport (SST) k-ω turbulence model was used [17]. This turbulent model was developed to effectively blend the robust and accurate formulation of the k-ω model in the near-wall region with free-stream independence of the k-ω model in the far field. The user-defined function (UDF) code of the ANSYS/FLUENT (Ver.14) [18] software based on C-language and systematic numerical process that is newly able to consider the extreme coherent gust (ECG) and the extreme operating gust (EOG) are originally developed to practically conduct viscous unsteady aerodynamic analysis of the wind turbine blades.

2.2   Structure Load Calculation

As presented in Fig.1(a), the wind turbine rotor inherently experience aerodynamic, inertial, and gravity forces. The blade coordinate system used in this study is also shown in Fig.1(b).

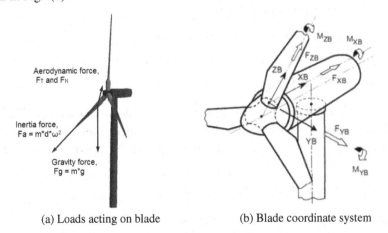

(a) Loads acting on blade   (b) Blade coordinate system

Fig. 1. Definition of rotor loads on blade coordinate system

Generated forces on the i-th blade root normally consist of aerodynamic, gravity, and inertial components:

$$F_{x,i} = F_{x,aero,i} + F_{x,grav,i} + F_{x,iner,i}$$
$$F_{y,i} = F_{y,aero,i} + F_{y,grav,i} + F_{y,iner,i} \qquad (7)$$
$$F_{z,i} = F_{z,aero,i} + F_{z,grav,i} + F_{z,iner,i}$$

where $F_{aero,i}$ means aerodynamic forces exerted on the i-th blade, $F_{grav,i}$ means gravitational forces, and $F_{iner,i}$ means inertia forces, respectively. Also, XB-, YB-, and ZB-directions are defined in the blade coordinated system as shown in Fig. 1(b). Gravitational and inertia forces without considering the elastic deformation effect of the blade can be simply calculated by the following equations. Tilt and precone angles of the rigid blade are considered herein.

$$F_{x,grav,i} = m_b g(\sin\theta_{tilt}\cos\theta_{cone} - \cos\theta_{tilt}\sin\theta_{cone}\cos\Psi_i)$$
$$F_{y,grav,i} = m_b g(\cos\theta_{tilt}\sin\Psi_i)$$
$$F_{z,grav,i} = m_b g(\sin\theta_{tilt}\sin\theta_{cone} - \cos\theta_{tilt}\cos\theta_{cone}\cos\Psi_i)$$
$$F_{x,iner,i} = m_b(R_h + z_{cm,b})\omega^2\sin\theta_{cone} \qquad (8)$$
$$F_{y,iner,i} = m_b(R_h + z_{cm,b})\dot{\omega}$$
$$F_{z,iner,i} = m_b(R_h + z_{cm,b})\omega^2\cos\theta_{cone}$$

where, θ_{cone}, θ_{tilt}, and Ψ are precone, tilt, and azimuth angle of blade rotor, respectively. m_b is blade mass, R_h is hub radius, and $z_{cm,b}$ is distance of blade center of mass from blade root along z-axis of blade coordinate system. The terms, $\dot{\omega}$ and ω are rotational acceleration and angular velocity, respectively.

Besides force components, torque components in the blade coordinate system are similarly defined as following equations.

$$M_{x,i} = M_{x,aero,i} + M_{x,grav,i} + M_{x,iner,i}$$
$$M_{y,i} = M_{y,aero,i} + M_{y,grav,i} + M_{y,iner,i} \qquad (9)$$
$$M_{z,i} = M_{z,aero,i}$$

where $M_{aero,i}$ means aerodynamic moments exerted on the i-th blade, $M_{grav,i}$ means gravitational moments, and $M_{iner,i}$ means inertia moments, respectively. Moment loads for rigid blade in the blade coordinate system can be analytically obtained by the following equations.

$$M_{x,grav,i} = m_b g(\cos\theta_{tilt}\sin\Psi_i)z_{cm,b}$$
$$M_{y,grav,i} = m_b g(\sin\theta_{tilt}\cos\theta_{cone} - \cos\theta_{tilt}\sin\theta_{cone}\cos\Psi_i)z_{cm,b}$$
$$M_{x,iner,i} = -J_b \dot{\omega} \qquad (10)$$
$$M_{y,iner,i} = m_b(R_h + z_{cm,b})\omega^2\sin\theta_{cone}z_{cm,b}$$

Here, the term, J_b is the blade mass moment of inertial about x-axis.

3   Results and Discussion

In this study, well known NREL 5-MW baseline wind turbine model has been considered for numerical simulations [13-14], [19]. Fig.2 shows verification results for the analytical solution of blade structural dynamic loads (inertia + gravity) with respect to azimuth angle in the blade coordinate system. Multi-body dynamic analysis program, SAMCEF software is used for this verification. The blade model considered herein is NREL 5-MW baseline model. Its blade specification is presented in Table 1. The analytical results considering tilt and precone angles show very good agreement with that by the multi-body dynamics computation.

Table 1. Blade specification of the NREL 5-MW Baseline model

Length	61.5 m
Overall mass	17,740 kg
First mass moment of inertia	363,231 kg m^2
Second mass moment of inertia	11,776,047 kg m^2
Center of mass location	20.475 m
Tilt angle	5 deg
Precone angle	2.5 deg

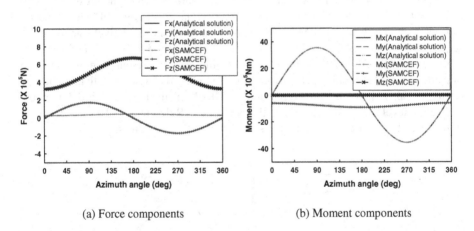

(a) Force components   (b) Moment components

Fig. 2. Comparison of one blade structural loads between analytical and computational solutions at 11.2 rpm

Table 2 shows the overall specification of the NREL 5-MW baseline wind turbine. The diameter of present rotor is 126 m. The cross sections of rotor blade are composed of series of DU (Delft University) and NACA 64xxx airfoils from the hub to the tip of the out-board section. Related aerofoil data can be found from Bazilevs et al.[14-15].

Table 2. Specification of the NREL 5-MW baseline wind turbine

Rating	5 MW
Rotor orientation	Upwind
Number of blade	3
Control	Variable Speed, Collective Pitch
Driventrain	High Speed, Multiple-State Gearbox
Rotor, Hub Diameter	126 m, 3 m
Hub Height	90 m
Cut-In, Rated, Cut-out Wind Speed	3 m/s, 11.4 m/s, 25 m/s
Cut-In, Rated Rotor Speed	6.9 rpm, 12.1 rpm
Rated Tip Speed	80 m/s
Overhang, shaft tilt, precone angles	5 m, 5 deg, 2.5 deg
Rotor Mass	110,000 kg
Nacelle Mass	240,000 kg
Tower Mass	347,000 kg
Coordinate Location of Overall CM	(-0.2m, 0.0m, 64.0m)

In order to show some comparison results for the present aerodynamic method, a three-bladed rotor model without tower and nacelle configuration has been considered first. Figure 3 shows the comparison results for the verification of the present aerodynamic analysis for different wind speeds. The present results by CFD and unsteady blade element moment (UBEM) aerodynamic analyses show relatively good agreement with the previous results by Sørensen et al.[12] using EllipSys3D and Chow et al.[13] using OVERFLOW2. Operational conditions for present computation are adopted from Chow et al.[13]. This condition is slightly different from that applied by Sørensen et al.[12]. As mentioned previously, UBEM analysis with stall delay model [3] is also developed and applied for this analysis. Referentially, Hansen et. al.[4] adopted Beddoes-Leishman's dynamic stall model in the previous study.

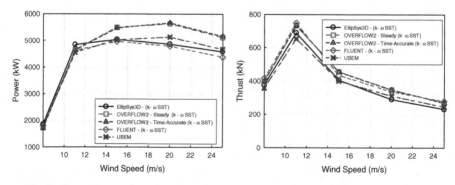

Fig. 3. Comparisons of predicted aerodynamic power and thrust for various wind speeds

Aerodynamic analyses for a full scale 5-MW wind turbine which includes nacelle and tower configurations have been also conducted. Unstructured grid topology is used to effectively generate complex mesh system for CFD computations as presented in Fig.4. More importantly, automatic grid generation procedure for the complex full wind turbine model with different blade pitching angles is practically developed in this study. The size of outer computational domain is 700 m (length) x 600 m (width) x 400 m (height) of which boundaries are extended as 3 and 8 times of rotor radius with respect to upstream and downstream directions, respectively. The total number of cells for full computational domain is approximately 2.8 million. Uniform and variable velocity conditions are prescribed as an upstream boundary in the inlet. A gauge pressure was assumed as sea level as a downstream boundary in outlet. The terrain was defined as a wall condition. Unsteady sliding mesh technique is applied to the interface boundaries between the inner flow domain of the rotating rotors and stationary outer flow domains.

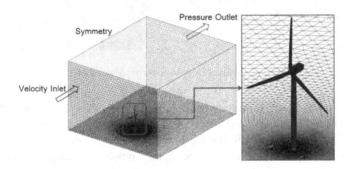

(a) Side view of geometry

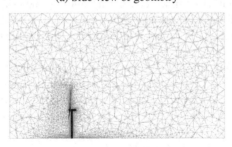

(b) Cross section at center plane

Fig. 4. Computational grid domain for CFD analysis of the full wind turbine model

Figure 5 shows typical flow streaklines around the present full wind turbine model at the wind speed of 15 m/s. It is shown in the figure that there are strong flow separations in the root area of the each blade and on the front section of the upper nacelle surface. It is physically well expected that the flow condition of the nacelle surface must be strongly affected by the position of the each blade. It was found in our previous study that this kind of interference effect tends to somewhat decrease the overall aerodynamic performance of huge wind turbine [21]. The present CFD analyses have been effectively carried out using clustered parallel server computers with Windows 7 operating system and its hardware specification of each machine is Intel i7-2600K @ 3.4 GHz (8-CPU) and 16 GB RAM. The computation time for each steady state solution is less than 1 hour. On the other hand, approximately 48 hours are required for each converged solution of unsteady solutions for the present 5-MW class wind turbine model.

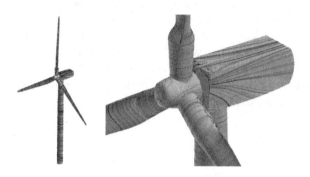

Fig. 5. Flow streaklines at freestream wind speed of 15 m/s

As mentioned previously, various critical design load cases (DLC) must be considered and analyzed based on wind turbine design guidelines such as IEC-61400-1 and GL Guidelines 2010 [20] etc. during the practical design process of a wind turbine. Typical design load cases of wind turbine include power production, power production plus occurrence of fault, start-up , normal shut-down, emergency shut-down, parked, parked plus fault conditions, etc. In this study, some critical DLC conditions for practical numerical demonstration are selected according to GL Guidelines 2010. Selected DLCs for numerical simulations are normal wind profile (NWP) and extreme operating gust (EOG) conditions.

According to GL Guidelines 2010, the normal wind profile (NWP) model defines the average wind speed as a function of height z above the ground. The assumed wind profile is used to define the average vertical wind shear across the rotor swept area. The mormal wind speed profile is calculated by the following power law:

$$V(z) = V_{hub} \left(\frac{z}{z_{hub}}\right)^{\alpha}$$ (11)

where $V(z)$ means the wind speed at the height z, z_{hub} means the hub height above ground, α means the power law exponent which is normally assumed as 0.2, and V_{hub} means the freestream wind velocity at the hub height. For the NREL 5-MW baseline model considered herein, z_{hub} is 90 m.

The extreme wind conditions are used to determine the extreme wind loads actiong on wind turbines. These conditions include peak wind speeds due to stroms and rapid changes in wind speed and direction. These extreme conditions include the potential effects of wind turbulence, with the exception of the extreme wind speed model (EWM), so that only the deterministic effects need to be considered in the design calculations. The extreme coherent gust (ECG) represents the transient coherent gust characteristics of natural wind speed. The coherent gust magnitude V_{cg} for design for the standard wind turbine classes is assumed as 15 m/s. The wind speed for the extreme coherent gust (ECG) condition is defined by the following equations

$$V(z,t) = \begin{cases} V(z) & \text{for } t < 0 \\ V(z) + 0.5V_{cg}\left(1 - \cos\left(\frac{\pi t}{T}\right)\right) & \text{for } 0 \le t \le T \\ V(z) + V_{cg} & \text{for } t > T \end{cases} \tag{12}$$

Here, $T = 10$ sec is the rise time and $V(z)$ is the normal wind profile (NWP).

Furthermore, the extreme operating gust (EOG) represents the transient gust characteristics of natural wind speed. The gust magnitude V_{gustN} at hub height for a recurrence period of N years shall be calculated for the standard wind turbine classes by the following relation ship [20]:

$$V_{gustN} = \beta\sigma_1 B \tag{13}$$

where V_{gustN} means the maximum value of the wind speed for the extreme operating gust, with an expected recurrence period of N years, σ_1 means standard deviation of the longitudinal wind velocity component at hub height, and B is size reduction factor. Related equations for σ_1 and B can be found in Ref.20.

The wind speed for the extreme operating gust (EOG) condition for a recurrence period of N years is defined by the following equations.

$$V(z,t) = \begin{cases} V(z) - 0.37V_{gustN}\sin\left(\frac{3\pi t}{T}\right)\left(1 - \cos\left(\frac{2\pi t}{T}\right)\right) & \text{for } 0 \le t \le T \\ V(z) & \text{for } t < 0 \text{ and } t > T \end{cases} \tag{14}$$

Here, the recurrence period, N is considered as 50 years, the extreme gust blowing time, T is considered as 14.0 sec. Also, the present wind turbine model is considered as Class I with turbulence category A in order to analyze the most severe extreme gust condition.

Figure 6 shows comparison resluts for aerodynamic thrust and power with respect to azimuth angle of the blade for a normal wind profile (NWP) condition at 11 m/s and 15 m/s. Unsteady CFD and UBEM methods are used to calculate those results. It is noted here that the blade pitch angles are different for 11 m/s and 15 m/s because of power regulation control. Aerodynamic interference effects for nacelle geometry is not considered in the present UBEM analysis. The results show overall good correlation but approximately 6% (11 m/s) and 10 % (15 m/s) differences are seen for the calculated thrust. One can see that the analysis results by the UBEM seems to predict higher values than those by the CFD. It is also clearly seen that there are strong interference effects between the blade and the tower at azimuth angles of 60, 180, and 300 degrees that exactly correspond to blade passing locations with respect to the tower.

In this study, 0 degree azimuth angle for number 1 blade is defined by the position of twelve o'clock. In addition, unsteady CFD theory can include all complex flow interference effects such as flow separation and viscous downstream wake interactions between the blade and the nacelle-tower geometries. As mensioned earlier, three-dimensional stall delay model [3] is adopted for the present UBEM analysis and it does not consider the interference effect with the nacelle.

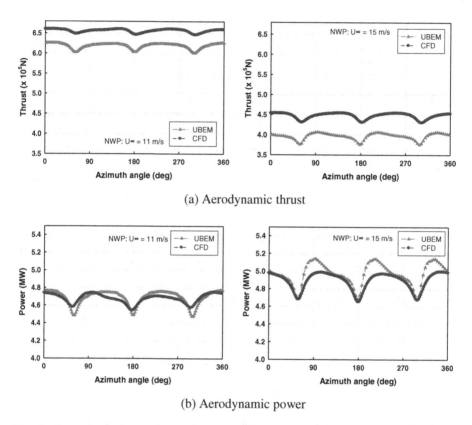

(a) Aerodynamic thrust

(b) Aerodynamic power

Fig. 6. Comparison of aerodynamic performances between CFD and unsteady BEM for a normal wind profile (NWP) condition

Figure 7 shows the comparison result of imposed vertical wind speed profile for an extreme coherent gust condition. The magnitude of the wind speed profile imposed in the CFD analysis shows very good agreement with the theoretical calculation by Eq.12. Figure 8 shows comparison resluts for aerodynamic thrust and power with respect to azimuth angle of the blade for an extreme coherent gust (ECG) condition at the average hub wind speed of 11 m/s and 15 m/s. Unsteady CFD and UBEM methods are also used to calculate those results. It is noted here that the blade pitch angles are different for 11 m/s and 15 m/s because of power regulation control. Aerodynamic interference effects for nacelle geometry is not considered in the present

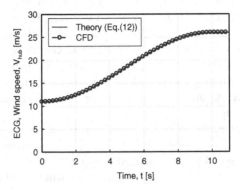

Fig. 7. Comparison of imposed wind velocity profile based on the extreme coherent gust (Example of an extreme coherent gust with $V_{hub} = 11 \text{m/s}$)

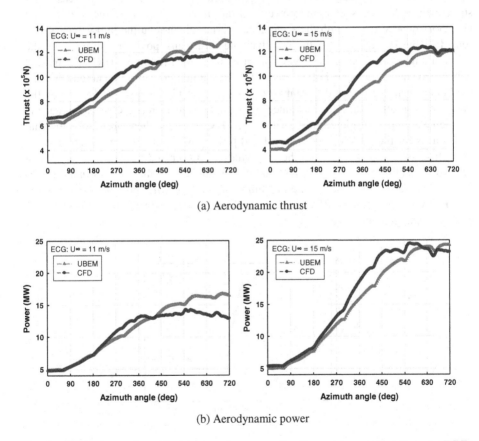

(a) Aerodynamic thrust

(b) Aerodynamic power

Fig. 8. Comparison of aerodynamic performances for the extreme coherent gust (ECG) condition

UBEM analysis. The results show overally good correlation. However, there are also somewhat differences for high wind speed region which highly dynamic characteristic of flow-field around wind turbine occurs. As a result, CFD computations predict much higher values of thrust and power for the ECG condition. The results clearly show that the predicted aerodynamic power considering transient wind gust tends to be dominantly changed during the period of extreme wind gust blowing. When the extreme wind gust occurs, there is large variation in the magnitude of power and thrust curve as seen in Fig. 8.

For the next simulation, the wind gust conditions based on an expected recurrence period of 50 years has been considered. Figure 9 shows the comparison between an imposed vertical wind speed profile obtained from CFD and Eq. 14 in case of an extreme operating gust condition. The magnitudes of the wind speed profile imposed in the CFD analysis are much closed to the data resulting from the theoretical calculation by Eq.14. When the EOG condition is considered for the wind turbine, imposed wind speed decreases in the beginning of wind gust duration and then quickly increases and subsequently decreases, thus the power and the thrust curves show the nearly same behaviors in accordance with the wind speed variation curve at the hub which is presented in Fig.10. They show the aerodynamic thrust and power resulting from the azimuth angle variation of the blade for an extreme operation gust (EOG) condition with the average wind speed of 11 m/s and 15 m/s at hub position. It is noted here that the blade pitch angles are different from the case for 11 m/s and for 15 m/s by the power regulation control logics. Since aerodynamic interference effects for nacelle geometry are not considered in the current UBEM analysis, there are also somewhat differences for the peak values of thrust and power. When the extreme wind gust occurs, there is a large variation in the magnitude of power and thrust curve as seen in Fig. 10. In addition, when the EOG condition is applied for the wind turbine, an imposed wind speed decreases at the beginning step of the wind gust duration. So, the transient wind gust clearly affects a predicted aerodynamic power during the period of extreme wind gust blowing.

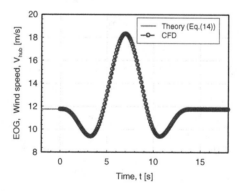

Fig. 9. Comparison of imposed wind velocity profile based on the extreme operating gust (Example of an extreme operating gust with a recurrence period of N = 50, category A, D = 126m, zhub = 90m, Vhub = 11m/s)

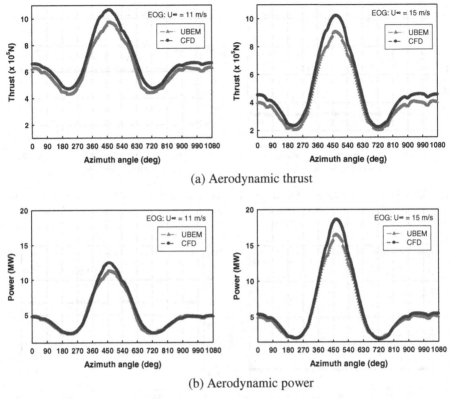

(a) Aerodynamic thrust

(b) Aerodynamic power

Fig. 10. Comparison of aerodynamic performances for the 50 year extreme operating gust (EOG) condition

Based on a coordinate system of blade in Fig.2. an aerodynamic load acting on a turbine rotor for normal wind profile condition in the case of 15 m/s is in Fig.11. An aerodynamic load acting on wind turbine rotor are closed to the data obtained from both CFD and UBEM code in Fig.11. In case of normal wind profile conditions, an atmosphere boundary layer is applied for CFD and UBEM solver. Load acting on each blade tend to slightly decrease from top position (blade is located at 0 o'clock position on clock) to tower position (blade is located at 6 o'clock position on clock) and vice versa. Loads such as the force and moment acting on the wind turbine blade during extreme coherent gust (ECG) with the hub velocity of 15 m/s are in Fig.12. Applied ECG as boundary conditions, the force and moment have a big variation for the first two revolutions. Also, there are the inflection points at azimuth angles of 180 deg and 540 deg due to an interference effect between the tower and the blade, where the tower is aligned with one blade in front point of view. It is located at 180 deg azimuth angle or six o'clock arm position on a clock. Loads acting on blade root for the 50 year extreme operating gust (EOG) condition in the case of 15 m/s are in Fig.13. The condition of gust blowing applied here results in a line of discontinuity on aerodynamic forces and moments. The conditions of extreme gust may be considered as an important factor in the design process of a wind turbine.

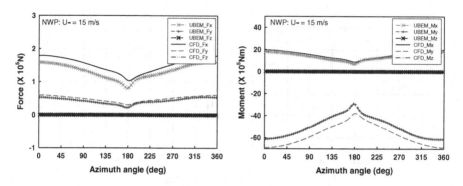

Fig. 11. Comparison of blade root loads for the normal wind profile (NWP) condition at 15 m/s (blade coordinate)

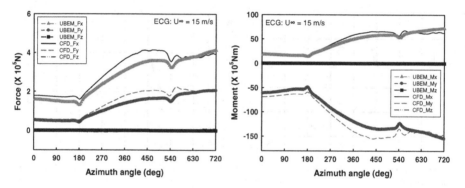

Fig. 12. Comparison of blade root loads for the extreme coherent gust (ECG) condition at 15 m/s (blade coordinate)

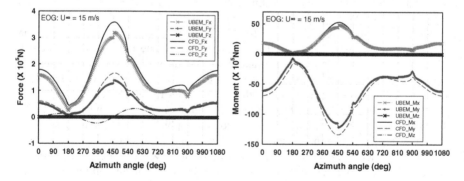

Fig. 13. Comparison of blade root loads for the 50 year extreme operating gust (EOG) condition at 15 m/s (blade coordinate)

4   Conclusions

The unsteady aerodynamic analysis methods considering the effects of wind profile and extremely operating gust using an unsteady blade element momentum theory (UBEM) and computational fluid dynamics (CFD) have been successfully developed. Analytical equations for the calculations of blade structural forces and moments including gravity and rotational inertia are applied here and validated against the published data with a good agreement. Also, the characteristics of unsteady aerodynamic performances and loads are investigated considering the interference effects between the blades and the tower in detail. The current study shows a unsteady blade element momentum theory (UBEM) has been still available to calculate loads acting on wind turbine blade nevertheless its inherent limitation. In other words, a UBEM method can be useful to determine wind turbine load preliminary as well as CFD work can be applied to get loads exactly required to the primary design.

Acknowledgments. This work was supported by the Human Resources Development program(No. 20124030200140) of the Korea Institute of Energy Technology Evaluation and Planning(KETEP) grant funded by the Korea government Ministry of Trade, Industry and Energy.

References

1. Hansen, M.O.L.: Aerodynamics of wind turbines. James and James. Science Publishers Ltd. (2000)
2. Hansen, M.O.L., Sorensen, J.N., Voutsinas, S., Sorensen, N., Madsen, H.A.: State of the art in wind turbine aerodynamics and aeroelasticity. Progress in Aerospace Sciences 42, 285–330 (2006)
3. Lindenburg, C.: Modeling of rotational augmentation based on engineering considerations and measurements. In: European Wind Energy Conference, London, November 22-25 (2004)
4. Hansen, M.H., Gaunaa, M., Madsen, H.A.: A Beddoes-Leishman type dynamic stall model in state-space and indicial formulations, Risoe-R-1354(EN), Roskidle, Denmark (2004)
5. Moriarty, P.J., Hansen, C.A.: AeroDyn Theory Manual, Tech. Rep. NREL/EL-500-36881, National Renewable Energy Laboratory, Golden, CO (December 2005)
6. Ivanell, S.S.A.: Numerical computations of wind turbine wakes, Technical Reports, Royal Institute of Technology, PhD Dissertation (2009)
7. Mikkelsen, R.: Actuator disc methods applied to wind turbines, Technical University of Denmark: PhD Dissertation (2003)
8. Carcangiu, C.E.: CFD-RANS Study of Horizontal Axis Wind Turbines, Università degli Studi di Cagliari, PhD Dissertation (2008)
9. Sarun, B.: Computational studies of horizontal axis wind turbines in high wind speed condition using advanced turbulence models. Doctor of Philosophy in the School of Aerospace Engineering Georgia Institute of Technology (2006)
10. Fernando, V., Marcelo, R., Adrian, I.: Assessment of Turbulence Models for Flow Simulation around a Wind Turbine Airfoil. Modeling and Simulation in Engineering, Article ID 714146, 8 (2011)

11. Sarun, B., Lakshmi, N.S.: Evaluation of turbulence models for the prediction of wind turbine aerodynamics. In: AIAA-2003-0517 (2003)
12. Sorensen, N.N., Johansen, J.: UPWIND, Aerodynamics and aero-elasticity rotor aerodynamics in atmospheric shear flow. Wind Energy Department, Risø National Laboratory & Department of Civil Engineering, Aalborg University (2006)
13. Chow, R., van Dam, C.P.: Inboard Stall and Separation Mitigation Techniques on Wind Turbine Rotors. In: 49th AIAA Aerospace Sciences Meeting, Orlando, FL (2011)
14. Bazilevs, Y., Hsu, M.C., Akkerman, I., Wright, S., Kakizawa, K., Henicke, B., Spielman, T., Tezduyar, T.E.: 3D simulation of wind turbine rotors at full scale Part I: Geometry modeling and aerodynamic. International Journal for Numerical Methods in Fluids 65, 207–235 (2011)
15. Bazilevs, Y., Hsu, M.C., Kiendl, J., Wüchner, R., Bletzinger, K.U.: 3D simulation of wind turbine rotors at full scale. Part II: Fluid–structure interaction modeling with composite blades. International Journal for Numerical Methods in Fluid 65, 236–253 (2011)
16. Bakker, A., LaRoche, R.D., Wang, M.H., Calabrese, R.V.: Sliding Mesh Simulation of Laminar Flow in Stirred Reactors. Trans. Ichem E 75, Part A (January 1997)
17. Menter, F.R.: Two-Equation Eddy-Viscosity Turbulence Models for Engineering Applications. AIAA Journal 32(8), 1598–1605 (1994)
18. ANSYS FLUENT Ver.14.0 User Documentations (2011)
19. Jonkman, J., Butterfield, S., Musial, W., Scott, G.: Definition of a 5-MW reference wind turbine for offshore system development. Technical Report NREL/TP-500-38060, National Renewable Energy Laboratory, Golden, CO (2009)
20. Guideline for the Certification of Wind Turbines, 2010th edn. Germanischer Lloyd, Hamburg (2010)
21. Tran, T.T., Ryu, G.J., Kim, Y.H., Kim, D.H.: CFD-based design load analysis of 5MW offshore wind turbine. In: 9th International Conference on Mathematical Problems in Engineering, Aerospace and Sciences: ICNPAA (2012)

Selecting LTE and Wireless Mesh Networks for Indoor/Outdoor Applications

Sahithi Chintapalli, Nishan Weragama, and Dharma P. Agrawal

School of Computing Sciences and Informatics
University of Cincinnati, Cincinnati, OH 45221-0008
{chintasi,weragans}@mail.uc.edu, dpa@cs.uc.edu

Abstract. The smart phone usage and multimedia devices have been increasing day by day and drastic increase of wireless technologies in these gadgets have been predicted for upcoming years. As data plans offered by the network service providers are expensive, users are inclined to utilize freely accessible and commonly available Wi-Fi networks indoors. LTE (Long Term Evolution) has been a topic of discussion in providing high data rates outdoors and various service providers are planning to roll out LTE networks all over the world. On the other hand, Wireless Mesh Networks (WMNs) are becoming popular for their flexibility and low cost in both indoors and outdoors applications. The objective of this presentation is to compare usefulness of these two leading wireless schemes based on LTE and WMN and bring forward their advantages for indoor and outdoor environments. Both these networks are heterogeneous in nature, employ cognitive approach and support multi-hop communication. We also investigate to see if a hybrid LTE-WMN network may be feasible. The main motivation behind this work is to utilize similarities in these networks, explore their capability of offering high data rates and generally have large coverage areas. In this work, we compare both these networks in terms of their data rates, range, cost, throughput, and power consumption. We also compare 802.11n based WMN with Femto cell in an indoor coverage scenario; while for outdoors, 802.16 based WMN is compared with LTE. The main objective is to help users select a network that could provide enhanced performance in a cost effective manner.

Keywords: Indoor, LTE, Outdoor, WiMAX, WMN.

1 Introduction

Wireless Mesh networks (WMN) have been in use from quite some time now. Mesh networks have been adopted to cover homes, enterprise buildings, etc. Due to the multi hop nature of WMN, they have been deployed to cover larger areas and also rural regions where it is rare to find Internet connection. Both indoor and outdoor deployments are possible in these networks. While there has been a lot of research already done in this area, no one looked into the possibility of comparing WMN with the latest technology of Long Term Evolution (LTE). In this paper, we compare WMN and LTE and pointed out few differences and similarities between them. We

B. Murgante et al. (Eds.): ICCSA 2013, Part V, LNCS 7975, pp. 143–153, 2013.
© Springer-Verlag Berlin Heidelberg 2013

keep our selves away from the discussions about which technology is better. We are trying to put forward advantages and disadvantages of each technology so that users can make a better choice in a particular scenario.

WMNs and WiMAX (World Wide Interoperability for microwave access) are mostly used for Home Networking, Enterprise Networking and follow open standard. On the other hand, LTE is a proprietary standard which has to be owned by the network operators. The spectrum bands are auctioned in every country and the operators try to buy as much spectrum as they can. The market strategies and the ongoing competition between both WiMAX and LTE are also indicated.

The rest of the paper is organized as follows: Section 2 gives the background and motivation for the work, section 3 describes the different scenarios that both LTE and WMN have been compared, and section 4 concludes the paper.

2   Background and Motivation

This work was motivated by the fact that we have been looking at the resemblances and variances between these two technologies. We have examined the major areas and summarized major differences and similarities.

2.1   Relaying

Multi-hop communication is adopted in a WMN so that coverage and quality of the network can be improved. The mesh routers (MRs) relay data from one MR to the other till it reaches the destination of the MR connected to Internet, commonly known as IGW (Internet Gateways). The MRs act as repeaters rather than making any changes to the data. MRs has been incorporated by 802.16j protocol to provide improved capacity and coverage.

LTE Release 10 supports implementation of relay nodes to improve the coverage and performance of the network. The relays are used to improve the performance at the edge of a cell where the signal to noise plus ratio is low. Such a multi-hop communication using relays is supported by LTE.

The main advantage of a WMN by multi-hop communication within the whole city or rural area is to have access to the Internet. In WMN, the mesh routers basically work as relays where they work as repeaters and relay the data to the next router if they are not the destination routers. LTE also supports multi-hop communication using relay technology in which relays are deployed to cover the entire cell site without any coverage holes, especially the border areas with weaker signal quality.

2.2   Heterogeneous Networks

In a WMN, MR's are equipped with multiple radios. As indicated in [1], "Using multiple heterogeneous radios offers tradeoffs that can improve robustness, connectivity, and performance". By using the wireless backhauls, different networks are interconnected in order to provide a better coverage. Heterogeneous WMN schemes can integrate networks following 802.11 and 802.16 protocols.

In LTE Advanced, heterogeneous networks are formed by deploying several low power nodes within a macro cell to improve the network capacity and the coverage. The low power nodes include micro-cell, pico-cell, femto-cell, and relay stations.LTE supports heterogeneous networks so as to optimize the performance of unequal distribution in terms of users and/or amount of traffic. Deploying low power nodes reduces the cost of organizing the network [2].

Both WMN and LTE support heterogeneity to increase the capacity of the network and to make the handover of calls much easier within the same macro cell and without the user being disconnected from the network. Each low power node in the network operates in different frequency range and has different bandwidth and coverage area.

2.3   Cognitive Networks

In any technology, the main bottleneck is the limited amount of allocated spectrum. In order to better utilize the available spectrum, cognitive radios have been suggested. In a Cognitive Radio WMN, two users share the same frequency spectrum efficiently by making use of cognitive radio technique. The primary user is the actual user to whom the spectrum is allocated. Secondary user is the user who uses another frequency to transmit its data. The secondary user keeps track of the frequency channel using cognitive radios and utilizes the available frequency until a primary user needs it. If a primary user wants to transmit data, then the secondary user leaves the frequency spectrum and again starts searching for an unused channel.

Cognitive radios are especially useful in the Heterogeneous network deployments. Several MRs implemented in different frequencies and with different coverage areas, will be looking for a spectrum to transmit its data. A cognitive radio is one of the solutions to avoid interference between the adjacent MRs. A lot of research is still being pursued in the area of cognitive radios. In the current scenario where finding a good spectrum has become a very difficult task, having cognitive radio is a definite advantage.

Although there are a few similarities between both the technologies, there are also few major differences that might give one technology an edge over the other.

2.4   Cost

The major advantage of the WMNs over LTE is its cost effectiveness. It is cheaper and easier to deploy Wi-Fi Access Points that are plug-and-play. The low cost of installing a Wi-Fi network has made all residential users and home buildings organize Wi-Fi networks inside the buildings to cover a larger area. In a WiMAX network, the majority of cost is incurred in setting up the base station and the other access and service network modules.

The cost of deploying LTE is comparatively larger than the WMNs. The towers have to be deployed at the cell site for outdoor coverage and femto cells, relays or distributed antennas to provide indoor coverage. While some of the deployment options might be easy; few of them might take a lot of time and also could cost a lot in placement of new infrastructure.

2.5   Range

Wi-Fi has a range of few meters whereas WiMAX has a range of few miles. The range of Wi-Fi is approximately 100 feet and the theoretical range of a single WiMAX base station is approximately 30miles. Practically, it covers 6miles with a high signal to noise ratio.

A single LTE tower covers up to 100Kms with a slight degradation of the signal after 30Kms.It has a longer range than any of the existing standards. In order to cover the entire region, it gears relays all over the cell site.

2.6   Spectrum Flexibility

WiMAX can be implemented in both licensed and unlicensed spectrum frequencies.The users don't have to have a permission to access the unlicensed spectrum whereas while using the licensed spectrum, the users should have the authorization to access it. The 2.5GHz spectrum has a lot of electronic applications using the frequency for transmission, so WiMAX implements its licensed spectrum in 2.5GHz and unlicensed spectrum in 5GHz.Wi-Fi uses both 2.5GHz and 5GHz.In 2.5GHz, there are 3 non-overlapping channels and thus there is a lot of interference. In 5GHz, there are about 12 non-overlapping channels and with less interference.

In LTE, different spectrums can be used to transmit data in the uplink and downlink. It supports both paired and unpaired transmission. In paired transmission, the uplink and downlink are operated on different frequency bands. In unpaired transmission, the uplink and downlink share the same frequency band and the transmissions are separated in the time domain.

3   Scenarios

We tried to compare both LTE and WMN in both indoor and outdoor scenarios. We considered Wi-Fi for indoor, WiMax for outdoor in the case of mesh networks; Femto cells for indoor, LTE macro cell for outdoor in the case of LTE networks.

3.1   Indoor Coverage Scenario

WMNs have been implemented inside a building using Wi-Fi. We considered 802.11n mesh routers. The technological specifications of 802.11n routers are: Orthogonal Frequency Division Multiplexing, Multiple Input Multiple Output, Maximal Ratio Combining, Spatial Multiplexing, Channel Bonding, and Frame Aggregation.A lot of data and voice usage is initiated inside the buildings. In order to provide handling inside the building, the LTE network operators explore the use of femto cells, distributed antennas systems, relays, and hotspots. Though each technique has its pros and cons, femto cells are considered as a cost effective solution that provides indoor coverage and improves the capacity of the network. In order to deploy distributed antenna systems inside the building, the network operators should install cables and run them inside the buildings and deploy repeaters. It takes a lot of time and cost to complete setting up the antennas.

Data Rate and Coverage

The coverage area of a Wi-Fi router is larger than a Femto Access Point. The coverage area of a Wi-Fi router is about 100m whereas of Femto Access Point it is 10-30m.The Wi-Fi routers have higher data rates than the Femto cells. A typical 802.11n router has a theoretical data rate of 300Mbps with a practical data rate of 54Mbps.The lower frequencies are better to get higher data rates so when deploying a Wi-Fi network using a 2.5GHz spectrum is better than 5GHz.802.11n is the first 802.11 standard to implement Multiple Input Multiple Output to improve the data rate.In MIMO, several antennas are placed at the transmitter and receiver. The signal is sent using multiple antennas and received at the receiver using multiple antennas. At the receiver, frequency equalization is performed to filter and get the best possible signal. The data rate of a femto cell is 6.7Mbps which is the average speed of a DSL link that is used to backhaul the data to the operator network [3].

Spectrum

WMN operates in two frequency spectrums 2.5GHz and 5GHz.The number of non overlapping channels in 2.4GHz is 3, and in 5GHz its 12.According to Bandaranayke in [4], the 802.11a 2.4GHz channels have a lot more interference among themselves. The spectrum is shared with other electronic devices like microwave oven, Bluetooth, etc. The upper band U-IIN channels perform worse than the lower band and the middle band, which makes it difficult to find channels that can have a good performance capability. But, when using an 802.11n routers by using channel bonding, two 20MHz channels are combined to form a single 40MHz channel so that more data can be transmitted through it. It has been shown that channel bonding degrades the performance in 2.5GHz by increasing the interference between the channels. So, it is advisable to implement a WMN in 5GHz than a 2.5GHz spectrum [5].

LTE Femto cell shares the spectrum with the macro cell. In the United States, LTE is implemented between 1.9 and 2.6GHz. This sharing of spectrum causes cross-tier interference between the macro cell and femto cell. In order to mitigate the interference problem; cognitive radios, fractional frequency reuse, and inter-cell interference coordination methods have been suggested.

Back Haul Access

When a Wi-Fi network is being constructed, the 802.11n routers are deployed in the given geographical area. The routers are placed so that the coverage area of one router doesn't overlap with it neighboring router and that could cause interference. The routers are plug-and-play and use broadband connection as the backhaul network. The MRsare connected to the core network through the IGW routers. WiMAX can also be used as the backhaul access for the Wi-Fi networks for improved coverage.

When femto cells are used to provide indoor LTE coverage, femto base stations are deployed in the user's home. The femto access points connect to the cellular networks using the user's DSL cable connection. The femto cells are cost effective solution for the network operators to provide improved capacity and coverage in indoor scenarios.

AT&T spokesman Seth Bloom told the DSL reports.com that "3G Microcell is primarily intended to enhance the voice call quality experience in your home. While it can carry mobile data traffic, that's not the primary objective. Wi-Fi is an optimal solution for home mobile data use. People take advantage of Wi-Fi capabilities and that's why all smart phones include Wi-Fi radios, and usage on Wi-Fi doesn't count against mobile data usage bucket" [6]. And according to AT&T's website, the 3G Microcell service agreement/terms and conditions states that the data used over the femto cell region will be charged from the data plan that the user has. So, the user ends up paying subscription fee for the femto cell service, monthly data plan fee, and the broadband connection provided by the networks operator or any other company.

Access Method

Wi-Fi networks are usually either password protected or unprotected. It depends on the Access Point owner to either have a password to connect to the network or not to have one. It is a good practice to have a password to your network so that an outsider doesn't connect to your network and degrade the performance. As a larger number of users try to connect to a single network, the bandwidth is spread between the users depending on the applications they are using.Normally, home users have a password protected Wi-Fi networks.Most of the public places like airports, coffee shops, malls, etc. have unprotected networks. It also depends on the coverage area that each Access Point serves. For example, an airport has a lot of user traffic and there has to be a lot of access points deployed to cover the entire area. The number of users connected to a single network can be very large, making the network very slow due to limited available bandwidth. So, even though the network is password protected to the user's traffic, the network's performance can be degraded. A coffee shop area is relatively small. So, having a password protected Wi-Fi connection doesn't make the network congested.The security of the network is improved by having a password, so that no one can access your network as an observer or simply involved in illegal activity.

Femto Access Points (FAP) can be a closed, open, or hybrid access. In the closed access, only the femto access point subscribers are allowed to connect to it. In open access, all the users having a contract with the network operator can access it. In hybrid access, both subscribers and non subscribers can connect to the access point, but have restricted rights to the resources. While closed access has advantage over the other access methods, there is a drawback of associated interference problem. When a non subscriber is in the coverage area of the femto access point, the signal from the AP will be stronger than the macro cell signal and will cause interference.In open access, as the user can connect to any femtocell, the number of handovers between the FAP's will increase, making the call transition painful for the service provider.Hybrid access is the tradeoff between the closed and the open access. The FAP placement should be taken in to consideration so that it doesn't interfere with the neighboring FAP's.

VoIP

802.11 standardnetworks support Skype calls but they don't support mobile phone calls. The VoIP capacity of wireless networks is improved by the enhancements in 802.11n MAC and PHY layers. The implementation of MIMO, Frame aggregation, Block ACK, and Reverse Direction improves the throughput, reduces packet loss and delay [7].According to [8], the packet loss in Wi-Fi networks during a VoIP call is 0% which gives it an advantage over Femto cells.

Femto cells support both mobile phone and also Skype calls. Usually, you don't get a proper mobile network signal when you are inside the building, especially in the basement. So, when you use a femto cell, it ensures that you have a strong coverage even when you are inside the buildings.You don't have to walk out of the building or stand in the corridor to make a phone call. With a femto cell deployed, you are always connected to the network. When you are out of the range of a femto cell, your call will be automatically handed over to the macro cell.According to [8], the packet loss in Femto cells during a VoIP call is 0.06% due to which Wi-Fi has a slight edge.

Power Consumption

The power consumption of the access point depends of the area it is covering and the number of users it is serving in a cell. Wi-Fi routers consume a lot of power than Femto cells because of its larger coverage area.These are summarized in Table 1[9].

Table 1. Femtocell and WiFi Specification

	Femtocell	WiFi
Data rates	7.2 – 14.4Mbps	11 and 54Mbps
Op. Frequency	1.9 – 2.6GHz	2.4 and 5GHz
Power	10, 100mW	100, 200mW
Range	20-30m	100-200m
Services	Primarily Voice, and data	Primarily Data, and voice

Video Streaming

Video streaming takes up a lot of bandwidth than any other mobile application. According to [8], the performance of the network decreases when streaming a video over Wi-Fi due to packet losses. As illustrated in Figure 1[8], femto cell out performs Wi-Fi in the case of video streaming.

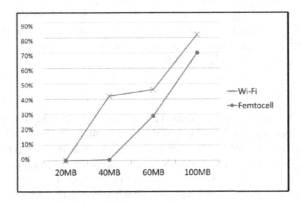

Fig. 1. Packet Loss in a Femtocell

LTE uses Wi-Fi and Femto cell as the offloading technique to meet the higher data rate needs of the users. As a majority of the users' traffic is originated inside the building, LTE offloads the data traffic to Wi-Fi or Femto cell, which are very well capable of handling the indoor coverage. WhileWi-Fi offers a higher data rate, is cost effective, has a good VoIP performance and takes an advantage of having unlicensed spectrum; it consumes more power and has a poor video streaming performance as compared to a femto cell. Femto cells consume less power, has a slightly degraded VoIP performance than Wi-Fi. Though the video streaming performance is good, users should consider the fact that accessing data, when connected to a femto access point, will be charged from the user's data plan. When using applications with high bandwidth requirements, it may cost more as compared to accessing the web browser, email, etc.

3.2   Outdoor Coverage Scenario

In the outdoor scenario, we considered WiMAX technology in the case of WMNs. It provides wireless broadband Internet access all over and is either a fixed, nomadic, or mobile network. WiMAX has been implemented even before LTE and it was considered as a tough competition for LTE. It basically has a performance similar to 802.11 networks but with a wider coverage comparable to the cellular networks. The WiMAX 802.16e Release 1.0 is comparable to 3GPP Rel-8 commercially known as LTE. The WiMAX 802.16m Release 2.0 is similar to 3GPP Rel-10 known as LTE-A. These are the two leading 4G technologies right now in the market and are trying to reach the IMT-Advances specifications.

The bandwidth of WiMAX Release 1.0 varies between 3.5, 5, 7, 8.75 and 10MHz, whereas the channel bandwidth of LTE Rel-8 varies between 1.4, 3, 5, 10, 15, 20MHz [10]. The bandwidth of WiMAX Release 2.0 is 5, 7, 8.75,10,20, and up to 100MHZ by using carrier aggregation.LTE-A has a bandwidth upto 100MHz.It uses contiguous and non-contiguous spectrum for carrier aggregationto provide higher data rate [11].

Data Rates and Coverage

According to Sprint, which has both LTE and WiMAX networks rolled out, the upload and download speed of a 4G LTE network is 2-3Mbps and 6-8 Mbps respectively, whereas the upload and download speed of a 4G WiMAX network is up to 1.5Mbps and 3-6Mbps respectively. The round trip latency of the LTE network is less than 60ms and of WiMAX is 120ms [12].

Spectrum

Mobile WiMAX mainly supports only TDD (Time Division Duplex) both in licensed and unlicensed spectrum while fixed WiMAX can maintain both FDD and TDD. In Line of sight access, the WiMAX base stations are operated in 10 to 66GHz band and in non-line-of-sight access, the WiMAX base stations are operated in 2 to 11GHz. The non-line-of-sight access bands let lot of users connect to a single base station, reducing the effective cost of the network which is the major advantage of WiMAX. The line-of-sight base stations operate at a higher bandwidth and are less prone to interference [13].In WiMAX, the spectrum ought to be synchronous because of the main motivation being the TDD.

LTE is operated in the 700MHz band in the United States. While LTE's main emphasis is on FDD, it supports both FDD and TDD. In fact, TD-LTE networks deployment have been initiated by few network operators. In FDD, the spectrum can be asynchronous. It can have two different spectrums assigned to uplink and downlink channels. The spectrum flexibility allows assignment of different bandwidth of approximately 1.25-20MHz in each frequency band.

Power Consumption

WiMAX implements OFDMA in both uplink and downlink. According to [14], the power consumed by a single WiMAX base station is 2.9kW.

LTE utilizes OFDMA in the downlink while SCFDMA is employed in the uplink. The main advantage of LTE to WiMAX is implementing SCFDMA which has a low Peak-Average-Power-Ratio. According to [14], the power consumed by a single LTE base station is 3.7 KW.

Latency

The frame duration in WiMAX is 20ms and is divided into 4 sub frames of 5ms each. Each sub frame is divided into several OFDM symbols.The frame duration in LTE is 10ms and is fixed. Each sub frame is divided into two slots of 0.5ms.The Transmission Time Interval is 1ms.The latency of WiMAX TDD is 30ms whereas LTE FDD is 10ms.The response time of LTE networks is faster than WiMAX [15].The lower latency of LTE helps in providing better performance for VoIP calls to the users.

VoIP

Both LTE Rel-8 and WiMAX 802.16e support VoIP calls. The number of users per sector each network supports is different. WiMAX 802.16e supports 20 users/sector where as LTE Rel-8 supports 80 users/sector. WiMAX 802.16m supports approximately >30 users/sector, where as LTE-A supports approximately >80 users/sector [11].

IndustrySupport

The major dis-advantage of WiMAX is its inability to be backward compatible with any of the existing UMTS networks. So, the network operator has to roll out the network from scratch.A lot of network operators and companies are backing out from the deployment of WiMAX and taking steps ahead in installing LTE networks. Cisco, Erricson, Alcatel-Lucent, Alvarion already started showing their support in utilizing LTE networks. LTE-A is interoperable with LTE and LTE is interoperable with the previous UMTS, CDMA networks. This gives LTE an advantage to the network operators to roll out the networks without much difficulty.

LTE is a 3GPP standard and is supported by most of the companies like Intel, Qualcomm, Cisco etc. WiMAX is an IEEE standard and Sprint, Clear wire are the only companies that support it. Sprint recently said it will stop supporting WiMAX and start deploying LTE networks. Thus, LTE has a lot of future scope to evolve though the deployment started 2 years after WiMAX's placement. Though there are a total of more than 500 deployments worldwide, WiMAX covers only 6% of the world population [16].

Most of the mobile companies support 3GPP standard. Samsung is the only company that produces WiMAX supported mobile phones. Apple's iPhones are all 3GPP standard supported which has the highest percentage of usage. This is also one of the factors that are affecting the popularity of WiMAX.

LTE-WiMAX Coexistence

A seamless integration of LTE and WiMAX networks has been proposed by Shyam S. Wagle, Minesh Ade, and M. GhazanfarUllah [17].The authors basically describe various functions that are designed to integrate both the networks.

LTE has clearly outperformed WiMAX in terms of data rates, VoIP performance, and popularity. The main disadvantage of LTE is the power consumption. Even the public safety department is planning to migrate its operation from Wi-Fi and WiMAX to LTE.

4   Conclusion

The objective of the paper is to help users select an appropriate network between the WMNs and LTEin indoor and outdoor scenarios. We started off by pointing out the similarities and differences between the two networks. In indoor coverage, 802.11n Wi-Fi has been compared with LTE Femtocell. In outdoor coverage, WiMAX 802.16e has been compared with LTE. We tried to parallel them in terms of their data rates, VoIP performance, video streaming, etc., and show the advantages and disadvantages of each network. While each network has associated pros and cons, we cannot clearly state that which network is better than the other. Depending on the application requirements, it is better to choose the cost effective solution.

References

1. Draves, R., Padhye, J., Zill, B.: Routing in multi-radio,multi-hop wireless mesh networks. In: MobiCom 2004 Proceedings of the 10th Annual International Conference on Mobile Computing and Networking, September 26-October 1, pp. 114–128 (2004)
2. Khandekar, A., Bhushan, N., Fang, J.T., Vanghi, V.: LTE-Advanced-Heterogeneous networks. In: 2010 European Wireless Conference, April 12-15, pp. 978–982 (2010)
3. http://www.dslreports.com/shownews/Average-US-Broadband-Connection-Now-67-Mbps-120700
4. Bandaranayake, A.U., Pandit, V., Agrawal, D.P.: Indoor Link Quality Comparison of IEEE 802.11a Channels in a Multi-radio Mesh Network Test bed. Journal of Information Processing Systems 8(1), 1–20 (2012)

5. WLAN Channel Bonding: Causing Greater Problems Than It Solves. Texas Instruments SPLY 2003 – White Paper (September 2003),
   http://www.ti.com/lit/wp/sply003/sply003.pdf
6. http://www.fiercebroadbandwireless.com/story/ts-3g-femtocell-service-counts-against-data-usage-caps/2010-06-20
7. Wang, C.-Y., Wei, H.-Y.: IEEE 802.11n MAC Enhancement and Performance Evaluation. ACM/Springer Mobile Networks and Applications Journal (MONET) 14(6), 760–771 (2009)
8. Bare, J.K.: Comparison of the Performance and Capabilities of Femtocell Versus Wi-Fi Networks. MS thesis, Naval Postgraduate School (September 2012),
   http://www.dtic.mil/dtic/tr/fulltext/u2/a567411.pdf
9. Hasan, S.F., Siddique, N.H., Chakraborty, S.: Femto cell versus Wi-Fi – A survey and comparison of architecture and performance. In: 1st International Conference on Wireless Communication, Vehicular Technology, Information Theory and Aerospace & Electronics Systems Technology (2009)
10. Abichar, Z., Chang, J.M., Hsu, C.-Y.: WiMAX vs. LTE: Who Will Lead the Broadband Mobile Internet? IEEE IT Professional 12(3) (May-June 2010)
11. Ahmadi, S.: Mobile WiMAX: A Systems Approach to Understanding IEEE 802.16m Radio Access Technology. Academic Press (2010)
12. http://www.sprint.com/landings/lte/index.html?ECID=vanity:4glte
13. Pareek, D.: WiMax – Taking wireless to a Max, May 30. Auerbach Publications (2006)
14. Vereecken, W., Van Heddeghem, W., Deruyck, M., Puype, B., Lannoo, B., Joseph, W., Colle, D., Martens, L., Demeester, P.: Power Consumption in telecommunications networks: overview and reduction strategies. IEEE Communications Magazine 49(6), 62–69 (2011)
15. Ball, C.: LTE and WiMAX Technology and Performance Comparison. In: Nokia Siemens Networks Radio Access, GERAN& OFDM Systems: RRM and Simulations, EW 2007 Panel (April 3, 2007), http://projects.comelec.enst.fr/EW2007/Documents/Comparison_LTE_WiMax_BALL_EW2007.pdf
16. http://www.researchandmarkets.com/reports/1194057/wimax_and_lte_the_case_for_4g_coexistence
17. Yahiya, T.A., Chaouchi, H.: On the Integration of LTE and Mobile WiMAX Networks. In: Proceedings of 19th International Conference on Computer Communicationsand Networks (ICCCN), August 2-5, pp. 1–5 (2010)

Improved Heuristics for Online Node and Link Mapping Problem in Network Virtualization

Hoang Viet Tran and Son Hong Ngo

School of Information and Communication Technology
Hanoi University of Science and Technology
sonnh@soict.hut.edu.vn
http://soict.hut.edu.vn/~sonnh

Abstract. Network virtualization is considered a promising way to improve flexibility and manageability for Internet. A major challenge in network virtualization is how to effectively allocate the underlying resources satisfying virtual network requests. Since the general network embedding problem is computationally unsolvable, past researches have followed various approaches to reduce the computational cost. For example, Minlan Yu et al. have introduced a greedy algorithm for online virtual network requests with low complexity. In this work, we further improve the performance of this algorithm by proposing three improvements: (1) sorting virtual nodes according to the amount of required resources, (2) using an adaptive function to evaluate available resources and (3) pre-checking invalid substrate links before link-mapping phase. Extensive simulation shows that our proposed algorithm gives better results in terms of acceptance ratio, revenue-to-cost ratio as well as computational cost.

Keywords: network virtualization, link and node mapping, future Internet.

1 Introduction

In the last few decades, while protocols and services have been evolving rapidly on the edge of the Internet, the core of the Internet still changes very slowly (Papadimitriou et al.[1]). For example, the new protocol IPv6 is deployed at an extremely slow pace (Google Statistics[2]). Network Virtualization is believed to be a solution to this issue by allowing multiple virtual networks (VNs) to run simultaneously on the same underlying infrastructure [3–6].

A major challenge in network virtualization is how to allocate the underlying resources to virtual networks and to satisfy the requirements in an efficient way (Haider et al. [7, 8]). This problem is similar to the Virtual Private Network design problem, which is NP-hard. Optimal resource allocation in network virtualization is even harder since the location of virtual nodes is unknown. Hence, heuristics are usually used to seek for a near optimal solution while the complexity of the problem is reduced.

B. Murgante et al. (Eds.): ICCSA 2013, Part V, LNCS 7975, pp. 154–165, 2013.

In this paper, we consider the online virtual network embedding (VNE) problem, i.e. the VN requests arrive dynamically and stay in the network for a period of time before departing. In practice, the embedding algorithm must handle VN requests as they arrive. In [9], Minlan Yu et al. have introduced conventional algorithm for online virtual network requests with low complexity. By dividing the VNE problem into node-mapping phase and link-mapping phase and using a greedy approach for each phase, the Yu's conventional algorithm can solve the VNE problem in a relatively small computational time.

In this work, we further improve the performance of Yu's conventional algorithm by proposing three improvements: (1) sorting virtual nodes according to the amount of required resources, (2) using an adaptive function to evaluate available resources and (3) pre-checking invalid substrate links before link-mapping phase. Extensive simulation shows that our proposed algorithm gives better results in terms of acceptance ratio, revenue-to-cost ratio as well as computational cost.

The rest of this paper is organized as follows: Section 2 describes the virtual network mapping problem and restates the greedy virtual mapping algorithm of Yu et al. [9]). Section 3 presents our improvement for this algorithm. The evaluation results are presented in Section 4. Conclusion and Future works are figured out in Section 5.

2   Problem Formulation and Related Works

In this section we describe the general virtual network embedding problem and some greedy schemes to solve this problem. The following description are derived mainly from Yu [9], Zhu [10] and Lischka [11].

2.1   Virtual Network Mapping Problem

We model the substrate network as an undirected graph $G^S = (N^S, L^S, A^S)$, where N^S is the set of substrate nodes and L^S is the set of substrate links. Every node or link $e|e \in N \cup L$ is associated with their attributes, denoted by A^S. Usually, CPU resource and location are the constraints of a node, while bandwidth and error rate are the constraints of a link. In this paper we consider CPU resource for node attribute and bandwidth for link attribute.

A virtual network (VN) is a logical network running on a substrate network. It is modeled as a graph $G^V = (N^V, L^V, C^V)$ where N^V, L^V are the set of virtual nodes and virtual links, respectively. C^V is the set of constraints, which commonly are required bandwidth and CPU resources.

A virtual network request $r = (G^V, t, l)$ is a request for creating the virtual network G^V at arrival time t. Duration l is the lifetime that the virtual network desires to exist in the system. A virtual network mapping (VNM) problem of a virtual network $G^V = (N^V, L^V, C^V)$ onto a substrate network $G^S = (N^S, L^S, A^S)$ is defined as a mapping of G^V to a subset of G^S, such that the constraints in G^V are satisfied:

$$M : G^V \mapsto (N^S, P^S)$$

where P^S is the set of all substrate paths in the substrate networks. The mapping M is valid if all constraints of G^V are satisfied.

2.2 Objectives Functions

Several objective functions have been used in the VNM problem. In [9, 11] the authors decided to allocate the network resources by economic interest. Following this, two main objectives are set:

1. Accept as many VN requests as possible, and
2. Serve the VN requests with highest economic benefit.

For the first one, Lischka et al. [11] try to maximize the VN request *Acceptance Ratio*, which is defined as the proportion of arriving VN requests which are accepted. For the second one, above works aim at maximizing the revenue and minimizing the cost per each VN request acceptance. In these works, revenue gained by accepting a VN request is evaluated by the total bandwidth of all virtual links, while the cost of a mapped VN request is determined by the amount of substrate resources reserved for that VN. Specifically, the revenue R and the cost C of a VN request $G^V = (N^V, L^V)$ are defined as [9]:

$$R = \sum_{l \in L^V} bw^v(l), \; C = \sum_{p \in P(G^V)} length(p) \times bw^s(p, G^V) \tag{1}$$

where:

- $bw^v(l)$ is the required bandwidth of the virtual link l,
- $length(p)$ is the hop count of path p on substrate network,
- $P(G^V)$ is the set of all substrate paths corresponding to virtual links in G^V,
- $bw^s(p, G^V)$ is the occupied bandwidth on path p by G^V.

2.3 Greedy Node Mapping and Link Mapping without Path-Splitting Algorithms

Yu et al. [9] use a greedy approach which involve two main stages: *node mapping* and *link mapping*.

At the first stage (Greedy Node Mapping), all requests in one time-window are collected and stored into a queue of requests in the order of decreasing revenue. For each request, they map its virtual nodes to substrate nodes with maximum available resource. The available resource (AR) of a physical node n is measured by Eq.2.

$$AR(n) = CPU(n) \times \sum_{l \in L(n)} bw(l) \tag{2}$$

where:

- $CPU(n)$ is the free computing resource of node n
- $L(n)$ is the set of all adjacent substrate links of node n
- $bw(l)$ is the spare bandwidth resource of link l

The pseudo code of the Greedy Node Mapping algorithm is described as follows.

Algorithm 1. Greedy Node Mapping

1: Sort the requests according to their revenues.
2: Take a request with the largest revenue.
3: Find the subset S of substrate nodes that satisfy available CPU capacity (larger than that specified by the request).
4: **for** each virtual node **do**
5:     Find the substrate node with the "maximum available resource" in S.
6: **end for**
7: **if** fail in **3.** or **4.** **then**
8:     Store this request and GOTO **2.**
9: **end if**

At the second stage, the selected nodes are connected following k-shortest paths algorithm to form a completed virtual network. A found path is accepted if it has enough bandwidth. Below is the link-mapping algorithm for requests without path-splitting. Note that in this paper, we only consider the case of un-splittable flows.

Algorithm 2. Link Mapping with k-Shortest Paths [9]

1: Sort the requests according to their revenues.
2: Take one request with the largest revenue.
3: **for** each virtual link **do**
4:     Search k-shortest paths p until p satisfies bandwidth constraints
5: **end for**
6: **if** fail in **3.** **then**
7:     Reject and store this request, GOTO 2.
8: **end if**

---

3   Enhanced Virtual Network Mapping Algorithm

The advantages of above greedy algorithm are the low computational cost which is preferred for large-scale applications. However, this approach has some drawbacks as follows.

Firstly, the virtual nodes of a VN request are mapped without considering about their required resource (step 4. of algorithm 1). Therefore, a "heavy" virtual node, i.e. node that requires more CPU resources than the others do, may be selected to map on a substrate node having low available CPU resource. Consequently, resource exhaustion may quickly happen on this node, thus reduces the performance of the VNE algorithm.

Secondly, the "available resource" is evaluated by using a fixed function (as in Eq. 1). Although this function already takes into account both available CPU and spare bandwidth resources, it still does not separate the case when substrate links are much more loaded than substrate nodes' CPU and vice versa. For example, a VN mapping on lightly-loaded nodes and very heavily loaded links may cause congestion on its corresponding substrate links, thus reduces the performance of this resource allocation strategy.

Finally, we also observe that the bandwidth constraints in the link mapping phase are checked only after k-shortest paths are found [12]. In many cases, all of these paths do not satisfy the bandwidth constraints, causing the mapping ends unsuccessfully (as illustrated in Fig. 1).

Based on the above observation, we propose the Enhanced Greedy Node Mapping scheme and the Link Mapping with Pre-Checking scheme for the virtual network mapping as below.

3.1   Enhanced Greedy Node Mapping (EGNM)

In this scheme, the virtual nodes are sorted according to their "required resource" (see step 4. of Algorithm 3.) before mapping them onto corresponding substrate nodes. By this way, virtual nodes with more "required resource" will be mapped to substrate nodes which have more available resource. This will help to avoid bottleneck at substrate nodes lacking of bandwidth resource. The "required resource" of a virtual node n^v is defined in a similar way to "available resource", as below:

$$RR(n^v) = CPU(n^v) \times \sum_{l^v \in L(n^v)} bw(l^v) \qquad (3)$$

To overcome the drawback of the fixed AR function, we adopt the adaptive optimization strategy [10] in which the AR value depends on the value of a variable T. In our model, threshold T is defined as the ratio of *average CPU load* on all substrate nodes to *average link load* on all substrate links:

$$T = \frac{\sum\limits_{n^s \in N^S} CPULoad(n^s)/|N^S|}{\sum\limits_{l^s \in L^S} LinkLoad(l^s)/|L^S|} \qquad (4)$$

where $CPULoad(n)$ is the ratio between occupied CPU resource of substrate node n and its total CPU resource; similarly, $LinkLoad(l)$ is the ratio between occupied bandwidth and total bandwidth of substrate link l. The AR value of node n is then defined as an adaptive function of the variable T:

$$AR(n) = \begin{cases} CPU(n) \times \sum\limits_{l \in L(n)} bw(l) & \text{if } T \geq 1 \\ \sum\limits_{l \in L(n)} bw(l) & \text{if } T < 1 \end{cases} \tag{5}$$

Algorithm 3. Enhanced Greedy Node-Mapping (EGNM)

1: Sort the requests according to their revenues.
2: Take one request with the largest revenue.
3: Find the subset S of substrate nodes that satisfy available CPU capacity (larger than that specified by the request.)
4: Sort the virtual nodes according to their "required resource".
5: **for** each virtual node **do**
6:    Find the substrate node n with the max "available resource" (Eq.5) in S.
7: **end for**
8: **if** fail in **3.** or **6. then**
9:    Store this request and GOTO **2.**
10: **end if**

In comparison with the conventional GNM algorithm, the additional computation time of the EGNM algorithm is devoted for finding the "required resource", sorting the virtual nodes and calculating the variable T. The time to find the "required resource" of all virtual nodes in virtual network $G^V = (N^V, L^V)$ is $O(|N^V| + |L^V|)$. The time to sort the virtual nodes is $O(|N^V|.\log|N^V|)$. Since we always keep track of available resources, the variable T can be calculated in amortized time $O(|N^S| + |L^S|)$. We also notice that T is evaluated only once per each time-window.

3.2 Link Mapping with Pre-checking

Starting from the observation that in many cases, all k-shortest paths do not satisfy the bandwidth constraints, we propose a pre-checking scheme that verifies the status of resource usage in the substrate network before the link mapping. Specifically, those substrate links that do not have enough available bandwidth will be marked and removed from the graph. Then, we will search for the single shortest path using Dijkstra's algorithm. This process improves the acceptance ratio by assuring that once we find out a path, it will satisfy the bandwidth constraints.

Fig. 1 illustrates the efficiency of link mapping with pre-checking. In this figure, the number on each substrate link represents the available bandwidth. Here, we need to map the virtual link ab onto the substrate network $\{A, B, C, D, E, F\}$, suppose that the virtual nodes a and b have been already mapped to substrate nodes A and B. In the left scenario, the conventional Yu's algorithm finds out two paths ACB and ADB. However, both of them do not satisfy bandwidth

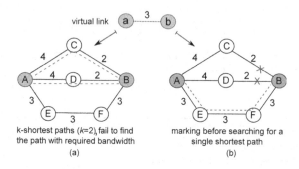

k-shortest paths (k=2), fail to find   marking before searching for a
the path with required bandwidth   single shortest path
(a)   (b)

Fig. 1. Comparing two link-mapping strategies

constraints. The right scenario, in contrast, temporarily remove two substrate links CB and DB and find a valid path $AEFB$ using Dijkstra's algorithm.

The pseudo code of the Link Mapping with Pre-Checking scheme is described as follows.

Algorithm 4. Link Mapping with Pre-Checking

1: Sort the requests according to their revenues.
2: Take one request with the largest revenue.
3: **for** each virtual link l^v **do**
4:     **for all** $l^s \in L^S$ **do**
5:         **if** spare bandwidth of l^s is less than the requirement of virtual link **then**
6:             $weight(l^s) = \infty$
7:         **end if**
8:     **end for**
9:     Find the single shortest path p.
10: **end for**
11: **if** fail in **3.** or **9.** **then**
12:     Reject and store this request, GOTO **2.**
13: **end if**

Given a virtual link, the time complexity of the k-shortest paths scheme to find a link mapping without path-splitting is $O(|N^S|log|N^S| + k.|N^S|)$ [12]. The link mapping with pre-checking scheme can be solved in $O(|L^S| + |L^S|.log|N^S|)$ time complexity with Dijkstra's algorithm.

4   Experimental Results

4.1   Simulation Environment

We implemented three algorithms (described in Table 1.) in C++ and using the OMNeT++ network simulator [13] to perform the simulation. The number of shortest paths for the GNM-kPaths and EGNM-kPaths is set to $k = 2$. Substrate networks and virtual networks are generated with random topologies by using the popular GT-ITM tool [14]. The number of nodes of the substrate network is 30 and 50, respectively. Each pair of substrate nodes is randomly connected with probability 0.2. The number of nodes of a virtual network request follows uniform distribution between 2 and 10, each pair of virtual nodes is randomly connected with probability 0.2. In fact, we adopted the similar simulation parameters as in the related work considering online problem [9, 15].

VN requests arrive as a Poisson process with an average rate at $\lambda = 5$ requests per time window and have exponentially distributed lifetime, with 10 time windows on average. The size of time window is 20 units by default. The CPU resources of substrate nodes and the bandwidth resources of substrate links follow uniform distribution between 50 and 100 units. *The CPU requirement of a virtual node and the bandwidth requirement of a virtual link follow are 50 units on average.*

Table 1. Algorithms to be evaluated

Algorithm	Description
GNM-kPaths	Yu's algorithm: Greedy Node Mapping and Link Mapping with k-shortest path
EGNM-kPaths	Enhanced Greedy Node Mapping, Link Mapping with k-*shortest* path
EGNM-preCheck	Enhanced Greedy Node Mapping, Link Mapping with Pre-checked

4.2 Result and Discussion

We evaluate the performance of these algorithms based on these criteria: *Acceptance Ratio, Cumulative Revenue, R/C-Ratio* and *Computational Time*. These objectives have been introduced in section 2.2.

Figs. 2 & 3 show the acceptance ratio and cumulative revenue over time, respectively. Cumulative revenue is the total revenue of all mapped VN requests from the beginning of simulation. It is observed from these figures that in terms of acceptance ratio and cumulative revenue, two modified algorithms provide much better result than that of the conventional Yu's algorithm. It is also noticed that the algorithm EGNM-preCheck is considerably better than the others. This observation suggests that our proposed improvements in both node-mapping phase and link-mapping phase allow substrate resources to be used more efficiently.

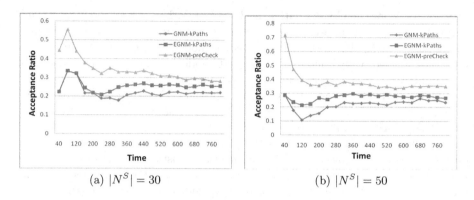

(a) $|N^S| = 30$ (b) $|N^S| = 50$

Fig. 2. Acceptance Ratio

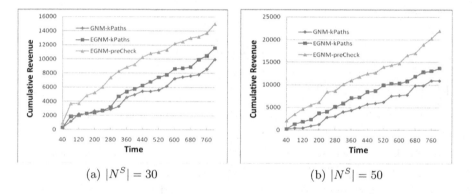

(a) $|N^S| = 30$ (b) $|N^S| = 50$

Fig. 3. Cumulative Revenue

Since the revenue alone does not express exactly the economic efficiency, we also use the revenue-to-cost ratio (R/C-ratio) as the unified evaluation of revenue and cost. The value of R/C-ratio varies between 0 and 1. If a VN request is not accepted, the revenue is set to 0. In the optimal case, R/C-ratio reaches 1 when each virtual link is mapped to a path that spans only one substrate link.

Fig. 4 shows the performance of three algorithms in terms of R/C-ratio. It is also observed that our proposed algorithm EGNM-kPaths outperforms the conventional GNM-kPaths. However, the EGNM-preCheck has the poorest result among these algorithms despite the fact that it has highest cumulative revenue and acceptance ratio. The reason is that the EGNM-preCheck always tries to find out a valid path by all means, even the path it found is much more longer than the shortest path, thus increase the cost and decrease the R/C-ratio. One possible solution to improve the performance of EGNM-preCheck is that we should reject those VN requests which have too long path, we will consider this issue in our future work.

In addition, we also observe that the value of acceptance ratio and R/C-ratio tends to be fluctuating at the beginning, then becomes steady after about 400

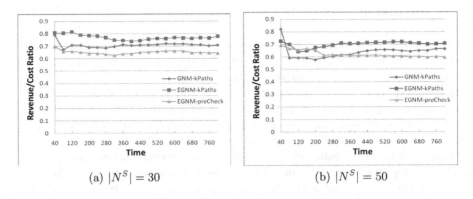

(a) $|N^S| = 30$ (b) $|N^S| = 50$

Fig. 4. Revenue/Cost-Ratio

Table 2. Average simulation running time (in *sec.*)

Scenario	GNM-kPaths	EGNM-kPaths	EGNM-preCheck
(1) $|N^S| = 30, \lambda = 5$	8.094	8.406	5.719
(2) $|N^S| = 50, \lambda = 5$	14.375	15.125	9.240
(3) $|N^S| = 50, \lambda = 10$	27.687	25.656	19.860

time units. This observation suggests that at this point the system has entered a balanced state where the amount of departing requests is approximately equal to the number of arriving requests.

To evaluate the computation time, we measure the simulation running time for each turn and compute the average running time. We also use two substrate networks with the number of nodes are 30 and 50, respectively and shows the results in Table 2. It is notable that the EGNM-kPaths and GNM-kPaths have almost similar execution time while the EGNM-preCheck has the lowest execution time. These results show that the enhanced node mapping scheme just slightly increases the computational complexity and therefore it is suitable for a greedy approach.

We also notice that in scenario (3), the conventional algorithm has the biggest computation time. When the VN arrival rate increases ($\lambda = 10$), the number of rejected VN requests which are re-inserted into the queue will significantly increase. This figure of GNM-kPaths algorithm is larger than that of the improved algorithms since this conventional algorithm has the lowest acceptance ratio.

5   Conclusion

In this paper, we presented our improved virtual network mapping algorithms with three enhanced schemes: sorting virtual nodes based on their required resources, using an adaptive evaluation of available resource in the node-mapping phase and applying a pre-check process in the link-mapping phase. We implemented the proposed schemes in two enhanced algorithms (EGNM-kPaths and EGNM-preCheck) and compared their performance with the conventional Yu's algorithm by means of simulation. The evaluation results show that our algorithms are better than conventional one in term of several economical objectives such as acceptance ratio, cumulative revenue, revenue-to-cost ratio, while the computational complexity is still reasonable. In future works, we will investigate more sophisticated heuristics using evolutionary computing for VN mapping problem as well as consider the survivability issue in network virtualization.

Acknowledgment. This work was supported by the Ministry of Science and Technology (MOST) under the Vietnam - Japan protocol project named "Models for Next Generation of Robust Internet", grant number 12/2012/HD-NDT.

References

1. Papadimitriou, P., Maennel, O., Greenhalgh, A., Feldmann, A., Mathy, L.: Implementing network virtualization for a future internet. In: 20th ITC Specialist Seminar, Hoi An, Viet Nam (May 2009)
2. Google. Global IPv6 statistics, http://www.google.com/ipv6/statistics.html (accessed December 20, 2012)
3. Feamster, N., Gao, L., Rexford, J.: How to lease the internet in your spare time. SIGCOMM Comput. Commun. Rev. 37(1), 61–64 (2007)
4. Chowdhury, N.M.M.K., Boutaba, R.: A survey of network virtualization. Comput. Network 54(5), 862–876 (2010)
5. Carapinha, J., Feil, P., Weissmann, P., Thorsteinsson, S.E., Etemoğlu, Ç., Ingpórsson, Ó., Çiftçi, S., Melo, M.: Network virtualization - opportunities and challenges for operators. In: Berre, A.J., Gómez-Pérez, A., Tutschku, K., Fensel, D. (eds.) FIS 2010. LNCS, vol. 6369, pp. 138–147. Springer, Heidelberg (2010)
6. Wang, A., Iyer, M., Dutta, R., Rouskas, G.N., Baldine, I.: Network virtualization: Technologies, perspectives, and frontiers. Journal of Lightwave Technology 31(4), 523–537 (2013)
7. Haider, A., Potter, R., Nakao, A.: Challenges in resource allocation in network virtualization. In: 20th ITC Specialist Seminar, Hoi An, Viet Nam (May 2009)
8. Belbekkouche, A., Hasan, M.M., Karmouch, A.: Resource discovery and allocation in network virtualization. IEEE Communications Surveys Tutorials 14(4), 1114–1128 (2012)
9. Yu, M., Yi, Y., Rexford, J., Chiang, M.: Rethinking virtual network embedding: substrate support for path splitting and migration. SIGCOMM Comput. Commun. Rev. 38(2), 17–29 (2008)
10. Zhu, Y.: Routing, Resource Allocation and Network Design for Overlay Networks. PhD thesis, Georgia Institute of Technology (December 2006)

11. Lischka, J., Karl, H.: A virtual network mapping algorithm based on subgraph isomorphism detection. In: Proceedings of the 1st ACM Workshop on Virtualized Infrastructure Systems and Architectures, VISA 2009, pp. 81–88 (2009)
12. Eppstein, D.: Finding the k shortest paths. SIAM J. Computing 28(2), 652–673 (1998)
13. The OMNeT++ network simulator, http://www.omnetpp.org (accessed December 20, 2012)
14. Zegura, E.W., Calvert, K.L., Bhattacharjee, S.: How to model an internetwork. In: Proceedings IEEE INFOCOM, vol. 2, pp. 594–602 (March 1996)
15. Chowdhury, N.M.M.K., Rahman, M.R., Boutaba, R.: Virtual network embedding with coordinated node and link mapping. In: IEEE INFOCOM 2009, pp. 783–791 (April 2009)

Accelerating Metric Space Similarity Joins with Multi-core and Many-core Processors

Shichao Jin, Okhee Kim, and Wenya Feng

School of Software and Microelectronics, Peking University, Beijing, China
{shichaojin.cs,anniekim.pku,pkuwenyafeng}@gmail.com

Abstract. The similarity join finds all pairs of similar objects in a large collection. This search problem could be successfully divided into many sub-problems by an algorithm called Quickjoin recently. Besides, this algorithm could be extended to a wide range of application areas as it is based on metric spaces instead of vector spaces only. When the volume of a dataset reaches to a certain degree or the distance measure of the similarity is complex enough, however, Quickjoin still takes much time to accomplish the similarity join task, which leads us to develop a parallel version of the algorithm. In this paper, we present two parallel versions of the Quickjoin algorithm exploiting multi-core and many-core processors respectively as well as evaluate them. Experiments show that the parallelization of this algorithm in our many-core processor yields speedup to 22 at most compared with its non-parallel version.

Keywords: Similarity Joins, Metric Spaces, Multi-core, Many-core.

1   Introduction

Similarity joins are useful in a variety of fields, such as medical imaging, multimedia database and geographic information systems. For example, users might use it to find the association rules, to deal with the data cleaning task, and even to identify data clusters. In brief, the responsibility of similarity joins is to find all pairs of similar objects in a large database (or find pairs of similar objects from two different databases), where the similarity is defined in advance. Here the similarity refers to the distance between two objects (ϵ denotes this distance in this paper).

Metric spaces could offer a general way to handle the similarity join problem. It means that not only objects relying on vector spaces could be processed, but also objects like strings including DNA sequences could be directly handled, provided that the distance measure used to calculate the distance between two objects meets the requirements of metric spaces. Specifically, in the object oriented programming, one only needs to define an interface to realize the problem, where two parameters are the class of the object and the type of the distance function respectively, while the remaining part of a metric algorithm could almost stay unchanged. The distance measure might vary depending on the type of objects tested. For example, Edit distance [1] is a way to calculate the distance between two string objects including

B. Murgante et al. (Eds.): ICCSA 2013, Part V, LNCS 7975, pp. 166–180, 2013.
© Springer-Verlag Berlin Heidelberg 2013

common texts and DNA sequences; the Jaccard coefficient [2] can be used to quantify the similarity between two sets including web documents.

Quickjoin [3] is an algorithm capable of processing metric data, and it employs the divide-and-conquer method to recursively partition the similarity join problem until the scale of the problem could be smoothly handled in a brute force way (comparing each pair of objects).   Although Quickjoin algorithm could substantially decrease the time costs in solving the similarity join problem compared to a brute force way, the computing complexity is still high, especially when the distance measure is CPU bound. For example, the Edit distance has $O(nm)$ runtime complexity, where n and m represent the lengths of two strings respectively.

Considering the reality of rapid development in multi-core and many-core processors, we propose two parallel versions of the Quickjoin algorithm. One of them makes use of the multi-core processor, while the other exploits the resources of the many-core processor, that is, the GPU. The reason of using the GPU is that its general purpose platform increasingly offers a great power in parallel processing. On the other hand, the multicore CPU is another powerful parallel platform without specific constraints of the GPU architecture, such as transferring between main memory and GPU global memory. Both platforms offer a high performance/price ratio, and we are interested in their comparative results in terms of parallelizing the Quickjoin algorithm. To the best of the authors' knowledge, this is the first paper in surveying for the problem of metric space similarity joins in the general purpose platform of GPUs, and a maximum speedup of 22 with GTX460 card is reported in our experimental results.

The rest of this paper is organized as follows. We first review the related work in Section 2. After we discuss the preliminaries, including overview of metric spaces, the GPU platform and the Quickjoin algorithm in Section 3, the way to parallelize the algorithm in two platforms will be described in Section 4. Section 5 reports the experimental results and the conclusion is made in Section 6.

2   Related Work

There have been plenty of works for effectively solving the similarity join problems. However, most of them focus on dealing with the objects relying on the vector format. For example ϵ-kdB [4] is an indexed approach, where each level splits on a different dimension into ϵ sized pieces. Once the index is built, a traverse is performed to find results. Without forming an index, EGO [5] partitions the search space into an ϵ sized gird, and then processes each cell in the grid. LSS [6] uses sorting and searching operations to cast a similarity join operation as a GPU sort-and-search problem. When objects are not multi-dimensional data, some embedding methods [7] are usually used to transform the objects into the vector format. Nevertheless, how to make the result of transformation satisfying is another question.

As for metric space techniques, an index called eD-index could be utilized to fulfill the self-similarity join task [8], which partitions the data with a ball partition strategy [12]. Then, an in-memory technique follows to find the final results. Quickjoin algorithm is another metric space approach to find the similar pairs, while it does not construct any index at all. Since Quickjoin is the cornerstone of our work, it will be described in detail in the next section.

3   Preliminaries

In this section we first present a brief introduction to similarity joins in metric spaces, and some concepts about GPU platform used in our implementation will also be stated here. As Quickjoin algorithm is the basis of our work, it will be explained in this section as well.

3.1   Metric Space

Formally, a metric space $\mathcal{M}$ is a pair $\mathcal{M} = (\mathcal{D}, d)$, where $\mathcal{D}$ is the domain of objects (or the objects' keys or indexed features) and d is the distance function $d: \mathcal{D} \times \mathcal{D} \to \mathbb{R}$ with the following properties for any object $x, y, z \in \mathcal{D}$:

- $d(x, y) \geq 0$ (non-negativity),
- $d(x, y) = d(y, x)$ (symmetry),
- $x = y \Leftrightarrow d(x, y) = 0$ (identity),
- $d(x, z) \leq d(x, y) + d(y, z)$ (triangle inequality).

The similarity join [2] between two datasets $X \subseteq \mathcal{D}$ and $Y \subseteq \mathcal{D}$ retrieves all pairs of objects $(x \in X, y \in Y)$ whose distance does not exceed a given threshold ϵ. Mathematically, the result of the similarity join $J(X, Y, \epsilon)$ can be defined as:
$$J(X, Y, \epsilon) = \{(x, y) \in X \times Y : d(x, y) \leq \epsilon\}.$$

3.2   GPU Computing

NVIDIA's CUDA (Compute Unified Device Architecture) [9] is a general purpose parallel computing platform, which exploits the massive computing power inside the GPU. As our many-core implementation is based on this platform, CUDA's memory hierarchy and thread hierarchy will be explained briefly as follows.

From the hardware aspect, the GPU consists of several streaming multiprocessors (SMs). In each SM, there are multiple streaming processors (SPs). The SMs share the global memory and L2 cache in the GPU, while SPs in a SM are tightly coupled and share the resources like shared memory. Unlike the way of instruction execution in the CPU, the GPU follows a specific design called single instruction multiple threads (SIMT) to execute hundreds of threads concurrently. As a result, all SPs execute the same instruction at the same time.

From the perspective of logical structure, a number of threads form a thread block, and many blocks constitute a grid. This double hierarchical parallel structure corresponds to GPU's hardware design (SMs and SPs) as well as its memory hierarchy as shown in Fig. 1. To be specific, each thread has a private local memory,

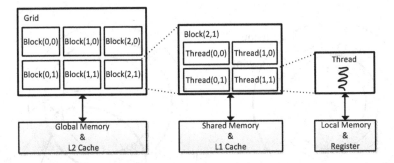

Fig. 1. CUDA's thread hierarchy and memory hierarchy

and each thread block has a shared memory available to all threads in the block, while the global memory is visible to all threads in a grid. CUDA offers the synchronization mechanism inside a thread block, while there is no natural synchronization method provided among blocks. Instead, this kind of synchronization should be implemented through accessing the global memory normally.

A general way to code a program running in the GPU is to allocate the GPU (device side) memory first, and then transfer the data from host to the device memory. After a function executed in the GPU that is called kernel finishes processing the data, the final result might be retransferred to the main memory in the host side.

3.3   The Quickjoin Algorithm

The Quickjoin algorithm recursively partitions the data until a partition is small enough to be handled in a brute force way, which is actually a nested loop comparing every pair of objects in the partition. Fig. 2 illustrates an example of a dataset partitioned through the Quickjoin algorithm. Note that since it is intuitive in the two-dimensional space, all the illustrations related to a dataset sample in this paper will be explained under the assumption that objects are 2-D data, and meanwhile the Euclidean distance is employed. However, methods to find similarity pairs are still applicable to metric spaces. In the algorithm, the data could be partitioned through a technique called ball partition, which partitions the data according to a distance r and a pre-selected central pivot. For example, in Fig. 2. (a) p_1 is the selected central pivot, and the data are partitioned into two parts according to $d(p_1, p_2)$, where p_2 is a randomly selected object in the dataset. One part (*partL*) contains objects that $d(o_i, p_1) \leq r$ holds, while the other part (*partG*) contains objects that $d(o_i, p_1) > r$ holds. The algorithm continues to make a partition in each subsection.

After partitioning into two parts, however, some pairs of similar objects might be missing as one object in *partL* might be similar to an object in *partG*. To correctly report all the similar pairs, two windows must be created. As shown in Fig. 2. (a), *winL* and *winG* are two new windows, where we must find similar pairs of objects. The width of each window is ϵ, which is just the quantified similarity. Each pair of similar objects (x, y) in windows consists of an object from each window, where

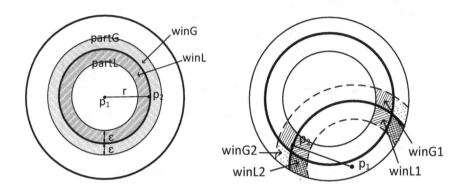

Fig. 2. Top level of the Quickjoin algorithm (a) main partition (b) partition in the windows

$x \in winL \wedge y \in winG$ or $y \in winL \wedge x \in winG$. The pair (x, y) that $x \in winL \wedge y \in winL$ and $x \in winG \wedge y \in winG$ has been calculated in *partL* and *partG* respectively. Again, the Quickjoin algorithm continues to partition the data in two windows.

When it comes to the partition in two windows, the situation is slightly different as four windows might be created instead of just two windows. Fig. 2. (b) shows the partition situation in *winL* and *winG*. Likewise, p_1 and p_2 are randomly selected in advance. Thus, both *winL* and *winG* are partitioned into two parts respectively according to $d(p_1, p_2)$ and p_1. These parts could be partitioned further or we can directly compare every pair of objects from each part if the scale is small enough. As a result, some missing pairs should be especially cared about like what happens in the main partition. Two dashed lines in Fig. 2. (b) depict the area where the missing pairs fall in. To be specific, the algorithm must find the missing pairs of similar objects between *winL1* and *winG2*, and between *winG1* and *winL2*.

4   Parallelization of the Quickjoin Algorithm

In this section, we present two parallelized versions of the Quickjoin algorithm. Since the mutli-core version of the algorithm is just slightly different from the original one, we introduce it first as it might shed light on some specifics of the serial algorithm. Then, the many-core version of the Quickjoin algorithm will be thoroughly discussed under the assumption that an outline of the Quickjoin algorithm in the CPU is well understood.

4.1   Multi-core Version

In essence, the multi-core version of the Quickjoin algorithm exploits the multithreading resources in CPU. Its implementation is based on OpenMP [10], which is an API that could be used to explicitly direct shared memory parallelism.

Function Quickjoin(objs)	
1.	**begin**
2.	**if** (objs.size < constSmallNumber) *// If the amount of objects is smaller than the threshold,*
3.	NestedLoop(objs); *// the brute force method could be used, which ends the*
4.	return; *// recursion function.*
5.	**end if**
6.	
7.	$p_1 \leftarrow$ randomObject(objs); *// Pivots are randomly selected and they should be different.*
8.	$p_2 \leftarrow$ randomObject(objs- p_1);
9.	(partL, partG, winL, winG) $\leftarrow$ Partition(objs, p_1, p_2);
10.	
11.	***#pragma omp task*** *// create a new task to calculate in winL $\cup$ winG*
12.	QuickjoinWin(winL, winG);
13.	***#pragma omp task*** *// create a new task to calculate in partL*
14.	Quickjoin(partL);
15.	***#pragma omp task*** *// create a new task to calculate in partG*
16.	Quickjoin(partG);
17.	**end**

Fig. 3. The pseudo code of the multi-core version of the Quickjoin algorithm: main partition

We incrementally parallelize the serial Quickjoin algorithm through directives provided by OpenMP. Fig. 3 shows the pseudo code of the main partition, and the one of the window partition is displayed in Fig. 4.

In order to adapt to the multi-core environment, the serial Quickjoin algorithm is parallelized with the following straightforward scheme:

- Each thread is responsible for a different part's computation because there is no intersection between any two partitioned parts at the same recursive depth.
- As the windows, in which the missing pairs of similar objects exist, are newly created by pointing to the corresponding objects, calculations in windows could be isolated from the one in the main partition. As a result, another thread could be allocated for computing in the windows.
- Since functions including *Quickjoin* and *QuickjoinWin* might be recursively invoked many times, the amount of the threads should be controlled within a certain threshold in case of performance degradation.

A new concept called task [11], which is introduced in OpenMP 3.0, provides a simple way to accomplish the parallelization goal meeting the three requirements listed above. As shown in Fig. 3 and Fig. 4, task directives are displayed in bold and italic style to be distinguished from other sentences. If these task directives are omitted in the pseudo codes, the multi-core version could be directly transformed into the serial one. In our case, a task refers to calculations in a part.

The execution flow of the parallelized algorithm is that before the function *Quickjoin* is executed, a bundle of threads will be created first. Once a task directive is executed, a thread might be allocated to execute the corresponding task immediately or not depending on the available resources. In Fig. 3, ideally calculations in *partL* and *partG* are executed by two different threads, and a third thread is responsible for calculations in the *winL∪winG*. Likewise, in Fig. 4, four threads are allocated to make calculations in *winL1 ∪ winG2*, *winG1 ∪ winL2*, *partL1∪partL2* and *partG1∪partG2* respectively. When there is no enough resource, a task might be delayed waiting for an available thread to execute it.

Function QuickjoinWin(objs1, objs2)	
1. **begin**	
2. totalLen ← objs1.size + objs2.size;	// If the amount of objects is smaller than the threshold,
3. **if** (totalLen < constSmallNumber)	// the brute force method could be used, which ends the
4. NestedLoopWin(objs1, objs2);	// recursion function.
5. return;	
6. **end if**	
7.	
8. allObjs ← objs1 ∪ objs2;	// Pivots are randomly selected from the union and they
9. p_1 ← randomObject(allObjs);	// should be different.
10. p_2 ← randomObject(allObjs - p_1);	
11. (partL1, partG1, winL1, winG1) ← Partition(objs1, p_1, p_2);	
12. (partL2, partG2, winL2, winG2) ← Partition(objs2, p_1, p_2);	
13.	
14. ***#pragma omp task***	// create a new task to calculate in winL1 ∪ winG2
15. QuickjoinWin(winL1, winG2);	
16. ***#pragma omp task***	// create a new task to calculate in winG1 ∪ winL2
17. QuickjoinWin(winG1, winL2);	
18. ***#pragma omp task***	// create a new task to calculate in partL1 ∪ partL2
19. QuickjoinWin(partL1, partL2);	
20. ***#pragma omp task***	// create a new task to calculate in partG1 ∪ partG2
21. QuickjoinWin(partG1, partG2);	
22.	
23. ***#pragma omp taskwait***	// join to wait for the forked threads, because the windows
24. **end**	// should be deleted after finishing calculations

Fig. 4. The pseudo code of the multi-core version of the Quickjoin algorithm: window partition

4.2   Many-Core Version

The many-core version of the Quickjoin algorithm is based on GPU in the main, with a small portion of the implementation on CPU. A rule followed throughout our many-core implementation is that enough workloads should be provided to the GPU to fully exploit its computing power, because currently a GPU owns hundreds of cores (streaming processors) at least. Otherwise, the performance in the GPU might be not comparable to the one executing in a peer CPU as utilizing a few of SPs in the GPU is not competitive at all in terms of the performance. Guided by this philosophy, we consistently deliver calculations in the parts to the GPU side, because in metric spaces many distance measures are complex, such as the string Edit distance with a $O(nm)$ complexity. In the following paragraphs, we would like to discuss the evolution of the many-core version of the Quickjoin algorithm.

Our first attempt to speed up the algorithm in the GPU is to employ the same idea as we do in the CPU that calculations in each part are executed by different threads. In order to implement it, commands issued from the host (CPU) and executed in the device (GPU) should be grouped into streams, which are capable of executing calculations of different parts concurrently in the GPU. The host side pseudo code of the main partition is listed in Fig. 5, and we omit the pseudo code of the window partition as they are similar to each other. Note that in this first version we set the configuration values of Grid and Block marked as both bold and italic in Fig. 5 to 1, which means only one thread in the GPU will be allocated to execute the kernel function *Quickjoin*. As shown in Fig. 5, it is necessary to allocate two stacks to support the recursion, one of which resides in the host side and the other in the device side memory, since currently few of GPUs support recursion with the allocation of new resources (blocks, threads). In the device side of this version, the implementation is similar to the serial algorithm in the CPU, apart from recording partition information (start and end position of a segment) into the GPU side stack.

Function QuickjoinLaunch(d_objs, d_stack)

```
1.   begin
2.      cudaStreamCreate(&s);        // create a stream to run in the GPU
3.      streams.Add(s);             // Positions of the pivots are generated randomly, and they are
4.      RandomPivots(&p₁, &p₂);     // the parameters transferred to the device side.
5.      Quickjoin<<<Grid, Block, 0, s>>> (d_objs, d_stack, 0, d_objs.size-1, p₁, p₂...);
6.
7.      while (1)
8.         for i ← 0 to streams.size     // wait for the completion of every stream
9.            cudaStreamSynchronize(streams[i]);
10.        end for
11.
12.        h_stack ← CopyStackFromDevice(d_stack);   // copy the stack from device to host
13.        if (h_stack.size == 0)
14.           break;
15.        end if
16.
17.        while (h_stack.size > 0)     // get start and end positions of every segment from the stack
18.           elem ← h_stack.Pop();
19.           cudaStreamCreate(&s);    // create a stream for each segment
20.           streams.Add(s);
21.           RandomPivots(&p₁, &p₂);   // generate random positions of pivots for each segment
22.           Quickjoin<<<Gird, Block, 0, s>>> (d_objs, d_stack, elem.start, elem.end, p₁, p₂...);
23.        end while
24.     end while
25.  end
```

Fig. 5. The pseudo code of the many-core version of the Quickjoin algorithm (host side)

However, this scheme works poorly in performance mainly due to the following three reasons:

- At first, there is only one thread executing the calculations, while the amount of threads doubles during every outer *while* loop. As a result, the utilization of the GPU is very low initially.

- Although the utilization of the GPU might grow after creating enough threads, there is still a toplimit of the amount of concurrently running streams that restrains the efficiency (for example, the toplimit is 32 on devices of compute capability 3.5 and 16 on devices of lower compute capability [9]).
- The algorithm cannot guarantee a balanced partition, and thus the execution time depends on the stream with the most objects during every outer *while* loop resulting from the synchronization strategy.

The first reason implies that computing resources in the GPU are not extracted to the full, and thus the algorithm should be modified to make use of those free resources as many as possible. The second reason indicates that a number of tasks with a few of distance calculation times are not favorable to the GPU. Instead, it suggests that the partition granularity has to be experimented, and tasks with heavy workloads should be delivered to the GPU side. The last reason could be easily solved in the devices of compute capability 3.5 as a grid management unit [13] is equipped to support the recursion in the GPU, and hence there is no need to synchronize in the host in our case. However, we use the device of compute capability 2.1 to run experiments, and the loss of performance due to this reason is already reckoned in.

Guided by these analyses, we introduce our second version of the many-core Quickjoin algorithm, which overcomes the deficiency of exploiting the GPU ineffectively. The main idea in this version is that each thread in the GPU is responsible for one distance calculation. Fig. 6 illustrates how this version of the many-core Quickjoin algorithm works. At first, each thread in the GPU gets its object. Then, in each block, objects are divided into two groups according to the separating distance r, and the amount of group members is recorded. As a result, group's location in the output array could be calculated by atomic operations in the global memory in the GPU. Inside the group, the position of each object is decided by atomic operations in the shared memory.

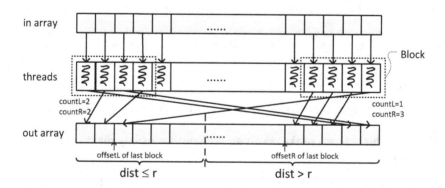

Fig. 6. The illustration of the many-core version of the Quickjoin algorithm

Kernel Quickjoin(objs, stack, start, end, p₁, p₂, in, out, offset)

```
1.    begin
2.        thread_id ← blockIdx.x × blockDim.x + threadIdx.x;  // Each thread gets its global id.
3.        size ← end – start + 1;
4.        if (thread_id ≥ size)          // The redundant threads should be returned as early as possible.
5.            return;
6.        end if
7.
8.        count ← 0;                     // used to synchronize among blocks
9.        data ← in[start + thread_id];  // Each thread gets its own object.
10.
11.       __shared__ r, pivotObj, countL ← 0, countR ← 0, posL ← 0, posR ← 0;
12.       if (threadIdx.x == 0)
13.           pivotObj ← LoadPivot(objs, p₁);  // load central object in position p₁
14.           r ← pivotObj.Distance(objs[p₂]);  // get the separating distance
15.       end if
16.       __syncthreads();       // synchronize in the block to initialize r
17.                                       // each thread calculates the distance between the
18.       dist ← pivotObj.Distance(objs[data]);  // central pivot and its own object
19.       if (dist ≤ r)
20.           atomicAdd(&countL, 1);       // count objects in this block whose distance is ≤ r
21.       else
22.           atomicAdd(&countR, 1);       // count objects in this block whose distance is > r
23.       end if
24.       __syncthreads();       // synchronize in the block to record countL and countR
25.
26.       __shared__ offsetL, offsetR;    // locate the left position (≤ r) and right position (> r) of this
27.       if (threadIdx.x == 0)           // block in the whole objects
28.           offsetL ← start + atomicAdd(&(offset→posL), countL);
29.           offsetR ← end - (atomicAdd(&(offset→posR), countR) + countR) + 1;
30.           count ← countL + countR;
31.       end if
32.       __syncthreads();       // synchronize in the block to initialize offsetL and offsetR
33.
34.       if (dist ≤ r)   // locate the left position (≤ r) and right position (> r) of each thread in this block
35.           out[offsetL + atomicAdd(&posL,1)] ←data;
36.       else
37.           out[offsetR + atomicAdd(&posR,1)] ←data;
38.       end if
39.
40.       if (threadIdx.x == 0 && atomicAdd(&(offset→count),count) + count == size)
41.           part ← end - offset→posR        // the block's first thread that adds the variable count
42.           if (start < part)               // last is responsible for pushing the partition results
43.               stack→Push(start, part);    // into the stack
44.           end if
45.           if (part+1 < end)
46.               stack→Push(part+1, end);
47.           end if
48.       end if
49.   end
```

Fig. 7. The pseudo code of the many-core version of the Quickjoin algorithm (device side)

The host side pseudo described in Fig. 5 is still applicable to this many-core version if configuration parameters (Grid and Block) vary according to the assigned workloads ($Grid = \lceil \frac{Objs.size}{Block} \rceil$) instead of allocating only one thread directly. Fig. 7 lists the device side pseudo code of the main partition, and the window partition is similar to it which is not shown here to avoid redundancy. According to Fig. 7, each thread finds its corresponding object through the unique *thread_id* on line 2. Then, the first thread in each block loads the central pivot into the shared memory and calculates the separating distance r from line 12 to 15. The synchronization on line 16 is necessary to let r be visible to all the threads in the block before calculating the distance between central pivot and the object carried by each thread on line 18. Line 19 to 23 record the amount of objects with *distance* $\leq r$, and the amount of objects with *distance* $>r$ in the block in order to compute their start addresses in the global output array from line 27 to 31 by taking the block as the unit. The address inside a block calculated from line 34 to 38 is the final place to accommodate an object. The first thread of a block that adds the *count* variable (used to synchronize among blocks) last should push the partition information into the GPU side stack.

When the amount of the workload (measured by the times of the distance calculation) in the GPU is less than *constSmallNumber*, the brute force approach could be used, where each thread in the GPU processes one distance calculation only.

To further accelerate this many-core version, some implementation specifics must be taken care of, especially the implicit synchronization [9, 14] as it might degrade the concurrency in the GPU. In summary, we list them as below:

- It is better to pre-allocate the device memory (stack, window and offset) in case of the implicit synchronization.
- In order to support the overlap of data transfer and kernel execution, page-locked host memory is required to replace the pageable memory (window), and it should also be pre-allocated in case of the implicit synchronization.
- To utilize the feature of concurrent execution between host and device and to avoid the implicit synchronization, it is suggested to use asynchronous APIs with respect to the host side.

5   Experimental Study

In this section, we evaluate experimental results of our parallel implementations of the Quickjoin algorithm. Serial, multi-core and many-core version of the brute force method are also implemented, and added to the experiment as a kind of a sanity test. The multi-core version is based on the OpenMP's work sharing construct, while in the many-core version of the brute force approach each distance calculation is processed by one thread in the GPU.

We use two real datasets listed in Table 1 to conduct experiments, both of which are publicly available[1]. It is worth mentioning that the distance calculation between two strings is relatively computation bound compared to the one in our vector

---

[1] Metric spaces library, http://www.sisap.org/Metric_Space_Library.html

database as being a feature of metric spaces many distance calculations are complex. All the results about the performance are averaged on 10 times.

Table 1. Details of datasets used in experiments

Dateset	Size	Type	Description
Color Histograms	112,682	vector, double	They are 112-dimensional vectors extracted from a real image database, where the Euclidean distance is chosen to measure the difference between two objects. The maximum distance is $\sqrt{2}$.
English Dictionary	69,069	string	It is an English dictionary with words disordered to avoid dependences on lexicographical ordering. The maximum length of the word in the dictionary is 21. The similarity is measured by the Edit distance.

Details of the platform where we conduct experiments are listed in Table 2. We intentionally take a GPU that is almost half of the price of the CPU in the market to explore its performance/price ratio. Since the shared memory is not heavily exploited in our many-core implementation, we set the configurable combination as 48KB L1 cache and 16KB shared memory in the GPU. As for the compilation, the serial and the multi-core version are compiled with ICC, while the host side code of the many-core version is compiled with the default compiler as CUDA 5.0 currently does not support ICC in Windows platform.

Table 2. Details of the experimental platform

Equipment	Description
CPU	Intel Xeon E3-1230 v2, 4 physical cores and 8 logical cores, clock speed 3.3GHz, max turbo frequency 3.7GHz
GPU	GeForce GTX460 v2, compute capability 2.1, 7 multiprocessors (336 cores), 1GB global memory, 48KB L1 cache, 384KB L2 cache, bus width 192 bits
Memory	DDR3-1600 8GB
Compiler	Intel C/C++ Compiler v11.0 (ICC) and the default compiler in Visual Studio 2008, compile codes with full optimization
Toolkit	CUDA 5.0

Before testing the execution time of all implementations, we firstly find out the most appropriate threshold (*constSmallNumber*) for the CPU based (serial) Quickjoin algorithm and the one for the GPU based (many-core) version respectively. It is crucial to present an ideal threshold, especially for the GPU based version, as a very small threshold might cause many transmissions between the host and the device with workloads divided into pieces. The serial and the multi-core version use the same threshold, where *constSmallNumber* is compared with the objects' amount, while in the GPU based version it is compared with the times of the distance calculation. To avoid failure in partition, the threshold should be larger than the amount of those objects that have the equal distance between each other in a dataset. In Fig. 8, the threshold in the color histograms dataset is experimented with $\epsilon = 5\%$ of the maximum distance, and the one in the English dictionary is attained with $\epsilon = 1$. According to Fig. 8, we find that in the vector dataset the performance of the CPU based version peaks at around 120, and 50,000 for the GPU based version. As for the string dataset, the *constSmallNumber* of the serial version should be 50, and 50,000 for the many-core version. As a result, we apply these values in the following experiments.

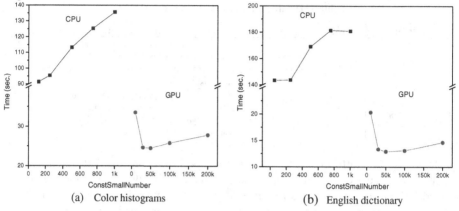

Fig. 8. Performance tests to find an appropriate value of *constSmallNumber* in both the CPU and the GPU based Quickjoin algorithm

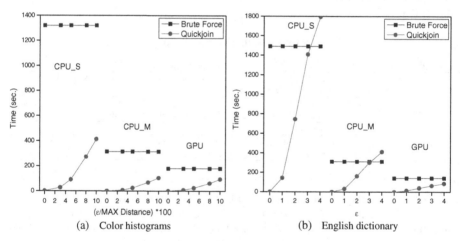

Fig. 9. Comparative results of similarity join costs in sequential CPU, multi-core CPU, and GPU platforms with two approaches: brute force and Quickjoin

Execution time (including data transmission between the device and the host) of all implementations is shown in Fig. 9. For the vector database, results are reported with $\epsilon=0$, 3%, 5%, 8%, and 10% of the maximum distance respectively, while results are observed with $\epsilon=0$, 1, 2, 3 and 4 in the string database. Within expectation, the performance of the Quickjoin algorithm in three platforms is better than the one of the brute force method in most cases. Another fact we could see is that the response time grows with the increasing ϵ in the Quickjoin algorithm in all platforms. In general, the GPU based implementations offer the best performance followed by the multi-core versions. Specifically, in the vector dataset, results of the many-core Quickjoin algorithm are almost the same as the ones of the multi-core version, where the average speedup is around 4.3. However, in the string dataset, the maximum speedup is 22 observed in the GPU based Quickjoin algorithm, compared to the improvement with 4.68 in the multi-core version. A main reason to explain this difference is the high measurement complexity with the Edit distance which might lessen the effects from other procedures such as transmission between the host and the device, repetition of a recursion.

6   Conclusions

In order to effectively and efficiently handle similarity joins in metric spaces, we present two parallel versions of the Quickjoin algorithm exploiting multi-core and many-core processors respectively, and evaluate them on both vector and string datasets. The best experimental result is shown in the many-core version, which obtains speedup around 22 in the string dataset, compared to the serial version in the CPU. Also, the multi-core implementation of the algorithm scales well according to the amount of physical cores in the CPU.

As future work, we would like to survey the performance of the many-core version of the Quickjoin algorithm in devices of compute capability 3.5 as a grid management unit is equipped to truly support the recursion in the GPU.

All of our implementations in this work are mainly based on C++, and the source codes used in our experiments are available to the public domain[2].

References

1. Levenshtein, V.I.: Binary Codes Capable of Correcting Deletions, Insertions and Reversals. Soviet Physics Doklady 10 (1966)
2. Zezula, P., Amato, G., Dohnal, V., Batko, M.: Similarity Search: The Metric Space Approach. Advances in Database Systems, vol. 32. Springer (2006)
3. Jacox, E.H., Samet, H.: Metric Space Similarity Joins. ACM Transaction on Database Systems 33(2), 1–38 (2008)
4. Shim, K., Srikant, R., Agrawal, R.: High-Dimensional Similarity Joins. IEEE Transactions on Knowledge and Data Engineering 14(1), 156–171 (2002)
5. Böhm, C., Braunmüller, B., Krebs, F., Kriegel, H.-P.: Epsilon Grid Order: An Algorithm for the Similarity Join on Massive High-Dimensional Data. In: SIGMOD, pp. 379–388 (2001)
6. Lieberman, M.D., Sankaranarayanan, J., Samet, H.: A Fast Similarity Join Algorithm Using Graphics Processing Units. In: ICDE, pp. 1111–1120 (2008)
7. Faloutsos, C., Lin, K.: FastMap: A Fast Algorithm for Indexing, Data-Mining and Visualization of Traditional and Multimedia Datasets. In: SIGMOD, pp. 163–174 (1995)

---

[2] Implementations by Authors,
http://sourceforge.net/projects/paralquickjoin/

8. Dohnal, V., Gennaro, C., Zezula, P.: Similarity Join in Metric Spaces Using eD-Index. In: Mařík, V., Štěpánková, O., Retschitzegger, W. (eds.) DEXA 2003. LNCS, vol. 2736, pp. 484–493. Springer, Heidelberg (2003)

9. CUDA C Programming Guide: CUDA Toolkit Documentation, `http://docs.nvidia.com/cuda/cuda-c-programming-guide/index.html`

10. OpenMP: An API for multi-platform shared-memory parallel programming in C/C++ and Fortran, `http://www.openmp.org/`

11. Ayguadé, E., Copty, N., Duran, A., Hoeflinger, J., Lin, Y., Massaioli, F., Teruel, X., Unnikrishnan, P., Zhang, G.: The Design of OpenMP Tasks. IEEE Transactions on Parallel and Distributed Systems 20(3), 404–418 (2009)

12. Yianilos, P.N.: Data Structures and Algorithms for Nearest Neighbour Search in General Metric Spaces. In: SODA, pp. 311–321 (1993)

13. Dynamic Parallelism in CUDA, `http://developer.download.nvidia.com/assets/cuda/docs/TechBrief_Dynamic_Parallelism_in_CUDA_v2.pdf`

14. CUDA C/C++ Streams and Concurrency, `http://developer.download.nvidia.com/CUDA/training/StreamsAndConcurrencyWebinar.pdf`

Using a Multitasking GPU Environment for Content-Based Similarity Measures of Big Data

Ayman Tarakji[1], Marwan Hassani[2], Stefan Lankes[1], and Thomas Seidl[2]

[1] Chair for Operating Systems, RWTH Aachen University, Aachen, Germany
{ayman,lankes}@lfbs.rwth-aachen.de
[2] Data Management and Data Exploration Group
RWTH Aachen University, Germany
{hassani,seidl}@cs.rwth-aachen.de

Abstract. Performance and efficiency became recently key requirements of computer architectures. Modern computers incorporate Graphics Processing Units (GPUs) into running data mining algorithms, as well as other general purpose computations. In this paper, different parallelization methods are analyzed and compared in order to understand their applicability. From multi-threading on shared memory to using NVIDIA's GPU accelerators for increasing performance and efficiency on parallel computing, this work discusses the parallelization of data mining algorithms considering performance and efficiency issues. The performance is compared on both many-core systems and GPU accelerators on a distance measure algorithm using a relatively big data set. We optimize the way we deal with GPUs in heterogeneous systems to make them more suitable for big data mining applications with heavy distance calculations. Moreover, we focus on achieving a higher utilization of GPU resources and a better reuse of data. Our implementation of the content-based similarity algorithm SQFD on the GPU outperforms by up to 50× CPU counterparts, and up to 15× CPU multi-threaded implementations.

Keywords: GPGPU, Similarity Measures, Data Mining, Heterogeneous Parallel Systems.

1 Introduction

General-Purpose Computation on Graphics Processing Units (GPGPU) is one of the current trends in the field of parallel processing [15]. The GPU became the most powerful discrete processor in a computer, only considering Floating-Point Operations Per Second (FLOPS). Hardware developers and research groups around the world started to simplify the rather complex programming of GPUs for non-graphic related computations. Nowadays, a single graphics card is able to provide over 4 TFLOPS for single-precision computations [20] and a variety of programming interfaces is available.

One important application area of the usage of GPUs as a powerful processing tool when dealing with huge datasets, is data mining. Most powerful mining

B. Murgante et al. (Eds.): ICCSA 2013, Part V, LNCS 7975, pp. 181–196, 2013.
© Springer-Verlag Berlin Heidelberg 2013

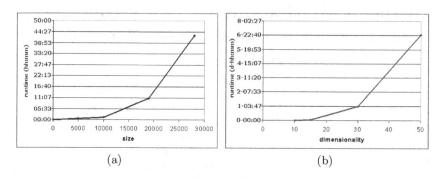

Fig. 1. The scalability of "SUBCLU" clustering algorithm [7] when implemented on CPU by: (a) increasing the size of the data, (b) increasing the dimensionality

techniques and information retrieval algorithms are expensive when considering their run times. This has limited the acceptance of such algorithms in many fields. Figure 1 depicts an example on the scalability of a data clustering algorithm called SUBCLU [7]. It is obvious that the complexity of this algorithm is quadratic in both the size and the dimensionality of the data. By minding the huge running times on the y-axis, it can be seen that it is sometimes too expensive to get meaningful data mining results out of big data. These results were gained by implementing the algorithm over CPUs merely. In modern parallel systems GPUs have proved their efficiency in solving many complex parallelizable computation problems when compared with the mere CPU performance. In this paper, we study their efficiency when using a quadratic distance function called SQFD (Signature Quadratic Form Distance) [2] that is widely used in the area of data mining and information retrieval.

Over the last few years, data mining applications have grown in size and complexity. When a query is sent to a data base, a huge amount of computations should be done on large data sets to return the search results for this query. Because of the long runtime of these applications on single core processors, data mining researchers are looking at running data mining computations on modern parallel architectures to accelerate their applications.

GPGPU reaches from the consumer sector, all the way to super computers in High-Performance Computing (HPC). However, enhanced performance could be ensured for a quite limited range of applications, if uncertainty about the essential features of the architecture exists when using GPUs for performing general purpose computations [18]. In the field of high performance computing, there exists a great distinction among parallel architectures, such as the multiprocessor, many-core and stream processing architectures.

The focus of this work lies on comparing issues among different parallelizing paradigms, optimization techniques, and maximizing of utilization when using graphics devices in general purpose computations. We mainly discuss different important aspects and identify different trade-offs of running data mining applications on the GPU stream processing units. In order to compare stream processors on GPUs with today's other trends regarding parallelization paradigms, multi-threading on shared memory is also introduced.

The reminder of this paper is organized as follows: In Section 2 we introduce the key concepts of parallel processing and applying it in the field of GPGPU using CUDA API, then we explain the SQFD distance measure used in this paper. In Section 3, we present the three different strategies followed in the context of this work to parallelize the SQFD. Later in Section 4, we present a thorough evaluation of the approach over a relatively big image dataset. Section 5 lists some related work, and Section 6 concludes the work and discusses future directions.

2 Background

The high level of demand for processing complex computations in the range of short-time dynamics steered the attention of many researchers in the field of data mining to parallelization techniques, which might accelerate the processes in data mining applications.

2.1 Multi-threaded Programming Model

In data mining applications, it often turns out that the same operations should be executed on a huge amount of data items. Since searches are computational independent operations which can be performed in parallel using multiple program threads, shared memory programming models provide an appropriate environment for them. These problems can benefit from the multi-threaded computational power of multi-core processor systems. The reached enhancement in performance using multi-threaded programming has paid off for programmer when compared to the required effort involved in the parallelization. Nowadays, the most widespread implementation of multi-threading nowadays, is called OpenMP (Open Multi-processing) [13]. This is a method of parallelizing programs by forking a specified number of threads. The multiple threads run concurrently and execute the program in a work-sharing fashion. In the context of this work, we used this parallelization technique for comparison purposes.

2.2 GPGPU Programming

Dedicated APIs like CUDA (Compute Unified Device Architecture) [11] and OpenCL (Open Computing Language) [10] have been released allowing algorithms across all fields of application to be coded and executed on the graphics processing units. Parallelism on highly programmable GPUs basically conforms to a data parallel computing model. The same operation is applied on the whole set of data, where one instruction after the other in the program comes to execution. This corresponds to the well known SIMD model (Single Instruction Multiple Data) [17].

A GPU programmer should be able to partition the computation problem into sub-problems that could be solved independently. An arbitrary number of threads will be created to do this work. Each of them performs the same operation on different data, i.e. on a vector (the maximum vector width can vary depending on the generation of the GPU and its vendor).

Since the performance of processing units is measured by both number of floating point operations and bandwidth per time, GPUs have shown a massive increase of performance compared to CPUs in the preceding years. This increase was measured and verified in [1] on one of the most advanced computing GPU architectures released by NVIDIA, called *Fermi*.

In the context of this work, we developed our approach using CUDA on two different modern graphics cards. The first platform uses the card C2050 from NVIDIA which is built on the basis of the *Fermi* architecture and provides the compute capability 2.0 [5]. For comparison purposes, we also performed the same tests on Tesla K20Xm, the newst generation of graphics cards from NVIDIA. This new hardware is based on the *Kepler* architecture [12], which was first released in 2012.

2.3   The SQFD Distance Function

In the field of content-based similarity search of multimedia objects, distance measures are the main part that decides the similarity of these objects. Smaller values of the distance function usually imply higher similarities, and vice-versa. Applying these distance measures usually consists of two main steps: 1) a pre-processing step that digitizes the contents of the image objects, which results in some value in the defined *feature space* and 2) a distance calculation step which utilizes these features values to decide about the similarity between the image objects.

The feature space usually includes among others: color, texture and position information. Rubner et al. [16] has introduced the following feature signature definition:

Definition 1. *Feature Signature* *A feature signature Q of length n^q is defined as a set of tuples: $Q := \{\langle c_i^q, w_i^q \rangle, i = 1, \ldots, n^q\}$ where: c_i^q implies the weighted centroid inside the defined feature space, and w_i^q represents its weight.*

Beecks et. al [2] has made use of the above feature signature to introduce SQFD:

Definition 2. *Signature Quadratic Form Distance* *Given two feature signatures $Q = \{\langle c_i^q, w_i^q \rangle | \ i = 1, \ldots, n\}$ and $P = \{\langle c_i^p, w_i^p \rangle | \ i = 1, \ldots, m\}$, the Signature Quadratic Form Distance $SQFD_{A_{fs}}$ between Q and P is defined as:*

$$SQFD_{A_{fs}}(Q, P) = \sqrt{(w_q| - w_p) \cdot A_{fs} \cdot (w_q| - w_p)^T},$$

where $A_{fs} \in \mathbb{R}^{(n+m) \times (n+m)}$ is the similarity matrix, $w_q = (w_1^q, \ldots, w_n^q)$ and $w_p = (w_1^p, \ldots, w_m^p)$ are weight vectors, and $(w_q| - w_p) = (w_1^q, \ldots, w_n^q, -w_1^p, \ldots, -w_m^p)$.

Fig. 2. Some results from our GPU-based implementation of the SQFD over the Wang dataset [19]. First row contains the queries, second row contains their best matches.

Figure 2 gives some examples that were collected after applying a random set of queries over the Wang image database [19]. The SQFD similarity measure was implemented within the GPU-based method which is presented in this paper.

3   Design

Using the SQFD algorithm [2] in order to return query results from the considered data set, the required computations are basically divided into two phases: First, the features similarity matrix must be created, this includes the Euclidean distance between the query object on the one hand, and each other object from the data set on the other. When the distance matrix is created, the similarity values for the query item could be defined. These values are normalized by scaling them between 0 and 1.

In the follow-up, we briefly present the used procedure of parallelizing the algorithm using multi-threaded processing with OpenMP, this will be used as a reference point for the performance measurements besides the single processing simplified version of the considered algorithm. After that, we introduce a straight-forward implementation of the SQFD algorithm using CUDA (*naive CUDA version*). Finally, we are going to demonstrate the advanced variant for developing a solution of the SQFD algorithm using the GPGPU extensively.

3.1   The Multi-threaded Implementation

As depicted in Algorithm 1, there are two levels in the calculations required in the SQFD algorithm. Multiple threads are created using the OpenMP paradigm

Algorithm 1. Multi-threaded Version

1: preprocessing of database
2: get query S^q
3: fork
4: *distribute the computations over available threads*
5: **for all** signatures S^o of database **do**
6:    Euclidean ground distance function L_2 between S^q and S^o
7: **end for**
8: **for all** signatures S^o of database **do**
9:    compute similarity matrix $A_{fs} \in \mathbb{R}^{5 \times 5}$
10: **end for**
11: **repeat**
12:    compute $SQFD_{A_{fs}}(S^q, S^o)$
13: **until** end of database
14: **return** results of query S^q
15: End

on shared memory, each of the available threads performs a part of the whole computation. At the end of the parallel area, the results are collected by the main thread and the processing of the query finishes.

3.2   The Straight-Forward CUDA Implementation

Algorithm 2 explains in details the implementation of our naive CUDA version. In the preprocessing phase, the data set will be read and the corresponding values of objects features will be imported in the framework. Since the calculation is outsourced to the GPU, the features information of all objects of the queried data set including the query object itself will be sent to the GPU. Also Figure 3 shows, which phases are included in the procedure of processing a received query. In the first phase after the preprocessing, a CUDA kernel program is used to create the Euclidean distance matrix. This is performed based on the feature data that already exist in the global memory of the GPU at this point. The feature similarity matrix will be created, this includes the Euclidean distance calculations for the query object considering the whole data set. As an intermediate result, the Euclidean-distance matrix remains in the GPU memory until the processing of the current query is finished. In the second phase, a further CUDA kernel program was developed to calculate the values of the similarity between the respective query object and all other objects in the considered data set. In order to perform this calculation, further information are required by the GPU to follow the description of the SQFD computation. These information include the weight vector of each signature and the number of features for each item in the data set, since this number can differ from one item to another. At the end of the second phase, the result of the query will be sent back to the CPU and will be represented in integral numbers ranging between 0 and 1. After each processed query, the reserved storage on the GPU side is released and the GPU process for the corresponding query is concluded.

Algorithm 2. Naive CUDA Version

1: preprocessing of database
2: get query S^q
3: *Phase 1 on GPU*
4: transfer centroid vectors c to GPU
5: **for all** signatures S^o of database **do**
6:     Euclidean ground distance function L_2 between S^q and S^o
7: **end for**
8: **for all** signatures S^o of database **do**
9:     compute similarity matrix $A_{fs} \in \mathbb{R}^{5\times5}$
10: **end for**
11: *Phase 2 on GPU*
12: transfer all weight vectors w_o to GPU
13: **repeat**
14:     compute $SQFD_{A_{fs}}(S^q, S^o)$
15: **until** end of database
16: send results to CPU
17: release storage on GPU
18: *any further queries*
19: if $(S_q) == 0$ **then**
20:     goto 4
21: **end if**
22: End

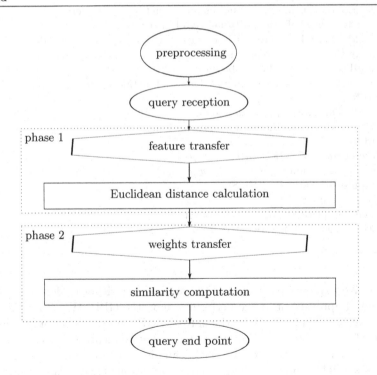

Fig. 3. Naive CUDA implementation

3.3   The Optimized CUDA Implementation

In order to enhance the performance of written programs using GPUs, a well-thought distribution of the whole work between the CPU and the GPU is required. In this context, complex issues like resource management and data transfer should be carefully considered when developing applications on GPUs, otherwise it ends with a jammed co-processor and results in wasted time.

Algorithm 3. Optimized CUDA Version

1: read database
2: create data structures
3: *Preprocessing of Data Set on GPU*;
4: transfer centroids vectors c to GPU
5: **for all** signatures S_o in database **do**
6:     Euclidean ground distance function L_2 *(one sided)*
7: **end for**
8: **for all** signatures S_o in database **do**
9:     generate "diagonal" similarity matrix
10: **end for**
11: **repeat**
12:     get query S^q
13:     *Preprocessing of Query*;
14:     calculate the Euclidean ground distance function L_2 of query
15:     generate "diagonal" similarity matrix of query
16:     transfer query data, data set information, weight vectors w_s to GPU
17:     update the reference to data set
18:     **for all** signatures S_o in database **do**
19:         calculate the Euclidean ground distance function L_2 *(two sided)*
20:     **end for**
21:     **for all** signatures S_o in database **do**
22:         generate "reverse diagonal" similarity matrix
23:     **end for**
24:     *Calculation of SQFD in two Steps*;
25:     create weight buffers $(w_q| - w_p)$, $(w_q| - w_p)^T$
26:     compute the $SQFD_{A_{f_s}}(S^q, S^o)$
27:     returns results to CPU
28: **until** end of queries to the same data set
29: transfer "diagonal" similarity matrix to be saved on CPU
30: End

Method. We explored each step of the studied algorithm in depth to ensure a maximum improvement in the performance on the GPU. We observed that a variety of actions can be taken on the naive GPGPU implementation. In the calculations of the SQFD algorithm, we noticed that the feature similarity matrix which is created on the basis of the Euclidean distance (as was explained in Section 2.3) essentially consists of four equal square sub-matrices. The sub-matrices can be calculated separately and each two of them depends on a certain part

$$A_{fs} = \begin{bmatrix} \boxed{A} & \boxed{C} \\ \boxed{D} & \boxed{B} \end{bmatrix}$$

Fig. 4. The similarity matrix A_{fs} divided into four equal square sub-matrices: A and B in the diagonal direction of A_{fs} and, C and D in the reverse diagonal

of the input data, as illustrated in Figure 4. The "diagonal" sub-matrix B is based on the Euclidean distance measurement between the features themselves, which belong to the same considered object from the data set (one-sided). This sub-matrix is used for the processing of each further query when the same object from the data set is considered, because the feature similarity matrix is created each time in the same way. In our approach, the sub-matrices B for all items of the queried data set will be calculated in the preprocessing phase on the GPU as it is shown in Algorithm 3, and thus it will be possible to reuse them for each further received query on the same data set. The other "diagonal" sub-matrix A represents the Euclidean distance measurement between the features of the query object themselves. This matrix will be calculated on the CPU and transfered to the GPU when a new query must be processed.

The other two "reverse diagonal" sub-matrices C and D of the feature similarity matrix could also be calculated separately. The sub-matrix D represents the Euclidean distance between the query object on the one hand, and each other object of the dataset on the other. The other sub-matrix C represents simply its transpose, thus it can be easily derived from it. For simplicity in our approach, the "diagonal" and the "reverse diagonal" sub-matrices of the feature similarity matrix are called "diagonal" similarity matrices and "reverse diagonal" similarity matrices, respectively.

In the follow up, when a query is received, the required information about the queried data set are sent to the GPU. An important optimization measurement can be observed in the algorithm, this is represented in updating the reference to the considered data set for each further query in a separate step. This is done for the purpose of switching between different data sets when it is required. At this step, if the data set that is currently used on the GPU, must be considered in processing the current query, no further action will be taken there, otherwise an update of the reference will mean sending the "diagonal" similarity matrix of the corresponding data set to the GPU.

The calculation of the SQFD values is also divided into two further separate kernel programs in our approach. This has a first advantage that a higher utilization of processing bandwidth on the GPU could be achieved. Additionally, since small kernels are preferred for applications in which different tasks run on the same GPU simultaneously, a higher suitability for running such applications in a multitasking multiprocessing environment on the GPU is achieved. Algorithm 3 explains in details the implementation of our optimized CUDA version.

Data Reuse. The preprocessing phase is basically divided into two separate parts. The distribution of the preprocessing between CPU and GPU brought the great advantage of reusing data on two levels. On the first level, when multiple queries are sent to the same data set repeatedly, in this case, the calculated part on the GPU during the preprocessing phase could be reused for arbitrary number of queries. The second level in data reuse is represented by the case of multiple datasets on the same GPU system. The separation of computations on the GPU enables the reconstruction of data structures, when different datasets must be considered for different queries correspondingly. For this purpose, the calculated part in the preprocessing phase must be saved on the host memory. In order to switch to another data on the GPU, this part must be transfered again to it. Since in new generations of GPUs, asynchronous transfers from CPU to GPU and vice versa are enabled, the costs of this transfer will be negligible since it occurs during the processing of other queries. Thus, further calculations could be proceeded until the processing of the current query is finished.

4   Experiments

The upcoming chapter aims at the critical analysis, evaluation and also the verification of the three different developed implementations of the Signature Quadratic Form Distance on the used data base. First, the testing environment is described, including the test algorithms used to capture different characteristics of problems. After that follows a verification section, where it is checked, if the requested features of the considered data mining application have been realized and are functioning correctly. The end is formed by a set of tests and measurements used for performance and design concept evaluation.

4.1   The Testing Environment

The both test platforms are equipped with NVIDIA computing GPUs, these are Tesla C2050 and Tesla K20Xm respektively.

The Fermi card is attached via PCIe 2.0 x16 to a server with 2 Quad-Core AMD Opteron Processors 2600GHz and 16 Gigabyte of main memory, running Fedora 16. The other platform with the Kepler card contains an Intel(R) Xeon(R) CPU E5-2650 0 @ 2.00GHz with 32 cores. This Platform was only used to run the optimized version of our implementation for comparison purposes. In the context of this work, all tests were performed using the double precision floating-point format.

4.2   Execution Models

In our approach, besides the single processing CPU implementation using the object programming language C++, we have developed three different implementations based on different models and strategies of optimization. The first version was developed using a multi-threaded parallel environment with means

Tesla C2050
. NVIDIA Fermi
. 14 Streaming Multiprocessors (SM), each hosting 32 CUDA cores (total number of cores available is 448)
. 3GB of ECC capable GDDR5 memory
. 2 Asynchronous DMA engines
. 515 GigaFLOPS of double precision peak performance
. 3 Hardware instruction queues
. Max. 4 concurrent kernels
. Compute capability 2.0

Tesla K20Xm
. NVIDIA Kepler
. 14 Streaming Multiprocessors (SM), each hosting 192 CUDA cores (total number of cores available is 2688)
. 8GB of ECC capable GDDR5 memory
. 2 Asynchronous DMA engines
. 1 TFLOP of double precision peak performance
. 32 Hardware instruction queues
. Max. 16 concurrent kernels
. Compute capability 3.5

of OpenMP. In this implementation, 8 threads were created on the two AMD Quad-Core processors, which run in the same process and share the same memory area. In the second and third implementations, a GPGPU programming model was used and developed in two different methods. Using CUDA, different kernel programs were developed to achieve the different parts of computations in the SQFD algorithm. Transfer and compute operations were performed in two different models and process orders, depending on the distribution of work between CPU and GPU, which is distinguished between the both versions. In the naive CUDA implementation, only two kernel programs were developed dividing the procedure in two stages of computation. The required data for kernel computations are sent to the GPU before each of the two phases. At the end, the results of the currently processed query are sent back to the CPU.

In the last implementation (optimized CUDA implementation), a quite different strategy was used when comparing to the previous implementation. The computations were divided into more small kernel programs which are run on the GPU. The corresponding transfer operations were done in such a way that data reuse and better utilization of processing units could be achieved.

4.3   Dataset

In our experimental part, we ran similarity measures on the Wang dataset [19], to verify the results of the SQFD algorithm using the three different strategies presented in this work. The Wang dataset contains 1 000 images classified into ten themes. The themes belong to different topics like: buses, dinosaurs, elephants, flowers, horses, mountain, beaches, etc. Table 1 shows a small sample of the results of random queries which were created and processed using the mentioned data set.

During the first phase of our experiments, a set of queries were processed using a different implementation version each time. In the second phase, we performed scalabilty tests on both test platforms used in the experimental part of this work. In this case, the data set used to process received queries consists

Table 1. Query results

Similarity Level ╲ Query Object	pic 1	pic 2	pic 3	pic 4	pic 5
First Level	0.188481	0.187199	0.148227	0.141532	0.157001
Second Level	0.203505	0.189603	0.156521	0.156579	0.182344

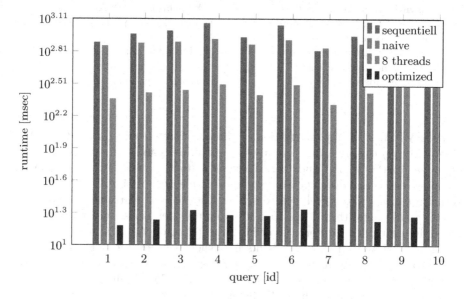

Fig. 5. Runtime of randomized queries

of a several instances of the considered data set. Each time the size of the data set was increased, the same query was executed on the enlarged data set.

4.4   Efficiency Results

Runtime results of performing a random set of queries using different strategies are shown in Figure 5. It demonstrates the obvious enhancement achieved when running the SQFD algorithm on the GPU. In particular, the optimized version has shown a sharp decline in the runtime which is measured on the same test platform. A speed-up factor of about 50 was achieved when running the optimized CUDA version on the GPU, compared to the sequential implementation on the CPU. Also using the multi-threaded programming model, we achieved a good speedup for the selected queries when comparing with the single processing CPU counterpart. A speed-up factor of about 4 was achieved on this way using 8 threads which run parallel and calculate a part of the similarity values. However, in our experiments we have observed that increasing the number of parallel threads over 8 could not produce any considerable improvement in the performance of our multi-threaded implementation.

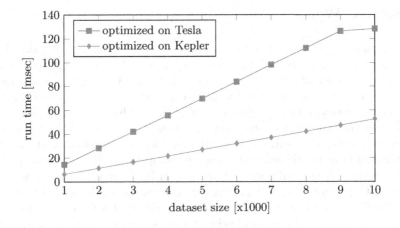

Fig. 6. Scalability tests

Considering the naive CUDA implementation of the SQFD algorithm, an enormous difference in runtime can be clearly seen in the measures. Also the multi-threading has done much better than the naive CUDA strategy, this could be explained by the high costs of the needed preprocessing and transfer operations if GPUs are used to perform general purpose tasks with different levels of computations naively.

A further important issue to be considered is the instability in the time measures in the most of strategies if different queries are compared with each other. However, the optimized CUDA implementation of the SQFD algorithm showed the most stable measures over the different random queries used on the data set. We believe that both the factors: The separation of different computation levels on the GPU and the data reuse, represent the most important factors which played a major role in the stability of the measures.

4.5   Scalability Results

As depicted in Figure 6, scalability tests were performed when executing one query on more than one instance of the considered data set repeatedly. Each time the data set size was enlarged, the run time has increased by less than two times of the value before. This shows that the speedup of our approach decreases linearly with the size of the used data set. Also, on the newst architecture from NVIDIA (Kepler), it exhibits a similar behaviour when using a progressively enlarged data set for repeated queries. In addition, the run time of processing these queries had been drastically decreased when compared to the first platform, which uses the older graphics card (Fermi). This can be explained by the ability of our approach to exhaust the huge resources of the new architecture.

5   Related Work

Using stream processing to enhance performance and efficiency of parallelized applications has been studied in many works, Abhishek Das and William J. Dally introduced an interesting work on this area [4]. A framework for scheduling of stream programs was introduced, in which kernel execution and memory transfers are overlapped to hide memory latency. They have performed high-level optimizations on stream programs, and they showed that stream schedulers make efficient utilization of resources. However, the focus of this work lied on using certain media benchmarks using the KernelC compiler [9]. Hans-Friedrich Papst and Jan Springer proposed another approach in [14]. They developed an iterative Beziér Clipping for exact trimming that runs on the GPU. In this approach, they showed that running out of memory is very likely if the GPU resources were not managed carefully. They could achieve a considerable acceleration using the GPU, however, much care had to be taken to keep the memory requirements low and to fit the program code on the GPU. Similar to our approach, Feng Cao and Anthony Tung developed a method for clustering of very large datasets using GPUs in [3]. An enhancement of performance was achieved using popular clustering algorithms. However, different to all of the mentioned proposals, our approach efficiently deals with heavy quadratic distance function calculations, an arbitrary number of queries can be answered and the runtime of processing them was reduced by a factor of 50 times comparing with CPU counterparts.

Another very similar approach to our work was introduced in [8], where an improved performace of the SQFD algorithm was presented by using the GPU. A parallel execution of multiple SQFD computations was achieved during the evaluation process of a query, and a parallelization on single SQFD computation level was also provided. Different from our approach, no data reuse was achieved in that work since both CPU and GPU were employed to a far extent. The data reuse feature is important to support running data mining applications in a multi-tasking GPU environment [18]. Additionally, [8] has employed more than one graphics card in order to get an increased performance of the algorithm.

6   Conclusion and Future Work

We have developed different techniques to partition the computations required in the SQFD distance function on the GPU. We presented a real application of our approach based on a relatively big image dataset. The numerous experiments have shown enormous profits in performance and efficiency using GPUs. The presented CUDA approach was compared with another parallelized alternative using a multi-threaded environment on a multi-core processor system. We showed that a well-thought distribution of the computations between the CPU and the GPU offers extremely good odds. Special advantages were obtained concerning data reuse and data amount reduction. We also verified the linear scalability of our method. A special strategy is currently under development, with the aim of running stream clustering algorithms on big *streaming* datasets in a multi-processing multitasking GPU environment. Such algorithms are much more complicated than a distance function, and the necessity for real time answers to user queries is very obvious there. Using a special GPU scheduling tool [18], the user will be able to run the different data mining application at the same time with an arbitrary number of different general purpose computation tasks on the same GPU system. This model is very suitable for the stream clustering algorithms (e. g. [6]) that we want to tackle in our future work. Additionally, multiple queries will be executed on the same data set repeatedly, when other general purpose computations are also running on the same device. This is currently in developement and will be presented in future works.

Acknowledgments. The authors would like to thank Christian Beecks for the useful discussion and for providing the feature signatures of the Wang dataset objects. This work has been partially supported by the UMIC Research Center, RWTH Aachen University, Germany.

References

[1] Abdalla, A.M.H.: Applications Performance on GPGPUs with the Fermi Architecture. MA thesis. The University of Edinburgh (2011)

[2] Beecks, C., Uysal, M.S., Seidl, T.: Signature Quadratic Form Distance. In: Proceedings of the ACM International Conference on Image and Video Retrieval, CIVR 2010, pp. 438–445. ACM (2010)

[3] Cao, F., Tung, A.K.H., Zhou, A.: Scalable clustering using graphics processors. In: Proceedings of the 7th International Conference on Advances in Web-Age Information Management, WAIM 2006, pp. 372–384. Springer (2006)

[4] Das, A., Dally, W.J., Mattson, P.: Compiling for Stream Processing. In: Proceedings of the 15th International Conference on Parallel Architectures and Compilation Techniques, PACT 2006, pp. 33–42. ACM (2006)

[5] Glaskowsky, P.N.: NVIDIA's Fermi: The First Complete GPU Computing Architecture. Tech. rep. NVIDIA Corporation (2009)

[6] Hassani, M., Spaus, P., Gaber, M.M., Seidl, T.: Density-Based Projected Clustering of Data Streams. In: Proceedings of the 6th International Conference on Scalable Uncertainty Management, SUM 2012, pp. 311–324. Springer (2012)

[7] Kailing, K., Kriegel, H.-P., Kroeger, P.: Density-Connected Subspace Clustering for High-Dimensional Data. In: Proceedings of the Fourth SIAM International Conference on Data Mining, SDM 2004, pp. 246–257 (2004)

[8] Krulis, M., Lokoc, J., Beecks, C., Skopal, T., Seidl, T.: Processing the signature quadratic form distance on many-core GPU architectures. In: Proceedings of the 20th ACM International Conference on Information and Knowledge Management, CIKM 2011, pp. 2373–2376. ACM (2011)

[9] Mattson, P., Dally, W.J., Rixner, S., Kapasi, U.J., Owens, J.D.: Communication Scheduling". In: Proceedings of the Ninth International Conference on Architectural Support for Programming Languages and Operating Systems. In: ASPLOS IX, pp. 82–92. ACM (2000)

[10] Munshi, A.: The OpenCL 1.2 Specification. Khronos OpenCL Working Group. Khronos Group. Khronos (2012)

[11] NVIDIA CUDA C Programming Guide. NVIDIA Corp. (2012),
http://www.nvidia.com

[12] NVIDIA Corp., ed. NVIDIA's Next Generation CUDA Compute Archi- tecture:
Kepler TM GK110. The Fastest, Most Efficient HPC Architecture Ever Built
(2012)

[13] OpenMP Architecture Review Board. The OpenMP API Speciffication For Parallel Programming (2011)

[14] Pabst, H.-F., Springer, J.P., Schollmeyer, A., Lenhardt, R., Lessig, C., Froehlich, B.:
Ray casting of trimmed NURBS surfaces on the GPU. In: Proceedings of the 2006
IEEE Symposium on Interactive Ray Tracing, pp. 151–160 (2006)

[15] Preis, T.: Econophysics complex correlations and trend switchings in financial
time series. The European Physical Journal Special Topics, 5–86 (2011)

[16] Rubner, Y., Tomasi, C., Guibas, L.J.: The Earth Mover's Distance as a Metric for
Image Retrieval. International Journal of Computer Vision, 99–121 (2000)

[17] Tanenbaum, A.S.: Parallel Computer Architectures. In: Structured Computer Organization. Pearson Studium (2001) isbn: 0130959901

[18] Tarakji, A., Marx, M., Lankes, S.: The Development of a Scheduling System
GPUSched for Graphics Processing Units. In: The International Conference on
High Performance Computing Simulation, HPCS (2013)

[19] Wang, J.Z., Li, J., Wiederhold, G.: SIMPLIcity: Semantics-Sensitive Integrated
Matching for Picture LIbraries. IEEE Transactions on Pattern Analysis and Machine Intelligence, 947–963 (2001)

[20] Wasson, S.: Nvidia Kepler powers Oak Ridge's supercomputing Titan. Tech. rep.
PC Hardware Eplored (2012)

Hierarchical Tori Connected Mesh Network

M.M. Hafizur Rahman[1], Asadullah Shah[1], Masaru Fukushi[2],
and Yasushi Inoguchi[3]

[1] Dept. of Computer Science, KICT, International Islamic University Malaysia
(IIUM), P.O. Box. 10, 50728, Kuala Lumpur, Malaysia
[2] Graduate School of Science and Engineering, Yamaguchi University, Tokiwadai
2-16-1, Ube, 755-8611, Japan
[3] Research Center for Advanced Computing Infrastructure,
Japan Advanced Institute of Science and Technology (JAIST),
Ishikawa 923-1292, Japan
{hafizur,asadullah}@iium.edu.my, mfukushi@yamaguchi-u.ac.jp,
inoguchi@jaist.ac.jp

Abstract. Hierarchical interconnection networks provide high performance at low cost by exploring the locality that exists in the communication patterns of massively parallel computers. A **Hierarchical Tori** connected **Mesh Network (HTM)** is a 2D-torus network of multiple basic modules, in which the basic modules are 3D-mesh networks that are hierarchically interconnected for higher-level networks. This paper addresses the architectural details of the HTM and explores aspects such as degree, diameter, cost, average distance, arc connectivity, bisection width, and wiring complexity. We also present a deadlock-free routing algorithm for the HTM using two virtual channels and evaluate the network's dynamic communication performance using the proposed routing algorithm under uniform traffic and bit-flip traffic patterns. We evaluate the dynamic communication performance of HTM, H3DM, mesh, and torus networks by computer simulation. It is shown that the HTM possesses several attractive features, including constant node degree, small diameter, low cost, small average distance, moderate (neither too low, nor too high) bisection width, small wiring complexity, and high throughput per link and very low zero load latency, which provide better dynamic communication performance than that of H3DM, mesh, and torus networks.

Keywords: Interconnection network, HTM network, Deadlock-free routing algorithm, Static network performance, Uniform traffic patterns, Bit-Flip traffic patterns, Dynamic communication performance.

1 Introduction

High-performance computing is necessary in solving the grand challenge problems in many areas such as development of new materials and sources of energy, development of new medicines and improved health care, strategies for disaster prevention and mitigation, weather forecasting, and for scientific research including the origins of matter and the universe. This makes the current supercomputer

B. Murgante et al. (Eds.): ICCSA 2013, Part V, LNCS 7975, pp. 197–210, 2013.

changes into massively parallel computer (MPC) systems with thousands of node (Kei, Cray XT5-HE), that satisfy the insatiable demand of computing power. In near future, we will need computer systems capable of computing at the petaflops or exaflops level. To achieve this level of performance, we need MPC with tens of thousands or millions of nodes. Interconnection networks play a crucial role in the performance of MPC systems [1]. Many recent experimental and commercial parallel computers use direct networks for low latency and high bandwidth of inter-processor communication. The hierarchical interconnection network (HIN) provides an alternative way in which several network topology can be integrated [2]. For future MPC with millions of nodes, the large diameter of conventional topologies is intolerable. Hence, HIN is an efficient way to interconnect the future MPC[2]. A variety of hypercube based HINs found in the literature, however, its huge number of physical links make it difficult to implement. To alleviate this problem, k-ary n-cube based HIN [3,4] is a plausible alternative way.

In our previous study, we have studied H3DM network. A Hierarchical 3D-Mesh (H3DM) Network [5] is a 2D-mesh network of multiple basic modules (BMs), in which the BMs are 3D-torus networks that are hierarchically interconnected for higher-level networks. BMs are 3D-torus ($m \times m \times m$) and they are hierarchically interconnected by 2D-mesh ($n \times n$). The limited connectivity of higher level links of 2D-mesh network become congested with the increase of packet in the network [6]. It has already been shown that a torus network has better dynamic communication performance than a mesh network [1]. This is the key motivation that led us to consider the higher level network as a 2D-torus network and to reduce the total number of links we have considered 3D-mesh as basic module instead of 3D-torus network, despite the fact that a 3D-torus has better performance than a 3D-mesh network.

A HTM network is a 2D-torus network of multiple basic modules (BMs), in which the BMs are 3D-mesh networks that are hierarchically interconnected for higher-level networks. BMs are 3D-mesh ($m \times m \times m$) and they are hierarchically interconnected by 2D-torus ($n \times n$). Wormhole routing [7] has become the dominant switching technique used in contemporary multicomputers. This is because it has low buffering requirements and it makes latency independent of the message distance. Deterministic, dimension-order routing is popular in MPC because it has minimal hardware requirements and allows the design of simple and fast routers. Wormhole routing relies on a blocking mechanism for flow control, deadlock can occur because of cyclic dependencies over network resources during message routing. Virtual channels (VCs) [9] are used to solve the problem of deadlock in wormhole-routed networks. Since the hardware cost increases as the number of VCs increases, the unconstrained use of VCs is not cost-effective in MPC systems.

The remainder of the paper is organized as follows. In Section 2, we briefly describe the network structure of the HTM. The dynamic routing algorithm is proposed in Section 3 and its freedom from deadlock is also proved. Static network performance and the dynamic communication performance are discussed in Section 4 and 5, respectively. Finally, we conclude this paper in Section 6.

2   Interconnection of the HTM Network

The *HTM* network is a HIN consisting of multiple BM that are hierarchically interconnected for higher level networks. The BM of the HTM network is a 3D-mesh network of size $(m \times m \times m)$, where m is a positive integer. m can be any value, however the preferable one is $m = 2^p$, where p is a positive integer. The BM of a $(4 \times 4 \times 4)$ mesh, as depicted in Figure 1 (a), has some free ports at the contours of the xy-plane. A $(m \times m \times m)$ BM has $4 \times m^2$ free ports for higher level interconnection. All free ports, typically one or two, of the exterior Processing Elements (PEs) are used for inter-BM connections to form higher level networks. Successively higher level networks are built by recursively interconnecting lower level sub-networks in a 2D-torus of size $(n \times n)$, where n is also a positive integer. As portrayed in Figure 1(b), a Level-2 HTM network can be formed by interconnecting 16 BMs as a (4×4) 2D-torus network. Similarly, a Level-3 network can be formed by interconnecting n^2 Level-2 sub-networks, and so on. Each BM is connected to its logically adjacent BMs. For each higher level interconnection of HTM network, a BM must use $4m(2^q)$ of its free links: $2m(2^q)$ free links for y-direction and $2m(2^q)$ free links for x-direction interconnections. Here, $q \in \{0, 1, \dots, p\}$, is the inter-level connectivity, where $p = \lfloor log_2^m \rfloor$. $q = 0$ leads to minimal inter-level connectivity, while $q = p$ leads to maximum inter-level connectivity. It is depicted in Figure 1(a) that the $(4 \times 4 \times 4)$ BM has $4 \times 4^2 = 64$ free ports. With $q = 0$, $(4 \times 4 \times 2^0 =)$ 16 free links are used for each level interconnection, 8 for y-direction and 8 for x-direction interconnections as portrayed in Figure 1(b). The highest level network which can be built from a $(m \times m \times m)$ BM is $L_{max} = 2^{p-q} + 1$. With $q = 0$, Level-5 is the highest possible level to which a $(4 \times 4 \times 4)$ BM can be interconnected. The total number of nodes in a network having $(m \times m \times m)$ BMs and $(n \times n)$ higher level is $N = \left[m^3 \times n^{2(L_{max}-1)} \right]$. Thus, the maximum number of nodes which can be interconnected by the HTM network is $N = \left[m^3 \times n^{2(2^{p-q})} \right]$. If $m = 4$, $n = 4$, and $q = 0$, then $N = 4^3 \times 4^8 = 4194304$, i.e, about 4.2 million.

The address of a PE at Level-L HTM network is represented by Eq. 1.

$$A^L = \begin{cases} (a_z)(a_y)(a_x) & \text{if } L = 1, \text{ i.e., BM} \\ (a_y^L)(a_x^L) & \text{if } L \geq 2 \end{cases} \tag{1}$$

More generally, in a Level-L HTM, the node address is represented by:

$$A = A^L A^{L-1} A^{L-2} \dots\dots\dots A^2 A^1$$

$$= a_\alpha \, a_{\alpha-1} \, a_{\alpha-2} \, a_{\alpha-3} \dots\dots\dots a_3 \, a_2 \, a_1 \, a_0$$

$$= (a_{2L} \, a_{2L-1}) \dots\dots\dots (a_4 \, a_3)(a_2 \, a_1 \, a_0) \tag{2}$$

Here, the total number of digits is $\alpha = 2L + 1$, where L is the level number. The first group contains three digits and the rest of the groups contain two digits. Groups of digits run from group number 1 for Level-1, i.e., the BM, to group number L for the L-th level. In particular, i-th group $(a_{2i} \, a_{2i-1})$ indicates the location of a Level-$(i - 1)$ sub-network within the i-th group to which the node

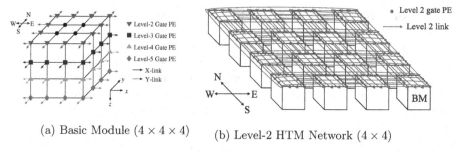

(a) Basic Module ($4 \times 4 \times 4$)   (b) Level-2 HTM Network (4×4)

Fig. 1. Interconnection of a HTM Network

belongs; $2 \leq i \leq L$. In a two-level network, for example, the address becomes $A = (a_4 \, a_3) \, (a_2 \, a_1 \, a_0)$. The last group of digits $(a_4 \, a_3)$ identifies the BM to which the node belongs, and the first group of digits $(a_2 \, a_1 \, a_0)$ identifies the node within that BM.

3   Routing Algorithm for HTM Network

3.1   Routing Algorithm

Routing of messages in the HTM network is first done at the highest level network; then, after the packet reaches its highest level sub-destination, routing continues within the sub-network to the next lower level sub-destination. This process is repeated until the packet arrives at its final destination [13]. When a packet is generated at a source node, the node checks its destination. If the packet's destination is the current BM, the routing is performed within the BM only. If the packet is addressed to another BM, the source node sends the packet to the outlet node which connects the BM to the level at which the routing is performed. We have considered a simple deterministic, dimension-order routing algorithm. Routing at the higher level HTM is performed first in the y-direction and then in the x-direction. In a BM, the routing order is z-direction, y-direction, and x-direction, respectively.

Routing in the HTM network is strictly defined by the source node address and the destination node address. Let a source node address be $s_\alpha, s_{\alpha-1}, s_{\alpha-2}, ..., s_1, s_0$, a destination node address be $d_\alpha, d_{\alpha-1}, d_{\alpha-2}, ..., d_1, d_0$, and a routing tag be $t_\alpha, t_{\alpha-1}, ..., t_1, t_0$, where $t_i = d_i - s_i$. The source node address of HTM is expressed as $s = (s_{2L}, s_{2L-1}), (s_{2L-2}, s_{2L-3}), ..., (s_2, s_1, s_0)$. Similarly, the destination address is expressed as $d = (d_{2L}, d_{2L-1}), (d_{2L-2}, d_{2L-3}), ..., (d_2, d_1, d_0)$. Figure 2 shows the routing algorithm for the HTM network.

Suppose a packet is to be transported from source node 0000000 to destination node 1131230. In this case, we see that routing should first be done at Level-3, therefore, the source node sends the packet to the Level-3 outlet node 0000130, whereupon the packet is routed at Level-3. After the packet reaches the $(1, 1)$ Level-2 network, then routing within that network is continued until the packet reaches the BM $(3, 1)$. Finally, the packet is routed to its destination node $(2, 3, 0)$ within that BM.

Routing HTM(s,d);
source node address:$s_\alpha, s_{\alpha-1}, s_{\alpha-2}, ..., s_1, s_0$
destination node address: $d_\alpha, d_{\alpha-1}, d_{\alpha-2}, ..., d_1, d_0$
tag: $t_\alpha, t_{\alpha-1}, t_{\alpha-2}, ..., t_1, t_0$
for $i = \alpha : 3$
    if ($i/2 = 0$ and ($t_i > 0$ or $t_i = -(n-1)$)), routedir = North; endif;
    if ($i/2 = 0$ and ($t_i < 0$ or $t_i = (n-1)$)), routedir = South; endif;
    if ($i\%2 = 1$ and ($t_i > 0$ or $t_i = -(n-1)$)), routedir = East; endif;
    if ($i\%2 = 1$ and ($t_i < 0$ or $t_i = (n-1)$)), routedir = West; endif;
     while ($t_i \neq 0$) do
       $N_z = outlet_z(s, d, L, \text{routedir})$
       $N_y = outlet_y(s, d, L, \text{routedir})$
       $N_x = outlet_x(s, d, L, \text{routedir})$
       BM_Routing(N_z, N_y, N_x)
       if (routedir = North or East), move packet to next BM; endif;
       if (routedir = South or West), move packet to previous BM; endif;
       if ($t_i > 0$), $t_i = t_i - 1$; endif;
       if ($t_i < 0$), $t_i = t_i + 1$; endif;
     endwhile;
  endfor;
  BM_Routing(t_z, t_y, t_x)
end
BM_Routing (t_2, t_1, t_0);
BM_tag t_2, t_1, t_0 = receiving node address (r_2, r_1, r_0) − destination (d_2, d_1, d_0)
  for $i = 2 : 0$
    if ($t_i > 0$), movedir = positive; endif;
    if ($t_i < 0$), movedir = negetive; endif;
    if (movedir = positive and $t_i > 0$), distance = t_i; endif;
    if (movedir = positive and $t_i < 0$), distance = $m + t_i$; endif;
    if (movedir = negative and $t_i < 0$), distance = t_i; endif;
    if (movedir = negative and $t_i > 0$), distance = $-m + t_i$; endif;
  endfor
  while($t_2 \neq 0$ or distance$_2 \neq 0$) do
    if (movedir = positive), move packet to $+z$ node; distance$_2$ = distance$_2$ − 1; endif;
    if (movedir = negetive), move packet to $-z$ node; distance$_2$ = distance$_2$ + 1; endif;
  endwhile;
  while($t_1 \neq 0$ or distance$_1 \neq 0$) do
    if (movedir = positive), move packet to $+y$ node; distance$_1$ = distance$_1$ − 1; endif;
    if (movedir = negetive), move packet to $-y$ node; distance$_1$ = distance$_1$ + 1; endif;
  endwhile;
  while($t_0 \neq 0$ or distance$_0 \neq 0$) do
    if (movedir = positive), move packet to $+x$ node; distance$_0$ = distance$_0$ − 1; endif;
    if (movedir = negetive), move packet to $-x$ node; distance$_0$ = distance$_0$ + 1; endif;
  endwhile;
end

Fig. 2. Dimension-Order Routing Algorithm of the HTM Network

3.2   Deadlock-Free Routing

A key component of reliability is its routing algorithm should be deadlock-free. A deadlock-free routing algorithm can be constructed for a wormhole routed interconnection network by introducing virtual channels [9]. Since the hardware cost increases as the number of virtual channels increases, the unconstrained use of virtual channels is prohibited for cost-effective parallel computers. A deadlock-free routing algorithm with a minimum number of virtual channels is preferred.

We have applied the dimension-order routing in each level of the HTMM network like hierarchical routing algorithms (HiRA) [13]. To prove the proposed routing algorithm for the H3DM network is deadlock-free, we divide the routing path into three phases, as follows:

- *Phase 1:* Intra-BM transfer path from source PE to the face of the BM.
- *Phase 2:* Higher level transfer path.
  - **sub-phase** $2.i.1$: Intra-BM transfer to the outlet PE of Level $(L - i)$ through the y-link.
  - **sub-phase** $2.i.2$: Inter-BM transfer of Level $(L - i)$ through the y-link.
  - **sub-phase** $2.i.3$: Intra-BM transfer to the outlet PE of Level $(L - i)$ through the x-link.
  - **sub-phase** $2.i.4$: Inter-BM transfer of Level $(L - i)$ through the x-link.
- *Phase 3:* Intra-BM transfer path from the outlet of the inter-BM transfer path to the destination PE.

The proposed routing algorithm enforces some routing restrictions to avoid deadlocks [8,14]. By using the following lemmas and theorem, we will prove that the proposed routing algorithm for the HTM network is deadlock-free using minimum number of virtual channels.

Lemma 1. *If a message is routed in the order $z \to y \to x$ in a 3D-mesh network, then the network is deadlock-free with 1 virtual channel.* [8]

Proof: If the channels are allocated according to Eq. 3 for a 3D-mesh network and the messages are routed according to the above mentioned phenomena, then cyclic dependency will not occur. Therefore, freedom from is proved.

$$C = \begin{cases} (l, a_2), \ z+ \ \text{channel}, \\ (l, m - a_2), \ z- \ \text{channel}, \\ (l, a_1), \ y+ \ \text{channel}, \\ (l, m - a_1), \ y- \ \text{channel}, \\ (l, a_0), \ x+ \ \text{channel}, \\ (l, m - a_0), \ x- \ \text{channel} \end{cases} \tag{3}$$

Here, $l = \{l_0, l_1, l_2, l_3, l_4, l_5\}$ are the links used in the BM, $l = \{l_0, l_1\}$, $l = \{l_2, l_3\}$, and $l = \{l_4, l_5\}$ are the links used in the z–direction, y–direction, and x–direction interconnection, respectively, m is the size of the BM, and a_0, a_1, and a_2 are the node addresses in the BM.

Lemma 2. *If a message is routed in the order $y \to x$ direction in a 2D-torus network, then the network is deadlock-free with 2 virtual channels.* [8]

Proof: If the channels are allocated as shown in Eq. 4 for the higher level 2D-torus network, and the messages are routed according to the above-mentioned phenomena, then cyclic dependency will not occur. Therefore, freedom from deadlock is proved. Initially, messages are routed over virtual channel 0 (lower). Then, messages are routed over virtual channel 1 (higher) if the message is going to use a wrap-around channel.

$$C = \begin{cases} (l, vc, a_{2L}), \ y+ \ \text{channel}, \\ (l, vc, n - a_{2L}), \ y- \ \text{channel}, \\ (l, vc, a_{2L-1}), \ x+ \ \text{channel}, \\ (l, vc, n - a_{2L-1}), \ x- \ \text{channel} \end{cases} \tag{4}$$

Here, $l = \{l_6, l_7\}$ are the links used for higher-level interconnection, l_6 is used in the interconnection of the east and west directions and l_7 is used in the interconnection of the north and south directions. $vc = \{VC_0, VC_1\}$ are the virtual channels, n is the size of the higher level networks, and a_{2L} and a_{2L-1} are the node addresses in the higher level, where L is the level number.

Theorem 1. *A HTM network is deadlock-free with 2 virtual channels.*

Proof: The BM of HTM is a 3D-mesh network and they are hierarchically interconnected by a 2D-torus network for higher level interconnection. In phase-1 and phase-3 routing, packets are routed in the source-BM and destination-BM, respectively. The BM of the HTM network is a 3D-mesh network. According to Lemma 1, the number of necessary virtual channels for phase-1 and phase-3 routing is 1. The routing of the message in source-BM and destination-BM is carried out separately. Thus, 1 virtual channel is shared in phase-1 and phase-3 routing. In phase-2 routing, packets are routed in the higher level networks. Intra-BM links between inter-BM communication on the xy-plane of the BM are used in sub-phases $2.i.1$ and $2.i.3$. These sub-phases utilize channels over intra-BM links, sharing either the channels of phase-1 or phase-3. The exterior links at the contours of the xy-plane of the BM are used in sub-phase $2.i.2$ and sub-phase $2.i.4$, and path of these links form the higher level networks. And the higher level networks of the HTM network is a 2D-torus network. According to Lemma 2, the number of necessary virtual channels for phase-2 routing is 2. The first virtual channel is used when the intra-node routing takes place and the second one is used when the wrap-around links is used in routing. The intra-node routing in phase-2 use the virtual channel of either phase-1 or phase-3. The only one additional virtual channel is required in phase-2 routing.

Therefore, the total number of necessary virtual channels for the whole network is 2.

4   Static Network Performance

Comparing performance, both static and dynamic communication, among different hierarchical interconnection networks such as HTM and H3D-mesh [6] alongwith conventional network topology is not an easy task, because each network has a different interconnection architecture, which makes it difficult to match the total number of nodes. For fair comparison we should have equal number of nodes for all the considered network. If $m = 4$, $n = 4$, and $L = 2$, then the total number of nodes in the HTM network is 1024. Again, Level-2 H3D-mesh with $n = 4$, 32×32 mesh, and 32×32 torus networks also have 1024 nodes. According to the structure of the 3D-mesh and 3D-torus network, it is not possible to construct a 1024-node 3D mesh and torus networks. This is why, in this paper, we have evaluated the performance of 1024-node networks.

The topology of an interconnection network determines many architectural features that affect several performance metrics. Although the actual performance of a network depends on many technological and implementation issues, several topological properties and performance metrics can be used to evaluate and compare different network topologies in a technology-independent manner. Most of these properties are derived from the graph model of the network topology. In this section, we discuss some of the properties and performance metrics that characterize the cost and performance of an interconnection network.

To evaluate the static network performance we have considered node degree, diameter, cost, average distance, arc connectivity, bisection width, and wiring complexity as performance metrics. The node degree is defined as the maximum number of links emanating from a node. The diameter and average distance are the maximum distance and mean distance, respectively, among all distinct pairs of nodes along the shortest path in a network. The product (*diameter* × *node degree*) is a good criterion for measuring the relationship between cost and performance of a MPC system [10]. The Bisection Width (BW) is defined as the minimum number of links that must be removed to partition the network into two equal halves. Arc Connectivity is defined as the minimum number of links that must be removed to partition the network into two disjoint parts.

We have considered Level-2 HTM and H3DM networks 32 × 32 mesh and torus networks for performance comparison. The static performance of these networks are tabulated in the Table 1. The node degree of the HTM is 6, and it is independent of network size. We have evaluated the diameter and average distances of H3DM and HTM networks by simulation, and conventional mesh and torus networks by their corresponding formulae. The HTM has a much smaller diameter and average distance than those of mesh and torus networks and slightly higher than that of H3DM network. It is shown in Table 1 that the cost of HTM is lower than that of mesh and H3DM networks, and slightly higher than that of torus network. We have calculated the bisection width of various networks by their respective formula, and it is shown in Table 1 that the the bisection width of the HTM is exactly equal to that of mesh network, lower than that of torus network, and higher than that of H3DM network.

Table 1. Comparison of static network performance of various networks

	Node Degree	Diameter	Cost	Average Distance	Arc Connectivity	Bisection Width	Wiring Complexity
2D-Mesh	4	62	248	21.33	2	32	1984
2D-Torus	4	32	128	16.00	4	64	2048
H3DM	8	21	168	9.23	6	16	3168
HTM	6	26	156	10.95	3	32	2432

The wiring complexity of a network refers to the total number of links required to form the network. The wiring complexity depends on the node degree and it has a direct correlation to hardware cost and complexity. It is shown that the wiring complexity of the HTM network is lower than that of H3DM network. Arc connectivity is a measure of connectivity. A network is maximally fault-tolerant if its connectivity is equal to the degree of the network. Arc Connectivity of the HTM network is higher than mesh and lower than torus and H3DM networks.

5   Dynamic Communication Performance

The overall performance of a MPC system is affected by the performance of the interconnection network as well as by the performance of the node. Low performance of the underlying interconnection network will severely limit the speed of the entire MPC system instead of high performance individual node is used. Therefore, the success of a MPC is highly dependent on the efficiency of their interconnection networks.

5.1   Performance Metrics

The dynamic communication performance of a MPC system is characterized by message latency and network throughput. Message latency refers to the time elapsed from the instant when the first flit (header flit) is injected into the network from the source to the instant when the last data flit of the message is received at the destination. Network throughput refers to the maximum amount of information delivered per unit of time through the network. For the network to have good performance, low latency and high throughput must be achieved.

5.2   Traffic Patterns

One of the most important factors influencing dynamic communication performance is the traffic pattern. In an interconnection network, sources and destinations for messages form the traffic pattern. Traffic characteristics such as message length, message arrival times at the sources, and destination distribution have significant performance implications. Message destination distributions vary a great deal depending on the network topology and the application's mapping onto different nodes. In order to evaluate the dynamic communication performance, we have considered uniform traffic pattern and bit-flip traffic pattern.

- **Uniform** – In the uniform traffic pattern, the source and the destination are randomly selected, i.e., every node sends messages to every other node with equal probability in the network [11].
- **Bit-flip** – The node with binary coordinates $b_{\beta-1}, b_{\beta-2} \ldots \ldots b_1, b_0$ communicates with the node $\left(\overline{b_0}, \overline{b_1}, \ldots \ldots \overline{b_{\beta-2}}, \overline{b_{\beta-1}}\right)$ [12].

5.3   Simulation Environment

We have developed a wormhole routing simulator using C programming language to evaluate the dynamic communication performance. We use a dimension-order routing and uniform traffic pattern. In the evaluation of performance, flocks of messages are sent through the network to compete for the output channels. Packets are transmitted by the request-probability r during T clock cycles and the number of flits which reached at destination node and its transfer time is recorded. Then the average transfer time and throughput are calculated and plotted as average transfer time in the horizontal axis and throughput in the vertical axis. The process of performance evaluation is carried out with changing the request-probability r. We have considered that the message generation rate is constant and the same for all nodes. Flits are transmitted at $20,000$ cycles i.e., $T = 20000$. In each clock cycle, one flit is transferred from the input buffer to the output buffer, or vice versa if the corresponding buffer in the next node is empty. Thus, transferring data between two nodes takes 2 clock cycles. The message length is considered as 16 flits; and the buffer length of each channel is 2 flits. For fair comparison of dynamic communication performance, two VCs per physical link are simulated, and the VCs are arbitrated by a round robin algorithm.

5.4   Dynamic Communication Performance Evaluation

We have evaluated the dynamic communication performance of several networks using deadlock-free dimension order routing with minimum number of virtual channels under the uniform traffic pattern and bit-flip traffic pattern. For fair comparison of dynamic communication performance, we have also considered same number of nodes, 1024 nodes, for all the considered network.

Uniform Traffic Patterns. The most frequently used, simplest, and most elegant traffic pattern is the uniform traffic pattern where the source and the destination are randomly selected. Uniform traffic patterns provide a benchmark for interconnection networks. Figure 3 show the average transfer time as a function of network throughput under uniform traffic pattern for different networks. The average transfer time at no load is called zero load latency. As shown in Figure 3, the zero load latency of the HTM network is significantly lower than that of the mesh and torus networks and a slightly higher than that of H3DM network. The throughput and latency of a network is increased with the increase of load. Because the links and virtual channels become congested and the message compete to each other for the network resources, links and channels. With the injection of more and more messages and in course of time, the network become saturated. After saturation, the message latency is increasing dramatically while the network throughput will not increase anymore. Up to saturation the trade-off between throughput and latency of the HTM network is better than that of mesh and torus networks and not as good as H3DM network as illustrated in Fig. 3.

Bit-Flip Traffic Patterns. Bit Permutation and Computation (BPC) traffic patterns is exhibited in many parallel applications such as computing multidimensional FFT, matrix problems, finite elements, and fault-tolerant routing. Bit-flip traffic patterns is very common in many scientific applications. It is also considered as benchmarks for interconnection networks. In a bit flip traffic pattern, a node with address Node $(b_{\beta-1}, b_{\beta-2} \ldots \ldots b_1, b_0)$ sends messages to Node $(\overline{b_0}, \overline{b_1}, \ldots \ldots \overline{b_{\beta-2}}, \overline{b_{\beta-1}})$. Figure 4 portrays the result of simulations under bit flip traffic pattern for the various networks. Like uniform traffic patterns,

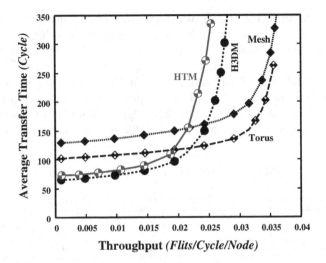

Fig. 3. Dynamic communication performance of various networks using dimension-order routing under uniform traffic pattern: 1024 nodes and 2 VCs

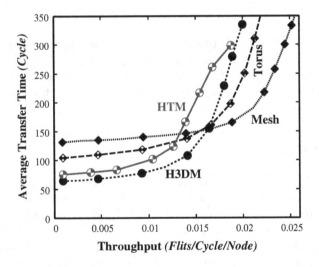

Fig. 4. Dynamic communication performance of various networks using dimension-order routing under bit-flip traffic pattern: 1024 nodes and 2 VCs

the zero load latency of the HTM network is significantly lower than that of the mesh and torus networks and a slightly higher than that of H3DM network. The maximum throughput of the HTM is lower than that of the mesh, torus, and H3D-mesh networks. However, as compared between Fig. 3 and 4, the difference of maximum throughput between H3D-mesh network and HTM is diminishing.

Maximum Throughput Per Link. The maximum throughput of H3DM network is higher than that of the HTM network, however, this high throughput of the H3DM network is yielded with the cost large number of links in the H3DM network as depicted in Table 1. If we compare the maximum throughput per link between H3DM and HTM, as illustrated in Table 2, we found that the maximum throughput per link of HTM is significantly higher than that of H3DM network. In percentage, it is 18% and 22% higher than that of H3DM network under uniform traffic pattern and bit-flip traffic pattern, respectively.

Table 2. Comparison of maximum throughput

Network	Number of links	Maximum throughput	Throughput per link
Uniform Traffic Pattern			
H3DM	3168	0.0282504	8.91×10^{-6}
HTM	2432	0.0254227	10.50×10^{-6}
Bit-Flip Traffic Pattern			
H3DM	3168	0.0199644	6.30×10^{-6}
HTM	2432	0.0187320	7.70×10^{-6}

The wrap-around links between end-to-end node in the basic module is removed to reduce the wiring complexity. However, due to the presence of wrap-around links between end-to-end nodes in the higher level networks, the congestion in the bisection is lessened. Thus, the maximum throughput per link of the HTM network is increased.

6   Conclusion

A new hierarchical interconnection network, called **H**ierarchical **T**ori connected **M**esh (**HTM**) network is proposed for the high performance massively parallel computer systems. The architecture of the HTM, addressing of nodes, routing of messages, and static network performance were discussed in detail. From the static network performance, it has been shown that the HTM possesses several attractive features, including constant node degree, small diameter, low cost, high connectivity, small average distance, moderate (neither too low, nor too high) bisection width, and low wiring complexity.

A deadlock-free routing algorithm using dimension order routing with 2 virtual channels has been proposed for the HTM. By using the routing algorithm described in this paper, and using uniform and bit-flip traffic patterns, we have evaluated the dynamic communication performance of the HTM, as well as that of several other interconnection networks. The average transfer time of the HTM is lower than that of the conventional mesh and torus networks and slightly higher than H3DM network. However, maximum throughput per link of the HTM is higher than that of H3DM network. A comparison of dynamic communication performance proves that the HTM achieves better performance than the mesh, torus, and H3DM networks. The HTM yields low latency and high throughput per link with reasonable cost, which are indispensable for high-performance massively parallel computers. Therefore, HTM would be a good choice of interconnection network for next generation massively parallel computers. The important issue of evaluating dynamic communication performance by other non-uniform traffic patterns and assessing the dynamic communication performance improvement of the HTM network by the adaptive routing algorithm remains a subject for further exploration.

Acknowledgment. This work is supported in part by KICT, IIUM and IIUM Endowment-B research fund EDW B11-169-0647, Research Management Center (RMC), IIUM, Malaysia. The authors are grateful to the anonymous reviewers for their constructive comments which helped to greatly improve the clarity of this paper.

References

1. Dally, W.J.: Performance Analysis of k-ary n-cube Interconnection Networks. IEEE Trans. on Computers 39(6), 775–785 (1990)
2. Abd-El-Barr, M., Al-Somani, T.F.: Topological Properties of Hierarchical Interconnection Networks: A Review and Comparison. Journal of Electrical and Computer Engineering 2011, 12 pages 2011)
3. Lai, P.L., Hsu, H.C., Tsai, C.H., Stewart, I.A.: A class of hierarchical graphs as topologies for interconnection networks. Theoretical Computer Science, Elsevier 411, 2912–2924 (2010)
4. Liu, Y., Li, C., Han, J.: RTTM: A New Hierarchical Interconnection Network for Massively Parallel Computing. In: Zhang, W., Chen, Z., Douglas, C.C., Tong, W. (eds.) HPCA 2009. LNCS, vol. 5938, pp. 264–271. Springer, Heidelberg (2010)
5. Horiguchi, S.: New Interconnection for massively Parallel and Distributed System. Research Report, 09044150, JAIST, pp. 47-57 (1999)
6. Rahman, M.M.H., Shah, A., Inoguchi, Y.: On Dynamic Communication Performance of a Hierarchical 3D-Mesh Network. In: Park, J.J., Zomaya, A., Yeo, S.-S., Sahni, S. (eds.) NPC 2012. LNCS, vol. 7513, pp. 180–187. Springer, Heidelberg (2012)
7. Ni, L.M., McKinley, P.K.: A Survey of Wormhole Routing Techniques in Direct Networks. IEEE Computer 26(2), 62–76 (1993)
8. Dally, W.J., Seitz, C.L.: Deadlock free Message Routing in Multiprocessor Interconnection Networks. IEEE Trans. on Computers C36(5), 547–553 (1987)
9. Dally, W.J.: Virtual-Channel Flow Control. IEEE Trans. on Parallel and Distributed Systems 3(2), 194–205 (1992)
10. Kumar, J.M., Patnaik, L.M.: Extended Hypercube: A Hierarchical Interconnection Network of Hypercube. IEEE Trans. Parallel Distrib. Syst. 3(1), 45–57 (1992)
11. Najaf-abadi, H.H., Sarbazi-Azad, H.: The Effects of Adaptivity on the Performance of the OTIS-Hypercube Under Different Traffic Patterns. In: Jin, H., Gao, G.R., Xu, Z., Chen, H. (eds.) NPC 2004. LNCS, vol. 3222, pp. 390–398. Springer, Heidelberg (2004)

12. Najaf-abadi, H.H., Sarbazi-Azad, H.: The Effects of Adaptivity on the Perfor-
    mance of the OTIS-Hypercube Under Different Traffic Patterns. In: Jin, H., Gao,
    G.R., Xu, Z., Chen, H. (eds.) NPC 2004. LNCS, vol. 3222, pp. 390–398. Springer,
    Heidelberg (2004)
13. Holsmark, R., Kumar, S., Palesi, M., Mekia, A.: HiRA: A Methodology for
    Deadlock Free Routing in Hierarchical Networks on Chip. In: Proc. of the 3rd
    ACM/IEEE NOCS, pp. 2–11 (2009)
14. Koibuchi, M., Anjo, K., Yamada, Y., Jouraku, A., Amano, H.: A Simple Data
    Transfer Technique using Local Address for Networks-on-Chips. IEEE Transactions
    on Parallel and Distributed Systems 17(12), 1425–1437 (2006)

Optimization of Sparse Matrix-Vector Multiplication for CRS Format on NVIDIA Kepler Architecture GPUs

Daichi Mukunoki[1] and Daisuke Takahashi[2]

[1] Graduate School of Systems and Information Engineering, University of Tsukuba
[2] Faculty of Engineering, Information and Systems, University of Tsukuba
1-1-1 Tennodai, Tsukuba, Ibaraki 305-8573, Japan
{mukunoki@hpcs.,daisuke@}cs.tsukuba.ac.jp

Abstract. Sparse matrix-vector multiplication (SpMV) is an important operation in scientific and engineering computing. This paper presents optimization techniques for SpMV for the Compressed Row Storage (CRS) format on NVIDIA Kepler architecture GPUs using CUDA. Our implementation is based on an existing method proposed for the Fermi architecture, an earlier generation of the GPU, and takes advantage of some of the new features of the Kepler architecture. On a Tesla K20 Kepler architecture GPU on double precision operations, our implementation is, on average, approximately 1.29 times faster than that the Fermi optimized implementation for 200 different types of matrices. As a result, our implementation outperforms the NVIDIA cuSPARSE library's CRS format SpMV in CUDA 5.0 on 174 of the 200 matrices, and the average speedup compared to the cuSPARSE SpMV routine across all 200 matrices is approximately 1.45.

Keywords: sparse matrix-vector multiplication, SpMV, Kepler architecture, GPU, CUDA.

1 Introduction

Sparse matrix-vector multiplication (SpMV), that performs $y = Ax$ (where x and y are vectors and A is a sparse matrix) is one of the most important operations in scientific and engineering computing. In general, in order to save memory, a sparse matrix is stored in two kinds of arrays: a data array which only stores the non-zero elements of the matrix and index arrays which store the addresses of the non-zero elements. Thus, sparse matrix operations require complex memory access patterns when compared to dense matrix operations. Moreover, various kinds of distributions of the non-zero elements are considered. Therefore, efficient implementation of SpMV requires a large number of optimization techniques.

This paper presents optimization techniques for SpMV for the Compressed Row Storage (CRS) format on NVIDIA Kepler architecture GPUs using CUDA. The CRS format is one of the most widely used storage formats for sparse

B. Murgante et al. (Eds.): ICCSA 2013, Part V, LNCS 7975, pp. 211–223, 2013.

$$\begin{bmatrix} 8\ 9\ 0\ 0\ 4\ 5 \\ 0\ 7\ 5\ 6\ 2\ 0 \\ 0\ 6\ 0\ 7\ 0\ 0 \\ 9\ 0\ 0\ 6\ 0\ 2 \\ 0\ 0\ 2\ 0\ 0\ 0 \\ 0\ 0\ 7\ 2\ 8\ 0 \end{bmatrix}$$

val = [8, 9, 4, 5, 7, 5, 6, 2, 6, 7, 9, 6, 2, 2, 7, 2, 8]

ind = [1, 2, 5, 6, 2, 3, 4, 5, 2, 4, 1, 4, 6, 3, 3, 4, 5]

ptr = [1, 5, 9, 11, 14, 15, 18]

Fig. 1. CRS format

matrices. In the CRS format, a sparse matrix is stored into the data array by scanning the matrix in the row direction using two index arrays: an index array, which represents the column number of the non-zero elements in the data array, and a pointer array, which points to the first non-zero element of each row (see Figure 1).

Various storage schemes, which perform better than the CRS format, have been proposed for GPUs. For example, Bell and Garland [2] proposed a new storage format, HYB (Hybrid), which combines the existing ELL (Ellpack) and COO (Coordinate) formats and was implemented on GPUs. The HYB format outperforms the CRS format. Many other storage schemes have been implemented on GPUs so far [13] [7] [10]. On the other hand, an auto-tuning method is effective for SpMV since the optimal storage scheme or the optimal implementation technique is dependent upon the distribution of the non-zero elements of the matrix. For instance, Kubota and Takahashi [9] presented an auto-tuning method which selects the optimal storage format using a percentage and variability of non-zero elements.

However, an SpMV routine for the CRS format that can perform well for a wide variety of matrices is still necessary, especially in numerical libraries. It may be necessary to convert the storage format from CRS to other formats in some cases. On auto-tuning methods, it may be necessary to scan the matrix to determine the optimal storage format in advance. In fact, the SpMV routine for the CRS format is still provided in various numerical libraries such as the NVIDIA cuSPARSE library [11] for sparse matrix operations on GPUs for CUDA environments.

In this paper, we will implement a fast SpMV routine for the CRS format for Kepler architecture GPUs. Our implementation is based on an existing method proposed for the Fermi architecture, which is an earlier generation of the GPU, and takes advantage of some of the new features of the Kepler architecture. This paper will be organized as follows: Section 2 will describe related work. Section 3 will briefly introduce the Kepler architecture. Section 4 will describe our implementation. In Section 5, we will show the effects of the optimization techniques and compare the performance of our SpMV routine to the NVIDIA cuSPARSE library. Finally, we will conclude this paper in Section 6.

2   Related Work

Two methods to implement SpMV for the CRS format on GPUs have been presented by Bell and Garland [2]. The first one, the CRS-scalar method, performs the calculation of each row of a matrix (calculation of one element of the vector y in $y = A \times x$) using one thread per row. The second method, the CRS-vector method, assigns multiple threads to calculate a single row. The CRS-scalar method can easily implemented with minimum changes from the CPU code, but it may be not suitable for efficient memory access on GPUs. On GPUs, several memory access transactions are coalesced into a single transaction when consecutive threads access consecutive memory addresses. The CRS-vector method can offer more efficient memory access patterns than the CRS-scalar method. Bell and Garland allocated 32 threads for the calculation of one row.

In the CRS-vector method, if the number of non-zero elements per row is less than 32, reducing the number of calculation threads per row may improve the performance. Baskaran and Bordawekar [1] used 16, instead of 32, threads to compute a row with CRS-vector. Guo and Wang [8] proposed a method that switches the number of threads to either 16 or 32 based on the characteristics of the input matrix. El Zein and Rendell [6] switched between the CRS-scalar and CRS-vector methods based on the number of non-zero elements per row. Reguly and Giles [12] improved the performance of the CRS-vector method by selecting the optimal number of threads from among 1, 2, 4, 8, 16 and 32 in proportion to the average number of non-zero elements per row. Furthermore, Yoshizawa and Takahashi [14] selected the optimal number of threads from among 1, 2, 4, 8, 16 and 32 based on the maximum number of non-zero elements per row. The strategy of varying the number of threads from 1–32 based on the number of non-zero elements per row is effective. The average-based approach is preferred because the average number of non-zero elements per row can be calculated without pre-scanning the input matrix. Therefore, our implementation is based on Reguly and Giles's method.

The Kepler architecture was launched by NVIDIA in 2012. Davis and Chung [4] compared the performance of the Kepler and the Fermi, which is an earlier generation of GPU, using the GeForce series of GPUs. Their report shows that the Kepler is slower than the Fermi, but they used the same program on the both GPUs. Most existing reports have focused on an earlier generation of GPU. As far as we know, there is no research regarding the implementation and evaluation of optimization techniques for the Kepler architecture of GPUs. In this paper, we will target a Tesla K20 GPU which is based on the Kepler architecture.

3   Kepler Architecture GPUs

An overview of the Kepler architecture can be found in the White Paper [3] by NVIDIA. The most major change from the previous generation of Fermi architecture GPUs is that the streaming multiprocessor, called SM on the Fermi architecture, has been replaced with an updated version called SMX. The SM

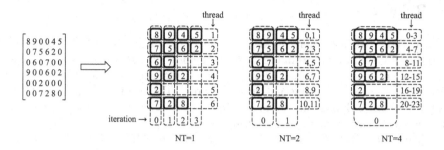

Fig. 2. Thread mapping for the cases of NT=1, 2 and 4 on the CRS-vector method

has 32 CUDA cores, but that number has been increased to 192 on the SMX. As a result, the maximum number of warps, threads, and thread blocks per multiprocessor have also increased. MaxGridDimX (the number of thread blocks in the x-direction that can be defined in a grid) has also increased from 65,535 to 2,147,483,647. In addition, the total number of registers has doubled to 65,536 and the total number of registers available to a thread has also increased from 63 to 255. Moreover, the execution efficiency of double precision operations has been improved from the Fermi architecture by improving the warp scheduler.

On the other hand, the Kepler architecture supports some new features. Among them, is a new 48KB read-only data cache and new shuffle instructions that we expect will improve the performance of SpMV. The 48KB read-only data cache can only be used to load data that does not change value during the kernel execution. This cache was accessible by using the texture unit on earlier generations of GPUs, but has seen major improvements on the Kepler architecture. The shuffle instructions are new instructions used to access a value different threads in the same warp. In this paper, we will utilize the 48KB read-only data cache, shuffle instructions and expansion of the MaxGridDimX to optimize SpMV on the Kepler architecture. We will explain the details of our three optimization techniques in the next section.

4   Implementation

This section describes our implementation. Our SpMV routine, implemented using double precision, computes $y = \alpha Ax + \beta y$, which is compatible with the SpMV routine of the cuSPARSE library. Our implementation is based on Reguly and Giles's method which is based on the CRS-vector method and selects the number of threads for the calculation of a single row (NT) from among NT = 1, 2, 4, 8, 16 and 32 in proportion to the average number of non-zero elements per row. The average number of non-zero elements per row is available in advance without pre-scanning the input matrix.

Figure 2 shows a conceptual diagram of the thread mapping for the cases of NT=1, 2 and 4 on the CRS-vector method. When NT=1, it is equivalent to the

```
int SpMV (char trans, int m, int n, int nnz, double alpha,
              double *a_val, int *a_ptr, int *a_idx, double *x,
              double beta, double *y) {
    int NT, ntx, nbx;
    float nnzrow = (float)nnz/(float)m;
    NT = max(1, min(32, (int)pow(2.,ceil(log2(nnzrow)))));
    ntx = 128;
    nbx = m / (ntx / NT) + ((m % (ntx / NT)) != 0);
    dim3 threads (ntx);
    dim3 grid (nbx);
    if (trans == 'N') {
        if (NT == 32) {
            cudaFuncSetCacheConfig(SpMV_kernel32, cudaFuncCachePreferL1);
            SpMV_kernel32 <<< grid, threads >>>
                          (m, alpha, a_val, a_ptr, a_idx, x, beta, y);
        } else if (NT == 16) {
            ....
        } else if (NT == 2) {
            ....
        } else {
            ....
        }
    }
}
```

Fig. 3. Host code of SpMV

CRS-scalar method. The CRS-vector method computes an inner product in the row direction using multiple threads. In Figure 2, "iteration" means a loop in the row direction. The NT can be up to 32 because thread synchronization is not required within a warp (=32 threads). Figure 3 shows the host code. In the host code, the NT is determined and the kernel codes for each NT are called.

Figure 4 shows the kernel code for the Fermi architecture. In this paper, we further optimized the implementation for the Kepler architecture by (1) using the 48KB read-only data cache, (2) avoiding the outermost loop, (3) using shuffle instructions. Figure 5 shows the kernel code optimized for the Kepler architecture. Note that, the second for-loop in the kernel codes (Figures 4 and 5) is unrolled since the number of iterations (=NT) is determined in advance of the kernel launch. We will explain the details of our three optimization techniques for the Kepler architecture in the following subsections.

4.1   48KB Read-Only Data Cache

The 48KB read-only data cache can be applied only to data that does not change value during the execution of a kernel. The cache can be accessed via a texture unit by mapping data in global memory to texture memory, which can also be

```
texture<int2, cudaTextureType1D, cudaReadModeElementType> tex_x;
static __inline__ __device__ double fetch_x (const int &i) {
    register int2  v = tex1Dfetch(tex_x, i);
    return __hiloint2double(v.y, v.x);
}


__global__ void SpMV_kernelNT_Fermi (int m, double alpha,
            double *a_val, int *a_ptr, int *a_idx,
            double *x, double beta, double *y) {
    int i;
    int tx = threadIdx.x;
    int tid = blockDim.x * blockIdx.x + tx;
    int rowid = tid / NT;
    int lane  = tid % NT;
    __shared__ double vals[128];
    while (rowid < m) {
        vals[tx] = 0.0;
        for (i = a_ptr[rowid] + lane; i < a_ptr[rowid + 1]; i += NT)
            vals[tx] += a_value[i] * fetch_x(a_index[i]);
        for (i = NT / 2; i > 0; i >>= 1)
            vals[tx] += vals[tx + i];
        if (lane == 0)
            y[rowid] = alpha * vals[tx] + beta * y[rowid];
        __syncthreads ();
        rowid += gridDim.x * blockDim.x / NT;
    }
}
```

Fig. 4. Kernel code of SpMV for the Fermi architecture (NT is one of 1, 2, 4, 8, 16 and 32. The second for-loop is unrolled)

done on the earlier Fermi generation architecture as shown in Figure 4, but before Kepler using the cache required complex programs and had many limitations. However, starting with the Kepler architecture, the cache can be accessed directly from the SM with general load operations. Reading through the read-only data cache is performed using an independent path of the L1 cache path. The read-only data cache is automatically managed by the CUDA compiler by using "const" and "__restrict__" qualifiers to direct the compiler to pass data in this read-only cache as arguments to kernel functions. We utilized the 48KB read-only data cache to read the vector x from the global memory in our implementation.

4.2   Avoid Outermost Loop

On the Kepler architecture, MaxGridDimX (the number of thread blocks that can be defined in the direction of dimension x in a grid) was extended from 65,535 to 2,147,483,647. As a result, we can avoid the outermost loop in the

```
__global__ void SpMV_kernelNT_Kepler (int m, double alpha,
              double *a_val, int *a_ptr, int *a_idx,
              const double * __restrict__ x, double beta, double *y) {
    int i, int val_hi, val_lo;
    int tx = threadIdx.x;
    int tid = blockDim.x * blockIdx.x + tx;
    int rowid = tid / NT;
    int lane  = tid % NT;
    double val;
    if (rowid < m) {
        val = 0.0;
        for (i = a_ptr[rowid] + lane; i < a_ptr[rowid + 1]; i += NT)
            val += a_val[i] * x[a_idx[i]];
        for (i = NT / 2; i > 0; i >>= 1) {
            val_hi = __double2hiint(val);
            val_lo = __double2loint(val);
            val += __hiloint2double(
                    __shfl_xor(val_hi, i, 32), __shfl_xor(val_lo, i, 32));
        }
        if (lane == 0)
            y[rowid] = alpha * val + beta * y[rowid];
    }
}
```

Fig. 5. Kernel code of SpMV for the Kepler architecture (NT is one of 1, 2, 4, 8, 16 and 32. The second "for" loop is unrolled)

CRS-vector method to calculate an index of a vector by using a thread ID and a thread block ID.

In the CRS-vector method, RowMax, the maximum dimension of a matrix that can be calculated, is obtained by setting RowMax = MaxGridDimX × BlockDim.x / NT. "BlockDim.x" means the number of threads for the x dimension in a thread block. In our implementation, the optimal size of the BlockDim.x is 128. RowMax becomes minimum when NT = 32. Thus on the Fermi architecture, RowMax = $65,535 \times 128/32 = 262,140$. In order to compute a vector longer than 262,140, it is necessary to recalculate the vector's address using a loop: recalculation of rowid and the outermost while-loop are required as shown in Figure 4 instead of the outermost if-statement in Figure 5. In addition, a thread synchronization instruction was required when using shared memory for reduction (we can avoid this thread synchronization by declaring the shared memory with the volatile suffix, but this method performs worse than using thread synchronization).

On the other hand on the Kepler architecture, RowMax has increased: RowMax = $2,147,483,647 \times 128/32 = 8,589,934,588$. The capacity of global memory on current GPUs is less than 10GB. RowMax is equivalent to a 32GB single-precision vector. Therefore, MaxGridDimX on the Kepler architecture is

Table 1. Evaluation Environment

CPU	Intel Xeon E3-1230 3.20GHz
RAM	16 GB (DDR3)
OS	CentOS 6.3 (kernel: 2.6.32-279.14.1.el6.x86_64)
GPU	Tesla K20 (5GB, GDDR5, ECC-enabled)
CUDA	CUDA 5.0 (Driver version: 304.54)
Compiler	gcc 4.4.6 (-O3), nvcc 5.0 (-O3 -arch sm_35)

sufficient to point to the index of any vector that can be loaded with current GPUs and the outermost loop is not required.

4.3   Shuffle Instruction

On the CRS-vector method, we compute a reduction at the second for loop shown in Figure 5. When NT ≥ 2 on the CRS-vector method, the NT threads perform the reduction within a single warp. On the earlier architectures, the operation must be performed using shared memory to exchange values among the threads in a warp. On the Kepler architecture, we can access a value on any other thread within the warp without shared memory by using shuffle instructions. There are 4 types of shuffle instructions that are supported starting with Kepler: indexed any-to-any (__shfl), shift right to N-th neighbour (__shfl_up), shift left to N-th neighbour (__shfl_down) and butterfly (XOR) exchange (__shfl_xor). Whereas shared memory requires separate load and store steps, shuffle instructions reduce this to a single step, and thus we expect that the shuffle instructions will outperform the equivalent shared memory instructions.

We used the butterfly exchange shuffle instruction for the reduction. Because the shuffle instructions support only a 32-bit value, moving 64-bit data requires two 32-bit movements. We converted a 64-bit double value into two 32-bit integer values and exchange the two 32-bit values with the shuffle instruction, and then reconverted the two 32-bit integer values to a single 64-bit double value.

5   Performance Evaluation

5.1   Evaluation Methods

We used an NVIDIA Tesla K20 Kepler architecture GPU. Our evaluation environment is shown in Table 1. The "-arch sm_35" compiler flag for nvcc is required in order to use the features of the Kepler architecture. We evaluated the GPU kernel execution time. The execution time does not include the time spent transferring data between the CPU and GPU over PCI-Express. To measure the performance accurately, we repeatedly executed a routine for at least one second at least 3 times, then computed the average execution time.

All input values other than the matrix A are composed of uniform random numbers. For our input sparse matrices, we randomly selected 200 matrices from

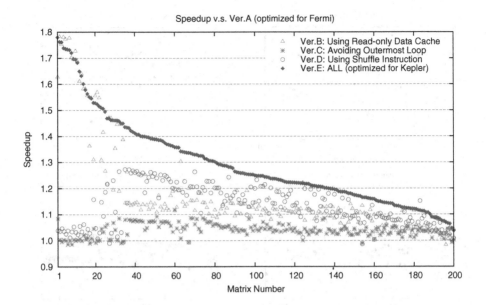

Fig. 6. Effects of the optimization techniques for the Kepler architecture

The University of Florida Sparse Matrix Collection[5]. The selected matrices are all real square matrices that have a different number of non-zero elements or a different number of rows. The number of rows varies between 1,813–5,558,326, and the number of non-zero elements varies between 4,257–117,406,044. We will show the details of the features of the matrices along with the performance evaluation results in the next section.

In order to investigate the effect of each optimization technique for the Kepler architecture used in this study, we implemented and evaluated the following five implementations:

- Ver. A: Optimized for the Fermi architecture (does not use Vers. B–D optimization techniques)
- Ver. B: Using 48KB read-only data cache
- Ver. C: Avoiding outermost loop
- Ver. D: Using shuffle instruction
- Ver. E: Optimized for the Kepler architecture (uses Vers. B–D optimization techniques)

Ver. A is optimized for the Fermi architecture and is the same implementation shown in Figure 4. For Vers. B–D, we applied each optimization one by one. All implementations except for Ver. B take advantage of the texture cache by mapping the data on global memory to texture memory for loading the vector x, like on the Fermi architecture. Ver. B uses the read-only data cache instead of the texture cache. Ver. E is the final version and is optimized for the Kepler architecture and is the same implementation shown in Figure 5.

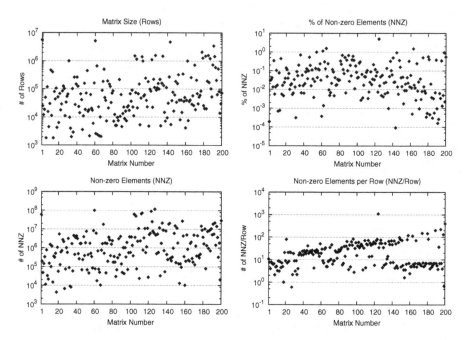

Fig. 7. Characteristics of the 200 matrices

5.2   Result

Figure 6 shows the relative performance of Vers. B–E to that of Ver. A. In the figure, data is sorted by the speedup of Ver. E from highest to lowest. The "Matrix Number" on the horizontal axis indicates which of the 200 matrices is being used and is numbered from the left to the right. The Matrix Number is common for all figures in this paper. For the 200 matrices, the average speedup of Ver. E is approximately 1.29. The minimum and maximum speedups are approximately 1.04 and 1.78, respectively. The worst case for any version was approximately 0.98. Figure 7 shows the details of the features of the 200 matrices: the number of rows (Rows), the number of non-zero elements (NNZ), the percentage of non-zero elements and the average number of non-zero elements per row (NNZ/Row).

Figure 8 shows the Flops performance of Ver. E and NVIDIA's sparse matrix numerical library cuSPARSE 5.0. The performance of our implementation for the 200 matrices is on average approximately 1.45 times faster than that of the cuSPARSE and it outperforms the cuSPARSE for 174 of the 200 matrices. The maximum speedup is approximately 7.60 times faster. On the other hand, the worst case was approximately 0.07 times of cuSPARSE.

5.3   Discussion

As we shown in Figure 6, we achieved performance improvement by using the three optimization techniques on most of the 200 matrices. However, strong

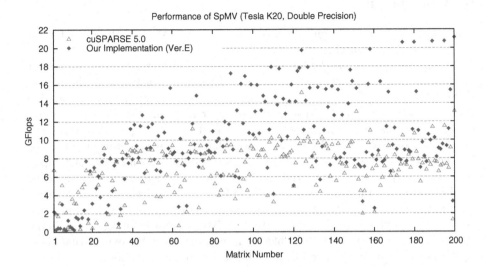

Fig. 8. Flops performance of our implementation and cuSPARSE 5.0

correlations between the effects of the optimization techniques shown in Figure 6 and the properties of the 200 matrices shown in Figure 7 are not indicated.

We infer there is a relation between the effect of the shuffle instruction and the NNZ/Row. Our implementation selects the number of threads for the calculation of a single row (NT) from among NT = 1–32 in proportion to the NNZ/Row. The speedup gained from using the shuffle instructions is relatively high when the NT is large and therefore the reduction in the number of iterations is large as well.

For matrices around Nos. 1–35, using the read-only data cache was quite effective. On these matrices, the Flops value of our implementation is relatively low and the number of non-zero elements per row (NNZ/Row) is relatively small. When the NT is large, memory access is well coalesced and the Flops value increases, but when the NT is small, the memory access efficiency is down. We infer in such cases that the performance depends strongly on the cache performance.

We note that the speedup of the Ver. B to the Ver. A shown in Figure 6 shows that on the Kepler architecture using the read-only cache improves the performance when compared to the implementation using the texture cache by accessing via a texture unit by mapping data in global memory to texture memory, like on the Fermi architecture. Thus, we can conclude that the read-only cache should be used instead of the texture cache on the other linear algebra operations on which the texture cache is effective.

6   Conclusion

This paper presented optimization techniques for SpMV for the CRS format on NVIDIA Kepler architecture GPUs using CUDA. Our implementation is based on the existing method proposed for the Fermi architecture, an earlier generation of GPUs, and takes advantage of three new features of the Kepler architecture: a 48KB read-only data cache, shuffle instructions and expanding the MaxGridDimX. On the Tesla K20 Kepler architecture GPU on double precision operations, our implementation optimized for the Kepler architecture is on average approximately 1.29 times faster than the implementation optimized for the Fermi architecture for the 200 matrices we used. As a result, our implementation outperforms the NVIDIA cuSPARSE library's implementation of the SpMV routine for the CRS format for 174 of the 200 matrices, and it is on average approximately 1.45 times faster than the SpMV routine. We conclude that the techniques shown in this paper are effective for implementing a fast SpMV routine for the CRS format on Kepler architecture GPUs. We expect that these methods will also be effective for other linear algebra operations.

Acknowledgment. This research was supported by JST, CREST.

References

1. Baskaran, M.M., Bordawekar, R.: Optimizing Sparse Matrix-Vector Multiplication on GPUs. IBM Research Report RC24704 (2009)
2. Bell, N., Garland, M.: Efficient Sparse Matrix-Vector Multiplication on CUDA. NVIDIA Technical Report NVR-2008-004 (2008)
3. NVIDIA Corporation: Whitepaper NVIDIAs Next Generation CUDA Compute Architecture: Kepler GK110. itepaper.pdf (2012), `http://www.nvidia.com/content/PDF/kepler/NVIDIA-Kepler-GK110-Architecture-Wh`
4. Davis, J.D., Chung, E.S.: SpMV: A Memory-Bound Application on the GPU Stuck Between a Rock and a Hard Place. Microsoft Technical Report MSR–TR–2012–95 (2012)
5. Davis, T., Hu, Y.: The University of Florida Sparse Matrix Collection, `http://www.cise.ufl.edu/research/sparse/matrices/`
6. El Zein, A.H., Rendell, A.P.: Generating Optimal CUDA Sparse Matrix Vector Product Implementations for Evolving GPU Hardware. Concurrency and Computation: Practice and Experience 24, 3–13 (2012)
7. Feng, X., Jin, H., Zheng, R., Hu, K., Zeng, J., Shao, Z.: Optimization of Sparse Matrix-Vector Multiplication with Variant CSR on GPUs. In: Proc. IEEE 17th International Conference on Parallel and Distributed Systems (ICPADS 2011), pp. 165–172 (2011)
8. Guo, P., Wang, L.: Auto-Tuning CUDA Parameters for Sparse Matrix-Vector Multiplication on GPUs. In: Proc. International Conference on Computational and Information Sciences (ICCIS 2010), pp. 1154–1157 (2010)
9. Kubota, Y., Takahashi, D.: Optimization of Sparse Matrix-Vector Multiplication by Auto Selecting Storage Schemes on GPU. In: Murgante, B., Gervasi, O., Iglesias, A., Taniar, D., Apduhan, B.O. (eds.) ICCSA 2011, Part II. LNCS, vol. 6783, pp. 547–561. Springer, Heidelberg (2011)

10. Matam, K., Kothapalli, K.: Accelerating Sparse Matrix Vector Multiplication in Iterative Methods Using GPU. In: Proc. International Conference on Parallel Processing (ICPP 2011), pp. 612–621 (2011)

11. NVIDIA Corporation: cuSPARSE Library (included in CUDA Toolkit), https://developer.nvidia.com/cusparse

12. Reguly, I., Giles, M.: Efficient sparse matrix-vector multiplication on cache-based GPUs. In: Proc. Innovative Parallel Computing: Foundations and Applications of GPU, Manycore, and Heterogeneous Systems (InPar 2012), pp. 1–12 (2012)

13. Xu, W., Zhang, H., Jiao, S., Wang, D., Song, F., Liu, Z.: Optimizing Sparse Matrix Vector Multiplication Using Cache Blocking Method on Fermi GPU. In: Proc. 13th ACIS International Conference on Software Engineering, Artificial Intelligence, Networking and Parallel/Distributed Computing (SNPD 2012), pp. 231–235 (2012)

14. Yoshizawa, H., Takahashi, D.: Automatic Tuning of Sparse Matrix-Vector Multiplication for CRS format on GPUs. In: Proc. 15th IEEE International Conference on Computational Science and Engineering (CSE 2012), pp. 130–136 (2012)

Parallel Two-Phase K-Means[*]

Cuong Duc Nguyen, Dung Tien Nguyen, and Van-Hau Pham

International University – VNU-HCM
{ndcuong,ntdung,pvhau}@hcmiu.edu.vn

Abstract. In this paper, a new parallel version of Two-Phase K-means, called Parallel Two-Phase K-means (Par2PK-means), is introduced to overcome limits of available parallel versions. Par2PK-means is developed and executed on the MapReduce framework. It is divided into two phases. In the first phase, Mappers independently work on data segments to create an intermediate data. In the second phase, the intermediate data collected from Mappers are clustered by the Reducer to create the final clustering result. Testing on large data sets, the newly proposed algorithm attained a good speedup ratio, closing to the linearly speed-up ratio, when comparing to the sequential version Two-Phase K-means.

Keywords: Data Clustering, K-means, Parallel Distributed Computing, MapReduce.

1   Introduction

The K-means algorithm is one of the most popular Data Mining algorithm. There are several parallel versions of the algorithm implemented on different programming frameworks, such as Parallel Virtual Machine (PVM), Message Passing Interface (MPI) or MapReduce. These parallel versions utilize the computing power of slave nodes to speed up the clustering process of K-means.

There are two drawbacks can be recognized in the current parallel versions of the K-means algorithm. Firstly, in some parallel algorithms [1-4], data are divided into equal subsets to compute on slave nodes. This assumes that the processing times on slave nodes are equivalent. In these algorithms, intermediate data are collected from slave nodes to update the global information by a Master node and then to broadcast this global data to slave nodes. Therefore, a synchronization is required the end of each iteration. Secondly, several parallel versions [1-5] require the data set is fully loaded, divided and sent to slave nodes. If the memory size of a slave node is smaller than the size of the data subset, the data will be swapped between memory and hard disk, so that this swapping will slow down the algorithm.

To overcome the limitations mentioned above, this paper introduces a new parallel version of K-means, called Par2PK-means, that is implemented on MapReduce. Par2PK-means reads the data set as a stream of data segments, independently

---

[*] The work is supported by DOST, Hochiminh City under the contract number 283/2012/HD-SKHCN.

B. Murgante et al. (Eds.): ICCSA 2013, Part V, LNCS 7975, pp. 224–231, 2013.
© Springer-Verlag Berlin Heidelberg 2013

processes each data segment on slave nodes to create intermediate data and finally processes intermediate data to create final clustering result. In Par2PK-means, each data segment is independently processed once on a Mapper so that the communication cost between Mappers and Reducer are reduced.

The remainder of the paper is organized as fellows. Section 2   aims at discussing about the related works. The new algorithm is described in Section 3. Evaluation of the new parallel version is   presented in Section 4. Section 5   concludes the paper

2   Related Works

2.1   The K-Means Algorithm

K-means (KM) was first   introduced by MacQueen in 1967 [6], and it has become one of the most widely used algorithms for Data Mining because of its efficiency and low complexity, $O(Knl)$, where n is the number of objects, K is the required number of clusters and l is the maximum number of iterations. However, K-means is often converge to a local optimum. To overcome this   drawback, the Incremental K-means algorithm (IKM) algorithm [7], an improved version of K-means, can empirically reach to the global optimum by stepping k from 1 to the required number of clusters. However, IKM has a higher complexity, $O(K^2nl)$.

The Two-Phase K-Means (2PKM) [8], is introduced to scale up K-means to process large data sets. The K-means algorithm requires several scans over data sets, so that, to speed up the data accessing, the data set has to be fully loaded to the computer memory. With large data sets having several TBs, this requirement is hard to fulfill. 2PKM is introduced to overcome this drawback. 2PKM has 2 phases. In Phase 1, 2PKM   loads and processes piece by piece of the dataset to produce the temporary cluster set which is stored for Phase 2. The K-means algorithm is used in Phase 1 due to its low complexity. In Phase 2, 2PKM clusters all the intermediate data to create final clustering result by IKM.

With the strategy of dividing the clustering process into two phases, 2PKM only requires one scan over the large data set and particularly this can be done by computers with limited memory. It can achieve approximate clustering result of the result that is created by K-means working the whole data set [8]. If 2PKM uses IKM in both phases, 2PKM can achieve approximate clustering result of the result that is created by IKM working the whole data set. Therefore, in this paper, IKM is used in both two phases of 2PKM.

2.2   Parallelizing the K-Means Algorithm

The parallel K-means of Kantabutra and Couch [5] is implemented in the master/slave model on the Message-Passing Interface (MPI) framework and executed on a network of workstations. This algorithm uses one slave to store all data objects of a cluster. It divides K subsets of the data set and sends each subset to a slave. In each iteration of K-means, a new center is re-calculated in each slave and then broadcasted to other slaves. After this center broadcasting, data are sent between slaves to make each subset on a slave only keeps data objects nearest to the center in that slave. This step of data re-arrangement requires a big data transmission between slaves and makes this strategy not suitable for big data sets.

The parallel K-means algorithm of Zhang *et al.* [1] is realized in the master/slave model based on the PVM framework and executed on a network of workstations. In the early state of the algorithm, the master reads the data and randomly initialize the cluster set. In each iteration, the master sends the cluster set to all slave nodes. The master divides the data set into S subsets (S can be larger than K) and consequently sends each subset to a slave node. A slave receives a subset of data, independently clusters this subset based on the cluster set and then sends its intermediate result back to the master node. The master node re-computes the position of the cluster centers based on the intermediate results received from slaves and then to starts a new iteration until the cluster set is stable. This parallel version of K-means requires the full load of the data set on the master node and a synchronization of data at the end of an iteration.

The parallel K-means algorithm of Tian *et al.* [2] same strategy as the one of Zhang et al. 1. It requires the data set fully loaded on the master node and divides the data set into m subset (m is the number of processors). The paper only estimate the complexity of the proposed algorithm. No practical implementation is executed to make any conclusion about the empirical performance of the proposed algorithm.

A parallel K-means algorithm, called ParaKMeans [3], is implemented in the multi-threading approach on a single computer to cluster biological genes. The parallel model in this algorithm is similar as the Tian's algorithm. The only difference is ParaKMeans uses Sufficient Statistics to measure the cluster's quality and in the stop condition.

A distributed K-means algorithm is introduced in [4]. It is designed to execute on multi-processor computers. Randomly split data subset is delivered to each processor before the algorithm starts. In the beginning, each processor randomly initialize the center of its K cluster centers. In each iteration, the cluster set of each processor is re-calculated based on its data subset, broadcast its cluster set to other processors and then re-calculated its cluster set again based on the received data. The process is repeated until the cluster set is stable. In general, the parallel strategy of this paper is similar to Zhang's algorithm but without using the master node, so that it has the same similar drawbacks.

With the introduction of MapReduce in 2004 [9], all following parallel versions of K-means in this section are implemented in this framework. The MapReduce framework uses a machine to play the role of the master node in the master/slave model. The Master node splits data into subsets, sends each subset to each Mapper (playing a role of a slave node), invokes the action of all Mappers. Each Mapper executes on its data subset to create intermediate data and then sends its intermediate data to a Reducer. Reducers collect intermediate data from Mappers to create final results. A node (Master, Mapper, Reducer) can be an independent computer.

Chu's work [10] proposed a framework that applied the parallel programming method of MapReduce on several Machine Learning algorithms, including K-means. In the parallel K-means algorithm in this framework, each Mapper works on different split data and then return its intermediate data (the sum of vectors in each data subgroup) to the Master node. After collecting all intermediate data from Mappers, the Master send received data to Reducer to compute the new centroids and return the final clustering result to Master. Comparing to the original K-means algorithm, the repeated converge process is not mentioned in the parallel version in this framework, so that the parallel version cannot reach to a good clustering result.

Another parallel K-means algorithm, called PKMeans, is introduced in [11] and implemented on MapReduce. This algorithm is also used in [12] for document clustering. Another type of nodes, called Combiner, is used in PKMeans. In each iteration, the Mapper, Combiner and Reducer are serially executed. A Mapper only assigns each sample to the closest center. A Combiner, that is executed on the same computer with the Mapper, partially sum the values of the data points assigned to the same cluster and then return an array of records. Each record stores the sum of values and the number of data points of a cluster. This array will be send to the Reducer to compute the new position of cluster centers, that is a global variable and can be accessed by Mappers. Several iteration is repeated until the cluster set is stable. Therefore, the model and the limits of this work is similar to Zhang's algorithm.

In conclusion, several parallel versions of K-means uses the data parallel strategy but the parallel strategies are different when using or not using framework MapReduce. When not using MapReduce, each slave node uses the initialized data from the master node,   processes its   sub data set and then synchronizes local data by sending the master node or broadcasting to other computing nodes before repeating the next iteration. This strategy has several drawbacks, such as the master node often has to load the full data set to deliver to computing nodes, a synchronization step is required in the end of each iteration, several scans over the data set is also required. When using MapReduce, each data object is processed by Mapper so that Mapper is called several times. In addition, when the algorithm requires several iterations, communication cost between nodes is much higher.

3   Parallel Two-Phase K-Means (Par2PK-Means)

The new proposed algorithm, called Parallel Two-Phase K-means (Par2PK-means), is developed based on the model of 2PK-means and implemented on Hadoop following the MapReduce model. The detailed algorithm of Par2PK-means is described in Figure 1. When there is an available Mapper, it reads a data segment and executes the IKM algorithm on that data segment to create an intermediate clustering result. Reducer retrieves the intermediate clustering result of all Mappers. When all Mappers finish their tasks, Master invoke Reducer to execute IKM on all received intermediate results from Mappers to create the final clustering result.

Parameter K_t decides the speed and the clustering quality of Par2K-means. Phase 1 of Par2PK-means plays a role of a compressor that reduces the number of data objects in a data buffer to K_t clusters. Parameter K_t decides the size of the intermediate data that enters Phase 2 so that controls the speed of the algorithm. The smaller K_t is, the faster Phase 1 and Phase 2 of Par2K-means are. However, K_t cannot be too small.

```
Parallel Two-Phase K-means:
Input:_
    K and K_t being the number of clusters and the number
    of clusters for Mappers
    L being the length of the data segment
Output: the cluster set
Algorithm:

Mapper: map (inputValue, inputKey, outputValue,
outputKey)
    1. Load a data segment from inputValue
    2. Execute IKM(K_t) on the loaded data segment
    3. Format outputValue as the clustering result
       (cluster centers with number of belonged number
       of data objects)
    4. Output (outputValue, outputKey)
    Note: outputKey has the same value for all Mappers

Reducer: reduce (inputValue, inputKey, outputValue,
outputKey)
    1. Load intermediate result from inputValue
    2. Execute IKM(K) on the received intermediate
       result
    3. Output the cluster set as the final result
```

Fig. 1. The Parallel Two-Phase K-means algorithm

Par2K-means uses the model of Two-Phase K-means, so that it only produces an approximate result of the sequential version of Two-Phase K-means. The larger K_t is, the higher the quality of clustering result is. Therefore, the selection of K_t is trade-off between the speed and the clustering quality. In Two-Phase K-means, K_t is selected correspondent to the size of a data buffer. With big data sets, K_t can be selected around one thousands to one percentage of the size of data buffer without much reduction in clustering result.

4   Evaluation

The proposed algorithm is evaluated its speed-up ratio when comparing to its serial version. Experiments are executed on a virtual computer cluster powered by Openstack [17] which is an open source cloud computing platform. It provides mechanisms to provision the virtual machines from resource pool. To provide the virtual machine, we need to have the hypervisor which allows to create a virtual machine out of a physical one. Openstack supports several hypervisor such as Xen[16], VMWare[14], KVM[15]. In our case, we use KVM. In order to create the cluster of virtual machines, we have created the machine image that has the Hadoop version 1.0.4 and Java 1.6.0_26 installed. Hadoop consists of two sub components: Hadoop MapReduce and Hadoop Distributed File System.

Data Distribution: HDFS is a distributed file system based on Master/Slave model, which prepares an environment for Hadoop work. In HDFS, the master node (also called NameNode) manages all events read/write activities of the system. It splits the input data into blocks whose size is specified by user (either 64MB or 128MB) and distributes these blocks to other slave nodes called DataNodes. The master node keeps the map of data blocks and DataNodes. For fault tolerance, user can specify the number of replications of a data block on the cluster. DataNode propagates the data blocks to the specified number of nodes.

Computing: The Hadoop MapReduce runs on top of HDFS. Hadoop MapReduce utilizes the capability of data awareness of HDFS to distribute the appropriate computing tasks to slave nodes. It also uses the Master/Slave architecture, which contains A JobTracker(master) and a number of TaskTracker nodes (slaves). JobTracker queries the locations of data on NameNode and deliver tasks to TaskTracker nodes. The results then are aggregated and reported to the user by the JobTracker.

As we show in the next paragraphs, we use several cluster size. We use the same virtual machine configuration though (Each virtual computer has a 2.8GHz CPU and 1GB of memory).

The first experiment uses data set CoverType from the UCI Repository [13]. This data set has 581,012 data objects with 52 attributes. In this experiment, only the first 4 attributes are used. The data set is enlarged by adding noise to the original data set. The enlarged data set has 2,324,048 data objects, 4 attributes, a size of 87.6MB. The number of slave nodes are 1, 2, 4 and 8. The number of data segments is selected as 100. Number of clusters K and K_t are selected as 10.

The speed-up ratios of Par2PK-means execution times on different number of slave nodes comparing to its execution time on one slave node are shown in Figure 2. From the figure, the speedup ratio of Par2PK-means approaches its limit (linear speedup). The lost percentage of Par2PK-means compared to the maximum ratio is around 10% due to the initialization of Hadoop.

In the second experiment, the CoverType data set is enlarged by adding noise to the original data set. The enlarged data set has 29,050,600 data objects, 4 attributes, a size of 1.23GB. The number of segments is also selected as 1000, but the number of clusters K and K_t are increased to 20 due to the larger data size. The evaluation results on the speedup ratio in this experiment is shown in Figure 3. With a larger data set and bigger number of clusters, the calculation of a Mapper is much longer, so that the percentage of time of initializing tasks and Phase 2 execution on the calculating time is reduced. This reduction means the speedup ratio of Par2K-means is closer to the linear situation (a perfect situation for parallelizing an algorithm).

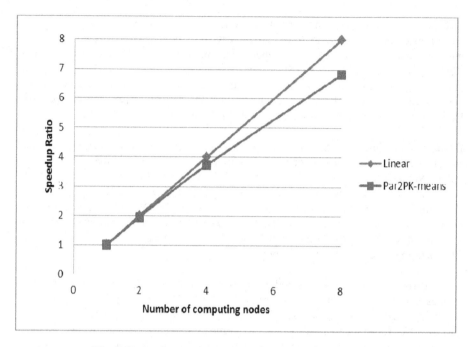

Fig. 2. Evaluation results on the original CoverType data set

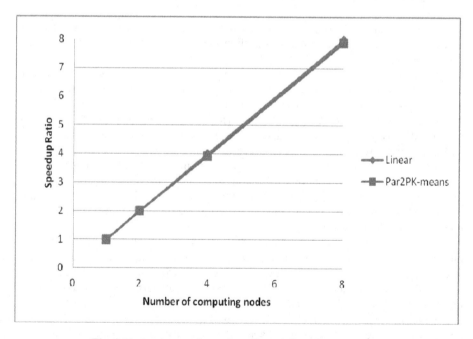

Fig. 3. Evaluation results on the enlarged CoverType data set

5   Conclusion

The paper introduces the parallel version of Two-Phase K-means. The proposed algorithm has achieved a good speedup ratio on tested data sets. However, its performance can be analyzed with other parallel strategies. Its incremental attribute (the algorithm can stopped on a number of data segments and give the best so-far clustering result) should be studied.

References

1. Zhang, Y., Xiong, Z., Mao, J., Ou, L.: The Study of Parallel K-Means Algorithm. In: Proceedings of the Sixth World Congress on Intelligent Control and Automation (WCICA 2006), vol. 2, pp. 5868–5871 (2006)
2. Tian, J., Zhu, L., Zhang, S., Liu, L.: Improvement and Parallelism of k-Means Clustering Algorithm. Tsinghua Science & Technology 10(3), 277–281 (2005)
3. Kraj, P., Sharma, A., Garge, N., Podolsky, R., McIndoe, R.A.: ParaKMeans: Implementation of a parallelized K-means algorithm suitable for general laboratory use. BMC Bioinformatics 9, 200 (2008)
4. Pakhira, M.K.: Clustering Large Databases in Distributed Environment. In: IEEE International Advance Computing Conference (IACC 2009), pp. 351–358 (2009)
5. Kantabutra, S., Couch, A.L.: Parallel K-means clustering algorithm on NOWs. NECTEC Technical Journal 1(6), 243–247 (2000)
6. MacQueen, J.B.: Some methods for classification and analysis of multivariate observations, Berkeley, California, vol. (1), pp. 281–297. University of California Press, Los Angeles (1967)
7. Pham, D.T., Dimov, S.S., Nguyen, C.D.: An Incremental K-means Algorithm. Proceedings of the Institution of Mechanical Engineers, Part C: Journal of Mechanical Engineering Science 218, 783–795 (2004)
8. Pham, D.T., Dimov, S.S., Nguyen, C.D.: A two-phase k-means algorithm for large datasets. Proceedings of the Institution of Mechanical Engineers, Part C: Journal of Mechanical Engineering Science 218(10), 1269–1273 (2004)
9. Dean, J., Ghemawat, S.: MapReduce: Simplified Data Processing on Large Clusters. In: OSDI 2004: Sixth Symposium on Operating System Design and Implementation, San Francisco, CA, pp. 137–150 (2004)
10. Chu, C.-T., Kim, S.K., Lin, Y.-A., Yu, Y., Bradski, G.R., Ng, A.Y., Olukotun, K.: Map-reduce for machine learning on multicore. In: NIPS, pp. 281–288 (2006)
11. Zhao, W., Ma, H., He, Q.: Parallel K-Means Clustering Based on MapReduce. In: Jaatun, M.G., Zhao, G., Rong, C. (eds.) Cloud Computing. LNCS, vol. 5931, pp. 674–679. Springer, Heidelberg (2009)
12. Zhou, P., Lei, J., Ye, W.: Large-Scale Data Sets Clustering Based on MapReduce and Hadoop. Journal of Computational Information Systems 7(16), 5956–5963 (2011)
13. Frank, A., Asuncion, A.: UCI Machine Learning Repository. University of California, School of Information and Computer Science, Irvine (2010), http://archive.ics.uci.edu/ml
14. VMware virtualization technology, http://www.vmware.com (accessed in May 2013)
15. Kernel based virtual machine, http://www.linux-kvm.org (accessed in May 2013)
16. Linux Foundation Collaborative Projects, http://www.xen.org/products/xenhyp.html (Last accessed in May 2013)
17. Openstack: Open source software for building private and public cloud, http://www.openstack.org/ (Last accessed in May 2013)

MobiPDA: A Systematic Approach to Mobile-Application Development

Khoi-Nguyen Tran and Hong-Quang Nguyen

International University, School of Computer Science and Engineering,
Quarter 6, Linh Trung Ward, Thu Duc District, Ho Chi Minh City, Vietnam
`gen.tran.1991@gmail.com, nhquang@hcmiu.edu.vn`

Abstract. The advent of mobile-computing platforms and profitable application-distributing channels has increasingly attracted a large number of non-professional people to mobile application development. They typically start with a simple idea about the applications of their interest; and to complete these applications, they often face two key problems: (*i*) the lack of a systematic method for *exploring ideas* and (*ii*) the lack of a method for *organizing activities* to keep balance between learning, designing and coding. Unfortunately, existing approaches have inadequately address these problems. In this paper, we propose a novel approach, named *MobiPDA*, to mobile application development targeting non-professional developers. Our systematic approach provides (*i*) conceptual tools for exploring an initial idea from different perspectives and (*ii*) an activities-organization process to apply these tools in software projects. We performed a comparative analysis with Mobile-D and Mobia, and evaluated MobiPDA on two pilot software projects.

1 Introduction

Non-professional developers are application developers who are not formally trained for software development. Prime examples of these developers include students, accountants and doctors who are motivated by their own passion to develop software products themselves for personal use or for sharing. They usually have good understanding, as a direct consumer, about the applications and features of their needs. Due to the explosive growth and low-entry barrier of the mobile application development, these non-professional developers have increasingly become a common force in the software market.

Many approaches to end-user development and mobile-application development have been proposed to enhance the work of non-professional developers on mobile platforms. Unfortunately, none of them directly addresses two major problems facing the non-professional developers. First, these developers tend to lack an effective method for *idea exploration*. They often start a software project with an initial idea. Since the project is of their interest and with no reference from any stakeholder (such as end-users, customers or investors), these developers need an effective way to explore their idea and generate requirements for their project. Second, the non-professional developers often lack a method for

B. Murgante et al. (Eds.): ICCSA 2013, Part V, LNCS 7975, pp. 232–247, 2013.

a balanced *activities organization* in a software project. Due to the absence of formal education on software development, they have to learn a lot of technical knowledge for developing the project. Thus, they require an effective way to organize various activities – including learning, designing and coding – in a balanced manner.

In this paper, we propose a new mobile-application development approach that aims to solve the above problems by answering two following questions:

(a) *How to effectively explore a product idea?*
(b) *How to organize activities in such a way that preserves balance between learning and other application-development activities?*

The remainder of this paper is structured as follows. Section 2 presents existing work on mobile application development for non-professional developers. An overview of MobiPDA is given in Section 3 while its two components – PKP and DORMI – are discussed in Sections 4 and 5. Section 6 presents a comparative analysis with current well-known approaches (such as Mobile-D [2] and Mobia [6]) and provides an evaluation on two pilot projects. Section 7 concludes with future work.

2   Related Work

In the past thirty years, along with explosive growth in information technology, we also witnessed the emergence of small software development firms and non-professional developers. The changing trend from "easy-to-use" to "easy-to-develop" in the field of Human-Computer Interaction – along with high-level programming languages, better Application Programming Interfaces (API) and abundance of online channels for learning technical skills – have increasingly created many opportunities for many non-professional people to participate in software development. Such a changing trend has resulted in a number of new software-development approaches, such as End-User Developments [9], Model-Driven Development [5,6]). The explosive growth of mobile computing platforms with computing power and functions that can rival traditional desktop computers [1] and profitable application distribution channels (also known as "appstores") had further increased such an opportunity of non-professional developers, especially students [15].

Non-professional developers share many similarities with end-user developers. They act more like professionals in the disciplines who create computer programs as pragmatic tools to solve work-related problems: they work with implicit requirements and specification, are overconfident about testing and verification, debug opportunistically, and have no plan about reuse [9]. Distinguished from the end-user developers, the non-professional developers, especially on mobile platforms, often have the motivation to share or sell their products instead of keeping for their personal use and often accept to learn new technical skills.

The stated traits of non-professional developers enable us to use some existing works on end-user development to further understand this type of developers.

Working process of non-professional developers also involves five main activities: requirements, design, reuse, testing-and-verification, and debugging [9]. The promotion of rigorous testing process and formal verification is often problematic [7]. Further, their development process can be supported by combining existing components [8] or Model-Driven development [5].

To create mobile applications from an initial idea, developers need a method for idea exploration and a process to activities organization in project. While the field of mobile computing is not as mature as other research fields in computer science, there are some works related to our stated issues. Mobile-D [2] is an Agile-based software-engineering process using test-driven development, pair programming, continuous integration and refactoring to help a team of developers to create quality mobile applications. In mobile end-user development, Mobia [6] is a Model-Driven Approach that allows developers to create mobile applications by specifying abstract designs. MobiMash [8] is also an End-User Development approach, but it is based on combining software components. Mobichart [3] provides visualization tools for mobile-application development. It extends the existing Objectchart with characteristics of mobile-computing environment such as migration, hoarding, cloning, and disconnected operations.

The range of mobile-application topics is broad. To provide an optimal solution, we need to focus on a specific subset of the topics, thus we need a taxonomy. In [11], author divided mobile application into two groups: highly goal-driven and entertainment-focus. The highly goal-driven applications are the ones that respond quickly to user's inquiries, whereas the entertainment-focus applications help users pass time. Six different categories of mobile application are proposed in [10], including standalone applications, office software, Internet applications, enterprise applications, location-aware applications and groupware applications. For mobile enterprise applications, five types including m-broadcast, m-information, m-transaction, m-operation and m-collaboration are presented in [16]. A combination of the stated taxonomy is given in [13].

3   MobiPDA: An Overview

In this paper, we propose a **P**erspective-**D**issolve **A**pproach for **Mobi**le application development (MobiPDA). It aims to guide a team of up to four non-professional developers to transform an initial idea into a complete application in four weeks. This can be achieved by investigating the product idea from different perspectives and keeping balance between technical learning and application development.

In MobiPDA, each software project is considered a problem-solving activity, whereby (i) a given topic is the problem, and (ii) the complete software product for that topic is the solution. Because of the variety of mobile-application topics, our proposed MobiPDA only focuses on one subset of the problems which we defined as a *"Process-related Problem"* (PrP).

PrP refers to application topics that center around a set of step-by-step activities to achieve a meaningful result. An example is task-management which consists of listing, estimating length, scheduling and tracking tasks. According to existing mobile-application topic taxonomies, PrP belongs to Goal-Driven group and encompasses Standalone category. PrP is the focus of MobiPDA because of several reasons. First, it can offer developers higher degree of freedom and creativity because is related to their daily-life activities. Second, PrP less technical demanding then other topics like enterprise or system utility applications. The last reason is that PrP is among the most common topics on mobile application markets.

MobiPDA can be used for exploring PrP without any restrictions, however for application development, certain level of technical expertise – "ability to create program code or the level of programming knowledge and experience" [6] – is required. Different levels of technical expertise are classified against three *personas* based on Goal, Skill and Attitude (Table 1).

Table 1. Personas of developers

	Goal	Skill
Novice	Create a software product that can be used for their needs, in their domains	No experience whatsoever in programming, has difficulty in expressing what he wants in a logical manner
Competent	Same as novice	Has knowledge about designing software applications and ability to write simple programs
Expert	Create a fully working software application to satisfy their needs or for others	Can design, develop, debug and deploy software products with any development technology. Might not be in the field, but know where to gather requirements

The competent level of technical expertise, which enables developers to learn new technology platforms and design techniques based on basic computer-science knowledge, is required for MobiPDA. This decision derived from the success of the research in [15] with competent technical expertise level on developing mobile-application projects in four weeks.

MobiPDA consists of two components: PKP, an idea exploration method, and DORMI, a software process that applies PKP. To illustrate our approach, throughout this paper, simplified examples are taken from our on-going project using MobiPDA called "task-management application" – a Windows Phone software to help learners manage their time, task assignments and some data related to their work.

4   PKP: Our Proposed Idea Exploration Method

Idea exploration consists of activities to investigate a given product idea for understanding and deriving requirement specifications. PKP is our proposed

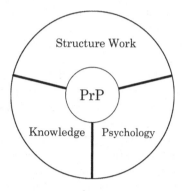

Fig. 1. Perspectives of Process-related Problems (PrP)

idea exploration method to investigate a PrP by performing research on its three perspectives: **P**rocess – **K**nowledge – **P**sychology. Process refers to a set of step-by-step activities defined by PrP to achieve a meaningful result. The set of listing, scheduling and tracking tasks activities is an example for process perspective of "task-management application" PrP. Knowledge refers to set of theories and experiences related to the process of PrP. Psychology perspective investigate internal problems of practitioners of the field that might degrade the performance of process. Together, these three perspectives form a complete view of the PrP that developers are trying to solve.

4.1   Process Perspective

Process covers the set of step-by-step activities of PrP. It is the most important perspective in defining a list of functionalities, or features, of the final software product.

The objective of a Process exploration is finding a list of activities related to the chosen PrP and put them into a relevant order of importance. The developers could adopt many different ways to perform this investigation, including:

(a) *Becoming members of a domain or potential users of the application.* For example, developers of a task-management application could try to take the role of manager in some small projects. This activity can force developers to put more effort in finding an optimal process to manage tasks and time.

(b) *Working with, observing, or interviewing the work of practitioners of the field.* For example, developers can interview a friend who is working as a manager. This approach is less demanding than the first one, but it can result in low-quality process if the practitioner is not reliable.

(c) *Reading reference books, journals and tutorials about working processes and best practices of the field.* A good article on task-management guidelines can provide developers a lot of information to identify an optimal task-management process. Developers have total control on the time and effort spending on this activity, but finding relevant document might be a challenge for some PrP.

(d) *Learning from products of a similar topic.* Influential software applications are results of continuous improvement based on users feedback, so to some degree, they represent a standard. These applications should be investigated to obtain an overview of activities related to chosen PrP, but the developers must avoid making their works copies of existing products.

To obtain the best results, the developers should try to combine the above methods in different ways, depending on their chosen PrP.

4.2   Knowledge Perspective

Around the process of PrP is related theories and experiences of the domain. Around task-management process are, for example, knowledge about factors of effective management and elements of a good task list. This type of knowledge helps developers obtain a better understanding about functionalities they are going to develop. In PKP, this type of knowledge is covered by *"Knowledge"* perspective.

The purpose of Knowledge perspective exploration is creating a knowledge base for chosen PrP. Several suggested ways to perform this investigation include:

(a) *Joining classes and workshops of the field,* like soft-skill classes on task-management when developing a task-management application. This approach usually results in good and structured knowledge.

(b) *Learning from practitioners of the discipline.* For example, developers can ask for experiences from a friend who can manage their work effectively to develop task-management application. This approach is not always accurate; hence, the developers should only use it as a preliminary study.

(c) *Learning from books and journals of the field or related to the chosen problem.* This method allows the developers to work at their own pace instead of having to follow a class schedule, but it can take more time, and for some PrP, it can be very difficult to find high-quality reading materials.

4.3   Psychology Perspective

The application of the process of PrP in real life scenarios is often effected by inner problems of the practitioners. For example, an optimal task-scheduling process might lose its effectiveness because of procrastination and irrational resistance of users. The *"Psychology"* perspective of PKP aims to explore these problems.

An investigation in the Psychology perspective results in a knowledge base. In PKP, we propose two methods for this exploration:

(a) *Reading books and journals.* The most difficult part of this approach is to find relevant documents. Basic guideline for this search is focusing on "Why" questions, for example, *why the procrastination occurs? Why is it so difficult for users to avoid procrastination?*

(b) *Consulting practitioners of the domain or potential users of the application.*
Practitioners of the domain, especially novice ones like university freshmen, usually experience many psychological issues. By observing, interviewing and analyzing their problems, the developers could identify solutions or at least the starting point for their research. However, the developers should be mindful about the accuracy of information being collected.

5   DORMI: Our Proposed Mobile-Application Development Process

The aim of our proposed PKP method is to mainly provide conceptual tools to perform *idea exploration*. To apply PKP in actual software projects, non-professional developers need a process to organize their development activities.

Define-**O**rganize-**R**esearch-**M**ap-**I**mplement (DORMI) is our proposed approach to organizing activities in a mobile-application project that preserve the balance between learning, designing and implementing. This process has two phases (Idea Exploration and Software Implementation) and five steps (Define, Organize, Research, Map and Implement). These phases and steps are built around theory of PKP, which will guide the non-professional developers in exploring their initial idea and turning it into a complete software product.

Phase 1, *Idea Exploration*, is composed of the first three steps: Define, Organize and Research. In this phase, the PKP method is applied to investigate the selected process-related problems, which can be done without any required technical expertise.

Phase 2, *Software Implementation*, is composed of the last two steps: Map and Implement. In this phase, the non-professional developers are required to learn technical skills, map software functionalities from papers to software-design models, and finally, implement these functionalities. At the end of each step, the developers will create a set of software artifacts or documents, including tables, lists and informal notes that can be archived and reused in other similar projects.

Working with DORMI mainly involves answering many questions through research. To make this activity simple, DORMI proposes a pair of working sessions: Brainstorm and Analysis Sessions. In brainstorm session, the developers will try to list as many ideas related to the working question as possible regardless the quality of ideas. In analysis session, the developers will attempt to group and link the brainstormed ideas to form a complete answer to the working questions. The brainstorm session is suggested to last around 25 minutes while the analysis session should only last up to 60 minutes.

DORMI can be used by a team of up to 4 members. However, Idea Exploration tasks should be performed individually. At the end of each task, a review session will be held as a chance for developers to discuss and combine individual results and incorporate new findings with their existing documents. The following subsections will present each step of DORMI in more detail.

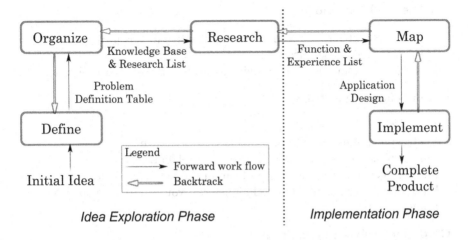

Fig. 2. DORMI process with backtrack steps

5.1   Step 1: Define the Problem

In this step, we aim to solve a PrP with PKP by defining the problem and evaluating its feasibility. This step takes the initial idea of the application as an input while its output is a *Problem Definition Table* presented in Table 2.

Table 2. Problem Definition Table

	Problem's Name
Given	- Field of the problem - List of acquired field knowledge - List of prepared technical skills and tools - Initial understanding about problem
Goal	- List of functions that should be provided - Traits and criteria that application should be able to satisfy
Guideline	- List of references for field knowledge - List of references for technical knowledge and training - Working schedule, if one is available

The main goal of Define step is to define the problem in a clear and concise manner in term of Given-Goal-Guidelines. *Given* describes known or given information at the beginning of problem solving. *Goal* gives information about the desired outcome of the problem solving. *Guidelines* describes what the developers need to do in order to reach the "Goal" from "Given". Result of this definition activity is better understanding and more accurate evaluation of developers about their chosen PrP. The focus of this stage is to complete the Problem Definition Table (Table 2). The following questions should be used as starting points for small brainstorm sessions to guide developers and prevent biases.

Goal-related Questions

– What is the problem that you are trying to solve in this project?
– Why is the project significant?
– What kind of solution will be provided to solve the selected problem?

Given-related Questions

– What is the field of problem?
– What expertise on the field of problem that you have?
– What experience can you use to solve this problem?
– How can technology solve such problem in the field?
– What kind of technology and techniques will you use to develop this application?

Guidelines-related Questions

– Where can you learn more about selected problem?
– What source of knowledge can be used to learn technical skills?

With above questions as starting points, developers can initialize some brainstorm sessions to find the answer. Through this process, developers will be able to identify weak points and feasibility of project and decide whether they should continue the development.

5.2   Step 2: Organize

The *"Organize"* step marks the beginning of applying PKP in the project. In this step, existing knowledge about PrP are organized into three perspectives according to PKP to create a "snap-shot" of current level of understanding about the chosen PrP, and identify weak points to do research on. Outcome of this step is stored in an important artifact called "Knowledge base and Research List" 3.

To fill in the table, the developers must extract all existing knowledge about PrP from their mind. Using several brainstorm and analysis session is one approach. Key questions for brainstorming are: *"What activities that people do in real life related to this PrP?"*, *"What knowledge and experience have we learned about this domain?"* and *"What have we learned about Psychology related to this PrP?"*.

During the process of extracting knowledge, some questions will appear. For example, when recording the fact that a learner can create a dictionary to store terms from courses, developers might discover that they do not know how to organize data in a dictionary. These questions indicate weak points in the knowledge base of the developers and are recorded in a Research List.

Developers can return to *Define* step if they find that their problem definition is incorrect or their project is not as feasible as they expected. When developers are satisfied with the knowledge extraction, they can move on to the next step.

Table 3. Initial Knowledge Base and Research List of a Management Application for learners

	Process	Knowledge	Psychology
Known	Time management involves creating schedule for repeated tasks and to-do lists for unexpected tasks - Study material management includes setting up a dictionary to record terms and a structure to organize books and slides	- Learners have to manage time, study materials and budget	- Most students are effected by procrastination
Required	- What activities involve in student budget management	- What types of information students usually record from materials	- Why procrastination happens? How to solve? - What psychological traits effects a learner's budgeting and spending?

5.3   Step 3: Research

The *"Research"* step of DORMI serves two purposes: enhancing knowledge base by answering questions in research list and creating a Functionality List (Table 4) and an Experience List (Table 5). These lists are the most important artifacts for creating the application.

Questions in the research list can be answered by examining all the three perspectives of PKP. Each completed question is removed from the list of questions and its answer is recorded into the corresponding perspectives in Knowledge Base. Long answers can be stored in separate documents and indexed into Knowledge Base.

With a better understanding about chosen PrP, the developers can begin to specify the solution with functionalities and experiences that a complete software product provides. Functionalities are tasks or actions that software product is intended to perform. A list of functionalities is usually drawn from the list of activities identified in the Process perspective. Functionalities are recorded using a natural language in the Function List (Table 4).

Experience concerns with how the application performs. Understanding about the required experience is drawn from the *Knowledge* and *Psychology* perspectives. For example, a user of task-management application needs his/her supporting tools to be easy to use, so the application must have simple user interfaces with at most 3 layers. Such knowledge is recorded with a natural language in the Experience List (Table 5).

Table 4. A simplified Function List of the management application for learner

Function Name	Description	Rationale
Create a To-do List	This functionality allows learners to add a task with name and deadlines, many system will automatically create alarms to remind the learners.	Besides repeated tasks like going to class, learners have many other tasks such as doing assignments. This functionality helps them to plan the time for these tasks.
Create a Term Dictionary	This functionality allows a learner to create dictionaries and record terms they learn in courses.	Learners face many new terms in a course, which can be confusing and hard to remember. Recording all these terms in a dictionary can be a great help for learners.

Table 5. A simplified Experience List of the management application for learner

Experience	Description	Rationale
Simple user interface	The hierarchy of interface i.e. number of steps that users have to follow before reaching required functionalities should not exceed 3.	System should help to ease the study of learners, not making it more complicated.

As shown in Figure 2, developers can return to Organize step to update the question list and re-organize the Knowledge Base, then continue working in the Research step. This common loop is called "Research Cycle", and the number of cycles is an important metric in Idea Exploration. The completion of the Research step marks the end of Phase 1 – Idea Exploration.

5.4   Step 4: Map

Phase 2, Software Implementation, starts with the *"Map"* step which enables non-professional developers to map the descriptions of the functionalities and experiences (written in a natural language) into an application design. The result of this step is recorded as various design models (or diagrams) that will be used for the later implementation.

Many different modeling methods could be adopted to design an application. Object-Oriented Modeling Methodology is a common approach to designing interactive applications. Unified Modeling Language (UML) is among the most popular modeling language. We leave the decision on which method to use on the developers.

5.5   Step 5: Implement

The *"Implement"* step is where all hard, technical work happens. This step takes as an input the application design created in the Map step, along with any required information from the results of the previous steps. The Develop step outputs a working application that is ready to solve the chosen PrP.

This step involves 3 main activities: (*i*) technical research and training, (*ii*) software coding and (*iii*) deployment.

As non-professional developers are not very well-versed in technical skills, learning and technical training are usually given most of project time which results in wasteful use of development effort. To balance between technical training and development tasks, we propose an arrangement with two main points:

(a) *Postponed technical training*, especially programming skills, until the Idea Exploration phase is accomplished, to ensure that proper amount of effort is put into understanding the application.

(b) *On-demand technical training*: technical skills that are only required by the development activities should be learned. When facing a new technology platform, the developers should only learn from introductory tutorials to obtain a basic understanding of such a new technology. Advanced functions of the platforms are learned only when needed for the implementation.

Software Implementation consists of main construction activities to create the final product. There are many ways to do coding and testing, so DORMI leaves the decision to developers. Test-driven development is highly recommended to enhance the overall quality of the final product [2,13,12].

Deployment is the most rewarding activity of a project, where the developers introduce their product to the world through distribution channels, such as Google Play Store [14] or Apple Appstore [4]. In this final step of the DORMI process, the developers should prepare an archive to record ratings and critical feedback from their customers. Such last artifact will be source of information to enhance the product and prepare for next projects.

6   Comparative Evaluation

To evaluate our approach, we performed two activities: (*i*) comparing MobiPDA with existing methods based on their documents and (*ii*) performing two evaluations on mobile-application development projects using MobiPDA.

6.1   Comparative Analysis

To identify strengths and limitations of our proposed MobiPDA against existing approaches in a software project of non-professional developers, we compare MobiPDA with an agile-based process called Mobile-D [2] and a model-driven approach called Mobia [6] on fourteen criteria as presented in Table 6.

The comparison results show that MobiPDA supports more types of application topics than Mobia, but it is still not as flexible as Mobile-D. It is not dependent on interaction with end-users like Mobile-D, as understanding about the application is gathered mainly through reading documents. In terms of simplicity, MobiPDA is better for the non-professional developers as it requires less

Table 6. Comparison between MobiPDA, Mobile-D and Mobia

Criteria	MobiPDA	Mobile-D [2]	Mobia [6]
1. Approach	Simplified software process with perspective dissolve theory	Agile software development for mobile	Model-driven development
2. Application domain	Process-related Problems	Any development problems for mobile application	Mobile health (mHealth)
3. Customer and end-users interaction	Not required	Important	End-users are developers in project
4. Number of phases	2	5	2 (from observation)
5. Number of steps	5	At least 18	2 (from observation)
6. Number of application specification steps	3	2 from first phase	Not specified in document
7. Number of development phases	2	At least 3 for each iteration	Not specified in document
8.Number of planning phases	0	At least 3	0
9. Proposed team size	1 to 4	Up to 10	Not specified, inferred to 1
10. Propose project time		10 weeks	Not specified.
11. Number of artifacts to manage	6 main artifacts with related documents	At least 4 sets of documents	Not specified, can be inferred to informal notes only
12. Development techniques	Up to developer's preferences	Mobile test-driven	No programming activities
13.Technical training guidelines	Supported	Not supported	Do not require technical expertise
14. Theory for idea exploration	Supported	Not supported	Not supported

steps, less artifacts, smaller teams and no management activities; however MobiPDA is outperformed by Mobia in this category. The main advantages of MobiPDA over other approaches are the freedom of developers to decide their own design and implementation methods, and the support in form of a detailed Idea Exploration theory and process that aims to preserve balance between learning and development.

In summary, each approach to mobile-application development has its own strengths and limitations, and is suitable for different scenarios. In case of small projects on topics that can be regarded as PrP, MobiPDA is more suitable for the non-professional developers as it offers an idea exploration method, freedom and balance between various development activities. However, larger projects on non-PrP problems will have high chance to succeed with Mobile-D, as it provides more team management and planning tools.

6.2   Pilot Projects

To test performance and acceptance of MobiPDA in real-life scenarios, we applied our proposed approach to an Android project and a Windows Phone 7.5 project. For the Android project, we use MobiPDA to develop an application that supports novel writing with management tools. As we develop this project directly, all of the project time is used in Idea Exploration and software development, so it represents idea condition to apply MobiPDA. For the Windows Phone project, a team of two computer-science students and two biotechnology students uses MobiPDA to create a learner-support application. Our role in this project is introducing MobiPDA to the non-professional developers. As a moderate amount of project time must be spent on learning to use MobiaPDA, this project is closer to real life scenarios.

At the time of writing this paper, both projects are in middle of the Implement step. The Android project has been running for six weeks, five hours a week. The project on Windows Phone has been running for four weeks, three hours a week for each team member. With the current schedule, each project are estimated to be completed in two more weeks.

Three key points have been drawn from a preliminary evaluation based on early feedback from those have participated in the experiments: (i) the activities of Idea Exploration provide useful results in real-life scenarios, (ii) finding PrP process, as well as the process concept, can be confusing for non-professional developers, and (iii) non-professional developers tend to skip reading materials to begin working directly with MobiPDA artifacts and create inaccurate results. This could be a result of resistance to working with processes that seem to be "technical".

7   Conclusion

Explosive growth and low entry-barrier of mobile-computing platforms has significantly increased the number of non-professional developers. They face two major problems: (i) the lack of an idea exploration method and (ii) the lack of a balanced activities-organization method. In this paper, we proposed a novel approach, *MobiPDA*, to mobile-application development. Our approach includes the PKP method for idea exploration that centers around dissolving a common type of topics called *"Process-related Problem"* (PrP) into different perspectives for analysis. The DORMI process provided by MobiPDA helps developers organize activities in project that preserve the balance between learning and development activities with a focus on designing application.

MobiPDA faces some limitations that can affect its operations in long term. It focuses on PrP. As mobile markets are a fast-changing environment, new demands may create a new kind of applications and reduce popularity of PrP. MobiPDA highlights the importance on learning from reference books and articles. This approach could be time-consuming and unattractive for some non-professional developers. Idea elicitation through brainstorm is also limited as some developers are not comfortable with the brainstorm methods. Further, MobiPDA concentrates only on the process of understanding the problem to create software products, while the success of a software project also depends on planning and management.

In the future, more pilot projects with different domain topics and team organization should be performed to gather more data, focusing on number of research cycles, technical training time and amount of time to learn MobiPDA. Using these results, a set of metrics for ranking difficulty levels of their problems and estimating the amount of required effort should be created. DORMI could be further enhanced with simple team-management tools and planning activities to support larger teams of non-professional developers.

References

1. Abowd, G.D., Iftode, L., Mitchell, H.: Guest editors' introduction: The smart phone–a first platform for pervasive computing. IEEE Pervasive Computing 4(2), 18–19 (2005)
2. Abrahamsson, P., Hanhineva, A., Hulkko, H., Ihme, T., Jäälinoja, J., Korkala, M., Koskela, J., Kyllönen, P., Salo, O.: Mobile-d: an agile approach for mobile application development. In: Companion to the 19th Annual ACM SIGPLAN Conference on Object-oriented Programming Systems, Languages, and Applications, OOPSLA 2004, pp. 174–175. ACM, New York (2004)
3. Acharya, S., Mohanty, H., Shyamasundar, R.K.: Mobicharts: A notation to specify mobile computing applications. In: Proceedings of the 36th Annual Hawaii International Conference on System Sciences (HICSS 2003) - Track 9, HICSS 2003, vol. 9, p. 298. IEEE Computer Society, Washington, DC (2003)
4. Apple Appstore, http://store.apple.com
5. Balagtas-Fernandez, F.T., Hussmann, H.: Model-driven development of mobile applications. In: Proceedings of the 2008 23rd IEEE/ACM International Conference on Automated Software Engineering, ASE 2008, pp. 509–512. IEEE Computer Society, Washington, DC (2008)
6. Balagtas-Fernandez, F., Tafelmayer, M., Hussmann, H.: Mobia modeler: easing the creation process of mobile applications for non-technical users. In: Proceedings of the 15th International Conference on Intelligent User Interfaces, IUI 2010, pp. 269–272. ACM, New York (2010)
7. Bareiss, R., Sedano, T.: Improving mobile application development. In: Proceedings of the FSE/SDP Workshop on Future of Software Engineering Research, FoSER 2010. ACM (2010)
8. Cappiello, C., Matera, M., Picozzi, M., Caio, A., Guevara, M.T.: Mobimash: end user development for mobile mashups. In: Proceedings of the 21st International Conference Companion on World Wide Web, WWW 2012 Companion, pp. 473–474. ACM, New York (2012)

9. Ko, A.J., Abraham, R., Beckwith, L., Blackwell, A., Burnett, M., Erwig, M., Scaffidi, C., Lawrance, J., Lieberman, H., Myers, B., Rosson, M.B., Rothermel, G., Shaw, M., Wiedenbeck, S.: The state of the art in end-user software engineering. ACM Comput. Surv. 43(3), 21:1–21:44 (2011)
10. Kunz, T., Black, J.P.: An architecture for adaptive mobile applications. In: Proceedings of Wireless 99, the 11th International Conference on Wireless Communications, pp. 27–38 (1999)
11. Oinas-Kukkonen, H., Kurkela, V.: Developing successful mobile applications. In: Proceedings of the International Conference on Computer Science and Technology, pp. 50–54 (2003)
12. Siniaalto, M., Abrahamsson, P.: A comparative case study on the impact of test-driven development on program design and test coverage. In: Proceedings of the First International Symposium on Empirical Software Engineering and Measurement, ESEM 2007, pp. 275–284. IEEE Computer Society, Washington, DC (2007)
13. Spataru, A.C.: Agile development methods for mobile applications. Master's thesis, University of Edinburge (2010)
14. Google Play Store, https://play.google.com/store (accessed on March 09, 2013)
15. Teng, C.-C., Helps, R.: Mobile application development: Essential new directions for it. In: Proceedings of the 2010 Seventh International Conference on Information Technology: New Generations, ITNG 2010, pp. 471–475. IEEE Computer Society, Washington, DC (2010)
16. Unhelkar, B., Murugesan, S.: The enterprise mobile applications development framework. IT Professional 12(3), 33–39 (2010)

On the Effectiveness of Using the GPU for Numerical Solution of Stochastic Collection Equation

Nikita O. Raba[1] and Elena N. Stankova[1,2]

[1] Saint-Petersburg State University
198504 St.-Petersburg, Russia, Petergof, Universitetsky pr., 35
[2] Saint-Petersburg Electrotechnical University "LETI",
197376, St.-Petersburg, Russia, ul.Professora Popova 5
no13@inbox.ru, lena@csa.ru

Abstract. Effective parallel algorithm is presented for numerical solution of stochastic collection equation on graphical processors. System of stochastic collection equations describes evolution of spectrum of drops and ice crystals due to the process of coalescence in the numerical models of natural convective clouds. They present the most computationally expensive part of such models aimed for forecasting dangerous weather phenomena such as thunderstorm, heavy rain and hail. Our results show that use of GPU can accelerate calculations in 15-20 times and thus provide operational forecast of such disaster events. Parallel algorithm is specially developed for calculations on GPU using CUDA technology. It is quadratic in time and allows using more than 1000 threads for effective parallelization. Special methods for optimization of memory access are also described.

Keywords: parallel algorithm, GPU, CUDA technology, stochastic collection equation, numerical modeling.

1   Introduction

Stochastic collection equation (SCE) is a complicated integro-differential equation describing the evolution of the spectrum (distribution function) of particles in dispersed medium through collision and subsequent coalescence. Analytical solution of such equation is possible only in case of a special type of kernel; in other cases it should be solved numerically. System of stochastic collection equations constitutes microphysical blocks of numerical models of natural convective clouds. They allow calculating evolution of the spectra of water droplets and ice particles through coagulation and thus describe the details of the process of precipitation that are very important for forecasting dangerous convective phenomena such as heavy rain, thunderstorm and hail.

Sequential algorithms that are used for SCE numerical solution, as a rule, have the computational complexity of the order of $O(N^3)$, where N is the total number of intervals of the partition range of cloud particle spectrum in size or mass. These algorithms are quite computationally expensive, even for calculating spectrum of one type of particles. And if you consider 7 types of them (the number which is usually

B. Murgante et al. (Eds.): ICCSA 2013, Part V, LNCS 7975, pp. 248–258, 2013.
© Springer-Verlag Berlin Heidelberg 2013

considered in the cloud models) and perform spectrum calculations at each time and space step, then computational costs are so substantial that the computations can last for several hours (in case of 2-D and 3-D models). This timing does not allow using the models in the operational practice for the prediction of dangerous convective phenomena such as thunderstorms, hail and heavy rain. The way out is to parallelize the most computationally expensive microphysical block, namely the algorithm for calculating the stochastic collection equation.

The parallel algorithm presented in the paper is developed on the base of GPU since it can be considered as a powerful SMP system placed in single, rather cheap computational device. Cheapness and availability are essential as the models have to be used in the regional meteorological centers and airports, where installation of expensive high-performance computational resources is not possible. CUDA technology was chosen because it is free and widely used platform, besides a lot of tutorials are available and C language can be used for programming.

Currently GPU have been actively used for solving the problems of fluid dynamics [1-3]. CUDA technology has been used in [1] for numerical integration of the three-dimensional, time-dependent, incompressible fluid dynamics equations in the Lagrangian representation. Special optimization procedure for GPU provides speed up to 57 times greater compared to the CPU version. Transferring the part of calculations to graphics accelerators (GPU) significantly improves performance of the OpenFOAM software package [3]. GPU has been used in [4] for the implementation of the parallel code developed to integrate the time-dependent Schrödinger equation that describes triatomic scattering reaction. The analysis of these and many other sources shows that the GPU is a powerful modern tool for accelerating the calculations and it can be effectively used to solve a wide range of scientific problems.

We use modified method of Kovetz and Olund [5] as a basic sequential algorithms for the numerical solution of the stochastic collection equation. The approach of Kovetz and Olund is known to be numerically very efficient. So we think that this is a proper choice for subsequent parallelization.

Our first tests show that parallel version of the algorithms in conjunction with GPU use can accelerate calculations in 5-10 times and thus is very perspective for being used in the cloud models for operational forecasting.

2   Numerical Calculation of Stochastic Collection Equation

The stochastic collection equation is given by Pruppacher and and Klett [6]:

$$\frac{\partial f(m)}{\partial t} = \frac{1}{2} \int_0^m K(m-m',m')f(m-m')f(m')dm'$$

$$-f(m)\int_0^\infty K(m,m')f(m')dm'. \tag{1}$$

where $f(m, t)$ is the drop mass distribution function at time t and $K(m, m')$ is the collection kernel describing the rate at which a drop of mass $(m-m')$ is collected by a drop of mass , m' thus forming a drop of mass m. The first integral on the right hand side of (1) describes the gain rate of drops of mass m by collision and coalescence of two smaller drops, while the second integral denotes the loss of drops with mass m due to collection by other drops.

The equation (1.1) can be rewritten in the following form [7]:

$$\frac{\partial f(m)}{\partial t} = \frac{1}{2} \int_0^\infty \int_0^\infty K(m',m'')\delta(m - m' - m'')f(m')f(m'')dm'dm''$$

$$-f(m)\int_0^\infty K(m,m')f(m')dm',$$

(2)

where $\delta(m - m' - m'')$ is the delta function. This expression is the continuous analogue of the formulas proposed by Kovetz and Olund [8].

Mass grid should be introduced for the numerical solution of (1) or (2). An equidistant grid m_i is chosen in case of modified Kovetz and Olund method, where m_i, $i = 0, 1,..., N$ are the discrete points along the m axis, $m_0 = 0$.

M_i is the mass of the droplets in the interval $[m_{i-1/2}, m_{i+1/2}]$. It is considered as a variable instead of distribution function $f(m)$

$$M_i = \int_{m_{i-1/2}}^{m_{i+1/2}} mf(m)dm \approx m_i f_i \Delta m_i,$$

$$f_i = f(m_i).$$

(3)

Then the equation for M_i can been written in the following form:

$$\frac{dM_i}{dt} = S_i^+ - S_i^-,$$

(4)

where S_i^- describes the loss of the particles from the interval i due to collection of the particles from the other intervals j.

$$S_i^- = \sum_{j=1}^{N-1} S_{ij}$$

(5)

$$S_i^+ = \sum_{j=1}^{i-1} \sum_{k=j+1}^{i} B_{jki} S_{jk}$$

(6)

where $S_{ij} = \dfrac{K(m_i, m_j)M_i M_j}{m_j}$. S_{jk} describes the appearance of new drops due to collection of the particles from interval j and interval k. Mass of the new particles may differ from the discretized mass points m_i. Instead it can be located between the mass intervals m_i and m_{i+1}. Thus the mass M_i has to be split up in grid boxes i and $i+1$. Redistribution is provided by means of coefficients B_{jki}.

$$B_{jki} = \begin{cases} (m_j + m_k - m_{i-1})/(m_i - m_{i-1}), if \ m_{i-1} \ < \ m_j \ + m_k \ < \ m_i, and \ i \ < \ N \\ (m_{i+1} - m_j - m_k)/(m_{i+1} - m_i), if \ m_i \ \leq \ m_j \ + m_k \ < \ m_{i+1}, and \ i \ < \ N-1 \\ 0 \quad in \ other \ cases \end{cases} \quad (7)$$

Computational complexity of Kovetz and Olund algorithm is of the order of $O(N^3)$, where N is the total number of grid points [9].

The algorithm has been optimized before parallelization in order to decrease its computational complexity. Optimization is based on the fact that the matrix B_{ijk} is very sparse. It is evident that the most part of its elements is equal zero and it does not make sense to summarize them (there are at most two nonzero elements of B_{jki} for any j and k).

Two 2-D arrays A_{jk} and I_{jk} have been introduced instead of 3-D array B_{jki}. A_{jk} presents matrix of weight coefficients and I_{jk} presents matrix of indices.

$$A_{jk} = (m_{i+1} - m_j - m_k)/(m_{i+1} - m_i) = B_{jki} \quad (8)$$

$$I_{jk} = i, if \ m_i \leq m_j + m_k < m_{i+1} \ and \ I_{jk} = 0 \ if \ there \ is \ no \ such \ i. \quad (9)$$

Note, that

$$B_{jki} = A_{jk}, if \ i = I_{jk};$$

$$B_{jki} = 1 - A_{jk}, if \ i = I_{jk} + 1, \quad (10)$$

$$B_{jki} = 0 \ in \ the \ other \ cases$$

Collision of the particles from the intervals j and k results in decrease of particle concentration in the intervals j and k and increase of particle concentration in the intervals i and $i+1$ if $m_i \leq m_j + m_k < m_{i+1}$ ($i = I_{jk}$). Newly formed particles are redistributed between the intervals i and $i + 1$ with the weight coefficients A_{jk} and $1-A_{jk}$. This process can be formulated in terms of time evolution of particle concentration in the following way:

$$\partial f_j/\partial t|_{jk} = -K_{jk} f_j f_k,$$

$$\partial f_k/\partial t|_{jk} = -K_{jk} f_j f_k, \quad (11)$$

$$\partial f_{i+1}/\partial t|_{jk} = K_{jk} f_j f_k (1 - A_{jk}),$$

$$\partial f_y/\partial t|_{jk} = 0, if \ y \ do \ not \ equal \ j, \ k, \ i \ and \ i + 1$$

where $\partial f_y/\partial t|_{jk}$ is variation in particle concentration in y interval due to collisions of the drops of the intervals j and k. The total mass of particles in a collision does not change.

Summarizing all collisions in all mass intervals we obtain total change of particle concentration in y interval due the process of collection (collision + coalescence):

$$\partial f_y/\partial t = \sum_{j=1}^{N-1} \sum_{k=j+1}^{N} \partial f_y/\partial t|_{jk}, \quad (12)$$

So sequential algorithm for numerical calculation of collection process for one type of particles can be described as follows:

1. Set initial values for S^- and S^+: $S^-_i := 0$, $S^+_i := 0$, for i from 1 to N.
2. Sort out all pairs of intervals (j, k) of colliding particles, $j < k$, and implement: $i := I_{jk}$, $F := K_{jk} f_j f_k$, $S^-_j := S^-_j + F$, $S^-_k := S^-_k + F$, $S^+_i := S^+_i + F A_{jk}$, $S^+_{i+1} := S^+_{i+1} + F (1 - A_{jk})$ for each pair on the condition that $I_{jk} \neq 0$.
3. Calculate $\partial f_i / \partial t := S^+_i - S^-_i$, for i from 1 до N.

Microphysical block of modern cloud models is usually not limited by description of one type of colliding particles. At least 7 types of hydrometeors should be taken into account: water drops, columnar ice crystals, plate ice crystals, ice dendrites, snowflakes, graupel and frozen drops. So system of seven equations of type (12) should be numerically solved describing variation of seven functions f_i due to collision of different types of hydrometeors with each other. This task is much more complex as collision may occur between the different types of particles, and the type of the resulting particles depends on both the type and mass of the colliding particles and the ambient temperature. The equation for the change in the concentration f_i of the particles of i-th type with mass m is as follows:

$$\partial f_i(m) / \partial t = \sum_{j=1}^{N_{PT}} \sum_{k=1}^{N_{PT}} C_{ijk}(T) \int_0^{m/2} K_{jk}(m - m', m') f_j(m - m') f_k(m') dm' - f_i(m) \sum_{j=1}^{N_{PT}} \int_0^\infty K_{ij}(m, m') f_j(m') dm', \tag{13}$$

where N_{PT} is the number of particle types, $C_{ijk}(T)$ is the coefficient, which is equal to 1, if a particle of i-th type is forming due to collision of particle of j-th type with the particle of k-th type at temperature T (mass of k-th type particle is less than mass of j-th type particle). $C_{ijk}(T)$ is equal to zero in the other cases. $K_{ij}(m, m')$ — is collection kernel (probability of collision and merging of i-th type particle with mass m and j -th type particle with mass m').

The following algorithm [9] have been developed for calculation of change of concentrations of all type of the particles due to the process of collection. (Note, that arrays S^- and S^+ have dimension $(N_{PT} \times N)$; f_{ni} is the concentration of particles of type n in the interval i; grid intervals are equal for all types of particles).

1. Set initial values for S^- and S^+: $S^-_{ni} := 0$, $S^+_{ni} := 0$, for n from 1 to N_{PT}, i from 1 to N.
2. Sort out all pairs of intervals (j, k) of colliding particles, and sort out types of colliding particles m (type of particles from j interval) and n (type of particles from k interval) $m \leq n$ for each pair when $I_{jk} \neq 0$. Define p — type of a particle, formed due collision for each four indices (m, n, j, k) if $j < k$, or $m \neq n$. Implement $i := I_{jk}$, $F := K_{mnjk} f_{mj} f_{nk}$, $S^-_{mj} := S^-_{mj} + F$, $S^-_{nk} := S^-_{nk} + F$, $S^+_{pi} := S^+_{pi} + F A_{jk}$, $S^+_{p\,i+1} := S^+_{p\,i+1} + F (1 - A_{jk})$.
3. Calculate $\partial f_{ni} / \partial t := S^+_{ni} - S^-_{ni}$, for n from 1 до N_{PT}, i from 1 до N.

Such an algorithm is universal: it can be applied to models of different dimensions and with different types of colliding particles (N_{PT}).

The developed algorithm has computational complexity of the order of $O(N^2)$ if we consider one type of colliding particles and of order of $O(N^2_{PT} N^2)$ if we consider several types of colliding particles.

3   Parallelization Method for Kovetz and Olund Algorithm

Modern graphics processors (GPU) have much more processing cores (hundreds) compared to CPU (2-8). To take full advantage of GPU, you need special "substantially parallel" algorithms (their implementation should be broken up into thousands of threads that perform the same operations on different data). There are additional restrictions: operations must be as simple as possible, number of possible branching should be minimal as well as thread synchronization and address to the global memory.

Currently, there are several techniques to use GPUs for general-purpose computation (GPGPU - General-Purpose computing on Graphics Processing Units): CUDA, ATI Stream, OpenCL, DirectCompute. One of the most popular technology is CUDA presented by NVidia company. It allows you to make full use of the GPU, and it is based on the language C / C $^{++}$ with some extensions and restrictions, making it easier to write programs. Program for CUDA uses both CPU and GPU. CPU executes sequential calculations, and GPU executes massive parallel calculations. These parallel calculations are presented in the form of concurrent threads.

The following method of calculation is presented. Each thread processes the collision of particles of type m from mass interval j with particles of different types ($n \geq m$) from the mass interval k, i.e. one thread handles with N_{PT} collisions maximum (or less, for example, when $k \leq j$). The values of j, m and k depend on the indices of thread ($threadIdx$) and block ($blockIdx$) (in CUDA threads are combined in blocks, the blocks are combined in a grid, threads and blocks have their multidimensional indices, threads within a block may be synchronized and have access to a fast shared memory).

Such type of parallelization comes across several problems.

Multiple threads may try to change the values in the same memory location (e.g,, in the same cell of the array S^+), and it is not possible to predict what particular value appears there as a result.

The following algorithm has been developed in order to prevent such type of record collisions. Four memory cells have been allocated separately for each pair of intervals (j, k) of the colliding particles of types m and n . The first memory cell is responsible for reduction of the concentration of particles of type m in j interval. The second cell is responsible for reduction of the concentration of particles of type n in k interval. The third and the forth cells are responsible for the increase of the concentration of particles of type p in the intervals i and $i + 1$ respectively, provided that $i = I_{jk}$ This procedure is realized by introducing the array MP(instead of arrays S^- and S^+.

Four matching indices for the array MP are stored for every four values (m, n, j, k): $ind0mnjk, ind1mnjk, ind2mnjk, ind3mnjk$. The indices are calculated in such a way that the memory cells related to the increase (or decrease) in the concentration of particles of the same type should be next to each other (i.e. form a coherent field in the array MP). Moreover, the memory cells related to an increase (or decrease) in the concentration of particles of the same type in the same mass range, too, must be close to each other. Some cells may be left unfilled, they may not have an appropriate index *ind*. Indices of all these successive areas: *pos0ni* - index of the last element in the array MP, related to a decrease in the concentration of particles of type *n* in the interval *i*, *pos1ni* - the index of the last element in the array MP, relating to the increase in the concentration of particles of type *n* in the interval *i*. The method of calculating index *ind* , *pos* and dimension of the array MP is presented in [10].

Calculation of the concentration variation is performed in 4 stages.

At the first stage, each thread calculates $F = K_{mnjk} f_{mj} f_{nk}$ for each corresponding collision and fills the four cells of the array *MP*: *MP [ind0mnjk] =-F, MP [ind1mnjk] =-F, MP [ind2mnjk] = F Ajk, MP [ind3mnjk] = F (1 - Ajk)*. The array MP is completely filled after the calculation of all collision. Initially array *MP* is filled with zeros, so those cells of MP for which index *ind* is not defined stay equal zero.

Summation of the separate array cells is provided at the second and the third stages of calculation. Note, that cells of the array under summation should belong to the same grid intervals.

The performance of such algorithms is limited mainly to the bandwidth of memory bus. Since bandwidth of memory bus of modern video cards is ten times greater than that of the computer's memory, the above algorithm can substantially (see below) speed up the calculations.

The pseudo-code of the algorithm is presented below:

```
//initialization of the number of hits and exceptions for the certain mass interval
for all iPT from 1 to NPT
    for all i from 0 to N − 1
        collisionM[iPT][i] = 0
        collisionP[iPT][i] = 0
//calculation of the number of hits and exceptions
/preliminary calculation of indM1, indM2, indP1, indP2 (just only local indices)
    for all i from 1 to N − 2
        for all j from 1 to N − 2
            k = I[i, j]
            if k ≠ 0, then
                for all iPT1 from 1 to NPT
                    for all iPT2 from iPT1 to NPT
                        newType = GetNewType(iPT1, iPT2, i, j)
                        if (j > i) or (iPT1 ≠ iPT2), then
                            indM1[iPT1, iPT2, i, j] = collisionM[iPT1][i]
                            indM2[iPT1, iPT2, i, j] = collisionM[iPT2][j]
                            indP1[iPT1, iPT2, i, j] = collisionP[newType][k]
                            indP2[iPT1, iPT2, i, j] = collisionP[newType][k+1]
                            collisionM[iPT1][i] = collisionM[iPT1][i] + 1
                            collisionM[iPT2][j] = collisionM[iPT2][j] + 1
                            collisionP[newType][k] = collisionP[newType][k] + 1
                            collisionP[newType][k+1] = collisionP[newType][k+1] + 1
//calculation of NM, NP, posM, posP, collisionSumM, collisionSumP
    sumM = 0
    sumP = 0
```

```
for all i_PT from 1 to N_PT
    for i from 0 to N − 1
        collisionSum_M[i_PT][i] = sum_M
        collisionSum_P[i_PT][i] = sum_P
        sum_M = sum_M + collision_M[i_PT][i]
        sum_P = sum_P + collision_P[i_PT][i]
        pos_M[i_PT][i] = sum_M − 1
        pos_P[i_PT][i] = sum_P − 1
//final calculation of ind_M1, ind_M2, ind_P1, ind_P2 (now they are global indices)
    for all i from 1 to N − 2
        for all j from 1 to N − 2
        k = I[i, j]
        if k ≠ 0, then
            for all i_PT1 from 1 to N_PT
                for all i_PT2 from i_PT1 to N_PT
                newType = GetNewType(i_PT1, i_PT2, i, j)
                if (j > i) or (i_PT1 ≠ i_PT2), then
                    ind_M1[i_PT1, i_PT2, i, j] = ind_M1[i_PT1, i_PT2, i, j] + collisionSum_M[i_PT1][i]
                    ind_M2[i_PT1, i_PT2, i, j] = ind_M2[i_PT1, i_PT2, i, j] + collisionSum_M[i_PT2][j]
                    ind_P1[i_PT1,  i_PT2,  i,  j]  =  ind_P1[i_PT1,  i_PT2,  i,  j]  +
collisionSum_P[newType][k]
                    ind_P2[i_PT1,  i_PT2,  i,  j]  =  ind_P2[i_PT1,  i_PT2,  i,  j]  +
collisionSum_P[newType][k+1]
```

4   Test Results

Test results have been performed using different processors and video cards. Their characteristics are as follows:

1. Intel Core 2 Duo E6400 - 2 cores, 2.13 GHz
2. Intel Core 2 Quad Q8200 - 4 cores, 2.33 GHz
3. Intel Core i3 2310M - 2 cores, 2.1 GHz Hyper-Threading technology
4. Intel Core i5 2400 - 4 cores, 3.1 GHz

Characteristics of video cards are as follows:

1. NVidia GeForce GTX460 - 336 cores (7 multiprocessors, each with 48 cores), GPU frequency - 700 MHz, video memory frequency - 3600 MHz, video memory bus width - 256 bit (video memory bus bandwidth - 115 GB / s);
2. NVidia GeForce GTX470 - 448 cores (14 multiprocessors, each with 32 cores), GPU frequency - 630 MHz, video memory frequency - 3348 MHz, video memory bus width - 320-bit (video memory bus bandwidth - 134 GB / s).

We compared the time need for calculating the process of collection with the help of CPU (E6400) and the GPU (GTX460 and GTX470) and assessed the benefits of using the latter. The results are presented in table 1,2 and fig.1.

Table 1. Calculation time (seconds) obtained with the help of the CPU (E6400) and the GPU (GTX460 and GTX470). N is the number of grid points

N	64	128	256	512
E6400	3.343	13.718	55.484	225.015
GTX460	0.766	2.484	8.578	32.375
GTX470	0.593	1.813	6.235	22.203

Data presented in table 1 show that parallelization with the help of GPU essentially decrease time of calculation. The more grid points we consider, the more evident is the effect of parallelization. Video card with the more number of cores (NVidia GeForce GTX470) gives slightly better results. It is interesting to note that calculation time is proportional to the number of GPU cores. The number of cores in GTX460 and GTX470 differs by about 25%. Calculation time differs also by about 25%.

Table 2. Speed up (S)obtained by using CPU (E6400) and the GPU (GTX460 and GTX470). N – is the number of grid points

N	64	128	256	512
E6400	1.00	1.00	1.00	1.00
GTX460	4.36	5.52	6.47	6.95
GTX470	5.64	7.57	8.90	10.13

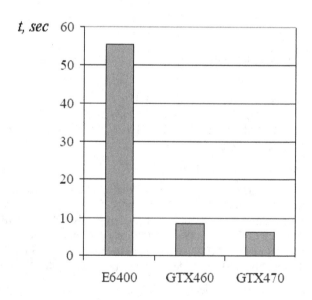

Fig. 1. Calculation time (seconds) obtained with the help of the CPU (E6400) and the GPU (GTX460 and GTX470). N = 256

Data presented in table 2 and Fig.1 show that use of GPU allow achieving a substantial acceleration of computations (up to 10 times).

Preliminary test results show also that the speed up could be more increased by using special procedure of parallel reduction [11].

In this algorithm, each block of threads is responsible for the summation of the corresponding element of the array. The elements of the array, which corresponds to some block, are split into pairs, parallel summation of each pair is implemented in a separate thread. Then the results of such a summation are again split into pairs, and parallel summation of these pairs is also conducted. A similar process is repeated until a single element is left, which will present the total sum of the initial elements. Since the number of items is reduced by half at each step, then this algorithm requires log_2M steps to sum M elements.

This algorithm can be applied only to the areas, the size of which is equal to a power of two. The size of the areas must also be equal (because all the blocks of threads that summarize these areas have the same dimension). Therefore it is impossible to apply this algorithm to sum the areas of variable sizes, the case we have with our parallel algorithm. Some efforts have been provided to adopt it for using procedure of parallel reduction. Dimension of the array MP is chosen equal to special RBS (reduction block size).

RBS is chosen to be a power of 2. Result of the summation is stored in the MP array in the cells with indexes multiple to RBS. In each grid interval only those elements are summing which indices are multiples of RBS, each thread performs summation in its own grid cell.

Results of using procedure of parallel reduction are presented in Fig.2 and are quite impressive.

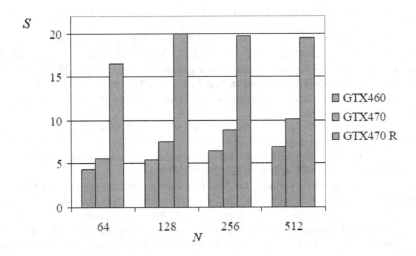

Fig. 2. Speed up achieved on GPU versus CPU («R» shows the results of the application of the procedure of parallel reduction)

As we can see from the figure using parallel reduction allow to increase values of speed up to 15 – 20 depending upon the number of grid intervals.

5   Conclusions

The multi-threaded parallel algorithm is developed for numerical solution of stochastic collection equation on graphical processors using CUDA technology. The way of splitting of computations on several stages is discussed; method of redistribution of calculations on different threads is presented and analyzed. We describe also some methods for preventing record collisions when multiple threads try to change the values in the same memory cell.

Results of the numerical tests show that the parallel algorithm developed for GPU allows reducing computation time in 5 – 10 times and even in 15-20 times when using special procedure of parallel reduction.

The developed algorithm is universal: it can be applied to models of different dimensions and with different types of colliding particles. So it is a prompt instrument for being used in the hardware and software systems intended for operational forecast of dangerous weather phenomena.

References

1. Boreskov AV Basics of CUDA technology / Boreskov AV, AA Kharlamov, 232 c. DMK Press, Moscow (2010) (in Russian) ISBN 978-5-94074-587-5
2. Sanders, D.: CUDA technology in the examples: An Introduction to GPU / D. Sanders, Kendrot A.- M., 231 p. DMK Press (2011) ISBN 978-5-94074-504-4
3. NVidia CUDA C Programming Guide / NVidia Corporation. Version 4.0, 2011.05.06, Santa Clara, cop., 187 p. (2011),
   http://developer.download.nvidia.com/compute/DevZone/docs/html/C/doc/CUDA_C_Programming_Guide.pdf
4. The use of video cards for calculations / Alex Tutubalin, http://www.gpgpu.ru
5. Stankova, E.N., Zatevakhin, M.A.: The modified Kovetz and Olund method for the numerical solution of stochastic coalescence equation. In: Proceedings 12th International Conference on Clouds and Precipitation, Zurich, August 19-23, pp. 921–923 (1996)
6. Pruppacher, H.R., Klett, J.D.: Microphysics of Clouds and Precipitation, p. 954. Kluwer Academic (1997)
7. Voloshchuk, V.M.: The Kinetic Theory of Coagulation. Leningrad, Gidrometeoizdat, 284 p. (1984)
8. Kovetz, A., Olund, B.: The effect of coalescence and condensation on rain formation in a cloud of finite vertical extent. J. Atm. Sci. 26, 1060–1065 (1969)
9. Raba, N.: Optimization algorithms for the calculation of physical processes in the cloud model with detailed microphysics. In: Applied Mathematics, Informatics, Control Processes. West. SPSU. Ser. 10, vol. 3, pp. 121–126 (2010) (in Russian)
10. Raba, N.: Development and implementation of the algorithm for calculating the coagulation in a cloud model with mixed phase using CUDA technology. In: Applied Mathematics, Informatics, Control Processes. Bulletin of St. Petersburg State University, Series 10, vol. 4, pp. 94–104 (2011) (in Russian)
11. Boreskov AV Basics of CUDA technology/ Boreskov Alexander Kharlamov, 232 c. DMK Press, AA - M (2010) (in Russian) ISBN 978-5-94074-587-5

Graph-Based Interference Avoidance Scheme for Radio Resource Assignment in Two-Tier Networks

Hsiu-Lang Wang and Shang-Juh Kao

Department of Computer Science and Engineering
National Chung Hsing University
250 Kuo Kuang Rd., 40227 Taichung, Taiwan
gary6319@gmail.com, sjkao@cs.nchu.edu.tw
http://www.cs.nchu.edu.tw

Abstract. In two-tier Orthogonal Frequency Division Multiple Access (OFDMA) networks, femtocells are commonly deployed to offload the traffic from the macrocell, but the dense deployment may result in severe interference. This paper presents an interference avoidance scheme by detecting the inter-interference relationship among femtocells graphically to avoid the co-tier interference. And, the cross-tier interference can also be reduced through the coordination of radio assignment. The traffic load among femtocells is taken account for the radio resource assignment so the average resource blocking rate can be decremented. The simulation results show the average data transfer rate of the proposed scheme can be increased in the range of 14% ~ 88% under low and high system capacity. The average blocking rate is decreased up to 41% and 70%, under a low system capacity, as compared with conventional graph coloring and graph-based dynamic frequency reuse schemes.

Keywords: two-tier OFDMA networks, graphic-based interference avoidance, radio resource assignment.

1 Introduction

Orthogonal Frequency Division Multiple Access (OFDMA) technology has the characteristics of a high data transfer rate, adjustable bandwidth requirement for each user and the ability to reduce the attenuation of multipath channel, and has been adopted by IEEE 802.16e [1] and Long Term Evolution (LTE) [3] as a downlink solution. In OFDMA networks, the frequency and time resources are allocated to users in an orthogonal manner, which improves the distribution of spectrum resources effectively.

Femtocell technology offers plug-and-play for configuration, low deployment cost, and provides traffic offloading from the macrocell. A femtocell consists of a small cellular base station, operates on a licensed spectrum the same spectrum with microcell and connects to a service provider's broadband system to offer wireless access service [15]. A sample deployment of a two-tier LTE network consisting of several femtocell base stations (FBSs) and a macrocell base station (MBS).

B. Murgante et al. (Eds.): ICCSA 2013, Part V, LNCS 7975, pp. 259–269, 2013.

FBS and MBS provide access service for femto user equipment (fUE) and macro user equipment (mUE), respectively. Femto gateway (FGW) is an intermediate node that controls and manages FBSs and performs traffic routing in the core network. In addition, It also supports femto-specific functionalities such as admission control, handover control and interference management. Since femtocells can be installed in an indoor environment by consumers, it can improve the macrocell's capacity and coverage in a simple and economical manner [2, 4].

Due to the radio resource is limited, macrocells and femtocells have to share the frequency spectrum. The sharing could lead to signal interference especially in a dense deployment of femtocells. Management of the interference problem can be either dedicated channel-based or co-channel-based deployment. As shown in Figure 1(a), a dedicated channel deployment reserves a given frequency for femtocells sharing, thereby is free from cross-tier interference. However, this approach is inefficiency because a reserved frequency is wasted in the case of less fUE connections. Oppositely, co-channel deployment, as shown in Figure 1(b), has the better spectrum utilization since the whole bandwidth is shared among macrocells and femtocells. Unfortunately, this strategy may result in significant interference.

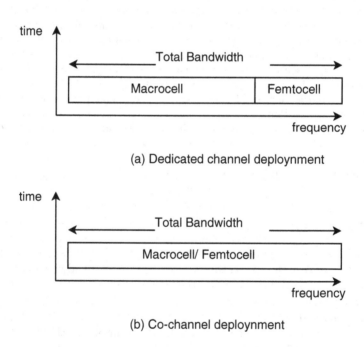

Fig. 1. The interference management schemes

Several interference management approaches based on Fractional Frequency Reuse (FFR) were proposed in [12, 11, 6, 5]. The frequency spectrum are divided into four sub-bands, with each sub-band assigned to a different area within the signal coverage of an MBS to resolve the cross-tier interference. However, co-tier interference still exists when FBSs are densely installed within the same sub-area.

As several downlink interference scenarios for two-tier networks were presented by Saquib *et al.* [9], the interference of downlink transmission, which is scheduled either by MBS or FBS, is expected to be avoided by an integrated scheduling method. Saha and Saengudomlert [8] proposed a scheduling algorithm with a resource reuse strategy among femtocells, and with preserved resource blocks for macro users. Although the cross-tier and co-tier interferences can be avoided through the proposed scheduling, it's hard to determine the amount of reserved resource blocks.

The graph-based spectrum allocation may be applied to the co-tier interference problem, as reported in [13, 14]. Based on the principle that neighboring cells cannot exploit the spectrum at the same time, where the frequencies allocated graphically. These approaches may solve the co-tier interference problem efficiently; however, they could suffer from inefficient and unfair spectrum utilization, especially in dense deployment scenarios. Moreover, the deterministic reserved spectrum does not take the users requirement into account, which degrades the performance as more users than expected are associated to the FBS.

In this paper, we focus on the downlink transmission in two-tier OFDMA networks. The co-tier interference among femtocells is avoided by a proposed non-conflict graph-based algorithm. All femtocells within a group have no adjacency with each other, hence any frequency can be reused among the femtocells without conflict. The cross-tier interference mitigation in-between macrocells and femtocells are compromised by the scheduler of the MBS and FGW. The average traffic loadings of both MBS and FBS are taken into account for frequency assignment in order to optimize the spectrum utilization.

2   System Model

A simple two-tier network is exemplified as follows. There is a MBS which provides macrocell access services for M macro user equipments (mUEs). Numbers of FBS are randomly deployed within the coverage of the MBS, with N femto user equipments (fUEs) being associated. The downlink OFDMA system with R resource blocks (RBs) in total is considered in this study.

Both average blocking rate and average data transfer rate are particularly interested for evaluation. The blocking rate indicates the percentage of unsatisfied resource fulfillment. It is defined as in the following, where a zero is assigned to ϵ_i^m(mUE) or ϵ_i^m(fUE), whenever it fails to obtain the RB_i.

$$Average\ blocking\ rate = \frac{\sum_{i=1}^{M}(1 - \epsilon_i^m) + \sum_{j=1}^{N}(1 - \epsilon_j^f)}{M + N}. \qquad (1)$$

The Shannon Capacity Formula [7] is used to determine the maximal data transfer rate over a communication channel. The formula is expressed below.

$$C = W \times \log_2(1 + \frac{Signal}{Noise}), \qquad (2)$$

where C is the data transfer rate (bits per second), W is the channel bandwidth in Hertz, $Signal$ is the received signal power and $Noise$ is the total noise or interference power.

The impact of interference is measured using the signal to interference and noise ratio (SINR), which is the product of $Signal$ divided by $Noise$.

$$SINR = \frac{Signal}{Noise} \tag{3}$$

3   Graph-Based Interference Avoidance Scheme and Resource Assignment

In this Section, we first introduce the interference graph construction among the fBSs, and then present a graph coloring algorithm for the purpose of segmenting fBSs into groups. The interference relation-ship among fBSs is further simplified, and a load-based radio resource assignment is also presented.

3.1   Non-conflict Group Discovery Algorithm

Two nearby FBSs having a shard frequency channel may result in co-channel interference. To avoid co-channel interference, the potential interference sources around each FBS has to be found out when the FBS is about to become active. Normally, a new installed FBS scans and listens to all available frequency channels. Scanning results include the identifiers of FBSs and a neighbor list, which in turn are forwarded to its serving FGW. FGW collects these messages and generates a corresponding interference graph, with each node presenting a FBS and an edge indicating an interference between two end nodes.

The spectrum allocation problem then can be modeled as a graph coloring problem where the color represents the available spectrum. This transforms the spectrum allocation problem into the vertex coloring problem, where adjacent nodes must not share the same color in order to avoid the interference. The conventional graph coloring algorithm employs a greedy strategy to color the nodes and thereby obtain the smallest chromatic number K. Similar to the approach presented in [13], we proposed a modified greedy-based coloring algorithm to segment the FBSs into K groups. Within each group, FBSs are free of interference from each other.

As shown in Algorithm 1, all nodes are assigned with the same color ($nc = 1$, where nc represents the node color.), initially. And $node\text{-}i$ iteratively getting its adjacent nodes through the adjacent matrix. Two adjacent nodes indicate they are mutual interfered. Then $node\text{-}i$ updates its nc to the conflict color list ($cclist$) of each adjacent $node\text{-}j$. The $cclist$ indicates the node colors have been assigned to neighbor nodes. Since two adjacent nodes cannot have the same color in order to avoid the interference, each $node\text{-}j$ then compares its nc with $cclist$, and selects the one as its nc, which is not present in $cclist$. As a result of exiting the $for\text{-}loop$ with $node\text{-}i$, there are K node colors required. And finally, all nodes are segmented into K groups.

Algorithm 1. non-Conflict Group Discovery

1: *Initially, FBSs are in same color (nc = 1)*
2: **for** $i = 1 \rightarrow number\ of\ FBSs$ **do**
3:   **for** $j = i + 1 \rightarrow number\ of\ FBSs$ **do**
4:     **if** $adjMatrix(i, j) == 1$ **then**
5:       $FBS(j).cclist \leftarrow FBS(i).nc$
6:       **if** $FBS(j).nc \in FBS(j).cclist$ **then**
7:         $FBS(j).nc = FBS(j).nc + 1$
8:       **end if**
9:     **end if**
10:   **end for**
11: **end for**

3.2 Load-Based Radio Resource Assignment

To access the radio resource of a MBS/ FBS, mUEs/ fUEs have to grant the access permission via some control channel as illustrated in [10]. Both MBS and FBS control the access requirement and schedule the resource blocks for resource allocation. In general, the MBS is preferable to schedule the radio resource, because of its carrier-grade quality. In order to avoid the cross-tier interference, the FGW is cooperated with the MBS for resource assignment.

A time period (T) is defined for each FBS to estimate and determine the RB requirement for the next time duration based on its traffic loading. The FGW collects information from each FBS after T interval, synchronizes with MBS for the fraction of radio resources, and determines the RBs for each FBS. Since after non-conflict group discovery algorithm, each FBS has been segmented. The θ_k represents the maximum RB requirement for the k_{th} group. The Θ is the summation of θ_k where $1 \leq k \leq K$. $N_{RB_M}^{DL}$ is the number of required downlink RBs for an MBS. The threshold τ is defined to prevent the resource starvation of FBSs, in case the MBS has become overloaded. A dynamic value δ is used to constrain the allocation of RBs between FBSs and the MBS. The load-based radio resource assignment algorithm is presented as Algorithm 2.

Once the FGW has synchronized with the MBS and collected the RB requirement of FBSs, the resource assignment for FBSs is determined based on the system capacity, $N_{RB_M}^{DL}$, and Θ. If $\Theta \leq R - \delta$, it indicates the resource can fulfill the requirement among MBS and FBSs, the MBS can assign $N_{RB_M}^{DL}$ RBs to mUEs while $R - \delta$ RBs are assigned to K groups according to θ_k. Since the FBSs are not adjacent within the same group, the RBs assigned to k_{th} group can be reused. Hence, the θ_k can fulfill the resource requirement among FBSs which are belonging to k_{th} group.

However, in the case of $\Theta > R - \delta$, the resource cannot fulfill the requirement. In order to assign as many as RBs to FBSs, the θ_k then be sorted by a descending order and $R - \delta$ RBs are assigned to K groups one by one. A group with higher θ_k is given higher priority to obtain downlink RBs. After the resource assignment, the FBSs are expected to have the lowest blocking rate.

Algorithm 2. Load-based radio resource assignment

1: **for** *each T* **do**
2:    **if** $N_{RB_M}^{DL} \geq \tau$ *and* $\Theta \geq R - \tau$ **then**
3:        $\delta = \tau$
4:    **else if** $N_{RB_M}^{DL} > \tau$ *and* $\Theta < R - \tau$ **then**
5:        $\delta = R - \Theta$
6:    **else**
7:        $\delta = N_{RB_M}^{DL}$
8:    **end if**
9:    **if** $\Theta \leq R - \delta$ **then**
10:        *assign* Θ *RBs to K groups directly.*
11:    **else**
12:        *sort* θ_k *by descending order.*
13:        **for** $k = 1$ *to* K **do**
14:            $\theta_k' = 0$
15:        **end for**
16:        $RB_{Counter} = R - \delta.$
17:        **while** $RB_{Counter} > 0$ **do**
18:            **for** $k = 1$ *to* K **do**
19:                **if** $\theta_k' \leq \theta_k$ **then**
20:                    *assign one RB to* $\theta_k'.$
21:                    $RB_{Counter} = RB_{Counter} - 1$
22:                **end if**
23:            **end for**
24:        **end while**
25:    **end if**
26: **end for**

4   Simulation and Results

A simulation using MATLB simulator is used to evaluate the performance for average signal quality, data transfer rate and blocking rate. We first show the simulation environment, then the simulation results are presented as compared with other interference management schemes.

4.1   Simulation Environment and Simulation Parameters

In our simulation, the network topology is deployed as similar the Figure 2. An MBS is located within the center of network topology, and parameter α is used to control the numbers of mUE which are associated with the MBS. There are numbers of FBS randomly deployed within the signal coverage of the MBS. Parameter β is used to control the numbers of fUE which are associated with a specific FBS. The numbers of fUE are randomly generated according to β parameter. We also assumed that the radio resource assignment is one RB for associated user equipment. The main parameters are listed in Table 1, which are referenced from [11, 14, 16].

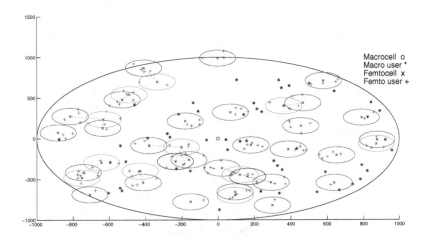

Fig. 2. The network deployment

Table 1. Main simulation parameters

Parameters	Value
System bandwidth	20MHz
Number of resource blocks	100
Indoor path loss model	$PathLoss = 127 + 30 \times log_{10}(\frac{d}{1000})$, d in meter
Outdoor path loss model	$PathLoss = 28 + 35 \times log_{10}d$, d in meter
Max Tx power of MBS	$40dBm$
Max Tx power of FBS	$10dBm$
Number of users per macrocell	$\alpha = 10 \sim 90$
Number of users per femtocell	$\beta = 6$

The performance metrics of SINR and data transfer rate are compared among full channel reuse (FCR), conventional graph coloring (CGC) and graph-based dynamic frequency reuse (GBDFR). With FCR, all RBs are shared and used among FBSs and MBS. In the case of CGC, RBs are reserved to FBSs according to the results of the graph coloring algorithm. Six RBs are reserved and reused among FBSs with GBDFR approach, according to [13]. The average RB blocking rate is compared among CGC and GBDFR. Since FCR utilizes and shares all available RBs, unless the RBs request is over the system capacity, there is no RB blocking. Here, our proposed load-based radio resource assignment is denoted as LRRA.

4.2 Simulation Results

The simulation results are shown in Figure 3- 7. Observed from Figure 3 and 4(β = 6), the fUEs' measured SINR are recorded and averaged, under a low (mUE=10) and high system capacity (mUE=90). The SINR degrees as more FBSs are deployed. The graph-based approaches, including CGC, GBDFR, and LRRA, have better SINR than FCR. According to formula 3, since FBSs share

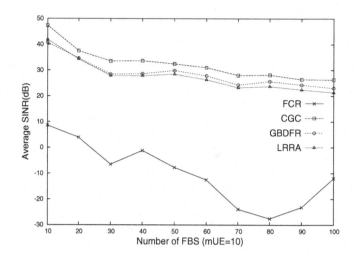

Fig. 3. The average fUE measured SINR under high system capacity

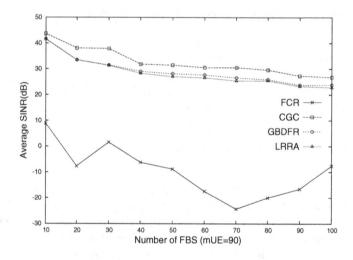

Fig. 4. The average fUE measured SINR under low system capacity

all RBs with MBS in FCR, once mUEs get close to FBSs, they would suffer the downlink interference. The distribution of SINR is more flat in graph-based approaches than FCR. This is because graph-based approaches will not use the same RBs which are assigned to mUEs. LRRA has a little lower SINR than CGC and GBDFR, since both CGC and GBDFR adopting a static RB reservation strategy, which also reduces the probability of co-tier interference.

The average data transfer rate of fUEs with various system capacities are shown in Figure 5 and 6. It is observed that even CGC and GBDFR have the higher SINR value then LRRA, while poor average data transfer rate then

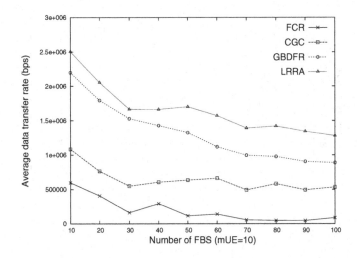

Fig. 5. The average data transfer rate under high system capacity

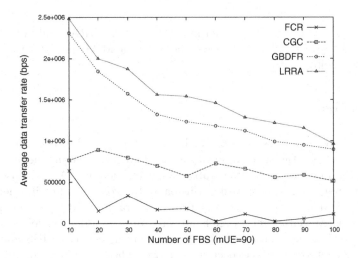

Fig. 6. The average data transfer rate under low system capacity

LRRA. This is because both CGC and GBDFR occupied a statically reserved resource. However, the reserved RBs may not fulfill the activation of fUEs, which results in poor performance. LRRA outperforms than others since it takes the traffic loading of each FBS for a dynamic resource allocation. The poor performance of FCR is caused by the co-tier and cross-tier interference.

The average RB blocking rate of graph-based approaches are shown in Figure 7. All graph-based approaches may block a request coming from mUEs or fUEs, once the system capacity has been reached. CGC has the highest blocking rate and is outperformed by GBDFR. This is because GBDFR has reserved more

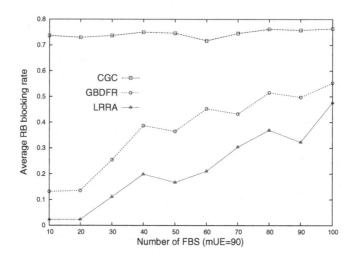

Fig. 7. The average RB blocking rate under low system capacity

RBs for FBSs, which can be reused more efficiently. LRRA would dynamically balance and adjust the demand for RBs between the MBS and FBSs to minimize the lowest blocking rate.

5   Conclusion

The limited frequency resource has to be reused or shared in two-tier networks in order to increase the spectrum utilization. However, the dynamic deployment of femtocells may result in unbearable signal interference. A load-based radio resource management is proposed in this paper to avoid downlink interference in two-tier OFDMA networks. We first present a graph-based coloring algorithm to classify disjoint interference-free femtocell groups in order to avoid the co-tier interference. Through the scheduling corporation in between MBS and FGW, the cross-tier interference is expected to be free. The traffic load among FBSs is taken into account for resource allocation. The assigned RBs among FBSs can be reused within the same group to increase the average resource utilization as well as to decrease the average resource blocking rate.

Matlab simulator is used to simulate the proposed scheme. The simulation results reveal that fUEs gains the better performances in data rate and blocking rate with our approach. As compared with other graph-based interference mitigation approaches, the average data rate of fUEs can be increased in the range of 14% ∼ 88% under low and high system capacity. Additionally, the cross-tier interference can also be effectively avoided by cooperative resource management. And, the average blocking rate of fUEs is successfully decreased up to 41% and 70%, under low system capacity, as compared with conventional graph coloring and graph-based dynamic frequency reuse schemes.

References

[1] IEEE Standard for Local and metropolitan area networks Part 16: Air Interface for Broadband Wireless Access Systems (2009)

[2] 3GPP TS 22.220 Service requirements for Home Node B (HNB) and Home eNode B (HeNB), Release 10 (2012)

[3] 3GPP TS 36.211 Evolved Universal Terrestrial Radio Access (E-UTRA) physical channels and modulation, Release 10 (2012)

[4] Chandrasekhar, V., Andrews, J., Gatherer, A.: Femtocell networks: a survey. IEEE Communications Magazine 46(9), 59–67 (2008)

[5] Lee, S., Ahn, C., Cho, T., Park, J., Shin, J.: Dynamic frequency allocation of femtocells for guaranteed macrocell throughput in ofdma networks. In: Proceedings of the 6th International Conference on Ubiquitous Information Management and Communication, ICUIMC 2012, pp. 3:1–3:5 (2012)

[6] Poongup, L., Taeyoung, L., Jangkeun, J., Jitae, S.: Interference management in lte femtocell systems using fractional frequency reuse. In: 2010 The 12th International Conference on Advanced Communication Technology (ICACT), vol. 2, pp. 1047–1051 (2010)

[7] Rappaport, T.S.: Wireless communications: principles and practice. Prentice Hall communications engineering and emerging technologies series. Prentice Hall PTR, Upper Saddle River (2002)

[8] Saha, R.K., Saengudomlert, P.: Novel resource scheduling for spectral efficiency in lte-advanced systems with macrocells and femtocells. In: 2011 8th International Conference on Electrical Engineering/Electronics, Computer, Telecommunications and Information Technology (ECTI-CON), pp. 340–343 (2011)

[9] Saquib, N., Hossain, E., Long Bao, L., Dong In, K.: Interference management in ofdma femtocell networks: issues and approaches. IEEE Wireless Communications 19(3), 86–95 (2012)

[10] Sesia, S., Toufik, I., Baker, M.: LTE, The UMTS Long Term Evolution: From Theory to Practice. Wiley Publishing (2009), 1611434

[11] Taeyoung, L., Hyuntai, K., Jinhyun, P., Jitae, S.: An efficient resource allocation in ofdma femtocells networks. In: 2010 IEEE 72nd Vehicular Technology Conference Fall (VTC 2010-Fall), pp. 1–5 (2010a)

[12] Taeyoung, L., Jisun, Y., Sangtae, L., Jitae, S.: Interference management in ofdma femtocell systems using fractional frequency reuse. In: 2010 International Conference on Communications, Circuits and Systems (ICCCAS), pp. 176–180 (2010b)

[13] Uygungelen, S., Auer, G., Bharucha, Z.: Graph-based dynamic frequency reuse in femtocell networks. In: 2011 IEEE 73rd Vehicular Technology Conference (VTC Spring), pp. 1–6 (2011)

[14] Yuyu, W., Kan, Z., Xiaodong, S., Wenbo, W.: A distributed resource allocation scheme in femtocell networks. In: 2011 IEEE 73rd Vehicular Technology Conference (VTC Spring), pp. 1–5 (2011)

[15] Zhang, J., Roche, G.D.L.: Femtocells: Technologies and Deployment. Wiley Publishing (2010), 1841737

[16] Zheng, K., Wang, Y., Lin, C., Shen, X., Wang, J.: Graph-based interference coordination scheme in orthogonal frequency-division multiplexing access femtocell networks. IET Communications 5(17), 2533–2541 (2011)

An Efficient Handover Mechanism by Adopting Direction Prediction and Adaptive Time-to-Trigger in LTE Networks

Fu-Min Chang[1], Hsiu-Lang Wang[2,*], Szu-Ying Hu[2], and Shang-Juh Kao[2]

[1] Department of Finance
Chaoyang University of Technology
168, Jifeng E. Rd., Wufeng District, 41349 Taichung, Taiwan
fmchang@cyut.edu.tw
http://www.cyut.edu.tw/~finance/cindex.htm
[2] Department of Computer Science and Engineering
National Chung Hsing University
250 Kuo Kuang Rd., 40227 Taichung, Taiwan
{gary6319,hszuying}@gmail.com, sjkao@cs.nchu.edu.tw
http://www.cs.nchu.edu.tw

Abstract. In this paper, we propose an efficient handover mechanism for Long Term Evolution (LTE) networks. Both moving direction prediction and Time-to-Trigger (TTT) adjustment are taken into consideration for processing to lower the unnecessary handover. By referencing previous locations, user equipment's (UE) moving direction is estimated by the cosine function to determine a better destined E-UTRAN NodeB (eNB). With adaptive TTT, we categorize UEs into 3 levels according to the velocity. Initially, TTT value is given for each level, and the value is decremented when the handover failure rate gets higher. In addition, we also combine both approaches of moving direction prediction and adaptive TTT into handover procedure. Simulation results show that the proposed method can effectively improve the handover efficiency. As compared with the standard handover procedure, an average reduction of 31% handovers can be obtained by applying the moving direction prediction. And, as compared with fixed TTT, up to the 36.5% of handover trigger failure rate can be expected by using the adaptive TTT handover triggering approach. The performance can be further improved when both approaches are taken in dealing with the handover procedure.

Keywords: handover, direction prediction, adaptive time-to-trigger, LTE.

1 Introduction

To meet the increasing demand of wireless data services, third generation partnership program (3GPP) has proposed UTRAN (UMTS Terrestrial Radio Access

---

* Corresponding author.

B. Murgante et al. (Eds.): ICCSA 2013, Part V, LNCS 7975, pp. 270–280, 2013.

Network) LTE (Long Term Evolution) [2]. LTE provides high speed communication within uplink and downlink, as well as aiming to provide more capacity along with less network complexity and low installation and maintenance cost. Two different radio access mechanisms are used in LTE, OFDMA (Orthogonal Frequency-Division Multiple Access) is used for downlink and SC-FDMA (Single Carrier-Frequency-Division Multiple Access) is used for uplink. OFDMA provides high spectral efficiency which is very immune to interference and reduces computation complexity in the terminal within larger bandwidths [3]. SC-FDMA has lower PAPR (Peak to Average Power Ratio) provisions, which leads towards longer battery life as well as larger radio coverage. Another advantage of using LTE is its capability to switch back to older legacy systems such as WCDMA (Wideband Code-Division Multiple Access) etc., when a user equipment (UE) gets out of coverage from an LTE network.

As shown in Figure 1, the LTE network architecture consists of evolved NodeBs (eNBs), mobility management entity (MME), and system architecture evolution gateways (S-GW) [3]. The eNBs are connected to the MME/S-GW by the S1 interface, and they are interconnected by the X2 interface. One of the important features within LTE networks is handover. In LTE networks, hard handover has been considered to minimize the radio resource requirement. Because hard handover is more sensitive to radio link failure than soft handover, it is required to minimize the radio link failure rate for maintaining the given Quality of Service (QoS) within a communication link in LTE networks. Hence, an efficient handover mechanism with lower radio link failure rate is necessary for LTE networks.

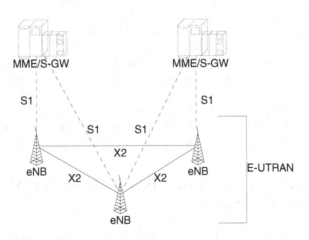

Fig. 1. E-UTRAN architecture

In LTE networks, the UE performs several downlink radio channel measurements, which are used to quantify network performance, on both serving cell and neighboring cells by using reference symbols (RS). Handover execution may

occur mainly due to poor cell coverage or poor QoS. The cell coverage is monitored using the Reference Signal Received Power (RSRP) from serving cell and
neighboring cells. On the other hand, QoS is measured using the Reference Symbols Received Quality (RSRQ) and other parameters. The RSRP and RSRQ are
measured during a designated measurement interval. The UE periodically checks
if certain conditions are met in order to trigger the handover procedure. In case
of coverage based handover, if the RSRP of one of the UE neighboring cells
is greater that the RSRP of the serving cell plus a designated hysteresis value
for at least a Time-To-Trigger period (TTT), the handover procedure triggering
occurs. Once triggering conditions are met, the UE reports back to the serving eNB the measurement report, indicating the triggering event and the target
cell with the highest RSRP level when compared with the current serving cell.
Based on the received measurement report, the serving eNB starts the handover
preparation phase.

During TTT period, if RSRP in the serving cell becomes higher again than
that in the target cell, handover would not be executed. Hence, by using TTT,
we can mitigate ping-pong handover effects which are wasteful switching of the
serving cell for a short time period. On the other hand, the radio link failure
rate can be increased because of delayed handover execution during TTT period.
Therefore, to achieve the lower ping-pong rate and the lower radio link failure
rate simultaneously, efficient TTT values in accordance with UE's speeds need to
be investigated. Besides, the prediction of UE's moving direction is useful technique for reducing the numbers of candidate target eNBs. By eliminating eNBs
located at the opposite of UE's moving direction, the number of unnecessary
handover can be reduced. Hence, in this paper, we propose an efficient handover
mechanism by adopting direction prediction and adaptive time-to-trigger in LTE
Networks. By referencing previous locations, UE's moving direction is estimated
by the cosine function to determine a better target eNB. With adaptive TTT,
we categorize UEs into 3 levels according to the velocity. In addition, we also
combine both approaches of moving direction prediction and adaptive TTT into
handover procedure.

This paper proceeds as follows. Section 2 describes the researches on prediction-
based handover algorithm and TTT. The proposed mechanism is described in section 3. The simulation results are shown in section 4. Finally we give a conclusion.

2   Researches on Prediction-Based Handover Algorithm and Time-To-Trigger

Different research approaches try to use movement predictions as an addition
to classical handover preparation and triggers. In [8], the authors proposed a
mobility management technique which uses simple handover prediction based
on cross-layer architecture. Their prediction technique uses both simple moving average for inertial movements and simple mobility pattern matching for
non-inertial movements. They also introduced a new reporting event for UE's
measurement reports and designed a handover prediction algorithm for eNodeB's

handover preparation. In [4], the authors proposed a handover approach with a simple handover prediction based on acts of users of mobile history. They used the user mobility database and simple mobility pattern matching as well. A valid update database approach is proposed, to ensure the database time in tracking the user mobility actions. However, both approaches reduce the evaluating cost of more candidate target eNBs, while increase the cost of mobility database.

For the researches on TTT, in [9], the authors examined the handover performance by using various TTT values in LTE networks, depending on UE speeds and cell configurations within an allowable radio link failure rate. In the first step, the adaptive TTT values for each UE speeds were selected in macro-macro handover and macro-pico handover scenarios. And then, the authors fitted the curve by using the result of selected TTT values for both neighboring cell configurations. From the TTT curve, the authors suggested the criteria for grouping UE speeds and the proper TTT values in accordance with the cell configurations. The simulation results showed that the performance was comparable with that of the case when applying the adaptive TTT values for an arbitrary UE speed. And the performance is significantly improved compared to that of the cases when a fixed TTT value was applied in both cell type configurations.

In [6], the authors evaluated the hard handover performance based on the different hand handover algorithms for LTE networks. With RSRP and RSRQ measurements in the different velocity environment, this research focused on the impact that two kinds of hard handover algorithms based on UE velocity and optimized TTT value, as compared with A3 event. The simulation results showed that the proposed both handover algorithms have gotten a better handover success rate than classical A3 via the change of hysteresis and TTT in the different velocity interval. This paper suggested that while the UE velocity increases, the handover decision rule may modify the handover parameter-hysteresis and TTT self-adaptively to avoid the increasing of handover failure rate.

3   Proposed Handover Mechanisms

As UEs move around the signal coverage of LTE networks, the terrain may influence the received signal power, especially when UEs locate far away its serving eNB. To maintain a certain communication quality, eNBs will inform UEs to trigger handover procedures as the feedback signal strength is weak of UEs. However, the handover may result in disconnection and data loss, where impact a real-time service. Therefore, a direction prediction scheme is proposed to predict the moving direction of UEs in order to reduce the number of handovers. We also propose a dynamic TTT scheme to avoid handover fail, which is archived by assigning the dynamic TTT value according to a predefined threshold.

3.1   Direction Prediction Scheme

The handover procedures in LTE can be divided into three distinct phases; there are the preparation, execution and completion phases. Before UEs enter in handover preparation phase, UEs have feedback measurement reports to its serving eNB periodically. The measurement reports include the signal strength of its serving and neighbor eNBs. The neighbor eNBs form as candidate eNBs for handover, and it takes time to select a target eNB for handover. In order to select a precise target eNB for handover, we adopt the Global Positioning System (GPS) to identify the position of each eNBS and UE.

The GPS is a satellite navigation system, and provides accurate positioning, velocity and precision standard time for most of the Earth's surface area. Since most of smart phones have been already embed the GPS service, it is convenient for devices to acquire its position. The position of each eNB can also be measured by GPS, because the deployment of eNBs is planned and designed by network operators. Therefore, with the help of GPS, the position of each eNB and UE can be identified separately for tracing.

Since the moving direction of an UE may toward a specific group of neighbor eNBs, the moving path of the UE has to be recorded at first for its direction prediction. The position of the UE can be retrieved from the GPS and represented by P_1, P_2, P_3, where P_3 is the position at the current time-stamp, P_2 is at previous and P_1 is the position at the before previous time-stamp as showing in Figure 2. Recall the trigonometric functions and inner product formula, giving three points, an θ angle is obtained by a function of angle calculator as shown in Algorithm 2.

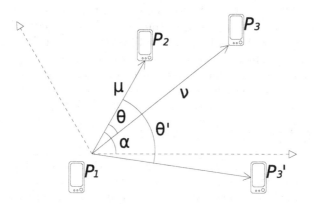

Fig. 2. The interference management schemes

In Figure 2, where μ is the vector of $\overrightarrow{P_1P_2}$, ν is the vector of $\overrightarrow{P_1P_3}$, and θ is the angle of $\angle P_2P_1P_3$. The θ is calculated along with the moving path of an UE. It is used to determine if the UE moves in a specific direction, and the location of eNBs ahead the UE are candidates for handover instead of all neighbor eNBs. The direction prediction for handover procedure is presented as Algorithm 1, and a function of angle calculator as Algorithm 2.

Initially, position P_1 and P_2 are obtained from GPS. Each time the new position P_3 is updated, the angle calculator is called with P_1, P_2, P_3 parameters to

---

Algorithm 1. Direction prediction assisted handover procedure

---

1: *Initially, P_1 and P_2 are obtained from GPS. S is a set contains eNBs. TTT_n is initialized by uE's velocity. HO_{FR} and uHO_R are continual counted .*

2: **while** P_3 **do**

3:     $\theta \leftarrow AngleCalculation(P_1, P_2, P_3)$

4:     **if** $\theta < \alpha$ **then**

5:         **for** *each eNB* $\in S$ **do**

6:             *Increase the weight of eNB by 1, if it locates within $\pm\alpha$ degree angle of $\overrightarrow{P_1 P_2}$*

7:         **end for**

8:     **else**

9:         $P_1 \leftarrow P_2$

10:         $P_2 \leftarrow P_3$

11:     **end if**

12:     **for** *each eNB* $\in S$ **do**

13:         **if** $eNB_{RSRP}^c > (eNB_{RSRP}^s + Hysteresis)$ *and target eNB has hightest weight.* **then**

14:             $TTT_n \leftarrow DynamicTTT Adjustment(TTT_n, HO_{FR}, uHO_R)$

15:             *While $eNB_{RSRP}^c > (eNB_{RSRP}^s + Hysteresis)$ is true and consist within TTT_n, then response a measurement report to its serving eNB.*

16:         **end if**

17:     **end for**

18: **end while**

Algorithm 2. Angle Calculator

1: *Input: P_1, P_2, and P_3.*

2: $\overrightarrow{\mu} \leftarrow \overrightarrow{P_1 P_2}$

3: $\overrightarrow{\nu} \leftarrow \overrightarrow{P_1 P_3}$

4: $\theta \leftarrow \arg\cos(\frac{\overrightarrow{\mu} \cdot \overrightarrow{\nu}}{|\overrightarrow{\mu}| \times |\overrightarrow{\nu}|})$

5: *return θ*

---

find out the θ value and return. The θ is compared with α value that is used to check if the directions of $\overrightarrow{P_1 P_3}$ and $\overrightarrow{P_1 P_2}$ are the same. And the α is a range where the location of a candidate eNB is considered to be associated with. The eNB is assumed to be equipped with three antennas to provide its serving area. Each antenna can provide signal coverage with the range of 120 degree angle, and form as a cellular network. Therefore, we defined α has 60 degree angle offset for a main direction boundary of an UE in this study, the moving path within the range of 120 degree angle totally is determined as forwarding to the same direction. In short, the moving path $\overrightarrow{P_1 P_3}$ has the same direction with $\overrightarrow{P_1 P_2}$ in the case of the angle $\angle P_3 P_1 P_2$ less than α as shown in Figure 2.

In the case of $\overrightarrow{P_1 P_3}$ has the same moving direction with $\overrightarrow{P_1 P_2}$, all neighbor eNBs located within $\pm\alpha$ degree angle of $\overrightarrow{P_1 P_3}$, are candidates for handover. The *weight* of these candidate eNBs is increased. In the case of $\overrightarrow{P_1 P_3'}$ has a different moving direction with $\overrightarrow{P_1 P_2}$, where θ' is not less than α, the main moving direction of the UE is re-defined. Then, P_2 is transferred to P_1, and P_3'

is transferred to P_2. As the result, the new direction of $\overrightarrow{P_1P_2}$ forms as a new moving direction of the UE.

After the comparison of moving direction procedure, the algorithm checks the handover condition. If the RSRP of a candidate eNB (eNB_{RSRP}^c) is larger than the RSRP of a serving eNB plus a hysteresis ($eNB_{RSRP}^s + Hysteresis$), and the candidate eNB has the highest *weight*, then the UE reports to its serving eNB for handover. The *weight* is reset to zero for all eNBs in order for the next handover procedure. To check the candidate eNB has the highest *weight* is used to ensure that the candidate eNB is a real target since the UE may pass-by a candidate eNB for a while. Handover to a pass-by candidate eNB will increase the number of handover and increase the risk of connection fail.

3.2   Dynamic Time-to-Trigger Adjustment Scheme

In LTE networks, UEs response the measurement report to its serving eNB for a handover event being triggered. A serving eNB constantly compares the signal quality of UEs in the serving cell with that among candidate eNBs during TTT once UEs enter in "entering condition". The serving eNB then executes handover procedures if a candidate eNB has higher signal quality during TTT period. Otherwise, UEs enter in "leaving condition" to stop reporting and abort handover procedures [3]. By using TTT, unnecessary handover is expected to be mitigated. However, an unsuitable TTT value may result in handover fail or unnecessary handover [9].

The LTE specification recommends the UE will adjust TTT depending on the UE speed. Here, the UE speeds are classified into three groups and giving an initial TTT value individually as presented in [7]. In order to reduce the number of handover fail, the value of TTT is proposed adjustable with a certain handover fail rate (HO_{FR}) and an unnecessary handover rate (uHO_R). The definitions are in the following.

$$HO_{FR} = \frac{HO_F}{HO_F + HO_S}. \tag{1}$$

$$uHO_R = \frac{uHO}{HO}. \tag{2}$$

Where HO_F is the number of failed handover, HO_S is the number of successes handover, uHO is the number of unnecessary, and HO is the number of handover. A handover back to its original serving eNB within 0.5s is defined as an unnecessary handover. The UE obtains an initial TTT value according to its velocity, and count HO_{FR} while it moves. Once HO_{FR} archives a certain threshold, a dynamic TTT adjustment is triggered as shown in Algorithm 3. The sixteen available TTT values specified by LTE recommends are indexed from TTT_0 to TTT_{15}. If HO_{FR} over $\tau1$, the TTT value is adjusted shorter to earlier trigger the handover procedure. In the case of uHO_R over $\tau2$, the TTT value is adjusted longer in order to prevent unnecessary handover.

Algorithm 3. Dynamic TTT adjustment

1: *Input: TTT_n, HO_{FR} and uHO_R.*
2: **if** $HO_{FR} > \tau 1$ **then**
3:     $TTT_n \leftarrow TTT_{n-1}$
4: **else if** $uHO_R > \tau 2$ **then**
5:     $TTT_n \leftarrow TTT_{n+1}$
6: **end if**
7: *return TTT_n*

---

4   Simulation and Results

To verify the proposed mechanism, a MATLAB-based LTE downlink system level simulation (SL Simulator) [5] was adopted to evaluate the performance. Both average handover rate and handover fail rate are evaluated with various velocities of UEs. We first introduce the simulation environment and related parameters, and then the simulation results are presented.

4.1   Simulation Environment and Parameters

The network topology is constructed as shown in Figure 3. There are nineteen eNBs, each eNB has three sectors, and each sector has one UE associated. There are 57 UEs totally deployed. Some main parameters are listed in Table 1, which

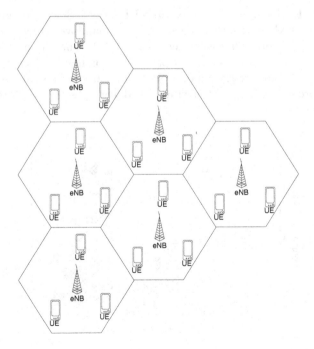

Fig. 3. Network topology

Table 1. Main simulation parameters

Parameters	Value
Frequency	$2GHz$
Bandwidth	$5MHz$
Thermal noise density	$-174dBm/Hz$
Receiver noise value	$9dB$
eNB's transmission power	$46dBm$
Resource Blocks per eNB	25
Bandwidth per eNB	$180KHz$
TTI(Subframe duration)	$1ms$
Numbers of UE	57
Cell layout	19 sites, 3 sector per site
Simulation time	$20s$
UE speed(km/hr)	$30, 60, 120km/hr$
Mobility model	Random waypoint

are referenced from [1]. UEs move randomly with predefined speed in a simulation time. The velocities are configured and classified into three levels to represent low, medium, and high speed, which are the $30km/hr$, $60km/hr$, and $120km/hr$. Each TTT is initiated with $512ms$, $256ms$, and $100ms$, respectively.

4.2   Simulation Results

Figure 4 shows the handover rate for various UE speeds. Since the signal strength of UEs' serving eNB is taken into account for a handover trigger only, an UE may response an incorrect measurement report. With our proposal, the moving direction of UEs is also adopted as a parameter for handover decision; it estimates numbers of unnecessary handovers. As compared with standard approach, proposed approach reduces the average handover rate to 43.5% at $30km/hr$

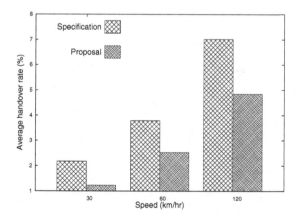

Fig. 4. Average handover rates with various velocities

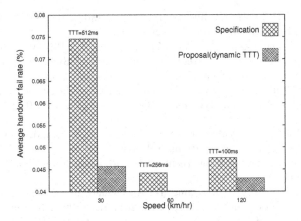

Fig. 5. Average handover fail rate with various velocities

(low velocity level), 33.3% at $60km/hr$ (medium velocity level), and 30.9% at $120km/hr$ (high velocity level), respectively.

Figure 5 shows the handover fail rate according to TTT values for various UE speeds. It is observed as UE moves at $30km/hr$ speed, it has the highest handover fail rate, and this is because the unnecessary handover. Too many unnecessary handover may result in a ping-pong situation. Our dynamic TTT can reduce the handover fail rate efficiently by adopting HO_{FR} and uHO_R to adjust a suitable TTT.

5 Conclusions

In this paper, an efficient handover mechanism for LTE networks was proposed. Both moving direction prediction and TTT adjustment were taken into consideration for processing to lower the unnecessary handover. Simulation results revealed that an average reduction of 31% handovers can be obtained by applying the moving direction prediction as compared with the standard handover procedure. And, up to 36.5% of the handover trigger failure rate can be expected by using the adaptive TTT handover triggering approach as compared with the fixed TTT value. The performance can be further improved when both approaches are taken in dealing with the handover procedure. The comparison with other prediction techniques, TTT adjustment policies and further comprehensive performance analysis on the handover preparation are now in progress for refining the proposed technique.

References

[1] 3GPP TR 36.942 Evolved Universal Terrestrial Radio Access (E-UTRA); Radio Frequency (RF) system scenarios (2012)
[2] 3GPP TS 36.211 Evolved Universal Terrestrial Radio Access (E-UTRA) physical channels and modulation, Release 10 (2012)

[3] 3GPP TS 36.331 Evolved Universal Terrestrial Radio Access (E-UTRA); Radio Resource Control (RRC); Protocol specification (2012)

[4] Huaining, G., Xiangming, W., Wei, Z., Zhaoming, L., Bo, W.: A history-based handover prediction for lte systems. In: International Symposium on Computer Network and Multimedia Technology, CNMT 2009, pp. 1–4 (2009)

[5] Ikuno, J.C., Wrulich, M., Rupp, M.: System level simulation of lte networks. In: 2010 IEEE 71st Vehicular Technology Conference (VTC 2010-Spring), pp. 1–5 (2010)

[6] Linlin, L., Muqing, W., Yuhan, C., Xian, H., Chaoyi, Z.: Handover parameter optimization of lte system in variational velocity environment. In: IET International Conference on Communication Technology and Application (ICCTA 2011), pp. 395–399 (2011)

[7] Shih-Fan, C., Hsi-Lu, C., Chia-Lung, L.: An efficient measurement report mechanism for long term evolution networks. In: 2011 IEEE 22nd International Symposium on Personal Indoor and Mobile Radio Communications (PIMRC), pp. 197–201 (2011)

[8] Tae-Hyong, K., Qiping, Y., Jae-Hyoung, L., Soon-Gi, P., Yeon-Seung, S.: A mobility management technique with simple handover prediction for 3g lte systems. In: 2007 IEEE 66th Vehicular Technology Conference, VTC-2007 Fall, pp. 259–263 (2007)

[9] Yejee, L., Bongjhin, S., Jaechan, L., Daehyoung, H.: Effects of time-to-trigger parameter on handover performance in son-based lte systems. In: 2010 16th Asia-Pacific Conference on Communications (APCC), pp. 492–496 (2010)

Constructing Elastic Scientific Applications Using Elasticity Primitives

Guilherme Galante and Luis Carlos Erpen Bona

Department of Informatics
Federal University of Paraná
Curitiba, PR – Brazil
{ggalante,bona}@inf.ufpr

Abstract. Elasticity can be seen as the ability of a system to increase or decrease the computing resources allocated in a dynamic and on demand way. Considering its importance, some mechanisms to explore elasticity have been proposed by public providers and by academy. However, these solutions are inappropriate to provide elasticity for scientific applications or are limited to a specific programming model. In this context, we present Cloudine, a platform for development of elastic scientific applications based in simple elasticity primitives. These primitives enable the dynamic allocation and deallocation of resources in several levels, ranging from nodes of a virtual cluster, to virtual processors and memory of a node. Using this basic building blocks it is possible to develop applications in different models. The Cloudine effectiveness is demonstrated in the experiments, where two elastic applications in different models were developed.

Keywords: Cloud computing, elasticity, scientific applications.

1 Introduction

Scientific computation is a broad field with applications in many domains of science and industrial settings. Evolution of scientific computation methodology is fast and the construction of infrastructures for computational science lies at the front edge of technical development. During the last decades, scientific applications have been executed over high performance infrastructures, including cluster and grid computing, and more recently cloud computing [1].

Cloud computing refers to a flexible model for on-demand access to a shared pool of configurable computing resources (such as networking, servers, storage, platforms and software) that can be easily provisioned as needed, allowing immediate access to required resources without needing to purchase any additional infrastructure [2]. It implies that the amount of resources used by an application may be changed over time, without any long-term indication about the future demands [3]. This flexibility in resources allocation is called *elasticity*.

Currently, cloud elasticity has been used for scaling traditional web applications in order to handle unpredictable workloads, and enabling companies

B. Murgante et al. (Eds.): ICCSA 2013, Part V, LNCS 7975, pp. 281–294, 2013.
© Springer-Verlag Berlin Heidelberg 2013

to avoid the downfalls involved with the fixed provisioning (over and under-provisioning) [4][5]. In scientific scenario, the use of cloud computing is discussed in several studies [6][7], but the use of elasticity in scientific applications is a subject that is starting to receive attention from research groups [8].

This interest is related to the benefits it can provide, that include, improvements in applications performance, cost reduction and better resources utilization. Improvements in the performance of applications can be achieved through dynamic allocation of additional processing, memory, network and storage resources. Examples are the addition of nodes in a master-slave application in order to reduce the execution time, and the dynamic storage space allocation when data exceeds the capacity allocated for the hosted environment in the cloud.

The cost reduction is relevant when using resources from public clouds, since resources could be allocated on demand, instead allocating all of them at the beginning of execution, avoiding over-provisioning. It could be used in applications that use the MapReduce paradigm, where is possible to increase the number of working nodes during the mapping and to scale back resources during the reduction phase. Elastic applications can also increase computational capabilities when cheaper resources became available. An example is the allocation of Amazon Spot Instances, when the price becomes advantageous [9].

In private clouds, the elasticity can contribute to a better resources utilization, since resources released by an user can be instantaneously allocated to another one. It allows an application to start with few resources and increasing them according to availability, without have to wait for the ideal number of resources to become free, as occurs in non-elastic applications.

Elasticity solutions have been proposed by public cloud providers, e.g., Amazon [10], Azure [11] and Rackspace [12], but most of this solutions are suitable for server-based applications, such as, HTTP, e-mail and database servers, which relies on the replication of virtual machines and uses load balancers to distribute the workload among the numerous instances [13]. In general, scientific applications do not adapt to this approach.

Some academic researches have addressed the development of elastic scientific applications. It is possible to find works focusing on workflows [14], MapReduce [9][15], MPI [16] and master-slave applications [17]. However, to the best of our knowledge, there are not frameworks or platforms that could be used independently from a specific application model.

In this context, we present Cloudine, a platform for development and deployment of elastic scientific applications in Infrastructure-as-a-Service (IaaS) clouds. Unlike current elasticity solutions, our platform provides a platform and a set of elasticity primitives for supporting the development of elastic applications in different programming models, e. g., multithread, master-slave, bag-of-tasks, mapreduce and single program multiple data (SPMD) applications.

In addition, Cloudine enables exploration of elasticity in a broad way, allowing (depending on the cloud used) the dynamic allocation and deallocation of resources in several levels, ranging from nodes of a virtual cluster, to virtual processors and memory of a node. The Cloudine effectiveness is demonstrated in the experiments, where two different elastic applications were developed: an OpenMP heat transfer problem and a bag-of-tasks file sorting application.

The remainder of the paper is organized as follows. Section 2 presents an overview of the current cloud elasticity mechanisms. Section 3 introduces the proposed platform. In Section 4 we present an example of how Cloudine works. In Section 5, the experiments are presented and the results are discussed. Finally Section 6, concludes the paper.

2   Cloud Elasticity: An Overview

Many elasticity solutions have been implemented by public cloud providers and by academy [8]. In general, IaaS public cloud providers offer some elasticity feature, from the basic, to more elaborate automatic solutions.

Amazon Web Services [10], one of the most traditional IaaS cloud providers, offers a mechanism based in virtual machine replication called Auto-Scaling. The solution is based on the concept of Auto Scaling Group (ASG), which consists of a set of instances that can be used for an application. Amazon Auto-Scaling uses a reactive approach, in which, for each ASG there is a set of rules that defines the instances number to be added or released. The metric values are provided by CloudWatch monitoring service, and include CPU usage, network traffic, disk reads and writes.

GoGrid [18] and Rackspace [12] also implement replication mechanisms, but unlike Amazon, they do not have native automatic elasticity services. Both providers offer an API to control the amount of virtual machines instantiated, leaving to the user the implementation of more elaborate automated mechanisms.

Other IaaS cloud providers also provide elasticity mechanisms but the features presented are not substantially distinct from presented above. Basically, the current elasticity solutions offer a virtual machine (VM) replication mechanism, accessed using an API or via interfaces, and in some cases, the resources allocation is managed automatically by a reactive (based in rules) controller.

In turn, applications developed in Platform-as-a-Service (PaaS) clouds have implicit elasticity. These clouds provide execution environments, called containers, in which users can execute their applications without having to worry about which resources will be used. In this case, the cloud platform automatically manages the resource allocation, thus, developers do not have to monitor the service status or interact to request more resources [19].

An example of PaaS platform with elasticity support is Google AppEngine [20], a platform for developing scalable web applications (Java, Python, and JRuby) that run on top of Google's server infrastructure. These applications are executed within a container (sandbox) and AppEngine take care of automatically scaling when needed.

Azure [11] is the solution provided by Microsoft for developing scalable .NET applications for clouds. Despite offering platform services, Azure does not provide an automatic elasticity control. In its approach, the user must configure the allocation of resources.

Several academic projects also developed elasticity mechanisms. They are similar to those provided by commercial providers, but include new techniques and methodologies for elastic provisioning of resources. Examples of elasticity solutions proposed in academy includes the works of Lim et al. [21], Roy et al. [22], Gong et al. [23], Sharma et al. [24] and several others.

Considering the elasticity solutions presented so far, most of them are appropriate for server-based applications, such as, http, e-mail and database, which relies on the replication of virtual machines and load balancers to distribute the workload among the numerous instances. In general, scientific applications do not adapt to this approach, since most of them need a higher level of coordination, requiring synchronizations and message exchanges. In this context, some works address the development of frameworks and platforms focusing scientific applications.

Aneka [25] is a .NET-based application development platform (PaaS), which offers a runtime environment and a set of APIs that enable developers to build applications by using multiple programming models such as task, thread and MapReduce, which can leverage the compute resources on either public or private clouds. In Aneka, when an application needs more resources, new container instances are executed to handle the demand, using local or public cloud resources.

In the works of Chohan et al. [9] and Iordache et al. [15] the elastic execution of MapReduce applications are presented. In the former, the authors investigate the dynamic addition of Amazon Spot Instances to reduce the application execution time. The latter presents a system that implements an elastic MapReduce API that allows the dynamic allocation of resources of different clouds.

Raveendran et al. [16] present a framework for the development of elastic MPI applications. The authors proposed the adaptation of MPI applications by terminating the execution and restarting a new one using a different number of virtual instances. Vectors and data structures are redistributed and the execution continues from the last iteration. Applications that do not have an iterative loop cannot be adapted by the framework, since it uses the iteration index as execution restarting point.

Rajan et al. [17] presented Work Queue, a framework for the development of elastic master-slave applications. Applications developed using Work Queue allow adding slave replicas at runtime. The slaves are implemented as executable files that can be instantiated by the user on different machines. When executed, the slaves communicate with the master, that on demand coordinates task execution and the data exchange. Based in Work Queue, Yu et al. [26] developed MWPC, an architecture for execution of elastic workflows on heterogeneous distributed computing resources.

Elastic workflows execution is also addressed in the work of Byun et al. [14], where the authors propose an architecture for the automatic execution of large-scale workflow applications on dynamically and elastically provisioned computing resources.

How we can observe, each elasticity solution was developed focusing a specific programming model, such as, message-passing (MPI), master-slave, workflow or mapreduce. There is not a generic framework that enables the construction of applications of any programming model. Other point to be observed is that the presented solutions support only horizontal elasticity, i. e., the allocation unit is a virtual machine. Vertical elasticity, in which processing, memory and storage resources can be added/removed from a running virtual instance, is not currently supported by the presented solutions.

In next section, we present our elasticity solution called Cloudine (*Cloud Engine*), whose objective is to provide mechanisms to the development of elastic applications in different programming models and supporting exploration of horizontal and vertical elasticity.

3   Cloudine: Elasticity via Programming Primitives

Cloudine is a platform for development and deployment of elastic scientific applications in computational clouds. The platform focuses parallel and distributed non-interactive batch applications that runs directly over the VM's operating system (IaaS clouds). Unlike the solutions proposed in Section 2, which are restrict to a programming model, our solution is based in low level elasticity primitives, with which is possible to construct applications in different models, according to the program structure.

These primitives enable the dynamic allocation and deallocation of resources in several levels, ranging from nodes of a virtual cluster, to virtual processors and memory of a node. Primitives for collecting information and monitoring data from the cloud system is also provided. Thus, the platform enables the development of elastic applications, by inserting basic building blocks into the source code, letting the application itself to control its elasticity.

In addition, Cloudine provides mechanisms to deployment and execution of applications using cloud resources in an automatic and easy way. The platform manages the creation of the virtual environment and coordinates the entire execution process, from source code submission to results collection phase.

To enable the development and deployment of elastic applications, Cloudine utilizes three main components, *Resources and Execution Manager*, *User Interface* and *Elasticity API*, as shown in Fig. 1, and described in next sections.

3.1   Resources and Execution Manager

The *Resources and Execution Manager* (REM) is the module that coordinates the applications execution and controls all resources allocation. This module is composed by five components, as presented in Fig. 2.

The *REM Server*, receives and analyses the requests sent by the other Cloudine modules and dispatches them to the appropriate component. These requests include the allocation of new virtual environments, allocation of complementary

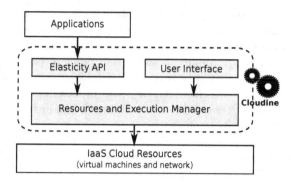

Fig. 1. Cloudine architecture

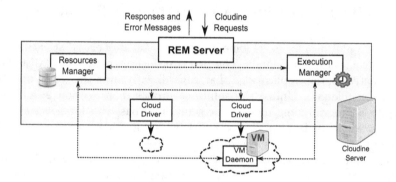

Fig. 2. Resources and Execution Manager modules

resources, deallocation of resources, status information and execution of applications. REM is also responsible for sending responses and error codes to other modules.

All requests involving resources are handled by the *Resources Manager*. It receives from REM Server the demands and allocates through *Cloud Drivers* the requested resources. Cloud Drivers are used to translate the Cloudine requests to a set of commands of a specific cloud, allowing the use of several providers since a driver is implemented for each one of them. Thus, it is possible to use a single interface internally, independently from the cloud used. To date, it is not possible to combine resources from different clouds.

The execution requests are processed by *Execution Manager*. This module is responsible for upload the application dependencies, build and execute the application. To perform these tasks on each VM, a *VM Daemon* is used. A VM Daemon is automatically started at the VM initialization and its role is to run the actions submitted by Execution Manager (e.g., copy file, compile, execute script) internally in the virtual machine, and collect the application status. VM Daemon is also used by Resources Manager to obtain the VM resources information.

3.2   User Interface

Cloudine provides a command-line *User Interface* (UI) that is used to request the initial virtual environment and to submit the applications. This module is also used to track the execution progress and receive error messages.

The virtual environment is configured in a file containing a set of attribute-value, that can vary according to the cloud (or driver) used. The attributes can include number of nodes, number of Virtual CPUs (VCPU), memory amount, virtual machine image or instance type, networking configurations and others. These settings are used to create a template that is sent to REM, which provides the virtual environment creation.

Similarly, the application execution is described by an attribute set that comprises script name, the applications dependencies (files to be uploaded) and the output directory, where results will be saved. The execution is requested to REM via command-line, which coordinates the process. An example of configuration file is presented in Fig. 3.

```
[resources]
cloud=OpenNebula
env_type=cluster
nodes=4
vcpu=2
memory=512
image=ubuntu.img

[app]
name=test
exec=exec.sh
exec_on_node=1
work_dir=/home/guilherme/test
```

Fig. 3. Example of a simple configuration file

In this example, a cluster of four nodes is requested in an OpenNebula cloud. Each node (VM) has two virtual CPUs (vcpu), 512 MB RAM, and uses an image called *ubuntu.img*. The rest of the file describes the application name, the script name (automatically executed on node 1), and the application work directory.

3.3   Elasticity API

The *Elasticity API* provides a set of primitives that enable the implementing of elastic applications for the Cloudine platform. The use of an API to develop elastic applications is interesting because the elasticity control is done by the application itself, which relies on internal information and monitoring data coming from the platform to request its own resources and does not require external mechanisms or user interaction.

Besides the creation of elastic applications, another possible use of the API is the construction or adaptation of parallel processing middleware that transparently supports elasticity. This feature could be used, for example, to adjust the processing capability when occur changes in the number of threads in an OpenMP application through `omp_set_num_threads()` function. A second example is the creation of a new virtual machines when a new MPI process is created with the `MPI_Comm_spawn()` primitive, defined in MPI-2 standard.

To date, the API supports C/C++ languages and offers 12 primitives, providing dynamic allocation of VCPUs, memory and virtual machines. Information about CPU and memory is also provided. As required, new primitives can be added in the future. Table 1 shows the functions implemented so far and their descriptions.

Table 1. Cloudine API functions

Function	Description
`int clne_add_vcpu(int N)`	Add N VCPUs to the current VM. Return 1 if successful, 0 if not.
`int clne_rem_vcpu(int N)`	Remove N VCPUs from the current VM. Return 1 if successful, 0 if not.
`int clne_add_node(int N)`	Add N nodes to the virtual environment (cluster). Return 1 if successful, 0 if not. This function also creates (or updates) a file in the VM containing the IPs of the cluster machines.
`int clne_rem_node()`	Turn off and remove the actual node from the virtual environment (cluster). Return 1 if successful, 0 if not.
`int clne_add_memory(long int N)`	Add N megabytes of memory to the current VM. Return 1 if successful, 0 if not.
`int clne_rem_memory(long int N)`	Remove N megabytes of memory from the current VM. Return 1 if successful, 0 if not.
`int clne_get_freemem()`	Returns the free memory amount of the VM host machine.
`int clne_get_maxmem()`	Returns the total memory amount of the VM host machine.
`int clne_get_mem()`	Returns the total memory amount of the current VM.
`int clne_get_freecpu()`	Returns the free CPU amount of the VM host machine.
`int clne_get_maxcpu()`	Returns the total CPU amount of the VM host machine.
`int clne_get_vcpus()`	Returns the CPU amount of the current VM.

To use the API the user must include the Cloudine library `clne.h` and compile passing the "`-lclne`" flag to compiler. The files `clne.h` and `libclne.so` are automatically provided in the VMs instantiated by Cloudine.

All primitives are implemented in libclne.so, that makes the requests to REM. When a primitive is invoked, a message including the operation and the arguments is assembled and sent to Resources Manager via TCP sockets. Resources Manager sends the request to Cloud Driver, that executes the operation in the cloud. According to operation result, a return code is sent back to application. Thus, depending on the cloud used or the operation, the response time can vary from few milliseconds (e.g., get operations) to minutes (VM request).

It is important to highlight that not all clouds support all functions. For example, Amazon EC2 does not support the dynamic allocation of memory or VCPUs, supporting only the allocation and deallocation of virtual machine instances. The information about the supported functions must be implemented in the corresponding Cloud Driver.

4   Cloudine Operation

To demonstrate how Cloudine works, this section presents an example of using the platform to instantiate a virtual machine and to run an elastic application. The process is illustrated in 14 steps as shown in Fig. 4 and explained in sequence.

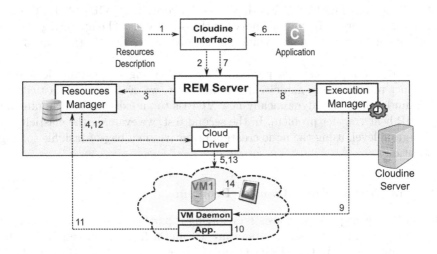

Fig. 4. Resources allocation and application execution steps

The process starts when (1) the user request a virtual machine via Cloudine Interface; REM Server (2) receives and (3) forwards the request to the Resource Manager that (4) chooses the correct Cloud Driver and sends an allocation command. (5) A new virtual machine (VM1, in the figure) is created in the cloud.

After the VM creation, (6) the application is submitted to execution, that is handled by (7) REM Server and (8) Execution Manager. (9) Execution Manager uses the VM Daemon to (10) start the elastic application on the VM1.

To exemplify the use of elasticity API, suppose that application request a new VCPU. This request is (11) sent to Resources Manager that (12) asks to Cloud Driver if this operation is supported by the cloud. If the allocation is possible, (13) the Cloud Driver sends the command to the cloud. Finally, (14) the VCPU is added to the VM. Note that the communication between application and Resource Manager is provided by the Cloudine API.

5   Tests and Results

In order to evaluate the proposed platform, a prototype was implemented using Python, C and Shell Script. An OpenNebula cloud [27] (version 3.4) using Xen as virtualization technology was used to perform the experiments. The choice of a private OpenNebula cloud was made based on the flexibility for VMs creation, which allows the configuration of resources according to application needs, unlike Eucalyptus and OpenStack, in which a pre-configured instance class must be chosen. The Xen Hypervisor was used due its capacity to handle the allocations of VCPUs and memory on-the fly, supporting all Cloudine API primitives presented in Table 1.

The hardware used is a 24-core workstation equipped with 3 AMD Opteron 6136 at 2.40 GHz and 96 GB RAM. We used 4 cores and 4 GB RAM for dom0 and the remaining CPUs and memory for the other VMs. The operating system in all tests, including virtual machines, is Ubuntu Server 12.04 with kernel 3.2.0-29.

To validate our proposal, we performed two tests using applications developed in distinct programming models. In the first test, we evaluate the elasticity in fine-grain level, adding dynamically new VCPUs to an existing VM running an OpenMP heat transfer problem. In the second test, we evaluate the elasticity in coarse-grain level, using the node creation primitives in a bag-of-task file sorting application.

5.1   OpenMP 2D Heat Transfer Problem

The heat transfer problem consists in solving a partial differential equation to determine the variation of the temperature within the heat conducting body. The code is implemented in C and uses OpenMP to provide the threads parallelism.

The application has an iterative loop that determines the temporal evolution of the simulation. At the beginning of the loop, a `clne_get_freecpu()` primitive was inserted in the source code to verify if there are idle CPUs. If CPUs were available, they are allocated to the VM using the `clne_add_vcpu()` primitive. The number of active OpenMP threads is set to the number of VCPUs of the VM.

In Fig. 5 is presented the results of the elastic allocation of resources. In the beginning of the simulation, the VM has only 2 VCPUs and 10 GB RAM. In this experiment, every 240 seconds 2 new VCPUs are available to allocation, as shown in Fig. 5.

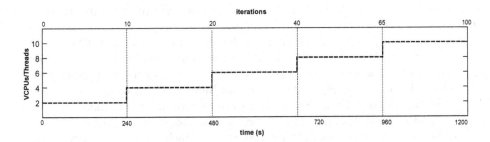

Fig. 5. Elastic allocations of VCPUs

Figure 6 shows the execution time and the speedup of normal and elastic versions. The average execution time (1,230 seconds) and the speedup (5.2) obtained by the elastic version are very close to those obtained by the normal version using 6 threads. The overhead caused by the insertion of the primitives is 4.6% or 56 seconds.

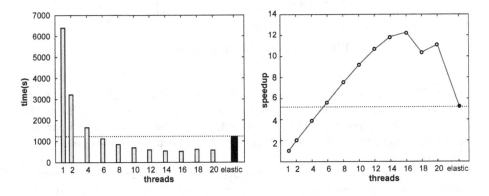

Fig. 6. Execution time and speedup

It is important to note that the inelastic version in the same scenario (with only 2 VCPUs available at the starting time) present average execution time of 3,220 seconds, i. e., 2.6 times slower than the elastic version. It is clear that the performance gains may vary according to the amount of available resources during execution, achieving more significant results when resources become available early or in greater quantities.

5.2 Bag-of-task Files Sorting

In second experiment, we tested a file sorting application implemented in bag-of-task model. The application consists of one server and N workers. The role of the server is to control the list of files to be sorted and to create dynamically the workers. The workers only sort the files.

Initially, the server is started and immediately instantiates a new VM to host a worker using `clne_add_node()`. Next, the worker application is upload and started in VM. The worker communicates with the server to get a file to sort, download the file and perform the operation. New workers are created (if there are available resources) while there are files to be sorted. At execution's end, all worker nodes are deallocated using the `clne_rem_node()` primitive.

In the tests, we used 400 files with 100,000 lines each. A timeline showing the workers allocation is presented in Fig. 7. The white balls represent the allocation request sent by the server and the black balls represent the moment when the worker operation is started.

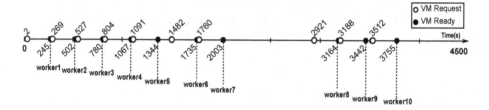

Fig. 7. Dynamic allocation of workers

At the beginning of the execution, 7 workers are sequentially allocated, taking advantage of available resources. After 900 seconds, new resources became available and the last three workers are allocated. The experiment ends at approximately 4,500 seconds, using 10 workers. The VM allocation time is on average 4 minutes.

As in the previous experiment, the platform proved to be efficient in allocating resources elastically, enabling the use of idle resources of the private cloud, and consequently improving application performance.

6   Conclusion and Future Work

In this paper we presented a platform for development and deployment of elastic scientific applications in IaaS clouds. The main contributions of our platform is the exploration of the elasticity using simple primitives, allowing the development of applications in several paradigms and models, and exploring horizontal and vertical elasticity.

The results show that Cloudine was successfully used to provide elasticity for different applications, being used to reduce the execution time and to increase the utilization of shared resources in a private cloud. Besides, the platform could be also employed in other programming models and scenarios, depending on the application and goals to be achieved.

As future work, we plan to develop elastic parallel middleware (e. g. Elastic OpenMP) with which users will be able to construct applications transparently, or to port applications with few modifications in source code. We also intend to develop cloud drivers to support different clouds, such as, OpenStack, Eucalyptus and public clouds.

Acknowledgments. This work is supported by CAPES and INCT-MACC (CNPq grant nr. 573710/2008-2).

References

1. Villamizar, M., Castro, H., Mendez, D.: E-Clouds: A SaaS Marketplace for Scientific Computing. In: Proceedings of the 2012 IEEE/ACM Fifth International Conference on Utility and Cloud Computing, UCC 2012, pp. 13–20. IEEE Computer Society (2012)
2. Badger, L., Patt-Corner, R., Voas, J.: DRAFT Cloud Computing Synopsis and Recommendations Recommendations of the National Institute of Standards and Technology. Nist Special Publication 146 (2011)
3. Leymann, F.: Cloud Computing: The Next Revolution in IT. In: Proc. 52th Photogrammetric Week, pp. 3–12. Wichmann (September 2009)
4. Chieu, T.C., Mohindra, A., Karve, A.A., Segal, A.: Dynamic Scaling of Web Applications in a Virtualized Cloud Computing Environment. In: Proceedings of the 2009 IEEE International Conference on e-Business Engineering, ICEBE 2009, pp. 281–286. IEEE (2009)
5. Armbrust, M., Fox, A., Griffith, R., Joseph, A.D., Katz, A., Konwinski, A., Lee, G., Patterson, D., Rabkin, A., Stoica, I., Zaharia, M.: A View of Cloud Computing. Commun. ACM 53(4) (April 2010)
6. Wang, L., Zhan, J., Shi, W., Liang, Y.: In Cloud, Can Scientific Communities Benefit from the Economies of Scale? IEEE Transactions on Parallel and Distributed Systems 23(2), 296–303 (2012)
7. Oliveira, D., Baio, F.A., Mattoso, M.: Migrating Scientific Experiments to the Cloud. HPC in the Cloud, http://www.hpcinthecloud.com/hpccloud/ 2011-03-04/migrating_scientific_expe
8. Galante, G., Bona, L.C.E.: A Survey on Cloud Computing Elasticity. In: Proceedings of the International Workshop on Clouds and eScience Applications Management, CloudAM 2012, pp. 263–270. IEEE/ACM (2012)
9. Chohan, N., Castillo, C., Spreitzer, M., Steinder, M., Tantawi, A., Krintz, C.: See Spot Run: Using Spot Instances for Mapreduce Workflows. In: Proceedings of the 2nd USENIX conference on Hot Topics in Cloud Computing, HotCloud 2010. USENIX Association (2010)
10. Amazon Web Services, http://aws.amazon.com/
11. Microsoft Azure, http://www.windowsazure.com/
12. Rackspace, http://www.rackspace.com/
13. Vaquero, L.M., Rodero-Merino, L., Buyya, R.: Dynamically Scaling Applications in the Cloud. SIGCOMM Comput. Commun. Rev. 41, 45–52 (2011)
14. Byun, E.K., Kee, Y.S., Kim, J.S., Maeng, S.: Cost Optimized Provisioning of Elastic Resources for Application Workflows. Future Gener. Comput. Syst. 27(8), 1011–1026 (2011)

15. Iordache, A., Morin, C., Parlavantzas, N., Riteau, P.: Resilin: Elastic MapReduce over Multiple Clouds. Technical Report RR-8081, INRIA (2012)
16. Raveendran, A., Bicer, T., Agrawal, G.: A Framework for Elastic Execution of Existing MPI Programs. In: International Symposium on Parallel and Distributed Processing Workshops and PhD Forum, IPDPSW 2011, pp. 940–947. IEEE (2011)
17. Rajan, D., Canino, A., Izaguirre, J.A., Thain, D.: Converting a High Performance Application to an Elastic Cloud Application. In: 3rd International Conference on Cloud Computing Technology and Science, CLOUDCOM 2011, pp. 383–390. IEEE (2011)
18. GoGrid, http://www.gogrid.com/
19. Caron, E., Rodero-Merino, L.F., Desprez, A.M.: Auto-Scaling, Load Balancing and Monitoring in Commercial and Open-Source Clouds. Technical Report 7857, INRIA (2012)
20. Google App Engine, http://code.google.com/appengine
21. Lim, H.C., Babu, S., Chase, J.S., Parekh, S.S.: Automated Control in Cloud Computing: Challenges and Opportunities. In: 1st Workshop on Automated Control for Datacenters and Clouds, ACDC 2009, pp. 13–18. ACM (2009)
22. Roy, N., Dubey, A., Gokhale, A.: Efficient Autoscaling in the Cloud Using Predictive Models for Workload Forecasting. In: 4th International Conference on Cloud Computing, CLOUD 2011, pp. 500–507. IEEE (2011)
23. Gong, Z., Gu, X., Wilkes, J.: PRESS: PRedictive Elastic ReSource Scaling for Cloud Systems. In: 6th International Conference on Network and Service Management, CNSM 2010, pp. 9–16. IEEE (2010)
24. Sharma, U., Shenoy, P., Sahu, S., Shaikh, A.: A Cost-Aware Elasticity Provisioning System for the Cloud. In: Proceedings of the 31st International Conference on Distributed Computing Systems, ICDCS 2011, pp. 559–570. IEEE (2011)
25. Calheiros, R.N., Vecchiola, C., Karunamoorthy, D., Buyya, R.: The Aneka Platform and Qos-Driven Resource Provisioning for Elastic Applications on Hybrid Clouds. Future Generation Computer Systems 28(6), 861–870 (2011)
26. Yu, L., Thain, D.: Resource Management for Elastic Cloud Workflows. In: Proceedings of the 2012 12th IEEE/ACM International Symposium on Cluster, Cloud and Grid Computing, CCGRID 2012, pp. 775–780. IEEE (2012)
27. OpenNebula, http://www.opennebula.org/

A Private Cloud-Based Architecture for the Brazilian Weather and Climate Virtual Observatory

Rafael Duarte Coelho dos Santos, Luiz Alberto R. Correa,
Eduardo Martins Guerra, and Nandamudi L. Vijaykumar

Brazilian National Institute for Space Research, São José dos Campos - SP, CEP
12227-010, Brazil
rafael.santos@inpe.br

Abstract. Brazil has a significant amount on a wide range of data about
weather and climate, collected from sensors or calculated by numerical
models, and are very important for historical reasons for the understanding of climate change and prediction of extreme weather events. This data
represent different physical measures, have different temporal and spatial scales and is stored in different formats; there is no unified way to
discover which data is available and under which conditions it can be
used.

In this paper we describe the architecture of the Brazilian Weather
and Climate Virtual Observatory, a set of software tools that works as a
Virtual Observatory (VO) to allow weather and data metadata discovery
and data access and processing. The VO will be partially deployed in a
private cloud; reasons and benefits of doing so will also be explained.

Keywords: Virtual Observatories, Private Cloud, Metadata,
Distributed Data Processing.

1 Introduction

Brazil now has a significant amount of information on a wide range of climate-
and weather-related measured variables and predictions. There is a growing understanding of the expected impacts of climate change and extreme weather
phenomena and a strong interest on better understanding it. Strategies to face
the effects of climate-related phenomena are important for all the government
sectors: national, state and local, for commercial enterprises (tourism, farms and
agricultural cooperatives, etc.) and for the community in general.

It is a fact that relevant data and information on climate and weather belong to several organizations that deal with different sectors of expertise. Each
organization has its own agenda of policies to operate and therefore such data
and information are stored in files and databases with formats and resolutions
(time and space) that most suit its needs and applications, often with different
public or private access policies and interfaces. That is, data and information are

B. Murgante et al. (Eds.): ICCSA 2013, Part V, LNCS 7975, pp. 295–306, 2013.

distributed and diverse in terms of completeness, formats, quality, etc., sometime with incomplete or unavailable metadata, making it hard to know which data with a specific time and/or spatial coverage is even available for queries. Therefore, it is more than natural that decision-makers, researchers, students and common citizens face difficulties to access such data. The difficulty is increased when there is need to use multiple data sources or combine data with different time and spatial scales, moreover when these data products must be kept up-to-date; and sometimes the most difficult task is to discover which type of data is available for a specific need.

Solutions to facilitate planning, policy-making, decision taking related to this kind of data must be made available. In order to achieve this, it is necessary to figure how to make available and accessible complete, reliable, good quality and easy to use information on climate and weather related data, without the need to reformulate the already existing systems, databases and interfaces and allowing the discovery of existing datasets and data processing operations. Along with the data itself, online applications that process this data (e.g. for classification, regression, summarization, prediction, visualization, etc.) could also be catalogued so users could combine applications and datasets to create their own workflows for weather and climate data analysis.

One approach to solve similar data dissemination and utilization problems was proposed by the astronomy and astrophysics community more than ten years ago: Virtual Observatories (VOs). Virtual Observatories are frameworks that use information technology (IT) to organize, maintain and explore information on large, distributed and dynamic datasets [1,2]. Within this framework it is possible to catalog data; process, visualize or cross-correlate it with tools both on the desktop and on the web; generate new data collections and create and use workflows and processing pipelines to automate new analysis and discoveries. The concept of Virtual Observatories is also being used in other science fields, such as Earth Sciences [3], Solar-terrestrial Physics [4], Environmental Sciences [5] and even Computer Science and IT itself [6]. Virtual Observatories are a natural extension of the paradigm of centralized middleware proposed to allow access to data and tool for specific domains [7,8], but allowing the inclusion of external data and tools, therefore increasing their usage and possibilities.

Figure 1 illustrates the outline of a Virtual Observatory. Users have access to portals that provide data catalogues, the data itself and processing tools, implemented as Web Services to ensure portability and flexibility. Users may also contribute with data and processing resources to the VO. The primary roles of a VO are to facilitate data discovery (through the portals or a registry of all catalogued data and services), data access (through web services or other methods, also allowing the use of local data) and data federation (combination of data from different sources) [1].

In this paper we describe the architecture of the Brazilian Weather and Climate Virtual Observatory, a set of software tools and data access tools that will enable users in different levels and with different skills to discover data, do basic analysis and visualization, using uniform data access protocols. This Virtual

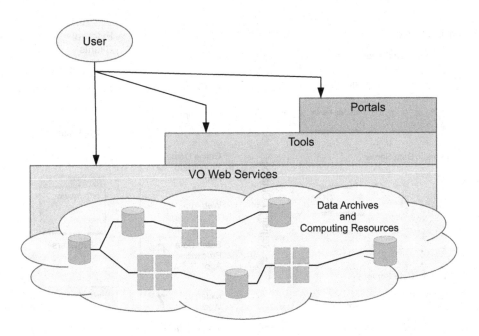

Fig. 1. Conceptual outline of a VO (adapted from [1])

Observatory will also enable users to include their own datasets in a catalog so other users can also access and use it.

This paper is organized as follows: Section 2 presents the general architecture for the Virtual Observatory, detailing its software components and the role of a private cloud for deploying the core functions of the Virtual Observatory. Section 3 comments on the present status of development and deployment of the VO and also on the future steps for the project and Section 4 present our conclusions.

2 The Weather and Climate VO Architecture

2.1 Introduction

In this section we describe the Brazilian Weather and Climate VO Architecture, in particular, its software components, their functions and how those components are integrated. We also explain the reasons to deploy the core part of the VO in a private cloud, and the expected problems and benefits of it. Finally, we describe a middleware that allows the execution of user-defined code inside the private cloud, effectively bringing the code "close" to the data (in the sense of reducing the need for sending the data through the network for processing), making possible the execution of algorithms that use large amounts of the data in an efficient way.

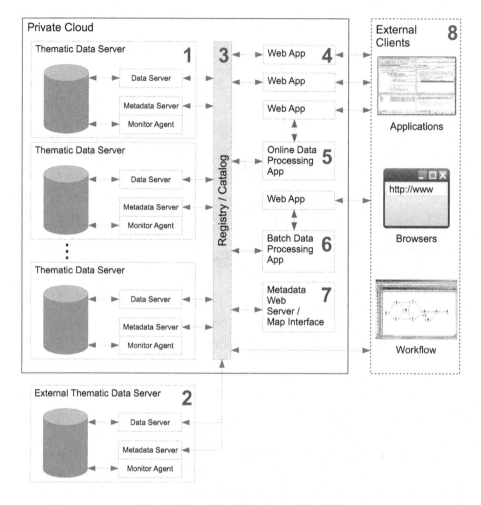

Fig. 2. General architecture of the VO

2.2 Architecture of the VO

Figure 2 presents the general architecture of the Virtual Observatory. The software components of the VO are grouped into three categories: private-cloud based server databases and applications (the VO core), external data server and external clients. Each major software component of the VO is labeled and will be commented in this section.

1: A **Thematic Data Server** contains data about a specific data collection or generation effort, usually with data related to a specific theme (e.g. rainfall, temperature, atmospheric electrical discharges, wind speed and direction, rainfall calculated from numerical models, etc.). The data on the server can also be spatially or temporally limited, e.g. there are databases that cover historic rainfall records for a particular state or only for a specific sensor which operated in a period of time.

The thematic data servers are self-contained systems composed of a database server with an associated data server that allows access to its data (optionally with constraints to avoid improper use). It is important to point out that under the VO architecture the databases are exposed to client applications only through their associated web services.

The thematic data servers also contain other two important software components: a Monitor Agent and a Metadata Server. The Monitor Agent regularly queries the database to extract metadata about it. In the VO context, metadata is information about the weather or climate data stored in the database, particularly, information about the data coverage (the spatial or temporal extent of the data on that database). This information is relayed to users and to the registry so users and applications can get information about the data before querying it.

2: **External Thematic Data Servers** are hosted outside of the private cloud, which is the expected scenario for collaborators who have their own databases already operational, possibly in other locations. In order to be able to share information with the VO these data servers must implement the Monitor Agent and Metadata Server, but this can be done without any real changes to the database itself, making possible the linking of legacy databases to the VO.

3: The **Registry** or Catalog is the most important component on the Weather and Climate Virtual Observatory. It is a central repository of information about data, metadata and tools for the VO, and can be searched by geographic coordinates, time intervals, data type, provider, keywords and combinations of those; returning a set of resources (usually web services) that can be used to get the data itself. Interaction with the registry can be done via a web interface or via web services, so they can be used directly by other applications.

The registry will be feed metadata from each thematic data servers' monitor agents through their metadata servers, ensuring that at any time the final user or application can find what data is available and its restrictions.

4: **Web Applications** are interfaces to the VO registry and to the databases associated to the VO. These applications are implemented as simple web services that perform queries on the databases or registry and return the results to the client applications. These web applications can also compose results from databases or other applications; distribute and aggregate queries, execute specific algorithms. The main difference between the data servers that are part of the thematic data servers and the web applications is that the former are designed to allow access to chunks of data with as little processing as possible (e.g. extrema and averages on time series), while the latter are applications that may be able to answer more complex queries that may involve more complex algorithms.

5: **Online data processing applications** are applications specifically designed to query and process one or more data servers and/or the registry to create results in almost real-time (i.e. being able to use the most recent records in the database). These applications, by their nature, must not be data- or CPU-intensive. These applications will interface with external users and applications through specific web applications that will work as portals to the data processing applications, i.e. interfaces that are able to call the applications, passing parameters and returning results to the final user.

Some possible examples of these applications are static visualization tools, for example, tools that overlay data points over a map.

6: Batch data processing applications are also applications that will access data on the data servers and registry, process this data and return it to the final users, through specific web applications that control the execution of the data processing applications. As the name of this software component points out, execution of applications will be done in batch, therefore it is possible to run more data-intensive and/or CPU-intensive algorithms but without guarantee that the results will be delivered in real time.

Since applications developed with this model will be executed in batch the web applications that control it must implement basic batch processing techniques: implementation of priority queues, running and monitoring processes, batch communication of results to users, basic authentication mechanisms, etc.

One important aspect of batch data processing applications is that they will use a framework that allows the execution of user-defined code in a sandbox. This will be described in subsection 2.4.

7: A **Metadata Web Server / Map Interface** that is a specific web application that allows the visual discovery of the available data. This application will present two ways to discover data: one by web services, proper for interaction with other applications, that will be able to list all data sources corresponding to specific constraints (e.g. data type, time and spatial limits, data quality/completeness, etc.). This application also will present to the users a visual interface, with the results for sources overlaid in a map, similar to SciScope (www.sciscope.org). With this application users will be able to quickly locate regions in space and time which contains the data of interest.

8: External clients that use the metadata and data available through the VO. We expect to have several types of clients of the VO data and metadata, such as applications developed by expert users that access the data and metadata through web services, simple clients like browsers and workflow management systems, that are able to visually compose the web services available to answer specific questions.

2.3   A Private Cloud for the VO

As described in Figure 2 the core functionality of the VO (some of its data servers, the registry, some applications) will be deployed in a private cloud, i.e. a cloud computing environment deployed in and operated by a single company or institution.

One could expect that the deployment of VO tools in a private cloud is a contradiction of the open, distributed nature of resources a VO is supposed to provide. There are several reasons for deploying the core of the VO infrastructure in a private cloud, which, in our opinion, more than justify its adoption:

- The web applications and the registry will be hosted in the same physical environment: hardware that makes part of the private cloud are connected through a high-performance internal network, ensuring fast access to the data servers.
- Thematic data servers may have lots of features in common; templates for the virtual servers can be created in order to facilitate the deployment of new data servers.
- The generic advantages of the cloud (sharing of resources, expected reduced costs, quick and easy deployment of servers, tools to increase the pool of resources, etc.) also apply.
- Monitoring the performance of the virtual data servers and web applications could give interesting insights about usage, which could lead, for example, to changes in the resources allocated to the servers. Since all those servers are hosted in a private cloud, the cloud manager itself could give information on the performance and loads on the servers.

Additionally it must be pointed that deploying the VO core infrastructure in a private cloud does not prevent or hinder the deployment of other tools or external data servers outside the cloud.

2.4   Running User-Defined Code on the VO Servers

Let's consider a simple use case of the VO: calculating simple statistics on a set of data with some constraints. For example, one could want to discover the largest difference between consecutive monthly averages of temperatures (i.e. greatest variation in consecutive months), constrained or not to a geographic region. Implementation of this algorithm is straightforward, and it can be made simpler if there are already web services to provide monthly temperature averages from some weather stations.

Expert users could implement this algorithm and query the appropriate web services to get all the data they need, but that would imply in running several web services and transferring their results over the Internet to the client application. It would be more convenient to prototype the algorithm in a small subset of the data, then transfer the algorithm implementation to the Batch Data Processing App (shown in Figure 2) so it would run "close to the data" improving its efficiency.

In order to achieve this we propose a middleware that will be one of the components of the VO and that allows the execution of user-defined code in a sandbox. By using this feature, a user will be able to develop his own processing algorithm and submit it to be executed in the server. The code submitted by the user should be described by custom metadata that will be used by the middleware to provide the right services to its classes [9]. The submission will be composed by an archive with classes and metadata descriptors as presented in Figure 3.

The code submitted to the server is able to define parameters to be received in each execution request as input. As output the class should have a result and

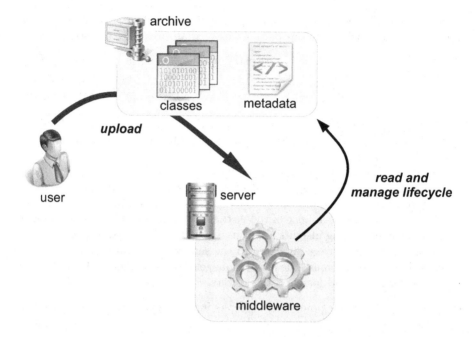

Fig. 3. Submission of user-defined code for execution on the server

optionally log entries. The class attributes used to store the processing inputs and outputs will be defined using metadata and can be used to dynamically generate graphical or programmable interfaces to a request submission. For a safe execution, this class will have restricted access to resources in the server. The middleware will be responsible to inject into the class the services necessary for its execution [10]. The instance injected is encapsulated with a proxy that is responsible to monitor the service access. The services retrieved by the middleware and inserted in the user class are also determined by metadata. Figure 4 presents a representation of this architecture.

The execution life cycle will be managed by the middleware. It should be able to ensure restrictions in the execution of external classes, being allowed to interrupt the execution if necessary. A configurable policy to limit the execution time or the number of services requests can be used to avoid the consumption of services resources by only one process. The metadata-based API used to define the classes allows a flexible definition of the services provided by the server [11]. This kind of solution also increases the decoupling between the framework and the submitted class, allowing each one of them to evolve independently. Consequently, a code submitted in an earlier version of the middleware will work on new versions, even if it evolve the existing metadata schema. The motivation of this architecture is to allow users to available new executions on the server. The middleware is used to ensure that the process will have access to the services that it needs and that its execution will not harm other processes spending excessive computational resources. As a result, it will provide a safe environment,

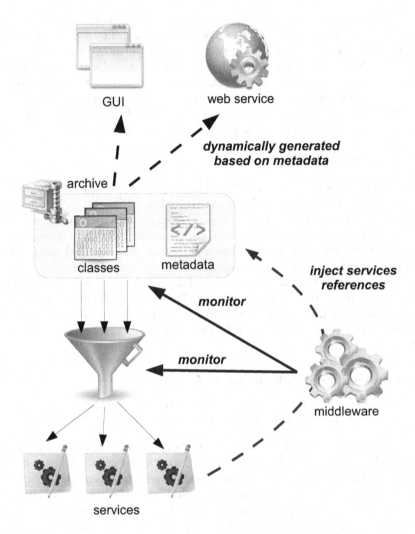

Fig. 4. Interaction between middleware and the submitted archive

in which the algorithms execution is performed close to the data, consequently avoiding excessive network invocations and having a better performance.

The middleware implementation would allow the execution of specific, user-defined algorithms with a more efficient access to the data. This would allow the implementation and deployment of data-intensive algorithms such as the used in data mining and static visualization applications [12].

3   Present Status

The Brazilian Weather and Climate Virtual Observatory is an ongoing research and development project – working results are, so far, not visible to the public. We already have three thematic data servers (one with precipitation data, one with atmospheric discharges and one with ground temperature). The next steps for those thematic data servers will be the implementation of the monitor agent and metadata server. After that we will able to deploy the registry and start publishing data and services, and after that the web applications. We expect to be able to get a first working implementation of the basic services and tools shown in Figure 2 by the end of 2013 – this step is called "first light", and it is borrowed from astronomy, meaning the first data collection from an astronomical instrument.

3.1   Next Steps

Future steps in this project will be based on users' demand. There is an existing "wish-list" of tools and services that was collected by CPTEC (INPE's Center for Weather Forecasting and Climate Research) from its data users and researchers. The online and offline processing tools will be chosen to satisfy some needs of the users of the data and to show the potential of using the VO paradigm for the development of new tools and solutions.

Another important future step is the validation of the web services included in the VO for use with workflow tools (e.g. Taverna [13]). These tools allow the visual definition of processing steps to find, collect and process distributed data and will prove very valuable for quick exploration and prototyping of additional tools for the VO.

3.2   Research and Implementation Challenges

Some noteworthy research and implementation challenges being considered at the moment for the Brazilian Weather and Climate Virtual Observatory are:

– Develop a metadata scheme that can deal with spatiotemporal data coverage in an efficient and compact way. Some of the data that will be stored in the thematic data servers can be represented as time series containing specific physical measures related to a geographic location (e.g. daily rainfall in a city). The metadata for this data server must represent the period in which the data was collected and its frequency, but also must somehow represent existing gaps in the data, so a potential user can know, beforehand, if that dataset will be suitable for his/her needs. The specific challenge is to represent the gaps in a way that will not make the metadata itself too large or complex.
– It is expected that the VO may collect redundant data or data that could be substituted by alternative data under certain conditions. For example, some large cities have their own meteorological stations, that may collect data with a different time frequency from other sources. Mechanisms to indicate alternative data sources could be implemented, considering spatiotemporal coverage and data quality indicators [14,15].
– Some of the tools for the VO (e.g. the ones implemented in the Batch Data Processing Application) may require a large amount of CPU cycles, but at the same time they may not need to be executed frequently – one example is the visual outliers map tool [12], that shows which time series are too divergent from

similar time series in a neighborhood, and can be used to detect problems in the data collection. These data- and CPU-intensive applications may demand additional resources from the private cloud, so we must investigate ways to automatically deploy additional resources when needed, and release them after computation is complete. As we mentioned in subsection 2.3, monitoring the performance of the virtual servers in the cloud may also give interesting insights on its operation, which could be used to optimize the resources available.

4   Conclusions

This paper presented the proposed architecture for a private cloud-based Virtual Observatory (VO) for Weather and Climate data, still under development. Some tools for real use will be created, and efforts to garner support from the scientific community and population in general will be done as part of the VO objectives.

Most of the resources (hardware, software and data) for the Brazilian Weather and Climate Virtual Observatory ill be initially developed by and hosted at INPE, the Brazilian National Institute for Space Research. INPE collects and generates, through its several scientific missions, a very large amount of earth observation data, including meteorological data from sensors and weather model simulations. Some of this data is presently available through web interfaces designed for human use [16], but without data access integration, full metadata or centralized information about it.

Besides the VO itself, some of the results we expect to achieve with the development and implementation of it are:

- Educational and outreach material for users of the VO tools (e.g. code samples, simple fully documented applications, tutorials for using the available data and including the users' own data on the VO, etc.).
- Know-how on data federation, curation, publication and distributed processing that can be applied to other data-related research at INPE and other institutions.
- Studies on how the Brazilian Weather and Climate Virtual Observatory can integrate with other existing frameworks that could benefit of data interchange (e.g. the Environmental Virtual Observatory, http://www.evo-uk.org/).

Acknowledgments. The authors would like to acknowledge the grants provided by the Brazilian Space Agency (AEB) and Brazilian Research Council (CNPq), process number 560188/2010-2.

References

1. Djorgovski, S., Williams, R.: Virtual observatory: From concept to implementation. In: Proceedings of the Astronomical Society of the Pacific Conference Series, vol. 345, pp. 1–14. Astronomical Society of the Pacific (2005)
2. Szalay, A.S.: The national virtual observatory. In: Astronomical Data Analysis Sofware and Systems X. ASP Conference Series, vol. 238, pp. 3–12 (2001)

3. Donnellan, A., Rundle, J., Fox, G., McLeod, D., Grant, L., Tullis, T., Pierce, M., Parker, J., Lyzenga, G., Granat, R., Glasscoe, M.: Quakesim and the solid earth research virtual observatory. In: Yin, X., Mora, P., Donnellan, A., Matsu'ura, M. (eds.) Computational Earthquake Physics: Simulations, Analysis and Infrastructure, Part II, pp. 2263–2279. Springer (2007)

4. Fox, P., McGuinness, D., Cinquini, L., West, P.G., Benedict, J.L., Middleton, D.: Ontology-supported scientific data frameworks: The virtual solar-terrestrial observatory experience. In: Computers and Geosciences, pp. 724–738 (2009)

5. Gurney, R., Emmett, B., McDonald, A., Blair, G., Buytaert, W., Freer, J.E., Haygarth, P., Rees, G., Tetzlaff, D.: EVO Science Team: The environmental virtual observatory: A new vision for catchment science. In: American Geophysical Union, Fall Meeting (2011)

6. Matray, P., Csabai, I., Haga, P., Steger, J., Dobos, L., Vattay, G.: Building a prototype for network measurement virtual observatory. In: Proceedings of the 3rd Annual ACM Workshop on Mining Network Data (MineNet 2007), pp. 23–28. ACM (2007)

7. Kiemle, S.: From digital archive to digital library - A middleware for earth-observation data management. In: Agosti, M., Thanos, C. (eds.) ECDL 2002. LNCS, vol. 2458, pp. 230–237. Springer, Heidelberg (2002)

8. Sinderson, E., Magapu, V., Mak, R.: Middleware and web services for the collaborative information portal of NASA's mars exploration rovers mission. In: Jacobsen, H.-A. (ed.) Middleware 2004. LNCS, vol. 3231, pp. 1–17. Springer, Heidelberg (2004)

9. Guerra, E., Oliveira, E.: Metadata-based frameworks in the context of cloud computing. In: Zaigham, M. (ed.) Cloud Computing - Methods and Practical Approaches, pp. 2263–2279. Springer (2013)

10. Fowler, M.: Inversion of control containers and the dependency injection pattern (2004), http://martinfowler.com/articles/injection.html

11. Guerra, E., Fernandes, C., Silveira, F.: Architectural patterns for metadata-based frameworks usage. In: Proceedings of Conference on Pattern Languages of Programs (2010)

12. Garcia, J.R.M., Monteiro, A.M.V., Santos, R.D.C.: Visual data mining for identification of patterns and outliers in weather stations' data. In: Yin, H., Costa, J.A.F., Barreto, G. (eds.) IDEAL 2012. LNCS, vol. 7435, pp. 245–252. Springer, Heidelberg (2012)

13. Hull, D., Wolstencroft, K., Stevens, R., Goble, C., Pocock, M.R., Li, P., Oinn, T.: Taverna: a tool for building and running workflows of services. Nucleic Acids Research 34, W729–W732 (2006)

14. Cruz, S.A.B., Monteiro, A.M.V., Santos, R.: Increasing Process Reliability in a Geospatial Web Services Composition. In: Proceedings of the 17th International Conference on Geoinformatics (2009)

15. Cruz, S.A., Monteiro, A.M., Santos, R.: Automated geospatial web services composition based on geodata quality requirements. Computers & Geosciences 47, 60–74 (2011)

16. Andrade, R.B., Nunes, L.H., Barbosa, E.B., Vijaykumar, N.L., Santos, R.D.C.: A web service-based framework for temporal/spatial environmental data access. In: Proceedings of the 12th International Conference on Computational Science and Its Applications, pp. 7–13. IEEE (2012)

Interactive Face Labeling System in Real-World Videos

Hai-Trieu Nguyen, Ngoc-Hien Nguyen, Thang Ba Dinh, and Tien Ba Dinh

Faculty of Information Technology, University of Science, Ho Chi Minh City, Vietnam
{nhtrieu,nnhien}@live.com, {dbthang,dbtien}@fit.hcmus.edu.vn

Abstract. We propose a robust semi-auto system of labeling faces for characters in video by combining face detection, tracking, and recognition. At the very first step, our system detects the faces automatically in each video frame. After that, the face detection responses will be linked together to form face sequences (raw tracklets) in each video shot by employing simple temporal-spatial constraints to associate. Then we apply a tracking algorithm to not only extend those raw tracklets bi-directionally to cover more appearance views of the object instead of focusing only the frontal view, but also to help fixing the "gaps" caused by missed detection. After being merged among the potential overlapped ones, in the next step, these extended non-overlapped face tracklets across the video are associated with each other by our proposed Heuristics clustering algorithm. In order to achieve high accuracy, we use both generative and discriminative appearance models of the faces and also the context information, which is the clothing color feature in our case. An extensive experiment is performed on approximately 3.5 hours of videos cut from two TV series "Friends", "How I Met Your Mother" to show the robustness of our system.

Keywords: Face labeling, interactive system, face detection, face tracking, face recognition, real-world videos.

1 Introduction

Face labeling in video is a process to identify the name of the characters in each frame of a video. Along with the dramatic increase in the amount of videos uploaded on the Internet, labeling faces is important for many applications such as providing training data for face recognition, and search engine. In fact, face tagging has been popular on Facebook and other social networks for several years. It helps users link their photos taken by their friends to their own account without need to download and re-upload them. However, labeling faces in video is a completely different story. To label them manually is an extremely tedious task, because, typically, a second of video contains 24 frames, which means an hour of it contains 86400 frames. That is a huge number enough to probably scare any manual labeler away. And naturally, manual processing on large data tends to get more errors when the operators get tired. Moreover, face recognition, which

B. Murgante et al. (Eds.): ICCSA 2013, Part V, LNCS 7975, pp. 307–320, 2013.
© Springer-Verlag Berlin Heidelberg 2013

Fig. 1. Example of Face Labeling in videos. (a) Friends; (b) How I Met Your Mother.

is the key factor to build an efficient naming system in videos, is a challenging problem since there are a lot of practical issues in real-world videos, where the environment is uncontrolled, such as large change of pose, illumination, expression, and occlusion.

To deal with these problems, naming characters in video by combining face sequences and textual information is one of the common trends [1,2,3]. However, this approach also faces with several complicated problems in correspondence between face-text, false positive detection, missed detection, ambiguous text information, especially when there is limited number of videos having subtitle. Later on, some researchers focused on video-based recognition using tracking and recognition with the purpose to create face models for training and recognizing faces of characters on videos [4,5,6,7]. Video-based recognition achieved many impressive results as using tracking to take the full advantage of both spatial and temporal information, which helps build effective appearance models for recognition. Although these frameworks improved remarkably the performance in naming characters, they still have not been applied to real-world videos widely since making a one-fit-all model or trained database is not trivial.

Recently, face detection and object tracking have become more and more mature, which motivates us to combine them together in order to build a complete interactive face labeling system to deal with real world videos. We employ a robust frontal face detector [8] to extract as many frontal faces in video as possible. After that, TLD (Tracking Learning Detection) [9], a recent robust tracking algorithm, is applied to help extend the face responses into tracklets by tracking them bidirectionally. Then, we employ temporal-spatial features to associate the tracklets together if they contain the same character. More than that, the context information, *i.e.* the appearance of the characters' clothes, is also exploited to build a robust similarity measurement between two tracklets. Finally, we cluster the tracklets into groups of same characters using our proposed Heuristics algorithm. Our contributions are: (i) build an interactive face labeling system for real world videos by combining detection, tracking and recognition algorithms effectively; (ii) propose a robust tracking framework to retrieve all face sequences in different viewpoints; (iii) Apply P-N learning model and clothing appearance for recognition; (iv) And propose an Heuristics algorithm for tracklet association.

2   Related Work

Labeling faces in unconstrained videos is a very challenging problem because of many difficulties such as variations in viewpoints, different illumination conditions, and changes in face expressions. Therefore, the performance of many model-based approaches completely depending on facial features such as facial motion [10] or active appearance model (AAM) [11], easily get affected when the conditions change. To improve the model-based systems, people exploited several additional cues such as hair appearance [12], global positioning information [13], or clothing appearance [14,15]. Some approaches used logical context information such as same person does not appear twice in the same photo to improve system performance [13,14,16].

Fortunately, in several typical types of videos such as movies, and TV shows there are many sources of information that can be used to improve the accuracy of recognition such as subtitles. Other than that, in most of videos, clothing appearance, and voice of the characters in the videos, are also great cues helping improve the recognition system. Name-it [1] was one of the earliest approaches using name and face association in news videos based on the co-occurrence between detected faces and transcripts. The similarity of the faces was defined as an eigenface while the names of the characters were extracted by natural language processing. However, this framework also faced with other challenges such as misalignment between face and text, miss face detection or false name extraction. Likewise, Yang et al. [2] prepared a set of named people, and extracted a variety of features and constraints from multiple videos before training a model which was used to predict the name of person. This approach strongly relies on the pre-trained model, which is hard to be applied in practice. To build a person identification for the sitcom Seinfeld, Li et al. [17] used both audio and visual data. They applied two strategies: audio-verify-visual and visual-verify-audio in the experiments. The first technique, which used faces to improve speaker identification, achieved recognition precision higher than the other one. Nonetheless, in most of the case, there is only one person speaking at time making it not feasible to label all of the characters in the same frame, which happens quite often in real-world videos.

Recently, Everingham et al. [5] proposed a robust approach for unsupervised labeling frontal face track in video, which combines several visual information cues such as faces, clothing, and textures, with subtitles . The frontal faces were detected in every frame and matched using a Kanade-Lucas-Tomasi tracker (KLT) [18] when they had not been continuously detected. In the recognition stage, voice detection and transcript are employed to help match the speaking person and their name. Later, all unlabeled point tracks are matched against with labeled point tracks using nearest neighbor algorithm. In addition, clothing cue is also used to support labeling decision when face descriptors are affected by lighting variation, face expression and so on. This approach achieved satisfactory results, which becomes to be an important premise for improvement later. However, this method does not provide a reliable tracker to help obtain face responses in every frame, which pushes all of the workload on the recognition module as the face tracklets are broken into a lot of small segments.

Based on the above method, Sivic *et al.* increased the coverage kind of labeled person to half-full profile faces in [6]. Köstinger *et al.* [19] renovated the recognition module by using a semi-supervised multiple instance learning algorithm to weakly incorporate face instances during the training phase, which improved the performance in the same datasets. Nonetheless, it is clear that these approaches depend heavily on subtitles and transcripts which are not usually available for a typical video daily uploaded by common users. Here, we target in solving the problem of naming the persons not only for movies or TV shows but also for any uploaded videos, given that the names of the persons are associated by some source. Our system retrieves and clusters all face tracklets into groups before the user manually tag the groups with any names they want, which is similar to the concept of Facebook tagging, except here, it is for videos, not for photos.

The rest of paper is organized as follow: Section 4 introduces overview of all components in our systems. Section 5 mentions details about tracking and recognition framework. Section 6 performs a number of comparative experiments in our dataset which built on 2 series "Friends", "How I met your mother". The paper finishes with our evaluations and suggestions for future research and practical applications.

3   Overview of Our Approach

Our system can be separated into 5 steps:

- Step 1 - Shot Boundary Detection: is implemented using a frame difference approach to split the video into shots.
- Step 2 - Face Detection: is applied to detect all frontal faces in every frame of the video.
- Step 3 - Tracklet Extraction: gathers face sequences of the characters in every video shot. Based on the detection responses, with some constraints about appearance and location, raw tracklets are formed by linking the face responses in consecutive. As soon as a raw tracklet, it is extended bidirectionally to form an extended tracklet by tracking. In a shot, local tracklet association using space-time constraints helps to merge similar tracklets to full tracklets.
- Step 4 - Global Tracklet Association: uses both TLD appearance model and clothing appearance to calculate the similarity between every pair of full tracklets before using our Heuristics algorithm to associate them and group them to the same character appearance in videos.
- Step 5 - Manual Refinement: the groups of tracklets can be labeled as well as easily modified, deleted, or merged by using some drag-drop operations in our interactive software.

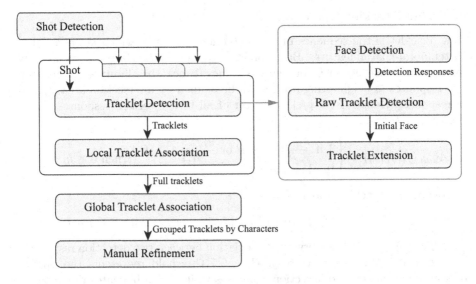

Fig. 2. System overview

4   Tracklet Extraction

The fact is that there is no perfect detector, which means miss detections, false alarms, and false positive exist. Even though, the faces are detected in every single frame of the video, there are "gaps" between frames, while non-face regions also show up as detected faces here and there. Hence, to achieve a complete face sequence in one shot, we have to use tracking to both link the incoherent face detection results, and remove the false alarms/positive. To extract all of face tracklets in a shot of video, our system follows two steps: Raw Tracklet Detection, and Tracklet Extension. The goal of Raw Tracklet Detection is to restrict false alarm caused by face detection. After that, they are extended into long tracklets by tracking to build robust appearance models which are extremely important for tracklet clustering in the recognition phase.

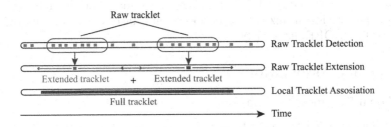

Fig. 3. Tracklet Extraction

4.1   Raw Tracklet Detection

Raw Tracklet is the sequence of face detection responses which are associated by temporal-spatial features. Based on the assumption that face targets are only move a small distance in L consecutive frames (time), the affinity score between two responses in neighboring frames can be defined by intersection area (space) of responses. Denote $F = \{f_i\}$ is the set of all face detection responses, S_{f_i} is the area of f_i region.

$$A_{link}\left(f_i, f_{i+m}\right) = \begin{cases} 1 \text{ if } \frac{S_{f_i \cap f_{i+m}}}{S_{f_i}} \geq P \text{ or } \frac{S_{f_i \cap f_{i+m}}}{S_{f_{i+m}}} \geq P \\ 0 \text{ otherwise} \end{cases} \text{ when } 0 < m \leq L$$

We define raw tracklet S as below:

$$S = \{f_{i_1}, f_{i_2}, ..., f_{i_n}\} \text{ when } A_{link}\left(f_{i_j}, f_{i_{j+1}}\right) = 1 \text{ with } 0 \leq j < n$$

To achieve an safe and conservative association between two detection responses, we use a dual-thresholds strategy. $P = 0.5$ threshold represents the rate of overlapping between two detection responses while $L = 3$ indicates the distance of them in time. It means two responses are linked if they are at most 3 frames far away from each other and their intersection area is bigger than a half of the area of each detection response.

4.2   Tracklet Extension - Tracklet Building

We employ TLD tracker [9] into our system to extend the face tracklets, as it is a robust and efficient algorithm, especially for face tracking [20]. It combines tracking, detection and learning to build a novel long-term tracking framework for an unknown object in a video stream. The tracker follows the object frame by frame, while the detector identifies all of the possible object regions, and the learning model is used to verify those responses to choose the best one as the tracking result. The object detector is updated using positive-negative (P-N) learning to improve the performance by identifying errors and avoid them in the future. It is robust in handling scale and illumination changes, background clutter, partial occlusions, and can run in real-time. In addition, the outcome of this approach is also a learned appearance model of the object built while tracking, which can be used for recognition. In our system, after a raw tracklet is detected, a TLD tracker is used to track forward and backward as far as possible. To retrieve the good face tracking results and to decline false tracking rate, this tracker is initialized by the most frontal face which is computed and evaluated by SHORE detector [8].

5   Tracklet Association

5.1   Local Tracklet Association

After being extended frame-by-frame, it is not guaranteed that we have a full tracklet for each character as it is not optimized. In the other words, the full set of images of a character can be split into several continuous segments which we call tracklets. Therefore, to assure that all tracklets are continuous and uninterrupted, we defined the standards to associate them to build full tracklets.

Time Period of Tracklet T_x: $t_x = \left[t_x^{begin}, t_x^{end}\right]$

Two tracklets T_x, T_y are associated together when

$$\frac{1}{|t_x \cap t_y|} \sum_{t \in (t_x \cap t_y)} Overlapped(r_x^t, r_y^t) > \theta_{st}$$

where r_x^t is the face region of tracklet x at time t, $\theta_{st} = 0.5$ is the acceptance threshold, and $Overlapped(r_x^t, r_y^t)$ is the intersection area of the detection responses between two tracklets at time t in space.

5.2 Global Tracklet Association

In the final step of our system, all tracklets are clustered into groups of the same character. We formulate the problem in graph and propose an Heuristics algorithm to solve it. Suppose that the video is a graph $G = (V, E)$, where V is the set of all nodes and E is the set of all edges connecting the adjacent nodes. The nodes are tracklets and the edges are the global distances between two tracklets, which is built by using TLD appearance model and clothing appearance. The clustering problem now can be formulated as the problem of how to separate the graph to multiple subgraphs which contain "close" nodes. In Algorithm 1, Disjoint Set Union-Find Data Structure is used for simplicity.

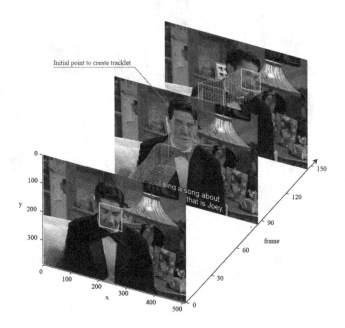

Fig. 4. Tracklet Extension in Space-Time

Algorithm 1. Heuristics clustering

Input: List of Tracklets
Output: Tracklet Groups
Operation:

- create-set(u): Create a set containing a single node u
- find-set(u): Find the set that contains a given node u
- union(u, v): Merge the set containing u and the set containing v into a common set
- size-set(u): return number element of the set that contains a given node u

Build graph $G = (V, E)$ from list of tracklets and set of global distances
for all node v in V **do**
    create-set(v)
end for
Sort E descending by weight
for edge (u, v) in E **do**
    if find-set(u) $\neq$ find-set(v) and [size-set(u)$< \theta$ or size-set(v)$< \theta$] **then**
        union(u, v)
    end if
end for
return collection of sets that mean groups of character

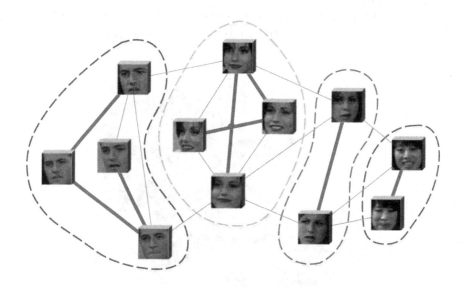

Fig. 5. Clustering by Heuristics Algorithm based on similarity of tracklets which is defined by P-N Learning Model and Clothing Feature

6   Appearance Model

6.1   P-N Learning Model

P-N learning method [21] includes two types of experts: P-expert estimates false negative before adding them to training set with positive label, otherwise N-expert estimates false positive to add training set with negative label.

In learning component of TLD, the appearance of object is represented by a model (M) which based on P-N learning method. It consists of positive patches and negative patches collections, $M = \{p_1^+, p_2^+, ..., p_m^+, p_1^-, p_2^-, ..., p_n^-\}$, where p^+, p^- represent object appearance and background information, respectively. In details, the patch is collected from an image in a specific bounding box and then it is normalized to resolution (15 x 15 pixels). According to TLD framework, similarity between two patches $S(p_i, p_j)$ is measured by *Normalized Correlation Coefficient* (NCC):

$$S(p_i, p_j) = 0.5(NCC(p_i, p_j) + 1)$$

Given an arbitrary patch p and an object model M:

- Similarity with the positive nearest neighbor: $S^+(p, M) = \max_{p_i^+ \in M} S(p, p_i^+)$
- Similarity with the negative nearest neighbor: $S^-(p, M) = \max_{p_i^- \in M} S(p, p_i^-)$
- Relative similarity S^r is defined the similar rate of patch p and positive patches collection of model M.

$$S^r(p, M) = \frac{S^+(p, M)}{S^+(p, M) + S^-(p, M)}$$

6.2   Clothing Appearance

In fact, the face recognition based on face descriptors have to face with many challenges caused by lighting variation, different face poses as well as relative similarity of character faces in some situations. Therefore, we decide to add clothing feature of characters to increase tracklet discrimination. Depending on the detection responses, the clothing bounding box is predicted based on the relative position and scale of face detection. As a result, every face detection response is attached by a clothing descriptor which is, basically, a color histogram in our case. Comparison between two clothing descriptors is defined by Bhattacharyya distance $S^c(f_1, f_2)$.

6.3   Global Model Appearance

In our approach, the similarity $S_{global}(T_1, T_2)$ of two tracklets is measured basing on TLD score S_{TLD} and clothing score $S_{clothes}$. To reduce the influence of variation expression or rotation face, we only choose a set of the frontal faces to compute this score. Let us denote E_i as the set of the frontal faces of tracklet i, M_i as the P-N learning model of tracklet i, this score is the basis for clustering the similar tracklets in character groups which works as follow:

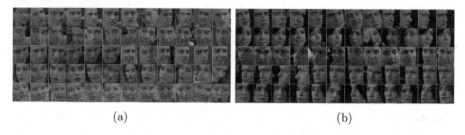

(a)   (b)

Fig. 6. Example of Tracklets in videos. (a) Friends; (b) How I Met Your Mother.

- All frontal faces E in tracklet T_1 are compared with the object model M_2 of tracklet T_2, and vice versa, by the relative similarity score:

$$S_{TLD}(T_1, T_2) = \frac{1}{2} \left(\frac{1}{|E_1|} \sum_{e \in E_1} S^r(e, M_2) + \frac{1}{|E_2|} \sum_{e \in E_2} S^r(e, M_1) \right)$$

- The clothing score is calculated:

$$S_{clothes}(T_1, T_2) = \frac{1}{|E_1||E_2|} \sum_{e_1 \in E_1} \sum_{e_2 \in E_2} S^c(e_1, e_2)$$

Then, the global score is calculated as following:

$$S_{global}(T_1, T_2) = \alpha S_{TLD}(T_1, T_2) + (1 - \alpha) S_{clothes}(T_1, T_2)$$

Empirically, we choose the weight $\alpha = 0.7$ represents for the priority to score TLD due to the fact that clothing of characters can be changed or different characters wear clothes with the same color.

7   Experiment Results

We choose several randomly videos in *Friends* and *How I met your mother* TV series to evaluate the performance of our system. The dataset consists of 60 videos with about 300,000 frames in total. All videos include many characters in various contexts. We compare the performance of face sequences retrieval using tracking in comparison with only face detection. Besides, we also present the performance of tracklet clustering.

7.1   Tracklet Retrieval Performance

Our system focuses on using TLD tracker to retrieve all face sequences, which are used in recognition phase as well as manual refinement. Therefore, we carry out this experiment to show the performance of our tracking framework for face retrieval in videos. It shows that number of detected faces in tracklets increases

Table 1. The results of Tracklet Detection. DF: Detected Faces, NT: Number of Tracklets, FT: Faces of Tracklets, FFT: False faces of Tracklets, ACRT: Average of Correct Rate of Tracklet

Videos	DF	NT	FT	FFT	ACRT
Friends	96427	1663	144513	5274	97.22%
How I Met Your Mother	125624	2273	165614	4392	98.05%

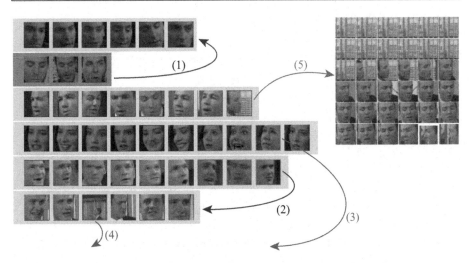

Fig. 7. Some mistakes in our system: **(1)** Separated Group; **(2)**,**(3)** False Groups Clustering; **(4)** False Alarm Tracklet; and **(5)** False Faces in a Tracklet caused by tracking. Tree based tracklet clustering helps us be able to sort these tracklets in depth-first pre-order, that parent node (□) is always on the left of child nodes (□), so these errors can be fixed by moving the root (very left node) of the wrong sub-tree.

significantly in comparison with only face detector. Besides, the percentage of true faces of the tracklets in our testing videos account is above 97% indicating the efficient combination of face detector and tracker in our tracking framework. Especially, the tracking process collected complete face sequences in videos which reduced remarkably the size of data for recognition. Particularly, face clustering only carried out 1663 tracklets in 30 videos of Friends and 2273 tracklets in 30 videos of How I Met Your Mother while it is 96427 faces and 125624 faces respectively if we use only face detector results.

7.2 Tracklet Clustering Performance

The groups of tracklets are the output of our system for a manual refinement step required users' operations. There are several mistakes in final output (Figure 7) which are caused by the false similarity evaluation or false tracklet clustering. To overcome this problem, we implement some basic drag-drop operations such as merging groups, changing group of tracklets, deleting groups or tracklets to

Table 2. The results of Tracklet Clustering. MMO: Minimum Manual Operations number; AP: Average of Precision of the system

Videos	Number of Tracklets	MMO	AP
Friends (30 videos)	1663	231	86.95%
How I Met Your Mother (30 videos)	2273	265	88.32%

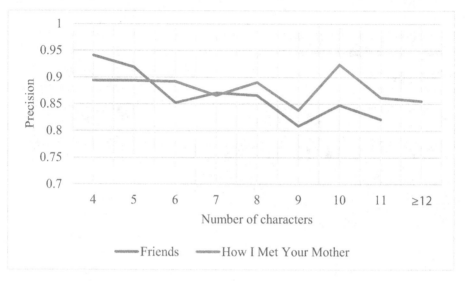

Fig. 8. The variation in performance of our system by number of characters in video

help users easily refine the clustering results. In this experiment, our system is evaluated by some metrics in the performance of clustering step. The metrics are involved:

- Minimum Manual Operation Number (MMO): the minimum number of manual operations to achieve the perfect result.
- Precision: the precision of a video is defined by the minimum number of operations divided by the total number of tracklets of the video.

The evaluation result is presented in Table 2. In details, average of precision in 30 videos of Friends is 86.95% with 231 operations to achieve best labeling data for these videos. Likewise, this figure is 88.32% with 265 operations in 30 videos in How I met your mother. In addition, we illustrates the variation in performance of our system by number of characters in video in Figure 8. As we can see, the number of manual operations completely depend on quantity of characters in video. In other words, if there are too many characters in video, the performance of our system degrade, so the users have to take more operations to achieve a perfect labeling result. On average, our system process a video containing 4500 frames in 15 minutes on a Intel Dual Core 2.0 GHz PC. Our previously successful short interactive videos can be found at the website [22].

8   Conclusion

In this paper, we proposed an interactive face labeling system in videos based on the combination of face detection, tracking and recognition. TLD tracker was combined with SHORE face detector to build an efficient tracking framework for face sequence extraction in videos. For recognition, we made use the object model of TLD tracker which is built in while tracking, and combined it with the clothing features to help making robust face associations. Finally, the system clusters tracklets into character groups by our Heuristics algorithm before user manually label the name groups. Our extensive experiments indicated very good results from our system. We also implemented user interface tools help users easily label the name of characters in videos and manually refine the results when needed. In the future, we plan to improve the tracklets recognition to achieve higher accuracy by using more context information of video such as social context as well as improving our clustering method to build a smarter and more robust interactive system.

Acknowledgments. We are very grateful to Fraunhofer Institute for providing us with the face detector SHORE [8], to Nam Vo and Loc Huynh for making a number of helpful comments. This research is supported by research funding from Advanced Program in Computer Science, University of Science, Vietnam National University - Ho Chi Minh City.

References

1. Satoh, S., Nakamura, Y., Kanade, T.: Name-it: naming and detecting faces in video by the integration of image and natural language processing. In: Proceedings of the Fifteenth International Joint Conference on Artifical Intelligence, IJCAI 1997, vol. 2, pp. 1488–1493. Morgan Kaufmann Publishers Inc., San Francisco (1997)
2. Yang, J., Hauptmann, A.G.: Naming every individual in news video monologues. In: Proceedings of the 12th Annual ACM International Conference on Multimedia, MULTIMEDIA 2004, pp. 580–587. ACM, New York (2004)
3. Houghton, R.: Named faces: Putting names to faces. IEEE Intelligent Systems 14(5), 45–50 (1999)
4. Sivic, J., Everingham, M., Zisserman, A.: Person spotting: Video shot retrieval for face sets. In: Leow, W.-K., Lew, M., Chua, T.-S., Ma, W.-Y., Chaisorn, L., Bakker, E.M. (eds.) CIVR 2005. LNCS, vol. 3568, pp. 226–236. Springer, Heidelberg (2005)
5. Everingham, M., Sivic, J., Zisserman, A.: "Hello! My name is... Buffy" – Automatic naming of characters in TV video. In: British Machine Vision Conference (2006)
6. Sivic, J., Everingham, M., Zisserman, A.: "Who are you?" – learning person specific classifiers from video. In: IEEE Conference on Computer Vision and Pattern Recognition (2009)
7. Aggarwal, G., Chowdhury, A.K.R., Chellappa, R.: A system identification approach for video-based face recognition. In: Proceedings of the 17th International Conference on Pattern Recognition (ICPR 2004), vol. 4, pp. 175–178. IEEE Computer Society, Washington, DC (2004)

8. Fraunhofer: SHORE - Sophisticated High-speed Object Recognition Engine (2009), http://www.iis.fraunhofer.de/en/bf/bsy/produkte/shore.html

9. Kalal, Z., Mikolajczyk, K., Matas, J.: Tracking-Learning-Detection. IEEE Transactions on Pattern Analysis and Machine Intelligence 34(7), 1409–1422 (2012)

10. Essa, I.A., Pentland, A.P.: Coding, analysis, interpretation, and recognition of facial expressions. IEEE Trans. Pattern Anal. Mach. Intell. 19(7), 757–763 (1997)

11. Cootes, T.F., Edwards, G.J., Taylor, C.J.: Active appearance models. IEEE Transactions on Pattern Analysis and Machine Intelligence, 484–498 (1998)

12. Yacoob, Y., Davis, L.: Detection, analysis and matching of hair. In: Proceedings of the Tenth IEEE International Conference on Computer Vision (ICCV 2005), pp. 741–748. IEEE Computer Society (2005)

13. Naaman, M., Yeh, R.B., Garcia-Molina, H., Paepcke, A.: Leveraging context to resolve identity in photo albums. In: Proceedings of the 5th ACM/IEEE-CS Joint Conference on Digital Libraries, JCDL 2005, pp. 178–187. ACM, New York (2005)

14. Zhang, L., Chen, L., Li, M., Zhang, H.: Automated annotation of human faces in family albums. In: Proceedings of the Eleventh ACM International Conference on Multimedia, MULTIMEDIA 2003, pp. 355–358. ACM, New York (2003)

15. Sivic, J., Zitnick, C.L., Szeliski, R.: Finding people in repeated shots of the same scene. In: British Machine Vision Conference (2006)

16. Anguelov, D., Lee, K.C., Göktürk, S.B., Sumengen, B.: Contextual identity recognition in personal photo albums. In: 2007 IEEE Computer Society Conference on Computer Vision and Pattern Recognition (CVPR 2007), June 18-23. IEEE Computer Society, Minneapolis (2007)

17. Li, D., Wei, G., Sethi, I.K., Dimitrova, N.: Person identification in tv programs. J. Electronic Imaging 10(4), 930–938 (2001)

18. Tomasi, C., Kanade, T.: Detection and tracking of point features. Technical report. International Journal of Computer Vision (1991)

19. Köstinger, M., Wohlhart, P., Roth, P.M., Bischof, H.: Learning to recognize faces from videos and weakly related information cues. In: AVSS, pp. 23–28 (2011)

20. Kalal, Z., Mikolajczyk, K., Matas, J.: Face-TLD: Tracking-Learning-Detection applied to faces. In: ICIP, pp. 3789–3792 (2010)

21. Kalal, Z., Matas, J., Mikolajczyk, K.: P-N learning: Bootstrapping binary classifiers by structural constraints. In: The Twenty-Third IEEE Conference on Computer Vision and Pattern Recognition, CVPR 2010, June 13-18, pp. 49–56. IEEE, San Francisco (2010)

22. Interactive Video (2013), http://interactive-video.appspot.com

Combining Descriptors Extracted from Feature Maps of Deconvolutional Networks and SIFT Descriptors in Scene Image Classification

Dung A. Doan[1], Ngoc-Trung Tran[1], Dinh-Phong Vo[1], Bac Le[1], and Atsuo Yoshitaka[2]

[1] University of Science, 227 Nguyen Van Cu street, District 5
Ho Chi Minh City, Viet Nam
[2] Japan Advanced Institute of Science and Technology, Japan
anhdungnt91@gmail.com, {tntrung,vdphong,lhbac}@fit.hcmus.edu.vn,
ayoshi@jaist.ac.jp

Abstract. This paper presents a new method to combine descriptors extracted from feature maps of Deconvolutional Networks and SIFT descriptors by converting them into histograms of local patterns, so the concatenation operation can be applied and ensure to increase the classification rate. We use K-means clustering algorithm to construct codebooks and compute Spatial Histograms to represent the distribution of local patterns in an image. Consequently, we can concatenate these histograms to make a new one that represents more local patterns than the originals. In the classification step, SVM associated with Histogram Intersection Kernel is utilized. In the experiments on Scene-15 Dataset containing 15 categories, the classification rates of our method are around 84% which outperforms Reconfigurable Bag-of-Words (RBoW), Sparse Covariance Patterns (SCP), Spatial Pyramid Matching (SPM), Spatial Pyramid Matching using Sparse Coding (ScSPM) and Visual Word Reweighting (VWR).

Keywords: Scene image classification, Deconvolutional Networks, Bag-of-Word model, Spatial Pyramid Matching.

1 Introduction

Scene classification is an essential and challenging open problem in computer vision with multiples of applications involved, for example: content-based image retrieval, automatic assigning labels to images and image grouping from given keywords. Because of some natural conditions of images such as the ambiguity, illumination, scaling, etc, scene classification is a difficult problem as well as there are many approaches proposed to overcome these challenges.

Specifically, early works on classifying scenes extract appearance features (color, texture, power spectrum, etc) [1][2][3] and use dissimilarity measures [4][5] to distinguish scene categories, but they can only be used in the case of

B. Murgante et al. (Eds.): ICCSA 2013, Part V, LNCS 7975, pp. 321–331, 2013.

classifying images into small number of categories such as indoor/outdoor and human-made/natural. In 2006, Svetlana Lazebnik et al. [6] presents Spatial Pyramid Matching (SPM), which is a remarkable extension of bag-of-features (BoF). SPM exploits descriptors inside each local patch, partitions the image into the segments and computes histograms of these local-features within each segment. After that, SPM framework uses histogram intersection kernel associated with Support Vector Machine (SVM) to classify images. According to Lazebnik [6], SPM gets high performance when using SIFT descriptors [7] or "gist" [8]. The method is also the major component of the state-of-the-art systems [9].

In 2010, Matthew Zeiler et al. [10] proposed Deconvolutional Networks (DN) which reconstructs images, but maintains stable latent representations and local information such as edge intersections, parallelism and symmetry. To be more specific, DN uses convolution operator principally, hence the networks can assist the grouping behavior and pooling operation on feature maps to extract descriptors (DN descriptors, briefly).

However, because edges in scene images are complex and DN cannot control the shape of learned filters, it is difficult for DN to represent edge information comprehensively. Figure 1 shows some bad reconstructed scene images that are done by 1 layer-DN. One way should be considered to overcome this disadvantage of DN is SIFT descriptors, because SIFT features totally provides directional information to enhance edge representation more powerful. Nevertheless, DN and SIFT descriptors are not the same representation, so we cannot ensure to get better performance if naively combining them together.

In this paper, we propose a new method to convert DN and SIFT descriptors into histograms of local patterns, then these histograms can be concatenated to produce a new one for classification step. Specifically, we first use K-means clustering algorithm to construct two codebooks, each word in these codebooks corresponds to a local pattern. Then, two spatial histograms are also built to represent the distribution of local patterns in an image. After that, histogram concatenation is carried out to make a new histogram that represents more local patterns than originals. Finally, SVM associated with Histogram Intersection Kernel [6] is used to classify images into appropriate labels. Our approach representing local patterns is similar to bag-of-features [11] [12]. However, note that to improve our method's performance, Spatial Pyramid Histograms are constructed by following the approach of Lazebnik [6]. To evaluate the performance of our method, we experimented the method in a large database of fifteen natural scene categories [6] and get an significant improvement of accuracy over some recent methods.

To express our method clearly, we organize the paper as follows: Section 2 introduces about recent related works. In section 3, we introduce our method in detail. Our results on Scene-15 Dataset are illustrated in section 4. Finally, in section 5, conclusions, discussion and future works are described.

Fig. 1. Original images and correspondence images reconstructed by the first layer of Deconvolutional Networks

2   Related Work

2.1   Deconvolutional Networks

Deconvolutional Networks proposed by Matthew Zeiler et al. [10] in 2010 has only a decoder which tries to reconstruct feature maps being expectantly close to the original image. To be more specific, from an input image and learned filters, DN sums over values which are generated by convolution of feature maps and filters in each layer, hopefully these values are proximate with the image. Sparseness constraint is also added to feature maps to encourage economical representation at each level of the hierarchy, thus more complex and high-level features are naturally produced. With both sparsity and convolution approaches, DN can preserve locality, mid-level concepts and basic geometric elements which can open the way for pooling operation and grouping behavior to extract descriptors. In practice, DN descriptor is particular successful when applying to object recognition and denoising images.

2.2   Image Classification by Spatial Pyramid Matching

Motivated by Grauman and Darrell's method [13] in pyramid matching, Spatial Pyramid Matching (SPM) is proposed by Svetlana Lazebnik [6] in 2006. First, SPM extracts local descriptors (for example: SIFT descriptors) inside each sub-region of images, quantizes these descriptors into vectors and then performs

K-means to construct dictionary. Secondly, with the constructed dictionary, SPM computes histogram of local descriptors and then these histograms are multiplied by appropriate weights at each increasing resolution. Finally, putting all the histograms together to get a pyramid match kernel for classification step.

3   The Proposed Method

3.1   Descriptor Extraction

Deconvolutional Networks Descriptors. In training filters of DN, given a set of I_u unlabeled images $y_u^{(1)}$, $y_u^{(2)}$, $y_u^{(3)}$,..., $y_u^{(I_u)}$. K_0 denotes the number of color channel, so we have a cost function for the first layer:

$$C_1(y_u) = \frac{\lambda_T}{2} \sum_{i=1}^{I_u} \sum_{c=1}^{K_0} || \sum_{k=1}^{K_1} \left(z_{k,1}^{(i)} \oplus f_{k,c}^1 \right) - y_{u,c}^{(i)}||_2^2 + \sum_{i=1}^{I} \sum_{k=1}^{K_1} |z_{k,1}^{(i)}|^p \qquad (1)$$

where, $z_{k,l}^{(i)}$ and $f_{k,c}^l$ respectively are feature map and filter k of layer l, K_l indicates the number of feature maps in layer l, obviously, $l = 0$ and $l = 1$ in equation (1). λ_T is a constant value that balance the the contribution of reconstruction of $y_{u,c}^{(i)}$ and the sparsity of $z_{k,1}^{(i)}$. We can also follow Matthew Zeiler [10] to form a hierarchy that means feature maps in layer l become inputs for layer $l+1$.

In reconstruction step, with the learned filters $f_{k,c}^1$ and a label/scene image y^{scene}, we infer $z_{k,1}$ by minimizing reconstruction error:

$$\min_{z_{k,1}} \frac{\lambda_R}{2} \sum_{c=1}^{K_0} || \sum_{k=1}^{K_1} \left(z_{k,1} \oplus f_{k,c}^1 \right) - y_c^{scene}||_2^2 + \sum_{k=1}^{K_1} |z_{k,1}|^p$$

We can also infer feature maps in higher layer if following [10].

Each feature map $z_{k,1}$ is split into overlapping $p_1 \times p_1$ patches with spacing of s_1 pixels. Each patch are pooled and then grouped to give local descriptors.

SIFT Descriptors. By conducting some experiments, we observe that because edges in scene images are very complicated, hence the feature maps z of 1 layer-DN are not enough (figure 1). Therefore, we utilize SIFT descriptors to support edge representation in scene recognition problem. Concretely, we densely exploit local SIFT descriptors in overlapping $p_2 \times p_2$ patches at a stride of s_2 pixels.

3.2   Building Histograms

Given a set of SIFT descriptors $X_{SIFT} = [x_{SIFT}^{(1)}, x_{SIFT}^{(2)}, ..., x_{SIFT}^{(N)}]^T \in R^{N \times 128}$ and DN descriptors $X_{DN} = [x_{DN}^{(1)}, x_{DN}^{(2)}, x_{DN}^{(2)}, ..., x_{DN}^{(M)}]^T \in R^{M \times D}$, we represent $x_{SIFT}^{(i)}$ and $x_{DN}^{(i)}$ being 128 and D−dimensional feature space respectively. With

B_{SIFT} and B_{DN} being codebooks of SIFT and DN descriptors, K-means method are applied to minimize the following cost functions:

$$\min_{\substack{V_{SIFT}, B_{SIFT} \\ V_{DN}, B_{DN}}} \sum_{i=1}^{N} ||x_{SIFT}^{(i)} - (B_{SIFT})^T . v_{SIFT}^{(i)}||_2^2$$
$$+ \sum_{i=1}^{M} ||x_{DN}^{(i)} - (B_{DN})^T . v_{DN}^{(i)}||_2^2$$

(2)

$$\text{subject to:} ||v_{SIFT}^{(i)}||_0 = 1; \ ||v_{SIFT}^{(i)}||_1 = 1; \ v_{SIFT,j}^{(i)} \geq 0, \forall i,j; \ ||v_{DN}^{(i)}||_0 = 1;$$
$$||v_{DN}^{(i)}||_1 = 1; \ v_{DN,j}^{(i)} \geq 0, \forall i,j$$

where, $v_{SIFT,j}^{(i)}$ and $v_{DN,j}^{(i)}$ are elements of vector $v_{SIFT}^{(i)}$ and $v_{DN}^{(i)}$ respectively. $V_{SIFT} = [v_{SIFT}^{(1)}, v_{SIFT}^{(2)}, ..., v_{SIFT}^{(N)}]^T$ and $V_{DN} = [v_{DN}^{(1)}, v_{DN}^{(2)}, ..., v_{DN}^{(N)}]^T$ are indexes vectors. In training phase, we minimize cost function (2) with respect to B_{DN}, B_{SIFT}, V_{DN} and V_{SIFT}, but in coding phase, with the learned B_{DN} and B_{SIFT}, we only minimize equation (2) with respect to V_{DN} and V_{SIFT}.

After obtaining V_{DN} and V_{SIFT}, we compute the histogram:

$$H_{SIFT} = \frac{1}{N} \sum_{i=1}^{N} v_{SIFT}^{(i)}$$
$$H_{DN} = \frac{1}{M} \sum_{i=1}^{M} v_{DN}^{(i)}$$

With the aim to improve our performance, Spatial Pyramid Histogram are made by following to approach of SPM [6].

3.3 Image Classification

In this stage, DN and SIFT descriptors are the same representation, so following equation may be applied:

$$H = H_{SIFT} \odot H_{DN}$$

where, $\odot$ denotes concatenation operation. The new histogram H, that is moved into SVM associated with Histogram Intersection Kernel [6], represents more local patterns than H_{SIFT} and H_{DN}.

Figure 2 shows all steps of our method.

4 Experiments

We adopt Scene-15 Dataset [6], which contains 15 categories (office, kitchen, living room, mountain, etc). Each category has from 200 to 400 images which have average size 300×250 pixels, the major image sources is COREL collection,

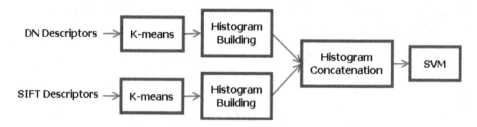

Fig. 2. Each step of our method. After extracting DN and SIFT descriptors, K-means is applied to build two codebooks. Then, the distribution of local patterns is represented in Histogram Building stage. Histogram Concatenation are carried out to produce a new histogram that represents more local patterns. Finally, SVM associated with Histogram Intersection Kernel is used to classify images.

Google image search and personal photographs. Example images of Scene-15 Dataset are illustrated in figure 5.

In our experiments, all images are converted into gray-scale and then contrast normalization before applying to DN. We train 8 feature maps of 1 layer-DN by using only 20 images which consist of 10 fruit and 10 city images, Scene-15 images are only used to train supervised classifier. Specifically, we follow experiment setup for Scene-15 Dataset of Lazebnik [6], training on 100 images per class and testing on the rest.

In classification step, multi-class SVM rule is 1-vs-All: a SVM classifier is trained to classify each class from the rest and a test image is assigned to the label having highest response.

4.1   Codebook Size

Selecting the codebook size influences the trade-off between discriminative and generalizable characteristics. Concretely, small codebook size can lead to the lack of discriminative characteristic, the dissimilar features may be assigned to same cluster/local patterns. On the other hand, a large codebook size is more discriminative but less generalizable, less tolerant to noises; because the similar features may be mapped to different local patterns.

Therefore, in this experiment, we would like to survey how the size of DN and SIFT codebook affects to the classification rate. Specifically, in the survey of SIFT codebook size, we keep the size of DN codebook fixed in 200 and gradually increase SIFT codebook size from 50 to 2000. Similarly, in the survey of DN codebook size, SIFT codebook is fixed in size of 200, and DN codebook is also raised between 50 and 2000. Note that spatial pyramid histogram is not used in this experiment and our detailed results are illustrated in figure 3.

On both situations, as the dictionary size increases from 50 to 1000, the performance rises rapidly, and then reaches the peak. When keeping on increasing the dictionary size, the classification rate decreases gradually.

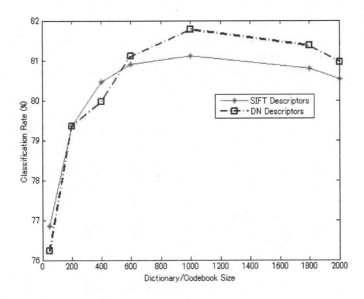

Fig. 3. The classification rates at different sizes of DN and SIFT codebook

4.2   Histogram Combination and Naive Combination

In this experiments, we compare histogram combination with naive combination. Concretely, in naive combination, the following equation is applied on DN and dense SIFT descriptors firstly:

$$x = x_{DN} \odot x_{SIFT} \tag{3}$$

where, x_{DN} and x_{SIFT} denote DN and dense SIFT descriptors, respectively. Then, codebook is constructed by using K-means, building local pyramid histograms in each sub-region of images at increasingly resolutions. Finally, SVM associated with Histogram Intersection Kernel are utilized to classify images.

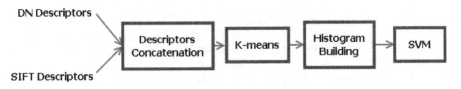

Fig. 4. In naive combination, DN and SIFT descriptors are concatenated firstly. Then, a codebook is constructed by using K-means. The distribution of local patterns are represented in Histogram Building. Finally, SVM associated with Histogram Intersection Kernel is utilized in classification step.

The detail of the steps are illustrated in figure 4. In naive combination approach, the parameters that we setup are $K_1 = 8$, $p_1 = 16$, $s_1 = 8$, $p_2 = 16$, $s_2 = 8$, $\lambda_T = \lambda_R = 10$ and the images are also resized to 150×150 before extracting dense SIFT descriptors to easily perform equation (3). Experiments are conducted 5 times with different randomly selected training and testing images, then mean and standard deviation are calculated.

Table 1. Histogram combination compares to naive combination, DN descriptors and SIFT descriptors

Method	Classification rate (%)
DN Descriptors	75.3 ± 0.9
SIFT Descriptors	81.5 ± 0.3
Naive combination	72.8 ± 0.9
Histogram combination	**84.3 ± 0.2**

Our mean and standard deviation results are shown in table 1, we compare histogram combination with not only naive combination approach but also DN and SIFT descriptors respectively. From table 1, the proposed histogram combination outperforms the others.

4.3 Comparison with Other Methods

In this experiment, we would like to compare the performance of our method with other recent methods, and the parameters that we use are $K_1 = 8$, $p_1 = 16$, $s_1 = 2$, $p_2 = 16$, $s_2 = 8$, $\lambda_T = \lambda_R = 10$.

The experimental processes are repeated 10 times with different randomly selected training and testing images. The final results are reported as the mean and standard deviation of the recognition rates, our results for 3-fold cross validation in SVM training are shown in table 2. As shown, our method outperforms some recent methods.

Table 2. Classification rate (%) comparison on 15-Scene Dataset

Method	Classification rate (%)	Year
SPM [6]	81.4 ± 0.5	2006
ScSPM [14]	80.3 ± 0.9	2009
VWR [15]	83.0 ± 0.2	2011
RBoW [16]	78.6 ± 0.7	2012
SCP [17]	80.4 ± 0.5	2012
Our method	**84.4 ± 0.4**	

Bedroom Suburb Industrial

Kitchen Living room Coast

Forest Highway Inside city

Mountain Open country Street

Tall Building Office Store

Fig. 5. Some example images of Scene-15 Dataset

5 Conclusion, Discussion and Future Works

Motivated by observations and experiments, we realize that because edges of scene images are very complicated and the learned filters of Deconvolutional Networks cannot be controlled, feature maps of DN are not enough to represent edge information comprehensively. In this paper, we propose a new method to use SIFT descriptors to support edge representation, both DN and SIFT descriptors are converted into histograms of local patterns, so we can concatenate these histograms together to make a new one that represents more local patterns than the originals. Consequently, our method makes data for classification step become more discriminative, experimental results on 15-Scene Dataset showed that our method has better performance than the recent methods.

However, there are some disadvantages in our method: it is still difficult to make a real-time application because the processing time is slow. Furthermore, by reason of using K-means in codebook construction, it is too restrictive in assigning each sample to only one local patterns.

In the future works, we would like to improve codebook construction step to be more flexible, not restrictive like K-means; supervised fashion in constructing codebook is also considered as well as implementing practical application is ongoing, too.

Acknowledgments. This research is supported by funding from Advanced Program in Computer Science, University of Science, Vietnam National University - Ho Chi Minh City.

References

1. Faloutsos, C., Barber, R., Flickner, M., Hafner, J., Niblack, W., Petkovic, D., Equitz, W.: Efficient and effective querying by image content. Journal of intelligent information systems (1994)
2. Hampapur, A., Gupta, A., Horowitz, B., Shu, C., Fuller, C., Bach, J., Gorkani, M., Jain, R.: Virage video engine. In: Electronic Imaging 1997, International Society for Optics and Photonics (1997)
3. Ma, W., Manjunath, B.: Netra: A toolbox for navigating large image databases. In: International Conference on Image Processing (1997)
4. Puzicha, J., Buhmann, J., Rubner, Y., Tomasi, C.: Empirical evaluation of dissimilarity measures for color and texture. In: The Proceedings of the Seventh IEEE International Conference on Computer Vision (1999)
5. Santini, S., Jain, R.: Similarity measures. IEEE Transactions on Pattern Analysis and Machine Intelligence (1999)
6. Lazebnik, S., Schmid, C., Ponce, J.: Beyond bags of features: Spatial pyramid matching for recognizing natural scene categories. In: IEEE Computer Society Conference on Computer Vision and Pattern Recognition (2006)
7. Lowe, D.G.: Towards a computational model for object recognition in IT cortex. In: Lee, S.-W., Bülthoff, H.H., Poggio, T. (eds.) BMCV 2000. LNCS, vol. 1811, pp. 20–31. Springer, Heidelberg (2000)
8. Torralba, A., Murphy, K., Freeman, W., Rubin, M.: Context-based vision system for place and object recognition. In: Proceedings of the Ninth IEEE International Conference on Computer Vision (2003)
9. Bosch, A., Zisserman, A., Muoz, X.: Image classification using random forests and ferns. In: IEEE 11th International Conference on Computer Vision, ICCV 2007 (2007)
10. Zeiler, M., Krishnan, D., Taylor, G., Fergus, R.: Deconvolutional networks. In: 2010 IEEE Conference on Computer Vision and Pattern Recognition (CVPR) (2010)
11. Yang, J., Jiang, Y., Hauptmann, A., Ngo, C.: Evaluating bag-of-visual-words representations in scene classification. In: Proceedings of the International Workshop on Multimedia Information Retrieval (2007)
12. Jiang, Y., Yang, J., Ngo, C., Hauptmann, A.: Representations of keypoint-based semantic concept detection: A comprehensive study. IEEE Transactions on Multimedia (2010)
13. Grauman, K., Darrell, T.: The pyramid match kernel: Discriminative classification with sets of image features. In: Tenth IEEE International Conference on Computer Vision, ICCV 2005 (2005)
14. Yang, J., Yu, K., Gong, Y., Huang, T.: Linear spatial pyramid matching using sparse coding for image classification. In: IEEE Conference on Computer Vision and Pattern Recognition, CVPR 2009 (2009)

15. Zhang, C., Liu, J., Wang, J., Tian, Q., Xu, C., Lu, H., Ma, S.: Image classification using spatial pyramid coding and visual word reweighting. In: Kimmel, R., Klette, R., Sugimoto, A. (eds.) ACCV 2010, Part III. LNCS, vol. 6494, pp. 239–249. Springer, Heidelberg (2011)
16. Parizi, S., Oberlin, J., Felzenszwalb, P.: Reconfigurable models for scene recognition. In: 2012 IEEE Conference on Computer Vision and Pattern Recognition (CVPR) (2012)
17. Wang, L., Li, Y., Jia, J., Sun, J., Wipf, D., Rehg, J.: Learning sparse covariance patterns for natural scenes. In: 2012 IEEE Conference on Computer Vision and Pattern Recognition (CVPR) (2012)

Tensor Field Visualization Using Eulerian Fluid Simulation

Marcelo Caniato Renhe*, José Luiz de Souza Filho, Marcelo Bernardes Vieira, and Antonio Oliveira

Universidade Federal de Juiz de Fora, DCC, Brazil
Universidade Federal do Rio de Janeiro, COPPE, Brazil
{marcelo.caniato,jsouza,marcelo.bernardes}@ice.ufjf.br,
oliveira@lcg.ufrj.br
http://www.gcg.ufjf.br

Abstract. In symmetric second order tensor fields, the colinearity and coplanarity of the represented structures are properties of major interest. In this paper, we present a method that induces the human perceptual system to extract these structures by using an Eulerian fluid simulation. Differently of previous approaches, our model explores the interaction between the particles to improve the perception of the underlying structures in the tensor field. The main contribution is the introduction of a tensor based advection step and a gradient based external force, suitable for viewing colinear and coplanar structures. Our experimental results show that fluids are suitable to reveal curves, surfaces and their connections.

Keywords: Tensor Field Visualization, Orientation Tensor, Fluid Simulation.

1 Introduction

Among the different tensor field visualization methods, those developed considering a dynamic approach usually present better results. This approach consists in using motion to stimulate the human visual system. Applying some sort of dynamics to the tensor field makes it easier for the observer to identify and analyze specific data of interest in the field. For example, we could define the motion as a function of characteristics of the tensors, in order to enhance structures that they represent.

A viable solution for introducing motion into the field is to use it as a medium for simulating fluid flow. Previous works related to tensor field visualization used advection mechanisms [1–3], which are an essential part of any fluid simulation. Many works in the computer graphics area explored techniques for controlling the fluid direction of flow in order to meet specific application requirements.

---

* Authors thank to Fundação de Amparo à Pesquisa do Estado de Minas Gerais and CAPES for financial support.

B. Murgante et al. (Eds.): ICCSA 2013, Part V, LNCS 7975, pp. 332–347, 2013.

Tensor field visualization means to turn visible some of their properties or somewhat continuous colinear and coplanar structures formed by the tensors. Showing this kind of structures can be challenging due to the huge amount of data. Using particle-tracing methods allows to enhance desired properties by analyzing the paths and arrangement of the particles along the field. Some of these methods are static and does not describe interaction between particles [4, 1].

We propose an interactive and dynamic method for tensor field visualization. It is based on the usage of fluid dynamics to rule the behavior of particles inside tensor fields. An Eulerian method based on Navier-Stokes equations was chosen for that. An external force was proposed to concentrate particles around areas of higher energy. A modification in the the advection is also proposed. This change consists on using properties of the acting tensors to make particles arrange in a way to highlight consecutive colinear structures.

2   Related Work

Most works related to tensor visualization are interested in developing more intuitive forms of visualization. Usually, the problem of analyzing a tensor field is its multivariate information. Methods encountered in literature fall into three categories. The discrete methods often employ some kind of glyph representation to visualize the field. Shaw *et al* [5] proposes using superquadrics to display the tensor data. The idea of superquadrics was later used by Kindlmann [6] to generate a tensor glyph that encodes the geometric shape of the tensor. A tensor shape can be determined based on coefficients introduced by Westin *et al* [7]. The main advantage of using these measures in a visualization is that they avoid ambiguities in the identification of the tensor shape.

The continuous methods, on the other hand, try to establish a continuous path along the tensor field. The hyperstreamlines method [8, 4], followed by the tensorlines method [1], are examples of them. They define paths that can be used to achieve a smooth visualization of the field in a global fashion. This kind of visualization is very useful for tracking linear features in the field, which are highly present in DT-MRI, for example. Vilanova *et al* [9] presents a thorough explanation of the methods mentioned above and other discrete and continuous methods. More recently, Mittmann *et al* [10] devised a real-time interactive fiber tracking method that can determine new paths automatically based on user feedback, while Crippa *et al* [11] proposed an interpolation method to avoid low-anisotropy regions during fiber tracking.

Finally, we have the dynamic visualization methods. These methods focus on exploring visual cues to stimulate the human perceptual system, making it easier for the observer to analyze the tensor field. Kondratieva *et al* [12] proposes the advection of particles through the directions of a generated tensor field. The goal of the advection is to induce motion in the field to enhance user perception. Leonel *et al* [2] used this idea to create an adaptive method, in which the position and orientation of the observer are taken into account in an interactive visualization of the field. Souza Filho *et al* [3] extended Leonel's work by using multiresolution data to improve the method, while reducing the number of required parameters and artifacts that were present in previous works.

This work proposes a dynamic visualization of tensor fields using fluid simulation. We use a simple and stable fluid simulator just like the one found in Stam's work [13]. Many other recent methods for fluid simulation exist, but they are focused on achieving physically accurate simulations, which is not our objective. We seek to use fluids as an alternative to introduce motion in the field. To direct the flow along a specific path, previous works used gradient fields to define an external force. In [14], this force is applied to the fluid in a morphing context, making the densities travel from an origin point to a predefined target. Similarly, we define a gradient field that is used in the calculation of the external forces. We also propose a modification in the advection mechanism, where the motion is distorted by the local tensors.

3   Fundamentals

3.1   Tensors

In this work, we are interested in visualizing tensor fields composed by second-order tensors. This type of tensor is represented by a 3x3 matrix, representing a linear transformation between vector spaces. Westin [7] introduces a special case of tensor, which is non-negative, symmetric and has rank 2. This tensor, used for estimating orientations in a field is called a local orientation tensor. It can be defined as:

$$\mathbf{T} = \sum_{i=1}^{n} \lambda_i e_i e_i^T$$

where $e_1, e_2, ..., e_m$ is a base of $\mathbb{R}^n$ and λ_i is the eigenvalue associated to the eigenvector e_i.

In $\mathbb{R}^3$, the orientation tensor can be decomposed into a sum of the contributions of its linear, planar and spherical features. The equation, then, becomes:

$$\mathbf{T} = (\lambda_1 - \lambda_2)\mathbf{T}_l + (\lambda_2 - \lambda_3)\mathbf{T}_p + \lambda_3 \mathbf{T}_s$$

The above decomposition has an important geometrical meaning. Analyzing the relation between the three eigenvalues, we can determine the approximate shape of the tensor. Basically, we have three possibilities:

- $\lambda_1 \gg \lambda_2 \approx \lambda_3$ indicates an approximately linear tensor
- $\lambda_1 \approx \lambda_2 \gg \lambda_3$ indicates an approximately planar tensor
- $\lambda_1 \approx \lambda_2 \approx \lambda_3$ indicates an approximately isotropic tensor

Usually we are interested in the main direction of the tensor. A tensor, when applied to a vector, distorts the space in accordance with its inherent shape.

Isotropic tensors have no main orientation, reason why we try to avoid them in the visualization process, as we will discuss in Section 4.

In order to identify the shape of the tensor, Westin defines three coefficients, each related to one of the types of tensors mentioned above. The so-called coefficients of anisotropy are defined as:

$$c_l = \frac{\lambda_1 - \lambda_2}{\lambda_1 + \lambda_2 + \lambda_3},$$
$$c_p = \frac{2(\lambda_2 - \lambda_3)}{\lambda_1 + \lambda_2 + \lambda_3},$$
$$c_s = \frac{3\lambda_3}{\lambda_1 + \lambda_2 + \lambda_3},$$

where each coefficient lies in the range $[0, 1]$. Besides, they have the following property: $c_l + c_p + c_s = 1$.

3.2   Fluid Simulation

Fluids have been extensively researched in the Computer Graphics field for the last decade. Simulations involving fluids evolved to the point of being able to reproduce very complex phenomena. Instead, even a simpler fluid simulation requires the use of many techniques from the Computational Fluid Dynamics field. All these techniques must be put together in order to solve the Navier-Stokes equations, which are responsible for describing the motion of fluids. The Navier-Stokes equations, in their vector form, are presented below:

$$\nabla \cdot \mathbf{u} = 0, \tag{1}$$
$$\frac{\partial \mathbf{u}}{\partial t} = -(\mathbf{u} \cdot \nabla)\mathbf{u} - \frac{1}{\rho}\nabla p + \nu \nabla^2 \mathbf{u} + \mathbf{f}, \tag{2}$$

where $\mathbf{u}$ represents the velocity vector, p is the pressure, ν is the fluid viscosity and f is the sum of external forces. The first equation is called the continuity equation, and accounts for the conservation of mass for incompressible flow. Equation 2 imposes momentum conservation [15].

Each term of the Navier-Stokes equations has its own meaning. The first term represents the advection part. The advection is the process of transporting properties through the fluid. These properties include the velocity itself. Other examples of properties or substances that can be advected through the velocity field are the density, the temperature or any other value which might be interesting for the simulation. The second term accounts for the pressure. It plays an important role in the projection step of the simulation. This step is responsible for enforcing Equation 1, i.e., the divergent of the velocity must always be zero, ensuring conservation of mass. The next term accounts for the viscosity, which smooths out the velocities, limiting the fluid motion. That's why numerical dissipation in the fluid simulation is also called numerical viscosity. It makes velocities dampen faster. We will discuss numerical issues in Section 4.3. Finally, the last term represents any external forces acting on the fluid. This term is often used as a means of controlling the motion direction of the fluid [14]. We could, for example, add a constant wind-like force in a certain portion of the fluid to shape its path.

Usually, in standard fluid simulations, we deal with non-viscous fluids. In these cases, the viscosity term can be suppressed. The resulting equation, suitable for inviscid fluids, is called the Euler equation. While inviscid means that density is constant throughout the fluid, we can still advect density, like any other property, to use it for visualization.

Two approaches are commonly employed with respect to the internal representation of the fluid. The Lagrangian approach focuses on the fluid itself, discretizing it in a number of particles. Thus, the properties of the fluid are represented in each particle. On the other side, the Eulerian approach, which we use in this work, takes the space surrounding the fluid as the reference, discretizing it into a grid. Each cell of the grid stores the values of the fluid properties at that point in the space.

In an Eulerian simulation, in order to solve the Navier-Stokes equations and actually simulate the fluid, we employ a technique called fractional steps, or operator splitting. It consists in breaking the target equation into parts, and solve each part with an adequate numerical method. The advantage is that we simplify the solution of the equation by allowing the conjugation of different methods for the solution of its terms. This means we can choose, for a particular term, the best suitable method to solve it. This technique can be used whenever the equation involves a sum of different terms that present different contributions to the final result. The solution of each term is then takes as input to the next term of the equation.

The first step in the simulation process consists in the application of the external forces. This stage produces a new velocity field, which is then taken as input for the advection step. In this step, we employ an unconditionally stable method presented in [13], which guarantees that no output velocity is bigger than the maximum velocity in the input field. This feature prevents the simulation from blowing off. The process is simple: we trace the advected property back in time to find what was its value in its previous position in the grid. Then, we use that value interpolated with its neighbors to update the current value of the property.

After the advection, we pass on to the last step, called projection. It takes the advected velocity field and makes it divergence free by subtracting the pressure from it. This is accomplished by observing a mathematical statement, known as the Helmholtz-Hodge Decomposition, which says that any vector field can be decomposed into a sum between a divergence free vector field and a scalar field.

4   Proposed Method

In this section, we present our method, which conjugates tensor field visualization with fluid simulation. The tensor field information is used to govern the fluid motion, allowing the observer to detect regions of interest in the field.

4.1   Fluid Representation

As stated in Section 3.2, we implemented an Eulerian fluid representation. We define a grid with a number of cells equal to the dimension of the tensor field. For simplicity, each fluid property is discretized at the center of each cell, as in [13]. Some works in the literature use a staggered arrangement for the velocities [16], in which they are discretized at the cell faces instead of its center. Trottenberg *et al* [17] analyzes the advantages of using the staggered grid, which are mainly related to avoidance of possible numerical instabilities that may arise during the simulation, creating a checkerboard pattern in the visualization (hence the name checkerboard instability). We were able to obtain, though, satisfactory visualizations even without using a staggered arrangement. However, it is important for non-staggered grids, when using central differences, to define pressure values at all boundaries of the grid to achieve second-order accuracy.

4.2   Extenal Force from the Tensor Field

In order to control the direction of the flow, we apply an external force based on tensor information. A more straightforward definition for this force would be to use the main eigenvector from the tensor matrix to set the direction of the force vector. Although this definition could produce rather good results, we decided to use a gradient field based on each tensor weight. The weight of a tensor is defined as:

$$w = \frac{1}{\sqrt{\lambda_1^2 + \lambda_2^2 + \lambda_3^2}} \tag{3}$$

We then set the external force to be the product between the gradient of the tensor weight at a specified and a parameter used to scale the force modulus. We discretize the gradient as follows:

$$\nabla w = \frac{1}{2} \cdot [w_{i+1,j,k} - w_{i-1,j,k} \quad w_{i,j+1,k} - w_{i,j-1,k} \quad w_{i,j,k+1} - w_{i,j,k-1}]^T \tag{4}$$

where $(i, j, k) \in \mathbb{N}^3$ are the coordinates of the neighbor tensors in the grid. The force from the gradient is applied to all cells in the grid. Since the tensor field is static, the applied force produces a velocity field which quickly converges to a stable configuration through the advection process. This makes the velocity field completely still, which is not interesting, since we want a dynamic visualization. Thus, we employed a different scheme for the application of the force. At each iteration, we apply the force only for a limited amount of tensors. Considering a 3-dimensional grid, these tensors are located in a single plane. So, in each iteration, forces are added to a specific plane in the grid. We sweep the all the planes through the iterations until we get to the first plane again. We then restart the cycle. With this mechanism, we are able to inject energy at varying portions of the grid, generating the illusion of motion in the velocity field. This immediatly allows us to visualize the tensor field simply by updating the velocity field over time, discarding the advection of densities, which is commonly used to visualize the fluid.

4.3   Advection with Tensors

The advection is the transport of properties along the flow path, as mentioned in Section 3.2. We solve the advection term in the Navier-Stokes equations through a semi-Lagrangian scheme. To compute the new velocity at a point x, we imagine a particle at this point and trace its path backwards in time, finding the position x_0 where it was located in the previous iteration. We then interpolate the value at x_0 with the values in its adjacent cells. This new value is used to update the velocity at the point x. Notice that, as we interpolate through already existent velocity values, the advection itself cannot cause the simulation to blow up.

In our case, however, the fluid path is determined by the tensor field. The tensors distort the space in which the fluid is enclosed. So, if in a previous iteration the supposed particle was at x_0, we must take into account the contribution of the tensors in the cells used for interpolation. In other words, we must find the direction determined by these tensors that led the particle to its current position. Thus, at each cell, we apply the tensor located at that site to its associated velocity vector, yielding a new velocity in a distorted space. This distorted velocity is then used to update the current one. This process is depicted in Figure 1.

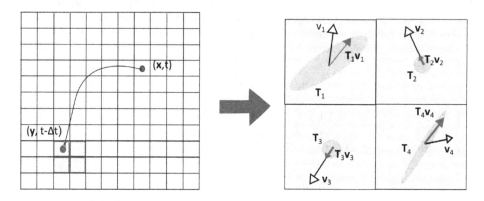

Fig. 1. Advection in the tensor field. The velocity of the cell at position **y** in time $t - \Delta t$ is used to define the position **x** in time t. The nearest tensors of **y** are applied on their respective velocities. The distorted velocities are summed to define the new velocity at **x**.

Generalizing the example in Figure 1 for threedimensional grids, suppose $\mathbf{y} \in \mathbb{R}^3$ in time $t - \Delta t$ is the point computed from the grid point $\mathbf{x} \in \mathbb{N}^3$ at time t using its current velocity backwards. Let $\{\mathbf{T}_1, \mathbf{T}_2, \mathbf{T}_3, \mathbf{T}_4, \mathbf{T}_5, \mathbf{T}_6, \mathbf{T}_7, \mathbf{T}_8\}$ be the tensors from the nearest neighbors of the point **y** in the grid, forming a 2x2x2 block of cells. We can then define the new velocity $\mathbf{v}_x$ at **x** as

$$\mathbf{v}_x = \sum_{i=1}^{8} \alpha_i \mathbf{T}_i \cdot \mathbf{v}_i, \tag{5}$$

where α_i are the weighting factors of a trilinear interpolation.

The semi-Lagrangian method for advection is highly subject to numerical dissipation. Because we are linearly interpolating properties every iteration, the numerical error tends to accumulate. The excessive averaging required by the linear interpolation acts as a low-pass filter, smoothing out values over time [18]. The result is that velocities dampen quite fast, and interesting visual effects, like swirling movements in the fluid, can be lost during the simulation. Possible overcomings include the utilization of sharper interpolation methods [16] and techniques like the vorticity confinement method, first presented by [19] and also used by [16]. As for the first solution, care must be taken to avoid overshooting, because it does not provide the same stability as a simple linear interpolation.

For our purpose of tensor field visualization, however, the usual dissipation that arises from the semi-Lagrangian advection is not exactly an issue. We are more interested in adding some level of fluid dynamics to the field than to accurately and realistically simulate the fluid. The fluid is just an interesting medium for the field visualization. The real problem is that the application of tensors to the velocities deeply aggravates the already present dissipation.

To work around this tensor dissipation issue, we adopted a simple solution. We defined a boost parameter to apply to each tensor in the field before it is used in the velocity calculation. Thus, we counteract the numerical dissipation with a simple scale in the tensor. If correctly adjusted, it is possible to create a balance between the boosting and the dissipation, producing controlled fluid simulations, which allow nice visualizations as the one shown in Figure 3 in the next section. We also can control which tensors receive the boost through checking of its weight or some of its anisotropy coefficients mentioned in Section 3.1. This is useful, for example, if we want to visualize just linear tensors. We then set the boost based on the linear anisotropy coefficient, letting the dissipation naturally smooth out the velocities in the cells with isotropic or planar tensors.

5   Results

In this section, we present some results from our visualization. All experiments were conducted in a computer with an Intel i3 processor running at 2.1GHz, 4GB RAM memory and an nVidia GeForce 540M graphics adapter. We generated a simple tensor field in order to demonstrate our method behavior. In a regular 2D grid of 64×64 elements, with integer coordinates $(i, j) \in [0, 63] \times [0, 63]$, we defined three main points $\mathbf{q}_1 = (16, 16)$, $\mathbf{q}_2 = (48, 16)$ and $\mathbf{q}_3 = (32, 48)$. For all points $\mathbf{p}_{ij} = (i, j)$ in the grid, a tensor is computed as

$$\mathbf{T}_{ij} = \sum_{k=1}^{3} \frac{(\mathbf{q}_k - \mathbf{p}_{ij}) \cdot (\mathbf{q}_k - \mathbf{p}_{ij})^T}{||(\mathbf{q}_k - \mathbf{p}_{ij})||},$$

where $\mathbf{q}_k \neq \mathbf{p}_{ij}$. The result is a tensor $\mathbf{T}_{ij}$ that captures the uncertainty of the direction between the grid point $\mathbf{p}_{ij}$ and all $\mathbf{q}_k$ points, weighted by their Euclidean distances. Figure 2a shows a representation of the tensor norms of this 2D field, varying from lowest (blue) to highest (red). In Figure 2b, the gradient of the tensor norms is shown.

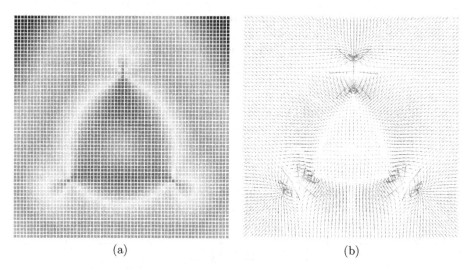

(a)   (b)

Fig. 2. (a) Norms of a 64 × 64 2D tensor field represented by a colored dot going from blue (lowest) to red (highest). (b) The corresponding gradient field of the tensor norms.

We employ a fluid simulation for the visualization as explained in Section 4. Observing the Figure 2b, we can see that the gradient norms are greater near the main points, since we use the tensor weight at each site to calculate them. The central region has the weakest gradients, which is in accordance with the fact that the tensors in that area are mainly isotropic. In our simulation, it means that, if we apply a force in that area, any velocity generated by this force will soon be reduced to zero.

We can see a screenshot of the aforementioned simulation in Figure 3. The velocity vectors are represented as two-colored lines, with the red edge indicating the direction. We interactively applied a series of external forces to the fluid in different points of the grid. These forces produced motion in the direction of the force, but the flow direction soon converged to the charges. As one can observe, the tensor-guided simulation produces continuous paths between the charges, leading the fluid from one charge to another. The velocities originated from the borders are directed towards the center of the field. In this simulation, we applied a tensor boost equals to 1.2, which avoided rapid velocity dissipation and created a fluid-like motion of the velocity field, with the velocities continuously moving around the charges. The velocities seen at the very center of the field in Figure 3 are not created there, because, as previously said, that area is full of isotropic tensors. Those velocities come from the linearly anisotropic tensors present in the diagonal line connecting the left (or the right) and the top charges.

This field represents a simple example where it is more clear to visualize the wanted behavior. There are two more challenging examples that are a 3D helical field and a DT-MRI brain. In these examples, we did not use interactive force placement. The details on how forces were applied are explained in the following discussion.

Fig. 3. Screenshot of the 3-point field visualization. Time step = 0.1; tensor boost = 1.2.

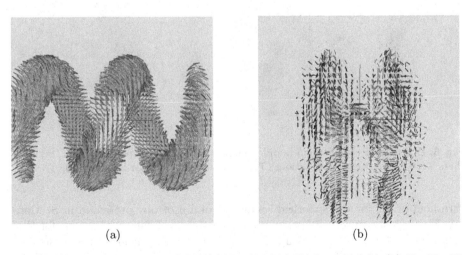

(a)   (b)

Fig. 4. Snapshots of the tensor fields: (a) $40 \times 39 \times 38$ helical tensor field. (b) $40 \times 38 \times 37$ DT-MRI brain.

The 3D helical field is an artificial tensor field, in which we have mainly linear tensors, composing the helix, surrounded by isotropic tensors. In order to better illustrate the tensor field, Figure 4 only shows the tensors below the

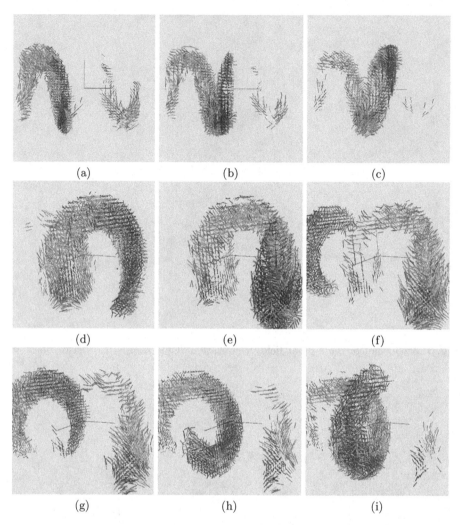

Fig. 5. Helical field visualization in different times and different points of view. Screenshots were taken at each 3 iterations. Time step = 0.001; tensor boost = 1.0.

threshold of 0.7 for the spherical coefficient of anisotropy presented in Section 3.1. Figure 5 shows screenshots of our simulation in different iterations for the helical field. As in the 2D example, we used the velocity field itself to visualize the tensor field. Since tensors do not inherently point in a particular direction, we used parallelepipeds as glyphs to represent the velocity vectors. Besides, to put the fluid into motion we need to apply an external force in the beginning of the simulation. This force starts the velocity advection process. If we do not apply additional forces, the velocities tend to dissipate at some point during the simulation. If we keep the tensor boost on, they will at most reach equilibrium, presenting no movement at all. We cannot arbitrarily scale the boost, because

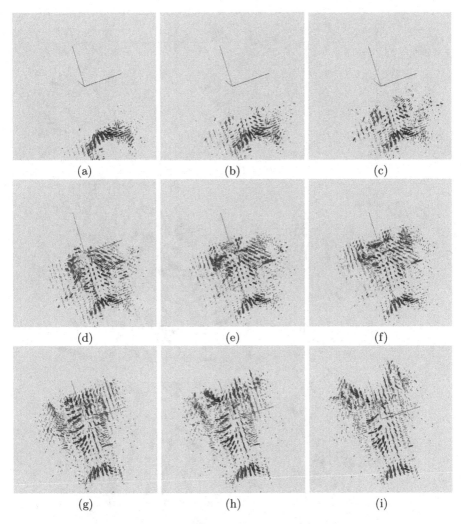

Fig. 6. Visualization of the DT-MRI brain field. Screenshots show 9 sequential frames, taken at subsequent iterations. Time step = 0.001.

the simulation can become unstable. On the other hand, if we apply a constant force in all the grid in every iteration, the velocities will be directed along the linear tensors path, as expected, but they will still not move. Thus, we employed a scheme in which we apply forces cyclically along varying planes of the 3D grid. In the helical field example, we applied forces, based on the tensor weight gradients, to every cell in the plane $x = k$, with $k = 0..N - 1$. Letting the tensor boost off, the velocities in the cells not subject to the gradient force tend to smooth out, while the affected cells create new velocities, which are advected through the tensor field. With this method, we were able to induce motion in the

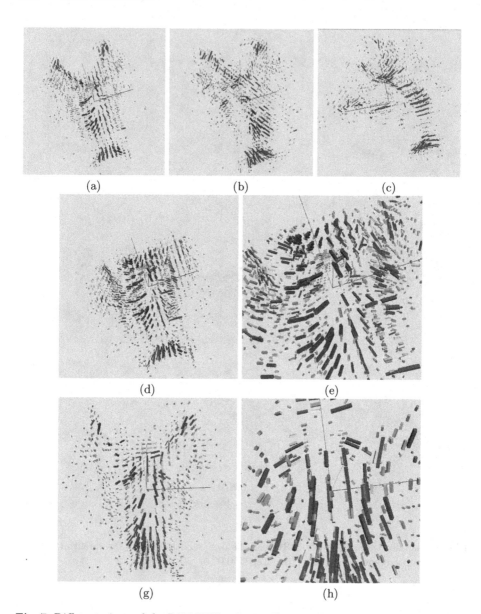

(a)   (b)   (c)

(d)   (e)

(g)   (h)

Fig. 7. Different views of the DT-MRI brain visualization depicted in Figure 6. Figure (c) shows a side view of the brain. Figure (g) shows it as seen from the top. Figures (e) and (h) show scaled views of the frames shown in Figures (d) and (g), respectively, to provide a detailed view of the tensor glyphs. Time step.

field, generating the visualization shown in Figure 5. The colors of the glyphs indicate the velocity norm, which means that the red velocities represent the places where the external force is being applied at the current iteration. This simulation did not run in real-time due to the grid resolution needed for this field. A downscaled version of the field may be used to work around this issue, although we lose some information, producing slightly poorer simulations.

Finally, we simulated the DT-MRI brain tensor field. For this simulation, we used a size-reduced version of the field. Since the brain field is quite large, we were also not able to achieve a real-time visualization, even with downscaling. Plus, the brain field is highly noisy. So, we used a filtered version of the original field. Figure 4 shows this field. Tensors with spherical anisotropy coefficient higher than 0.7 were omitted, as in the previous example. Again, as in the helical example, we applied forces by sweeping the grid planes $x = k$, with $k = 0..N - 1$. Also, we let the tensor boost in each cell be set by a gaussian function with respect to the distance between the cell plane and the plane $x = k$. Figure 6 shows 9 subsequent frames from a simulation, with a time step of 0.001. The size of each tensor glyph is defined as a function of its weight w, as defined in Section 4.2, and the velocity norm at the cell. The brain central region, which has many colinear fibers, contains tensors with higher weight, resulting in the red glyphs seen in the figure. The difference between the glyphs can be better seen in Figure 7, which zooms in the mentioned region in the brain. Also, Figure 7 shows alternative views of the brain, like its side and top views, for instance.

6   Conclusion

We presented a new visualization method based upon fluid simulation. We use the fluid as a means for visualizing tensor fields. The velocity field generated by the simulation is used to define the paths along linear tensors in the field. We apply external forces to the field in such a way that the velocities produce a continuous movement, enhancing the visualization process. Our method works both for 2D and 3D fields. We have shown our results for two artificial fields and for a DT-MRI brain field.

As mentioned in Section 5, we were not able to achieve real-time simulations for high resolution grids. Even with downscaling, the brain field visualization had to be generated offline. In an Eulerian simulation, fluid properties are calculated for every grid cell. The original brain field has dimensions 80x95x74. Further research must be done in order to work around this limitation and achieve real-time visualization. Parallelization may be a feasible solution, since it is not difficult to parallelize the stages of the simulation. Another possible experiment includes the utilization of particles through a hybrid fluid simulation method.

We used velocities to visualize the field, which led us to find a way to induce motion due to the reasons presented in Section 5. The use of fluid densities or other property like temperature would be a better solution for the movement issue. However, we still need to evaluate the best way to represent the density. More importantly, we would need to determine the best places to insert fluid density into the grid in such a way that the visualization produced can actually enhance the observer experience. Also, it would be interesting to explore the possibilities created by tweaking other fluid properties, like viscosity for instance.

References

1. Weinstein, D., Kindlmann, G., Lundberg, E.: Tensorlines: advection-diffusion based propagation through diffusion tensor fields. In: VIS 1999: Proceedings of the conference on Visualization 1999, pp. 249–253. IEEE Computer Society Press, Los Alamitos (1999)
2. de Almeida Leonel, G., Peçanha, J.P., Vieira, M.B.: A viewer-dependent tensor field visualization using particle tracing. In: Murgante, B., Gervasi, O., Iglesias, A., Taniar, D., Apduhan, B.O. (eds.) ICCSA 2011, Part I. LNCS, vol. 6782, pp. 690–705. Springer, Heidelberg (2011)
3. de Souza Filho, J.L.R., Renhe, M.C., Vieira, M.B., de Almeida Leonel, G.: A viewer-dependent tensor field visualization using multiresolution and particle tracing. In: Murgante, B., Gervasi, O., Misra, S., Nedjah, N., Rocha, A.M.A.C., Taniar, D., Apduhan, B.O. (eds.) ICCSA 2012, Part II. LNCS, vol. 7334, pp. 712–727. Springer, Heidelberg (2012)
4. Delmarcelle, T., Hesselink, L.: Visualizing second-order tensor fields with hyper streamlines. IEEE Computer Graphics and Applications 13(4), 25–33 (1993)
5. Shaw, C.D., Hall, J.A., Blahut, C., Ebert, D.S., Roberts, D.A.: Using shape to visualize multivariate data. In: NPIVM 1999: Proceedings of the 1999 Workshop on New Paradigms in Information Visualization and Manipulation in Conjunction with the Eighth ACM Internation Conference on Information and knowledge Management, pp. 17–20. ACM, New York (1999)
6. Kindlmann, G.: Superquadric tensor glyphs. In: Proceedings of IEEE TVCG/EG Symposium on Visualization 2004, pp. 147–154 (May 2004)
7. Westin, C., Peled, S., Gudbjartsson, H., Kikinis, R., Jolesz, F.: Geometrical diffusion measures for mri from tensor basis analysis. In: Proceedings of ISMRM, vol. 97, p. 1742 (1997)
8. Delmarcelle, T., Hesselink, L.: Visualization of second order tensor fields and matrix data. In: VIS 1992: Proceedings of the 3rd Conferencse on Visualization 1992, pp. 316–323. IEEE Computer Society Press, Los Alamitos (1992)
9. Vilanova, A., Zhang, S., Kindlmann, G., Laidlaw, D.: An introduction to visualization of diffusion tensor imaging and its applications. Visualization and Processing of Tensor Fields, 121–153 (2006)
10. Mittmann, A., Nobrega, T., Comunello, E., Pinto, J., Dellani, P., Stoeter, P., von Wangenheim, A.: Performing real-time interactive fiber tracking. Journal of Digital Imaging 24(2), 339–351 (2011)
11. Crippa, A., Jalba, A., Roerdink, J.: Enhanced dti tracking with adaptive tensor interpolation. Visualization in Medicine and Life Sciences II, 175–192 (2012)
12. Kondratieva, P., Kruger, J., Westermann, R.: The application of gpu particle tracing to diffusion tensor field visualization. In: IEEE Visualization, VIS 2005, pp. 73–78. IEEE (2005)
13. Stam, J.: Stable fluids. In: Proceedings of the 26th Annual Conference on Computer Graphics and Interactive Techniques, pp. 121–128. ACM Press/Addison-Wesley Publishing Co. (1999)
14. Fattal, R., Lischinski, D.: Target-driven smoke animation. ACM Transactions on Graphics (TOG) 23, 441–448 (2004)
15. Ferziger, J.H., Perić, M.: Computational methods for fluid dynamics, 3rd edn. Springer (2002)

16. Fedkiw, R., Stam, J., Jensen, H.: Visual simulation of smoke. In: Proceedings of the 28th Annual Conference on Computer Graphics and Interactive Techniques, pp. 15–22. ACM (2001)
17. Trottenberg, U., Oosterlee, C.W., Schüller, A.: Multigrid. Academic Press (2001)
18. Bridson, R.: Fluid Simulation For Computer Graphics. Ak Peters Series. A K Peters (2008)
19. Steinhoff, J., Underhill, D.: Modification of the Euler equations for "vorticity confinement": Application to the computation of interacting vortex rings. Physics of Fluids 6(8), 2738–2744 (1994)

Interactive Photographic Shooting Assistance Based on Composition and Saliency

Hiroko Mitarai, Yoshihiro Itamiya, and Atsuo Yoshitaka

Japan Advanced Institute of Science and Technology,
1-1 Asahidai, Nomi, Ishikawa, Japan
{hmitarai,itamiya0408-jyht,ayoshi}@jaist.ac.jp
http://awabi.jaist.ac.jp:8000/yoshitaka_lab/index_e.html

Abstract. When taking photographs, compositions support them to clarify the subject. The preliminary survey indicated that the professional photographers apply 1.7 types of composition on average; they tend to apply multiple types of compositions, such as triangular, diagonal and contrasting compositions in one photo. The proposed method considers co-occurrence of recognized compositions and candidate proposing compositions. We propose a novel photo shooting method which suggests one or more types of compositions and superimposes the suggestions on the photo being taken. Semantic differential method revealed that the proposed method increases compositional effects on the photograph.

Keywords: Composition, rule of thirds, photography.

1 Introduction

Digital cameras have been widely used among people. Performance and functionality of the camera have been improving year by year. Nowadays, photos of high image quality can be taken only by pressing the shutter release button. However, photos of high image quality are not always "good"; professional photographers exploit photographic techniques to take good photos, such as focusing on the subject, adjusting exposure, and considering the object arrangements in the photo to achieve better compositions [1].

Recently, common digital cameras have a feature to superimpose grid lines on the viewfinder to help shooting with an awareness of the rule of thirds, which is a popular and traditional compositional device. However, it is advised not to place the subject rigidly on the displayed lines [1]. Since the digital cameras on the market do not have features to display other types of reference lines as a guide to place elements in the photos, the amateur photographers do not have the opportunities to be conscious of other types of compositions. Less variation of composition causes narrower range of expression.

Good composition can be achieved by placing the elements of a picture in a harmonious, interesting, and even unusual way to capture attention [2].

This paper proposes a framework for digital still cameras to support the amateur photographers by detecting basic compositions in the photograph being

B. Murgante et al. (Eds.): ICCSA 2013, Part V, LNCS 7975, pp. 348–363, 2013.

taken, displaying the photographers compositional suggestions as hints for better compositions. Such framework can support novice photographers to consider exploiting compositions and experienced photographers to discover potential composition choices they might not have found in the frame. The acquisition of judgmental skills for the application of knowledge and the composition are difficult.

In the literature, there are studies such as recognizing if compositions are applied to photos or digitally alter the photo to apply compositions; however, only the rule of thirds was mainly considered as the target composition and compositions such as diagonal composition or triangular composition were not considered well.

In order to discover importance on the other compositions, the author carried out a preliminary survey by finding tendencies of compositional application of professional photographers and the application ratio. The results revealed that the professional photographers apply 1.7 types of composition on average; they tend to apply multiple types of compositions, such as triangular, diagonal and contrasting compositions, in one photo; however, previous studies are only capable of suggesting only one type of composition. The author proposes a novel photo shooting method which suggests one or more types of compositions and superimposes the suggestions on the photo being taken.

The author targets the basic five compositions, namely central composition, rule of thirds, triangular composition, diagonal composition and contrasting composition, which were applied frequently in the preliminary survey. When shooting photos, positions of prominent lines are extracted by the implemented system using face and saliency detection. After composition recognition is carried out considering the prominent area and lines in the frame, the proposed compositions are calculated and displayed on the viewfinder. The proposed method considers co-occurrence of recognized compositions and candidate proposing compositions. The system is implemented with a digital camera, a laptop computer and a USB-driven LCD panel with touch screen.

The proposed method using the implemented system was examined to measure its effectiveness. The examinees were directed to photograph subjects in three different ways: without any guidance, with grid lines, and with the photographic guidance. The author carried out a subjective evaluation on the photos taken by the examinees using a semantic differential method. It revealed that there is some effect. While questionnaires on the examinees indicated that there are still remaining issues on the proposed method, most of the examinees indicated that it is better to have compositional guidance in the camera.

The first contribution of this paper is that as we surveyed on the compositional use in professional photography, we discovered that the professionals exploit multiple compositions. The second contribution is that based on the discovery, we presented a interaction model and implemented a system and it can be a unique yet one of promising solutions for better photographs.

The rest of the paper is organized as follows. Types of photographic compositions are introduced in in Section 2. Related works in post-shooting guidance and in-shooting guidance are explained in Section 3. In Section 4, we describe the preliminary survey on the compositional use in professional photography. An interaction model is proposed in Section 5. Based on the model, the system implementation and the guidance algorithm are explained in Section 6. The implemented system is evaluated in Section 7, and concluding remarks are made in Section 8.

2   Types of Compositions

In [3], eleven different types of compositions are introduced (Fig. 1). Central composition arranges the subject in the center. Rule-of thirds composition is one of the most traditional compositions and subjects are arranged on the lines which divide the frame in three or on the intersection (also referred to as power points). Dual partitioning composition arranges the frame in two, such as the sky and the sea. Triangular composition arranges the subject of triangular shape, such as mountains. Diagonal composition arranges the subject on the diagonal line. Radiating composition arranges the subject in a radial fashion. Perspective composition emphasizes the distance using a natural law of perspective. Contrasting composition arranges two subjects for comparison. Curve composition arranges the subject in a curve, such as river. Symmetry composition arranges subject(s) in symmetry, horizontally or vertically. Tunnel composition emphasizes the subject using a frame within a frame.

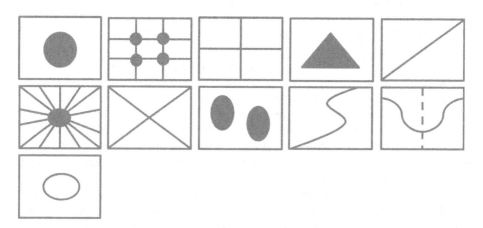

Fig. 1. Types of compositions. From left up to right, central composition, rule-of-thirds composition, dual-partitioning composition, triangular composition, diagonal composition, radiating composition, perspective composition, contrasting composition, curve composition, symmetry composition and tunnel composition.

3 Related Work

Widely classified, there are studies on post-shooting guidance (guidance after photoshoot) and in-shooting guidance (guidance during photoshoot).

3.1 Studies on Post-shooting Guidance

Mai et al. suggested a recognition method if the rule-of-thirds composition is applied on a shot photograph [4]. The method estimates the object position using Saliency Map and Objectness map and recognizes if the rule-of-thirds composition is applied. However, only the power points in the rule-of-thirds composition are considered. Horizontal or vertical lines of the rule-of-thirds composition and other compositions are not mentioned.

Bhattacharya et al. proposed an evaluation method on aesthetic quality based on a composition and a method to rearrange the subject [5]. When the user indicates the position of the main subject, the system segments the main object area and rearrange it to the appropriate position using Surface Layout Recovery method [6]. Note that it only exploits the rule-of-thirds composition.

Liu et al. also proposed an evaluation method on aesthetic quality based on composition [7]. The compositional elements considered in this paper is the rule-of-thirds composition, diagonal dominance, visual balance and size region. The author mentions that applying only the rule-of-thirds composition does not improve impression or attractiveness of the photograph.

Chang et al. proposed a system which finds favorable compositions out of panoramic photographs [8]. They defined photographs from National Geographic as "good photographs" and created a learning dataset by extracting compositions from the photographs. It extracts the rule-of-thirds, texture, and repeating patterns as photographic elements. The issue is that it only accepts panoramic photographs and only considers learned professional shooting styles.

Su et al. proposed a method called Bag of Aesthetics Preserving Features (BoAP) which does not depend on the formalized rules of photography [9]. BoAP creates feature vectors from color, texture, saliency and edges from the patched images. Using BoAP, a library is created from the collection of "beautiful" photographs. However, placement of the subject is not considered since compositions are not considered in this method.

Ieda et al. proposed a method to correct images by automatically selecting compositions and applying onto them [10]. Faces, prominent areas, triangles, horizontal and diagonal lines, and vanishing points. Central composition, rule-of-thirds composition, triangular composition, diagonal composition and perspective composition are considered. However, it does not consider saliency in this research.

After photographs are taken, it is possible to correct them however there are limitations. If the photographs are digitally altered, the semantic information may change and causes an unclear shooting intentions. Photography assistance during shooting reduces the artificiality of the photo, maintains the semantic information and therefore increase the possibility to express the shooting intention of the photographer. It also contributes to a shorter editing time.

3.2 Studies on In-shooting Guidance

There are several studies on the guidance during the shooting. In-camera automation of composition rules was proposed [11]. This system supports the user to locate the main object on the intersection points of the rule-of-thirds composition based on the previously-introduced main object detection method [12]. However, since a single object is considered and the background information cannot be used because it uses a special camera which blurs the background to detect the main object.

Real-time photographic recommendation system was proposed [13] based on the previous work [9]. The system recognizes a wider area than the user recognizes. The proposed method uses both bottom-up method using BoAP and top-down method using composition and scores the aesthetic characteristic of an image and recommends framing.

Cheng et al. propose a method to calculate the appropriate shooting frame automatically from a large input image [14]. They consider not only the co-occurrence of the neighboring area features but also features in a distance. In order to extract features related to colors and orientations, they used the color moment function and texture function using the method to calculate the similarity of color images [15]. The system segments the image, detects the face area and extract features. The extracted features are then compared with the image database consisted of professional photographs and the appropriate shooting frame is culculated. However, the authors automatically assume that each image area contains uniform color and texture and do not consider the compositions.

Different methods considering various types of features, photographic elements and compositions are proposed.

4 Preliminary Survey on Compositional Use

We carried out a survey on compositional use of professional photographers to find the frequency and types of the compositions used in professional photography. The target compositions are the eleven compositions described in the previous section. (Fig. 1) We created a dataset of 481 photographs taken by professional photographers with worldwide reputation[16][17][18]. Landscape, nature, and portrait photographs are selected as compositions are often carefully applied due to stability of the subjects. Since professional photographers sometimes shoot photographs slightly out of composition, we set the acceptable range for the rule-of-thirds composition, dual-partitioning composition and diagonal composition. The acceptable range of the compositions (as in Fig. 2, 3 and 4) are measured using the differences in the distance between positive examples and negative examples, both used to explain about the compositions in [19].

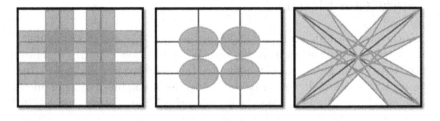

Fig. 2. Lines   **Fig. 3.** Points   **Fig. 4.** Diagonal

The result indicated that the acceptable range of rule-of-thirds(line) composition was ±11.4%, rule-of-thirds(point) composition was ±15.8%, and diagonal composition was ±8.5° (Table.1).

Table 1. Measured acceptable range

Composition	Positive examples	Negative examples	Acceptable range
Rule-of-thirds (line)	26	8	±11.4%
Rule-of-thirds (point)	16	7	±15.8%
Diagonal	10	4	±8.5°

Using the acceptable range, we manually counted the compositions in the photographs. The minimum unit of photographic element is measured according to [19] when we measured the range. When there are a multiple number of compositions in one photograph, the combination of the applied compositions. From the dataset analysis, it indicated that there are 1.7 compositions per photograph on average, and when two compositions are applied to a photograph, we found the co-occurrence of prominent area and prominent line (Table. 2).

Table 2. Composition co-occurrence

Photos that two compositions are applied	Ratio
Rule-of-thirds (point + horizontal line)	32.8%
Rule-of-thirds (horizontal + vertical line)	20.8%
Rule-of-thirds (point) + diagonal line	7.7%
Rule-of-thirds (point) + contrasting	4.9%
Rule-of-thirds (point + vertical line)	4.9%
Other	28.9%

5   Interaction Model

Fig. 5 is the proposed photographic shooting assistance model. In the model, the system supports proactive capture by user. When the user captures a photo for analysis, the system automatically estimates the main subject and calculate the proposing composition. It then calculates the corresponding guidance method and displays on the screen so that the user can check the capture method. If the user is satisfied with the composition of the photo, considering the suggested capture method, the user captures the photo.

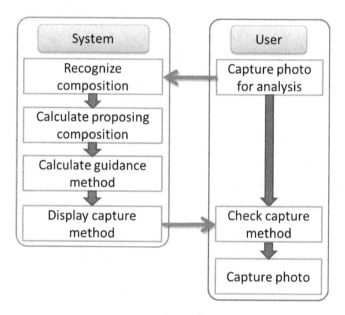

Fig. 5. Interaction model

6   System Implementation

6.1   User Interface

The user interface of the prototype system is shown in Fig. 6. It is implemented with Visual C++ and Intel Open Computer Vision Library. There are mainly four functions: normal capture, rule-of-thirds display, composition guide and composition display. Normal capture is a "shutter button". Rule-of-thirds display displays the grid lines indicating the rule of thirds (Fig. 7). Composition guide displays the target object placement superimposed on the image being shot (Fig. 8). Composition display detects the corresponding elements of the five main compositions (Central, rule-of-thirds, triangular, diagonal, and contrasting composition). When one of the composition is selected from the pull-down menu,

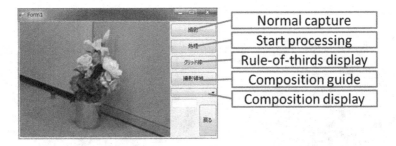

Fig. 6. Initial screen

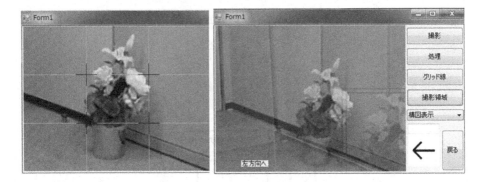

Fig. 7. Rule of thirds display   **Fig. 8.** Compositional navigation

corresponding photographic elements are highlighted. For example, the prominent area is highlighted when the central composition is selected (Fig. 9), or the diagonal line is highlighted when the diagonal composition is selected (Fig. 10).

6.2 Detection Algorythm on Compositional Elements

When the input image is entered to the system, it detects the photographic elements, which are prominent regions and prominent lines. When a composition or compositions are recognized, the system searches for compositions that are likely to co-occur with the existing composition(s). Finally, the system calculates the distance between the recognized composition and the proposing composition and guides the user to achieve the proposing composition. Fig. 12 describes the flow of the system.

Note that the system recognizes a wider area than the viewfinder so that it is able to detect the photographic elements that are not yet found by the user.

(1) Face Region Detection. After detecting the face using the method [20], the color information is extracted from the detected rectangular face area C_{rect}.

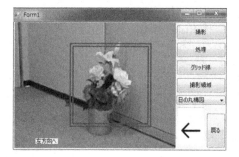

Fig. 9. Guidance on central composition

Fig. 10. Guidance on diagonal composition

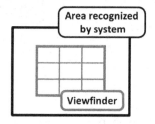

Fig. 11. System evaluation environment

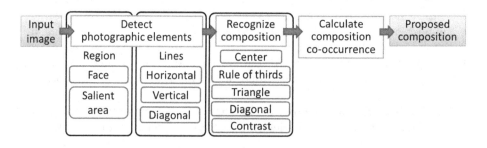

Fig. 12. Composition recognition algorithm

If the skin color region C_{skin} divided by C_{rect} is the same or more than Th_{skin}, C_{face} is true (Face is detected) . Skin color hue is determined at $6°$ to $38°$ according to [21].

$$C_{face} = \begin{cases} true, & (Th_{skin} \geq \frac{C_{skin}}{C_{rect}}) \\ false, & otherwise \end{cases} \tag{1}$$

(2) Salient Area Detection. The salient region of the photograph is detected using Saliency Map [22] and Graph Cut [23]. Saliency map indicates the salient region which attracts human attention based on low-level features such as brightness and color difference. Graph cut is a segmentation method which defines energy function from each segmented region and find the global minimum. It uses characteristics of both regions and borders. We set the region, which is calculated by Saliency Map to be the most salient, as a label of the graph cut. The label(seed) is necessary to learn foreground and background colors.

(3) Prominent Line Detection. The most frequent prominent lines utilized for better compositions are horizontal lines, vertical lines and diagonal lines. Smoothing using Gaussian filter removes the noises and emphasizes the edges of the captured image. Sobel and Laplacian filters are separately applied onto the smoothed image to detect the edges. Hough transform with angular limitation then detects the prominent lines. The angular limitations are as follows.

Table 3. Angle limitations of Hough transform

Type of line	Limitation
Vertical lines	$-8° < \theta < 8°$
Horizontal lines	$85° < \theta < 95°$
Diagonal lines (to the right)	$20° < \theta < 54°$
Diagonal lines(to the left)	$110° < \theta < 144°$

(4) Composition Detection. Compositions are detected based on the combinations of photographic elements as shown in Table. 4.

6.3   Hardware Organization

A prototype system(Fig. 13) was implemented. A digital camera, (CASIO EX-H10) is connected to the laptop PC(Panasonic CF-R8EW6AAS) via RCA connector (Princeton PCA-DAVP) for video imput. LCD panel with touch screen (SOUND GRAPH FINGERVU436W) accepts manipulations of the system and displays the picture being taken. The laptop PC has Intel Core2 Duo 1.20GHz and a 3GB memory, and processes still images captured by the system.

Table 4. Combination of compositional elements

Composition	Combination of elements
Central	One salient area
Rule of thirds	Combination of salient area, horizontal, and vertical lines
Triangular	Diagonal line (to the left) AND diagonal line (to the right)
Diagonal	Diagonal line (to the left) OR diagonal line (to the right)
Contrast	Two salient areas

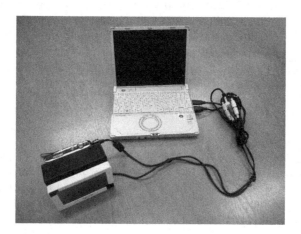

Fig. 13. Prototype system

7   System Evaluation

7.1   Condition of Experiment

In order to evaluate effectiveness of the photographic composition support system, we carried out a system performance evaluation on ten male graduate students, who had never been trained in photography. Before the evaluation, the examinees were requested to read prepared documents explaining about compositions. The evaluation is consisted of two parts: photo shooting and photo evaluation.

Photo Shooting. In this process, examinees were requested to carry out a photo shoot on three different subjects: one subject (a flower vase, Fig. 15), two subjects (a flower vase and a man, Fig. 16), and landscape (Fig. 17) with three different methods: no guidance, with the rule-of-thirds display, with the guidance of the system. There will be total nine photos per examinee. Prior to shooting with the system, the examinees were instructed how to use the system.

Fig. 14. System evaluation environment

Fig. 15. One subject   **Fig. 16.** Two subjects   **Fig. 17.** Landscape

Photo Evaluation. We carried out a subjective evaluation of the photos taken by the examinees using Semantic Differential Method. In order to unify the evaluation environment, the evaluation was carried out under the same environment using the monitor FlexScan EV2116W. When the examinees evaluated the photos, we changed the display order to avoid the order effect, and only presented them the photos of other examinees.

7.2 Analysis of Experiment

Misalignment Ratio. We first analyzed the distance of salient regions to consider how much the system affected the examinees to take compositions into account. Fig. 18 indicates the difference in distance of prominent regions between standard compositional position and detected position in photographs that examinees took. There is no notable difference, however, it indicates that with the grid line (rule-of-thirds) and the system, the examinees could reduce the misalignment ratio.

Likewise, Fig. 19 indicates the difference in distance of prominent lines. Even though there is no notable difference, it indicates that the misalignment ratio is lower especially when two objects are shot.

Difference in Impression. Using Semantic Differential Method, we analyzed the impression of the photos taken during the experiment. Five pairs of adjectives were selected as they represent general expressions when compositions are

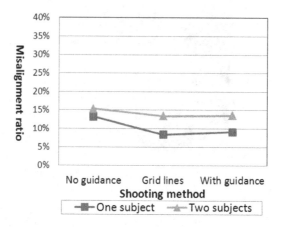

Fig. 18. Distance of prominent regions

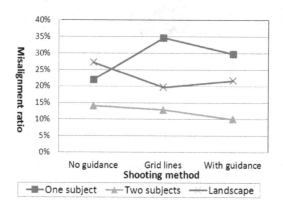

Fig. 19. Distance of prominent lines

applied. Fig. 20 is the factor analysis of the photos taken with the system guidance. It indicates that the factor loading is higher when there are two subjects or is a landscape photo. It implies that the system guides better if the frame is more complicated.

Fig. 21 is the factor analysis of the landscape photos. It indicates that the grid lines and the system have a similar effect on the photographic composition, however, the system has a very high factor loading in being "unified" and "complicated".

Result on Questionnaire. We also requested examinees to fill out a questionnaire after the experiment. It indicated that the examinees had difficulties using

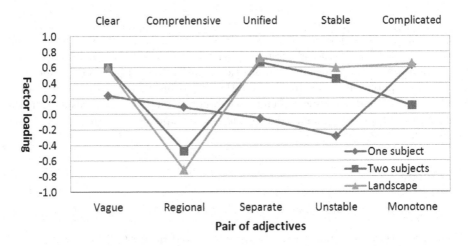

Fig. 20. Factor analysis of the photos taken with system

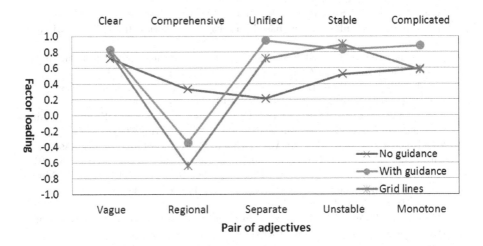

Fig. 21. Factor analysis of the landscape photos

the system. In terms of system operability, the result indicated 2.9 on average. 5 is the most operable and 1 is the least operable.

When the examinees were asked if they had been taking photographs considering compositions, the result was 2.3, 5 the highest consciousness in composition and 1 the lowest. Finally, the examinees indicated 3.7 for the question if the compositional guidance is necessary.

8   Conclusion

In this paper, we proposed a shooting guidance method which calculates a proposing composition considering multiple compositions. It is based on the survey that the professional photographers tend to take photographs applying 1.7 compositions on average. Compositions clarify the subject and balance the space in a photo. In order to evaluate the proposing method, we carried out a subjective evaluation using Semantic Differential Method. As a result, certain psychological quantity was recognized in photograph taken with the proposing method.

References

1. London, B., Upton, J., Kobre, K., Brill, B.: Photography, 7th edn. Prentice Hall (2001)
2. Cohen, D.: How to take good pictures, Kodak and Harper Collins (1995)
3. Ueda, K.: For better photos: Encyclopedia of Digital SLR Photography Techniques 101. Impress Japan Corp., Tokyo (2012) (in Japanese)
4. Mai, L., Le, H., Niu, Y., Liu, F.: Rule of Thirds Detection from Photograph. In: IEEE International Symposium on Multimedia, pp. 91–96 (2011)
5. Bhattacharya, S., Sukthankar, R., Shah, M.: A Framework for Photo-Quality Assessment and Enhancement based on Visual Aesthetics. In: ACM Multimedia, pp. 271–280 (2010)
6. Hoiem, D., Efros, A.A., Hebert, M.: Recovering Surface Layout from an Image. International Journal on Computer Vision 75(1), 151–172 (2007)
7. Liu, L., Chen, R., Wolf, L., Cohen-Or, D.: Optimizing Photo Composition. Computer Graphics Forum 29(2), 469–478 (2010)
8. Chang, Y., Chen, H.: Finding Good Composition in Panoramic Scenes. In: International Conference on Computer Vision, pp. 2225–2231 (2009)
9. Su, H., Chen, T., Kao, C., Hsu, W.H., Chien, S.: Scenic Photo Quality Assessment with Bag of Aesthetics-Preserving Features. ACM Multimedia, 1213–1216 (2011)
10. Ieda, A., Keum, J., Hagiwara, M.: Photo Quality Improvement System by Modification of Composition Reflecting Kansei. The Journal of the Society for Art and Science 9(4), 163–172 (2011)
11. Banerjee, S., Evans, B.L.: In-camera automation of photographic composition rules. IEEE Transactions on Image Processing 16(7), 1807–1820 (2007)
12. Banerjee, S., Evans, B.L.: Unsupervised Automation of Photographic Composition Rules in Digital Still Cameras. In: SPIE Conference on Sensors, Color, Cameras, and System for Digital Photography VI, pp. 364–373 (2004)
13. Su, H., Chen, T., Kao, C., Hsu, W.H., Chien, S.: Preference-Aware View Recommendation System for Scenic Photos Based on Bag of Aesthetics-Preserving Features. IEEE Transactions on Multimedia 14(3), 833–843 (2012)
14. Cheng, B., Ni, B., Yan, S., Tian, Q.: Learning to Photograph. In: ACM Multimedia, pp. 291–300 (2010)
15. Stricker, M.A., Orengo, M.: Similarity of color images. In: SPIE 2420, Storage and Retrieval for Image and Video Databases III, pp. 381–392 (1995)
16. Landscape Photographer of the Year, http://www.take-a-view.co.uk/

17. National Geographic International Photography Contest,
    http://www.nationalgeographic.com/
18. Magnum Photos, http://www.magnumphotos.co.jp/index.html
19. Digital Photo Editorial Department (ed.): Digital SLR Questions 300 Shooting techniques. SoftBank Creative Corp., Tokyo (2010) (in Japanese)
20. Lienhart, R., Maydt, J.: An Extended Set of Haar-like Features for Rapid Object Detection. In: IEEE International Conference on Image Processing, vol. 1, pp. 900–903 (2002)
21. Sherrah, J., Gong, S.: Skin Colour Analysis (2001),
    http://ist.ksc.kwansei.ac.jp/kono/Lecture/
    CV/Protected/Rejume/Skin.pdf
22. Itti, L., Koch, C., Niebur, E.: A model of Saliency-Based Visual Attention for Rapid Scene Analysis. IEEE Transactions on pattern analysis and machine intelligence 20(11), 1254–1259 (1998)
23. Boykov, Y., Funka-Lea, G.: Graph Cuts and Efficient N-D Image Segmentation. International Journal of Computer Vision 70(2), 109–131 (2006)

Development of an Endocrine Genomics Virtual Research Environment for Australia: Building on Success

Richard O. Sinnott, Loren Bruns, Christopher Duran, William Hu, Glenn Jayaputera, and Anthony Stell

Melbourne eResearch Group
University of Melbourne
Melbourne, Australia
rsinnott@unimelb.edu.au

Abstract. The $47m Australian National eResearch Collaboration Tools and Resources (NeCTAR - www.nectar.org.au) project has recently funded an initiative to establish an Australia-wide endocrine genomics virtual laboratory (endoVL – www.endovl.org.au) covering a range of disorders including type-1, type-2 diabetes, rare diabetes-related disorders, obesity/thyroid disorders, neuroendocrine/adrenal tumours, bone disorders and disorders of sex development. This virtual laboratory will establish a range of targeted databases, clinical registries and support a range of genetically targeted clinical trials leveraging a body of international projects and experiences garnered over many years through a range of EU and MRC funded initiatives. This paper focuses on the plans for endoVL and especially, the systems it leverages in supporting large-scale clinical, collaborative environments.

Keywords: endocrine disorders, virtual research environment, security.

1   Introduction

The endocrine system is comprised of a system of glands, each of which secretes different types of hormone into the bloodstream to regulate the body. At present across Australia and indeed internationally, communities of clinicians and biomedical researchers are working on aspects of clinical care and biomedical research associated with particular disorders of the endocrine system, however the infrastructure to support these networks and to facilitate international endocrine-wide research interactions does not exist. Furthermore, following the sequencing of the human genome and unprecedented growth and availability of genomic (and other –omics) data, the research opportunities and challenges across the post-genomic life sciences for personalised e-Health is growing at an exponential rate and "big data" is an increasingly common challenge that must be overcome. Obviously infrastructures for dealing with "big data" especially in the health context require far more than just large-scale data storage systems. Rather the systems need to be designed, developed and deployed to cope with the specific concerns of health-related data regarding information governance, privacy and confidentiality. Experience has shown that these

B. Murgante et al. (Eds.): ICCSA 2013, Part V, LNCS 7975, pp. 364–379, 2013.

challenges are enormous when considered in the large, e.g. consider the UK Connecting for Health initiative [1], however they can be successfully addressed when targeted to specific clinical and biomedical endeavours. Endocrine-related biomedical research represents a domain where a targeted endocrine genomics virtual laboratory (endoVL) for clinical and biomedical research would provide cohesion across many efforts across Australia. Establishing and operating such a facility on behalf of Australia-wide endocrine communities is the purpose of the endoVL project (www.endovl.org.au) The endoVL project itself commenced in early 2013, however it builds upon and leverages an extensive portfolio of software systems and experiences in development and provisioning of security-oriented collaborative platforms dealing with a wide variety of clinical and biomedical data and associated networks/research communities. Amongst others, these include major international European (EU FW7) and UK MRC funded projects including:

- The €6m EU funded ENSAT-CANCER (www.ensat-cancer.eu), which began in 2011 and has a focus on adrenal tumours [2];
- The £625k MRC funded International-DSD project (www.i-dsd.org), which began in 2011 (following on from the EU funded EuroDSD project www.eurodsd.eu). International-DSD has a focus on disorders of sex development [3];
- The €1m EU funded EuroWABB (www.euro-wabb.org), which began in 2010 and has a focus on the rare diabetes-related diseases Alstrom, Bardet-Biedl and Wolfram [4];
- The $200k Australian Diabetes Data Network (ADDN) (www.addn.org.au), which began in 2013 and has a focus on child type-1 diabetes [5].

Many of these systems have grown over time and have been used for a wide range of major clinical trials and studies including full 4-phase genetically targeted clinical trials. Whilst all of these projects are endocrine-related, the disorders themselves differ in their prevalence. Thus whilst some of these diseases are common, e.g. type-1 diabetes, others are quite rare. Adrenal tumors are typical of this. For example around 1-2 cases of adreno-cortical carcinomas (ACC) - one particular adrenal tumor type are diagnosed for every million individuals. About 600 cases of ACC are diagnosed for the whole of Germany, population of approximately 81m [6]. Given the sparseness of such information, it is difficult to establish how best to treat these patients and indeed what treatment regimes work based for which cancer types at particular points in their evolution. The need to aggregate such data and ideally make it available with analytical tools is compelling.

The Internet provides a ubiquitous data-sharing platform. However this of course has major challenges that must be addressed for such information sharing to occur: security, ethics and importantly the standardisation of the information that is to be shared. The endoVL project builds upon a portfolio of projects, systems and lessons learnt in development and delivery of such web-based infrastructures. This paper describes the goals of the endoVL project and highlights some of the key lessons learnt in the associated projects upon which it is being built.

The rest of this paper is paper is structured as follows. Section 2 focuses on the goals of the endoVL project and highlights the communities it serves. Section 3 illustrates the data challenges and kinds of web-based databases that are currently being built to service those needs. Section 4 demonstrates the utility of these databases focusing in particular on a major international clinical trial that is currently on going. Finally section 5 draws some conclusions on the work and outlines the challenges that remain to be solved.

2   EndoVL Needs and Requirements

The endoVL project has been established to serve the clinical/biomedical needs of a wide range of researchers, groups, networks and societies across Australia. These include the:

- Endocrine Society of Australia (ESA) - a national not-for-profit organisation of scientists and clinicians who conduct research and practice in the field of endocrinology. The ESA currently has over 900 members spread over all of Australia and New Zealand.
- Australian and New Zealand Bone and Mineral Society (ANZBMS) has over 450 members spread over all of Australian and New Zealand. ANZBMS brings together clinical and experimental scientists and physicians actively involved in the study of bone and mineral metabolism in Australia and New Zealand.
- Australian Diabetes Society (ADS) is the peak medical and scientific body in Australia working towards improved care and outcomes for people with diabetes. The society has over 500 members.
- Australasian Paediatric Endocrine Group (APEG) is a not-for-profit organization representing health professionals caring for children with diabetes across Australasia. APEG currently has 400 members distributed across Australasia.
- Australasian Disorders of Sex Development network (DSDnetwork) is a newly formed network of clinical and biomedical researchers across Australasia focused upon understanding the aetiology of inter-sex disorders.
- Australian Thyroid Foundation (ATF) is a national not-for-profit organisation that supports and educates its member base and promotes good thyroid health. There are currently 250 members of the ATF.
- Clinical Oncology Society of Australia (COSA) is the peak national body representing health professionals working in cancer control and treatment. COSA provides a national perspective on cancer control activity from those who deliver treatment and care services across all forms of cancer. COSA currently has 150 members.

The above societies represent the *direct* communities and societies that will be the beneficiaries of endoVL, however it is important to recognise that every hospital and every GP practice in Australia deal with patients with endocrine disorders. It s intended that the infrastructure that endoVL will provide will allow for knowledge exchange beyond the immediate groups and societies identified above.

As described in section 1, the most urgently demanded capability to support these networks and societies is for secure access to a range of detailed (disease/domain specific) data. For many groups this is a largely phenotypic data, but for others support for genetics and other –omics analysis capabilities is needed. A major focus of endoVL is on providing a seamless environment where phenotypic <u>and</u> genotypic information can be uniformly presented for clinicians and biomedical researchers alike.

It is important to note that each disorder has its own specific phenotypic data that is of interest and has been agreed as <u>the</u> data to be collected by that community. It is planned that all of these endocrine diseases and related information/clinical trials will be made available through a unified endocrine-wide virtual research environment (VRE) where endocrine disease-specific collaborations can occur, but also inter-disease collaborations can occur also. Thus the endocrine system is not a single system, but a collection of systems that interact with each other to regulate the body: adrenal tumours can cause hypertension; type-2 diabetes and thyroid conditions can give rise to chronic obesity; disorders of sex development is often associated with a range of other complications, e.g. stunted growth and bone development. A range of different endocrine axes has been identified which collectively have been shown to interact with the regulation of the body. In a similar vein, endocrine research communities need an infrastructure that offers more than a web-based database for "their disorder" or "their clinical trial", but a platform for exchange of information across a range of disorders (subject to security considerations). Such a system-of-systems does not currently exist and is the focus of endoVL. This will provide a broader perspective of patient information and their associated management regimes that goes beyond any given endocrine specialty.

An overview of the endoVL infrastructure to be developed is highlighted in Fig. 1. This is being delivered through a user-oriented research environment that is made available through the Internet2 Shibboleth-based Australian Access Federation (www.aaf.edu.au) but extended as described in section 3.

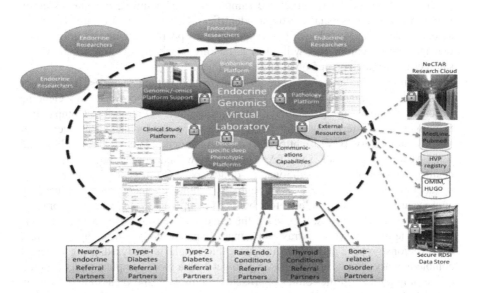

Fig. 1. High-level schematic of the endoVL Infrastructure

In Fig. 1, a range of clinical hospitals (referral partners) associated with the identified endocrine conditions act as the sources of clinical information. The endoVL infrastructure itself will support:

- A portfolio of disease-specific (deep) phenotypic databases that will subsequently allow much needed contextualisation of the –omics analysis of those patients. Each of these databases will allow screening (searching) for patients with certain phenotypic traits, on certain treatments, or a variety of other disease-specific criteria (particular karyotype, where a particular genetic screening using a given – omics approach has found (or not found) a particular mutation etc). These databases will be populated both through security-oriented forms that define and enforce data validation and structure, as well as live data feeds from hospital systems (as outlined below). Targeted bulk upload facilities for each endocrine disease will be supported (through processing of clinical data exported hospital databases to Excel spreadsheets for example). In this model, clinicians will be returned a list of auto-generated patient identifiers. This model has been realized through the I-DSD project.
- Clinical study platforms including development and support of new clinical study databases (typically populated with those patients screened from the given phenotypic databases subject to ethical approval and patient consent) together with electronic case report forms (eCRFs) targeted to the specific requirements of the clinical trials and studies.
- Biobanking support including Australia-wide biosample labelling and tracking between clinical referral partners and biomedical research groups. This should use international protocols for structured sample naming including the State, Centre, Patient-Id, and an associated sample identifier (such systems are already in use across Europe and work with a range of site-specific laboratory information management systems). This will allow a tracking capability to show where all the samples are for a given person at any given time.
- -omics support including a range of analyses of patient samples from major bioinformatics organisations across Australia including the Australian Genome Research Foundation (AGRF – www.agrf.org.au), Centre for Comparative Genomics in Western Australia (CCG - http://ccg.murdoch.edu.au/), Genomics Virtual Laboratory (GVL - https://genome.edu.au/wiki/GVL) and the National ICT Australia e-Health (NICTA - http://www.nicta.com.au/business/health/e-health). These groups cover the whole –omics space (from genomics; proteomics; metabolomics; transcriptomics through to whole exome/genome sequencing approaches). This will allow for both a comparative analysis of common bioinformatics approaches, e.g. where the same sample is analysed using the same –omics approach (microarray or mass spectroscopy etc) by different organisations/groups, and where the same biosample is analysed using different complementary –omics approaches, e.g. to get both a more complete understanding of the biology and on the -omics approaches themselves and their suitability in an applied clinical context. The raw data; stages of derived/processed data, and the final resultant (analysed) data together with all associated metadata from the - omics approaches for a given patient should be made available through the endoVL research environment.

- Pathology platform including standardised information on cellular information and associated pathological analysis of samples (including imaging). For tumours this will include information such as the Weiss score, number of mitoses, necrosis and information on when the biopsy/surgery was undertaken. Targeted databases will be developed for this purpose.
- Community forums for communication between the clinical and biomedical groups involved and for wider outreach efforts including to patient information support groups and networks. This will provide summary data on the number of patients included in the clinical databases, information on the clinical trials that are ongoing, and the process for access to and use of the endoVL resources more generally. Linkage / mining of the relevant research results from the international community through facilities such as PubMed and MedLine will be offered.
- Finally, where appropriate relevant data/metadata will be periodically published into the Human Variome Project Australia Node hosted at the University of Melbourne (HVPA - http://www.hvpaustralia.org.au/).

In the above figure, many of the images leverage pre-existing data models and services for these disorders, e.g. DSD and adrenal tumours developed in European projects. Other systems are currently ongoing, e.g. the ADDN national type-1 diabetes infrastructure. However as noted the seamless interplay across all of these systems is essential.

3   EndoVL Data and Security Challenges

Numerous hurdles must be overcome in the development and maintenance of such a complex environment. The main challenges are outlined in this section.

3.1   EndoVL Patient Linkage and Security

Central to the information model used in endoVL is the notion of a unique patient identity. A unique patient index is typically used to encapsulate the patient identity – typically within a particular hospital centre. It is the case however that each hospital has different approaches and solutions to tracking patients internally, and national systems are far from ubiquitous. Establishing the identity of individuals is critical to maintaining the consistency of information held within the endoVL databases but it must also be decoupled from any actual identity information in use locally, e.g. the name, data of birth, their address, Medicare number and potentially other local institutional identifiers that may be in place.

It should also be noted that for almost all of the disease areas under consideration within endoVL, an <u>independent</u> standalone secure database was required, i.e. a database without the requirement for interacting with existing hospital IT systems. For many of the rarer-disorders such as DSD this was quite satisfactory, however for some disorders such as type-1 diabetes this model was unrealistic. There are of the order of 120,000 children with type-1 diabetes across Australia and this figure is increasing dramatically [7]. Given this, automating data feeds from hospital settings was mandatory – to avoid the need for double data entry. In this model, patient specific information (name, address, Medicare number...) are kept in the hospital and a unique identifier generated (using a heartbeat counter specific to each centre). Thus the first patient from Westmead Hospital, Sydney, New South Wales, has identifier NSW-SYD-WM0001, the second NSW-SYD-WM0002 etc. This unique identifier is kept locally to the hospital and associated with other identifying information. This can be realised in many ways, e.g. a simple hash table or local database, depending upon the needs and experiences of the local hospital IT staff and technologies that they have in place. Lightweight clients are used to run local queries that extract the core identified data and associate them with the unique identifier NSW-SYD-WM0001, before pushing them out through outgoing-only hospital firewalls to the corresponding endoVL server where its digital signature is checked; it is decrypted and subsequently pushed into the database (validating the data as part of this process). This model of pushing data and not allowing incoming connections from the Internet appeases many of the security concerns of the hospital staff involved. A range of technical solutions to secure data linkage is described in [8] built on experiences in development of research infrastructures working with security-focused organisations.

An important aspect in this is the continual updates to data feeds where/when patients come to clinics. That is, the endoVL system needs to cope with updates to patient information and the synchronisation issues that this raises. In this case, a small subset of the core type-1 diabetes data needs to be updated based on information collected during the clinics. This includes such as the body-mass index and a range of typically fluctuating measurements and assessments that are captured on patient visits.

3.2   EndoVL Portal Security

The endoVL infrastructure needs to adhere to the highest standards of data security. The endoVL portal is to be provisioned with the Australian Access Federation (AAF – www.aaf.edu.au), which provides federated *authentication* – broadly speaking across Australian academic institutions. However guided by on-going efforts across Europe and Australian projects such as the Australian Urban Research Infrastructure Network (AURIN – www.aurin.org.au), users of the endoVL will also have specific privileges depending on their role within the endocrine disorder(s) efforts. These roles will be used to restrict access to data, tools and resources more generally. The default model will be to deny access, i.e. only individuals with authentic and valid credentials (digitally signed credentials recognised by the endoVL service components), will be able to access and use these resources. A targeted attribute authority will be set up for this purpose. These roles and privileges will be allocated under strict management of the endoVL staff, working in close cooperation with the clinical research communities on the assignment of these roles and privileges to the wider research community. Examples of the security policies that will be adopted (based on projects like I-DSD and ENSAT-CANCER) are that data can only be accessed by individuals from the same hospital; only by individuals from the same State (Victoria, NSW etc), or from the same country. Furthermore, certain individuals will be allocated roles that only allow read access to the database(s) whilst others will be allocated read/write access. Each contributing disorder centre will typically have a chief investigator responsible for dealing with the entry of patient data and acquisition of consent. These individuals are responsible for the data that is going into the endoVL databases by their staff or in the case of type-1 diabetes from the live data feeds.

As noted above, endoVL will not hold the names or addresses of any individuals or any direct information that can be used to identify any patient. Instead cases are identified for the purposes of communication with an automatically generated structured identifier that is unique within the clinical databases and an email address of the associated clinician responsible for the patient. This will leverage international standards and protocols that have been defined through projects such as ENS@T-CANCER on patient and biosample identification and tracking. At no time will a clinician ever be asked to reveal the identity of an individual to any researcher outside of their immediate clinical care environment. Furthermore, the linkage between the patient identifiers in the endoVL and identifiers used for particular patients within a live clinical setting will be both distinct and completely separated, i.e. it will never be possible to directly identify (through a given software query or direct observation of a particular software system) that a particular sample comes from a particular patient through use of the endoVL or other related IT system.

3.3   EndoVL Cloud Security

In addition to security related to portal-based access, the endoVL project will be hosting a wide range of genomic data including whole genome data sets derived using next generation sequencing technologies. To deal with the encryption and decryption of these data sets in virtualized environments and especially Cloud-based environments as offered through the National eResearch Collaboration Tools and Resources (NeCTAR) project (www.nectar.org.au), and especially their Research Cloud, the project is adopting the CSIRO TrustStore technology [9]. This allows users to create secure storage spaces on the Cloud by defining which storage, key management, and integrity services should be used and subsequently providing credentials to access these services. User profiles are used to support this process. With TrustStore, drag and drop ability is offered to copy files from a local machine to TrustStore storage. The files themselves are fragmented, encrypted, and hashed/signed with the encrypted fragments uploaded to the storage providers, and the keys stored in a TrustStore Key Management Service and signed digests stored in an Integrity Management Service.

It is planned that the genomic data itself is itself will be stored in the Leiden Open Variation Database (LOVD3 - http://www.lovd.nl/3.0/home). This system has already been established in endoVL (http://lovd.endovl.org.au). It is noted that this system has been used in other endoVL related projects including EuroWABB – https://lovd.euro-wabb.org.

3.4   EndoVL Data Flow Functions

The seamless movement of patient and/or genomic information is essential in developing systems such as endoVL. At the heart of the data flows is patient and sample tracking. Knowing that a patient is involved in a given clinical study should be used to inform for example whether they should be excluded from other studies. Data collected in a central registry should be used (where ethically approved) to filter and feed data into clinical trials. The interconnectedness of patients and their associated information is a key element of endoVL.

At the heart of this scheme are patient and study identifiers. Central identifiers used for patient tracking in the endoVL system, e.g. for the core data registries, need to be augmented with identifiers used within particular clinical trials and indeed for tracking of biosamples associated with those patients. Furthermore core registries established as part of endoVL, e.g. for adrenal tumours, should allow both the identification and recruitment of patients for particular clinical trials and studies, as well as the delivery of all relevant information on those patients for those trials and studies. In a similar manner data that is collected throughout the course of a particular clinical trial or study, should in principle (subject to ethics and information governance arrangements associated with the clinical trial) be available to wider collaborating researchers.

4   EndoVL Case Study Based on Adrenal Tumours

As identified, to realise the vision of a seamless interconnected virtual research environment for endocrine research across Australia, the endoVL project leverages a body of on going efforts and resources. The ENSAT-CANCER project is a significant undertaking that focuses on adrenal tumours (one form of neuroendocrine tumour). The ENSAT-CANCER project focuses specifically on four main types of adrenal tumour:

- aldosterone producing adenomas (APA);
- pheochromocytomas and paragangliomas (Pheo/PGL);
- non-aldosterone cortical adrenal adenomas (NAPACA),
- and adrenocortical carcinomas (ACC).

Each of these subtypes has different manifestations; involves different molecular mechanisms and ultimately requires different treatment regimes for optimal patient care. Given the rarity of the above kinds of adrenal tumours, the availability of a large collection samples, with associated clinical, biomedical/-omic data and accompanying treatment information is essential to better understand the differentiators of these tumour types; their aetiology and their associated molecular mechanisms.

The ENSAT-CANCER project commenced in 2011 and has since grown to include data on over 4000 patients with one of the above four tumour types; over 27,000 clinical annotations (treatments, surgeries, clinical visit information), as well as offering a source of over 5500 physical biosamples for a range of biomedical research and –omics data analysis (see Fig. 2). The system is currently used by 37 different centres/hospitals from around Europe. Given the rarity of the conditions identified in section 1, the critical mass of clinical and biomedical data to conduct statistically relevant research is now possible, and indeed is currently taking place. A key goal of endoVL and this paper is to build on the success of projects like ENSAT-CANCER.

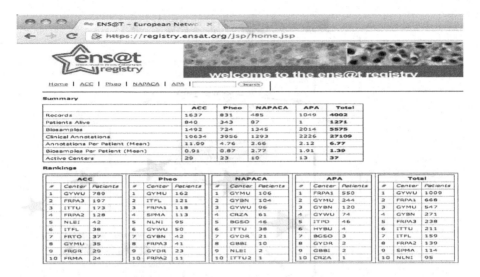

Fig. 2. Summary Data from ENSAT-CANCER

To illustrate how this system has evolved to be far more than simply a set of databases, but a truly collaborative research environment, we focus on one specific clinical trial that is currently on-going associated with Pheo/PGL patients. It is noted that a multitude of other genetically targeted clinical trials are now supported through ENSAT-CANCER including ADIUVO (ACC), FIRSTMAPPP (Pheo/PGL), FAMIAN (an imaging study) and a prospective study on recurrence of Pheo for patients with hypertension. It is envisaged that this latter study will run for many years (potentially over 25 years!).

4.1   PMT Study Background

The PMT study (Prospective Monoamine Tumour study) is a four-phase clinical trial targeted to patients who exhibit clinical indications of suspected pheochromocytoma through one or more of the following criteria:

- signs and symptoms;
- therapy-resistant hypertension;
- incidental finding on imaging for related condition;
- routine screening due to known mutation or hereditary syndrome;
- routine screening due to previous history of pheochromocytoma.

Thus far patients have been admitted to the study from a range of specialist clinics across Europe (including currently Dresden, Munich, Wurzburg, Nijmegen and Warsaw). Information on these patients is collected and tracked over time over the full four phases of the trial. These phases include initial screening of patients, clonidine tests for patients who meet the screening criteria, biochemical characterisation of patients and the subsequent follow-up of patients as shown in Fig. 3. This information is to be collected up to five years after the study completes.

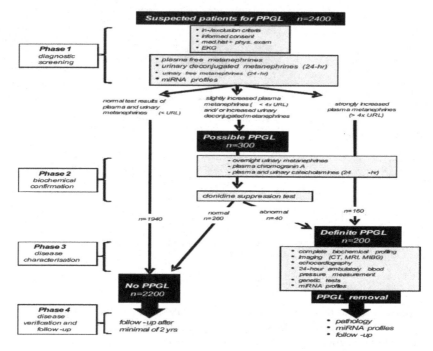

Fig. 3. Clinical Path of Patients in the PMT study

At present 865 patients have been recruited to the PMT study (see Fig. 4).

Pheo PMT Study

Home | Phase 1 | Phase 2 | Phase 3 | Phase 4A | Phase 4B | (Search)   Welcome, **Richard** | Sign Out | Account Details

	Full Study	Phase 1 (Screening)	Phase 2 (Clonidine)	Phase 3 (Pheo Characterization)	Phase 4 A (Excluded Follow-Up)	Phase 4 B (Pheo Follow-Up)
Patients	865	804	19	12	7	23

Fig. 4. PMT Recruitment Status

4.2   Pheo/PGL and PMT Data Flow

To support the recruitment (data flow) from the Pheo/PGL database, many of the required data points have been deliberately designed into the PMT data model directly. As a result, data from patients who match the criteria for recruitment to the PMT can be automatically fed into the PMT phase 1 recruitment/screening database (subject, amongst other things, to agreement/consent to participate in the study). It is also important to realise that many clinical trials (such as PMT) a whole range of additional information is required to be captured, e.g. the medications that are being received and the dosages, that would not generally be found in a disease registry like the ENSAT-CANCER Pheo/PGL register.

The actual clinical path (protocol) of patient progress through the PMT study is shown in Fig. 5. The patient identification naming scheme is similar to that found in the Pheo/PGL, e.g. Country code – Centre code – patient code. With this information, tracking of patients in the registry and the PMT study can thus be achieved. It is also worth knowing that additional data that is found in the PMT system can be directly fed back to the Pheo/PGL system, i.e. it is not always the case that the Pheo/PGL registry is used solely as a source of data/patients for clinical trials. Rather the information collected throughout the course of the clinical trials can be used to augment the data found in the registry.

Pheo PMT Study

Home | Phase 1 | Phase 2 | Phase 3 | Phase 4A | Phase 4B | (Search) | Welcome, **Richard** |
                                                                                      Sign Out | Account
                                                                                              Details

Create New Record (enter patient in phase 1)

1 2 3 4 5 6 7 8 9 10 11 12 13 14 15 16 17 18 19 20 21 22 23 24 25 26 27 28 29 30 31 32 33 34 35 36 37 38
39 40 41 Next Page

NOTE: starred patient IDs have skipped Phase 2

Phase 1 (Screening)				Phase 2 (Clonidine)	Phase 3 (Pheo Characterization)	Phase 4A (Excluded Follow-Up)	Phase 4B (Pheo Follow-Up)
Study ID	Record Date	Sex	Year of Birth				
GYDR-0011	2011-02-24	M	1943	GYDR-0014 GYDR-0016 GYDR-0028 GYDR-0031	PLWW-0047* GYMU-0274* GYMU-0014* PLWW-0128*	GYDR-0001* GYDR-0002* GYDR-0003* GYDR-0008*	GYDR-0004* GYDR-0005* GYDR-0006 GYDR-0007
GYDR-0012	2011-02-25	F	1939	GYDR-0045 GYDR-0048 GYMU-0311	PLWW-0139* PLWW-0172* PLWW-0213*	GYDR-0009* GYDR-0010* GYMU-0302*	PLWW-0010* GYDR-0018* GYDR-0042*
PLWW-0001	2011-03-01	F	1976	GYDR-0059 GYWU-0691 GYDR-0071	PLWW-0253* GYDR-0089* GYMU-0515*		PLWW-0088* GYDR-0052 GYMU-0273*
PLWW-0002	2011-03-01	M	1964	GYDR-0072 GYWU-0796	GYDR-0093* GYDR-0103		GYWU-0685* GYWU-0686*
PLWW-0003	2011-03-01	M	1954	GYWU-0797 GYDR-0081 GYDR-0078			GYWU-0695* GYWU-0770* GYDR-0069*
PLWW-0004	2011-03-01	F	1947	GYDR-0080 GYMU-0469 GYDR-0084			GYWU-0823* GYWU-0834 GBBI-0020
PLWW-0005	2011-03-01	F	1947	GYDR-0096			GBBI-0021* GYWU-0836* GYWU-0847*
PLWW-0006	2011-03-01	F	1960				GYWU-0869* GYDR-0092*

Fig. 5. PMT Phase Status

4.3   Biobanking Support

As identified in section 2, value added services that should be supported by a virtual research environment include broader data tracking – including tracking of physical biosamples between partner sites. In terms of biobanking, the production of labels for physical samples that are stored in laboratory fridges is of great significance. Whilst some centres have their own laboratory information management systems (LIMS), many others do not. Furthermore, there is considerable heterogeneity of LIMS systems in place and no commonality across naming and identification. The naming and tracking of physical samples is essential when dealing with large multi-centre studies like PMT. To accommodate this, biomaterial forms have been developed that allow the specification of samples stored, which studies they are primarily collected for, and the number of aliquots stored (this is stored permanently with the option for modification when it comes to the time of printing). The information captured includes the canonical identifier (including the center code), the biomaterial form number, the aliquot number, and the sample name. This is also captured in barcodes for readers that wish to capture the information electronically. There are many barcode standards to choose from – QR codes were initially tried but rejected due to the two-dimensional nature being unreadable on the curved surface of a typical sample tube. As a result the project has now adopted Interleaved 2 of 5 [10] as the agreed standard (a typical label example is shown in Fig, 6).

GBBI-0008
bio-ID 380
Study: EURINE-ACT
Date: 2012-01-06
Tumor Tissue (DNA)

Fig. 6. Biomaterial label output with barcode, form and study information (in this case EURINE-ACT study)

4.4   PMT Form Implementation and Adaptation

Many of the PMT study forms have relatively complicated user interface requirements: signs and symptoms have a matrix of symptoms versus frequency, duration and severity. Similarly, a common specification throughout the PMT study is the use of multiple units. Units can be input and converted between SI and imperial measures (e.g. pg/mL versus nmol/L). Originally these values were stored in one format – converted from what was originally input and rendered in both units when returned. However, the accuracy loss during conversion in this formatting was deemed unacceptable for clinical purposes and the information has now been stored as separate data-points, interchangeable if precision loss is tolerable, but losing none of the accuracy of the original input data units. It should be noted that the range of numbers is large – with values of three significant numbers to three decimal places for some measurements, e.g. 0.003 nmol/L, to five significant numbers for other measurements, e.g. 50,000 nmol/L.

Phase 1 - Screening Biochemical Tests   `Print`

Update the screening biochemical tests for this patient by entering the data manually in the fields below

PLEASE NOTE: only local Dresden users should be entering information on this form, for the correct processing of LC-MS/MS data. All other users will not have access - for further information please contact Prof Graeme Eisenhofer.

Summary			
	Plasma Free Metanephrines	Urine Free Metanephrines	Urine Deconj. Metanephrines
Date of test	2010-11-25	2010-11-25	
Normetanephrine	73.7 0.402 (pg/mL)(nmol/L)	21.5 117.3 (µg/day)(nmol/day)	314.0 1713.9 (µg/day)(nmol/day)
Metanephrine	29.8 0.151 (pg/mL)nmol/L)	14.7 74.5 (µg/day)(nmol/day)	110.2 558.8 (µg/day)(nmol/day)
Methoxytyramine	18.3 0.109 (pg/mL)(nmol/L)	25.9 154.8 (µg/day)(nmol/day)	188.0 1124.3 (µg/day)(nmol/day)
Creatinine (urine only)		1549 (mg/day)13.697 (mmol/day)	

Fig. 7. PMT Phase 1 Screening Biochemical Test Matrix

Other significant features used in the VRE include the ability to track medication inputs throughout the study – when they are added and when they are removed. This is especially important for the interpretation and contextualisation of biochemical results. These are listed in three categories of anti-hypertensives, other prescribed and "over-the-counter" supplements. It is typical that the storage of such information is vast and can quickly become unmanageable. However, the system allows for such tracking and capturing of dosages of medications.

A major challenge in clinical trials like PMT is the flow of information across the phases and when a patient may progress from Phase 1 to Phase 2 etc. Genetic information is captured and analysed in each phase. Patients are also able to move from phase 1 straight to phase 3 – where a patient has a Pheo/PGL confirmed without the requirement for supplemental clonidine testing. Even with this phase jump, it is important for the biochemical testing to still be performed. If the patient has gone through phase 2 already then this information is presented as complete in phase 3, but the clinical trial personnel have the option for manual input of data for patients that missed phase 2. In PMT, data entry to the biochemical pages, are restricted to only those users in Dresden, the central biochemical processing center for PMT. This allows the mass spectrometry information to be standardised to the lead center settings to ensure consistent quality of analysis and results.

A completeness function has also been implemented for each of the forms indicating what information still needs to be input for each phase. The method follows a simple method of green for complete, red for incomplete. The background information that provides this is a survey of the relevant database table, taking into consideration the points that have been marked as optional. These systems have been developed working in close consultation with the clinicians.

4.5   Data Output and Analytics

Of paramount importance is the ability to use clinical and biomedical information for advanced statistical analysis. Whilst it is possible to embed statistical tools within a browser, many researchers (and those of ENSAT-CANCER and PMT) prefer to directly download the data sets and analyse them on their local desktops. To support this, it is necessary to translate a relational database structure into a two-dimensional spreadsheet. A critical point here as identified by the biostatisticians was to have all the information for a single patient on each row. Whilst information is captured that exists in a one-to-many relationship in a database, in a flat file this must be flattened / rendered into one line as illustrated in Fig. 8.

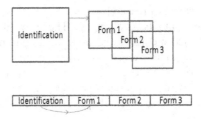

Fig. 8. Rendering one-to-many forms to a single spreadsheet line

The output of such data to these spreadsheets can often be "inelegant" as the size of the single line column width, must be pre-calculated from the size of the single patient that has the most table entries for that entity. The column number, and programmable features of output also depend on the format selected, with comma-separated value (.csv) files being the simplest with the ability to render in most simple text editors. The ENSAT-CANCER and PMT systems also allow export directly to .xls Excel files (with a maximum column width of 65536) and to the more advanced Excel .xlsx format which can accommodate a much greater number of columns. It is noted that many patients have over 10,000 columns (data points) in the PMT system.

The use of this formatting style is particularly important when tracking longitudinal information about treatment and follow-up summaries. For the ACC section of the ENSAT-CANCER registry, instances of recurrence, surgery resection status and treatments received, can be summarised in a single page. To export this summary to a similarly formatted spreadsheet, then to track this over time, is a critical function for progressing research into ACC treatments. Often this requires programmatic interfaces that calculate form dates to assess features such as time to recurrence or patient death. The algorithms to complete these are being updated as the study requirements develop.

5   Conclusions

The endoVL project is very much ramping up across Australia, however it is based upon established platforms that are used for a wide range of biomedical research endeavours. ENSAT-CANCER has become the central resource for a wide range of major multi-center clinical trials that are currently ongoing across Europe. These include ADIUVO (ACC clinical trial), FIRSTMAPPP (Pheo trial), EURINE-ACT, PMT (Pheo trial) with future studies including FAMIAN (imaging trial) and AVIS in the offing. Similarly, the DSD platform is now well established as the major disorders of sex development platform with a 5-year MRC platform grant to extend the EU EuroDSD system that was build originally. The security-oriented platforms that have been established have found a vibrant community of researchers willing to coordinate their efforts and these platforms have been hardened based on practical experiences from demanding consumers of IT infrastructures. It is important to note that before projects such as I-DSD and ENSAT-CANCER existed, the infrastructures and data sets in place across Europe can be best described as fragmented. Different researchers in different countries collected similar but not quite the same data; their IT systems used similar but not quite the same standards. This is currently the situation in Australia. Through endoVL we expect to revolutionise the way post-genomic research is undertaken across a spectrum of research areas.

This work is not without its challenges. The growth of –omics data and opportunities for personalised medicine and e-Health is putting increased challenges on data processing and storage. Across Australia major investments in national research data storage efforts is occurring. The $50m Research Data Storage Infrastructure (RDSI – www.rdsi.uq.edu.au) project is rolling out a 100Petabyte data storage system for Australia-wide researchers. The processing of such volumes of data is also challenging many communities. Across Australia the NeCTAR project has rolled out a national research Cloud that is being used already in endoVL.

A major challenge that is often non-technical is one of engagement with the researchers themselves and their organisations. Convincing them that their data is safe and meets their information governance concerns is an ever-present necessity. Similarly, whilst infrastructures like endoVL can be built in an agile manner, it is often extremely challenging to get consensus from the community on what data should be collected. We build on a body of work and many years of effort in this space however. The challenge of ethics and its associated processes are also non-trivial to deal with, but lie at the heart of successful biomedical collaborations. Whilst projects such as ENSAT-CANCER and I-DSD have ethics approval for all countries involved, the processes for obtaining approval for endoVL are different again in Australia. The endoVL team are currently working with endocrine-disorder research communities on the ethics application for storage of disease-specific phenotypic and genotypic data sets.

References

1. Doward, J.: Chaos as £13bn NHS computer system falters. The Guardian, London (retrieved August 11, 2008)
2. Sinnott, R.O., Stell, A.J.: Towards a Virtual Research Environment for International Adrenal Cancer Research. In: International Conference on Computational Science, Tsukuba, Japan (June 2011)
3. Ahmed, S.F., Rodie, M., Jiang, J., Sinnott, R.O.: The European DSD Registry – A Virtual Research Environmen. International Journal on Sexual Development, Special issue on Disorders of Sex Development, "New concepts for human disordersofsexdevelopment", Sex Dev. 4, 192–198 (2010), doi:10.1159/000313434).
4. Farmer, A., Ayme, S., Lopez de Heredia, M., Maffei, P., McCafferty, S., Mlynarski, W., Nunes, V., Parkinson, K., Paquis, V., Rohayem, J., Sinnott, R.O., Tillmann, V., Tranebjærg, L., Barrett, T.: EURO-WABB: An EU Rare Diseases Registry for Wolfram Syndrome, Alstrom Syndrome andBardet-Biedl Syndrome. To appear in Journal BMC Paediatrics (2013)
5. Colagiuri, S.B., Gomez, M., Fitzgerald, B., Buckley, A., Colagiuri, R.: DiabCo$t Australia Type 1: Assessing the burden of Type 1 Diabetes in Australia. Diabetes Australia. Canberra (2009)
6. Allolio, B., Fassnacht, M.: Clinical review: Adrenocortical carcinoma: clinical update. J. Clin. Endocrinol. Metab. 91(6), 2027–2037 (2006), doi:10.1210/jc.2005-2639.
7. Australian Institute ofHealthandWelfare (AIHW), Incidenceof Type 1 diabetes in Australianchildren 2000-2008. Cat. no. CVD 51. Canberra, AIHW
8. Sinnott, R.O., Stell, A.J., Jiang, J.: Classifying Architectural Data Sharing Models for e-Health Collaborations. In: HealthGrid 2011, Bristol, UK (June 2011)
9. Yao, J., Chen, S., Nepal, S., Levy, D., Zic, J.: TrustStore: MakingAmazon S3 Trustworthywith Services Composition. In: Proceedings of the CCGRID 2010, pp. 600–605 (2010)
10. Interleaved 2 of 5, http://en.wikipedia.org/wiki/Interleaved_2_of_5 (accessed March 2013)

Using MaxGWMA Control Chart to Monitor the Quality of TFT-LCD

Bi-Min Hsu[1], Thanh-Lam Nguyen[2], Ming-Hung Shu[3,*], and Ying-Fang Huang[3]

[1] Department of Industrial Engineering and Management, Cheng Shiu University,
Kaohsiung 83347, Taiwan
bmshu@csu.edu.tw
[2] Graduate Institute of Mechanical and Precision Engineering, National Kaohsiung University
of Applied Sciences, Kaohsiung 80778, Taiwan
green4rest.vn@gmail.com
[3] Department of Industrial Engineering and Management,
National Kaohsiung University of Applied Sciences, Kaohsiung 80778, Taiwan
{workman,winner}@kuas.edu.tw

Abstract. While the traditional Shewhart control charts are widely used to monitor and detect large shifts in the mean and variability of a manufacturing process respectively, the exponentially weighted moving average (EWMA) control charts are well employed to identify small ones. Its various extensions have been proposed. Among them, the Maximum generally weighted moving average (MaxGWMA) control chart is the superior. In the industry of the Thin Film Transistor Liquid Crystal Display (TFT-LCD), as TFT-LCD has become increasingly attractive and developed, it has received a high production volume. In practice, though Xbar – R control charts are widely used, they can't quickly detect the shift in the manufacturing process. Therefore, using the control charts may still lead to a risk of a high increase in production cost due to mass production of non-conformities. Hence, in this study, MaxGWMA control chart was found as a good replacement owing to its alarming ability to small shifts in the process.

Keywords: MaxGWMA, Control chart, Liquid Crystal Display.

1   Introduction

Nowadays, due to the high need of mass production in many different industries, controlling the quality of the industrial products has become a vital task to the survival of the industrial manufacturers because any inefficiency in performing the task will lead to a high increase of production cost, which will also reduce their competitive power on the market. Quality is well recognized as a competitive advantage [1]. Therefore, building and implementing an effective quality control program has been received a special attention these days. As a result, quality control has become a powerful

---

* Corresponding author.

B. Murgante et al. (Eds.): ICCSA 2013, Part V, LNCS 7975, pp. 380–390, 2013.

management tool in dealing with the key quality characteristics of visible products [2]. When the key quality characteristic lies within a small variability around its target value, the quality is said to be in-control and the products are believed to be consistently manufactured under a stable process. A stable process usually has a very low percentage of nonconformities and defects in its outputs, resulting in not only the decrease in the manufacturing cost but also the increase in customer satisfaction through using consistent products. In quality control, statistical process control (SPC) approach has been widely employed due to its effectiveness in reducing variability, achieving process stability as well as improving production capability [1].

Among the seven major tools in SPC, the most technically sophisticated one is the control charts pioneered by Shewhart. With the ability to quickly detect process shifts and identify abnormal conditions in on-line manufacturing processes, the control charts can provide enough analytical information and the values of important process parameters to preclude defects, lower scrap and rework, and avoid unnecessary process adjustment, meaning that they can play a significant role in both enhancing the productivity and production capacity and cutting down the production cost [1], [3]. A typical control chart is illustrated in Figure 1 on which the center line (CL) represents the estimated process target value and two control lines- the upper control limit (UCL) and lower control limit (LCL) are the boundaries of the normal variability used to test if the majority of the observations are in statistical control. A process can be classified as stable and in statistical control only if the entire statistic points of collected sample data fall within the limits and do not exhibit any systematic pattern on the control chart. Otherwise, the process indicates the presence of some assignable causes which need carefully investigated and further correct actions should be implemented. This stage is usually referred as phase I of control chart application.

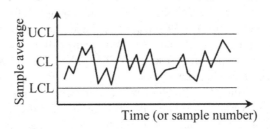

Fig. 1. A typical control chart

In phase I, Shewhart control charts are very effective in not only being easily constructed and interpreted but also detecting both large, sustained shifts in the process parameters, measurement errors, data recording and/or transmission errors, etc. which are to be settled in order to bring the process into a state of statistical control [1]. When the process is reasonably stable, it is then of great importance to continuously monitor the process performance; which is now referred as phase II of control chart application.

Since assignable causes resulting in large shifts are systematically eliminated in the first phase, smaller shifts must be detected in the phase II to make the products consistent. Due to the fact that the Shewhart control charts are relatively insensitive to small shifts, they are obviously incompatible with the second phase. Instead, other two control charts, including the cumulative sum (CUSUM) control chart and the exponentially weighted moving average (EWMA) control chart, are their good substitutes in the phase. Comparing the performance between the two alternative control charts, Vargas et al. [4] concluded that EWMA chart outperforms CUSUM charts in detecting small shifts of one standard deviation or less. Moreover, EWMA charts can quickly detect the small shifts at the beginning of the changes and do effective forecasting for the next period [4] as well as be set up and operated easier [1]. Consequently, EWMA chart is more preferred in practice. Many scholars have paid much effort in developing this traditional chart. Among its various extensions, a control chart named the Maximum generally weighted moving average (MaxGWMA) control chart has recently been introduced by Sheu et al. [5] which was claimed to be superior in diagnostic abilities. Therefore, in this paper, we propose using MaxGWMA control chart to monitor the quality of thin-film-transistor liquid crystal displays (TFT-LCDs).

This paper is organzied as the following. A review of MaxGWMA control chart is discussed in Section 2. Section 3 presents an application of the chart in monitoring quality of TFT-LCD. Also in this section, the chart is compared to the traditional Xbar – R charts to prove of its better performance. The conclusion is made up the last section.

2   Review of MaxGWMA Chart

The EWMA control chart was first introduced by Roberts [6] to detect a small shift in the process mean. Since then, it has successfully attracted the unceasing attention of many scholars as in the reviews by Xie [7], Han and Tsung [8], Eyvanzian et al. [9], Sheu et al. [10], Li and Wang [11], Zhang et al. [12] and Sheu et al. [5].

In monitoring the process mean and variability, many different methods have been suggested. Sweet [13] proposed the use of two EWMA charts to detect the shifts in process. Some other researchers have made a lot of effort in employing only a single chart. Chan et al. [14] introduced a chart which can identify the shift but it requires plotting the mean and variability separately, resulting in the difficulty in constructing. Whereas, the omnibus EWMA charts proposed by Domangue and Patch [15] are sensitive to shifts; but it cannot point out whether the shifts occur due to the process mean or variability or both [16]. Gan [17] recommended a joint scheme consisting of a two-sided EWMA mean chart and a two-sided EWMA variance chart which was found to perform well for several out-of-control circumstances. Chao and Cheng [18] came up with semicircle chart for variable data but it also fails to track the time sequence of the plotted pointed. Gan [16] presented a method using a two-dimensional chart of elliptical in-control region to plot the EWMA of logarithm of sample variance (s2) against the EWMA of sample mean (xbar). Chen and Cheng [19] first discussed about a new control chart called Max-chart by combining xbar-chart and s-chart under the maximum values of their relevant statistics. The Max-chart can effectively exhibit both process mean and variability on a single chart. Besides, in term of exponentially weighted moving average, Max-Min EWMA chart was proposed by Amin et al. [20]. Among other types of EWMA charts, such as MaxEWMA, SS-EWMA, EWMA-Max and EWMA-SC charts demonstrated by Xie [7], the MaxEWMA chart was proved to be the most effective in terms of ARL performance, and diagnostic ability and have high capability of detecting small shifts in the process mean and variability as well as identifying the source and the direction of an out-of-control signal [7].

The EWMA chart was first generalized by Sheu and Lin [21] to create a new chart named the generally weighted moving average (GWMA) control chart. The GWMA chart can detect small shifts much quicker than the EWMA can. However, while the EWMA chart is almost a perfectly distribution-free procedure, meaning that EWMA chart is robust to non-normality, GWMA chart is designed under the assumption that the data set comes from a normal distribution and it is sensitive to the departure from normality than EWMA chart when process shifts [20]. This assumption has been used by Sheu and Yang [22], Sheu and Tai [23] and Sheu et al. [5]. If the assumption of normality is violated, the number of false alarms caused by GWMA chart is more than that of EWMA chart; however, this problem can be fixed by increasing the control limit width in the construction of GWMA chart [21].

By combing the advantages of the MaxEWMA chart and GWMA chart, Sheu et al. [5] proposed a new chart called the Maximum generally weighted moving average (MaxGWMA) control chart. The MaxGWMA chart is more sensitive than the MaxEWMA chart [5]. The MaxGWMA chart is constructed based on the following procedure with five basic steps.

2.1   Estimate Unknown Process Parameters $\left(\mu_0, \sigma_0\right)$

Let x be the key quality characteristic following a normal distribution $N(\mu_0, \sigma_0^2)$, where μ_0 is process mean and σ_0 is process standard deviation. The process is said to be shifted to δ standard deviation in mean if its new mean μ_1 is $\mu_1 = \mu_0 \pm \delta\sigma_0$ where $\delta \neq 0$; similarly, it is said to be shifted of ρ standard deviation in variability if its new standard deviation σ_1 is $\sigma_1 = (1 \pm \rho)\sigma_0$ where $\rho \neq 0$. In practice, because μ_0 and σ_0 are usually unknown, they must be estimated by using sample data, at least 20 to 25 in-control samples recommended. Assume that we have collected the sample data $A = \left\{ A_i \middle| A_i = \left(x_{i1}, x_{i2}, \ldots, x_{in_i} \right), i = \overline{1, m} \right\}$ which consist of m random subgroups and each subgroup contains n observations of x.

Let $\overline{x}_i$ be the sample average of the ith sample.

$$\overline{x}_i = \frac{1}{n} \sum_{j=1}^{n} x_{ij}, \text{ for } i = \overline{1, m} \tag{1}$$

Then the grand sample average $\overline{x}$ is obtained by

$$\bar{x} = \frac{1}{m}\sum_{i=1}^{m}\bar{x}_i \,, \text{ for } i = \overline{1,m} \qquad (2)$$

Let s_i be the standard deviation of the ith sample.

$$s_i = \sqrt{\frac{\sum_{j=1}^{n}\left(x_{ij}-\bar{x}_i\right)^2}{n-1}} \qquad (3)$$

The average of the m standard deviations is

$$\bar{s} = \frac{1}{m}\sum_{i=1}^{m}s_i \qquad (4)$$

The unbiased estimators of the process mean μ_0 and standard deviation σ_0 are given by

$$\begin{cases} \mu_0 = E(\bar{x}) = \bar{x} \\ \sigma_0 = E(\bar{s}) = \bar{s}/c_4 \end{cases} \qquad (5)$$

where the values of c_4 for sample sizes $2 \le n \le 25$ are available in the textbooks [1], [2], [24].

2.2   Compute Two Transformed Statistics

Since $\bar{x}_i$ and s_i^2 are independent with different distributions, they can be transformed into two independent statistics

$$\begin{cases} M_i = \dfrac{(\bar{x}_i - \mu_0)}{\sigma_0/\sqrt{n}} \\ S_i = \Phi^{-1}\left\{F\left[\dfrac{(n-1)s_i^2}{\sigma_0^2}, n-1\right]\right\} \end{cases} \qquad (6)$$

where $F(a,b)$ denotes the chi-square distribution function of a with b degrees of freedom and $\Phi^{-1}(.)$ denotes the inverse standard normal distribution function.

2.3   Compute MaxGWMA Statistic

Let A be an event of interest. t denotes the counting number of samples between two adjacent occurrences of A. $P_j = P(t > j)$ is the probability that A does not occur in the first j samples. Assume that $\forall j \ge 1$ and $j < i$, we have $P_j > P_i$; then, the probability p_j of the occurrence of A at the jth sample can be obtained by

$$p_j = P(t > j-1) - P(t > j) = P_{j-1} - P_j \tag{7}$$

Let p_j be the weight in the weighted moving average; then, the GWMA statistics for the ith subgroup are obtained by

$$\begin{cases} U_i = \sum_{j=1}^{i} p_j M_{i+1-j} \\ V_i = \sum_{j=1}^{i} p_j S_{i+1-j} \end{cases} \tag{8}$$

For the ease of computation, Sheu and Lin [18] chose $P_j = q^{j^{\alpha}}$, where q is called a design parameter which is a constant with the value in [0,1] and α is called an adjustment parameter determined by the practitioner. So, the traditional EWMA chart is just a special case of GWMA chart when $\alpha = 1$ and $q = 1 - \lambda$. The probability in equation (7) is now rewritten as

$$p_j = q^{(j-1)^{\alpha}} - q^{j^{\alpha}} \tag{9}$$

As M_i and S_i are mutually independent, if the process is not shifted, U_i and V_i are also mutually independent and follow the same standard normal distribution. Their variances are determined by

$$\sigma_{U_i}^2 = \sigma_{V_i}^2 = \eta_i = \sum_{j=1}^{i} p_j \tag{10}$$

The MaxGWMA statistic, denoted by MG, for MaxGWMA chart is defined as

$$MG_i = \max\left(|U_i|, |V_i|\right) \tag{11}$$

A small value of MG_i indicates that the process mean and process variability are close to their respective targets and vice versa. With this approach, the MaxEWMA chart proposed by Xie [7] is just a special case of MaxGWMA chart at $\alpha = 1$ and $q = 1 - \lambda$.

2.4 Compute MaxGWMA Control Limits

Because MG_i is non-negative, Sheu et al. [5] proposed using only upper control limit (UCL) to monitor the MG_i. The UCL_i for the ith subgroup is determined by

$$UCL_i = E(MG_i) + L\sqrt{Var(MG_i)} \tag{12}$$

where L denotes the width of control limit.

Based on desired in-control ARL0 and sample size n, and optimal values of parameters q, α, and L for an initial state of the MaxGWMA chart, the approximate UCL_i in equation (12) is given by

$$UCL_i = (1.12838 + 0.60281 \times L)\sqrt{\eta_i} \qquad (13)$$

2.5   Compare MG_i with UCL_i and Construct MaxGWMA Chart

By comparing MG_i values with the control limit UCL_i, we can judge if a certain subgroup is in control or out of control.

- If $MG_i \leq UCL_i$, the ith subgroup is said to be in control.
- If $MG_i > UCL_i$, the ith subgroup is said to be out of control. In order to identify the situations if there is a change in the process mean and/or the process variability, for brevity, the symbols listed in Table 1 are used.

Table 1. Symbols for out-of-control points with their indications

No.	Situation	Symbol	Indication
1	$MG_i = U_i$ and $\|V_i\| \leq UCL_i$	m+	An increase in process mean
2	$MG_i = -U_i$ and $\|V_i\| \leq UCL_i$	m−	A decrease in process mean
3	$MG_i = V_i$ and $\|U_i\| \leq UCL_i$	v+	An increase in process variability
4	$MG_i = -V_i$ and $\|U_i\| \leq UCL_i$	v−	A decrease in process variability
5	$U_i > UCL_i$ and $V_i > UCL_i$	++	An increase in both process mean and variability
6	$-U_i > UCL_i$ and $-V_i > UCL_i$	−−	A decrease in both process mean and variability
7	$U_i > UCL_i$ and $-V_i > UCL_i$	+−	An increase in mean and a decrease in variability
8	$-U_i > UCL_i$ and $V_i > UCL_i$	−+	A decrease in mean and an increase in variability

3   A Practical Application

Thin Film Transistor Liquid Crystal Display (TFT-LCD) has become increasingly attractive and popular as a thin, light and low power consumption display device. A wide range of applications of TFT-LCD has been developed and commercialized, including notebook and desktop PCs, video camcorders, cameras, televisions, and public displays. The core components of a TFT-LCD are a Thin Film Transistor (TFT) panel and a Color Filter (CF) panel which should be accurately aligned for better performance of the display. In order to ensure the display quality of LCD panels, the inspection of surface defects becomes a critical task in LCD manufacturing.

Surface defects of a TFT-LCD panel will cause both visual and electrical failures. Therefore, monitoring the surface defects on TFT-LCDs will help the manufacturer detect any abnormal issues in the manufacturing process in order to have proper actions to not only improve the quality but also reduce the manufacturing cost by lowering the scrap and rework.

One of the key quality characteristics of a TFT-LCD is the Total Pitch (TP) which is actually the alignment error between the designed and the actual pitch values of the CF and TFT panels [25], [26]. A negative TP value means that the pattern is smaller than a design specification; and vice versa. For monitoring the TP, TFT-LCD manufacturers usually use the traditional Xbar – R control charts [25]. However, since the control charts are not able to quickly detect the ˋshift of process mean and variability, in the mass production, any uncontrolled shift will lead to mass manufacturing cost, accordingly reducing the manufacturer's benefit. Therefore, in this paper, we propose employing the MaxGWMA control chart as an alternative.

The TP data used in this paper were collected from 25 samples of size 5 of TFT-LCDs supplied by an LCD manufacturer in Taiwan. Based on these data, we will construct appropriate traditional Xbar – R control charts and MaxGWMA chart in order to compare the applicability of the proposed control chart used in this study.

For the Xbar chart, the upper and control limits are found with the values of 1.998 and -1.720, respectively. Whereas, the upper and lower control limits for the R chart are correspondingly 6.793 and 0. The Xbar – R control charts are shown in Figure 2.

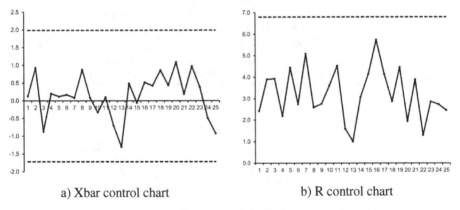

a) Xbar control chart   b) R control chart

Fig. 2. Xbar – R control charts

Based on the traditional Xbar – R control charts, there is no out-of-control signal detected, which leads to a conclusion that the quality of the TFT-LCDs is in control. Therefore, no intervention to the manufacturing process is needed.

However, suppose that based on the desired in-control average run length ARL0=370, we want to detect the shift of one standard deviation in the process mean and half standard deviation in the process variability. The parameters (q,α,L) for the constructing of MaxGWMA chart are (0.95, 1.10, 2.759) [5]. The results are shown in Table 2 with the corresponding MaxGWMA control chart shown in Figure 3.

Table 2. The control limit and relevant statistics for MaxGWMA control chart

Sample#	UCL_i	V_i	U_i	MG_i	Status	Symbol
1	0.140	-0.001	0.011	0.011	In control	
2	0.206	0.072	-0.027	0.072	In control	
3	0.254	-0.015	-0.089	0.089	In control	
4	0.293	-0.017	-0.068	0.068	In control	
5	0.325	-0.019	-0.130	0.130	In control	
6	0.351	-0.016	-0.139	0.139	In control	
7	0.373	-0.020	-0.241	0.241	In control	
8	0.392	0.048	-0.247	0.247	In control	
9	0.409	0.050	-0.251	0.251	In control	
10	0.422	0.007	-0.294	0.294	In control	
11	0.434	-0.001	-0.371	0.371	In control	
12	0.444	-0.079	-0.317	0.317	In control	
13	0.453	-0.218	-0.194	0.218	In control	
14	0.461	-0.196	0.196	0.196	In control	
15	0.467	-0.207	0.209	0.209	In control	
16	0.473	-0.167	0.323	0.323	In control	
17	0.478	-0.130	0.382	0.382	In control	
18	0.482	-0.053	0.393	0.393	In control	
19	0.486	-0.012	0.464	0.464	In control	
20	0.489	0.084	0.425	0.425	In control	
21	0.491	0.100	0.438	0.438	In control	
22	0.494	0.180	0.501	0.501	Out of control	m+
23	0.496	0.209	0.520	0.520	Out of control	m+
24	0.497	0.150	0.512	0.512	Out of control	m+
25	0.499	0.042	0.540	0.540	Out of control	m+

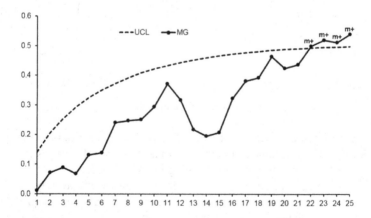

Fig. 3. MaxGWMA control chart for monitoring the TP of TFT-LCD

According to Table 2 and Figure 3, the manufacturing process is found to have an increase in the mean of the last four samples, indicating that compared to the designed alignment, the actual errors between the CF and TFT panels tend to become larger than one standard deviation. The trend was found in the 21st sample and therefore, a careful investigation on the manufacturing of the last five samples should be conducted to find out the assignable causes. If the causes are found, it is then of great importance to have proper corrective actions undertaken to either eliminate them from the process or reduce the variability induced by them so as to improve the performance quality of the TFT-LCDs.

With the above findings, in monitoring the Total Pitch of TFT-LCDs, MaxGWMA control chart outperforms the traditional Xbar – R control charts which are currently widely used in practice. Hence, it is strongly suggested that the former should be employed as a good substitute of the latter.

4   Conclusion

Today, building and implementing an effective quality control program has been a vital task to the survival of various industrial producers due to the need of mass production. Once the quality of the industrial products is not fully and effectively controlled, they usually face up with a high increase of production cost, which will also reduce their benefit and competitive power on the market. In the industry of manufacturing TFT-LCDs, to better monitor the quality of the displays, it is suggested that the traditional Xbar – R control charts should be replaced with the MaxGWMA chart because it is more sensitive to the shift in the process mean and variability.

References

1. Montgomery, D.C.: Introduction to Statistical Quality Control, 6th edn. Wiley, Asia (2009)
2. Walpole, R.E., Myers, R.H., Myers, S.L., Ye, K.: Probability and Statistics for Engineers and scientists, 9th edn. Pearson (2012)
3. Evans, J.R., Lindsay, W.M.: The Management and Control of Quality, 8th edn. South-Western College Publishing, Ohio (2011)
4. Vargas, V.D.C.C.D., Lopes, L.F.D., Souza, A.M.: Comparative study of the performance of the CuSum and EWMA control charts. Comput. Ind. Eng. 46, 707–724 (2004)
5. Sheu, S.H., Huang, C.J., Hsu, T.S.: Extended maximum generally weighted moving average control chart for monitoring process mean and variability. Comput. Ind. Eng. 62, 216–225 (2012)
6. Roberts, S.W.: Control chart test based on geometric moving averages. Technometrics 1, 239–250 (1959)
7. Xie, H.: Contributions to qualimetry, Ph.D. thesis, University of Manitoba, Winnipeg, Canada (1999)
8. Han, D., Tsung, F.: A generalized EWMA control chart and its comparison with the optimal EWMA, CUSUM and GRL Schemes. Anal. Stat. 32, 316–339 (2004)

9. Eyvazian, M., Naini, S.G.J., Vaghefi, A.: Monitoring process variability using exponentially weighted moving sample variance control charts. Int. J. Manuf. Technol. 39, 261–270 (2008)
10. Sheu, S.H., Tai, S.H., Hsieh, Y.T., Lin, T.C.: Monitoring process mean and variability with generally weighted moving average control charts. Comput. Ind. Eng. 57, 401–407 (2009)
11. Li, Z., Wang, Z.: An exponentially weighted moving average scheme with variable sampling intervals for monitoring linear profiles. Comput. Ind. Eng. 59, 630–637 (2010)
12. Zhang, J., Li, Z., Wang, Z.: A multivariate control chart for simultaneously monitoring process mean and variability. Comput. Stat. Data Anal. 54, 2244–2252 (2010)
13. Sweet, A.L.: Control charts using coupled exponentially weighted moving averages. IIE Trans. 18, 26–33 (1986)
14. Chan, L.K., Cheng, S.W., Spiring, F.A.: Alternative variable control chart, Technical report, University of Manitoba (1990)
15. Domangue, R., Patch, S.C.: Some omnibus exponentially weighted moving average statistical process monitoring schemes. Technometrics 33, 299–313 (1991)
16. Gan, F.F.: Joint monitoring of process mean and variance. Nonlinear Anal. 30, 4017–4024 (1997)
17. Gan, F.F.: Joint monitoring of process mean and variance using exponentially weighted moving average chart. Technometrics 37, 446–453 (1995)
18. Chao, M.T., Cheng, S.W.: Semicircle control chart for variables data. Qual. Eng. 8, 441–446 (1996)
19. Chen, G., Cheng, S.W.: Max-chart: Combining X-bar and s-chart. Statistica Sinica 8, 263–271 (1998)
20. Amin, R.W., Wolff, H., Besenfelder, W., Baxley Jr., R.: EWMA control charts for the smallest and largest observations. J. Qual. Technol. 31, 189–206 (1999)
21. Sheu, S.H., Lin, T.C.: The generally weighted moving average control chart for detecting small shifts in the process mean. Qual. Eng. 16, 209–231 (2003)
22. Sheu, S.H., Yang, L.: The generally weighted moving average control chart for monitoring the process mean. Qual. Eng. 18, 333–344 (2006)
23. Sheu, S.H., Tai, S.H.: Generally weighted moving average control chart for monitoring process variability. Int. J. Adv. Manuf. Technol. 30, 452–458 (2006)
24. Scheaffer, R.L., Mulekar, M.S., McClave, J.T.: Probability & Statistics for Engineers, 5th edn. Brooks/Cole, Cengage Learning (2011)
25. Tseng, Y.H., Tsai, D.M.: Using independent component analysis based process monitoring in TFT-LCD manufacturing. J. Chin. Inst. Ind. Eng. 23, 262–267 (2006)
26. Li, D.C., Chen, W.C., Liu, C.W., Lin, Y.S.: A non-linear quality improvement model using SVR for manufacturing TFT-LCDs. J. Intell. Manuf. 23, 835–844 (2012)

An Interactive System Building 3D Environment Using a Moving Depth Sensor

Duy Pham, Nghia Huu Doan, Thang Ba Dinh, and Tien Ba Dinh

Advanced Program in Computer Science, Faculty of Information Technology
VNU - University of Science, Ho Chi Minh City, Vietnam
{pduy,dhnghia}@apcs.vn, {dbthang,dbtien}@fit.hcmus.edu.vn

Abstract. We present an interactive system helping reconstruct a 3D indoor environment and manipulate it using a moving 3D sensor. Our system takes the depth stream from a depth sensor and converts it into point clouds. After that, the pose of the 3D sensor is tracked in real-time by matching the current point cloud to those from all previous frames; in the next step, it is mapped to a unique world point cloud in order to build the 3D model of that environment. Tracking the 3D sensor in real-time helps automatically fill the holes from the model, where the previous frames has not covered, making the model complete. Once the 3D model of the environment is ready, our system allows us to either add more 3D objects on it or remove an existing one. Here, we propose two different 3D object segmentation methods, which is the core module of our object removal function, for evaluation: a *K-means* based algorithm for simple models, and a graph-based algorithm for complex models. The system is tested and evaluated on various indoor environments.

Keywords: virtual reality, 3D reconstruction, KinectFusion, Large Scale, Segmentation, Graph-cut, sensor, K-means.

1   Introduction

Constructing 3D models from images is an interesting research topic in computer vision. It plays an important role in both scientific research and real-world applications. Having a 3D model helps solving many critical problems, which are extremely hard to tackle in 2D, such as self-occlusion, and confusion in object segmentation. Recently, 3D models are widely used in many applications such as 3D face recognition, 3D printing, 3D movies, and 3D virtual reality systems.

In the past decades, there were several efforts to construct corresponding 3D models from a set of 2D images, such as [1]. However, this approach is slow when processing tons of images, and the accuracy highly depends on the feature matching algorithm between the images. It means when the algorithm cannot find the good feature matching, the whole model will be affected. Recently, the emergence of depth cameras creates new opportunities for researchers in 3D reconstruction. The depth streams from such depth sensors provide the distance from each pixel to the camera. There are several sensors which has been released

B. Murgante et al. (Eds.): ICCSA 2013, Part V, LNCS 7975, pp. 391–406, 2013.

on the market such as SoftKinectic's DepthSense [2], Microsoft Kinect [3], and Mesa Swiss Ranger [4]. The price is varied from a few hundred dollars to several thousand dollars. Among them, Microsoft Kinect developed by PrimeSense [5] gets much more exposure than its competitors because of several advantages: has cheap price (around 150 dollars), integrated to the popular Microsoft XBox360 gaming system [6], and included Software Development Kit (SDK) with skeleton tracking. Those nice features bring big excitement to the research community, where everyone can easily get a sensor, hook it up, and test their ideas.

In 2011, Richard A. Newcombe et al. [7] proposed KinectFusion, a system which takes the depth stream from a moving Kinect and reconstructs the 3D model of a small region in real time. After that, KinectFusion is implemented as a project of Point Cloud Library [8] and extended by [9]. In their work, they used a cyclic buffer to continuously enlarge the 3D model when the camera moves to new positions. This method is then extended in PCL[8] in a project named Kinfu Large Scale to reconstruct a large area. The break-through of KinectFusion is its practical GPU implementation making the 3D model reconstruction real-time, which makes lots of applications now not as hard as before, such as virtual reality, and real-time 3D face recognition.

Being inspired by KinectFusion and Kinfu Large Scale, we propose a 3D virtual reality system which takes the depth streams from the Kinect and reconstructs a 3D model of an indoor environment such as an office room. After that, the system allows the users to interact with the model by adding or removing 3D objects.

2   Related Work

Microsoft Research published KinectFusion [10,7] in November, 2011. It takes the depth frames from the Kinect, and converts them into point clouds by putting the pixels' 2D coordinates and their depth data into a 3D coordinate system. Different point clouds from different frames are matched with each other using *Iterative Closest Point* algorithm (ICP) [1]. The general idea of the algorithm is to register the point clouds together by finding the best match. The result of this process is a transformation describing how one point cloud of the same model is transformed to match the other one from a different view. The whole system is implemented using CUDA [11]. CUDA architecture allows the computer to share calculations to the graphics cards, which have many processors, for parallel computing. This technology reduces the workload on CPU, dramatically; hence, boost up the speed of the whole system by a large factor.

In this system, we have to solve two problems: adding a pre-built 3D object to our 3D world model, and removing an existing 3D object out of the built model. While the first task requires an understanding about the planes/surfaces in the 3D environment to make the added objects look real, the other task requires a full 3D segmentation which is a classical problem in computer vision. There is a variety of methods to segment the objects in a 2D image such as applying clustering algorithms [12,13,14], region growing [15] based methods, and RANSAC

[16] based methods. *Graph-cut* [17,18,19] is also a very useful algorithm applied widely in image segmentation. In 2D image segmentation, the color distances among pixels is often used to build the energy functions [18] [20]. Meanwhile, in 3D point clouds, Aleksey Golovinskiy and Thomas Funkhouser [19] use Euclidean distances to build the graph, and applied max-flow algorithm by Boykov and Kolmogorov [21] to minimize these functions and get the minimum cut. Besides, there were also some efforts to use clustering algorithms in 3D point cloud segmentation [22].

3   Our Interactive System

We apply KinectFusion to set up a system which allows users to construct a 3D room using a moving Kinect. The 3D room is saved as a point cloud. Some operations to decompose analysis and manipulate the model, such as a polygon mesh construction method, a filtering technique, and segmentation, are also available. The users can add some other objects or decorations into their room, as well as remove the existing objects in the original model.

3.1   3D Reconstruction Using a Depth Sensor

Our system uses a Kinect sensor to construct the 3D model of a room. We applied the real-time KinectFusion [10] to build it. This system requires a computer with a high-end graphics card supporting Nvidia CUDA to fully execute the parallel computations. The reconstruction works as follow:

- Step 1: Move the Kinect sensor around the room, the system takes the depth stream from the Kinect
- Step 2: For each depth frame taken, the system converts it into a point cloud by putting every pixel into a 3D coordinate.
- Step 3: Each new point cloud is aligned to the previous data by running ICP. ICP takes two point clouds as input arguments and return a transformation matrix. For more details about how ICP work, please refer to [1] and [23].
- Step 4: Repeat the process from step 1 until users stop the camera

The detailed description of how KinectFusion works can be found in the papers [7] and [10].

3.2   Point Cloud Pre-processing

The cloud of points received from the KinectFusion will be sampled and analyzed. The Kinect sensor provides VGA resolution 3D *images,* each of which has 640×480 pixels. Although not all these pixels are converted into 3D points, the complete point cloud still contains hundred thousands of points. For our purpose, we do not need all the points, while they greatly slow down the later processes. One solution is to apply a uniform sampling method.

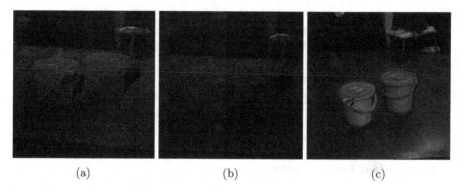

(a)   (b)   (c)

Fig. 1. Point Cloud from KinectFusion and *VoxelGrid filter*. (a) Raw Point Cloud before filtering. (b) Point Cloud after filtering, where visualization quality is strongly decreased. (c) Generated Polygon Mesh from filtered Point Cloud

Fig. 2. Plane segmentation using RANSAC. Each color represents a unique plane.

We perform a *VoxelGrid filter* on the point cloud. The voxel size can be set by users or by experiment. In each voxel, we keep the center point only. It is obvious that the number of points decreases when the voxel size increases. In our experiment, we remove approximately 75% of the points. For instance, the size of a model of 300000 points is reduced to 75000 after it is down-sampled. Based on our observation, the trade-off between this sampling process and the quality of the model is acceptable, while the improvement in processing time is noticeable. Figure 1 shows a sample point cloud taken from KinectFusion and the sampled one.

After the sampling has completed, we apply RANSAC [16] to detect all the planes in the model. In an indoor office environment, plane is the most basic component to form the structure. The algorithm divides the point cloud into 2 sets: *Inliers* and *Outliers*. While *Inliers* contains the points in the plane, *Outliers* contains the rest. Each loop of this algorithm gives us a plane and its equation in the model. We get all the planes by executing the algorithm to the sub-clouds recursively. Figure 2 shows the plane segmentation result on a model constructed by KinectFusion.

Finally, Greedy Projection Triangulation [24] is applied to construct a Polygon Mesh. It is used to improve the visualization, which is affected by the point density reduction caused by *VoxelGrid filter*, as shown in Figure 1b.

3.3   Object Insertion

The directions of objects in the model are expressed by their normal vectors. We define the original normal of all objects as $(0,0,1)$, and the original position of all object as $(0,0,0)$. That means we manually transform all objects to the origin so that their ground planes contain the origin, and their directions are all the same and along the z-axis before applying the transformations.

Surface Picking. The plane segmentation process gives us a set of points in each plane and a set of all plane equations. The coefficients in the equation are exactly the x, y, and z elements of the plane normal. The desired surface is specified by picking a point (x_1, y_1, z_1) in the point cloud. Then the system automatically gets the corresponding plane and its normal.

Transformation. Every object is put into the model by rotation and translation steps.

Rotation: What we need to do is to rotate the object from its direction to the destination planes direction. In other words, we need a rotation calculation to transforms the normal of the object to that of the plane.

The rotation angle is

$$\theta = \arccos \frac{\mathbf{n}_{object} \cdot \mathbf{n}_{plane}}{\|\mathbf{n}_{object}\| \cdot \|\mathbf{n}_{plane}\|}$$

where:

- $n_{object} = (0,0,1)$ is the normal vector of the object.
- $n_{plane} = (a, b, c)$ is the normal vector of the destination plane.

The rotation axis is the cross product between the two normals. After normalizing the axis, we get:

$$\mathbf{u} = \frac{\mathbf{n}_{object} \times \mathbf{n}_{plane}}{\|\mathbf{n}_{object} \times \mathbf{n}_{plane}\|}$$

We build the rotation matrix based on the rotation angle and rotation axis, which is:

$$R = I \cos\theta + \sin\theta \lfloor \mathbf{u} \rfloor_\times + (1 - \cos\theta)\mathbf{u} \otimes \mathbf{u} \tag{1}$$

where:

- I is the identity matrix
- $\lfloor \mathbf{u} \rfloor_\times$ is the cross product matrix
- $\otimes$ is the tensor product.

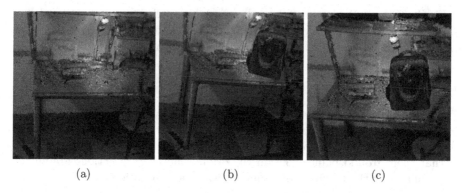

(a)   (b)   (c)

Fig. 3. Object Insertion example. (a) Original model taken from KinectFusion. (b) A bag is put on the table using affine transformation. (c) Rotating the object to find the best standing.

When the rotation matrix is constructed using equation (1), the point cloud is rotated to the angle of the destination surface by multiplying the coordinates of each point coordinates with the rotation matrix.

$$\mathbf{p_{new}} = R.\mathbf{p_{old}}$$

Where $\mathbf{p_{new}}$ and $\mathbf{p_{old}}$ are the new coordinates and old coordinates of each point, respectively.

Translation: We need to translate the object from its position to the position picked by the user. We are using a normal Cartesian coordinate, so the translation is interpreted by adding a translation vector $\mathbf{d}$ to every point. Let $\mathbf{p_{object}}$ be the object position and $\mathbf{p_{user}}$ be the position picked by the user, then we have:

$$\mathbf{p_{new}} = \mathbf{p_{old}} + \mathbf{d}$$

where $\mathbf{d} = \mathbf{p_{user}} - \mathbf{p_{object}}$

Self-Rotation After putting an object into a surface, we can fix its position by rotating it around its normal until we find the most suitable standing.

3.4   Point Cloud Segmentation

While adding more objects into the 3D room is straightforward by applying several transformation steps, modifying the existing structure of the room is more challenging. It now requires a robust 3D object segmentation algorithm rather than a plane one described in the previous section. To explore the feasibility of the system, we apply two different types of methods: a clustering method (using K-means) and a graph-based method (using minimum-cut).

Segmentation Using K-Means. Segmentation based on clustering algorithms is simple. Even though it does not produce excellent results on every model,

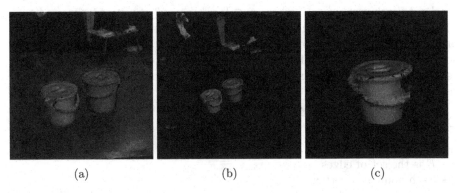

(a)   (b)   (c)

Fig. 4. Plane removal and K-means example. (a) Original model. (b) Model after removing the largest ground plane. (c) K-means result with $(n, k) = (1, 21)$

it works well on simple ones. An advantage of this approach is that it does not require much manual specification from the users, comparing to the other method in our system.

Pre-processing: To implement this module, we make two assumptions. First, in an indoor environment, every object in the room is put on a ground plane. Hence, in a simple model (*i.e.*, model with few and distinct objects), if we remove a ground plane, then any object connected to it will be disjoint to the rest of the model. Therefore, a clustering algorithm can take that object out. Second, the ground planes are always bigger than those of the objects put on it. Given that, a ground plane can be eliminated by removing n biggest planes, where n is a user-defined value.

Clustering using K-means: At the first stage, users are required to choose an object by picking a point, then the system execute the pre-processing. After n biggest planes have been removed, the point cloud is divided into k sub-clouds, in which the one containing the specified point is returned, by any clustering algorithm. In this paper, we apply *K-means*, one of the most popular clustering techniques. For how *K-means* work in details, see [13]. The pair (n, k) is initialized to be $(1, 2)$, the minimum value, and is updated manually after each run of *K-means*, until we get the desired object.

In the example shown in Figure 4, the value assigned to k is quite large because there are two objects standing near each other. If there is only one object on the floor, we can get the object with *(n,k)=(1,2)*. However, the value of *(n,k)* in this example shows that *K-means* is hardly effective on the more complex models. Therefore, we propose another method: Segmentation via *Graph-cut*. In this paper, we interpret a *Minimum-cut* approach.

Interactive Segmentation Using Minimum-Cut. Our work is mainly inspired by [18] and [19].

In computer vision, segmentation using graph-cut is usually formulated as a binary labeling problem. That is, given a graph $G(V, E)$, the algorithm assigns a label x_i to each vertex $i \in V$, where $x_i \in \{background(x_i = 0), foreground(x_i = 1)\}$ [18]. In some graph-cut projects on 2D images, a set of background seeds and a set of foreground seeds are defined manually before running graph-cut segmentation [18] [20] . Following the same fashion, our system allows users to scribble on the foreground and background to get the foreground and background seeds.

Graph Construction: After getting the seeding points, we construct a graph $G=(V,E)$ from our point cloud as follow:

- V is the set of vertices
- E is the set of edges
- Each point is a vertex
- There are two additional vertices: *source* and *sink*, or s and t. The *source* vertex belongs to the foreground, while the *sink* vertex belongs to the background. Each node is connected to *source* and *sink* by two extra edges. The weights of these edges are called the unary potentials, which are calculated by:

$$weight_s = \frac{d_b}{d_b + d_f}$$

and

$$weight_t = \frac{d_f}{d_b + d_f}$$

If only foreground seeds are available, then the unary potentials are a little bit difference, that is:

$$weight_t = d_f/r$$

In this case, the source weight $weight_s$ is a user-defined value, and is set by experiments. It is called the penalty, which determine whether a point belongs to the background or the foreground.

Here is the explanation of the terms used in the above equations:
- d_f is the minimum distance from the point to the foreground
- d_b is the minimum distance from the point to the background
- r is the object radius, and is also a user-defined value
- Each point is connected to k nearest neighbors by k edges respectively. The weights assigned to these edges are called the binary potential, and are calculated by:

$$weight_{ij} = \exp -(\frac{d_{ij}}{\sigma})^2$$

where:
- d_{ij} is the distance between two points i and j
- σ is the scaling factor. It is used to normalize the distance.

Energy function: In graph theory, a cut is a set of removed edges, which divides the graph into disjoint sets. In this problem, we need to have a method to choose a cut which absolutely divides the graph into two sub-graphs. One contains the *source*, the other contains the *sink*. Such cut is called an *s-t* cut.

Taking a detailed look at the unary and the binary potentials, we can see that:

- A unary potential of a node is the likelihood that the node belongs to the foreground or the background. If the source weight is larger, then the node is put in the foreground and its sink edge is cut, and vice versa. Hence, the role of unary potentials is to divide the point clouds into a background set and a foreground set.
- A binary potential is the similarity of 2 nodes. 2 similar nodes are likely to be in the same segment after the graph is cut. It means two adjacent vertices which distance is small are less likely to be separated by the object boundary than the two having a larger distance. Hence, the role of binary potentials is to sharpen the edge of the segmented object (the foreground set).

Therefore, we need to determine a set of edges which sum of weights is minimal. In other words, we minimize the following function:

$$E(X) = \sum_{i \in V} E_1(x_i) + \sum_{(i,j) \in E} E_2(x_i, x_j) \tag{2}$$

where:

- $X = x_1, x_2, , x_n$ is the solution set of the labeling problem.
- $E_1(x_i)$ is the unary potential of vertex i.
- $E_2(x_i, x_j) = \|x_i - x_j\| weight_{ij}$ is the binary potential of (i, j). The term $\|x_i - x_j\|$ indicates that E_2 represents a weight at the object boundary.

In graph-cut theory, E_1 and E_2 are called the energy terms, and function (2) is called an energy function. The minimum cut problems are solved by minimizing such functions. Here, we apply *max-flow algorithm*[21] proposed by Boykov and Kolmogorov.

Min-cut segmentation implementation in the virtual 3D room system: The segmentation works as follows:

- **Step 1**: Users are allowed to choose foreground seeds and background seeds (optional) simply by scribbling on the model.
- **Step 2**: We build the graph, as mentioned above, using Boost Library [25]. We choose *k=8* as the number of connections from each node to 8 nearest neighbors.

   The common issue when applying graph-cut in image/point cloud segmentation is choosing the materials to calculate the distance used in the energy function (2). In 2009, Aleksey Golovinskiy and Thomas Funkhouser [19] applied graph-cut to 3D point cloud using Euclidean distance to calculate the energy terms. Meanwhile, many graph-cut segmentation methods for 2D images, such as [18] and [20], use color distance. In our work, we make use of both color distance and Euclidean distance, which is different from previous

Fig. 5. *Graph-cut* example, with both background and foreground seeds. **Left:** Original model with scribbles from users. **Right:** The stair is segmented and filled with a different color. The object is obtained after 2 levels of revising.

graph-cut implementation. It is based on our observation that the two adjacency objects, as well as an object and its background floor, are often distinguished by color more than by Euclidean distance, which can be seen in our figures in this paper. And as we have mentioned above, the binary potentials are used to sharpen the edge of the object. Hence, we use color distances to calculate the binary potentials $E_2(x_i, x_j)$, while the unary potentials $E_1(x_i)$ are still constructed via Euclidean distance.

Before calculating the energy terms, we need to perform an additional processing, which is distance normalization. The distances are normalized by their scaling factors σ to avoid the difference between the scales of different point clouds and the imbalance between color distances and Euclidean distances, which can affect the segmentation results. In particular, we use $\sigma_{euclidean} = 0.01$ and $\sigma_{color} = 7$ because they are the common spacing in most of our data.

- **Step 3**: Finally, when the graph is ready, min-cut segmentation is executed to extract the object (i.e. the foreground) out of the model. We apply Boykov and Kolmogorov max-flow algorithm [21], which is implemented in Boost Library [25], to minimize the energy function and retrieve the object.

Interactive Process: If the segmented object is not satisfying, we can improve it by adding more points to both foreground seeds and background seeds and the update process will take in place by executing min-cut segmentation again. Based on our experiments, the satisfied object is often achieved after 2 to 3 rounds of min-cut segmentation.

Object Removal: After the segmentation is done, the object can be removed from the model easily. For example, in Figure 5, the stair is segmented after 2 seeding revisions. The detailed performance of our segmentation methods is shown in section 4.

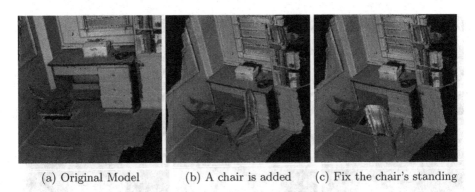

(a) Original Model   (b) A chair is added   (c) Fix the chair's standing

Fig. 6. Object Insertion Test Scene

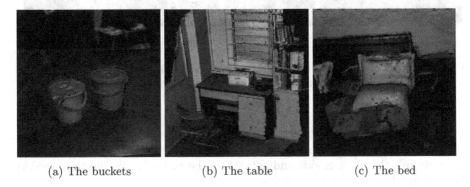

(a) The buckets   (b) The table   (c) The bed

Fig. 7. Segmentation Test Scenes

4   Experiment

We test our system on a computer system with an Intel Core i5 2.5 GHz processor, 4GB of RAM, and Geforce G630M graphics card with 96 cores. The whole system is implemented in C++ using Visual Studio 2010. All the experiments are executed in *Release mode*.

4.1   KinectFusion Performance

In our system, the 3D reconstruction usually operates at a frame rate of only 4fps, approximately. It is understandable because our testing GPU has less than 100 processors. The number of GPU's cores strongly affects the quality of parallel computing using CUDA. Because of the low frame rate, it is time-consuming when we reconstruct a virtual 3D environment. For example, it took approximately from 20 minutes to more than half an hour to build a model we use in this paper. However, all these models are still not complete rooms. Hence, to save time, we build only parts of the rooms to get the test scenarios for our interactive mechanisms.

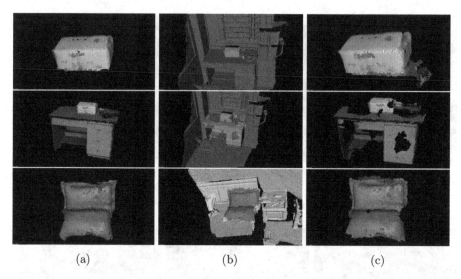

(a)   (b)   (c)

Fig. 8. Min-cut result: Segment the objects **without background seeds**.(a) Ground truth. (b) Segmentation result in the whole model. (c) The extracted object

After building the 3D model, some pre-processing is interpreted as described in section 3.2. In our experiments, it takes approximately from 0.5 to 3 seconds to down-sample the point clouds and extract all the planes, while the time needed for triangulation is from 3 to 8 seconds.

4.2   Object Insertion Performance

The core of object insertion is the affine transformations. In the example shown in section 3.3, the bag point cloud contains 6532 points, and the transformation time is 0.07 seconds. In Figure 6, another object is added into an virtual environment taken from KinectFusion. In this case, the extra chair contains 10070 points, and the time needed to put it beside the table is 0.08 seconds. The operation can be faster for smaller objects, such as a box, a flowerpot or the stationery.

4.3   Object Segmentation Performance

In this section, we make a more detailed analysis about the performance and results of our segmentation methods. We define 3 test scenarios, which are *The buckets*, *The table*, and *The bed*, as shown in Figure 7. In section 3.4, we have demonstrated the operation of *K-means* using the first model. As our initial expectation, the *K-means* method is used on simple model only. On complex models, we have to make a lot of trials to find a suitable pair (n, k) which can take the desired object out. Hence, in these cases, *Graph-cut* is always the better

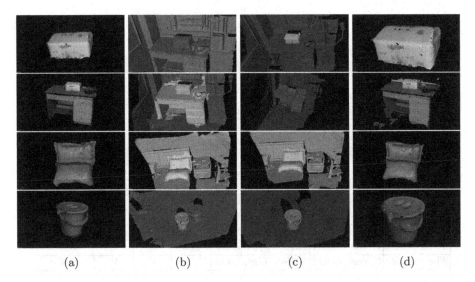

(a)   (b)   (c)   (d)

Fig. 9. Min-cut result: Segment the objects with **both foreground and background seeds**. (a) Ground Truth. (b) First min-cut run. (c) Second min-cut run. (d) The extracted object

choice, undoubtedly. This is the reason why the *K-means* is evaluated and compared to the *Graph-cut* on only the simplest model among our data. The other scenarios are more complex and are not suitable for *K-means*. The segmentation results are illustrated in Figure 8 and 9, while the detailed performance of the *K-means* example in section 3.4, as well as the other test scenes, is displayed in Table 1.

We can see from the Figures and Table 1 in section 4.3 that, in general, *Graph-cut* segmentation with background seeds gives a better result than the one with foreground seeds only; and an update on the seeds often help improve the results. However, for the model which is simple enough, such as the box in our test scenes, *Graph-cut* can be executed without background seeds, and the pre-defined penalty can remove the background from the foreground very well. In practice, if the seeds are not good (i.e. the users make mistakes when choosing the seeds), then the *Graph-cut* performance with background seeds can be even worse than the one without them. Strong dependence on the seed quality is an limitation of our *Graph-cut* segmentation. This is the reason why sometimes we update the seeds, re-execute *Graph-cut* and get a worse result than the previous run, as shown in Table 1.

For the *K-means*, Table 1 shows that it is seven-times faster than 2 rounds of *Graph-cut* on *The buckets* test scene, and the accuracy is acceptable (94.1 %). In this case, we set $k = 21$ as mentioned in section 3.4. Although such a large value of k may cause some inconvenience to find, the experiment data shows that this clustering-based method is not a bad choice on a simple model.

Table 1. Segmentation performance and results on the test scenes

Model	Object	Size	Method	Loops	F.size	B.size	Runtime
The table	Box	141264	Graph-cut	1	15	0	1.95
The table	Box	141264	Graph-cut	1	15	17	1.96
The table	Box	141264	Graph-cut	2	27	36	3.98
The table	Table	141264	Graph-cut	1	41	0	1.95
The table	Table	141264	Graph-cut	1	45	32	1.96
The table	Table	141264	Graph-cut	2	75	67	3.97
The bed	Pillow	53615	Graph-cut	1	30	0	0.69
The bed	Pillow	53615	Graph-cut	1	37	30	0.74
The bed	Pillow	53615	Graph-cut	2	77	69	1.49
The buckets	Green bucket	101909	K-means	100	0	0	0.38
The buckets	Green bucket	101909	Graph-cut	2	49	46	2.67

- **Size** is the size of the model
- **Loops** is the number of loops inside the *K-means* algorithm, or the number of *Graph-cut* executions.
- **F.size**(Foreground size) and **B.size**(Background size) are the total sizes of the seeds.
- **Runtime** is calculated in seconds

Object	Methods	Background	G.T	Segment	Correct	Accuracy	Error
Box	Graph-cut	No	1999	2416	1967	98.4	18.6
Box	Graph-cut	1	1999	2478	1992	99.6	19.6
Box	Graph-cut	2	1999	2042	1962	98.1	4.0
Table	Graph-cut	No	29438	19124	17904	60.8	6.4
Table	Graph-cut	1	29438	35217	28350	96.3	19.5
Table	Graph-cut	2	29438	33483	28997	98.5	13.4
Pillow	Graph-cut	No	7448	6997	6646	89.2	5.0
Pillow	Graph-cut	1	7448	7634	6762	90.8	11.4
Pillow	Graph-cut	2	7448	7005	6718	90.2	4.1
Green bucket	Graph-cut	2	2106	2195	2085	99.0	5.0
Green bucket	K-means	N/A	2106	1990	1982	94.1	0.4

- **Background** values: "No": no background seeds, 1 and 2: number of graph-cut executions with background seeds
- **G.T** is the size of the ground truth
- **Segment** is the size of the segment
- **Correct** is the number of correct points in the segment
- **Accuracy** is the percentage of the correct points comparing to the ground truth
- **Error** is the percentage of the wrong detected points comparing to the extracted segment

5 Conclusion

By integrating the new technologies and classical algorithms in computer vision with some improvements, we present a system to construct a 3D indoor office and provide tools to manipulate this virtual model. The system uses a depth sensor with KinectFusion support to do the 3D reconstruction in real-time, and applies some affine transformations, clustering techniques, and graph-based segmentation to build the interactive mechanisms. Extra objects can be added to the model in any direction, or existing ones can be removed out. The system is able to be applied in interior design to help people test their ideas before implementing them in real-life architectures.

There are still some limitations needed to be improved in the further research. First, the interactive mechanisms can be upgraded, such as adding more physical constraints to the objects arrangement, and building an object library so that the users can choose which one to add into their virtual room. Second, the object removal function creates the holes and thus damages the surfaces. Restoring them is another challenge. Finally, the segmentation methods can still be extended and completed. For instance, the structured light from the depth sensor can be used to segment the objects while reconstructing the 3D models; or the existing graph-based method can be completed by some post-processing techniques to remove noise.

Acknowledgment. This research is supported by research funding from Advanced Program in Computer Science, University of Science, Vietnam National University - Ho Chi Minh City.

References

1. Chen, Y., Medioni, G.: Object modelling by registration of multiple range images. Image Vision Comput. 10(3), 145–155 (1992)
2. SoftKinetic: Softkinetic solutions,
   http://www.softkinetic.com/fr-be/solutions.aspx/
3. Eisler, C.: Use the power of kinect for windows to change the world,
   http://blogs.msdn.com/b/kinectforwindows/archive/2012/01/09/
   kinect-for-windows-commercial-program-announced.aspx
4. Oggier, T., Lustenberger, F., Blanc, N.: Miniature 3D TOF camera for real-time imaging. In: André, E., Dybkjær, L., Minker, W., Neumann, H., Weber, M. (eds.) PIT 2006. LNCS (LNAI), vol. 4021, pp. 212–216. Springer, Heidelberg (2006)
5. Microsoft: Primesense supplies 3-d-sensing technology to project natal for xbox 360, http://www.microsoft.com/en-us/news/press/2010/mar10/
   03-31PrimeSensePR.aspx/
6. Microsoft: Kinect for xbox360,
   http://www.xbox.com/en-US/xbox360/accessories/kinect/KinectForXbox360/
7. Newcombe, R.A., Davison, A.J., Izadi, S., Kohli, P., Hilliges, O., Shotton, J., Molyneaux, D., Hodges, S., Kim, D., Fitzgibbon, A.: KinectFusion: Real-time dense surface mapping and tracking. In: 2011 10th IEEE International Symposium on Mixed and Augmented Reality, pp. 127–136. IEEE (October 2011)

8. Rusu, R.B., Cousins, S.: 3D is here: Point Cloud Library (PCL). In: 2011 IEEE International Conference on Robotics and Automation (ICRA), pp. 1–4. IEEE (May 2011)

9. Whelan, T., Kaess, M., Fallon, M., Johannsson, H., Leonard, J., McDonald, J.: Kintinuous: Spatially extended KinectFusion. In: RSS Workshop on RGB-D: Advanced Reasoning with Depth Cameras, Sydney, Australia (July 2012)

10. Izadi, S., Kim, D., Hilliges, O., Molyneaux, D., Newcombe, R., Kohli, P., Shotton, J., Hodges, S., Freeman, D., Davison, A., Fitzgibbon, A.: Kinectfusion: real-time 3d reconstruction and interaction using a moving depth camera. In: Proceedings of the 24th Annual ACM Symposium on User Interface Software and Technology, UIST 2011, pp. 559–568. ACM, New York (2011)

11. Nickolls, J., Buck, I., Garland, M., Skadron, K.: Scalable parallel programming with cuda. Queue 6(2), 40–53 (2008)

12. Ester, M., Kriegel, H.P., Sander, J., Xu, X.: A Density-Based algorithm for discovering clusters in large spatial databases with noise. In: Simoudis, E., Han, J., Fayyad, U. (eds.) Second International Conference on Knowledge Discovery and Data Mining, pp. 226–231. AAAI Press, Portland (1996)

13. Lloyd, S.: Least squares quantization in pcm. IEEE Trans. Inf. Theor. 28(2), 129–137 (2006)

14. MacQueen, J.: Some methods for classification and analysis of multivariate observations. In: Proc. Fifth Berkeley Symp. on Math. Statist. and Prob., vol. 1, pp. 281–297. Univ. of Calif. Press (1967)

15. Adams, R., Bischof, L.: Seeded region growing. IEEE Trans. Pattern Anal. Mach. Intell. 16(6), 641–647 (1994)

16. Fischler, M.A., Bolles, R.C.: Random sample consensus: a paradigm for model fitting with applications to image analysis and automated cartography. Commun. ACM 24(6), 381–395 (1981)

17. Felzenszwalb, P.F., Huttenlocher, D.P.: Efficient graph-based image segmentation. Int. J. Comput. Vision 59(2), 167–181 (2004)

18. Li, Y., Sun, J., Tang, C.K., Shum, H.Y.: Lazy snapping. In: ACM SIGGRAPH 2004 Papers. SIGGRAPH 2004, pp. 303–308. ACM, New York (2004)

19. Golovinskiy, A., Funkhouser, T.: Min-cut based segmentation of point clouds. In: IEEE Workshop on Search in 3D and Video (S3DV) at ICCV (September 2009)

20. Lempitsky, V.S., Kohli, P., Rother, C., Sharp, T.: Image segmentation with a bounding box prior. In: IEEE 12th International Conference on Computer Vision, ICCV 2009, Kyoto, Japan, September 27-October 4, pp. 277–284. IEEE (2009)

21. Boykov, Y., Kolmogorov, V.: An experimental comparison of min-cut/max-flow algorithms for energy minimization in vision. IEEE Trans. Pattern Anal. Mach. Intell. 26(9), 1124–1137 (2004)

22. Klasing, K., Wollherr, D., Buss, M.: A clustering method for efficient segmentation of 3d laser data. In: 2008 IEEE International Conference on Robotics and Automation, ICRA 2008, May 19-23, pp. 4043–4048. IEEE (2008)

23. Szymon Rusinkiewicz, M.L.: Efficient variants of the ICP algorithm. In: Third International Conference on 3D Digital Imaging and Modeling (3DIM) (June 2001)

24. Marton, Z.C., Rusu, R.B., Beetz, M.: On Fast Surface Reconstruction Methods for Large and Noisy Datasets. In: Proceedings of the IEEE International Conference on Robotics and Automation (ICRA), Kobe, Japan, May 12-17 (2009)

25. Siek, J., Lee, L.Q., Lumsdaine, A.: Boost random number library (June 2000), http://www.boost.org/libs/graph/

Analysis of the Linear Complexity
in Pseudorandom Sequence Generators*

Amparo Fúster-Sabater

Information Security Institute, C.S.I.C.
Serrano 144, 28006 Madrid, Spain
amparo@iec.csic.es

Abstract. In this paper, binary sequences generated by nonlinearly filtering maximal length sequences are studied. Specifically, the parameter linear complexity of the filtered sequences has been considered and analyzed. In fact, a method of computing all the nonlinear filters that generate sequences with a cryptographically large linear complexity has been developed. The procedure is based on the concept of equivalence classes of nonlinear filters and on the addition of filters from different classes. Three distinct representations of nonlinear filters have been systematically addressed. The method completes the class of nonlinear filters with guaranteed linear complexity found in the cryptographic literature.

Keywords: pseudorandom sequence, linear complexity, encryption function, nonlinear filter, cryptography.

1 Introduction

Sequence generators based on Linear Feedback Shift Registers (LFSR) are very commom procedures to generate pseudorandom sequences for multiple applications: computer simulation, circuit testing, error-correcting codes or cryptography (stream ciphers).

The encryption procedure in stream ciphers tries to imitate the mythic *one-time pad cipher* [1] that remains as the only known perfectly secure or absolutely unbreakable cipher. This encryption procedure is designed to generate from a short key a long sequence (*keystream sequence*) of seemingly random bits. Some of the most recent designs in stream ciphers can be found in [2, 3]. Typically, a stream cipher consists of a keystream generator whose output sequence is XORed with the plaintext (in emission) in order to obtain the ciphertext or with the ciphertext (in reception) in order to recover the original plaintext. References [4–7] provide a solid introduction to the study of stream ciphers.

Most keystream generators are based on maximal-length LFSRs [8] whose output sequences, the so-called *m*-sequences, are combined in a nonlinear way

---

* Research partially supported by CDTI (Spain) under Project Cenit- HESPERIA as well as by Ministry of Science and Innovation and European FEDER Fund under Project TIN2011-25452/TSI.

B. Murgante et al. (Eds.): ICCSA 2013, Part V, LNCS 7975, pp. 407–420, 2013.

(by means of nonlinear filters, nonlinear combinators, irregularly decimated generators, typical elements from block ciphers, etc) to produce sequences of cryptographic application. Desirable properties for such sequences can be enumerated as follows:

1. Long Period
2. Good statistical properties
3. Large Linear Complexity (LC).

One general technique for building a keystream generator is to use a nonlinear filter, i.e. a nonlinear function applied to the stages of a single maximal-length LFSR. That is the output sequence is generated as the image of a nonlinear Boolean function F in the LFSR stages. Period and statistical properties of the filtered sequences are characteristics deeply studied in the literature, see [9], [10] and the references above mentioned. In addition, such sequences have to pass all 19 DIEHARD tests [11] to be accepted as cryptographic sequences.

Regarding the third requirement, linear complexity of a sequence is defined as the amount of known sequence necessary to reconstruct the entire sequence. In cryptographic terms, LC must be as large as possible in order to prevent the application of the Berlekamp-Massey algorithm [12]. A recommended value for LC is about half the sequence period. Although several contributions to the linear complexity of nonlinearly filtered sequences can be found in [6], [13] and [14], the problem of determining the exact value of the linear complexity attained by any nonlinear filter is still open [15]. For an efficient calculation of Vandermonde matrices, the interested reader is referred to [16–18]

In this paper, a method of computing all the nonlinear filters applied to a LFSR with $LC \geq \binom{L}{k}$ (where L is the LFSR length and k the order of the filter) has been developed. The procedure is based on the concept of equivalence classes of nonlinear filters and on the handling of such filters from different classes. No restriction is imposed on the parameters of the nonlinear filtering function. The method completes the families of nonlinear filters with guaranteed LC given in [6].

The paper is organized as follows. Basic concepts and specific notation is introduced in Section 2. Three different representations of nonlinear filters are given in Sections 3 as well as an equivalence relationship for nonlinear filters is defined in Section 4. The construction of all possible filters preserving the cosets of weight k is developed in Section 5. Discussion on numerical features and an example is given in Section 6. Finally, conclusions in Section 7 end the paper.

2   Basic Concepts and Notation

Specific notation and different basic concepts are introduced as follows:

A *m-sequence*. Let $\{s_n\}$ be the binary output sequence of a maximal-length LFSR of L stages, that is a LFSR whose characteristic polynomial $P(x) = \sum_{j=0}^{L} p_j x^j$ with $p_j \in \{0, 1\}$ is primitive of degree L, see [6], [8]. In that case, the output sequence is a m-sequence of period $2^L - 1$. Moreover, $\{s_n\}$ is completely determined by the LFSR initial state and the characteristic polynomial $P(x)$. The sequence $\{s_n\}$ satisfies the linear recursion:

$$\sum_{j=0}^{L} p_j \, s_{n+j} = 0,$$

that allows one to express any term of the sequence as a linear combination of the previous L terms.

The roots of $P(x)$ are α^{2^i} $(i = 0, 1, \ldots, L-1)$ where α is a primitive element in $GF(2^L)$ that is an extension of the binary field $GF(2)$ with 2^L elements [19]. Any generic element of the sequence, s_n, can be written in terms of the roots of $P(x)$ as:

$$s_n = Tr(C\,\alpha^n) = \sum_{j=0}^{L-1} (C\alpha^n)^{2^j}, \quad n \geq 0 \tag{1}$$

where $C \in GF(2^L)$. Furthermore, the $2^L - 1$ nonzero choices of C result in the $2^L - 1$ distinct shifts of the same m-sequence. If $C = 1$, then $\{s_n\}$ it is said to be in its *characteristic phase*.

Nonlinear filter. It is a Boolean function $F(x_0, x_1, \ldots, x_{L-1})$ in L variables of degree k. For a subset $A = \{a_0, a_1, \ldots, a_{r-1}\}$ of $\{0, 1, \ldots, L-1\}$ with $r \leq k$, the notation $x_A = x_{a_0} x_{a_1} \ldots x_{a_{r-1}}$ is used. The Boolean function can be written as [20]:

$$F(x_0, x_1, \ldots, x_{L-1}) = \sum_A c_A \, x_A, \tag{2}$$

where $c_A \in \{0, 1\}$ and the summation is taken over all subsets A of $\{0, 1, \ldots, L-1\}$.

Filtered sequence. The sequence $\{z_n\}$ is the keystream or output sequence of the nonlinear filter F applied to the L stages of the LFSR. The keystream bit z_n is computed by selecting bits from the m-sequence such that

$$z_n = F(s_n, s_{n+1}, \ldots, s_{n+L-1}).$$

Cyclotomic coset. Let Z_{2^L-1} denote the set of integers $[1, \ldots, 2^L - 1]$. An equivalence relation R is defined on its elements $q_1, q_2 \in Z_{2^L-1}$ such as follows: $q_1 R q_2$ if there exists an integer j, $0 \leq j \leq L - 1$, such that

$$2^j \cdot q_1 = q_2 \bmod 2^L - 1.$$

The resultant equivalence classes into which Z_{2^L-1} is partitioned are called the *cyclotomic cosets* mod $2^L - 1$, see [8]. All the elements q_i of a cyclotomic coset have the same number of 1's in their binary representation; this number is called the *coset weight*. The leader element, E, of every coset is the smallest integer in such an equivalence class. Moreover, the cardinal of any coset is L or a proper divisor of L.

Characteristic polynomial of a cyclotomic coset. It is a polynomial $P_E(x)$ defined by $P_E(x) = (x + \alpha^E)(x + \alpha^{2E}) \ldots (x + \alpha^{2^{(r-1)}E})$, where the degree r ($r \leq L$) of $P_E(x)$ equals the cardinal of the cyclotomic coset E.

Characteristic sequence of a cyclotomic coset. It is a binary sequence $\{S_n^E\}$ defined by the expression $\{S_n^E\} = \{\alpha^{En} + \alpha^{2En} + \ldots + \alpha^{2^{(r-1)}En}\}$ with $n \geq 0$. Recall that the previous sequence $\{S_n^E\}$ satisfies the linear recurrence relationship given by $P_E(x)$, see [8], [19]. Moreover, $\{S_n^E\}$ is a decimation of the m-sequence $\{s_n\}$ obtained from such a sequence by taking one out of E terms.

3   Different Representations of Nonlinear Filters

According to the previous section, nonlinear filters can be characterized by means of different representations:

3.1   Algebraic Normal Form (ANF)

The equation (2) describes the ANF of a nonlinear filter $F(s_n, s_{n+1}, \ldots, s_{n+L-1})$. That is F is represented as the sum of distinct products in the variables $(s_n, s_{n+1}, \ldots, s_{n+L-1})$. For each nonlinear filter the ANF representation is unique. The algebraic degree, k, of the Boolean function F is the highest degree of a monomial in F. This representation of Boolean functions is currently used by the designer of nonlinear filters.

3.2   Bit-Wise Sum of the Characteristic Sequences

Now, if all the variables s_{n+j} ($0 \leq j \leq L - 1$) in the ANF representation of F are substituted by their corresponding expressions in (1) and the resulting terms grouped, then the generic element z_n of the filtered sequence $\{z_n\}$ can be written as:

$$z_n = F(s_n, s_{n+1}, \ldots, s_{n+L-1}) =$$
$$C_1 \alpha^{E_1 n} + (C_1 \alpha^{E_1 n})^2 + \ldots + (C_1 \alpha^{E_1 n})^{2^{(r_1 - 1)}} +$$
$$C_2 \alpha^{E_2 n} + (C_2 \alpha^{E_2 n})^2 + \ldots + (C_2 \alpha^{E_2 n})^{2^{(r_2 - 1)}} +$$

$$\vdots \tag{3}$$

$$C_N \alpha^{E_N n} + (C_N \alpha^{E_N n})^2 + \ldots + (C_N \alpha^{E_N n})^{2^{(r_N - 1)}},$$

where r_i is the cardinal of coset E_i, the subindex i ranges in the interval $1 \leq i \leq N$ and N is the number of cosets of weight $\leq k$.

Thus a nonlinear filter $F(s_n, s_{n+1}, \ldots, s_{n+L-1})$ can be represented in terms of the N characteristic sequences $\{S_n^{E_i}\}$ that appear in this sequential decomposition in cosets shown in equation (3).

At this point different features can be pointed out. Note that the i-th row of (3) corresponds to the nth-term of the sequence $\{C_i \alpha^{E_i n} + (C_i \alpha^{E_i n})^2 + \ldots + (C_i \alpha^{E_i n})^{2^{(r_i - 1)}}\}$, where the coefficient $C_i \in GF(2^{r_i})$ determines the starting point of such a sequence. In fact, as long as C_i ranges in its corresponding extension field we shift along the sequence $\{S_n^{E_i}\}$. If the corresponding characteristic polynomial $P_{E_i}(x)$ is a primitive polynomial, then the characteristic sequence $\{S_n^{E_i}\}$ is a m-sequence.

If $C_i = 0$, then $\{S_n^{E_i}\}$ would not contribute to the filtered sequence $\{z_n\}$. In that case, the cyclotomic coset E_i would be degenerate. Linear complexity of the filtered sequence is related to the number of coefficients C_i different from zero as the contribution to LC of any nondegenerate coset equals the cardinal of such a coset.

3.3 A N-tuple of Coefficients

This is a representation very close to the previous one. In fact, a nonlinear filter $F(s_n, s_{n+1}, \ldots, s_{n+L-1})$ can be represented in terms of a N-tuple of coefficients $(C_1, C_2, \ldots, C_N)$ with $C_i \in GF(2^{r_i})$ where each coefficient determines the starting point of the sequence $\{S_n^{E_i}\}$ with reference to its characteristic phase and N denotes, as before, the number of cosets of weight $\leq k$.

In this work, the three representations will be indistinctly used.

4 Equivalence Classes for Nonlinear Filters

The idea of grouping nonlinear filters in equivalence classes for their handling has been already developed in the literature, see [21]. This is the technique followed in this section to design filters with specific properties.

Let G be the set of the kth-order nonlinear filters applied to a LFSR of length L. We are going to group the elements of G producing the filtered sequence $\{z_n\}$ or a shifted version of $\{z_n\}$, notated $\{z_n\}^*$. From equation (3), it is clear that if we substitute C_i for $C_i \cdot \alpha^{E_i}$ $\forall i$, then we will obtain $\{z_{n+1}\}$. In general,

$$C_i \rightarrow C_i \cdot \alpha^{jE_i} \ \forall i \Rightarrow \{z_n\} \rightarrow \{z_{n+j}\}.$$

This fact enables us to define an equivalence relationship $\sim$ on the set G as follows: $F_0 \sim F_1$ with $F_0, F_1 \in G$ if

$$\{F_0(s_n, \ldots, s_{n+L-1})\} = \{F_1(s_n, \ldots, s_{n+L-1})\}^*.$$

Therefore, two different nonlinear filters F_0, F_1 in the same equivalence class will produce shifted versions of the same filtered sequence. In addition, it is easy to see that the relation defined above is an equivalence relationship. Making use of the third representation for nonlinear filters (N-tuple of coefficients) in the previous section, we see that the coefficients associated with F_0, F_1, notated $(C_{E_i}^0)$ and $(C_{E_i}^1)$ respectively, satisfy

$$C_{E_i}^1 = C_{E_i}^0 \cdot \alpha^{j\,E_i} \ \ \forall i. \tag{4}$$

Clearly, the number of elements in every equivalence class equals the period of the filtered sequence, T, so that in (4) the index j verifies $1 \leq j \leq T - 1$.

Definition 1. *Two nonlinear filters F_0 and F_1 in the same equivalence class are consecutive if they satisfy the equation (4) with $j = 1$ or equivalently*

$$F_1(s_n, \ldots, s_{n+L-1}) = F_0(s_{n+1}, \ldots, s_{n+L}).$$

Let $E_1, E_2, \ldots, E_M$ be the leaders of the nondegenerate cosets of weight at most k in $\{z_n\}$ and $r_1, r_2, \ldots, r_M$ their corresponding cardinals. Several results can be pointed out.

Lemma 1. *If p nonlinear filters in the same equivalence class are chosen*

$$(C_{E_i}), (C_{E_i} \cdot \alpha^{q_1 E_i}), (C_{E_i} \cdot \alpha^{q_2 E_i}), \ldots, (C_{E_i} \cdot \alpha^{q_{p-1} E_i}) \tag{5}$$

($q_1, q_2, \ldots, q_{p-1}$ being integers) in such a way that no characteristic polynomial $P_{E_i}(x)$ $(1 \le i \le M)$ divides the polynomial

$$Q(x) = (1 + x^{q_1} + \ldots + x^{q_{p-1}}), \tag{6}$$

then the nonlinear filter characterized by the coefficients

$$\tilde{C}_{E_i} = C_{E_i}(1 + \alpha^{q_1 E_i} + \ldots + \alpha^{q_{p-1} E_i}) \quad (1 \le i \le M) \tag{7}$$

preserves the same cosets E_i as those of the filters defined in (5).

Proof. The result follows from the fact that the coefficients of the new nonlinear filter verify

$$\tilde{C}_{E_i} = C_{E_i}(1 + \alpha^{q_1 E_i} + \ldots + \alpha^{q_{p-1} E_i}) \ne 0 \quad (1 \le i \le M)$$

as no α^{E_i} is a root of $Q(x)$. □

Therefore an easy way to guarantee the presence of all the cosets E_i in the new filter is just summing $p \le r_{min}$ consecutive nonlinear filters in the same equivalence class (r_{min} being the least cardinal of all the cosets E_i) as $deg\, Q(x) < deg\, P_{E_i}(x)$ $(1 \le i \le M)$.

Lemma 2. *The sum of nonlinear filters satisfying the conditions of Lemma 1 gives rise to a new nonlinear filter in a different equivalent class.*

Proof. We proceed by contradiction. Suppose that the new filter belongs to the same equivalence class. Then,

$$\tilde{C}_{E_i} = C_{E_i}(1 + \alpha^{q_1 E_i} + \ldots + \alpha^{q_{p-1} E_i}) = C_{E_i} \cdot \alpha^{q_1 E_i} \quad \forall i. \tag{8}$$

For simplicity reasons, assume that coset E_i=coset 1. Therefore, according to (8)

$$(1 + \alpha^{q_1} + \ldots + \alpha^{q_{p-1}}) = \alpha^j$$

and

$$(1 + \alpha^{q_1 E_i} + \ldots + \alpha^{q_{p-1} E_i}) = \alpha^{j E_i} \quad (2 \le i \le M).$$

Thus, it follows that

$$(1 + \alpha^{q_1} + \ldots + \alpha^{q_{p-1}})^{E_i} = (1 + \alpha^{q_1 E_i} + \ldots + \alpha^{q_{p-1} E_i}) \quad (2 \le i \le M).$$

Nevertheless, it is a well known fact that in $GF(2^L)$ this equality only holds for E_i of the form 2^m (i.e. the elements of coset 1) but not for the leaders of any coset $E_i \ne$ coset 1. □

4.1   Practical Design of Nonlinear Filters with Guaranteed Linear Complexity

According to the previous results, a method of constructing nonlinear filters with large linear complexity can be stated as follows:

1. Start from a nonlinear filter F_0 whose number of nondegenerate cosets is known to be large, for instance, a nonlinear filter with a unique term of order k and equidistant stages [6], which preserves all the cosets of weight k.
2. Sum two consecutive nonlinear filters in this class $F_0 + F_1$ in order to jump into a different equivalence class preserving all the cosets E_i of the previous class.
3. Repeat step 2 in order to generate as many different equivalence classes as desired.

If at least one of the nondegenerate cosets has as characteristic sequence a m-sequence, then we can jump into $2^L - 1$ different equivalence classes before coming back to the original class. In this way, the sum operation $F_0 + F_1$ is a simple source of generation of nonlinear filters that preserve the k-weight cosets.

5   Construction of All Possible Nonlinear Filters with Cosets of Weight k: A Specific Algorithm

In order to generate all the nonlinear filters with guaranteed cosets of weight k, we start from a filter with a unique term product of k equidistant phases of the form:

$$F_0(s_n, s_{n+1}, \ldots, s_{n+L-1}) = s_n s_{n+\delta} \cdots s_{n+(k-1)\delta} \tag{9}$$

with $1 \leq k \leq L$ and $gcd(\delta, 2^L - 1) = 1$. According to [6], the sequence obtained from this type of filters includes all the k-weight cosets.

Given F_0 in ANF, the computation of its N_k-tuple is carried out via the root presence test described in [6]. That is the computation of Vandermonde determinants for which there is a simple formula. Next, the N_k-tuple representations for $F_1 = \mathcal{S}(F_0)$ and $F_0 + F_1$ are easily computed too. The key idea in this construction method is shifting the filter $F_0 + F_1$ through its equivalence class and summing it with F_0 in order to cancel the successive components of its N_k-tuple.

The final result is:

1. A set of N_k basic filters of the form $(0, 0, \ldots, d_i, \ldots, 0, 0)$ $(1 \leq i \leq N_k)$ with $d_i \in GF(2^L), d_i \neq 0$.
2. Their corresponding ANF representations.

The combination of all these basic filters with d_i $(1 \leq i \leq N_k)$ ranging in $GF(2^L)$ (with the corresponding ANF representations) gives rise to all the possible terms of order k that preserve the cosets of weight k. Later, the addition of terms of order $< k$ in ANF permits the generation of all the nonlinear filters of order k that guarantee a linear complexity $LC \geq \binom{L}{k}$.

An algorithm for computing the basic nonlinear filters with cosets of weight k is depicted in Fig. 1. The employed notation is now introduced:

```
Input: One nonlinear filter with guaranteed k-weight cosets,
       F₀(sₙ,...,s_{L-1}) → (C₁⁰,...,Cᵢ⁰,...,C_{Nₖ}⁰),
```

$$F_0(s_n, \ldots, s_{L-1}) \to (C_1^0, \ldots, C_i^0, \ldots, C_{N_k}^0),$$

```
Compute F₁ = S(F₀(sₙ,...,s_{L-1})) → (C₁¹,...,Cᵢ¹,...,C_{Nₖ}¹),
for j = Nₖ to 2 do
    Step 1: Addition of the two filters: F₀ + F₁ = F₀₁ →
            (Cᵢ⁰) + (Cᵢ¹) = (Cᵢ²)
    Step 2: Comparison F₀ : F₀₁
            (C₁⁰,...,Cᵢ⁰,...,C_{Nₖ}⁰) : (C₁²,...,Cᵢ²,...,C_{Nₖ}²)
    Step 3: Shifting of (C₁²,...,Cᵢ²,...,C_{Nₖ}²) through its equivalence class
            until Cⱼ² = Cⱼ⁰
            (C₁²,...,Cⱼ²,...,C_{Nₖ}²) → (C₁³,...,Cⱼ⁰,...,0)
    Step 4: Addition
            (Cᵢ⁰) + (Cᵢ³) = (Cᵢ⁴) = (C₁⁴,...,0,...,0)
            keep (Iᵢ^{j-1}) = (Cᵢ⁴)
    Step 5: Substitution
            (Cᵢ⁰) ← (Cᵢ⁴)
end for

(Bᵢ¹) = (Iᵢ¹); Display the ANF.
for j = 2 to Nₖ do
    Step 6: Comparison (Bᵢ¹),...,(Bᵢ^{j-1}) : (Iᵢ^j)
        for l = 1 to j - 1 do
            Step 7: Shifting of (Bᵢˡ)
                    until Bₗˡ = Iₗ^j
        end for
    Step 8: Addition
            ∑_{l=1}^{j-1} (Bᵢˡ)' + (Iᵢ^j) = (Bᵢ^j)
            Display the ANF.
end for

Output: Nₖ basic filters (Bᵢ^j) = (0,0,...,dⱼ,...,0) to generate
        all the nonlinear filters preserving the k-weight cosets
        and their ANF representations.
```

Fig. 1. Pseudo-code of the algorithm to generate N_k basic filters

- F_0 is the initial filter with guaranteed cosets of weight k. Its N_k-tuple coefficient representation can be written as:

$$F_0 = (C_1^0, C_2^0, \ldots, C_{N_k}^0) = (C_i^0) \quad (1 \le i \le N_k).$$

- $F_1 = \mathcal{S}(F_0)$ is the consecutive filter in the same equivalence class. Its N_k-tuple coefficient representation can be written as:

$$F_1 = (C_1^1, C_2^1, \ldots, C_{N_k}^1) = (C_i^1) \quad (1 \le i \le N_k).$$

- $F_{01} = F_0 + F_1$ is a new filter in a different equivalence class whose N_k-tuple coefficient representation is:

$$F_{01} = (C_1^2, C_2^2, \ldots, C_{N_k}^2) = (C_i^2) \ (1 \le i \le N_k).$$

- The filter (C_i^2) ranges in its equivalence class until the j-th component $(C_j^2) = (C_j^0)$. The resulting filter is:

$$(C_1^3, C_2^3, \ldots, C_j^0, \ldots, 0) = (C_i^3) \ (1 \le i \le N_k),$$

where $C_l^3 = 0$ for $(j + 1 \le l \le N_k)$.
- The filter (C_i^4) is the sum of:

$$(C_i^0) + (C_i^3) = (C_i^4) = (C_1^4, C_2^4, \ldots, 0, \ldots, 0) \ (1 \le i \le N_k),$$

where $C_l^4 = 0$ for $(j \le l \le N_k)$.
- (I_i^j) is an intermediate filter where (C_i^4) is stored for the corresponding value of the index j.

$$(I_1^j, I_2^j, \ldots, I_{N_k}^j) = (I_i^j) \ (1 \le i \le N_k).$$

- (B_i^j) is a basic filter whose components are 0 except for the j-th component $d_j \ne 0$.

$$(B_1^j, B_2^j, \ldots, B_{N_k}^j) = (B_i^j) = (0, 0, \ldots, d_j, \ldots, 0) \ (1 \le i \le N_k).$$

The symbol $(B_i^j)'$ means that the initial filter (B_i^j) has been shifted through its equivalence class.

Next a pseudo-code of the programmed algorithm is given in Fig. 1.

6   Discussion

Regarding the previous sections, distinct considerations must be taken into account.

Recall that the construction method described in the previous section to compute the basic filters $(0, 0, \ldots, d_i, \ldots, 0, 0), d_i \ne 0$ involves very simple operations:

- Sum operation: that is reduced to a logic sum of filters for the ANF representation or to a sum of elements of the extended field $GF(2^L)$ that expressed in binary representation is just an exclusive OR operation.
- Shifting operation through an equivalence class: that means an increment by 1 in all the indexes in the ANF representation or the multiplication of powers of α by their corresponding factors α^{E_i} in the N-tuple representation that just means the addition of exponents.

Consequently, the efficiency of the computation method is quite evident. In brief, we provide one with the complete class of nonlinear filters with $LC \ge \binom{L}{k}$ at the price of minimal computational operations.

In the case that the presence of more cosets of weight $\ne k$ were guaranteed, the procedure here described continues being applicable just enlarging the coefficient vector to new components corresponding to those new guaranteed cosets in the N-tuple representation.

Let us now see an illustrative example.

No.	A.N.F.	Coeff.
0	$s_0s_1s_2 \oplus s_0s_1s_3 \oplus s_0s_1s_4 \oplus s_0s_2s_3 \oplus s_0s_2s_4 \oplus$ $s_0s_3s_4 \oplus s_1s_2s_3 \oplus s_1s_2s_4 \oplus s_1s_3s_4$	$(\alpha^5,0)$
1	$s_0s_1s_2 \oplus s_0s_1s_3 \oplus s_0s_1s_4 \oplus s_0s_2s_3 \oplus s_0s_2s_4 \oplus s_1s_2s_4$	$(\alpha^{12},0)$
2	$s_0s_1s_2 \oplus s_0s_1s_3 \oplus s_0s_2s_3 \oplus s_1s_2s_4 \oplus s_1s_3s_4$	$(\alpha^{19},0)$
3	$s_0s_2s_3 \oplus s_0s_2s_4 \oplus s_1s_2s_3 \oplus s_1s_2s_4 \oplus s_1s_3s_4 \oplus s_2s_3s_4$	$(\alpha^{26},0)$
4	$s_0s_1s_3 \oplus s_0s_2s_3 \oplus s_0s_2s_4 \oplus s_0s_3s_4 \oplus s_1s_3s_4$	$(\alpha^2,0)$
5	$s_0s_1s_3 \oplus s_0s_1s_4 \oplus s_0s_2s_4 \oplus s_1s_2s_4 \oplus s_2s_3s_4$	$(\alpha^9,0)$
6	$s_0s_1s_2 \oplus s_0s_1s_3 \oplus s_0s_2s_3 \oplus s_0s_3s_4 \oplus s_1s_2s_3 \oplus s_1s_2s_4$	$(\alpha^{16},0)$
7	$s_0s_1s_4 \oplus s_0s_2s_3 \oplus s_1s_2s_3 \oplus s_1s_2s_4 \oplus s_2s_3s_4$	$(\alpha^{23},0)$
8	$s_0s_1s_2 \oplus s_0s_2s_3 \oplus s_0s_3s_4 \oplus s_1s_2s_3 \oplus s_1s_3s_4 \oplus s_2s_3s_4$	$(\alpha^{30},0)$
9	$s_0s_1s_4 \oplus s_0s_2s_4 \oplus s_0s_3s_4 \oplus s_1s_2s_3$	$(\alpha^6,0)$
10	$s_0s_1s_2 \oplus s_0s_1s_3 \oplus s_0s_1s_4 \oplus s_1s_2s_3 \oplus s_1s_3s_4 \oplus s_2s_3s_4$	$(\alpha^{13},0)$
11	$s_0s_1s_2 \oplus s_0s_2s_4 \oplus s_0s_3s_4 \oplus s_1s_2s_4$	$(\alpha^{20},0)$
12	$s_0s_1s_3 \oplus s_0s_1s_4 \oplus s_0s_2s_3 \oplus s_1s_2s_3 \oplus s_1s_3s_4$	$(\alpha^{27},0)$
13	$s_0s_1s_2 \oplus s_0s_2s_4 \oplus s_1s_2s_3 \oplus s_1s_2s_4 \oplus s_1s_3s_4$	$(\alpha^3,0)$
14	$s_0s_1s_3 \oplus s_0s_2s_3 \oplus s_0s_2s_4 \oplus s_1s_2s_3$	$(\alpha^{10},0)$
15	$s_0s_1s_3 \oplus s_1s_2s_4 \oplus s_1s_3s_4 \oplus s_2s_3s_4$	$(\alpha^{17},0)$
16	$s_0s_2s_3 \oplus s_0s_2s_4 \oplus s_0s_3s_4 \oplus s_1s_2s_4 \oplus s_2s_3s_4$	$(\alpha^{24},0)$
17	$s_0s_1s_3 \oplus s_0s_1s_4 \oplus s_0s_2s_3 \oplus s_0s_2s_4$	$(1,0)$
18	$s_0s_1s_2 \oplus s_0s_1s_4 \oplus s_1s_2s_3 \oplus s_1s_2s_4$	$(\alpha^7,0)$
19	$s_0s_1s_2 \oplus s_0s_2s_3 \oplus s_2s_3s_4$	$(\alpha^{14},0)$
20	$s_0s_3s_4 \oplus s_1s_2s_3 \oplus s_1s_3s_4$	$(\alpha^{21},0)$
21	$s_0s_1s_4 \oplus s_0s_2s_4 \oplus s_1s_3s_4$	$(\alpha^{28},0)$
22	$s_0s_1s_2 \oplus s_0s_1s_3 \oplus s_0s_2s_4 \oplus s_1s_2s_3 \oplus s_2s_3s_4$	$(\alpha^4,0)$
23	$s_0s_1s_3 \oplus s_0s_3s_4 \oplus s_1s_2s_3 \oplus s_1s_2s_4 \oplus s_2s_3s_4$	$(\alpha^{11},0)$
24	$s_0s_1s_4 \oplus s_0s_2s_3 \oplus s_0s_3s_4 \oplus s_1s_2s_4 \oplus s_1s_3s_4 \oplus s_2s_3s_4$	$(\alpha^{18},0)$
25	$s_0s_1s_2 \oplus s_0s_1s_4 \oplus s_0s_2s_3 \oplus s_0s_2s_4 \oplus s_0s_3s_4 \oplus s_1s_2s_3 \oplus s_2s_3s_4$	$(\alpha^{25},0)$
26	$s_0s_1s_2 \oplus s_0s_1s_3 \oplus s_0s_1s_4 \oplus s_0s_3s_4 \oplus s_2s_3s_4$	$(\alpha,0)$
27	$s_0s_1s_2 \oplus s_0s_1s_4 \oplus s_0s_3s_4 \oplus s_1s_2s_4 \oplus s_1s_3s_4$	$(\alpha^8,0)$
28	$s_0s_1s_2 \oplus s_0s_1s_4 \oplus s_0s_2s_3 \oplus s_0s_2s_4 \oplus s_1s_3s_4 \oplus s_2s_3s_4$	$(\alpha^{15},0)$
29	$s_0s_1s_2 \oplus s_0s_1s_3 \oplus s_0s_2s_4 \oplus s_0s_3s_4 \oplus s_1s_3s_4 \oplus s_2s_3s_4$	$(\alpha^{22},0)$
30	$s_0s_1s_3 \oplus s_0s_1s_4 \oplus s_0s_2s_4 \oplus s_0s_3s_4 \oplus s_1s_2s_3 \oplus s_1s_2s_4 \oplus s_1s_3s_4 \oplus s_2s_3s_4$	$(\alpha^{29},0)$

Fig. 2. Class of nonlinear filters $(B_i^1) = (\alpha^5,0)$

6.1 A Numerical Example

Let $(L,k) = (5,3)$ be a nonlinear filter of third order applied to the stages of a LFSR of length $L = 5$ and primitive characteristic polynomial $P(x) = x^5 + x^3 + 1$ where α is a root of $P(x)$ so that $\alpha^5 = \alpha^3 + 1$. We have $N_3 = 2$ cyclotomic cosets of weight 3: coset 7= $\{7, 14, 28, 25, 19\}$ and coset 11= $\{11, 22, 13, 26, 21\}$. The initial filter with guaranteed cosets of weight 3 is $F_0(s_0, s_1, s_2) = s_0s_1s_2$. The algorithm described in Fig. 1 is applied.

No.	A.N.F.	Coeff.
0	$s_0 s_2 s_4 \oplus s_0 s_3 s_4 \oplus s_1 s_2 s_4$	$(0, \alpha^{13})$
1	$s_0 s_1 s_3 \oplus s_0 s_1 s_4 \oplus s_0 s_2 s_3 \oplus s_1 s_3 s_4$	$(0, \alpha^{24})$
2	$s_0 s_1 s_2 \oplus s_0 s_2 s_4 \oplus s_1 s_2 s_3 \oplus s_1 s_2 s_4 \oplus s_1 s_3 s_4 \oplus s_2 s_3 s_4$	$(0, \alpha^{4})$
3	$s_0 s_1 s_3 \oplus s_0 s_2 s_3 \oplus s_0 s_2 s_4 \oplus s_0 s_3 s_4 \oplus s_1 s_2 s_3$	$(0, \alpha^{15})$
4	$s_0 s_1 s_3 \oplus s_0 s_1 s_4 \oplus s_1 s_2 s_4 \oplus s_2 s_3 s_4$	$(0, \alpha^{26})$
5	$s_0 s_1 s_2 \oplus s_0 s_2 s_3 \oplus s_0 s_3 s_4 \oplus s_1 s_2 s_3 \oplus s_1 s_2 s_4$	$(0, \alpha^{6})$
6	$s_0 s_1 s_4 \oplus s_0 s_2 s_3 \oplus s_1 s_2 s_3 \oplus s_2 s_3 s_4$	$(0, \alpha^{17})$
7	$s_0 s_1 s_2 \oplus s_0 s_3 s_4 \oplus s_1 s_2 s_3 \oplus s_1 s_3 s_4 \oplus s_2 s_3 s_4$	$(0, \alpha^{28})$
8	$s_0 s_1 s_4 \oplus s_0 s_2 s_4 \oplus s_0 s_3 s_4 \oplus s_1 s_2 s_3 \oplus s_1 s_3 s_4$	$(0, \alpha^{8})$
9	$s_0 s_1 s_2 \oplus s_0 s_1 s_3 \oplus s_0 s_1 s_4 \oplus s_0 s_2 s_4 \oplus s_1 s_2 s_3 \oplus s_1 s_3 s_4$	$(0, \alpha^{19})$
10	$s_0 s_1 s_2 \oplus s_0 s_1 s_3 \oplus s_0 s_2 s_4 \oplus s_1 s_2 s_4$	$(0, \alpha^{30})$
11	$s_0 s_1 s_3 \oplus s_0 s_2 s_3 \oplus s_1 s_2 s_3 \oplus s_1 s_2 s_4$	$(0, \alpha^{10})$
12	$s_0 s_2 s_3 \oplus s_1 s_2 s_4 \oplus s_1 s_3 s_4 \oplus s_2 s_3 s_4$	$(0, \alpha^{21})$
13	$s_0 s_2 s_3 \oplus s_0 s_2 s_4 \oplus s_0 s_3 s_4 \oplus s_1 s_3 s_4 \oplus s_2 s_3 s_4$	$(0, \alpha^{1})$
14	$s_0 s_1 s_3 \oplus s_0 s_1 s_4 \oplus s_0 s_2 s_4 \oplus s_0 s_3 s_4 \oplus s_2 s_3 s_4$	$(0, \alpha^{12})$
15	$s_0 s_1 s_2 \oplus s_0 s_1 s_3 \oplus s_0 s_1 s_4 \oplus s_0 s_3 s_4 \oplus s_1 s_2 s_3 \oplus s_1 s_2 s_4 \oplus s_1 s_3 s_4$	$(0, \alpha^{23})$
16	$s_0 s_1 s_2 \oplus s_0 s_1 s_4 \oplus s_0 s_2 s_3 \oplus s_0 s_2 s_4 \oplus s_1 s_2 s_4 \oplus s_1 s_3 s_4$	$(0, \alpha^{3})$
17	$s_0 s_1 s_2 \oplus s_0 s_1 s_3 \oplus s_0 s_2 s_3 \oplus s_0 s_2 s_4 \oplus s_1 s_3 s_4 \oplus s_2 s_3 s_4$	$(0, \alpha^{14})$
18	$s_0 s_1 s_3 \oplus s_0 s_2 s_4 \oplus s_0 s_3 s_4 \oplus s_1 s_2 s_3 \oplus s_1 s_2 s_4 \oplus s_1 s_3 s_4 \oplus s_2 s_3 s_4$	$(0, \alpha^{25})$
19	$s_0 s_1 s_3 \oplus s_0 s_1 s_4 \oplus s_0 s_2 s_3 \oplus s_0 s_2 s_4 \oplus s_0 s_3 s_4 \oplus s_1 s_2 s_4 \oplus s_1 s_3 s_4$	$(0, \alpha^{5})$
20	$s_0 s_1 s_2 \oplus s_0 s_1 s_3 \oplus s_0 s_1 s_4 \oplus s_0 s_2 s_3 \oplus s_0 s_2 s_4 \oplus s_1 s_2 s_3 \oplus s_1 s_2 s_4 \oplus s_2 s_3 s_4$	$(0, \alpha^{16})$
21	$s_0 s_1 s_2 \oplus s_0 s_1 s_3 \oplus s_0 s_2 s_3 \oplus s_0 s_3 s_4 \oplus s_1 s_2 s_4 \oplus s_1 s_3 s_4 \oplus s_2 s_3 s_4$	$(0, \alpha^{27})$
22	$s_0 s_1 s_4 \oplus s_0 s_2 s_3 \oplus s_0 s_2 s_4 \oplus s_0 s_3 s_4 \oplus s_1 s_2 s_3 \oplus s_1 s_2 s_4 \oplus s_2 s_3 s_4$	$(0, \alpha^{7})$
23	$s_0 s_1 s_2 \oplus s_0 s_1 s_3 \oplus s_0 s_1 s_4 \oplus s_0 s_2 s_3 \oplus s_0 s_3 s_4 \oplus s_1 s_2 s_3 \oplus s_2 s_3 s_4$	$(0, \alpha^{18})$
24	$s_0 s_1 s_2 \oplus s_0 s_1 s_4 \oplus s_0 s_3 s_4 \oplus s_1 s_2 s_4 \oplus s_2 s_3 s_4$	$(0, \alpha^{29})$
25	$s_0 s_1 s_2 \oplus s_0 s_1 s_4 \oplus s_0 s_2 s_3 \oplus s_0 s_3 s_4 \oplus s_1 s_3 s_4$	$(0, \alpha^{9})$
26	$s_0 s_1 s_2 \oplus s_0 s_1 s_4 \oplus s_0 s_2 s_4 \oplus s_2 s_3 s_4$	$(0, \alpha^{20})$
27	$s_0 s_1 s_2 \oplus s_0 s_1 s_3 \oplus s_0 s_3 s_4$	$(0, 1)$
28	$s_0 s_1 s_4 \oplus s_1 s_2 s_3 \oplus s_1 s_2 s_4 \oplus s_1 s_3 s_4$	$(0, \alpha^{11})$
29	$s_0 s_1 s_2 \oplus s_0 s_2 s_3 \oplus s_0 s_2 s_4 \oplus s_1 s_2 s_3$	$(0, \alpha^{22})$
30	$s_0 s_1 s_3 \oplus s_1 s_2 s_3 \oplus s_1 s_3 s_4 \oplus s_2 s_3 s_4$	$(0, \alpha^{2})$

Fig. 3. Class of nonlinear filters $(B_i^2) = (0, \alpha^{13})$

INPUT: The nonlinear filter $F_0(s_0, s_1, s_2) = s_0 s_1 s_2 \rightarrow (C_i^0) = (\alpha^{20}, \alpha^{13})$
Compute: $F_1(s_0, s_1, s_2) = s_1 s_2 s_3 \rightarrow (C_i^1) = (\alpha^{20} \cdot \alpha^7, \alpha^{13} \cdot \alpha^{11}) = (\alpha^{27}, \alpha^{24})$
Initialize: $(I_i^2) = (C_i^0) = (\alpha^{20}, \alpha^{13})$
for $j = N_3$ to 2

– Step 1: Addition of two filters $F_0 + F_1 = F_{01}$

$$(C_i^0) + (C_i^1) = (C_i^2)$$
$$(\alpha^{20}, \alpha^{13}) + (\alpha^{27}, \alpha^{24}) = (\alpha^5, \alpha^5)$$

- Step 2: Comparison $F_0 : F_1$

$$(C_i^0) : (C_i^2)$$
$$(\alpha^{20}, \alpha^{13}) : (\alpha^5, \alpha^5)$$

- Step 3: Shifting of (C_i^2) until $(C_2^2) = (C_2^0)$

$$(C_1^2, C_2^2) \to (C_1^3, C_2^0)$$
$$(\alpha^5, \alpha^5) \to (\alpha^{27}, \alpha^{13}) = (C_i^3)$$

- Step 4: Addition

$$(C_i^0) + (C_i^3) = (C_i^4)$$
$$(\alpha^{20}, \alpha^{13}) + (\alpha^{27}, \alpha^{13}) = (\alpha^5, 0)$$
$$(I_i^1) = (C_i^4)$$

end for
Introduce $(B_i^1) = (I_i^1) = (\alpha^5, 0)$
for $j = 2$ to N_3

- Step 6: Comparison $(B_i^1) : (I_i^2)$

$$(\alpha^5, 0) : (\alpha^{20}, \alpha^{13})$$

- Step 7: Shifting of (B_i^1) until $(B_1^1) = (I_1^2) = \alpha^{20}$

$$(B_i^1) \to (B_i^1)'$$
$$(\alpha^5, 0) \to (\alpha^{20}, 0)$$

- Step 8: Addition

$$(B_i^1)' + (I_i^2) = (B_i^2)$$
$$(\alpha^{20}, 0) + (\alpha^{20}, \alpha^{13}) = (0, \alpha^{13})$$
$$(B_1^2, B_2^2) = (0, \alpha^{13})$$

end for
OUTPUT: $N_3 = 2$ basic nonlinear filters and their corresponding ANF
representations.

1. $(B_i^1) = (\alpha^5, 0)$
   ANF: $s_0 s_1 s_2 \oplus s_0 s_1 s_3 \oplus s_0 s_1 s_4 \oplus s_0 s_2 s_3 \oplus s_0 s_2 s_4 \oplus s_0 s_3 s_4 \oplus s_1 s_2 s_3 \oplus s_1 s_2 s_4$
   $\oplus s_1 s_3 s_4$.
2. $(B_i^2) = (0, \alpha^{13})$
   ANF: $s_0 s_2 s_4 \oplus s_0 s_3 s_4 \oplus s_1 s_2 s_4$.

Basic filters (B_i^1) and (B_i^2) range in their corresponding equivalence class (with
$2^5 - 1$ filters per class) as it is shown in Fig. 2 and Fig. 3, respectively. Filter (B_i^1)
includes a unique coset of weight 3 that is (coset 7) as so does (B_i^2) with (coset
11). None of the filters depicted in the previous figures attains the lower bound
$LC \geq \binom{L}{k}$. Nevertheless, summing up each one of the ANF representations in
Fig. 2 with every one of the ANF representations in Fig. 3, we get the 31×31
possible combinations of terms of order 3 that guarantee the cosets of weight
3 (coset 7 and coset 11). Next, the addition of terms of order < 3 in ANF
representation permits us the generation of all the nonlinear filters of order 3
applied to the previous LFSR that guarantee a linear complexity $LC \geq \binom{5}{3}$.

7 Conclusions

In this paper, a method of computing all the nonlinear dynamical filters applied to a LFSR that guarantee the cosets of weight k has been developed. The procedure is based on the handling of nonlinear filters belonging to different equivalence classes. The method not only includes the nonlinear filters (e.g. filters obtained from equidistance phases or combination of equidistance phases) found in the literature but also it formally completes the class of filters with a guaranteed linear complexity. In brief, an easy way of designing keystream generators for stream cipher purposes has been provided.

References

1. Nagaraj, N.: One-Time Pad as a nonlinear dynamical system. Commun Nonlinear Sci. Numer Simulat. 17, 4029–4036 (2012)
2. eSTREAM, the ECRYPT Stream Cipher Project,The eSTREAM Portfolio (2012), http://www.ecrypt.eu.org/documents/D.SYM.10-v1.pdf
3. Robshaw, M., Billet, O. (eds.): New Stream Cipher Designs. LNCS, vol. 4986. Springer, Heidelberg (2008)
4. Menezes, A.J., et al.: Handbook of Applied Cryptography. CRC Press, New York (1997)
5. Paar, C., Pelzl, J.: Understanding Cryptography. Springer, Heidelberg (2010)
6. Rueppel, R.A.: Analysis and Design of Stream Ciphers. Springer, New York (1986)
7. Tan, S.K., Guan, S.U.: Evolving cellular automata to generate nonlinear sequences with desirable properties. Applied Soft Computing 7, 1131–1134 (2007)
8. Golomb, S.: Shift-Register Sequences, revised edn. Aegean Park Press, Laguna Hills (1982)
9. Fúster-Sabater, A., Caballero-Gil, P., Delgado-Mohatar, O.: Deterministic Computation of Pseudorandomness in Sequences of Cryptographic Application. In: Allen, G., Nabrzyski, J., Seidel, E., van Albada, G.D., Dongarra, J., Sloot, P.M.A. (eds.) ICCS 2009, Part I. LNCS, vol. 5544, pp. 621–630. Springer, Heidelberg (2009)
10. Fúster-Sabater, A., Caballero-Gil, P.: Chaotic modelling of the generalized self-shrinking generator. Appl. Soft Comput. 11, 1876–1880 (2011)
11. A. Marsaglia, Test of DIEHARD (1998), http://stat.fsu.edu/pub/diehard/
12. Massey, J.L.: Shift-Register Synthesis and BCH Decoding. IEEE Trans. Information Theory 15(1), 122–127 (1969)
13. Caballero-Gil, P., Fúster-Sabater, A.: A wide family of nonlinear filter functions with large linear span. Inform. Sci. 164, 197–207 (2004)
14. Limniotis, K., Kolokotronis, N., Kalouptsidis, N.: On the Linear Complexity of Sequences Obtained by State Space Generators. IEEE Trans. Inform. Theory 54, 1786–1793 (2008)
15. Kolokotronis, N., Limniotis, K., Kalouptsidis, N.: Lower Bounds on Sequence Complexity Via Generalised Vandermonde Determinants. In: Gong, G., Helleseth, T., Song, H.-Y., Yang, K. (eds.) SETA 2006. LNCS, vol. 4086, pp. 271–284. Springer, Heidelberg (2006)
16. Lee, K., O'Sullivan, M.E.: List decoding of Hermitian codes using Gröbner bases. Journal of Symbolic Computation 44(12), 1662–1675 (2009)

17. Respondek, J.S.: On the confluent Vandermonde matrix calculation algorithm. Applied Mathematics Letters 24(2), 103–106 (2011)
18. Respondek, J.S.: Numerical recipes for the high efficient inverse of the confluent Vandermonde matrices. Applied Mathematics and Computation 218(5), 2044–2054 (2011)
19. Lidl, R., Niederreiter, H.: Finite Fields. In: Enciclopedia of Mathematics and Its Applications, 2nd edn., vol. 20. Cambridge University Press, Cambridge (1997)
20. Ronjom, S., Helleseth, T.: A New Attack on the Filter Generator. IEEE Trans. Information Theory 53(5), 1752–1758 (2007)
21. Rønjom, S., Cid, C.: Nonlinear Equivalence of Stream Ciphers. In: Hong, S., Iwata, T. (eds.) FSE 2010. LNCS, vol. 6147, pp. 40–54. Springer, Heidelberg (2010)

Communication-Efficient Exact Clustering of Distributed Streaming Data

Dang-Hoan Tran and Kai-Uwe Sattler

Department of Computer Science & Automation
Ilmenau University of Technology, Germany
{dang-hoan.tran,kus}@tu-ilmenau.de

Abstract. A widely used approach to clustering a single data stream is the two-phased approach in which the online phase creates and maintains micro-clusters while the off-line phase generates the macro-clustering from the micro-clusters. We use this approach to propose a distributed framework for clustering streaming data. Every remote-site process generates and maintains micro-clusters that represent cluster information summary from its local data stream. Remote sites send the local micro-clusterings to the coordinator, or the coordinator invokes the remote methods in order to get the local micro-clusterings from the remote sites. Having received all the local micro-clusterings from the remote sites, the coordinator generates the global clustering by the macro-clustering method. Our theoretical and empirical results show that the global clustering generated by our distributed framework is similar to the clustering generated by the underlying centralized algorithm on the same data set. By using the local micro-clustering approach, our framework achieves high scalability, and communication-efficiency.

1 Introduction

Mining distributed data streams has been increasingly received great attention [18,20]. Data in nowadays data analysis applications is too big to gather and analyze in a single location or data is generated by the distributed sources. Design and development of data stream mining algorithms in high-performance and distributed environments thus should meet system scalability yet assure the quality of mining results. This paper studies the problem of clustering distributed streaming data.

Clustering is considered an unsupervised learning method which classifies the objects into groups of similar objects [12]. Clustering data stream continuously produces and maintains the clustering structure from the data stream in which the data items continuously arrive in the ordered sequence [11]. There are two types of problems of clustering data streams: (1) Clustering streaming data is to classify the data points that come from a single or multiple data streams into groups of similar data points [1,6,15]; (2) Clustering multiple data streams is to classify the data streams into groups of data streams of similar behavior or trend [4,9]. The basic requirements for clustering data streams [3] includes

B. Murgante et al. (Eds.): ICCSA 2013, Part V, LNCS 7975, pp. 421–436, 2013.
© Springer-Verlag Berlin Heidelberg 2013

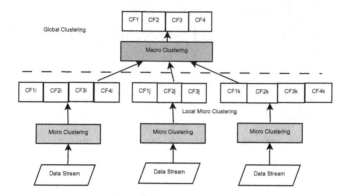

Fig. 1. A distributed two-staged framework for clustering streaming data

a compact representation of clustering structures, fast, incremental processing of recently incoming data items, and clear and fast identification of "outliers". Clustering distributed streaming data is to classify the data points that come from multiple data streams generated by the distributed sources. Clustering distributed streaming data can be used to solve many real world applications. Such an example is a sensor network that continuously monitors the changes in the environment such as a building. In a naive approach, every sensor continuously sends its stream of observations (temperature, humidity, light) to the base station for processing in order to give a global view of the building [13]. However, the challenges facing this approach include: (1) the limited communication bandwidth; (2) the overlapping of observations sensed by nearby sensors that generates data redundancy. Clustering distributed streaming sensor data may be a solution to this problem. There are four approaches to the problem of clustering distributed data streams: (1) *Centralization:* all the raw data is transmitted to the coordinator site at which a clustering algorithm is applied to the received data. This approach is impossible in the distributed context because of the resource constraints. First, it is impractical to send all the data to the coordinator for processing and mining due to the limited network bandwidth. The communication constraint even becomes more critical when the energy consumption is considered in the sensor network with nodes of the power-limited batteries [14]. Second, the transmission of all the raw data over a network can cause privacy and security concerns while the objective we want is only to understand and monitor the global view of the environment. (2) *Data summary:* First, each remote site sends the data representatives constructed from data stream by sampling method or synopsis construction method to the coordinator site. Second, the coordinator site combines all the received data representatives into the global representative. Third, the clustering algorithm is performed on this global representative [8,19,21]. Compared to the centralized approach, this approach considerably reduces the volume of data to be transmitted. However, a major drawback of the summary-based clustering is that the clustering quality is affected due to the loss of data during the sampling or synopsis construction process. (3) *Local clustering:* each remote site clusters its local data stream. The local clustering is sent to the coordinator site [7]. The coordinator site generates the global clustering by using some cluster ensembles method. The important advantage of this approach is that it is fault tolerant. That means the framework works well despite of the malfunction of some nodes in the network.(4) *P2P stream clustering:* An important distinction and this approach with the first three approaches is that each remote site can directly communicate with each other to generate the global clustering. As each remote site must maintain the same global clustering [2], this approach may incur high communication overhead.

The specific contributions of this paper are as follows.

- We propose a distributed framework for clustering data streams. Every remote-site process generates and maintains micro clusters that represent cluster information summary, from its local data streams. Remote sites send the local micro-clusterings to the coordinator, or the coordinator invokes the remote methods in order to get the local micro-clusterings from the remote sites. Having received all the local micro-clusterings from the remote sites, the coordinator generates the global clustering by the macro-clustering method.
- We achieve the global clustering as exact as the underlying centralized algorithm. Since the local micro clusterings that are received at the coordinator from the remote sites are the same as the local micro clusterings produced by the micro-clustering method of the centralized clustering algorithm and the macro-clustering algorithms are similar in both the distributed clustering algorithm and the centralized clustering algorithm.
- Our framework achieves high scalability, and communication-efficiency by using the local micro-clustering approach.

In comparison with the previous work, our framework has the following advantages. First, as local clusterings are much smaller than the local raw data, transmission of local clusterings significantly reduces the communication cost. Second, the strongest point in our framework is that no approximation is required when producing global clustering.

We formulate the problem in Section 2. Section 3 describes the algorithms in the framework in detail. The experimental results, as well as evaluations of our framework are presented in Section 4. Section 5 further discusses the research direction of local micro-clustering method in distributed stream mining. Related work is given in Section 6. Finally, we conclude and propose the future work in Section 7.

2   Problem Formulation

We consider a network of one coordinator site and N remote sites. Let $\{S_1, ..., S_N\}$ be the incoming data streams to the remote clustering modules at N remote sites. A data stream is an infinite sequence of elements

$S_i = \{(X_1, T_1), .., (X_j, T_j), ...\}$. Each element is a pair (X_j, T_j) where X_j is a d-dimensional vector $X_j = (x_1, x_2, ..., x_d)$ arriving at the timestamp T_j. We assume that S_i^t is a block of M data items that arrives at the remote site i during each update epoch. We also assume that all the data streams arrive at the remote sites with the same data speed, that means at each update epoch t, every remote site receives the same M data items. We assume that each site knows its own clustering structure but nothing about the clustering structures at other sites. Let $LC_i^t = \left\{ CF_i^1, .., .., CF_i^j, .., CF_i^{K_i} \right\}$ be a local micro-clustering generated by a clustering algorithm at remote site i at the update epoch t, where CF_i^j is the $j - th$ micro-cluster, K_i is the number of micro-clusters, for $i = 1, .., N$.

One of the fundamental properties of distributed computational systems is locality [17]. A distributed algorithm for clustering streaming data should meet the locality. A local algorithm is defined as one whose resource consumption is independent of the system size. Local algorithms can fall into one of two categories [10]: (1) exact local algorithms are defined as ones that produce the same results as a centralized algorithm; (2) approximate local algorithms are algorithms that produce approximations of the results that a centralized algorithm would produce. The objective here is to create and maintain the global clustering which is similar to the clustering created by the centralized stream clustering algorithm. In other words, our algorithm is an exact local algorithm.

A distributed streaming data clustering algorithm derived from a given underlying algorithm $\mathcal{A}$ is called $Dis\mathcal{A}$. The micro-clustering $Mic\mathcal{A}$ creates and maintains micro-clusters at the remote sites. The macro-clustering algorithm $Mac\mathcal{A}$ is used to produce meaningful macro-clustering at the coordinator.

3   Algorithm Description

Our distributed framework for clustering streaming data includes two fundamental processes: remote-site process and coordinator-site process. On the basis of type of query, we design the appropriate communication protocols as follows. First, if the clustering query is push-based query, the remote sites send the local micro-clusterings to the coordinator. As the coordinator process is a passive process, it continuously listens for connection requests from the remote sites and receives the local micro clusterings that are sent by the remote sites. Push-based query is widely used for querying sensor network. Second, if the clustering query is pull-based query, the coordinator invokes the remote methods on the local data streams from the remote sites in order to get the local micro-clusterings. Such a pull-based query is used in the well-known protocol HTTP. In order for the coordinator to communicate with many remote sites concurrently, the global clustering process is organized as a multi-threading process.

Our algorithm uses the same data structure *Clustering* for both local clustering and global clustering. An object *Clustering* consists of many micro-clusters where each cluster is a cluster representative. The first truly scalable clustering algorithm for data stream called BIRCH [22] constructs a clustering structure in a single scan over the data with limited memory. The underlying concept behind BIRCH called the cluster feature vector is defined as follows

Definition 1. *Given d-dimensional data points in a cluster:* $\{X\}$ *where* $i = 1, 2, .., N$, *the Clustering Feature (CF) vector of the cluster is a triple:* $CF = (N, LS, SS)$ *where* N *is the number of data points in the cluster,* $LS = \sum_{i=0}^{N-1} X_i$ *is the linear sum of the data points in the cluster, and* $SS = \sum_{i=0}^{N-1} X_i^2$ *is the squared sum of the* N *data points. The cluster created by merging two above disjoint clusters* CF_1 *and,* CF_2 *has the cluster feature is defined as follows*

$$CF = (N_1 + N_2, LS_1 + LS_2, SS_1 + SS_2) \tag{1}$$

The advantages of this CF summary are: (1) it does not require to store all the data points in the cluster; (2) it provide sufficient information for computing all the measurements necessary for making clustering decisions [22].

Micro-cluster extends the cluster feature vector by adding the temporal components [1].

Definition 2. *A micro-cluster for a set of* $d-$*dimensional points* $X_{i_1}, ..., X_{i_N}$ *with time stamps* $T_{i_1}, ..., T_{i_n}$ *is the* $(2d+3)$*-tuple* $(\overline{CF2^x}, \overline{CF1^x}, CF2^t, CF1^t, N)$, *wherein* $\overline{CF2^x}$ *and* $\overline{CF1^x}$ *each corresponds to a vector of d entries. The definition of each of these entries is as follows*

- *For each dimension, the sum of the squares of the data values is maintained in* $\overline{CF2^x}$. *Thus,* $\overline{CF2^x}$ *contains d values. The* $p-th$ *entry of* $\overline{CF2^x}$ *is equal to* $\sum_{j=1}^{N} \left(X_{i_j}^p\right)^2$.
- *For each dimension, the sum of the data values is maintained in* $\overline{CF1^x}$. *Thus,* $\overline{CF1^x}$ *contains d values. The* $p-th$ *entry of* $\overline{CF1^x}$ *is equal to* $\sum_{j=1}^{N} \left(X_{i_j}^p\right)$.
- *The sum of the squares of the timestamps* $T_{i_1}, ..., T_{i_n}$ *is maintained in* $CF2^t$.
- *The sum of the time stamps* $T_{i_1}, ..., T_{i_N}$ *is maintained in* $CF1^t$.
- *The number of data points is maintained in* N.

Definition 2 is the first definition of micro-cluster. In the later work, variants of micro-clusters are proposed to meet the specific requirements of the stream clustering algorithms. For example, in DenStream [6], micro-clusters fall into types such as core-micro-clusters, potential-c-micro-clusters, or outlier micro-clusters, which are used for density-based clustering.

3.1   The Remote Process

This subsection describes how a local micro-clustering algorithm at the remote site works. As we assumed in Section 2, a remote site processes the incoming data from its local stream in an update epoch t: S_i^t. For $i = 1, .., N$, the local clustering process for the push-based type of query at the remote site i works as follows.

1. make the connection between the remote site i and the coordinator.
2. while not at the end of the data stream S_i
   (a) initialize the stream learner $MicA(S_i^t)$.
   (b) read block of data items at update epoch t
   (c) create the local micro-clustering LC_i^t by calling micro-clustering learner $MicA(S_i^t)$ on block S_i^t.
   (d) transmit the local micro-clustering to the coordinator.

The clustering algorithm for push-based query, at each update epoch, every remote site generates and transmits the local micro-clustering to the coordinator. For $i = 1, .., N$, the local clustering process for the pull-based type of query at the remote site i works as follows.

1. make the connection between the remote site i and the coordinator.
2. while not at the end of the data stream S_i
   (a) initialize the stream learner $MicA(S_i^t)$.
   (b) read block of data items at update epoch e, let S_i^t denote the block of data items.

In contrast to the clustering algorithm for push-based type of query, in a clustering algorithm for pull-based query, the coordinator calls the local micro-clustering on the block of data items at epoch t from the remote-site process.

3.2   The Coordinator Process

This subsection describes how the coordinator process works. For each update epoch t, the coordinator works as follows.

1. for each remote site i, perform the following steps
   (a) make the connection between the remote site i and the coordinator.
   (b) receive the local micro-clustering LC_i^t from the remote site (for the push-based query), or remotely calls the local-micro clustering method from the remote site i on the block S_i^t (for the pull-based query).
   (c) add LC_i^t to a buffer buf.
2. if all the local micro-clusterings are received, the coordinator creates the global clustering GC^t by calling $MacA$ on the micro-clusters in the buffer.
3. process the global clustering GC_t.

We note that step 3 is executed on the basis of each specific context. For example, it may return the global clustering to the user as requested, or the global clustering can be stored in the history of global clusterings for some computation purpose, or it may be sent back to all the remote sites.

3.3   Algorithm Analysis

In this section we present some theoretical results including the global clustering quality, communication complexity, and computational complexity.

Global Clustering Quality. We shows that the global clustering produced by $DisA$ in an update epoch is similar to the clustering created by the underlying algorithm $DisA$ on the same data set. We first consider the case in which the clustering is generated by the centralized clustering algorithm A. Let S_i^t be the streaming data block at the remote site i in an update epoch t, for $i = 1, .., N$, where N is the number of remote sites. We consider the A and $DisA$ in an update epoch. We also assume that all the data streams arrive at the remote sites with the same data speed, that means at each update epoch t, every remote site receives the same data items. As such, the streaming data that the coordinator receive from all the remote sites in an update epoch is given by $S^{t+\triangle t} = \overset{N}{\underset{i=1}{\cup}} S_i^t$ where $\triangle t$ is the time for the coordinator to receive all the streaming blocks from the remote sites. Algorithm A works in two stages as follows

- *Micro clustering:* the output of the micro-clustering method is given by

$$LC_{centralized} = MicA\left(S^{t+\triangle t}\right) = \overset{N}{\underset{i=1}{\cup}} LC_i^t \qquad (2)$$

- *Macro clustering:* the output of the macro-clustering method is given by

$$GC_{centralized}^t = MacA\left(\overset{N}{\underset{i=1}{\cup}} LC_i^t\right) \qquad (3)$$

We now consider how our distributed stream clustering framework $DisA$ works. However, a distinction between the centralized algorithm A and the distributed algorithm $DisA$ is that while the both micro-clustering method and macro-clustering method of A are executed in the same process, the micro-clustering method of $DisA$ takes place at the remote sites, the macro-clustering method of $DisA$ occurs at the coordinator. The framework $DisA$ works as follows

- *At the remote site:* Each remote site generates the local macro-clustering LC_i, for $i = 1, .., N$. If the clustering query is pushed-based query, remote sites send its local micro-clustering to the coordinator.
- *At the coordinator:* If the clustering query is pull-based query, the coordinator invokes remote clustering methods in order to get all the local micro-clusterings. The list of local clusterings received from remote sites

$$LC_{distributed}^{t+\triangle t_1} = \overset{N}{\underset{i=1}{\cup}} LC_i^t \qquad (4)$$

where $\triangle t_1$ is the time needed to transmit all the local clusterings to the coordinator

The global clustering $GC^t_{distributed}$ is created by the macro clustering method $\mathcal{M}ac$ up to time $t + \triangle t_1 + \triangle t_2$ is given by

$$GC^t_{distributed} = \mathcal{M}ac\mathcal{A}\,(LC_{distributed}) = \mathcal{M}ac\mathcal{A}\left(\bigcup_{i=1}^{N} LC^t_i\right) \qquad (5)$$

where $\triangle t_2$ is the time needed to produce global macro-clustering (global clustering).

From the equations 2, 3 and 4, 5, we can conclude that the global macro-clustering produced by the distributed framework $Dis\mathcal{A}$ in an update epoch t is similar to the macro-clustering produced by the centralized version of the two-phased stream clustering algorithm $\mathcal{A}$.

Communication Complexity. One of the fundamental issues of a distributed computing algorithm is to evaluate the communication complexity. The communication complexity of a distributed framework for clustering is the minimum communication cost so that it produces the global clustering. Let b^t_i be the total bits sent between remote site i and the coordinator. The size of a local micro-clustering $|LC_i|$ is defined as the product of the size of each micro-cluster $|CF|$ by the number of clusters K_i in the local clustering structure $|LC_i| = |CF|K_i$. The number of bits used to transmit a local micro-clustering produced by the remote site i to the coordinator site is $\log_2 |LC^t_i|)$. As such, the communication cost of all local clusterings from the remote sites to the coordinator is given by.

$$\sum_{i=1}^{N} \log_2 |CF|K_i = \sum_{i=0}^{N} \log_2 K_i + \frac{(N+1)N}{2}\log_2 |CF| \qquad (6)$$

Let DisClustream be the distributed version of Clustream. If the DisClustream is used, and the number of clusters at all the remote sites and the coordinator are the same, the communication cost needed to answer a clustering query at an update epoch t is given by

$$N\log_2 K + \frac{(N+1)N}{2}\log_2 |CF| \qquad (7)$$

where K is the number of micro-clusters in a local micro-clustering, and $|CF|$ is the size of a micro-cluster.

Computational Complexity. This section answers the question how much time is needed to compute the global clustering. Let $\mathcal{T}_{mic}$ and $\mathcal{T}_{mac}$ denote the time needed to produce a micro-clustering and a macro-clustering respectively. The time needed to generate the clustering by an underlying two-phased stream clustering $\mathcal{T}_{centralized}$ is given by

$$\mathcal{T}_{centralized} = N\mathcal{T}_{mic} + \mathcal{T}_{mac} \qquad (8)$$

The time needed to generate the global clustering is computed from the time at which all the local micro-clustering algorithms start until the time at which the global macro-clustering is generated. Therefore, the time needed to compute the global clustering $\mathcal{T}_{distributed}$ is given by

$$\mathcal{T}_{distributed} = \mathcal{T}_{mic} + \mathcal{T}_{transmit}^{allLC} + \mathcal{T}_{mac} \tag{9}$$

Let $\mathcal{T}_{transmit}^{LC_i}$ denote the time needed for transmitting the local clustering from the remote site i to the coordinator. As our framework is multi-threaded, the total time needed to transmit all the local clusterings $\mathcal{T}_{transmit}^{allLC}$ is less than or equal to the sum of time needed to transmit each local clustering $\mathcal{T}_{transmit}^{LC_i}$, that means $\mathcal{T}_{transmit}^{all\,LC} \leq \sum_{i=1}^{N} \mathcal{T}_{transmit}^{LC_i}$. From the equation 9, we can reduce the time to produce the global clustering at the coordinator by the following ways: (1) speed up the local clustering algorithm (reduce $\mathcal{T}_{clust}^{LC}$); (2) reduce the time to send the local clustering $\mathcal{T}_{send}^{LC}$; (3) speed up the cluster ensemble algorithm at the coordinator site. The first task depends on the selection of stream clustering algorithm at remote sites. For the second task we reduce the size of local clusters transmitted. The third task depends on the selection of the ensemble cluster approach. The time to send a local clustering to the coordinator site depends on the network bandwidth, and the size of local clustering. As shown in the experiment part, the local micro-cluster is small. The speedup of our framework is given by

$$speedup = \frac{\mathcal{T}_{centralized}}{\mathcal{T}_{distributed}} \tag{10}$$

where $\mathcal{T}^{centralized}$ is the time needed to generate the clustering by the underlying two-phased stream clustering algorithm and $\mathcal{T}_{distributed}$ is the time needed to generate the global clustering by our framework.

From the equations 8, 9, and 10, we have

$$speedup = \frac{N\mathcal{T}_{mic} + \mathcal{T}_{mac}}{\mathcal{T}_{mic} + \mathcal{T}_{transmit}^{allLC} + \mathcal{T}_{mac}} \tag{11}$$

Based on the experiment in Subsection 4.3, the average values of $\mathcal{T}_{mic}$, $\mathcal{T}_{mac}$, and $\mathcal{T}_{transmit}^{LC_i}$ were computed. From these parameters, the function speedup was approximately given by $speedup = 0.98N$. In conclusion, the speed of our framework scales up with the increasing number of remote sites.

4   Empirical Results

All the experiments were performed on a Intel Pentium (R) $2.00GHz$ computer with $1.00GB$ memory, running Windows XP professional. We implemented an instance of the proposed framework, which we call DisClustream (Distributed Clustream). DisClustream was developed on the basis of the underlying stream clustering algorithm Clustream [5]. The streaming data sets (Sensor Stream, Powersupply Stream, and Kddcup99) that were used in our experiments were from the Stream Data Mining Repository [24]. Depending on the purpose of each group of experiments, we selected some data sets of the above data sets.

We sought to answer two empirical questions about our framework: (1) How accurate is our clustering framework for distributed data streams in comparison with the central clustering approach on the same data set? (2) How scalable is our framework? All the empirical evaluations were done in an update epoch.

4.1   Evaluation on Global Clustering

This group of experiments evaluated aspects of the global clustering in terms of clustering quality compared with the centralized clustering algorithm, time needed to create the global clustering. The goal of these experiments is to compare the clusterings produced by centralized and distributed stream clustering algorithms. We expect that the global clustering produced by distributed stream clustering will be almost the same as the clustering produced by centralized stream clustering algorithm. For ease of comparison, we chose a data set in which the number of attributes and the number of classes are small sufficient to visualize clusterings. The appropriate data set is for this task is Powersupply used to predict which hour the current supply belongs to. This data set includes 29928 records, each record consists of 2 attributes. These records can fall into one of 24 classes. To obtain clusterings, we ran the centralized version of ClusStream, and the distributed version DisClustream. For the centralized CluStream, the entire data set can be seen as the incoming data stream. For the distributed DisCluStream, we divided the data set into two data sets of the same size. There were two remote sites, each remote site produced a local micro clustering from one of two above data sets.

Figure 2(a) illustrates two clusterings produced by centralized (marked by square) and distributed (marked by circle) stream clustering algorithms.

As we expect, the clustering produced by distributed clustering algorithm DisClustream was almost the same as the clustering produced by centralized one CluStream (Figure 2(a)). In conclusion, the global clustering produced by our distributed clustering framework is almost similar to the clustering produced by centralized one. The quality of an underlying local stream clustering affects the overall quality of the global clustering.

We should distinguish between the time needed to produce the global clustering and the time needed to run the macro-clustering algorithm. The time needed to generate the global clustering in the distributed framework is the duration from the time at which all the local micro-clustering algorithms start until the time at which the global macro-clustering is generated. The time needed to generate micro clusters at the remote site is much greater than the time needed to run macro clustering algorithm in order to generate the global clustering at the coordinator. This is a direct consequence of the two-phased stream clustering approach in which the online phase consumes more time than the off-line one [1]. Our the experiment with KDDCup99 data set shows that time to produce micro clusters at the remote site (14021 instances, and number of micro-clusters 100, time needed is 25701 ms) is much greater than the time needed to produce

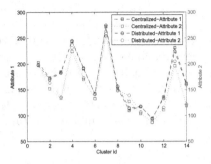 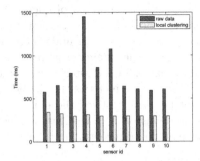

(a) A comparison of two clusterings that were created by distributed stream clustering algorithm and the centralized one on Power Supply data set

(b) Time to transmit block of streaming data vs. time to transmit local-micro clustering

Fig. 2. Comparison of DisClustream and Clustream

macro-clustering (391 ms) at the coordinator. Therefore, in order to speed up the algorithm, we should select, or develop the fast and high-quality micro-clustering approach.

4.2   Evaluation on Communication Efficiency

The goal of this subsection is to evaluate the time needed to transmit block of streaming data with the time needed to transmit the local micro-clustering built from the corresponding the block of streaming data. As shown in Section 3.3, the communication overhead needed to transmit local micro clusterings would be negligible compared to the raw data transmission involved. In other word, the time needed to transmit local micro-clustering would be much smaller than the time needed to transmit the raw streaming data.

We execute this groups of experiments with DisClustream on the Intel data set. We divided this data set into many data sets based on the sensor id. Without loss of generality, the sensor data set from sensor 1 to 10 was used as data streams at the remote sites. The number of micro-clusters was set to default number (100 kernels, in Clustream). For simplicity, we set up 10 experiments with 10 sensor data sets from 1 to 10 separately, that means each experiment include one coordinator and one remote site. Figure 2(b) shows that, the time needed to transmit raw streaming data depends on the size of streaming data block. The time needed to transmit micro-clusters is almost invariant as the number of micro-clusters were set to the same number 100 in all experiments. Our experiments shows that the micro-clusters produced by remote sites are small sufficient for meeting the requirement of communication efficiency.

The goal of this experiment is to determine the size of local clustering. The DisCluStream was used in this experiment. We used KDD Cup99 for this experiment. The framework consists of one remote site and one coordinator. At the remote site, the number of micro-clusters was fixed to 100. The coordinator received the local clustering and wrote it to file. We changed the number of instances in the window (an update epoch) in the range 1000, 2000, 3000, 4000, 5000. The size of local clustering file was invariant and equals to 13.2KB. In fact, the actual size of local clustering in memory may be smaller than 13.2 KB as the local clustering file includes the size of file format. The size of local clustering is invariant because we fixed the number of micro-clusters to 100. Therefore, the size of local clustering in DisClustream is independent of the number of instances that a remote site receives in an update epoch.

4.3   Effect of Block Width and Number of Micro-clusters

The success of a distributed framework for clustering streaming data depends on the scalability of the local stream clustering algorithm and the scalability in term of number of remote sites. The scalability of the micro-clustering algorithm used at the remote sites is mandatory because it must process the large volume of incoming data. The scalability of the micro-clustering algorithm Clustream in the number of data dimensions, and the number of clusters was thoroughly evaluated [1] in term of the time needed to produce clustering. In contrast, we evaluated the scalability of DisClustream in terms of communication time by increasing the number of micro-clusters, and the number of instances in the update block.

Effect of Block Width. This group of experiments studied the scalability in the increasing number of instance in the block. The underlying stream clustering algorithm was used to evaluate the scalability of our proposed framework is Clustream. As such, the micro-clustering algorithm at the remote sites the micro-clustering method of Clustream while the macro-clustering algorithm at the coordinator is K-means. Figure 3(a) shows how our distributed stream clustering framework scale with the number of instances in an update epoch. As we experimented with the sensor data set in which each record consisted of two attributes, we can observe the time to transmit the local clustering to the coordinator in Figure 3(a) . We also tested the scalability in window size with KDD Cup data set. The result of this experiment demonstrated that, the time needed to create local clustering scales with the number of instances in an update epoch. However, the time needed to transmit the local clustering is hardly varied. For example, with KDD Cup data set, the time needed to create the local clustering (18,11 seconds) was much larger than the time needed to transmit the local clustering (46 miliseconds) with the number of instances in an update epoch 10000. As the number of instances in an update epoch increases 10 times (100000), the time needed to create local clustering was almost 3 (minutes) while the time needed to transmit local clustering was only 47 (milliseconds).

Effect of Number of Micro-clusters. To evaluate the scalability of our algorithm in the increasing number of micro clusters. Evaluation on the scalability

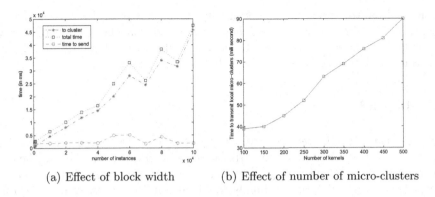

(a) Effect of block width   (b) Effect of number of micro-clusters

Fig. 3. Effect of block width and number of micro-clusters

of our algorithm was done on two data sets Intel sensor data set and KDD-Cup 99 data set. We set up experiment as follows. The number of instances was fixed to 10000 instances. There were nine remote sites connected to the coordinator sites. The number of micro clusters was selected from the range 100, 150, 250, 350, 400, 450, 500 as shown in Figure 3(b). An increase in the number of micro-clusters resulted in the increasing time to transmit micro-clusters In our experiments, time required to produce micro-clusters is much larger than the time required to transmit local micro-clusters.

Our experiments on the scalability with DisClustream showed that the time needed to create local clustering in an update epoch scales with the increasing number of instances in the window, and the increasing number of micro-clusters while the time needed to transmit a local clustering is almost invariant. The data stream clustering algorithms that are fast sufficient to deploy in resource-limited applications such as sensor networks are still open questions.

5   Further Discussion

Micro-clustering approach is used in many tasks of data mining. For instance, micro-clustering method is used to summarize cluster information for the data stream classification [16]. Our theoretical and empirical results show that we can develop a general framework for mining distributed streaming data by using micro-clusters. The remote sites create and maintain micro-clusters or the variants of micro-clusters. The coordinator can generate and maintain the global pattern from the micro-clusters that are received from the remote sites by using the micro-clusters based data mining such as classification, clustering, and frequent pattern mining. The reasonable foundation behind this local micro-clustering approach to data stream mining is that micro-clusters maintain statistics at a sufficiently high level of granularity (temporal and spatial)[1]. As such, micro-clustering algorithms can be seen as a data summarization method. For the problem of clustering distributed streaming data, there are some open questions such as: How to create and maintain the global clustering by cluster ensembles method on the micro-clusters that are received from the remote sites? How to build and maintain the global clustering in sliding window in a distributed environment by using the local micro-clustering algorithm?

6 Related Work

The first two-phased stream clustering algorithm Clustream introduced by Aggarwal et al [1] has two separate parts: online and off-line components. The role of the online component is to process and to extract summary information from data stream while the off-line builds the meaningful clustering structure from the extracted summary information.

Clustream can efficiently maintain a very large number of micro-clusters, which are then used to build macro-clusters. Further, clustering structure can be observed in arbitrary time horizons by using pyramidal time model. Hence, Clustream is capable of monitoring the evolution of clustering structure. However, one of the main drawbacks of Clustream is that it requires the number of clusters in advance. To overcome this drawback, Cao et al. [6] have proposed a density-based clustering data streams. This advantage of this approach is that it can handle outliers in data streams, and adapt to the changes of clusters. However, the major limitation of DenStream is that it frequently runs the off-line component in order to detect the changes of clusters. As such, the off-line component consumes the cost of computing clusters much. In the experimental evaluation, we use two well-known algorithms Clustream [1], and DenStream [6] as the local stream clustering algorithms at the remote sites.

Cormode et al. [7] presented a suite of k-center-based algorithms for clustering distributed data streams. The communication topology consists of N remote sites, and one coordinator. Any two remote sites can directly communicate with each other. Their approach can provide a global clustering that is as good as the centralized clusterings. However, this approach provide a approximate clustering by using the underlying k-centers stream clustering algorithm. Zhou et al. [23] considered the problem of clustering distributed data stream by using EM-based approach. The advantage of this approach is that it can deal with the noisy and incomplete data streams. To deal with the evolving data streams, they used the reactive approach to rebuild the local clustering and global clustering when it no longer suits the data. However, it is not scalable in the large number of remote sites such as in a sensor network of thousands of nodes. Although this drawback can be overcome by model-merging technique, the global clustering quality considerably reduces. Furthermore, the memory consumption at the remote site increases with the increasing number of data dimensionality and the increasing number of models. In contrast to the work introduced by Zhou et al. [23], Zhang et al.[21] introduced a suite of k-medians-based algorithms for clustering distributed data streams, which work on the more general topologies: topology-oblivious algorithm, height-aware algorithm, and path-aware algorithm. To deal with the evolving data streams, they selected periodic approach in which the global clustering at the root site is continuously updated every period called update epoch. Similar to [7], the global clustering is approximately computed on the summaries that are received from the internal and leaf nodes.

7 Conclusions

We have proposed a distributed framework for clustering streaming data by extending the two-phased stream clustering approach that is widely used to cluster a single data stream. While the remote sites create and maintain micro-clusters by using the micro-clustering method, the coordinator creates and maintains the global clustering by using the macro-clustering algorithm. We have theoretically analyzed our framework in the following aspects. The global clustering created in an update epoch is similar to the clustering created by the centralized stream clustering algorithm. We also estimated the communication cost as well as proved that the speed of our framework scales with the number of remote sites. Our empirical results demonstrated that the global clustering generated by our distributed framework was almost the same as the clustering generated by the underlying centralized algorithm on the same data set with low communication cost. In conclusion, by using the local micro-clustering approach, our framework achieves scalability, and communication-efficiency while assuring the global clustering quality. Our work along with other work can be seen as one of the first steps towards a novel distributed framework for mining streaming data by using micro-clustering method as an efficient and exact method of data summarization.

Acknowledgments. We thank anonymous reviewers for many helpful comments. This work was partly funded by the Vietnam Ministry of Education and Training, TU Ilmenau, and the DAAD.

References

1. Aggarwal, C., Han, J., Wang, J., Yu, P.: A framework for clustering evolving data streams. In: Proceedings of the 29th International Conference on Very Large Data Bases, vol. 29, pp. 81–92. VLDB Endowment (2003)
2. Bandyopadhyay, S., Gianella, C., Maulik, U., Kargupta, H., Liu, K., Datta, S.: Clustering Distributed Data Streams in Peer-to-Peer Environments (2004)
3. Barbará, D.: Requirements for clustering data streams. ACM SIGKDD Explorations Newsletter 3(2), 23–27 (2002)
4. Beringer, J., Hullermeier, E.: Online clustering of parallel data streams. Data & Knowledge Engineering 58(2), 180–204 (2006)
5. Bifet, A., Holmes, G., Kirkby, R., Pfahringer, B.: Moa: Massive online analysis. The Journal of Machine Learning Research 11, 1601–1604 (2010)
6. Cao, F., Ester, M., Qian, W., Zhou, A.: Density-based clustering over an evolving data stream with noise. In: Proceedings of the 2006 SIAM International Conference on Data Mining, pp. 328–339 (2006)

7. Cormode, G., Muthukrishnan, S., Zhuang, W.: Conquering the divide: Continuous clustering of distributed data streams. In: IEEE 23rd International Conference on Data Engineering, ICDE 2007, pp. 1036–1045. IEEE (2007)

8. Da Silva, A., Chiky, R., Hebrail, G.: Clusmaster: A clustering approach for sampling data streams in sensor networks. In: 2010 IEEE 10th International Conference on Data Mining (ICDM), pp. 98–107. IEEE (2010)

9. Dai, B., Huang, J., Yeh, M., Chen, M.: Clustering on demand for multiple data streams. In: Fourth IEEE International Conference on Data Mining, ICDM 2004, pp. 367–370. IEEE (2004)

10. Datta, S., Bhaduri, K., Giannella, C., Wolff, R., Kargupta, H.: Distributed data mining in peer-to-peer networks. In: IEEE Internet Computing, pp. 18–26 (2006)

11. Guha, S., Meyerson, A., Mishra, N., Motwani, R., O'Callaghan, L.: Clustering data streams: Theory and practice. IEEE Transactions on Knowledge and Data Engineering 15(3), 515–528 (2003)

12. Jain, A., Murty, M., Flynn, P.: Data clustering: a review. ACM computing surveys (CSUR) 31(3), 264–323 (1999)

13. Karnstedt, K., Sattler, D., Quasebarth, J.: Incremental mining for facility management. In: LWA 2007 Lernen–Wissen–Adaption, p. 183 (2007)

14. Klan, D., Karnstedt, M., Hose, K., Ribe-Baumann, L., Sattler, K.: Stream engines meet wireless sensor networks: Cost-based planning and processing of complex queries in anduin, distributed and parallel databases. Distributed and Parallel Databases 29(1), 151–183 (2011)

15. Kranen, P., Assent, I., Baldauf, C., Seidl, T.: Self-adaptive anytime stream clustering. In: Ninth IEEE International Conference on Data Mining, ICDM 2009, pp. 249–258. IEEE (2009)

16. Masud, M., Gao, J., Khan, L., Han, J., Thuraisingham, B.: A practical approach to classify evolving data streams: Training with limited amount of labeled data. In: Eighth IEEE International Conference on Data Mining, ICDM 2008, pp. 929–934. IEEE (2008)

17. Naor, M., Stockmeyer, L.: What can be computed locally? pp. 184–193 (1993)

18. Sun, J., Papadimitriou, S., Faloutsos, C.: Distributed pattern discovery in multiple streams. In: Advances in Knowledge Discovery and Data Mining, pp. 713–718 (2006)

19. Yin, J., Gaber, M.: Clustering distributed time series in sensor networks. In: Eighth IEEE International Conference on Data Mining, ICDM 2008, pp. 678–687. IEEE (2008)

20. Zaki, M., Pan, Y.: Introduction: recent developments in parallel and distributed data mining. Distributed and Parallel Databases 11(2), 123–127 (2002)

21. Zhang, Q., Liu, J., Wang, W.: Approximate clustering on distributed data streams. In: ICDE, pp. 1131–1139 (2008)

22. Zhang, T., Ramakrishnan, R., Livny, M.: BIRCH: an efficient data clustering method for very large databases. ACM SIGMOD Record 25(2), 103–114 (1996)

23. Zhou, A., Cao, F., Yan, Y., Sha, C., He, X.: Distributed data stream clustering: A fast em-based approach. In: IEEE 23rd International Conference on Data Engineering, ICDE 2007, pp. 736–745. IEEE (2007)

24. Zhu, X.: Stream data mining repository (2010), http://www.cse.fau.edu/~xqzhu/stream.html

Towards a Flexible Framework to Support a Generalized Extension of XACML for Spatio-temporal RBAC Model with Reasoning Ability

Tuan Ngoc Nguyen, Kim Tuyen Le Thi, Anh Tuan Dang, Ha Duc Son Van,
and Tran Khanh Dang

HCMC University of Technology, VNUHCM, Ho Chi Minh City, Vietnam,
{tuyenltk,khanh}@cse.hcmut.edu.vn

Abstract. XACML is an international standard used for access control
in distributed systems. However, XACML and its existing extensions
are not sufficient to fulfil sophisticated security requirements (e.g. access
control based on user's roles, context-aware authorizations, and the ability of reasoning). Remarkably, X-STROWL, a generalized extension of
XACML, is a comprehensive model that overcomes these shortcomings.
Among a large amount of open sources implementing XACML, HERAS-
AF is chosen as the most suitable framework to be extended to implement
X-STROWL model. This paper mainly focuses on the architecture design of proposed framework and the comparison with other frameworks.
In addition, a case study will be presented to clarify the work-flow of this
framework. This is the crucial contribution of our research to provide a
holistic, extensible and intelligent authorization decision engine.

Keywords: Access Control Model, GIS Database, HERAS-AF, Security Engineering, Spatio-temporal Data, XACML, X-STROWL.

1 Introduction

The huge success of the Internet as the platform for accessing data with the
rapid growth of sensor technology and mobile devices has made the Location
Based Service (LBS) and Geographic Information System (GIS) Applications
become more and more widespread. As a result, there is an emerging demand of
controlling data access, especially for spatial and temporal data. Moreover, it is
necessary for organizations to define, maintain and integrate the security policies
easily. Therefore, it could be more convenient to archive these two demands by
a common standard.

Extensible Access Control Markup Language (XACML) is widely accepted
because of its outstanding advantages (e.g. standard, generic, distributed and
powerful) [5]. It is a standard to describe both policy language and access control decision request/response language. The policy language is used to describe
general access control requirements with standard extensibility points including functions, data types, combining logic, etc. Meanwhile; the request/response

B. Murgante et al. (Eds.): ICCSA 2013, Part V, LNCS 7975, pp. 437–451, 2013.
© Springer-Verlag Berlin Heidelberg 2013

language allows users to form a query to ask if a given action should be allowed or not. With the use of XACML, access control policies can be expressed flexibly due to the extensibility of data types, functions and the ability of combining multiple decisions into a single decision. Besides, OASIS also defines an XACML Profile to meet the requirements of Role Based Access Control (RBAC) [6]. However, this profile suffers significant drawbacks such as ignoring the process of user-role assignment and not complying with the NIST standard of RBAC. Another limitation is that the spatial access restrictions are not addressed.

Although there are many models addressing the above disadvantages [25][18] [28], X-STROWL model is the most prominent model. It is an extension of XACML and supports the RBAC model according to the NIST standard as well as spatial-temporal conditions. Besides, a new kind of policy, the context policy, is also indicated in this extension. This policy defines a set of conditions generalized into a context profile and shared between access control policies. In addition, semantic reasoning on hierarchical roles is also taken into account in the X-STROWL model [29]. This paper is the result of building a flexible framework to implement X-STROWL model. Holistic Enterprise-Ready Application Security - Architecture Framework (HERAS-AF) is a well-established open source that is worth being chosen for the implementation. The extended framework supports XACML model, meets the requirements for RBAC as well as the ability of reasoning for hierarchical roles. The implementation of this framework which applies the X-STROWL access control model in the real world is the major contribution of this paper.

Besides, the solutions to optimize the query processing time and handle the partial-deny queries are also considered [19][20] (the partial-deny queries are queries that have the "Deny" evaluation response, but user can access the partial data according to rules of system). However, these issues will not be focused in this paper.

The remainder of the paper is organized as follows. In Section 2, we present a generalized extension of XACML. Section 3 describes a holistic framework supporting XACML. Then our framework is discussed in detail in Section 4. In Section 5, we compare the proposed framework with others to conclude its outstanding features. Finally, the last section represents the conclusion and further work on this topic.

2   A Generalized Extension of XACML for Spatio-temporal RBAC Model

As mentioned above, our framework is based on X-STROWL model, so we will examine its characteristics before going further. This model is an extension of XACML to support RBAC in conformity with the NIST standard. Moreover, it supports context-driven access restrictions, including spatial and temporal aspects. The reasoning ability is also an outstanding advantage of this model. Next subsections will go into detail of these key features.

2.1   Role Based Access Control

To support RBAC, XACML uses four specific types of policy: `Role<PolicySet>`, `Permission<PolicySet>`, `RoleAssignment<Policy>`, and `HasPrivilegesOf-Role<Policy>`. These types of policy are represented in more detail in [6]. With this approach, the role-permission relationship is statically defined by `PolicySetIdReference` in `Role<PolicySet>` without Static/Dynamic Separate of Duty (S/DSOD) constraints. That does not reach the third level of the NIST Standard for RBAC model [27]. This limitation is overcome in X-STROWL with two new policy sets: `PermissionAssignment<PolicySet>` and `SOD<PolicySet>` [29]. Namely, `PermissionAssignment<PolicySet>` indicates which permissions are assigned to which roles under which contextual conditions. Meanwhile, `SOD` `<PolicySet>` is used to define the S/DSOD constraints of the user-role and role-permission relationships. Table 1 represents the necessary functions for checking S/DSOD constraints. These functions, which get historical information about assigning and enabling roles for users, are inspired by [24].

Table 1. Extended functions for checking S/DSOD

Function	Result
getAssignedRoleSet(user)	All roles assigned to the given user
getAssignedUserCounter(role)	Number of users are assigned to the given role already
getAssignedRoleCounter(user)	Number of roles are assigned to the given user already
getEnabledRoleSet(user)	All roles enabled to the given user
getEnabledUserCounter(role)	Number of users are enabled to the given role already
getEnabledRoleCounter(user)	Number of roles are enabled to the given user already

In this table, the *assign* concept is to define user-role relationships statically. Meanwhile, the *enable* concept is to assign user to role in runtime. The functions, which get historical information about assigning and enabling permissions for roles are defined in a similar way.

2.2   Contextual Condition

Context-aware access control models are more and more necessary and important in advanced systems. In these systems, access restrictions not only depend on the subject's roles and permissions which are pre-assigned in traditional access control models like DAC or MAC but also depend on the context of the requests. In which, context is a combination of conditions based on attributes, state of subjects, resources, environment, historical access (e.g. role, position of the subject, access time, resource content, etc.). In X-STROWL, a context is defined in a

`Context<PolicySet>` policy which describes a set of conditions for a certain circumstance. Then these contexts can be used for defining constraints explicitly on user-role assignments and/or role-permission assignments. `Context<PolicySet>` policies are evaluated by the *ContextEA* to determine if the current context of the request meets the conditions in the `Context<PolicySet>` policies. Spatial and temporal conditions, two important aspects in X-STROWL, are supported by declaring new data types (e.g. Point, LineString, Polygon, DayTime, Period, Interval, etc.) and functions (e.g. Disjoint, Touches, Crosses, During, After, Before, etc.).

These data types and functions can be used in the `<Condition>` tag in XACML policies to define complex policies, including spatial and temporal conditions for fine-grained data access control systems. With this approach, administrators can define access restrictions based on state of subjects, resources, actions as well as information about environment, historical access and spatial-temporal relationships. Such restrictions are called context-driven access restrictions.

2.3   Reasoning Ability

In order to reduce the cost of administration and the storage space of policies, access control systems should have the capability of reasoning on the role hierarchy to inherit positive permissions from junior roles or negative permissions from senior roles [29]. The basic idea is to define role hierarchies by using Web Ontology Language (OWL) [9] and then define new XACML functions, which determine the relationship between roles based on these hierarchies.

For example, considering a hospital, there are following roles: Doctor, Nurse, Patient, Specialist Nurse, and Surgeon. Between two roles, it is able to have two kinds of relationships:

Is-a relationship: based on the definition of roles. For example, a Surgeon *is a* Doctor who does operations in a hospital.

Senior/Junior relationship: prescribed by organizations. For instance, the permissions of a Doctor may contain the permissions of a Nurse. In this case, Doctor can be considered as a *senior role* of Nurse. In other words, Nurse is a *junior role* of Doctor.

In [21,22], the relationships between roles are presented by one hierarchical tree. However in this paper, we use two trees for the two kinds of relationship. Fig. 1 is an example for these trees. With this approach, it is easier to maintain the hierarchies, and the policies are more expressive. The detailed analysis of these points is nevertheless beyond the scope of this paper. We can easily define the first hierarchy by the OWL built-in property named *subClassOf* (Fig. 1.(a)). For the second one, we have to use user-defined properties called *juniorRoleOf* and *seniorRoleOf* (Fig. 1.(b)).

For example, role Doctor can be defined as follows:

```
<rbac:Role rdf:ID="Doctor">
  <rdfs:label>Doctor</rdfs:label>
  <rdfs:subClassOf rdf:resource="Personnel"/>
  <rbac:seniorRoleOf>
    <rbac:Role rdf:resource="Nurse"></rbac:Role>
  </rbac:seniorRoleOf>
  <rbac:juniorRoleOf>
    <rbac:Role rdf:resource="Surgeon"></rbac:Role>
  </rbac:juniorRoleOf>
</rbac:Role>
```

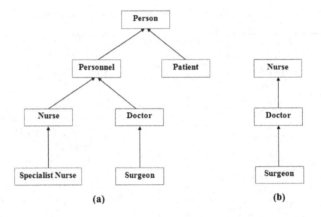

(a)   (b)

Fig. 1. Role Hierarchies

We need to define XACML extended functions which call reasoning services provided by an ontology reasoner. The PDP will use these functions to evaluate hierarchical relationships. Necessary functions for ascertaining the relationship between roles are listed in Table 2.

3 A Holistic Framework Supports XACML

There are many sources implement XACML; however, in the scope of this research, only open sources are considered. Table 3[1] provides some open sources implementing XACML. To compare these open sources, we consider the following criteria: (1) the components that the implementations support (PDP component has to be supported to evaluate the access request); (2) the version

---

[1] ? is unclear, _ is does not have and the Manual column represents the level that manual supports (Pdf – Web – txt is descending order of the completeness).

Table 2. XACML extended functions for supporting reasoning ability

Function	Return TRUE if
role-is-a(role1, role2)	*role1* is a sub class of *role2* in the hierarchy 1.(a)
role-is-senior(role1, role2)	*role1* is one of the senior roles of *role2* (hierarchy 1.(b))
role-is-senior-or-equal(role1, role2)	*role1* is *role2* or *role1* is one of the senior roles of *role2* (hierarchy 1.(b))
role-is-junior(role1, role2)	*role1* is one of the junior roles of *role2* (hierarchy 1.(b))
role-is-junior-or-equal(role1, role2)	*role1* is *role2* or *role1* is one of the junior roles of *role2* (hierarchy 1.(b))

Table 3. A general view about the open sources implement XACML

Released year	Name	XACML	RBAC	Manual	Components	Ref.
2004	Sun's XACML	1.1	-	Web	PDP, PEP	[11]
2010	OpenAz	2.0	-	Pdf	PEP	[8]
2010	XEngine	-	-	txt	PDP	[16]
2008	PAM_XACML	1.1	-	Web	PDP	[10]
2008	XACML Studio	2.0	-	Web	manage policies	[13]
2010	HERAS-AF	2.0	-	Pdf	PDP, PEP, PIP, PolicyRepository	[2]
2008	XACMLight	2.0	-	Web	PDP, PAP	[14]
2009	Enterprise Java XACML	2.0	-	?	PDP	[1]
2005	XACML.NET	1.0	-	?	PDP	[15]
2005	Margrave	2.0	?	Web	?	[4]
2011	WSO2	3.0	yes	Web	PIP	[12]

of XACML that the implementations support (2.0 and above); (3) whether the implementations support RBAC or not; and (4) whether it is fully documented or not.

Most of these open sources do not support RBAC, except WSO2 and Margrave. However, WSO2 does not support PDP and Margrave does not have clear information. Notable implementations are Sun's XACML, HERAS-AF, XACMLight and Enterprise Java XACML. Nevertheless, Sun's XACML only supports XACML version 1.1; the implementation of Enterprise Java XACML has not been updated for a long time, and it does not have a detailed manual. Compared with XACMLight, HERAS-AF supports more components and is more fully documented. For these reasons, HERAS-AF is chosen to implement the X-STROWL model.

Holistic Enterprise-Ready Application Security - Architecture Framework (abbreviated as HERAS-AF) is a well-established open source project hosted and supported by the University of Applied Science Rapperswil in Switzerland. The HERAS-AF XACML Core has become a comprehensive XACML engine repre- senting the de-facto reference implementation of XACML 2.0 [2]. HERAS-AF is a framework that supports the whole access control process. Particularly, the access request is sent to *Policy Enforcement Point (PEP)* and then forwarded to *Context Handler* (including the information about subject, resource, action and environment). *Policy Decision Point (PDP)* is responsible for the evaluation based on this information and the defined policies in *Operative Policies*. In case of missing attributes, *Policy Information Point (PIP)* will be called to provide these attributes for *PDP*. If the decision is "Permit", *PEP* contacts the *Resource* to get the result; otherwise, the decision may be "Deny", "Indeterminate" or "NotApplicable", *PEP* will perform some operations according to the pre-setup of the administrator. Fig. 2 represents the general architecture of the first ver- sion of HERAS-AF. This architecture is similar to the data flow of the XACML evaluation process.

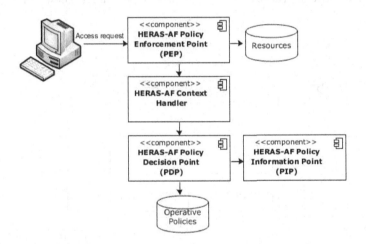

Fig. 2. General architecture of HERAS-AF framework

The current version of HERAS-AF (1.0.0-M2) has just focused on the Core component which makes the evaluation decision (e.g. the *Context Handler* and *PDP*). Besides the advantages inherited from the XACML language (standard, generic, distributed, powerful, and easy to extend), HERAS-AF also has other advantages. It supports most of the main functions of XACML 2.0 and is easy to extend by editing or adding packages and components. However, HERAS-AF framework does not support the following features:

- Spatial aspect (includes spatial data and spatial conditions),
- The reasoning ability,
- The access control based on the role of user,

– The *PEP*, *PIP* components in HERAS-AF are only implemented with basic functions and need to be improved.

These features are also the main points in X-STROWL model (as mentioned in section 2). Therefore, to use this framework to implement the model, HERAS-AF need to be extended to address these above disadvantages.

4   Towards a Flexible Framework

As mentioned above, to extend HERAS-AF to support X-STROWL model, new packages and components need to be added to the framework, more details will be represented in the following subsections. This framework called HERAS-AF+

4.1   Extended Architecture

HERAS-AF+ is constructed from five main components as shown in Fig. 3.

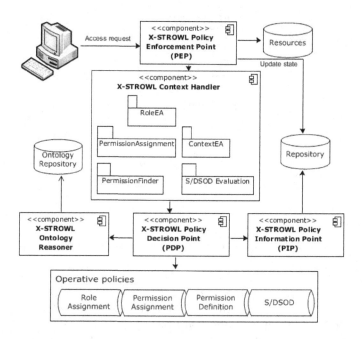

Fig. 3. Extended HERAS-AF for X-STROWL

Context Handler consists of five packages corresponding to five types of context handler. There is a general context handler to handle these types, but it is not represented in the architecture. Clearly, *RoleEA* is used to find the selected roles according to the request's subject and the current context.

PermissionFinder is responsible for finding the permissions according to the specific action and resource. *PermissionAssignment* defines which permissions is assigned to which roles in the specific context. These contextual conditions are evaluated by the *ContextEA*. Finally, *S/DSOD Evaluation* is used to evaluate the separate of duty constraints.

Operative Policies contain the necessary policies for the evaluation. Each type of context handler above will call *PDP* with the corresponding policies. Namely, *RoleAssignment* policies define which user is assigned to which role; *PermissionAssignment* policies determine which permission are assigned to which role; *PermissionDefinition* policies are the definition of which action can perform on which resource; and *S/DSOD* policies define the separate of duty constraints.

Ontology Reasoner is used for the reasoning ability. The *OntologyRepository* provides the basic knowledge for *Ontology Reasoner* to evaluate the hierarchical relationships between roles, permissions, or contexts. In this paper, we just consider the role hierarchy.

Repository provides the necessary attributes for the *PIP* in case of missing attributes. Besides, after receiving the result from *Context Handler*, *PEP* will update the state of attributes in the *Repository* to ensure that the current context is updated for the next access requests.

In general, the access request is sent to the *PEP* and then forwarded to the *Context Handler*. All the information about subject's role, permissions of role, the current context and the separate of duty constraints will be evaluated. Before using this information, *PDP* calls the *Ontology Reasoner* to evaluate the hierarchical relationships. In case of missing attributes, *PIP* will be called to provide attributes for *PDP*. The result will be returned to *PEP* after *PDP* evaluates the access request based on all of the information and the existent operative policies. Finally, the *PEP* updates the states of attributes in the *Repository* and gets the *Resource* (in case of returning "Permit").

With this extension, we need to add new packages (RoleEA, ContextEA, PermissionAssignment, PermissionFinder, and S/DSOD Evaluation), new components (OntologyReasoner, OntologyRepository), and some extended components (PIP, Repository, and Operative policies) to the HERAS-AF framework. In addition, we also have to define new data types and functions. Next section will represent this issue in detail.

4.2   New Data Types and Functions

Because the X-STROWL model supports both spatial and temporal aspects, we need to add spatial data types and functions to HERAS-AF framework. The Java Topology Suite (JTS) framework is used to solve this issue. JTS framework supports all of the necessary spatial types and functions [3]. Hence, the main work is to convert between JTS type and spatial type of X-STROWL. Table 4 and Table 5 represent the basic spatial data types and functions proposed by GeoXACML [7] that are used in the implementation.

Table 4. Basic spatial data types for the X-STROWL model

Geometry Type	Description
Point	0-dimensional geometric object and represents a single location
LineString	Curve with linear interpolation between Points
Polygon	planar Surface defined by 1 exterior boundary and 0 or more interior boundaries
MultiPoint	0-dimensional GeometryCollection
MultiLineString	MultiCurve whose elements are LineStrings
MultiPolygon	MultiSurface whose elements are Polygons

As mentioned above, most of the current GIS applications just provide the user functions and do not support the access control aspect. Next two subsections will represent how to apply our access control framework into real world.

Table 5. Basic spatial functions for the X-STROWL model

Spatial Functions	Return TRUE if
Contains (g1, g2)	the geometry g2 lies in the closure (boundary union interior) of geometry g1
Crosses(g1, g2)	the geometries g1 and g2 share some but neither is contained in the other, and the dimension of the intersection is less than that of both of the geometries
Disjoint(g1, g2)	the geometries g1 and g2 have no point in common
Equals(g1, g2)	the geometries g1 and g2 are equal (geometrically contain exactly the same points)
Intersects(g1, g2)	the geometries g1 and g2 have at least one point in common
Overlaps(g1, g2)	the geometries g1 and g2 share some but not all points in common, and the intersection has the same dimension as the geometries themselves
Touches(g1, g2)	the geometries g1 and g2 have at least one boundary point in common, but no interior points
Within(g1, g2)	the geometry g1 is spatially within geometry g2; that is if every point on g1 is also on g2

4.3   A Particular Application of Extended Framework

We introduce the architecture of a GIS application that integrates the extended framework mentioned in section 4.1, as shown in Fig. 4. This architecture is developed based on GIS N-Tier Architecture [23]. The middle tier is divided into many subsystems (layers). To guarantee security, the *Access Control Layer* is added to middle tier, the front of *Services and Business Logic Layer*. *Enforcement Service*, which can be built by our proposed framework, is the core of *Access Control Layer*. The communication between *Client Tier* and services then is intercepted by *Enforcement Service*, which is responsible for enforcing access

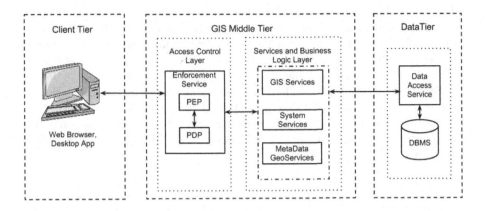

Fig. 4. Extended GIS Architecture

control. As a consequence, the work-flow of extended GIS architecture will be as follows. After receiving an access request from *Client Tier*, *PEP* contacts *PDP* to get access decision. If the decision is "Permit", *PEP* forwards request to *Services and Business Logic Layer*. Finally, the service response is sent back to *Client Tier* by *Enforcement Service*. Otherwise, *PEP* will deny the access request and return the result to client.

Integrating the proposed framework into this application makes it able to handle sophisticated security requirements related to spatial and temporal context. Therefore, in terms of security, the quality of service which this application provides will be enhanced significantly. Next part will represent a case study to provide a better understanding of this system.

4.4   Case Study

Case study is the application of our proposed framework into a hospital information system. This extended system provides support tools for business management and medical services. It is used by managers, personnel and patients. The system helps in monitoring personnel and managing patients, medical records more easily. For the sake of simplicity, we just focus on the following basic roles: Manager, Doctor, Surgeon, and Patient. In addition, data is classified into non-spatial data (such as personal information of patients and personnel, patients' medical records, department information, etc.), and spatial data (such as the positions of patients and personnel, the shapes and positions of departments, etc.) for spatial services.

Access Rules

The Manager has the highest privilege to access the database, including all information of patients, personnel, even their current positions in the hospital. However, he cannot add new medical records of patients or view private notes of doctors.

The Doctor has permission to view all medical information of patients who he is treating. Nonetheless, in emergency cases, Doctor can view medical records of a patient who is not normally his patient with a notification sent to the manager. In office hours, he can also add new medical records and private notes about his patient on doctor-patient confidentiality if he is in the hospital.

A Surgeon is also a Doctor. Therefore, he has all permissions which a Doctor has. Moreover, he is allowed to do operations in the hospital.

According to privacy rules, the Patient can view all stored information about him. For the sake of brevity, we will not discuss in detail the access rights of all other roles.

Work-flow

To clarify the work-flow of this framework, we consider the following example. In case of emergency of patient Peter, he needs to have surgery. Surgeon Daniel takes responsibility for doing this operation. Hence, he would like to view Peter's medical records. The access request of Daniel is transferred to HERAS-AF+ access control system. The request information includes the access user (*Daniel*), the action (*read*), the resource (*Peter's medical record*). HERAS-AF+ will process this request as follows:

Step 1: The request is separated into two parts which are processed in parallel. The first part is sent to *RoleEA* to determine the roles of user Daniel. The second part is sent to *PermissionFinder* to find the permissions which define the action of reading on patient's medical records. *RoleEA* gets all roles (*Manager, Doctor, Surgeon*, and *Patient*) in this system. For each role, *RoleEA* asks *PDP* if assigning the role to user Daniel is valid or not. If *PDP* returns "Deny", *RoleEA* will skip that role and continue checking the next role. If *PDP* returns "Permit", *RoleEA* will add that role into the list of valid roles of user Daniel. In this example, he has only one role *Surgeon*. In a similar way, *PermissionFinder* gets all permissions in this system and asks *PDP* to get valid permissions for action of reading patient's medical records. The returned result contains only one permission called *ReadingPatientRecord*. After that, *Context Handler* receives the result including role *Doctor* and permission *ReadingPatientRecord*.

Step 2: *Context Handler* then contacts *S/DSOD Evaluation* to evaluate whether assigning role *Surgeon* to user *Daniel* and permission *ReadingPatientRecord* to role *Surgeon* violate any SOD constraints or not. In this case study, there is no SOD constraint, so the result from *S/DSOD* is "no violation".

Step 3: *Context Handler* sends a request to *PermissionAssignment* to check whether assigning permission *ReadingPatientRecord* to role *Surgeon* is valid or not. *PermissionAssignment* passes this question to *PDP*. In this case, we just have the rule that *Doctor* has permission *ReadingPatientRecord* in emergency context. Therefore, *PDP* needs to ask *OntologyReasoner* if *Surgeon* is a *Doctor*. The result is "True". As a consequence, *Surgeon* can have permission *ReadingPatientRecord* in emergency context. *PDP* then contacts *PIP* to check current context.

Step 4: *PIP* asks *ContextEA* to check if the current context is emergency. In this example, the result is "True". The request of surgeon *Daniel* is permitted. After this step, *PEP* needs to do some obligations such as logging this request and sending an email to the manager for notification.

This case study is a simple example to clarify our framework work-flow. In this framework, the third level of NIST standard for RBAC is supported by determining user's roles through *RoleEA* and checking SOD constraints in the user-role and role-permission relationships through *S/DSOD*. HERAS-AF+ also supports context-driven access restrictions, which is evaluated by *ContextEA*. In which, spatial and temporal contexts are expressed through using spatial/temporal functions and types in policies which are evaluated by PDP. Finally, the reasoning ability of this framework is based on the inference of the *Ontology Reasoner* from hierarchical relationships defined by OWL. HERAS-AF+ will be evaluated by comparing with the relevant frameworks in next section.

5   Evaluation

The proposed framework, which is described in detail in previous sections, is extended from HERAS-AF to support a generalized access control model. In fact, the framework can deal with complex policies which require spatial, temporal information and the capability of reasoning.

In this section, we evaluate the features of the proposed framework compared with others. Considered criteria, which are given in Table 6, including supported data/function types, the ability of reasoning for hierarchical roles, XACML standard and NIST standard for RBAC model. The last criterion includes Flat RBAC (R0), Hierarchical RBAC (R1), Constrained and Symmetric RBAC (R2, R3).

Table 6. Supporting features of frameworks

Feature / Framework	Data/Function Type		Reasoning	XACML Standard [17,26]	NIST Standard RBAC [27]			
	Spatial	Temporal			R0	R1	R2	R3
Sun's XACML [11]		x		x				
XACMLight [14]		x		x				
Enterprise Java XACML [1]		x		x				
HERAS-AF [2]		x		x				
HERAS-AF+	x	x	x	x	x	x	x	

Each framework, which is surveyed, has its own pros and cons based on the model that each one implements. Indeed, Sun's XACML, HERAS-AF, XACMLight, and Enterprise Java XACML are open source implementations of the OASIS XACML standard. Hence, they can only support XACML standard and provide necessary data/function types, e.g. primitive types, temporal types.

Meanwhile, the proposed framework meets most of the given criteria, except for Symmetric RBAC model of NIST Standard. However, by adding components to manage role-permission assignments, our framework can easily fulfil the requirement of permission-role review as described in the Symmetric RBAC model. Furthermore, supporting spatial and temporal context and ability of reasoning for hierarchical roles are striking features that make the proposed framework able to meet more sophisticated security requirements in comparison with others.

6   Conclusion

In this paper, we proposed a framework based on X-STROWL model, which has three main advantages: conformity to the third level of NIST standard for RBAC, support for context-aware authorizations, and the reasoning ability. The architecture of this framework is also introduced. It is extended from HERAS-AF by adding spatial and temporal data types and functions, integrating particular components to manage and handle different aspects of the authorization process. To support reasoning ability, a new approach to define role hierarchies, together with functions used to evaluate hierarchical relationships, was presented. We also described the work-flow of evaluating access requests and represented functions of each component through a case study. Finally, we assessed our framework by comparing with well-known relevant frameworks.

In the future, we will complete implementation of the framework and assess its performance as well as propose solutions to optimize the request evaluation process. Besides, the spatial aspect of our framework will be verified with the test cases suggested by Open Geospatial Consortium for GeoXACML. The abnormal detection and resolution in policy sets are also taken into account in the long term. The goal is to reduce redundant and non-logical polices to assure the stability and performance of the system.

Acknowledgements. This paper was partially funded by the Ho Chi Minh City Department of Science & Technology, according to the Scientific Research Contract No. 121/HD-SKHCN, August 11, 2011 between Department of Science & Technology and University of Technology, Ho Chi Minh City, Vietnam.

References

1. Enterprise Java XACML (January 2013),
   http://code.google.com/p/enterprise-java-xacml/
2. HERAS-AF (January 2013), http://www.herasaf.org/
3. JTS (January 2013), http://www.vividsolutions.com/jts/jtshome.htm
4. Margrave (January 2013),
   http://www.cs.brown.edu/research/plt/software/margrave/index.html
5. OASIS Brief Introduction to XACML (January 2013),
   http://www.oasis-open.org/committees/download.php/2713/
   Brief_Introduction_to_XACML.html
6. OASIS XACML 2.0 Core Specification (January 2013),
   http://docs.oasis-open.org/xacml/2.0/access_control-xacml-2.0-core-
   spec-os.pdf

7. OGC GeoXACML 1.0.1 (January 2013),
   http://portal.opengeospatial.org/files/?artifact_id=42734
8. OpenAzine (January 2013),
   http://www.openliberty.org/wiki/index.php/OpenAz_Main_Page
9. OWL (January 2013), http://www.w3.org/TR/owl-features/
10. PAM_XACML (January 2013), http://pamxacml.sourceforge.net/
11. Sun's XACML (January 2013), http://sunxacml.sourceforge.net
12. WSO2 (January 2013), http://wso2.com/products/identity-server/
13. XACML Studio (January 2013), http://xacml-studio.sourceforge.net/
14. XACMLight (January 2013), http://sourceforge.net/projects/xacmllight/
15. XACML.NET (January 2013), http://mvpos.sourceforge.net/index.html
16. XEngine (January 2013), http://xacmlpdp.sourceforge.net/
17. Parducci, B., Lockhart, H., Rissanen, E.: Extensible access control markup language (XACML) version 3.0. OASIS Standard (2010)
18. Cuppens, F., Cuppens-Boulahia, N.: Modeling contextual security policies. International Journal of Information Security 7, 285–305 (2008)
19. Dang, T.K.: A Practical Solution to Supporting Oblivious Basic Operations on Dynamic Outsourced Search Trees. International Journal of Computer Systems Science and Engineering (CSSE) 21(1), 53–64 (2006) ISSN 0267-6192
20. Dang, T.K.: Solving Approximate Similarity Queries. International Journal of Computer Systems Science and Engineering (CSSE), 22(1-2), 71–89 (2007) ISSN 0267-6192
21. Ferrini, R., Bertino, E.: Supporting RBAC with XACML+OWL. In: Proceedings of the 14th ACM Symposium on Access Control Models and Technologies, Stresa, Italy, pp. 145–154 (2009)
22. Finin, T., Joshi, A., Kagal, L., Thuraisingham, B.: ROWLBAC: Representing Role Based Access Control in OWL. In: Proceedings of the 13th ACM Symposium on Access Control Models and Technologies, Estes Park, CO, USA, pp. 73–82 (2008)
23. Govorov, M., Khmelevsky, Y., Ustimenko, V., Khorev, A.: Security for GIS N-tier Architecture. In: Developments in Spatial Data Handling, pp. 71–83 (2005)
24. Helil, N., Rahman, K.: RBAC Constraints Specification and Enforcement in Extended XACML. In: Multimedia Information Networking and Security (MINES), Nanjing, China, pp. 546–550 (2010)
25. Kumar, M., Newman, R.E.: STRBAC - An approach towards spatio-temporal role-based access control. In: Communication, Network, and Information Security, Cambridge, Massachusetts, USA, pp. 150–155 (2006)
26. Rissanen, E.: Core and hierarchical role based access control (RBAC) profile of XACML version 3.0. OASIS Standard (2010)
27. Sandhu, R., Ferraiolo, D., Kuhn, R.: The NIST model for role-based access control: towards a unified standard. In: Proceedings of the Fifth ACM Workshop on Role-based Access Control, Berlin, Germany, pp. 47–63 (2000)
28. Thi, K.T.L., Dang, T.K., Kuonen, P., Drissi, H.C.: STRoBAC Spatial Temporal Role Based Access Control. In: Proceedings of the 4th International Conference on Computational Collective Intelligence. Technologies and Applications, Ho Chi Minh, Vietnam, pp. 201–211 (2012)
29. Thi, Q.N.T., Dang, T.K.: X-STROWL: A generalized extension of XACML for context-aware spatio-temporal RBAC model with OWL. In: Proceedings of the 7th International Conference on Digital Information Management, Macau, pp. 253–258 (2012)

Automatic Triggering of Constraint Propagation

Eric Monfroy[1], Broderick Crawford[2,3], and Ricardo Soto[2,4]

[1] LINA, Université de Nantes, France
`FirstName.Name@univ-nantes.fr`
[2] Pontificia Universidad Católica de Valparaíso, Chile
`FirstName.Name@ucv.cl`
[3] Universidad Finis Terrae, Chile
[4] Universidad Autónoma de Chile, Chile

Abstract. A constraint satisfaction problem requires a value, selected from a given finite domain, to be assigned to each variable in the problem, so that all constraints relating the variables are satisfied. The main feature of any constraint solver is constraint propagation, it embeds any reasoning which consists in explicitly forbidding values or combinations of values for some variables of a problem because a given subset of its constraints cannot be satisfied otherwise. It is very important to apply constraint propagation as efficiently as possible. In this paper, we present a hybrid solver based on a Branch and Bound algorithm combined with constraint propagation to reduce the search space. Some rules trigger constraint propagation based on some observations of the solving process. The results show that is possible to make reasonable use of constraint propagation.

Keywords: Constraint solving, Constraint propagation, Constraint Satisfaction Problems.

1 Introduction

Over the past decades, significant improvements have been developed for solving complex combinatorial optimization problems issued from real world applications. To tackle large scale instances and intricate problem structures, sophisticated solving techniques have been developed, combined, and hybridized to provide efficient solvers. Combinatorial problems are often modeled as Constraint Satisfaction Problems (CSP) or Constraint Optimization Problems (COP).

A Constraint Satisfaction Problem is defined by a set of variables, a set of values for each variable, and a set of constraints. A solution is an instantiation of variables satisfying all constraints. These problems are widely studied mainly because a lot of real-life problems can de modeled as CSPs. Complete solving techniques carry out a depth-first search by interleaving constraint propagation and enumeration. Constraint propagation prunes the search tree by eliminating values that cannot be in any solution. Enumeration creates a branch by instantiating a variable (e.g., $x = v$) and another one (e.g., $x \neq v$) for backtracking

B. Murgante et al. (Eds.): ICCSA 2013, Part V, LNCS 7975, pp. 452–461, 2013.
© Springer-Verlag Berlin Heidelberg 2013

when the first branch is proved unsatisfiable. All enumeration strategies preserving solutions are valid, but they have a huge impact on resolution efficiency. Moreover, no strategy is the best for all problems.

Constraint solving or hybridization of solvers become more and more complex: the user must select various solving and hybridization strategies and tune numerous parameters. Moreover, it is well-known that an a priori decision concerning strategies and parameters is very difficult since strategies and parameters effects are rather unpredictable and may change during resolution. The selection and the correct setting of the most suitable algorithm for solving a given problem has already been investigated many years ago [14]. The considerations of [14] are still valid and can indeed be considered at least from two complementary points of view: 1) selecting solving techniques or algorithms from a set of possible available techniques, and 2) tuning an algorithm with respect to a given instance of a problem. Moreover, setting should be changed to adapt themselves to the process during solving.

An Autonomous Search [10,7,12,5,17,6,8] system should provide the ability to advantageously modify its internal components when exposed to changing external forces and opportunities. It corresponds to a particular case of adaptive systems with the objective of improving its problem solving performance by adapting its search strategy to the problem at hand.

In [4], a framework for dynamic adaptation of enumeration strategies of a constraint propagation-based solver was proposed. In this paper, we use a subsequent instantiation of the framework [11]. The instantiation manages dynamically some strategies and components of a hybrid solver based on Branch and Bound (B&B) combined with Constraint Propagation in order to reduce the search space. To this end, the framework observes the resolution process, analyses the observations, and makes some decision to possibly dynamically adapt strategies and trigger some functions of the solver.

With our cooperative search we don't focus on improving the resolution of a single problem, but we are interested in quickly finding solutions on average for a set of problems. The experimental results are more than promising.

This paper is organized as follows. We present our proposal in Section 2. Experimental results are discussed in Section 3 and we conclude in Section 4.

2 The Dynamic Strategy Framework

The framework is based on four components exchanging information, see Figure 1: the first component runs a solver or a solver cooperation/hybridization based on some tunable modules/functions and on some solving strategies; the second one observes resolution and takes snapshots, i.e., some kind of quantitative summaries of the observations; the third one analyses snapshots and draws some indicators (that may be qualitative) about strategy and function quality; and the fourth one makes decisions to update strategy priorities and trigger functions (in [12,13] we used rule-based constraint propagation).

The SOLVE component is a solver, or an hybrid solver for solving constraint problems. This (hybrid) solver has at disposal several possible strategies

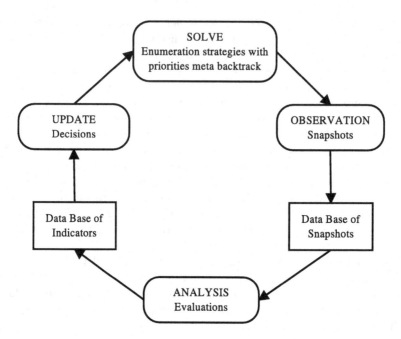

Fig. 1. The Dynamic Strategy Framework

and several functions that can be changed or triggered using some rules of the UPDATE component. For the experimentations, we consider a branch and bound algorithm: while there remain sub-problems, choose one (depth-first selection in our case) and treat it. Compared to an usual B&B, this algorithm is hybrid (Algorithm 1): each sub-problem P may be reduced by a propagation phase (e.g., [1]) to reduce the search space by eliminating values of variables that cannot participate in a solution. This technique is rather common (e.g., [15]), thus, we just give an overview of the method. We consider two models of the problem to be solved. The first one, M_{CSP} is based on CSP and uses finite domain variables (i.e., $X_{CSP} \in \{v_1, \ldots, v_n\}$), and the second one, M_{IP}, is a linear programming model (i.e., based on continuous variables $v_1 < X_{IP} < v_n$).

The two models are equivalent: at each node of the search tree, the problem is modeled by both techniques. Information discovered in a model is communicated to the other one (e.g., if X_{CSP} is reduced to the domain $\{v_1, \ldots, v_n\}$, then the corresponding variable X_{IP} can be constrained by $v_1 < X_{IP} < v_n$, and so on for objective functions and in the opposite way).

The (conditional) propagation phase is applied to the M_{CSP} model to reduce the search space by removing values of variables that cannot satisfy some constraints. The obtained information (reduction of domains) is communicated to the M_{IP} model; the relaxation phase is applied to the relaxation of the M_{IP} model (i.e., M_{IP} without integrity constraints). In case there is no propagation phase, the algorithm acts as a usual B&B: the relaxation is solved to determine the unsatisfiability of the problem or an integer solution.

Algorithm 1. Sketch of the hybrid algorithm

```
 1: procedure HYBRID_B&B
 2:     Optimum ← −∞
 3:     Nodes ← (P₀, −∞)
 4:     while Nodes ≠ ∅ do
 5:         P ← getSubproblem(Nodes)
 6:         PropagationPhase(P) (conditional)
 7:         RelaxationPhase(P)
 8:         checkSolution(P)
 9:         branching(P, (P₁, ..., Pₖ))
10:         for (1 ≤ i ≤ k) do
11:             Nodes ← Nodes ∪ (Pᵢ, bound(P))
12:         end for
13:         OptimumSolution ← Solution
14:         OptimumValue ← Optimum
15:     end while
16: end procedure
```

Then, when checking solution, the algorithm evaluates if the solution is feasible. If the solution is an integer with a better value than the actual one, the optimum value and the optimal actual solution are updated. If the solution is integer, but not better, it is discarded. If it is not integer, the branching is done on non integer variables, generating the respective subproblems w.r.t. the enumeration strategy (in this case, after some preliminary tests, we fixed the lexicographic order to select the variable, and the largest value of the variable).

In the following, we use our framework to (de)activate the propagation phase: by just changing the updating rules we obtain different types of hybridization.

The OBSERVATION component aims at observing and recording information of the current search tree, i.e., it spies the solving process of the SOLVE component. These observations (called **snapshots**) can be seen as an abstraction of the resolution state at a time t. Taking a snapshot consists in extracting (since search trees are too large) some information from a solving state.

In our experimentations, snapshots are taken at each node of the search tree and they contain the following data:

- Characteristic of the problem:
  - V_{total} : total number of variables
  - R_{eq} (R_{leq}, R_{geq} resp.): number of constraints of type $= (\leq, \geq$ resp.)
- Measures of the search tree
  - N_{expl} : number of explored nodes
  - N_{failed} : failed nodes (not leading to a solution)
  - N_{sol} : nodes with solutions
  - $Backs$: number of backtrackings
  - N_{prop} : nodes in which the propagation phase happened
  - $Depth$: current depth of the tree search
- Measures of the solving process

- T_{prop} : CPU time used in the propagation phase
- T_{relax} : CPU time used in the relaxation phase
- V_{fixen} : variables fixed by enumeration
- V_{fprop} : variables fixed by propagation
- VF : total of fixed variables (i.e., $V_{fixen} + V_{fprop}$)

The ANALYSE component analyses the snapshots taken by the OBSERVA-TION: it evaluates the strategies and functions, and provides **indicators** to the UPDATE component. They can be extracted, computed, or deduced from one or several snapshots. Numeric indicators (δn) are quantitative results computed from snapshots.: e.g., the depth of the search (δn_{depth}), the number of fixed variables (δn_{fix}), or fixed by enumeration (δn_{fixen}). Indicators can be more complex: e.g., the difference of depth between 2 snapshots to give information on the progress in the search tree, or the difference between the depth (δn_{depth}) of the search and the variables fixed by enumeration (δn_{fixen}) which gives the indicator (δn_{gap}) on how many unsuccessful enumerations were performed on the last variable (see [4]).

Boolean indicators (δb) reflect properties. Simple ones can be related to problems (e.g., there is a univariate constraint or a hard variable was fixed). More complex properties can be related to a quantitative or qualitative analysis of the snapshots, such as thrashing in the case of a constraint propagation based solver ([4]). Indicators that we used for our experimentations are described later on.

The UPDATE component makes decisions using the indicators: it makes interpretations of the indicators, and then updates the strategies priorities and/or triggers some functions of the SOLVE component.

The knowledge of the UPDATE component is contained in a set of rules. The head of such a rule is a conjunction of conditions on the indicators (disjunctions can be handled by several rules). There are two types of rules: for priority update rules ($\Rightarrow$rules), the body is a conjunction of updates of strategies priorities:

$$\bigwedge_{i=1}^{l}(\sum_{j \in J_i} \omega_j \times \delta n_j)\ op\ c_j \wedge \bigwedge_{i=1}^{k} \delta b_i \Rightarrow \bigwedge_{i=1}^{l} p_i = p_i + f_i(\delta n_1, ..., \delta n_l)$$

where:
- the ω_j are the weights of each numeric indicator δn_j in the condition, the c_j are constants, the J_i are subsets of all the indicators, and $op \in \{\leq, \geq, =\}$;
- the δb_i are some Boolean indicators;
- the f_i are functions over the indicators that returns real numbers to increase or decrease the priority p_i of the strategy i;
- and the $\bigwedge_{i=1}^{l}$ in the body of the rule is an abuse of language to mean that the l priorities can be updated.

whereas for function rules ($\rightarrow$rules) the body requests the application of a function, possibly with some parameters:

$$\bigwedge_{i=1}^{l}(\sum_{j \in J_i} \omega_j \times \delta n_j)\ op\ c_j\ \wedge\ \bigwedge_{i=1}^{k} \delta b_i \quad \rightarrow \quad function(...))$$

When the head of a rule is fulfilled (i.e., conditions are verified), its body is executed: for ⇒rules, the priorities of the strategies (e.g., enumeration strategies) are updated in the SOLVE component. Whereas for →rules, functions are triggered in the SOLVE component. The rules we used for experimentations are described in the next section.

3   An Hybrid B&B + Constraint Propagation Solver

We now experiment with a B&B + Constraint Propagation solver: snapshots, indicators, and rules enable us to simply change the strategies of hybridization. Here, we do not use priority of strategies. See [12,13]) for other strategies and rules.

In this work, we use rules of the second type to activate (*propagation*()) or deactivate (*nopropagation*()) propagation phases. Our goal is not to design the best solver, but 1) to show how simple it is to change hybridization strategies in our framework, 2) to observe the role of propagation on various problems, and 3) to give some hints on how to manage propagation.

The solving process was written in Gecode [16]; the propagation is achieved by the propagators of Gecode while the components of the B&B algorithm (e.g., relaxation) are achieved by *lp_solve* [3] is a linear (integer) programming solver based on the revised simplex method and the B&B method for the integers; it can be used as a library. The tests were run on an Intel Xeon 1.6GHz computer, with 4GB of memory. Each run was stopped after a time out of 1200 seconds.

The measures to evaluate the performance of the various hybridizations are: **v**alue of the **f**irst-found integer solution and **C**PU time (in s.) to obtain it; the same measures for the **b**est-found integer solution; and **C**PU time (in s.) to prove optimality. The solvers and their hybridizations were tested with various instances of 4 types of problems: the Multiple Balanced Academic Curriculum Problem (mbacp), the set covering problem (scp), the set partitioning problem (spp), the multidimensional knapsack problem (mknap). The instances and model of MBACP can be found in [9], and of the other problems in [2]. The size of the instances, in terms of variables and constraints, together with the best known solution and the best known CPU time are given in Table 1.

As a reference, Table 2 shows the results obtained with *lp_solve* alone. The columns show the value of the first solution and the time required to compute it; the best solution; the time required for proving optimality (”-” when optimality is not proven before the 1200 s. of timeout). With the same model, Gecode is faster to find a first solution, but of worse quality; it thus must explore a larger search space to find the optimum which it never reached before the timeout.

3.1   Propagation at Each Node

Here we do not consider any updating rule: propagation is triggered at each node of the search tree and the snapshots and indicators are not used. See Table 2 for results and in the following we give few comments since this basic hybridization

Table 1. Problem instances

Problem	Variables	Constraints	Best known sol.	Time
mbacp1	572	473	1	0.81
mbacp2	714	656	0	8.44
mbacp3	856	346	-	3.05
mbacp12	717	570	-	-
mbacp13	859	578	-	-
mbacp23	859	650	-	-
mbacp123	862	882	-	-
scp65	1000	200	161	-
scpa1	3000	300	253	-
scpb1	3000	300	69	-
sppaa01	8904	823	56138	238.76
sppaa03	8627	825	49649	12.47
sppaa04	7195	426	26402	319.19
sppnw18	10757	124	340160	2.19
mknapcb4a	250	5	59312	-
mknapcb4b	100	10	23064	-
mknapcb4c	100	10	22801	-
mknapcb2-1	100	10	22131	-

should be seen as a reference for the next hybridizations that make use of our updating rules.

MBACP. Times for proving optimality are significantly better than with the individual solvers (up to 29 times quicker for *mbacp2*). The cost of propagation is rather small, between 2% to 4% (depending on the instance) of the total execution time. However, propagation is quite inefficient.

SCP. The basic hybridization worsen the CPU times to get both the first and best solution. The propagation cost is around 5%: however, the reason is that propagation is quite inefficient for these problems and thus stops quickly.

SPP. Propagation is here more efficient, but costs between 26% to 40% of the total time.

MKNAP. These are the most difficult problems for the solvers. It seems that numerous valid solutions must be evaluated and the search tree is much larger.

3.2   Rule-Based Propagation

This strategy penalizes propagation when it is inefficient. The hybridization starts performing propagation; it carries on with propagation as long as it brings relevant information. If at a given node, propagation does not fix any variable, then propagation is not performed in the following next d nodes, d being a distance given a priori. The larger the value for d, the less propagation phases are achieved. The indicators only use the last taken snapshot (F):

Table 2. Results for lp_solve and the basic hybridization

Problem	lp_solve						Basic hybridization					
	First solution		Best solution		Opt.		First solution		Best solution		Opt.	
	Value	Time	Value	Time	Time		Value	Time	Value	Time	Time	
mbacp1	4	4.2	1	17.6	326.2		4	2.8	1	6.1	212.5	
mbacp2	4	15.1	0	315.5	320.1		4	2.2	0	10.4	11.7	
mbacp3	4	2.2	1	20.4	-		4	1.0	1	86.3	-	
mbacp12	4	34.9	1	213.0	594.2		4	9.5	1	133.1	340.8	
mbacp13	4	32.4	1	210.9	222.6		4	3.2	1	35.1	39.2	
mbacp23	4	3.4	1	175.4	191.7		4	2.3	1	79.3	84.3	
mbacp123	4	157.5	4	157.5	-		4	56.9	3	395.6	-	
scp65	166	0.2	161	9.0	23.4		166	0.2	161	10.4	26.1	
scpa1	271	0.8	253	107.0	167.5		271	0.9	253	118.6	184.7	
scpb1	75	0.9	69	11.4	33.1		75	1.2	69	13.9	38.8	
sppaa01	56363	25.1	56363	25.1	-		56363	33.4	56363	33.4	-	
sppaa03	49734	11.8	49649	54.3	71.7		49734	17.1	49649	97.2	127.0	
sppaa04	27802	10.6	27507	121.6	-		27802	16.8	27507	150.6	-	
sppnw18	342950	1.1	340160	28.5	28.5		342950	1.7	340160	39.3	39.5	
mknapcb4a	21277	0.1	23057	956.6	-		21277	0.0	23050	466.3	-	
mknapcb4b	20189	0.1	22704	878.8	-		20189	0.0	22704	1182.8	-	
mknapcb4c	20978	0.1	22131	858.6	1115.7		20978	0.0	22131	1124.3	-	
mknapcb2-1	57318	0.2	59015	1140.7	-		57318	0.1	59013	1018.1	-	

- δb_1 this indicator is true if at least one snapshot was taken; false otherwise.
- $\delta n_1 = (V_{fprop})_F$: the number of variables fixed by the last propagation phase.

$dist$ is the remaining distance before applying propagation again. r_1 means that propagation must be triggered if no snapshot has been taken, or propagation fixed variables in the father node, or $dist = 0$ (propagation is not currently penalized). When propagation does not fix variable, r_2 penalizes propagation with a distance d and r_3 decrements the penalty of the already penalized propagation.

$$r_1 : not(\delta b_1) \lor dist = 0 \lor \delta n_2 > 0 \rightarrow dist = 0; Prop()$$
$$r_2 : \delta n_2 = 0 \land dist = 0 \qquad\qquad \rightarrow dist = d; noProp()$$
$$r_3 : 0 < dist < d \qquad\qquad \rightarrow dist = dist - 1; noProp()$$

We made experimentations (see Table 3) with d ranging from 0 to 20, with an increment of 1. With $d = 0$, this hybridization is equivalent to the basic hybridization. With $d = 20$, propagation is nearly never used.

MBACP. This strategy improves the results of the basic hybridization and of lp_solve in various cases. In 3 instances, the best parameter d enables to reduce of 30% the time needed for proving optimality w.r.t. the basic hybridization.

SCP. Increasing d leads to results similar to lp_solve. However, with d larger than 6 there is no more improvement.

SPP. For proving optimality, there is no clear tendency: for sppaa03, increasing d we reach the efficiency of lp_solve; but, performing propagation and observing the

Table 3. Rule-based Propagation

Problem	Best d	First solution Value	First solution Time	Best solution Value	Best solution Time	Opt. Time
m1	1	4	2.8	1	7.5	213.7
m2	1	4	2.2	0	16.6	17.8
m3	4	4	1.0	1	18.0	-
m12	4	4	9.9	1	590.3	792.1
m13	9	4	2.2	1	23.6	27.6
m23	2	4	2.2	1	58.9	64.3
m123	1	4	80.8	3	533.9	-
scp65	13	166	0.2	161	9.1	23.3
scpa1	5	271	0.8	253	111.0	173.4
scpb1	2	75	1.0	69	12.3	35.4
sppaa01	12	56363	26.6	56363	26.6	-
sppaa03	3	49734	20.5	49649	78.9	114.7
sppaa04	0	27802	10.9	27507	121.9	-
sppnw18	6	342950	1.2	340160	22.6	22.6
mknapcb4a	0	21277	0.1	23057	993.8	-
mknapcb4b	2	20189	0.0	22704	1004.5	-
mknapcb4c	5	20978	0.0	22131	924.5	-
mknapcb2-1	4	57318	0.1	59013	826.0	-

solving process, we degrade the performance compared to *lp_solve*. For sppnw18, with $4 \leq d \leq 8$ the results are better than *lp_solve*.

MKNAP. On average, this hybrid is faster than *lp_solve* for finding the same "best" solution. However, for proving optimality of mknapcb4c, we see that propagation is an overhead. We could observe that using a larger d, we obtained better results than *lp_solve*, and thus, propagation effectively benefits to this type of problems. However, tuning correctly d for all the these instances is rather complicated.

4   Conclusion

In this paper, we presented a rule-based approach to constraint propagation. We experimented on a cooperative B&B + Constraint Propagation solver in which propagation can be triggered to respond to some observations of the solving process. The experimental results prove that our hybrid method often perform better than pure methods and branch and bound can be benefited from constraint propagation.

The benefits of constraint propagation emerge when it was embedded in B&B. The embedding suggested new ways of interleaving propagation and search and new ways of varying the propagation according to the particular features of the problem at hand. It could also be interesting to combine our method with learning or prediction techniques to select the hybridization w.r.t. the problem. The next step would be to change dynamically the hybridization strategy in order to respond to the change of the search state adaptively changing the rules.

References

1. Apt, K.R.: Principles of Constraint Programming. Cambridge Univ. Press (2003)
2. Beasley, J.E.: Or-library: distributing test problems by electronic mail. JORS 41(11), 1069–1072 (1990)
3. Berkelaar, M.: lpsolve—simplex-based code for linear and integer programming
4. Castro, C., Monfroy, E., Figueroa, C., Meneses, R.: An approach for dynamic split strategies in constraint solving. In: Gelbukh, A., de Albornoz, Á., Terashima-Marín, H. (eds.) MICAI 2005. LNCS (LNAI), vol. 3789, pp. 162–174. Springer, Heidelberg (2005)
5. Crawford, B., Castro, C., Monfroy, E., Soto, R., Palma, W., Paredes, F.: A hyperheuristic approach for guiding enumeration in constraint solving. In: Schütze, O., Coello, C.A.C., Tantar, A.-A., Tantar, E., Bouvry, P., Del Moral, P., Legrand, P. (eds.) EVOLVE - A Bridge between Probability, AISC, vol. 175, pp. 171–188. Springer, Heidelberg (2012)
6. Crawford, B., Soto, R., Castro, C., Monfroy, E.: Extensible CP-based autonomous search. In: Stephanidis, C. (ed.) Posters, Part I, HCII 2011. CCIS, vol. 173, pp. 561–565. Springer, Heidelberg (2011)
7. Crawford, B., Soto, R., Monfroy, E., Palma, W., Castro, C., Paredes, F.: Parameter tuning of a choice-function based hyperheuristic using particle swarm optimization. Expert Systems with Applications 40(5), 1690–1695 (2013)
8. Crawford, B., Soto, R., Montecinos, M., Castro, C., Monfroy, E.: A framework for autonomous search in the ecl^{i}ps^e solver. In: Mehrotra, K.G., Mohan, C.K., Oh, J.C., Varshney, P.K., Ali, M. (eds.) IEA/AIE 2011, Part I. LNCS, vol. 6703, pp. 79–84. Springer, Heidelberg (2011)
9. Gent, I., Walsh, T.: Csplib: a benchmark library for constraints. Technical report, APES-09-1999 (1999), http://csplib.cs.strath.ac.uk/
10. Hamadi, Y., Monfroy, E., Saubion, F. (eds.): Autonomous Search. Springer (2012)
11. Monfroy, E., Castro, C., Crawford, B., Figueroa, C.: Adaptive hybridization strategies. In: ACM Symposium on Applied Computing, pp. 922–923 (2011)
12. Monfroy, E., Castro, C., Crawford, B., Soto, R., Paredes, F., Figueroa, C.: A reactive and hybrid constraint solver. Journal of Experimental and Theoretical Artificial Intelligence 25(1), 1–22 (2013)
13. Monfroy, E., Crawford, B., Soto, R.: Interleaving constraint propagation: An efficient cooperative search with branch and bound. In: Blesa, M.J., Blum, C., Festa, P., Roli, A., Sampels, M. (eds.) HM 2013. LNCS, vol. 7919, pp. 52–61. Springer, Heidelberg (2013)
14. Rice, J.: The algorithm selection problem. Technical Report CSD-TR 152, Purdue Univ. (1975)
15. Rodosek, R., Wallace, M., Hajian, M.: A new approach to integrating mixed integer programming and clp. Baltzer Journal (1998)
16. Schulte, C., Tack, G., Lagerkvist, M.: Gecode: Generic constraint development environment. In: INFORMS Annual Meeting (2006)
17. Soto, R., Crawford, B., Monfroy, E., Bustos, V.: Using autonomous search for generating good enumeration strategy blends in constraint programming. In: Murgante, B., Gervasi, O., Misra, S., Nedjah, N., Rocha, A.M.A.C., Taniar, D., Apduhan, B.O. (eds.) ICCSA 2012, Part III. LNCS, vol. 7335, pp. 607–617. Springer, Heidelberg (2012)

Semantic Annotation and Publication of Linked Open Data

Serena Sorrentino[1], Sonia Bergamaschi[1],
Elisa Fusari[2], and Domenico Beneventano[1]

[1] DIEF - University of Modena and Reggio Emilia
via Vignolese 905, 41100 - Modena, Italy
{sonia.bergamaschi,serena.sorrentino}@unimore.it
[2] Graduate Student at DIEF - University of Modena and Reggio Emilia,
via vignolese 905,41100 Modena, Italy
57915@studenti.unimore.it

Abstract. Nowadays, there has been an increment of open data government initiatives promoting the idea that particular data produced by public administrations (such as public spending, health care, education etc.) should be freely published. However, the great majority of these resources is published in an unstructured format (such as spreadsheets or CSV) and is typically accessed only by closed communities. Starting from these considerations, we propose a semi-automatic experimental methodology for facilitating resource providers in publishing public data into the Linked Open Data (LOD) cloud, and for helping consumers (companies and citizens) in efficiently accessing and querying them. We present a preliminary method for publishing, linking and semantically enriching open data by performing automatic semantic annotation of schema elements. The methodology has been applied on a set of data provided by the Research Project on Youth Precariousness, of the Modena municipality, Italy.

1 Introduction

Nowadays, the availability of freely accessible information on the Web is constantly growing. In particular, recently, there has been an increment of open data government initiatives (e.g., data.gov for US and data.gov.uk for UK, dati.gov.it for Italy etc.) promoting the idea that certain data produced by public administrations (such as public spending, health care, education etc.) should be freely published in order to allow companies and citizens to browse, analyze and reuse them [12].

As a result, numerous open data sources are available on public organization's web sites. However, the great majority of these resources is published in an unstructured format (such as spreadsheets or CSV) and is typically accessed only by closed communities. Indeed, even if freely available on the Web, there are no connections among them and their structural and semantic heterogeneity makes it difficult to perform automatic or semi-automatic cross-data analysis, thus preventing to obtain high value information.

B. Murgante et al. (Eds.): ICCSA 2013, Part V, LNCS 7975, pp. 462–474, 2013.

The Linked Open Data (LOD) paradigm represents the key solution to improve and enrich the use of open data and to help consumers (citizens and companies) to access their integrated information. In the Semantic Web research area, the term Linked Data refers to a set of best practices for publishing and connecting structured data on the Web [4]. LOD extends the linked data paradigm by publishing data which are freely available to everyone and for any purpose. LOD data sets are represented and published in the RDF[1] standard format and any resource (e.g. things, persons, etc.) has a dereferenceable URI (Uniform Resource Identifier)(i.e., a string of characters used to identify a name or a resource) as global identifier. In this way, open data sets can be exposed on the Web and data consumers could use the current Web infrastructure to obtain relevant information about any resource. The RDF data sets, then, can be queried by using the standard SPARQL query language.

Nevertheless, providing a standard way to represent and query public data is not enough: a full and easy access to open data sources requires the opportunity to integrate multiple LOD data sets belonging to the same knowledge domain. However, while the LOD cloud is rich of instance links (e.g., *owl:sameAs* relationships), schema level mappings (e.g., *rdfs:subClassOf* relationships) which are fundamental for performing dataset integration are almost absent. In this context, *semantic annotation* of schemas, i.e. the explicit association of one or more *meanings* to a schema element with respect to a reference lexical thesaurus is a key tool. Its effectiveness has been proved in the task of discovering schema and ontology mappings, i.e. semantic correspondences at the schema-level [3].

Starting from these considerations, we present a preliminary and experimental semi-automatic methodology to perform: RDF-ization (i.e., RDF translation) of open data sets; semantically annotation of their schema elements; publication on the Web; linking in the unified LOD cloud. In particular, during the process we make use of different already developed methods and tools.Our methodology represents a first step towards an automatic LOD integration system allowing users to publish any kind of public data, dynamically integrating two or more LOD data sets by exploiting semantic information and querying them without any pre-configured statistic analysis.

The methodology has been applied on a real case: the data we used were provided by the Research Project on Youth Precariousness, of the Modena municipality, Italy, which is carrying out an investigation about the precarious situation of young people living in the Modena district[2].

The rest of the paper is organized as follows. In Section 2, we give a step-wise description of the methodology by illustrating its main features, functionalities and the tools employed. Section 3 describes and analyzes related work. Finally, in Section 4, we give our concluding remarks and describe future work.

---

[1] http://www.w3.org/RDF/

[2] The project is carried out by the Councillor for Youth Policies Fabio Poggi, in collaboration with Prof. Claudio Baraldi and Dr. Federico Farini of the Department of Language and Culture, University of Modena, Italy.

2   Annotation and Publication of Linked Open Data

Our goal was to provide a standard methodology to facilitate both source providers and consumers in publishing, semantically enriching and querying LOD data sets. To this aim, we studied a general methodology consisting of four main steps:

1. *RDF-ization* for modeling the data in a structured format and convert them into RDF;
2. *Semantic Enrichment* for understanding the semantics of source schema elements;
3. *Web Publishing* for making data accessible through a SPARQL query endpoint;
4. *Linking and mapping* for discovering instance level-links and semantic mappings between the public data sets and other LOD resources.

Figure 1 shows the process of annotation and publication of the Youth Precariousness data set, and the interaction with the tools and resources exploited during the methodology. In the following, we describe each of these steps in details. We start by briefly describing the data set we used.

2.1   The Youth Precariousness Data Set

Our methodology has been applied on a set of data collected within the Modena (Italy) district project *Youth Precariousness*. This project aims to analyze the actual situation of job and emotional insecurity that young people are living in Modena. The data about the Youth Precariousness were collected by means of a paper questionnaire and were stored in Excel spreadsheets. The questionnaire were filled up by young people with age between 20 and 35 years. The questionnaires were anonymous and composed by 29 questions including the personal data of the interviewed such as age, birth place, the actual employment situation (i.e., employed, unemployed or student), school and university career and questions about parents job, family and friends in order to assess their social environment. Moreover, it included psychological questions about what the interviewed expects from the future: uncertainties, difficulties, expectations etc (see [10] for more details).

At present, the data collection phase is still in progress. However, for our purpose, it was enough to apply our methodology on the first 315 collected questionnaires. The questionnaire data were stored within an Excel spreadsheet, which simply map each question with the corresponding answer (e.g., question Q3 answer a2).

2.2   RDF-ization

In [4], Sir Tim Berners Lee introduces a "'5-star rating system'" for open data. This system can be summed up as follow:

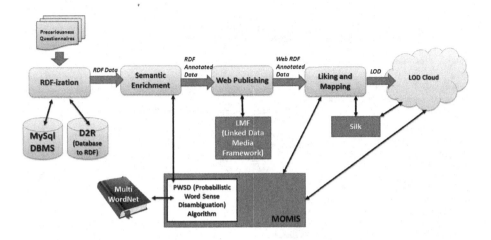

Fig. 1. The Annotation and publication process

1. one star: the resource is available on the web (whatever format);
2. two stars: the resource is available as structured data (e.g. excel instead of the scan of a table);
3. three stars: the resource is available as a non-proprietary structured data format (e.g. csv instead of excel);
4. four stars: the resource is available in the RDF format and use URLs to identify things;
5. five stars: the resource is linked to other open data.

Our methodology aims to make available the Youth Precariousness data set as a five stars open data. Thus, first of all we need to convert the data set into the RDF standard knowledge representation language. To represent data into RDF, we need first to convert them in a relational database and then to exploit one of the several freely available open source automatic tools for relational-RDF translation. The relational database has been realized in a semi-automatic way: starting from the Excel spreadsheet, we analyzed the data in order to design the corresponding Entity/Relationship diagram [19]. This process is fundamental and it has been performed manually with the support of Entity/Relationship editors. In particular, during this step, we identified the main concepts and the relationships among them.

Then, by using the open source MySQL Workbench tool[3], we automatically generated and populated a relational database storing the public data. From the collected questionnaires we created a database composed by 18 table for a total of approximately 5400 records.

Finally, we employed the D2R (Database To RDF)[4] open source software to convert the database into RDF. D2R is an HTTP server that allows us to convert

---

[3] http://www.mysql.com/products/workbench/
[4] http://d2rq.org/

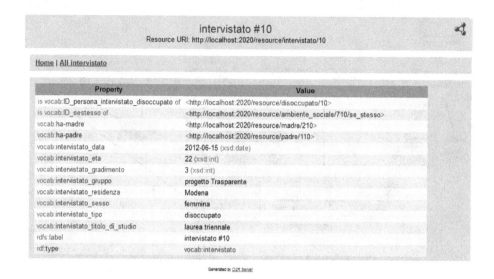

Fig. 2. Example of RDF data representation by using D2R

relational data in RDF triples through its specific internal language called D2RQ: a mapping file is created which maps each database table in a corresponding RDF class and each column in a RDF property. During this phase, the only information to be provided is the resource URI, which is mandatory for creating globally unique resource identifiers.

Figure 2 shows the RDF representation of an excerpt of the Youth Precariousness data set schema where each class and attribute has been converted into RDF by using the D2R vocabolary "*vocab*".

2.3   Semantic Enrichment

In order to efficiently use LOD data sets, consumers need to deeply understand the semantics of source schemas. Moreover, the hidden meanings associated to schema elements can be exploited for discovering semantic mappings and thus performing integration of different LOD data sets. Indeed, semantics facilitate and speed up the recognition of correspondences between data, contextualizing and enriching the information available.

To add semantics at the schema level, we use semantic annotation. *Semantic annotation* is the process of explicit alignment of one or more meanings to schema element labels (classes and attributes names). Manual semantic annotation is a time consuming and not scalable task. However, automatic semantic annotation is difficult due to the problem of term ambiguity. Thus, to perform automatic or semi-automatic annotation, a method for *Word Sense Disambiguation* (WSD), i.e. for identifying the sense of a term in a context [16], has to be devised.

We decided to utilize the PWSD (Probabilistic Word Sense Disambiguation) algorithm [18] developed in the MOMIS data integration system [2], which annotates schema elements with one or more meanings.

PWSD performs annotation with respect to the lexical reference database WordNet [15]. The strength the WordNet is that it provides a set of possible meanings, called synsets, for each term (nouns, adjectives, verbs and adverbs) and includes a wide network of semantic relationships among these meanings (e.g., hypernymy/hyponymy relationship defines between two concepts where one is more general/specific of the other) which can be used to infer semantic mappings among schema elements.

PWSD is composed by five different algorithms: the Structural Disambiguation algorithm which exploits terms that are related by a structural relationship (e.g., is-a relationships); the WordNet Domains Disambiguation algorithm which tries to disambiguate terms by exploiting domains information supplied by WordNet Domains [11]; the Gloss Similarity and the Iterative Gloss Similarity algorithms based on string similarity techniques; WordNet first sense heuristic rule selecting the first WordNet meaning (that is the more used in English) for a term. Before performing annotation, all the schema labels are preprocessed by using the normalization techniques described in [20], in order to expand abbreviations, remove stop words, and identify compound terms.

However, we cannot directly apply PWSD to the schema labels of our target data set which are in Italian, while the original version of WordNet includes English terms only. Therefore, we extended and modified the original implementation of PWSD in order to deal with Italian terms.

First of all, we needed to select an Italian thesaurus: we decided to use Multi-WordNet [5] which is a multilingual lexical database containing an Italian version of WordNet strictly aligned with the English WordNet (e.g., the Italian synset {corte, tribunale} is aligned with the English synset {court, tribunal, judicature}). Moreover, it includes the access to other versions of WordNet in several languages as well as all the relationships that exist between the various translations of the same word.

Then, as the PWSD algorithm has been designed for annotating English terms, we need to verify its applicability for annotating Italian terms in the context of Liked Open Data. We experimentally verified that the Structural, Gloss Similarity, Iterative Gloss Similarity and the First Sense algorithms can be directly applied to Italian terms as they exploit term features (like the structural relationships and the glosses) that do not depend on the language. As regards the WordNet Domains algorithm, it can be easily adapted to Italian terms: indeed, in MultiWordNet, the domain information has been automatically transferred from English to Italian, resulting in an Italian version of the resource WordNet Domains: for instance, as the English synset {court, tribunal, judicature} was associated with the domain LAW, also the corresponding Italian synset {corte, tribunale}, results automatically associated with the LAW domain.

---

[5] http://multiwordnet.fbk.eu/english/home.php

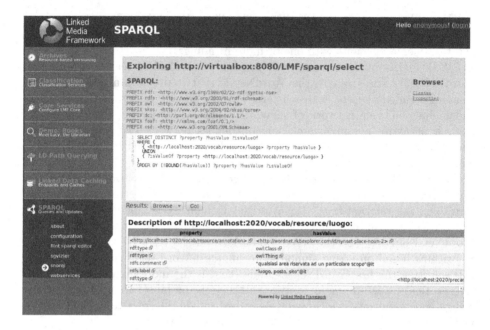

Fig. 3. Screen shot of the SPARQL service of LMF showing the annotation tag for the class label "luogo"

PWSD is a probabilistic algorithm, i.e. it associates to each annotation a probability value indicating the reliability of the annotation itself. The probability is used to filter annotations having reliability under a given threshold. An evaluation of PWSD, is out of our scope and can be found in [20]. However, to give an idea of its performance in the case of Italian schemas, we evaluated precision and recall of the annotation process on our resources: by using a probability threshold of 0.30, it obtained 0.71 in precision and 0.57 in recall. The recall value is not so high due to the application of a threshold greater than the one usually used in [20] (i.e., 0.15). However, this helps to reduce the risk of wrong annotations that might propagate errors in all the derived mappings in the LOD cloud.

To represent the semantic annotations in RDF, we used the class properties *owl:AnnotationProperty*, *rdfs:label* and *rdfs:comment*. In particular, we added as label the element name and its synonyms terms and as comment the Italian gloss (i.e., the definition of the meaning) taken from MultiWordNet.

As the English WordNet is available in the LOD cloud in RDF/OWL format[6] by using MultiWordNet, we can further enrich schema elements by linking Italian annotations to the corresponding English WordNet URI.

The RDF/OWL WordNet schema has three main classes: Synset, WordSense and Word. Each instance of Synset, WordSense and Word has its own URI. There

---

[6] http://www.w3.org/TR/wordnet-rdf/

is a pattern for the URIs so that it is easy to determine from the URI the class to which the instance belongs. The navigable URI provides some information on the meaning of the entity it represents. For example, the following URI

www.w3.org/2006/03/wn/wn20/instances/synset-bank-noun-2

is an instance of the class Synset representing the second meaning of the noun "bank". We used a custom class of *owl:AnnotationProperty* in order to link an *rdfs:Class* in our public schema with an instance of the class Synset in WordNet. The custom property *<owl:AnnotationProperty rdf:about="&vocab;annotation"/>* was called *vocab:annotation*, where the namespace "vocab" refers to the vocabulary that is automatically created by D2R during the RDF database conversion. By using this property, we can insert the link to the RDF WordNet synset in a navigable way. For instance, the semantic enrichment of the Italian schema element "luogo" is represented in the following way:

```
<owl:Class rdf:about="&vocab;luogo">
    <rdfs:label xml:lang="it">
        luogo, posto, sito
    </rdfs:label>
    <vocab:annotation rdf:resource="&wordnet;synset-place-noun-2/>
    <rdfs:comment xml:lang="it">
        qualsiasi area riservata ad un particolare scopo
    </rdfs:comment>
</owl:Class>

<!--http://wordnet.rkbexplorer.com/id/synset-place-noun-2 -->
<owl:Thing rdf:about="&wordnet;synset-place-noun-2">
    <rdf:type rdf:resource="&rdfs;Resource"/>
</owl:Thing>
```

where the annotation tags, actually, link the meaning from the WordNet thesaurus to the schema elements (in the form of URIs).

2.4   Web Publishing

The set of thesaurus-based annotation tags previously obtained represent the semantics of the schema. The following step is to make the data public on the Web. To this aim, we need an RDF repository exposing an HTTP de-referenceable SPARQL endpoint, so that the published data set can be referred and linked to other resources from the LOD cloud. To this aim, we evaluated two open source tools providing both the functionalities of RDF data storing and RDF querying: Fuseki[7] and LMF (LinkedData Media Framework)[8].

Fuseki is a SPARQL Server implemented by Apache-Jena, which allows us to query and analyze RDF data through a query engine called ARQ. Its main advantage is that it provides several query functionalities, such as grouping operators or counting functions, which can be used to perform statistic queries (e.g.,

---

[7] http://jena.apache.org/documentation/serving_data/index.html
[8] http://code.google.com/p/lmf/

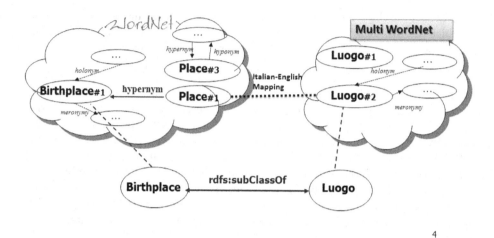

Fig. 4. An example Mapping discovery from semantic annotations: for the label "Luogo"

"how many interviewed are students? how many are employed and how many are unemployed?").

LMF is an application server employing the query service Snorql[9]. With respect to Fuseki, it permits to compose only simple SPARQL queries (e.g., it does not support grouping operators). However, it provides navigation functionalities that allow us to explore the WordNet URIs and see all the information related to its resource (see Figure 3).

For our purpose, we decided to employ LMF for its navigation functionalities. However, a composition of both the tools represent an interesting future work.

2.5   Linking and Mapping

After having published the data on the Web, we need to link the data set to other LOD resources. Creating links and mappings between the published resources is a key part of the Linked Open Data (LOD) paradigm [7]. We can identify two kinds of connections:

– *Instance-Level links* which are established between LOD data set instances (e.g., *owl:sameAs* established between two instances representing the same real world objects);
– *Schema-level mappings* which are established between schema concepts (e.g., *rdfs:subClassOf* used to state that all the instances of one class are instances of another);

In the LOD cloud, instance-links represent the great majority of connections. In our case, thanks to the annotation tags, our schema is automatically linked to the

---

[9] http://data.semanticweb.org/snorql/

RDF version of WordNet. To link the data set to other LOD resources, we can use Silk [14], a popular semi-automatic framework providing several patterns and property based techniques for helping data set providers in discovering instance links. We used Silk to discover links between our public data and DBPedia [5] which essentially, makes the content of Wikipedia available in RDF.

As observed before, schema-level mappings are almost absent in the LOD cloud even if, as previously described, they represents a fundamental means to integrate different LOD resources. Silk does not provide the semantic techniques needed to discovery semantic mappings among LOD schemas. We can discover semantic mappings among different LOD schemas exploiting the techniques developed for ontology and schema matching systems [9].

As will be pointed out in Section 4, this problem represents a core challenge and a future work of our research area. However, in this section, we want to present our approach to discover mappings starting from the previous obtained semantic annotations.

We can automatically discover RDF relationships by exploiting the wide semantic network provided by WordNet. In particular, let c_1 and c_2 be classes of two different schemas and $m(c_1)$ and $m(c_2)$ their meanings in WordNet, we consider the following possible RDF relationships:

- c_2 *rdfs:subClassOf* c_1, defined if $m(c_1)$ is a *hypernym* of $m(c_2)$ in WordNet;
- c_1 *rdfs:subClassOf* c_2, defined if $m(c_1)$ is a *hyponym* of $m(c_2)$ in WordNet;
- c_1 *owl:equivalentClass* c_2, defined if $m(c_1)$ is a *synonym* of $m(c_2)$ in WordNet.

Figure 4 shows an example of mapping discovery starting from semantic annotations: let us suppose that we want to discover the mappings between our Youth Precariousness data set and another English LOD schema. First of all we annotate the elements of both schemas with respect to MultiWordNet (in case of Italian terms) and WordNet (in case of English terms). By using the direct correspondence between Italian and English WordNet synsets, we can discover that there exists a hyponym relationship between the meaning associated to "luogo" (i.e., "place" in English) and the annotation of "Birthplace". Thus, we can automatically infer that there exists also an rdfs:SubClassOf relationship between these two RDF classes.

To perform semantic driven mapping discovery, we will use the open-source MOMIS data integration system which has been designed and tested by our research group. An open-source version of MOMIS is actually delivered and maintained by the academic Spin-Off DataRiver[10]. However, MOMIS will need to be reviewed and adapted in order to deal with RDF and LOD resources.

3   Related Work

As we have previously seen, the creation of semantic mappings at schema level has a fundamental role in the integration and alignment of LOD resources from

---

[10] http://www.datariver.it/

different domains. Several research groups have developed tools, frameworks and platforms for the integration of Linked Open datasets at the semantic level.

BLOOMS (Bootstrapping-based Linked Open Data Ontology Matching System)[13] is a system for the alignment of LOD ontologies at schema level. It utilizes a bootstrapping approach based on the Wikipedia category hierarchy. Essentially, BLOOMS for each matching candidate ontology class C identifies all its corresponding Wikipedia articles and construct a forest (i.e., a set of trees, one for each article) T_C, by selecting the Wikipedia super categories of each selected Wikipedia article. Then they compare each couple of forests (e.g., T_C and T_B for the classes C and B) in order to evaluate whether or not two classes should be aligned. The main drawback of this approach is that, it considers all the possible meanings (i.e., Wikipedia articles) for each ontology class, thus increasing the complexity of the method and the risk to discover wrong mappings. On the contrary, in our approach we address the problem of term ambiguity by performing automatic WSD. Moreover, they only present a system for LOD mapping discovery while we propose a complete methodology for translating, publishing and mapping LOD datasets.

Another system for publishing LOD resources is AGROPub [17], which facilitates integration of LOD agro-environmental resources. AGROPub comprises services and tools that enable resource providers to semantically annotate their resources by relevant concepts from selected agro-environmental domain ontologies, to generate and publish RDF descriptions of the resources to LOD and to link the published resources to related resources from LOD. Moreover, it provides services and tools that enable consumers of the agro-environmental resources to search and annotate published resources by adding their own annotations. However, the semantic annotations have to be added manually by the users (resource provides or consumers) by using the GUI, and no automatic or semi-automatic annotation method is provided.

Stratosphere[11] is an open-source cluster/cloud computing framework for Big Data analytics. In [12] this system has been extended for the integration of large data sets belonging to the Linked Open Data. In particular, it has been applied to integrate open governmental data with other popular LOD resources, such as DBpedia[12] and Freebase[13]. It addresses the problem of semantic and structural heterogeneity among different LOD resources by developing data cleansing operators for the Stratosphere framework. The integration methodology is mainly based on the analysis of the instances of the data sets and on the use of entity (e.g., persons, cities etc.) extraction and record linkage (i.e., identify the real world entities across the different data sources) techniques. However, in our case this method could not be applied as the great majority of data are numeric or do not correspond to entities.

Finally, WebSmatch [6] is a flexible environment for Web data integration with a service oriented architecture. It has been applied on the real scenario of Data

---

[11] https://www.stratosphere.eu/

[12] http://it.dbpedia.org/

[13] http://www.freebase.com/

Publica, a French company, providing added value over public data sets they crawl, such as visualization of data source or data integration. WebSmatch has been employed all over the process of metadata extraction, matching and visualizing data sources. For the matching phase, it exploits YAM++ (Yet Another Matcher) [8], a tool for pattern matching and alignment of ontologies, which combines different matching techniques mainly based on the string similarity, dictionary and thesauri like WordNet and instance based techniques. However, also in this case it does not make us of WSD techniques.

4   Conclusion and Future Work

In this paper, we presented an experimental and preliminary methodology to publish and link public open data to the LOD cloud. Moreover, we propose an automatic and multilingual method to semantically enrich LOD data sets by performing semantic annotation of schema elements with respect to the Multi-WordNet lexical thesaurus. The process has been applied to public data coming from the Research Project on Youth Precariousness of the district of Modena, Italy. However, it might be easily adapted for any other public data set.

As previously described, during the process we employed different open source software. The main drawback in using different tools is the need of creating custom interfaces in order to allow the automatic communication among them. Future work will be devoted to implement and integrated system providing all the functionalities supplied by the different tools used during the process in order to allow data providers and consumers to interact with them by using an integrated GUI.

Furthermore, we will investigate the application of traditional data integration and schema matching systems in the context of Linked Open Data: we will extend the MOMIS data integration system by adapting its schema matching method in dealing with RDF data; moreover, by using the MOMIS provenance module [1], we will add to the public data further RDF metadata describing the provenance of data (i.e., where data came from and how they were derived and modified over time) in order to provide consumers with valuable information that can be exploited during the LOD navigation.

Acknowledgments. Our sincere thanks to the Councillor for Youth Policies Fabio Poggi[14] and to Dr. Sergio Ansaloni, Head of Stradanove[15], Studies and Documentation Centre on Youth, for providing us the public data. This work is partially supported by the BIOGEST-SITEIA laboratory (www.biogestsiteia.unimore.it), funded by Emilia-Romagna (Italy) regional government.

---

[14] http://www.comune.modena.it/politichegiovanili/info/assessorato
[15] http://www.stradanove.net/

References

1. Beneventano, D.: Provenance based conflict handling strategies. In: Yu, H., Yu, G., Hsu, W., Moon, Y.-S., Unland, R., Yoo, J. (eds.) DASFAA Workshops 2012. LNCS, vol. 7240, pp. 286–297. Springer, Heidelberg (2012)
2. Bergamaschi, S., Castano, S., Vincini, M.: Semantic integration of semistructured and structured data sources. SIGMOD Record 28(1), 54–59 (1999)
3. Bergamaschi, S., Po, L., Sala, A., Sorrentino, S.: Data source annotation in data integration systems. In: Fifth International Workshop on Databases, Information Systems and Peer-to-Peer Computing, DBISP2P (2007)
4. Bizer, C., Heath, T., Berners-Lee, T.: Linked data - the story so far. Int. J. Semantic Web Inf. Syst. 5(3), 1–22 (2009)
5. Bizer, C., Lehmann, J., Kobilarov, G., Auer, S., Becker, C., Cyganiak, R., Hellmann, S.: Dbpedia - a crystallization point for the web of data. J. Web Sem. 7(3), 154–165 (2009)
6. Coletta, R., Castanier, E., Valduriez, P., Frisch, C., Ngo, D., Bellahsene, Z.: Public data integration with websmatch. CoRR, abs/1205.2555 (2012)
7. Cruz, I.F., Palmonari, M., Caimi, F., Stroe, C.: Towards "on the go" matching of linked open data ontologies. In: LDH, pp. 37–42 (2011)
8. Duchateau, F., Coletta, R., Bellahsene, Z., Miller, R.J. (not) yet another matcher. In: Cheung, D.W.-L., Song, I.-Y., Chu, W.W., Hu, X., Lin, J.J. (eds.) CIKM, pp. 1537–1540. ACM (2009)
9. Euzenat, J., Shvaiko, P.: Ontology matching. Springer, Heidelberg, DE (2007)
10. Fusari, E.: Linked open data: pubblicazione, arricchimento semantico e linking di dataset pubblici attraverso il sistema momis. Master Degree Thesis (2012), http://www.dbgroup.unimore.it/tesi/FusariElisa-tesi2012.pdf
11. Gliozzo, A.M., Strapparava, C., Dagan, I.: Unsupervised and supervised exploitation of semantic domains in lexical disambiguation. Computer Speech & Language 18(3), 275–299 (2004)
12. Heise, A., Naumann, F.: Integrating open government data with stratosphere for more transparency. J. Web Sem. 14, 45–56 (2012)
13. Jain, P., Hitzler, P., Sheth, A.P., Verma, K., Yeh, P.Z.: Ontology alignment for linked open data. In: Patel-Schneider, P.F., Pan, Y., Hitzler, P., Mika, P., Zhang, L., Pan, J.Z., Horrocks, I., Glimm, B. (eds.) ISWC 2010, Part I. LNCS, vol. 6496, pp. 402–417. Springer, Heidelberg (2010)
14. Jentzsch, A., Isele, R., Bizer, C.: Silk - generating rdf links while publishing or consuming linked data. In: ISWC Posters&Demos (2010)
15. Miller, A.: Wordnet: A lexical database for english. Communications of the ACM 38(11), 39–41 (1995)
16. Navigli, R.: Word sense disambiguation: A survey. ACM Comput. Surv. 41(2) (2009)
17. Nešić, S., Rizzoli, A.E., Athanasiadis, I.N.: Publishing and linking semantically annotated agro-environmental resources to LOD with aGROPub. In: García-Barriocanal, E., Cebeci, Z., Okur, M.C., Öztürk, A. (eds.) MTSR 2011. CCIS, vol. 240, pp. 478–488. Springer, Heidelberg (2011)
18. Po, L., Sorrentino, S.: Automatic generation of probabilistic relationships for improving schema matching. Inf. Syst. 36(2), 192–208 (2011)
19. shan Chen, P.P.: The entity-relationship model: Toward a unified view of data. ACM Transactions on Database Systems 1, 9–36 (1976)
20. Sorrentino, S., Bergamaschi, S., Gawinecki, M.: Norms: An automatic tool to perform schema label normalization. In: ICDE, pp. 1344–1347 (2011)

A Novel Approach to Developing Applications in the Pervasive Healthcare Environment through the Use of Archetypes

João Luís Cardoso de Moraes[1], Wanderley Lopes de Souza[1], Luís Ferreira Pires[2], Luciana Tricai Cavalini[3], and Antônio Francisco do Prado[1]

[1] Federal University of São Carlos, Computer Science, São Carlos, Brazil
{joao_moraes,desouza,prado}@dc.ufscar.br
[2] University of Twente, Software Engineering Group, Enschede, The Netherlands
l.ferreirapires@utwente.nl
[3] Fluminense Federal University, Department of Epidemiology, Niterói, Brazil
lutricav@vm.uff.br

Abstract. Pervasive Healthcare focuses on the use of new technologies, tools, and services, to help patients to play a more active role in the treatment of their conditions. Pervasive Healthcare environments demand a huge amount of information exchange, and specific technologies has been proposed to provide interoperability between the systems that comprise such environments. However, the complexity of these technologies makes it difficult to fully adopt them and to migrate Centered Healthcare Environments to Pervasive Healthcare Environments. Therefore, this paper proposes an approach to develop applications in the Pervasive Healthcare environment, through the use of Archetypes. This approach was demonstrated and evaluated in a controlled experiment that we conducted in the cardiology department of a hospital located in the city of Marília (São Paulo, Brazil). An application was developed to evaluate this approach, and the results showed that the approach is suitable for facilitating the development of healthcare systems by offering generic and powerful approach capabilities.

Keywords: Pervasive Healthcare, Ubiquitous Computing, *open*EHR, Archetypes, Domain Specific Language.

1 Introduction

Most countries face many problems related to healthcare, such as expensive care and the low quality of health services. These problems began to appear as a result of population increase and the lack of healthcare professionals, which is worsening due to the increasing number of the elderly, more chronic disease, and the growing demand for new treatments and technologies. For example, in the United States the number of graduated doctors was 15,632 in 1981 and 15,712 in 2001 (increasing by only 0.5%), while in the same period its population rose from 226 million to 281 million (increasing by 24%) [1].

B. Murgante et al. (Eds.): ICCSA 2013, Part V, LNCS 7975, pp. 475–490, 2013.

The current healthcare model is centered on employing highly specialized people, located in large hospitals and focusing on acute cases for treatment but needs to change into a distributed model, to produce faster responses and to allow patients to better manage their own health. The centralized healthcare model implies that patients and caregivers have to visit the same facility (a hospital or clinic) for the healthcare services to be delivered, which is often expensive and inefficient. A distributed healthcare model that pervades the daily lives of the citizens is more able to provide less expensive and more effective and timely healthcare, and characterizes Pervasive Healthcare. According to [2], the goal of Pervasive Healthcare is to enable the management of health and well-being by using information and communication technologies to make healthcare available anywhere, at anytime, and to anyone.

Ubiquitous Computing [3] has been considered to represent the new age of Computing and aims to enable the user to easily access and process information from anywhere, at anytime, and using any kind of device. Ubiquitous Computing environments whether in communities, homes, or hospitals can be extremely useful to build a Pervasive Healthcare model. Particularly, for the "soul" of this model to become directed toward the patient, the efficient care of that patient is fundamental. In this manner, it is necessary that the information exchange among the various professionals, doctors, nurses, health experts responsible for patients, is fast, efficient and safe.

The exchange of health information among heterogeneous Electronic Healthcare Record (EHR) systems in pervasive healthcare environments requires communication standards that enable interoperability between such systems. Although Health Level Seven (HL7)[1] is a widely-used international standard for message exchange between heterogeneous Healthcare Information Systems (HIS), it has some well-known limitations for representing clinical knowledge, such as its combined use of structured components and coded terms, which can result in inconsistent interpretations of clinical information [4]. openEHR[2] is a foundation dedicated to research into interoperable EHR, which defined an open architecture based on two-level modeling that separates information from knowledge, therefore addressing some of the limitations of HL7.

This paper proposes an approach to developing applications in the Pervasive Healthcare environment, by using archetype and Domain Specific Languages (DSLs) [5]. DSLs allow the creation of an infrastructure to reuse and generate most of the applications code of the target domain. The remainder of this paper is structured as follows. Section 2 introduces the main concepts about the openEHR standard. Section 3 describes the domain specific languages concept. Section 4 presents the proposed approach illustrated with a case study in the Healthcare domain. Section 5 shows the evaluation of the application developed using the proposed approach. Section 6 discusses some related work. Finally, section 7 presents concluding remarks and proposes further work.

---

[1] http://www.hl7.org
[2] http://www.openehr.org

2   *open*EHR Dual Model

The *open*EHR architecture was developed based on the two-level modeling paradigm, as shown in Fig. 1. At the first level, a common *Reference Model* (RM) was defined in terms of a predefined set of classes that model the structure of an electronic record; and on the second level, specific concepts were defined by restricting the RM classes in terms of so-called *archetypes*, expressed in the Archetype Definition Language (ADL) [6].

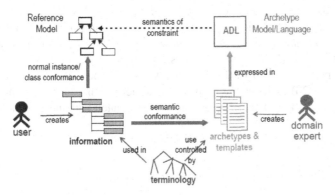

Fig. 1. *open*EHR two-level modeling paradigm

An archetype constitutes a formal model of a domain concept and is expected to be easily understandable by a domain expert. At the implementation level, archetypes can be translated into any language. In accordance with the two-level modeling, data input from users are stored according to the RM, but should also comply with the concepts expressed by the archetypes. The archetypes are designed by the domain experts, and not by information technologies professionals. This split should facilitate the interpretation of the knowledge extracted from the messages exchanged by health systems in various applications.

In the *open*EHR RM, the COMPOSITION class refers to one or more instances of the SECTION class, each containing ENTRY objects. The ENTRY class represents the actual recording of clinical content during a patient Observation, Examination, Assessment, or Intervention. ENTRY is defined as an abstract type with four concrete subtypes: (1) OBSERVATION, which can be used to represent clinical observations, such as blood pressure; (2) EVALUATION, which can be used to represent assessments made after a clinical observation is completed, such as risk assessment; and (3) INSTRUCTION and ACTION, which can be typically used to represent surgical procedures, medication, and other clinical interventions and actions taken. The ACTION subclass describes what was done and committed to the EHR as the result of an INSTRUCTION.

The Clinical Knowledge Manager (CKM) is the archetype repository proposed by the *open*EHR Foundation. This repository contains a set of archetypes that represent clinical concepts and can be reused in various health applications. We have reused some available archetypes provided by CKM, such as *Device, Device Details* and *Clinical Synopsis*, but we have also developed new archetypes to represent clinical concepts of the cardiology domain, such as *Pacemaker Implantation, Vascular Cardiac Surgery, Coronary Cardiac Surgery, Angioplasty Cardiac* and *Pacemaker Evaluation*.

The *openEHR* archetypes were defined for general reuse, and they can accommodate any number of natural languages and terminologies. An archetype consists of three sections: header, definition and ontology. Fig. 2 shows an extract of an ADL archetype for *Pacemaker Implantation* defined according to the *openEHR* standard. The *header* includes the name of the archetype (line 1). The *definition* section contains the structure and restrictions associated with the clinical concept defined by the archetype. *Pacemaker Implantation* specializes the ACTION class (line 4) of the RM. The CLUSTER part (line 8) refers to the stimulations mode for pacemaker implantation, and consists of an ELEMENT with a value of type DV_TYPE. The *ontology* section (line 14-20) includes the terminological definitions. In this example, the linguistic expression 'Pacemaker Implantation' is associated with the code 'at0000'.

```
1   archetype (adl_version=1.4) openEHR-EHR-ACTION.pacemaker_implantation.v1
2   [...]
3   definition
4     ACTION[at0000] matches {--Pacemaker Implantation
5       description matches {
6         ITEM_TREE[at0001] matches {   -- Tree
7           items cardinality matches {2..*; unordered} matches {
8             CLUSTER[at0008] occurrences matches {0..1} matches {--Stimulations Mode
9               items cardinality matches {1..2; unordered} matches {
10                ELEMENT[at0003] occurrences {0..1} matches {--Stimulation Mode
11                  value matches {
12                    DV_TEXT matches {*}}}
13  [...]
14  ontology
15    term_definitions = <
16      ["en"] = <
17        ["at0000"] = <
18          text = <"Pacemaker Implantation">
19          description = <"The Pacemaker implant is indicated for patients[...]">
20  [...]
```

Fig. 2. Archetype for the Pacemaker Implantation concept

The two-level modeling approach applied in *openEHR* allows designers to develop software systems separately from the domain modeling, by specializing and instantiating RM classes. Domain experts can introduce new concepts by defining new archetypes, so that the software systems do not have to be redesigned and redeployed whenever a new concept is defined. Therefore, the two-level modeling approach tends to make the systems easier to maintain and extend with new applications. Archetypes play the role of semantic gateways to terminologies, classifications and computerized clinical guidelines. Archetypes were introduced into *openEHR* to improve the level of semantic interoperability between various healthcare information systems.

3   Domain Specific Languages

A DSL is a language designed to be useful to a specific set of tasks within a given domain [7]. It can be defined by a meta-model, which represents the domain know-ledge of the target problem. Restricted to a specific domain, DSLs are usually small, consisting of just a set of abstractions and notations, which are closed to real terms known by the experts of this domain [8]. Thus, DSLs express solutions in the lan-guage and abstraction level of the problem domain, reducing the translation efforts of the concepts of this domain into the solution domain.

The use of DSLs in applications modeling, rather than general-purpose modeling languages (e.g., Unified Modeling Language – UML), allows the creation of more specific and complete models. Resources, such as frameworks, design patterns and components, may be included in models, creating an infrastructure that allows the execution of Model-to-Code (M2C) transformations, so as to generate a greater amount of code from the modeling.

4   Proposed Approach

In our approach, we applied *open*EHR dual-level modeling, as a model-driven metho-dology that has proved in software the achievement of semantic interoperability. The approach combines archetypes and templates in order to create communication inter-faces, using ADL as the domain specific language (DSL) that expresses the healthcare domain knowledge.

4.1   Overview

Using the Structured Analysis and Design Technique (SADT) [9] diagram notation, Fig. 3 shows an overview of the approach, which consists of two stages: Domain Engineering (DE) and Application Engineering (AE). In DE a DSL, represented by a

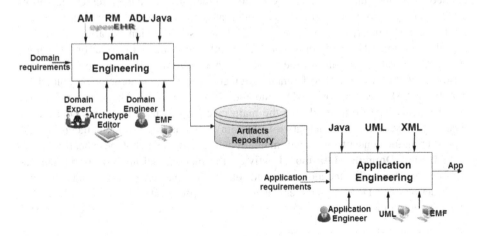

Fig. 3. Overview of the proposed approach

healthcare domain meta-model, and M2C transformations, to code generation are built. In AE, ubiquitous applications are developed through the reuse of the artifacts developed in the DE stage. The meta-model is used to support the application modeling, and the M2C transformations are employed to generate the code that handles *open*EHR messages using archetype.

4.2   Domain Engineering

Fig. 4 illustrates the three activities of the Domain Engineering stage: *Specify Domain Meta-model*, *Design Domain Meta-model*, and *Implement Domain Meta-model*.

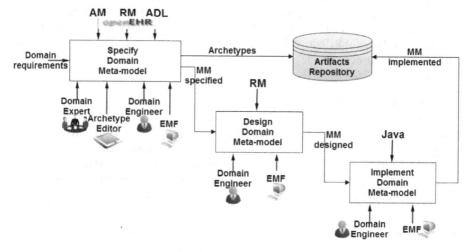

Fig. 4. Overview Domain Engineering

In **Specify Domain Meta-model** activity, the healthcare domain requirements are elicited, specified, analyzed and represented in a meta-model that expresses the knowledge about this domain. In the Eclipse Integrated Development Environment (IDE), the Domain Engineer, guided by the *open*EHR RM specifications, analyses the meta-model. The *open*EHR RM represents the healthcare domain information at a high abstraction level, which allows the use of a range of terminology in the concept description of this domain. The Domain Engineer specifies the archetypes, guided by *open*EHR AM specifications, ADL and the CKM repository. The Domain Experts and the Archetype Editor tool[3] are the main mechanisms used to support the archetypes specification. The outputs of domain analysis are domain meta-models, representing the concepts used in domain modeling, and the archetypes specified.

In **Design Domain Meta-model** activity, the meta-model analysis is refined according to standards, technologies and platforms that enable the meta-model's construction. The Domain Engineer, using the Eclipse Modeling Framework (EMF) [10], develops the architecture for the domain meta-model, showing the functional

---

[3] http://www.oceaninformatics.com/

decomposition of the domain. The process is guided by medical terminology coding, such as the Systematized Nomenclature of Medicine – Clinical Terms[4] (SNOMED-CT). SNOMED-CT is a scheme for identifying clinical terms and concepts, and are used by computer systems to support semantic interoperability. Fig. 5 shows part of a designed meta-model with a flexible structure which satisfies all important requirements and still leaves a large degree of freedom for its implementation.

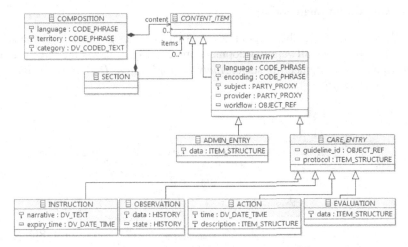

Fig. 5. Portion of *open*EHR RM meta-model

In **Implement Domain Meta-model** activity, based on the designed meta-model, the implementation is achieved. Supported by the EMF framework, the Domain Engineer automatically generates the Java code of the meta-model and of a model editor. Fig. 6 shows the Java code generation (2) for the ACTION class, based on the designed meta-model (1). This model editor will assist the Domain Engineer during

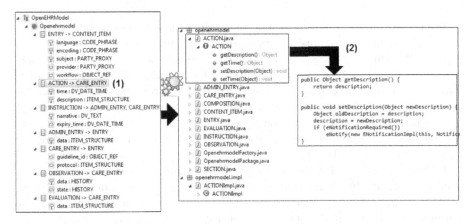

Fig. 6. Sketch of meta-model with Java code generation

[4] http://www.ihtsdo.org/snomed-ct/

the application model specification, during the AE stage. Both the model editor and the meta-model are available as a plugin in Eclipse IDE, ensuring that during AE the meta-model is instantiated to modeling the application. Once the meta-model is based on *open*EHR RM, the models created using DSL expressed by this meta-model, also correspond to refinement that should be performed to create the *open*EHR messages.

4.3   Application Engineering

The AE stage involves the disciplines of Analysis, Design, and Implementation, which are part of the software development process. Fig. 7 shows the three activities of the Application Engineering stage: *Analyzer Application*, *Design Application*, and *Implement Application*.

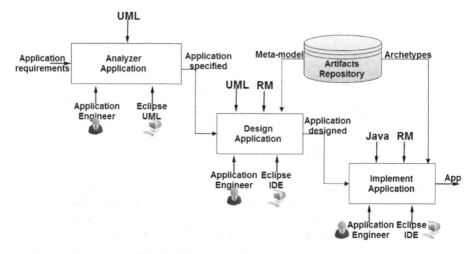

Fig. 7. Overview Application Engineering

The use of DSL expressed by healthcare domain meta-model to application modeling in AE, besides facilitating the application development for this domain, allows mapping the defined process for the *open*EHR messages development into the traditional activities of Software Engineering. Thus, the *open*EHR RM can be performed directly on models that correctly specify the application requirements. Moreover, the code generation automates many of the Application Engineer tasks associated with *open*EHR messages organization, complying with the concepts expressed by the archetypes. The healthcare meta-model also allows the reuse of domain knowledge in various projects during the AE.

To investigate the feasibility of the proposed approach, a case study was developed, building an ubiquitous application that allows evaluation of the implanted pacemaker in patients. The patient receives a notification on his mobile device to schedule an appointment. When the patient arrives in the clinic, the physician receives a message on his device containing information about the patient's pacemaker implantation. The messages received by the physician and patient are based on *ope-nEHR* archetypes related to the pacemaker implantation concept. The AE activities are detailed below, presenting examples based on the application described as the case study.

In the **Analyzer Application** activity, the application is specified from its requirements, which have been elicited, specified, analyzed and verified. For the specification, the Application Engineer uses UML techniques such as class, sequence and use cases diagrams. An example is the use case diagram shown in Fig. 8, which specifies the requirements to evaluate pacemaker implantation during an appointment. The clinical information was sent to the physician using the *open*EHR standard with archetypes.

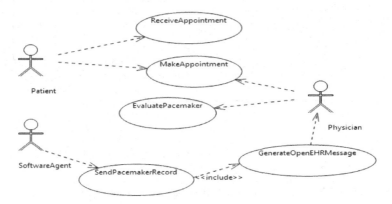

Fig. 8. Application Use Cases

In the **Design Application** activity, the application specification is refined using the technologies of hardware and software platforms, which enable the implementation of the application, such as Java Micro Edition (Java ME). Based on the use case diagrams, in this activity, the Application Engineer also performs the modeling of terms and concepts about the target problem domain as an instance of the meta-model built in DE, selecting the classes and attributes that are relevant to the application domain.

Fig. 9 shows a model built using the model editor plugin available in the Eclipse IDE. This model describes the information associated with the domain applications in accordance with the pacemaker implantation archetype. The *PacemakerImplantation* class is a subtype of the *ACTION* class from *open*EHR RM, as shown in Fig. 9 (1). The attribute *stimulationMode* represents the mode of stimulation used during the pacemaker implantation, according to the at0003 node in the *pacemaker_implantation* archetype, as shown in Fig. 9 (2).

Adopting the coding terminology used in the meta-model (SNOMED-CT), the Application Engineer combines the archetypes that define the clinical concepts represented in the models using the corresponding SNOMED-CT code. For example,

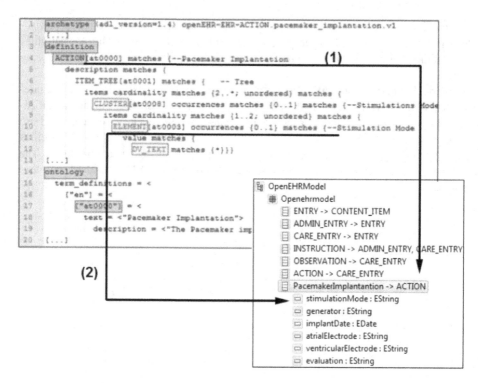

Fig. 9. Meta-model of the Pacemaker Implantation concepts

Fig. 9 shows the *stimulationMode* attribute of the *PacemakerImplantation* class, which was applied the SNOMED-CT code that identifies the pacemaker implantation concept on the archetype ontology section. Thus, the application becomes able to transmit understandable clinical data in an interoperable way.

In the **Implementation Application** activity, the application is implemented, including its communication interface, via integration of archetypes to *open*EHR messages. In the Eclipse IDE, the Application Engineer is assisted by the JET framework to perform the M2C transformations to generate partial code. This code handles the *open*EHR messages using archetypes, which will carry the data related to pacemaker implantation, according to specifications contained in the model. After the partial code generation, the Application Engineer performs the required supplements. Moreover, the other components that comprise the application are developed, such as the user integration interfaces and data persistence. At the end of implementation, tests are performed to provide feedback that indicates whether it is necessary or not to return to the previous activities of AE. Fig. 10 shows the *pacemaker_implantation* archetype and the user interface that contain clinical information related to the case study.

Fig. 10. Pacemaker evaluation – use case

5 Results and Discussion

We conducted a controlled experiment [11] in the cardiology department of the Santa Casa Hospital of Marília (São Paulo, Brazil), using the scenario discussed in Section 4. Table 1 shows the number of participants. The evaluation period was from 1 January 2012 to 30 July 2012.

Table 1. Participants (respondents)

Participants / Scenario	Physician	Medical Student	Nurse	Patient	Total
Scenario: Pacemaker Evaluation	3	2	2	15	22

In our study we used the Technology Acceptance Model (TAM) [12], which assumes that *perceived usefulness (PU)* and *perceived ease of use (PEU)* can predict attitudes towards usage and *intentions to use (IU)* technology. TAM is known to be a suitable model for explaining the technology acceptance process in the healthcare sector [13]. To test the effect on technology acceptance, two groups were selected for analysis, involving a limited but still relevant set of people: caregivers (physicians, medical students and nurses) and patients. We decided to group the caregivers together because they work as a team in the scenario.

After the implementation of the scenario, a structured questionnaire based on TAM was sent to the caregivers involved in the scenario, and a questionnaire was also distributed to patients after the medical procedure. Respondents were asked to indicate their agreement or disagreement with several statements on a five-point Likert scale

[14], ranging from 1=strongly disagree to 5=strongly agree. Many empirical studies have demonstrated that the psychometric properties of measurement scales can be affected by the ordering of items within the questionnaire. To avoid this bias, the questionnaire randomly intermixed items across constructs (PU, PEU, IU), and we conducted a group pretest to ensure that the scales were suitable. About 80% and 73% of the questionnaires distributed to caregivers and patients respectively were completed.

Our analysis consisted of two parts: (1) we tested the validity and reliability of the measurement model using Cronbach's Alpha [15]; (2) the data were analyzed using Structural Equation Modeling (SEM) [16] to test both the research model and the hypotheses. SEM is a statistical method that combines factor analysis and path analysis, enables theory construction and analyses the strengths of relationships among variables.

Table 2 shows the results of the validity tests for all scenarios, in which the internal consistency of the constructs were further evaluated for their reliability. The constructs had Cronbach's Alpha values at least close to the limit of 0.700, which is considered very good. Therefore, we concluded that these constructs are reliable for use in our data analysis. Based on this analysis, we concluded that the mean values showed good validity.

Table 2. Statistics and reliability of constructs

Participants	Constructs	Number of Items	Mean	Cronbach's Alpha
Caregivers	Perceived Ease of Use	4	4.05	0.813
	Perceived Usefulness	5	4.50	0.851
	Intention to Use	3	4.39	0.799
Patients	Perceived Ease of Use	4	4.13	0.790
	Perceived Usefulness	5	4.06	0.862
	Intention to Use	3	3.94	0.764

Next, the validated data were analyzed using SEM. The users' intentions to use the communication systems in each scenario can be explained or predicted based on their perception of ease of use and usefulness. Fig. 11 summarizes the research model used in this study. The labels on the arrows show the hypotheses and the path coefficients that show the relative strength and together indicate the causal relationships among variables, whereas R^2 is the percentage of total variance of the independent variables and indicates the predictability of the research model.

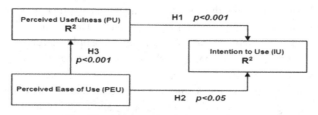

Fig. 11. Research model and hypotheses

The following three hypotheses were proposed and analyzed in each scenario:

H1. Perceived Usefulness positively affects Intention to Use the system.
- *Null hypothesis:* H_{1n}: $\mu_{PerceivedUsefulness} = \mu_{IntentionUse}$.
- *Alternative hypothesis:* H_{1a}: $\mu_{PerceivedUsefulness} \neq \mu_{IntentionUse}$.

H2. Perceived Ease of Use positively affects Intention to Use the system.
- *Null hypothesis:* H_{2n}: $\mu_{PerceivedEaseUse} = \mu_{IntentionUse}$.
- *Alternative hypothesis:* H_{2a}: $\mu_{PerceivedEaseUse} \neq \mu_{IntentionUse}$.

H3. Perceived Ease of Use positively affects Perceived Usefulness.
- *Null hypothesis:* H_{3n}: $\mu_{PerceivedEaseUse} = \mu_{PerceivedUsefulness}$.
- *Alternative hypothesis:* H_{3a}: $\mu_{PerceivedEaseUse} \neq \mu_{PerceivedUsefulness}$.

To verify these three hypotheses, we chose to apply *statistical regression analysis* [17] to the data collected from caregivers and patients in each scenario. The results are summarized in Table 3. For example, in the scenario (participants=caregivers), we test H_{3a} to verify if Perceived Usefulness is determined by Perceived Ease of Use. The details of the regression analysis are: $R^2=0.61$, p=0.0005 (very high significance with $p<0.001$ and $\alpha=0.05$). From the result of the regression we could reject the null hypothesis (H_{3n}), meaning that we empirically corroborate that Perceived Usefulness is determined by Perceived Ease of Use, and that H_{3a} was strongly confirmed. R^2 indicates that Perceived Ease of Use explains 61% ($R^2=0.61$) of the variance in Perceived Usefulness. Hypotheses H_{1a} and H_{2a} were confirmed, and Perceived Usefulness strongly determined Intention to Use, and Perceived Ease of Use was a significant secondary determinant.

Table 3. Statistic regression analysis

	Hypotheses	H1 (PU→IU)		H2 (PEU→IU)		H3 (PEU→PU)	
Scenario	Participants	R^2	p<0.001	R^2	p<0.05	R^2	p<0.001
Scenario:	Caregivers	0.72	0.0001	0.72	0.009	0.61	0.0005
Pacemaker Evaluation	Patients	0.35	0.0000	0.35	0.007	0.55	0.0002

Therefore, we could conclude that all three hypotheses were confirmed at all points of measurement. Based on this evaluation, the caregivers concluded that the system would be very useful for daily tasks and was very easy to use. Most patients identified some usability benefits, such as the efficient message exchange.

However, our data analysis approach has some limitations. First, the questionnaire model is not completely free of subjectivity for each respondent. Each respondent reacts in a particular way to a questionnaire. Second, we grouped all caregivers together and generalized the results, while we could have split them in several groups. Third, other factors may affect the decision of people to use a given technology, such as their prior experience and job relevance [18], which we did not take into consideration in our work. However, in this study we considered perceived usefulness and perceived ease of use as the most important factors to explain future intention to use the system.

6   Related Work

Due mainly to the complexity involves in the *open*EHR message development process and the lack of comprehensive support tools, various proposals to facilitate the use of the standard can be found in the literature.

In [19], there is a discussion of the implementation of a modeling tool for HL7 messages developed from Eclipse IDE. To handle the artifacts, the developed tool allows serializing of the messages in the XML Metadata Interchange (XMI) standard for later use.

In [20], an approach is presented for developing laboratory information system software by using archetypes to improve the software development process.

In [21], there is a discussion about the implementation of a software factory focused in the healthcare domain. The goal of developing the software factory was to automate the creation of communication interfaces, termed "collaboration ports", in order to facilitate message exchange.

In [22], a solution is presented to assist developers in handling messages. The authors' proposal is based on the use of a programmatic DSL to among other features support the messages creation, transmission, interpretation and validation.

The approach proposed in this paper is based on several characteristics of the works described above.It offers its own contributions through the evolution and adaptation of concepts described in related works. Among the contributions of this work, there is the use of archetypes to represent clinical concepts, which are integrated with *open*EHR messages. A DSL is also used that is based on a healthcare domain metamodel, which enables modeling and automatic code generation of the communication interfaces of the applications in that domain.

7   Conclusions

This paper has presented an approach to develop applications in the Pervasive Healthcare environment, and that approach integrates archetypes to *open*EHR messages via DSLs. The integration of the archetypes to *open*EHR messages achieves agility and efficiency in communication between heterogeneous HIS, once the communication becomes interoperable and clinical concepts can be easily identified. DSLs allow the creation of an infrastructure for code generation of these messages, and their reuse in different application of target domain, which reduces the necessary application development effort.

We performed a comprehensive study to investigate the usefulness and ease of use of pervasive healthcare technologies within healthcare environments. We presented the scenario that shows the acceptance of our pervasive healthcare approach by both caregivers and patients, who reacted positively with respect to the usefulness of the architecture. In our approach, the partial code for handling *open*EHR using archetypes messages was generated with the application of suitable M2C transformations.

In future work, we will further evaluate the performance of our approach, especially its scalability, which is a crucial non-functional requirement for realistic applications and for the simultaneous support of multiple scenarios. We will also measure any improvement in patient safety during the medical procedure. We plan future experiments to extend the TAM models to investigate the effect of other variables - such as, user experience, job relevance and output quality - on perceived usefulness, perceived ease of use, and intention to use.

Acknowledgment. We thank the National Council of Technological and Scientific Development (CNPq) for sponsoring our research in the context of the INCT-MACC.

References

1. Varshney, U.: Pervasive Healthcare Computing - EMR/EHR, Wireless and Health Monitoring. Springer, New York (2009)
2. Hansmann, U., Merk, L., Nicklous, M., Stober, T.: Pervasive Computing: The Mobile World. Springer (2003)
3. Weiser, M.: Some Computer Science Issues in Ubiquitous Computing. Commun. ACM 36, 75–84 (1993)
4. Browne, E.: openEHR Archetypes for HL7 CDA Documents. Ocean Informatics (2008)
5. Mernik, M., Heering, J., Sloane, A.M.: When and how to develop domain-specific languages. ACM Comput. Surv. 37, 316–344 (2005)
6. Beale, T., Heard, S.: Archetype Definition Language. openEHR Foundation (2007)
7. Gronback, R.: Eclipse Modeling Project: A Domain-Specific Language (DSL) Toolkit (2009)
8. van Deursen, A., Klint, P., Visser, J.: Domain-specific languages: an annotated bibliography. SIGPLAN Not 35, 26–36 (2000)
9. Ross, D.T.: Structured Analysis (SA): A Language for Communicating Ideas. IEEE Transactions on Software Engineering SE-3, 16–34 (1977)
10. Steinberg, D., Budinsky, F., Paternostro, M., Merks, E.: EMF: Eclipse Modeling Framework 2. Addison-Wesley Professional (2009)
11. Hevner, A.R., March, S.T., Park, J., Ram, S.: Design Science in Information Systems Research. MIS Quarterly 28, 75–105 (2004)
12. Davis, F.D.: Perceived Usefulness, Perceived Ease of Use, and User Acceptance of Information Technology. MIS Q 13, 319–340 (1989)
13. Chau, P.Y.K., Hu, P.J.H.: Investigating Healthcare Professionals' Decisions to Accept Telemedicine Technology: An Empirical Test of Competing Theories. Information & Management 39, 297–311 (2002)
14. Likert, R.: A Technique for the Measurement of Attitudes, New York (1932)
15. Nunnally, J.C.: Psychometric Theory. Current Contents/Social & Behavioral Sciences (1979)
16. Henseler, J., Chin, W.W.: A Comparison of Approaches for the Analysis of Interaction Effects Between Latent Variables Using Partial Least Squares Path Modeling. Structural Equation Modeling-a Multidisciplinary Journal 17, 82–109 (2010)
17. Goldin, R.F.: Review: Statistical Models-Theory and Practice. The American Mathematical Monthly 117, 844–847 (2010)
18. Davis, F.D., Venkatesh, V.: Toward Preprototype User Acceptance Testing of New Information Systems: Implications for Software Project Management. IEEE Transactions on Engineering Management 51, 31–46 (2004)

19. Banfai, B., Ulrich, B., Torok, Z., Natarajan, R., Ireland, T.: Implementing an HL7 version 3 modeling tool from an Ecore model. Studies in Health Technology and Informatics 150, 157–161 (2009)
20. Piho, G., Tepandi, J., Parman, M., Perkins, D.: From archetypes-based domain model of clinical laboratory to LIMS software. In: MIPRO, 2010 Proceedings of the 33rd International Convention, pp. 1179–1184 (2010)
21. Regio, M., Greenfield, J.: Designing and Implementing an HL7 Software Factory (2005)
22. Ohr, C., Václavík, M.: Using HL7 Processing Capabilities of the Open Ehealth Integration Platform in the Implementation of IHE Profiles. In: ICW Developer Conference, p. 11 (2010)

Observing Community Resiliency in Social Media

Robert M. Patton, Chad A. Steed, Chris G. Stahl, and Jim N. Treadwell

Oak Ridge National Laboratory, P.O. Box 2008, Oak Ridge, TN, USA, 37831
{pattonrm,steedca,stahlcg,treadwelljn}@ornl.gov

Abstract. In spite of social media's lack of structural integrity, accuracy, and reduced noise with respect to other forms of communication, it plays an increasingly vital role in the observation of societal actions before, during, and after significant events. In October 2012, Hurricane Sandy making landfall on the northeastern coasts of the United States demonstrated this role. This work provides a preliminary view into how social media could be used to monitor and gauge community resilience to such natural disasters. We observe, evaluate, and visualize how Twitter data evolves over time before, during, and after a natural disaster such as Hurricane Sandy and what opportunities there may be to leverage social media for situational awareness and emergency response.

Keywords: social media, temporal analysis, community resilience.

1   Introduction

Originally developed for entertainment purposes, social media systems have rapidly evolved to provide valuable benefits for such areas as business intelligence, national security, and disaster management. In combination with the development and adoption of smart phone technology, people have become mobile sensors providing "eyes" and "ears" as events unfold. Leveraging this capability provides significant advantages for situational awareness.

Unfortunately, this technology contains significant noise and error in the data as well as the inability to position the "sensors" in critical areas. Consequently, leveraging this capability can be quite challenging depending on the application purposes. Research into this issue has opened up several opportunities to overcome this problem. The work described here is a preliminary investigation into resolving the data noise issue as it relates to disaster management.

This work narrowly focuses on resolving a specific problem: identifying and characterizing the community resilience to natural disasters relating to physical infrastructure and social behavior using information obtained via Twitter. Community resilience is a measure of a community's ability to prepare, respond, withstand, and recover from disasters both natural and man-made. In the unfortunate event of a disaster, communities that have low resilience will suffer significant loss of life and damage to critical infrastructure. In addition, communities with low resilience require longer periods of time to recover to pre-disaster operations and quality of life. Communities with higher resilience suffer fewer losses and return more quickly to

B. Murgante et al. (Eds.): ICCSA 2013, Part V, LNCS 7975, pp. 491–501, 2013.
© Springer-Verlag Berlin Heidelberg 2013

pre-disaster operation and quality of life. Unfortunately, when a disaster occurs, the effects of any weakness in a community's resiliency will be amplified. Any ability for disaster management personnel and leadership to observe in real-time the effects of a disaster on a community provides a significant advantage and opportunity to respond more quickly. Consequently, the work described here is focuses on utilizing social media, specifically the Twitter platform, as a means of providing a real-time view into the impacts of a disaster on the community.

2   Related Works

Investigating the impacts of natural disasters on society is not a new research area. There are many works relating to the identification of different impacts as well as different studies that have been performed [1][11][12]. Many of these works and studies, however, have identified impacts that are difficult to measure quantitatively either before, during, or after the event. In addition, some works have relied on remote sensing techniques for assessing and monitoring disaster situations [4][9][13]. For our work, the goal is to develop an approach that would provide a more quantitative view of the event impact as well as be founded on field reporting about the event.

More recently, research has shifted into utilizing social media for disaster management. In [5], a system for crowdsourcing the collection of information for disaster management is described. Leveraging social networks, information is received and exchanged between volunteers in the field and disaster management command and control (DMCC). The DMCC can request the volunteers to visit specific locations and report on what is seen and heard, thereby making the DMCC much more adaptive to the circumstances. Unlike the work of [5] which relies on some structured data (e.g., automated geo-tagging of the volunteer's mobile device) to reduce the noise, the work of [10] uses natural language processing and data mining to extract situational awareness from Twitter data. Utilizing Twitter's location based search API, tweets are collected for specific areas and then processed and mined to provide a view into the current situation of a specific area. Results are then displayed as a combination of tag clouds and map-based user interfaces.

The primary problem with the two previous works is that they are very granular in detail as it relates to situational awareness. In disaster management, there are various "layers" of detail that allow disaster response personnel to view the event at different levels of abstraction. The two previous works provide little to no layers of abstraction. In the work of [2], Twitter data is processed in increasingly higher levels of abstraction such that individual tweets are eventually grouped into areas such as Health, Transport, Communication, and Lodging. While their work provides a critical connection of tweets to higher-level abstractions, what is lacking is a visualization or user interface to support a "top-down" view of the different layers of abstraction. The work described here attempts to address this deficit.

Others have also investigated the impact of events from news reports, but in different ways. In the work of [8], information from blogs was analyzed with respect to the actual sales of a book. The authors discovered that there was a direct relationship between blog chatter prior to a book release and the actual volume of book sales after the book release. The more people talked about a book on their blogs prior to its release, then the higher the volume of sales for the book after its release. The authors show clearly that blog chatter is a good indicator of the impact that a book release will create on sales volume. In the work of [6-7], the relationship between financial news and stock prices was investigated. The foundation of these works is the premise that financial news can have either a positive, negative, or natural impact on the price of a stock, and that the time lag between the news and the stock price was minimal, if any. The authors observed that, in fact, some of the financial news could provide indications as to the direction of the stock price (up, down, unchanged). However, their results are based on market simulations using real data. Regardless, their work provides evidence that events and event impacts can be monitored and gauged using news reports.

Finally, the work described here is an extension of our previous work in this area. In [14], news media was investigated in order to better understand various impacts of a disaster on society. In that work, the data source tended to be less noisy in terms of the language that was used. News media tend to be more factual and use more structured language as well as correct spellings. In contrast, the Twitter data uses abbreviated or incorrect syntax and spelling, and can be less factual and more biased to the individual needs and wants of a person. As addressed in [2] work, a bottom up approach was developed to identify trends in a larger population of people based on the tweets that depict individual wants and needs. The approach described here supports visualizing the higher-level trends and enables the ability to dive into details for specific areas of concern.

3   Approach

The primary focus is to detect, analyze, and characterize community resiliency metrics from events impacting society as observed in Twitter data. Inspiration for this work partially originated from the concept of decomposing a time-series into sub-series as discussed in [15], where time-series are decomposed into three sub-series referred to as seasonal, trend, and noise. The goal is to find complete or partial periodic patterns in a time-series with trends. Their work provides an approach to observe both short and long-term periodic patterns. They demonstrate their approach on atmospheric $CO2$ levels as well as stock prices. Figure 1 shows the conceptual view of our approach.

The original data begins as a time-series of all of the word and phrase counts that are observed in the Twitter fire hose. Next, a "textual prism" is created that is comprised of a set of taxonomies that describe the words and phrases related to the topics of interest by the user. The taxonomies are defined by the user and applied to the

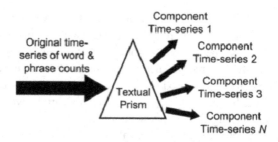

Fig. 1.

original time-series in order to produce a component time-series. This component time-series shows how the words and phrases of the taxonomy change over time.

As discussed previously, there have been investigations into the various impacts caused by natural disasters. For this work, the research performed by [3] provides the foundation for the impacts to be investigated. This work was chosen for its simplicity and extensibility. In [3], impacts can be observed in 4 areas: Technical, Organizational, Social, and Economic. Technical refers to the infrastructure (roads, bridges, power grid, water systems, etc.) of a society. Organizational refers to areas such as service crews that maintain or respond to the Technical aspects. Social refers to areas related to housing, shelters, provisions for human needs, etc. Economic refers to the impacts on the economy. This work focuses on the Technical and Social impacts of a natural disaster.

To begin, a set of taxonomies were developed for both the Technical and Social categories as shown in Table 1. These taxonomies were developed manually by analyzing news reports from the time period of August 25, 2005 to September 5, 2005. This is the time period when Hurricane Katrina made landfall in the Gulf Coast of the United States. This natural disaster created significant and widespread damage and resulted in extensive impacts on U.S. society. The news reports during this time period were clustered. The clusters were then analyzed to determine the most popular words and phrases that were used to describe specific conditions. These words and phrases were then categorized according to the framework defined in [3] as shown in Table 1.

As a specific example, the Shelters taxonomy consists of word and phrases such as: taking cover, taking shelter, seek refuge, shelters. The Movement taxonomy consists of words and phrases such as: mandatory evacuation, dawn curfew, flights canceled, etc. The Power taxonomy consists of phrases such as: downed power lines, backup power, emergency generator, and emergency power. The Roads taxonomy consists of phrases such as: alternate route, traffic backups, evacuation route, major streets, and interstates.

Table 1.

Technical	Social
Communications	Crime
Fuel	Death
Light	Health Hazards
Outage	Medical
Power	Movement
Rail	Shelter
Roads	Water (Health)
Structures	
Water (Infrastructure)	

After creating these taxonomies, they are then applied to the Twitter data as a form of "textual prism" to create component time-series as described in Figure 1. Every word and phrase observed from these categories is counted each day. For this particular investigation, the data was focused on the Hurricane Sandy event, and was collected using the Twitter location-based search to identify tweets in the geographic area of impact. When visualizing the volume of tweets for each taxonomy on the same chart, trends in numerically larger scales can obscure significant trends in smaller scales. To account for the volume of tweets for a particular taxonomy, the total counts of each day for each category was converted to be the percentage of total tweets for that taxonomy across the entire time that tweets were collected. For example, the Shelters taxonomy may have a total of 10,000 tweets over a 10-day period while the Medical taxonomy may have a total of 50 tweets over the same time period. On day 1, the Shelters total tweet count may have been 1,000 while the Medical may have been 5. For each taxonomy, the number of tweets on day 1 is 10% of the total number of tweets over the 10-day period, respectively. Thus, the data is normalized to a scale of 0 to 100.

4   Results

The results from the approach described previously are shown in Figures 2 through 5. Figures 2 and 3 show the trends of the Social taxonomies of Table 1. Figures 4 and 5 show the trends of the Technical taxonomies. Figures 2 and 4 show the total volume of tweets collected for the respective taxonomies, while Figures 3 and 5 show the normalized tweet counts, respectively.

As Hurricane Sandy makes landfall in New Jersey around midnight of October 29th, the trends in nearly every area begin moving upward. Of particular note, the Shelter and Movement taxonomies begin trending upward considerably and prior to the upward trends Death and Medical taxonomies. Unfortunately, most of the tweets in the Shelter and Movement taxonomies are not particularly useful. Many of the tweets are either originating from news agencies, retweets from news agencies, or tweets that contain information that could more easily be obtained from news articles. Furthermore, many of the tweets are simply wishing or praying for people to stay safe and seek shelter. Table 2 shows some example tweets during that time period.

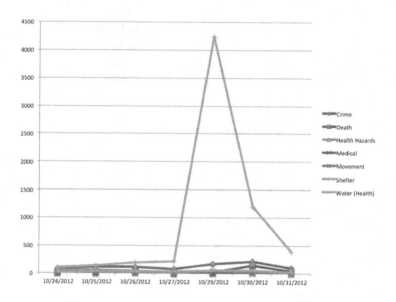

Fig. 2. Total number of tweets for Social taxonomies

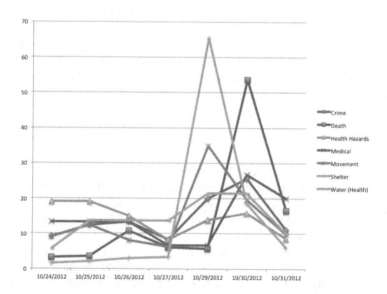

Fig. 3. Normalized number of tweets for Social taxonomies

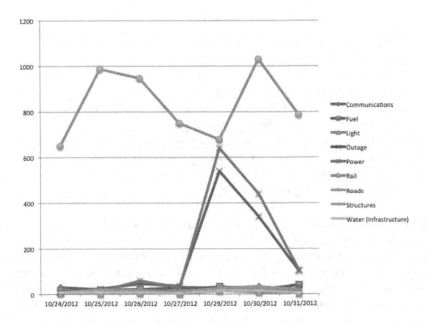

Fig. 4. Total number of tweets for Technical taxonomies

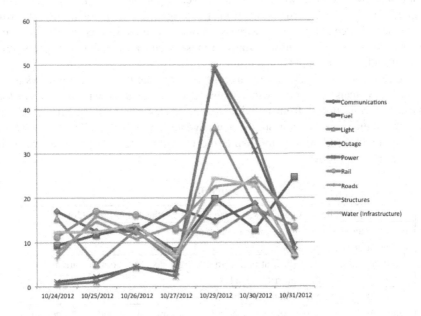

Fig. 5. Normalized number of tweets for Technical taxonomies

Table 2.

rt @sapendeleds: philadelphia to open shelters at 2 pm tomorrow for evac of flood-prone areas of city.
rt @corybooker: animal lovers: our jfk center shelter will have facilities to care for people displaced by storm who also have pets.
those without power: as of 11:30 pm the active living center in jim barnett park will be an emergency shelter for those without power.
all's mostly quiet in #cherrydale. still no power. headed to bed. stay safe all! #fb
please stay safe people who are on the east coast! #prayforeastcoast
... no we are staying put. as of now it's not mandatory in our town. plus we would have to find a dog friendly shelter :/

Table 3 shows example tweets from the Death and Medical taxonomies. Like the Shelter and Movement taxonomies, the tweets in Table 3 are mostly news oriented, retweet of news, or contain information more easily obtained from other sources. However, in two example tweets that departs from this pattern. One tweet is requesting information on where to receive medical attention in New York City (NYC), while another tweet states that a specific Twitter user needs medical attention. In the latter example, one would hope that 911 were called to assist the person if, in fact, it was an emergency and not a sarcastic remark.

Table 4 shows example tweets from the Power, Light, and Fuel taxonomies. While still consisting of tweets originating from news agencies, more of the tweets from these taxonomies consisted of eyewitness accounts such as downed power lines and trees and shortages at gas stations. Many of these tweets also tended to express more sentiment such as fear, anxiety, and relief.

These results highlight the discrepancy in the language that is used by news agencies and individuals who use Twitter. News agencies tend to express more factual information, while tweets from individuals tend to describe more specific details or express more emotion. In addition, the news agencies often include http links for followers to obtain more information, while the individuals tend to produce tweets directly to a specific follower.

Table 3.

death toll continues to rise after sandy steamrolls northeast http://t.co/2opfeg2b
rt @stormchaser4850: developing: update: ap - death toll from hurricane #sandy is now at least 33 via @abc7chicago
if ppl are in need of medical attention in nyc right now what options are available if any? #sandy cc. @hurricanehackrs
@\<Twitter ID> needs medical attention!
sandy death toll climbs; millions remain without power. thousands of homes are flooded others destroyed.

Table 4.

current power outages in monmouth and ocean counties http://t.co/j4zai70n via @asburyparkpress
power lines down and trees in nb
and the valley view power outages begin
backup generator just literally blew up
light pole fell and smashed a parked car on my street. #sandy is real!
people are lining up at gas stations where there isn't even gas just waiting for gas to come this is scary.
wtf all these gas stations are packed ????
just went on the road ..so dark no houses have light in them long lines at gas stations omg ??

In regard to trends, each taxonomy experiences an increasing trend as Hurricane Sandy approaches landfall and through the initial hours of landfall, but then quickly start declining. One exception of particular notice is the Fuel taxonomy as shown in Figure 5. While the other taxonomies are declining, the Fuel taxonomy makes a significant increase on the last day of our collection. Analyzing the individual tweets from this taxonomy reveals that a number of the tweets are from news agencies that are reporting shortages at gas stations. Further, there are an increasing number of tweets by individuals expressing frustration at the gas shortages and long lines at gas stations. For example, one tweet stated "people are lining up at gas stations where there isn't even gas just waiting for gas to come this is scary." Another tweet stated, "what gas stations are open and are not packed with hundreds of people. i highly doubt any." In the days and weeks after Hurricane Sandy, this gas shortage becomes a significant problem resulting in near riots and citizens venting their anger in various ways. Consequently, there is an opportunity here to supplement this preliminary work with sentiment analysis as well as additional taxonomies to capture social unrest in order to further capture the dynamics of the community as an event transpires.

5   Summary

This work is a preliminary investigation into the use of community resilience taxonomies applied to Twitter data. Visualizing these taxonomies over time provides insight into how communities respond to events in a variety of different ways. Our approach helps curb the effects of noise and errors in the Twitter data and provides a "top-down" visual analytics approach to leveraging this data for situational awareness purposes.

Additional work will incorporate sentiment analysis and additional taxonomies to capture more detail in the data as well as additional filters to further reduce noise and errors in the data. Furthermore, additional visualization involving parallel coordinates and a more sophisticated user interface to enable the user to thoroughly analyze the data will be developed.

Acknowledgements. Research sponsored by the Laboratory Directed Research and Development Program of Oak Ridge National Laboratory, managed by UT-Battelle, LLC, for the U. S. Department of Energy. LDRD #6427.

Prepared by Oak Ridge National Laboratory, P.O. Box 2008, Oak Ridge, Tennessee 37831-6285; managed by UT-Battelle, LLC, for the U.S. Department of Energy under contract DE-AC05-00OR2225. This manuscript has been authored by UT-Battelle, LLC, under contract DE-AC05-00OR22725 for the U.S. Department of Energy. The United States Government retains and the publisher, by accepting the article for publication, acknowledges that the United States Government retains non-exclusive, paid-up, irrevocable, worldwide license to publish or reproduce the published form of this manuscript, or allow others to do so, for United States Government purposes.

References

1. Alexander, D.: The Study of Natural Disasters, 1977-97: Some Reflections on a Changing Field of Knowledge. Disasters 21(4), 284–304 (1997)
2. Birregah, B., Top, T., Perez, C., Chatelet, E., Matta, N., Lemercier, M., Snoussi, H.: Multi-layer Crisis Mapping: A Social Media-Based Approach. In: 2012 IEEE 21st International Workshop on Enabling Technologies: Infrastructure for Collaborative Enterprises (WETICE), pp. 379–384 (2012)
3. Bruneau, M., et al.: A Framework to Quantitatively Assess and Enhance the Seismic Resilience of Communities. Earthquake Spectra 19(4), 733–752 (2003)
4. Chi, T., et al.: Research on information system for natural disaster monitoring and assessment. In: Proc. of the 2003 IEEE International Geoscience and Remote Sensing Symposium (2003)
5. Chu, E.T., Chen, Y.-L., Lin, J.-Y., Liu, J.W.-S.: Crowdsourcing support system for disaster surveillance and response. In: 2012 15th International Symposium on Wireless Personal Multimedia Communications (WPMC), pp. 21–25 (2012)
6. Fung, G.P.C., et al.: Stock prediction: Integrating text mining approach using real-time news. In: Proc. of the 2003 IEEE International Conference on Computational Intelligence for Financial Engineering, pp. 395–403 (2003)
7. Fung, G.P.C., et al.: The predicting power of textual information on financial markets. IEEE Intelligent Informatics Bulletin 5(1) (June 2005)
8. Gruhl, D., et al.: The predictive power of online chatter. In: Proceeding of the Eleventh ACM SIGKDD International Conference on Knowledge Discovery in Data Mining, Chicago, Illinois, pp. 78–87 (2005)
9. Hussain, M., et al.: Emerging geo-information technologies (GIT) for natural disaster management in Pakistan: an overview. In: Proc. of the 2nd International Conference on Recent Advances in Space Technologies (2005)
10. Jie, et al.: Using Social Media to Enhance Emergency Situation Awareness. Intelligent Systems 27(6), 52–59 (2012)

11. Kreimer, A., Arnold, M., Carlin, A.: Building Safer Cities: The Future of Disaster Risk. World Bank Publications (2003)
12. Little, R.G.: Toward more robust infrastructure: observations on improving the resilience and reliability of critical systems. In: Proc. of the 36th Annual Hawaii International Conference on System Sciences (2003)
13. Llinas, J.: Information fusion for natural and man-made disasters. In: Proc. of the 5th International Conference on Information Fusion, pp. 570–576 (2002)
14. Patton, R.M., et al.: Discovery, analysis, and characteristics of event impacts. In: 2008 11th International Conference on Information Fusion, pp. 1–8 (2008)
15. Yu, J.X., Ng, M.K., Huang, J.Z.: Patterns discovery based on time-series decomposition. In: Cheung, D., Williams, G.J., Li, Q. (eds.) PAKDD 2001. LNCS (LNAI), vol. 2035, p. 336. Springer, Heidelberg (2001)

Robust Secure Dynamic ID Based Remote User Authentication Scheme for Multi-server Environment

Toan-Thinh Truong[1], Minh-Triet Tran[1], and Anh-Duc Duong[2]

[1] University of Science, VNU-HCM
{ttthinh,tmtriet}@fit.hcmus.edu.vn
[2] University of Information Technology, VNU-HCM
daduc@uit.edu.vn

Abstract. Dynamic ID based authentication scheme is more and more important in wireless environments such as GSM, CDPD, 3G and 4G. One of important properties of such authentication scheme is anonymity. It must be guaranteed to defend the privacy of mobile users against outside attacks, and the scheme of Cheng-Chi Lee, Tsung-Hung Lin and Rui-Xiang Chang satisfies that requirement. However, another important property that should be considered is impersonation. The scheme must have capability to resist this kind of attack to protect legal users from illegal adversaries. In this paper, we demonstrate that Lee et al.'s scheme is still vulnerable to masquerade attack and session key attack with stolen smart card. Then we present an improvement of their scheme in order to isolate such problems.

Keywords: Authentication, Password, Dynamic identity, Smart card, Impersonation, Session key.

1 Introduction

Authentication is the important problem in the information systems. With the development of applications deployed in the internet, there are many transactions performed between users and service providers. Customers can access those services from many different devices such as, personal computer, tablet, mobile phone...Besides, communications can be set on diverse network technologies, for instances, 3G, 4G, wireless network...Therefore, to protect the security for such ambiances, two-way authentication is necessary to online services, especially financial ones.

Password authentication is a simple, efficient, and convenient authentication mechanism allowing a legal user to login to remote systems. A number of papers have put forward many ideas to improve password authentication schemes for safe login of legal users [5][10][11][12][13]. Nevertheless, all these schemes are based on static login identity, which is easy to leaking some information about a user's login message to an adversary. One solution to ID-theft is making identity of users vary for each login.

B. Murgante et al. (Eds.): ICCSA 2013, Part V, LNCS 7975, pp. 502–515, 2013.

In 2007, Liao and Wang proposed a secure dynamic ID-based remote user authentication scheme for multi-server environment [2]. The scheme has the following advantages: (1) It allows the users to choose and change their passwords freely. (2) It does not maintain any verifier tables in the remote system. (3) The remote user authentication scheme is secure against replay attacks, server spoofing attack, insider and stolen verifier attack.

In 2009, Han-Cheng Hsiang and Wei-Kuan Shih [3], however, showed that Liao and Wang's scheme is indeed completely vulnerable to insider's attack, masquerade attack, server spoofing attack, registration center spoofing attack and is not reparable. And they proposed a improved scheme to resist such problems. They claimed that their proposed scheme not only inherits Liao and Wang's advantages but it also enhances Liao and Wang's security by removing the security weaknesses.

In 2011, Cheng-Chi Lee, Tsung-Hung Lin and Rui-Xiang Chang showed that Hsiang and Shih's scheme is still vulnerable to a masquerade attack, server spoofing attack, and is not easily reparable. Moreover, it can not provide mutual authentication. So, they proposed an improved scheme and claimed that their scheme is more secure and efficient, compared with older schemes.

Unlike Lee et al.'s claimed, in this paper, we demonstrate that their scheme is still vulnerable to insider's attack, session key security problem, impersonation and stolen smart card attack. Then, we present an revised scheme to resist such problems.

The remainder of this paper is organized as follows: section 2 quickly reviews Lee et al.'s scheme; then section 3 discusses its weaknesses. Our proposed scheme is presented in section 4, while section 5 discusses the security and efficiency of the proposed scheme. Our conclusions are presented in section 6.

2   Review of Lee, Lin and Chang's Scheme

In this section, we review Lee, Lin and Chang's scheme. Their scheme includes four phases: registration phase, login phase, verification phase and password change phase. Some important notations in this scheme are listed below.

- U_i: ith user.
- ID_i: Unique identification of U_i.
- PW_i: Unique password of U_i.
- S_j: jth server.
- RC: Registration center.
- SC: Smart card.
- SID_j: Unique identification of S_j.
- CID_i: Dynamic ID of U_i.
- $h(.)$: A one-way hash function.
- x: The secret key maintained by registration center.
- y: A secret number maintained by registration center.
- $\oplus$: The exclusive-or operation.
- $\|$: The concatenation operation.

2.1   Registration Phase

When the user U_i wants to access the system, he/she has to submit his or her identity ID_i and PW_i to RC. Figure 1 illustrates the steps of the registration phase.

- Step R1. $U_i \Rightarrow RC$: ID_i, $h(b \oplus PW_i)$. U_i freely chooses his or her identity ID_i and PW_i, and computes $h(b \oplus PW_i)$, where b is a random number generated by U_i. Then U_i sends ID_i and $h(b \oplus PW_i)$ to the registration center RC for registration through a secure channel.
- Step R2. RC computes $T_i = h(ID_i \parallel x)$, $V_i = T_i \oplus h(ID_i \parallel h(b \oplus PW_i))$, $B_i = h(h(b \oplus PW_i) \parallel h(x \parallel y))$ and $H_i = h(T_i)$.
- Step R3. $RC \Rightarrow U_i$: RC issues a smart card to U_i, and the card contains $(V_i, B_i, H_i, h(.), h(y))$.
- Step R4. U_i keys b into his or her smart card, then the smart card contains $(V_i, B_i, H_i, b, h(.), h(y))$.

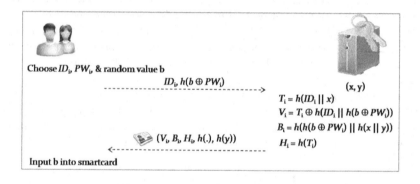

Fig. 1. Lee, Lin and Chang's registration phase

2.2   Login Phase

After receiving the smart card from RC, U_i can use it when he or she wants to login to S_j. Fig 2 illustrates the steps of login phase.

- Step L1. U_i inserts his or her smart card into the smart card reader and then inputs ID_i and PW_i. Then the smart card computes $T_i = V_i \oplus h(ID_i \parallel (b \oplus PW_i))$ and $H_i^* = h(T_i)$, and then checks if the H_i^* is the same as H_i. If they are the same, U_i continues to the next step. Otherwise the smart card rejects this login request.
- Step L2. The smart card generates a nonce N_i and computes $A_i = h(T_i \parallel h(y) \parallel N_i)$, $CID_i = h(b \oplus PW_i) \oplus h(T_i \parallel A_i \parallel N_i)$. $P_{ij} = T_i \oplus h(h(y) \parallel N_i \parallel SID_j)$ and $Q_i = h(B_i \parallel A_i \parallel N_i)$.
- Step L3. $U_i \Rightarrow S_j$: CID_i, P_{ij}, Q_i, N_i.

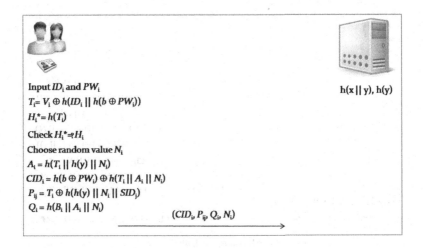

Input ID_i and PW_i

$T_i= V_i \oplus h(ID_i \parallel h(b \oplus PW_i))$

$H_i^*= h(T_i)$

Check $H_i^* =? H_i$

Choose random value N_i

$A_i = h(T_i \parallel h(y) \parallel N_i)$

$CID_i = h(b \oplus PW_i) \oplus h(T_i \parallel A_i \parallel N_i)$

$P_{ij} = T_i \oplus h(h(y) \parallel N_i \parallel SID_j)$

$Q_i = h(B_i \parallel A_i \parallel N_i)$

$h(x \parallel y), h(y)$

$(CID_i, P_{ij}, Q_i, N_i)$

Fig. 2. Lee, Lin and Chang's login phase

2.3   Verification Phase

After receiving the login request sent from U_i, S_j performs the following tasks to authenticate the user's login request. Fig 3 illustrates the steps of verification phase.

- Step V1. On receiving the login request $(CID_i, P_{ij}, Q_i, N_i)$, S_j computes $T_i = P_{ij} \oplus h(h(y) \parallel N_i \parallel SID_j)$, $A_i = h(T_i \parallel h(y) \parallel N_i)$, $h(b \oplus PW_i) = CID_i \oplus h(T_i \parallel A_i \parallel N_i)$ and $B_i = h(h(b \oplus PW_i) \parallel h(x \parallel y))$ by using received message (CID, P, N), $h(y)$ and $h(x \parallel y)$.
- Step V2. S_j computes $h(B_i \parallel A_i \parallel N_i)$ and checks it with Q_i. If they are not equal, S_j rejects the login request and terminates this session. Otherwise, S_j accepts the login request and generates a nonce N_j to compute $M'_{ij} = h(B_i \parallel N_i \parallel A_i \parallel SID_j)$. Finally, S sends the message (M'_{ij}, N_j) to U.
- Step V3. Upon receiving these message (M'_{ij}) from S_j, U_i computes $h(B_i \parallel N_i \parallel A_i \parallel SID_j)$ and checks it with received message M'_{ij}. If they are not equal, U_i rejects these messages and terminates this session. Otherwise, U_i authenticates successfully S_j and computes $M''_{ij} = h(B_i \parallel N_j \parallel A_i \parallel SID_j)$. Finally, U_i sends back the message (M''_{ij}) to S_j.
- Step V4. On receiving this message (M''_{ij}) to S_j computes $h(B_i \parallel N_i \parallel A_i \parallel SID_j)$ and checks it with received message (M''_{ij}). If they are equal, S_j authenticates successfully U_i. After finishing verification phase, U_i and S_j can compute $SK = h(B_i \parallel N_i \parallel N_j \parallel A_i \parallel SID_j)$ as the session key for securing communications with authenticator.

2.4   Password Change Phase

In this phase, U_i can change his or her password anytime when he/she wants. The steps of the password change phase are as follows:

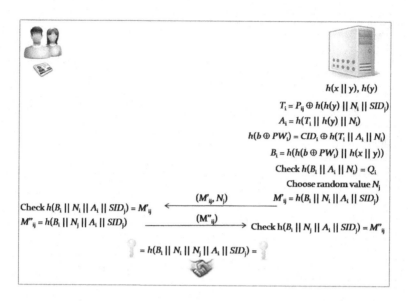

Fig. 3. Lee, Lin and Chang's verification phase

- Step P1. U_i inserts his or her smart card into the smart card reader and then inputs ID_i and PW_i.
- Step P2. The smart card computes $T_i = V_i \oplus h(ID_i \parallel h(b \oplus PW_i))$ and $H_i^* = h(T_i)$ and then checks if the H_i^* is the same as H_i. If they are the same, U_i chooses a new password and a new random number b new to compute $h(b_{new} \oplus PW_{new})$ and $V_{new} = T_i \oplus h(ID_i \parallel h(b_{new} \oplus PW_{new}))$. Finally, U_i sends ID_i and $h(b_{new} \oplus PW_{new})$ to RC via secure channel.
- Step P3. RC computes $B_{new} = h(h(b_{new} \oplus PW_{new}) \parallel h(x \parallel y))$. RC sends back B_{new} to U_i.
- Step P4. Finally, the smart card replaces V_i and B_i with V_{new} and B_{new}.

3   Cryptanalysis of Lee, Lin and Chang's Scheme

In this section, we present our results on Lee, Lin and Chang's scheme. We will show that their scheme is vulnerable to insider attack, impersonation attack, stolen smart card attack and session key attack. Besides, their scheme is not practical when only registration center is in charge of changing password for users. In our scheme, this task is shared with other servers.

3.1   Insider and Impersonation Attacks

In Lee, Lin and Chang's scheme, we see that any legal user can fake another legal user. If an attacker A is a legal user, A will have $h(y)$. Following are some steps A performs to steal important user's information.

- When other users send $(CID_i, P_{ij}, Q_i, N_i)$, A can block this package.
- Then, A can compute $T_i = P_{ij} \oplus h(h(y) \parallel N_i \parallel SID_j)$, $A_i = h(T_i \parallel h(y) \parallel N_i)$, $h(b \oplus PW_i) = CID_i \oplus h(T_i \parallel A_i \parallel N_i)$.
- With $h(b \oplus PW_i)$ in hand, A will re-register with RC by sending $(ID_{new}, h(b \oplus PW_i))$.
- Then, RC will perform steps in registration phase. One of them is $B_i = h(h(b \oplus PW_i) \parallel h(x \parallel y))$. So, when RC sends back $(V_i, B_i, H_i, h(.), h(y))$ to A, A will pick B_i.
- Finally, A has B_i, T_i and $h(b \oplus PW_i)$ of one legal user, attacker A is easy to fake that legal user.

Following are some steps A performs to fake another user.

- Attacker A will compute $A_i = h(T_i \parallel h(y) \parallel N_i)$, $CID_i = h(b \oplus PW_i) \oplus h(T_i \parallel A_i \parallel N_i)$.
- A continues to compute $P_{ij} = T_i \oplus h(h(y) \parallel N_i \parallel SID_j)$ and $Q_i = h(B_i \parallel A_i \parallel N_i)$, where N_i is A's random value, T_i is computed from A's previous blocking package of another legal user, and B_i is from A's re-registration phase.
- A fakes another legal user successfully by sending above parameters to server.

3.2   Stolen Smart Card and Session Key Attacks

In Lee, Lin and Chang's scheme, we see that stealing smart card of another legal user will cause a bad result for other legal users. Because attacker A will have $h(y)$ in the smart card, attacker A can perform following steps to achieve another user's $h(b \oplus PW_i)$.

- A captures any login message $(CID_i, P_{ij}, Q_i, N_i)$.
- Then, A computes $T_i = P_{ij} \oplus h(h(y) \parallel N_i \parallel SID_i)$ and $A_i = \mathrm{h}(T_i \parallel h(y) \parallel N_i)$.
- Finally, A achieves $h(b \oplus PW_i) = CID_i \oplus h(T_i \parallel A_i \parallel N_i)$.

With $h(b \oplus PW_i)$ in hand, A also re-registers to have B_i of another legal user. So, with B_i, A can impersonate that legal user or steal session key of that user by collecting random values of that legal user and S_j to compute $SK = h(B_i \parallel N_i \parallel N_j \parallel A_i \parallel SID_j)$. Clearly, all information encrypted with SK will be revealed.

3.3   Ineffectiveness in Password Change Phase

In Lee, Lin and Chang's scheme, we see that changing password of users must communicate with RC. So, when many users wants to change password due to some reasons, RC cannot process simultaneously. When a number of users increase, this problem will be worse. Clearly, their scheme is not practical.

4   Proposed Scheme

In this section, we will propose an revised scheme of Lee, Lin and Chang's scheme that removes the security problems described in the previous section. Our improved scheme not only inherits the advantages of their scheme, it also enhances the security of it. Our scheme is also divided into the four phases of registration, login, mutual verification and session key agreement and password change.

4.1   Registration Phase

When one user U_i wants to register to the service provider S_j, he or she has to submit his or her identity ID_i and password PW_i to RC. Then, RC performs the following steps:

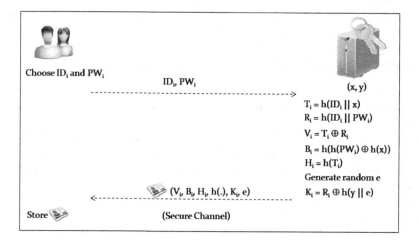

Fig. 4. Proposed registration phase

1. Compute $T_i = h(ID_i \parallel x)$, $R_i = h(ID_i \parallel PW_i)$, $V_i = T_i \oplus R_i$, $B_i = h(h(PW_i) \oplus h(x))$ and $H_i = h(T_i)$. Then, RC generates a random number e to compute $K_i = R_i \oplus h(y \parallel e)$.
2. Issue the smart card with the secret parameters $(V_i, B_i, H_i, h(.), K_i, e)$ to the user U_i through a secure channel.

4.2   Login Phase

The user U_i types his or her identity ID_i^*, password PW_i^* and the server identity SID_j to login the service provider S_j, and then the smart card performs the following steps:

1. Compute $R_i^* = h(ID_i^* \parallel PW_i^*)$, $T_i^* = V_i \oplus R_i^*$ and $H_i^* = h(T_i^*)$. Checks whether H_i and H_i^* is equal or not. If yes, the legality of the user can be assured and proceeds to the next step; otherwise, reject the login request.
2. Generate nonce N_i and compute (CID_i, P_{ij}, Q_i) in accordance with the following equations: $CID_i = h(PW_i) \oplus h(T \parallel R_i^* \parallel N_i)$, $P_{ij} = T_i \oplus h(R_i^* \parallel N_i \parallel SID_j)$ and $Q_i = h(B_i \parallel R_i^* \parallel N_i)$.
3. Send the login request message $(CID_i, P_{ij}, Q_i, N_i, K_i, e)$ to the service provider S_j.

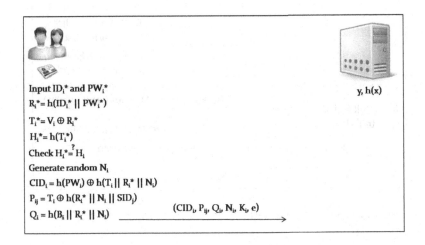

Fig. 5. Proposed login phase

4.3   Mutual Verification and Session Key Agreement Phase

In this session, the service provider S_j will receive the login request message $(CID_i, P_{ij}, Q_i, N_i, K_i, e)$ from U_i in the login phase, S_j authenticates the user U_i with the following steps:

1. Compute $R_i = h(y \parallel e) \oplus K_i$, $T_i = P_{ij} \oplus h(R_i \parallel N_i \parallel SID_j)$, $h(PW_i) = CID_i \oplus h(T_i \parallel R_i \parallel N_i)$ and $B_i = h(h(PW_i) \oplus h(x))$.
2. Compute $h(B_i \parallel R_i \parallel N_i)$, and then compare it with Q_i. If they are not equal, the server S_j rejects the login request and terminates this session.
3. Generate nonce N_j and compute $M_{ij1} = h(B_i \parallel N_i \parallel R_i \parallel SID_j)$, and then send back the message (M_{ij1}, N_j) to the user U_i.
4. On receiving the message (M_{ij1}, N_j), the user U_i continues to perform.
5. Compute $h(B_i \parallel N_i \parallel R_i^* \parallel SID_j)$ and compare it with M_{ij1}. If they are equivalent, it indicates that the legality of the service provider S_{ij} is authenticated; otherwise, the connection is terminated.
6. Compute $M_{ij2} = h(B_i \parallel N_j \parallel R_i^* \parallel SID_j)$, and then send back M_{ij2} to the service provider S_j.

7. Upon receiving the message M_{ij2}, the service provider S_j responds.
8. Compute $h(B_i \| N_j \| R_i \| SID_j)$ and compare it with M_{ij2}. If it is hold, the identity of U_i can be assured.

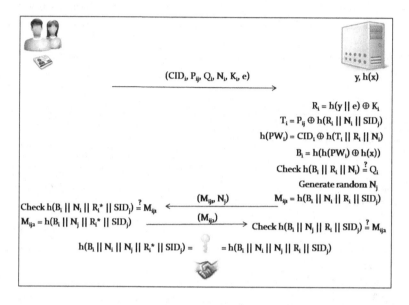

Fig. 6. Proposed mutual verification and session key agreement phase

After finishing above phase, the user U_i computes $SK_1 = h(B_i \| N_i \| N_j \| R_i^* \| SID_j)$ and the service provider S_j computes $SK_2 = h(B_i \| N_i \| N_j \| R_i \| SID_j)$. We see that $SK_1 = SK_2$.

4.4   Password Change Phase

When one user U_i wants to update his or her password without the help of RC but of Server S_j, he or she inserts his/her smart card to card reader and keys (ID_i^*, PW_i^*) corresponding to the smart card. To avoid the adversary updating password freely by way of stealing the smart card, the smart card first works as the step 1 of login phase. After assuring the legality of the card holder, the smart card performs some steps to update password

1. Smart card allows the card holder to re-submit a new password PW_i^{new}.
2. The smart card recomputes $R_i^{new} = h(ID_i \| PW_i^{new})$, V_i stored in the smart card can be updated with $V_i^{new} = T_i \oplus R_i^{new}$.
3. Similarly, K_i stored in the smart card can be replaced with $K_i^{new} = K_i \oplus R_i \oplus R_i^{new}$.

4. Then smart card computes $h(PW_i{}^{new})$ and sends $h(PW_i{}^{new})$ to S_j via secure channel.

5. S_j computes $B_i{}^{new} = h(h(PW_i{}^{new}) \oplus h(x))$. Finally S_j sends $B_i{}^{new}$ to U_i via secure channel.

6. U_i updates $B_i = B_i{}^{new}$.

5 Security and Efficiency Analysis

In this section, we analyze proposed scheme on two aspects: security and efficiency

5.1 Security Analysis

In this subsection, we analyzes our authentication scheme's security based on some kinds of attacks and standards which many previous schemes follows.

Two-Factor Security. We consider two-factor problem because the smart card cannot prevent the information stored in it being extracted. This is the important result which we review from [21] and [22]. Our scheme satisfies two-factor security[4] perfectly. Unlike Lee et al.'s scheme, when smart card of another legal user loss, attacker can know $h(y)$ to infer $h(b \oplus PW_i)$, and then can impersonate that legal user or re-register with RC to obtain B_i of other users. In our scheme, if legal user loss smart card and PW_i, attacker can not impersonate that legal user and other legal users. In our scheme, when attacker has smart card (V_i, B_i, H_i, $h(.)$, K_i, e) and PW_i, attacker can not compute R_i without ID_i. So, attacker can not obtain T_i. Clearly, attacker has no way to compute CID_i, P_{ij}, and Q_i to impersonate user.

Replay Attack. The replay attack is replaying the same message of the receiver or the sender again. Our scheme uses nonce instead of time stamp to withstand replay attacks. After incepting the ($CID_i{}^*$, $P_{ij}{}^*$, $Q_i{}^*$, $N_i{}^*$, $K_i{}^*$, e^*) from U_i, the attacker can replay the same message to the service provider S_j. Then, attacker can receive message (M_{ij1}, N_j) from S_j. Nevertheless, attacker can not compute M_{ij2} to respond to the server S_j without R_i and B_i. Similarly, when we assume that attacker replies a previous message ($M_{ij1}{}^*$, $N_j{}^*$) to U_i. On receiving ($M_{ij1}{}^*$, $N_j{}^*$), U_i computes $h(B_i \parallel N_i \parallel R_i \parallel SID_j)$ and checks if it is equal to $M_{ij1}{}^*$. It is clear that the equality can not hold since N_i is not equal to $N_i{}^*$.

Server Spoofing Attack. Our scheme resists server spoofing attacks because attacker has no way to have secret number y and $h(x)$ of S_j. So, attacker can not compute B_i to fool legal user.

Insider and Stolen Verifier Attack. In our scheme, we see that users have no way to compute y or $h(x)$ of S_j from information of their smart card. This is because RC generated random number e for each user. So, users can compute $h(y \parallel e)$ of their own by performing $K_i \oplus h(ID_i \parallel PW_i)$, but can not have $h(y \parallel e)$ of other users.

Stolen Smart Card Attack. Our scheme resists stolen smart card attack. Attacker can not compute anything to fake user or server. Attacker has no way to have $h(x)$ and y of server, and attacker also can not compute R_i to fake user without PW_i and ID_i.

Known-Key Attack. The known-key security means that compromise of a past session key can not derive any further session key. In our scheme, the session key SK is associated with B_i and R_i, which are unknown to the adversary. Even though the past session key SK is disclosed, the attacker can not derive B_i and R_i based on the security of one-way hash function. Thus, the attacker cannot obtain any further session key.

User Anonymity Protected. The user U_i will send the login request(CID_i, P_{ij}, Q_i, N_i, K_i, e) to the server S_j in each login. Thus, the attacker might incept and analyze the login message. It is infeasible to know ID_i from the login message. Furthermore, the login message is dynamic in each login. Among the parameters of login message, CID_i is associated with nonce N_i, R_i and dynamically changed. Similarly, the values of P_{ij} and Q_i are also related to nonce N_i and R_i. Consequently, an adversary can not identify the person who is trying to login. In other words, our scheme can protect user's anonymity.

Reparability. If the user U_i finds that $B_i = h(h(PW_i) \oplus h(x))$ has been compromised, he/she can re-register with RC via secure channel. U_i chooses a new password and send (ID_i, PW_{new}) to RC via secure channel. After receiving ID_i and PW_{new}, RC computes $R_{new} = h(ID_i \parallel PW_{new})$, $V_{new} = T_i \oplus R_{new}$, $B_{new} = h(h(PW_{new}) \oplus h(x))$ and $K_{new} = R_{new} \oplus h(y \parallel e_{new})$. Finally we see that a new smart card for U_i, (V_{new}, B_{new}, H_i, $h(.)$, K_{new}, e_{new}). Therefore, the improved scheme is easily reparable.

5.2 Efficiency Analysis

To analyze the computational complexity of our scheme, we introduce the notation T_h, which is the time for computing one-way hash function.

In table 1, there are three previous schemes. Liao and Wang's scheme needs 5 $\times$ T_h in registration phase and 15 $\times$ T_h in login and verification phase. Hsiang and Shih's scheme needs 6 $\times$ T_h in registration phase and 20 $\times$ T_h in login and verification phase. Lee, Lin and Chang's scheme, we see that we need 6 $\times$ T_h in registration phase and 16 $\times$ T_h in login and verification phase. Our scheme is the same as their scheme. It does not add any additional computational costs and the proposed scheme also enhances security.

Table 1. A comparison of computation costs

Phases / Schemes	Login-verification	Registration
Liao[2]	$15 \times T_h$	$5 \times T_h$
Hsiang[3]	$20 \times T_h$	$6 \times T_h$
Lee[1]	$16 \times T_h$	$6 \times T_h$
Ours	$16 \times T_h$	$6 \times T_h$

In table 2, we list the functionality comparisons between our improved scheme and others. According to Lee et al[1], the reparability of Liao[2] and Hsiang[3] is poor. Besides, all related schemes cannot resist to server spoofing, masquerade, and insider attacks. It can be seen that functionality comparisons of our scheme is more secure against various attacks.

Table 2. The functionality comparison between our scheme and the others

Schemes / Kinds of Attacks	Liao[2]	Hsiang[3]	Lee[1]	Ours
Reparability	No	No	Yes	Yes
Server spoofing	No	No	No	Yes
Masquerade	No	No	No	Yes
Insider	No	No	No	Yes
Mutual authentication	No	No	Yes	Yes
User anonymity	Yes	Yes	Yes	Yes
Replay attack	Yes	Yes	Yes	Yes
Known-key attack	Yes	Yes	Yes	Yes
Two-factor	No	Yes	No	Yes
No verification table	Yes	Yes	Yes	Yes

6   Conclusions

In this paper, we review secure dynamic ID based remote user authentication scheme for multi-server environment using smart cards of Lee et al. Although their scheme is equipped with anonymity and secure against to server and registration center spoofing attacks, we see that their scheme is vulnerable to insider, impersonation, session key, stolen smart card attacks and is not practical. So, we propose an improved scheme to eliminate such problems.

Compared with related schemes, the proposed scheme has the following main advantages; (1) User can choose the password freely. (2) It provides secure user anonymity. (3) It does not hold the password verification table for mobile users. (4) It provides mutual authentication and safe session key establishment. (5) Its password change phase is more reasonable than related schemes. As a result, the proposed scheme is able to provide greater security and be practical in wireless communication systems.

In the future, however, we will research a remote mutual authentication scheme for multi-server environment on elliptic curve cryptosystem (ECC) using smart card to enhance security more and apply to more applications in electronic transactions.

References

[1] Lee, C.-C., Lin, T.-H., Chang, R.-X.: A secure dynamic ID based remote user authentication scheme for multi-server environment using smart cards. Expert Systems with Applications 38(11), 13863–13870 (2011)

[2] Liao, Y.-P., Wang, S.-S.: A secure dynamic ID based remote user authentication scheme for multi-server environment. Computer Standards & Interfaces 31(1), 24–29 (2009)

[3] Hsiang, H.-C., Shih, W.-K.: Improvement of the secure dynamic ID based remote user authentication scheme for multi-server environment. Computer Standards & Interfaces 31(6), 1118–1123 (November 2009)

[4] Hwang, M.S., Lee, C.C., Tang, Y.L.: Improved efficient remote user authentication schemes. Int. J. Netw. Secur. 4(2), 149–154 (2007)

[5] Lee, C.C., Hwang, M.S., Yang, W.P.: Flexible Remote User Authentication Scheme Using Smart Cards. ACM Operating Systems Review 36(3), 46–52 (2002)

[6] Das, M.L., Saxena, A., Gulati, V.P.: A Dynamic ID-based Remote User Authentication Scheme. IEEE Transactions on Consumer Electronics 50(2), 629–631 (2004)

[7] Yoon, E.-J., Yoo, K.-Y.: Improving the Dynamic ID-Based Remote Mutual Authentication Scheme. OTM Workshops (1), 499–507 (2006)

[8] Chen, T.-H., Chen, Y.-C., Shih, W.-K., Wei, H.-W.: An efficient anonymous authentication protocol for mobile pay-TV. Advanced Topics in Cloud Computing 34(4), 1131–1137 (2011)

[9] Menezes, A.J., Oorschot, P.C., Vanstone, S.A.: Handbook of Applied Cryptograph. CRC Press, New York (1997)

[10] Lamport, L.: Password Authentication with Insecure Communication. Communications of the ACM 24, 770–772 (1981)

[11] Hwang, M.S., Lee, C.C., Tang, Y.L.: A Simple Remote User Authentication Scheme. Mathematical and Computer Modelling 36, 103–107 (2002)

[12] Li, L.H., Lin, I.C., Hwang, M.S.: A Remote Password Authentication Scheme for Multiserver Architecture Using Neural Networks. IEEE Transactions on Neural Network 12(6), 1498–1504 (2001)

[13] Shen, J.J., Lin, C.W., Hwang, M.S.: A Modified Remote User Authentication Scheme Using Smart Cards. IEEE Transactions on Consumer Electronics 49(2), 414–416 (2003)

[14] Xu, J., Zhu, W.-T., Feng, D.-G.: An efficient mutual authentication and key agreement protocol preserving user anonymity in mobile networks. Computer Communications 34(3), 319–325 (2011)

[15] Wang, R.-C., Juang, W.-S., Lei, C.-L.: Robust authentication and key agreement scheme preserving the privacy of secret key. Computer Communications 34(3), 274–280 (2011)

[16] Islam, S.H., Biswas, G.P.: A more efficient and secure ID-based remote mutual authentication with key agreement scheme for mobile devices on elliptic curve cryptosystem. Journal of Systems and Software (2011) (In Press); Corrected Proof, Available online (July 7, 2011)

[17] Vaidya, B., Park, J.H., Joel, S.-S.Y., Rodrigues, J.P.C.: Robust one-time password authentication scheme using smart card for home network environment. Computer Communications 34(3), 326–336 (2011)

[18] Liaw, H.-T., Lin, J.-F., Wu, W.-C.: An efficient and complete remote user authentication scheme using smart cards. Mathematical and Computer Modelling 44(1-2), 223–228 (2006)

[19] Boyd, C., Choo, K.: Security of Two-Party Identity-Based Key Agreement. In: Dawson, E., Vaudenay, S. (eds.) Mycrypt 2005. LNCS, vol. 3715, pp. 229–243. Springer, Heidelberg (2005)

[20] Shim, K.: Effient ID-based authenticated key agreement protocol based on the Weil pairing. Electron. Lett. 39(8), 653–654 (2003)

[21] Kocher, P.C., Jaffe, J., Jun, B.: Differential power analysis. In: Wiener, M. (ed.) CRYPTO 1999. LNCS, vol. 1666, pp. 388–397. Springer, Heidelberg (1999)

[22] Messerges, T.S., Dabbish, E.A., Sloan, R.: Examining smart-card security under the threat of power analysis attacks. IEEE Transactions on Computers 51(5), 54152 (2002)

Detection of Probabilistic Dangling References in Multi-core Programs Using Proof-Supported Tools

Mohamed A. El-Zawawy[1,2]

[1] College of Computer and Information Sciences,
Al Imam Mohammad Ibn Saud Islamic University (IMSIU)
Riyadh, Kingdom of Saudi Arabia
[2] Department of Mathematics, Faculty of Science, Cairo University
Giza 12613, Egypt
maelzawawy@cu.edu.eg

Abstract. This paper presents a new technique for detection of probabilistic dangling references in multi-core programs. The technique has the form of a simply structured type system and provides a suitable framework for proof-carrying code applications like mobile code applications that have limited resources. The type derivation of each individual analysis serves as a proof for the correctness of the analysis. The type system is designed to analyze parallel programs with structured concurrent constructs: fork-join constructs, conditionally spawned cores, and parallel loops.

For a given program S, a probabilistic threshold p_{ms}, and a probabilistic reference analysis for S, if S is well-typed in our proposed type system then all computational paths with probabilities greater than or equal to p_{ms} will contain no dangling pointers at run time. The soundness of the presented type system is proved in this paper with respect to a probabilistic operational semantics to our model language.

1 Introduction

Multi-core (multithreading) [36] is one of the main programming styles today. The use of multiple cores (threads) has many advantages; simplifying the process of structuring huge software systems, hiding the delay caused by commands waiting for resources, and boosting the performance of applications executed on multiprocessors. However the interactions between different cores complicate the compilation and analysis of multi-core programs.

One of the vital and attractive attributes of multi-core programs is memory safety (mainly including dangling-references detection) [9]. The importance that memory safety enjoys is justified by several facts including the fact that the absence of memory safety can cause the execution of programs to abort. This absence can be maliciously used to cause security breaches like in many recent cases. However low-level parallel programming languages, used to write most existing parallel-software applications, scarify safety for the sake of improving performance. Violating memory safety takes several forms including memory leaks, buffer overflows, and dangling pointers. Among causes for memory safety violations are explicit allocation and deallocation, pointer arithmetic, casting, and the interactions between multiple cores (threads).

B. Murgante et al. (Eds.): ICCSA 2013, Part V, LNCS 7975, pp. 516–530, 2013.
© Springer-Verlag Berlin Heidelberg 2013

Memory safety [9] is a critical compiler analysis used to decide whether a given piece of code contains memory violations (basically including dangling references). A conventional memory-safety analysis deduces whether a given program (i) is definitely safe (all program execution paths are safe), (ii) is definitely not safe (all program paths contain memory violations), or (iii) maybe safe (some paths are not definitely safe). A probabilistic memory-safety [30] analysis is a program analysis that decides for a given program S and a probability ϵ whether all execution paths with probabilities greater than or equal to ϵ are definitely memory-safe. On the one hand most traditional compiler optimizations count on precise memory-safety checks, and to ensure correctness cannot optimize in the "maybe" case which is the prevalent case. But on the other hand new speculative optimizations [34] can aggressively take advantage of the prevalent "maybe" case, especially in the presence of a probabilistic memory-safety (dangling-references) analysis.

Reference analysis [11,9,10,13,14,12] of a program calculates for each program point a reference (points-to) relationship that captures information about the memory addresses that may be referenced by (pointed-at) by program references (pointers). A probabilistic reference analysis [34,10] statically anticipates the likelihood of every reference relationship at each program point. An absolute memory-safety analysis follows an absolute reference analysis i.e. builds on the result of an absolute reference analysis. It is also the case that probabilistic memory-safety analysis follows or builds on the result of a probabilistic reference analysis.

This paper presents a new approach for detecting dangling references in multi-core programs. The proposed technique is probabilistic in its nature. For a given program S, probability threshold ϵ, and the result *pts* of a probabilistic reference analysis for S (like that in [10]), the proposed technique decides wether execution paths of S with probabilities greater than or equal to ϵ are memory safe (dangling-references free) with respect to *pts*. The proposed technique is flow-sensitive.

The algorithmic style [5,1], which relies on data-flow analysis, is typically used to present static analysis and optimization techniques of multi-core programs. Another framework for program analyses and optimizations is provided by type systems [11,9,20,10,13,14,12,21]. While the type-systems style works directly on the phrase structure of programs, the algorithmic style works on control-flow graphs (intermediate forms) of programs. One advantage of type-systems approach over the algorithmic one is that the former provides communicable justifications (type derivations) for analysis results. Certified code is an example of an area where such machine-checkable justifications are required. Another advantage of type-systems style is the relative simplicity of its inference rules. The technique presented in this paper for memory safety of multi-core programs has the form of a type system. The key to the proposed approach is to compute a post-type starting with the trivial type as a pre-type. Then a program that has this post-type is guaranteed to be memory safe over all computational paths whose probabilities greater than or equal to a given probabilistic threshold.

$$n \in \mathbb{Z}, \; x \in \textit{Var}, \; and \oplus \in \{+, -, \times\}$$

$$e \in \textit{Aexprs} ::= x \mid n \mid e_1 \oplus e_2$$

$$b \in \textit{Bexprs} ::= \textit{true} \mid \textit{false} \mid \neg b \mid e_1 = e_2 \mid e_1 \le e_2 \mid b_1 \wedge b_2 \mid b_1 \vee b_2$$

$$S \in \textit{Stmts} ::= x := e \mid x := \&y \mid *x := e \mid x := *y \mid \textit{skip} \mid S_1; S_2 \mid \textit{if } b \textit{ then } S_t \textit{ else } S_f \mid$$
$$\textit{while } b \textit{ do } S_t \mid \textit{par}\{\{S_1\}, \ldots, \{S_n\}\} \mid \textit{par-if}\{(b_1, S_1), \ldots, (b_n, S_n)\} \mid \textit{par-for}\{S\}.$$

Fig. 1. Our programming language model

Figure 1 presents the programming language that we study. The set *Var* is a finite set of program variables. The language is the simple *while* language enriched with basic commands for parallel computations; fork join, conditionally spawned cores, and parallel loops.

Motivation

Figure 2 presents a motivating example for a probabilistic memory-safety analysis. For this program we suppose that the condition of *if* statement in line 1 is true with probability 0.8. This program has four possible execution paths;

1. the *then* statement (line 2) followed by the first core (thread) (line 5) followed by the second core (line 6).
2. the *then* statement followed by the second core followed by the first core.
3. the *else* statement (line 3) followed by the first core followed by the second core.
4. the *else* statement followed by the second core followed by the first core.

The probability of each of the first two paths is 0.4 and that of each of the last two paths is 0.1. The last two paths are not memory safe as they contain dangling pointers (dereferencing of a in line 5). However the first two paths are memory safe. The motivation of our work is to design a technique that for a program like this one and a probabilistic threshold (for example 0.4) decides whether the program paths with probabilities greater than or equal to 0.4 are memory safe. The desired technique is also required to associate its decision with a correctness proof.

```
1.  if (a > b)
2.     then a := &b
3.     else a := 2;
4.  par{
5.        {b := *a}
6.        {b := &c}
7.     };
```

Fig. 2. A motivating example for a probabilistic dangling-reference analysis

Contributions

Contributions of this paper are the following:

1. An original type system carrying a probabilistic analysis for memory safety (mainly dangling-references detection) of multi-core programs.
2. A formal proof for the soundness of the proposed type system with respect to a probabilistic operational semantics.

Organization. The rest of the paper is organized as follows. Section 2 presents a type system for probabilistic memory safety of multi-core programs. This section also presents a formal correctness proof for the proposed type system with respect to a probabilistic operational semantics presented in an appendix to the paper. Future work and a survey of related work to memory safety (including the used of type systems in program analysis, and the analysis of multi-core programs) are discussed in Section 3.

2 Memory Safety

This section presents a new technique for memory safety (including dangling-references detection) of multi-core programs. The proposed technique is both forward and static (to be used during compilation time). The technique is also probabilistic in the sense that for a given program S and a probabilistic threshold (denoted by p_{ms} in this paper) the technique decides whether all computation paths (Definition 4) of S with probability greater or equal to p_{ms} are memory safe. This sort of information is required and intensively used in speculative optimizations that are parts of most modern compilers.

Memory safety of programs is a forward program analysis that is typically built on the result of a reference analysis. For probabilistic memory safety, the underlying reference analysis has to be probabilistic [34,10] as well. Hence we assume that our input program S is associated with the probabilistic threshold p_{ms} and the result of a probabilistic reference analysis[1] for S. To be more precise, we assume that the underlying probabilistic reference analysis associates each program point with a reference type pts drawn from a set of probabilistic reference types PTS. A natural formalization of the set PTS together with a subtyping relation on its types is introduced in [10] and reviewed in Definition 1. To build the memory safety analysis on robust ground, surely the underlying reference analysis has to be sound with respect to a robust semantics; in our case the operational semantics of this paper appendix. Suppose that for a statement S, the reference analysis associates a pre-reference type pts and a post-reference type pts', i.e. $S : pts \rightarrow pts'$. The soundness has the intuition that if the execution of S from a state (γ, p) of type pts ends at a state (γ, p'), then this final state has to be of type pts'.

Definition 1. *1. $Addrs = \{x' \mid x \in Var\}$ and $Addrs_p = Addrs \times [0, 1]$.*

2. $Pre\text{-}PTS = \{pts \mid pts : Var \rightarrow 2^{Addrs_p} \text{ s.t. } (y', p_1), (y', p_2) \in pts(x) \Longrightarrow p_1 = p_2\}$.

3. For $pts \in Pre\text{-}PTS$ and $x \in Var$, $\sum_{pts} x = \sum_{(z', p) \in pts(x)} p$.

---

[1] The reference analysis results (reference information) are typically assigned to program points of S.

4. *For pts $\in$ Pre-PTS and $x \in$ Var, $A_{pts}(x) = \{z' \mid \exists p > 0. (z', p) \in pts(x)\}$.*
5. *PTS $= \{pts \in$ Pre-PTS $\mid \forall x \in$ Var. $\sum_{pts} x \leq 1\}$.*
6. *pts $\leq$ pts' $\overset{\text{def}}{\Longleftrightarrow} (\forall x, y \in$ Var. $(y', p) \in pts(x) \Longrightarrow \exists p'. p \geq p' \,\&\, (y', p') \in pts'(x))$.*
7. *$\gamma \models pts \overset{\text{def}}{\Longleftrightarrow} (\forall x \in$ Var. $\gamma(x) \in$ Addrs $\Longrightarrow \exists p > 0. (\gamma(x), p) \in pts(x))$.*

2.1   Types

Our proposed approach for memory safety has the form of a type system. The types of this type system are enrichments of that of the underlying reference types (*PTS*). Therefore each memory-safety type is a triple (pts, v, p_s) where v is a set of variables that are guaranteed to contain addresses at the program point assigned this type and p_s is a lower bound for probabilities of reaching the program point assigned this type. The following definition gives a precise formalization for the set of memory-safety types (called safety types). The formal interpretation of assigning a safety type to a state is also introduced in the following definition.

Definition 2.   *– A safety type is a triple (pts, v, p_s) such that*
- *$pts \in$ PTS,*
- *$v \subseteq$ Var such that for every $x \in v$, there exists a pair $(z', p) \in pts(x)$ with $p > p_{ms}$, and*
- *$p_s \in [0, 1]$.*

– $(pts, v, p_s) \leq (pts', v', p_s') \overset{\text{def}}{\Longleftrightarrow} pts \leq pts', v \supseteq v'$, and $p_s \geq p_s' \geq p_{ms}$.
– A state (γ, p) has type (pts, v, p_s) with respect to the probability p_{ms}, denoted by $(\gamma, p) \models_{p_{ms}} (pts, v, p_s)$, if $\gamma \models pts, \forall x \in v(\gamma(x) \in$ Addrs$)$, and $p_{ms} \leq p_s \leq p$.

The key to our technique for probabilistic memory safety of multi-core programs is the following. Suppose that we have a statement S and a reference analysis for S in the form $S : pts \rightarrow pts'$. Then for a safety pre-type $(pts, v, p_s)^2$, a post-type derivation is attempted for S in the memory-safety type system. If such post-type exists then Theorem 1 below guarantees the following. It is memory-safe to execute S starting from a state (γ, p) that is of type (pts, v, p_s) and that is positive (Definition 5; has no execution paths with probability less than or equal to p_{ms}).

2.2   Inference Rules

The inference rules of our type system for probabilistic memory safety are as follows:

$$\frac{y \in v}{y : (x, pts, v) \rightarrow v \cup \{x\}} \; (y_1^m) \qquad \frac{\sum_{pts} y < p_{ms}}{y : (x, pts, v) \rightarrow v \setminus \{x\}} \; (y_2^m) \qquad \frac{}{n : (x, pts, v) \rightarrow v \setminus \{x\}} \; (n^m)$$

$$\frac{\forall y \in FV(e_1 \oplus e_2). \sum_{pts} y = 0}{e_1 \oplus e_2 : (x, pts, v) \rightarrow v \setminus \{x\}} \; (\oplus^m) \qquad \frac{x := e : pts \rightarrow pts' \quad e : (x, pts, v) \rightarrow v'}{x := e : (pts, v, p_s) \rightarrow (pts', v', p_s)} \; (:=^m)$$

---

^{2} Typically $(pts, v, p_s) = (pts, \emptyset, 1)$.

$$\frac{x \in v \quad pts(x) = \{(z'_1, p_1), \dots, (z'_n, p_n)\} \quad \forall z'_i \in A_{pts}(x). \, z_i := e : (pts, v, p_s) \to (pts_i, v', p_s)}{*x := e : (pts, v, p_s) \to (\Upsilon(pts, pts_1, \dots, pts_n), v', p_s)} \quad (* :=^m)$$

$$\frac{y \in v \quad pts(y) = \{(z'_1, p_1), \dots, (z'_n, p_n)\} \quad \forall i. \, x := z_i : (pts, v, p_s) \to (pts_i, v', p_s)}{x := *y : (pts, v, p_s) \to (\Upsilon(pts, pts_1, \dots, pts_n), v', p_s)} \quad (:= *^m)$$

$$\frac{x := \&y : pts \to pts'}{x := \&y : (pts, v, p_s) \to (pts', v \cup \{x\}, p_s)} \quad (:= \&^m) \qquad \overline{skip : (pts, v, p_s) \to (pts, v, p_s)}$$

$$\frac{S_1 : (pts, v, p_s) \to (pts'', v'', p_s'') \qquad S_2 : (pts'', v'', p_s'') \to (pts', v', p_s')}{S_1; S_2 : (pts, v, p_s) \to (pts', v', p_s')} \quad (seq^m)$$

$$\frac{\forall y \in FV(b)(\sum_{pts} y = 0) \quad \begin{array}{l} S_t : (pts, v, p_s) \to (pts_t, v_t, p_t) \\ S_f : (pts, v, p_s) \to (pts_f, v_f, p_f) \end{array} \quad \begin{array}{l} p_{ms} \le p_t \times p_{if} \\ p_{ms} > p_s \times (1 - p_{if}) \end{array}}{if \; b \; then \; S_t \; else \; S_f : (pts, v, p_s) \to (\Upsilon(pts_t, pts_f), v_t, p_t \times p_{if})} \quad (if_1^m)$$

$$\frac{\forall y \in FV(b)(\sum_{pts} y = 0) \quad \begin{array}{l} S_t : (pts, v, p_s) \to (pts_t, v_t, p_t) \\ S_f : (pts, v, p_s) \to (pts_f, v_f, p_f) \end{array} \quad \begin{array}{l} p_{ms} > p_s \times p_{if} \\ p_{ms} \le p_f \times (1 - p_{if}) \end{array}}{if \; b \; then \; S_t \; else \; S_f : (pts, v, p_s) \to (\Upsilon(pts_t, pts_f), v_f, p_f \times (1 - p_{if}))} \quad (if_2^m)$$

$$\frac{\forall y \in FV(b)(\sum_{pts} y = 0) \quad \begin{array}{l} S_t : (pts, v, p_s) \to (pts_t, v_t, p_t) \\ S_f : (pts, v, p_s) \to (pts_f, v_f, p_f) \end{array} \quad \begin{array}{l} p_{ms} \le p_t \times p_{if} \\ p_{ms} \le p_f \times (1 - p_{if}) \end{array}}{if \; b \; then \; S_t \; else \; S_f : (pts, v, p_s) \to (\Upsilon(pts_t, pts_f), v_t \cap v_f, \min\{p_t \times p_{if}, p_f \times (1 - p_{if})\})} \quad (if_3^m)$$

$$\frac{S_i : (\Psi(pts, \dots, pts_j, \dots \mid j \ne i), v \cap \cap_{j \ne i} v_j, \min\{p_s, p_j \mid j \ne i\}) \to (pts_i, v_i, p_i) \qquad p_{ms} \le \dfrac{\min_i p_i}{n!}}{par\{\{S_1\}, \dots, \{S_n\}\} : (pts, v, p_s) \to (\Upsilon(pts_1, \dots, pts_n), \cap_i v_i, \dfrac{\min_i p_i}{n!})} \quad (par^m)$$

$$\frac{par\{\{if \; b_1 \; then \; S_1 \; else \; skip\}, \dots, \{if \; b_n \; then \; S_n \; else \; skip\}\} : (pts, v, p_s) \to (pts', v', p_s')}{par\text{-}if\{(b_1, S_1), \dots, (b_n, S_n)\} : (pts, v, p_s) \to (pts', v', p_s')} \quad (par\text{-}if^m)$$

$$\frac{S : (\Psi(pts, pts'), v \cap v', \min\{p_s, p_s'\}) \to (pts', v', p_s')}{par\text{-}for\{S\} : (pts, v, p_s) \to (pts', v', p_s')} \quad (par\text{-}for^m)$$

$$\frac{\forall i \in [1, n], \forall y \in FV(b)(\sum_{pts_i} y = 0) \qquad (pts_1, v_1, p_{s_1}) \xrightarrow{S_t} (pts_2, v_2, p_{s_2}) \xrightarrow{S_t} \dots \xrightarrow{S_t} (pts_{n+1}, v_{n+1}, p_{s_{n+1}})}{while \; b \; do \; S_t : (pts_1, v_1, p_{s_1}) \to (\Upsilon(pts_{n+1}), v_{n+1}, p_{s_{n+1}})} \quad (whl^m)$$

$$\frac{(pts'_1, v'_1, p'_{s_1}) \le (pts_1, v_1, p_{s_1}) \qquad S : (pts_1, v_1, p_{s_1}) \to (pts_2, v_2, p_{s_2}) \qquad (pts_2, v_2, p_{s_2}) \le (pts'_2, v'_2, p'_{s_2})}{S : (pts'_1, v'_1, p'_{s_1}) \to (pts'_2, v'_2, p'_{s_2})} \quad (csq^m)$$

Judgments produced by the type system above has two forms. The judgment of an arithmetic expression has the form $e : (x, pts, v) \to v'$. The existence of such judgment for an expression e guarantees that calculating e in a state (γ, p) of type (pts, v, p_s) w.r.t. p_{ms}, i.e. $(\gamma, p) \models_{p_{ms}} (pts, v, p_s)$, does not fail. The judgment also guarantees that if the execution of the statement $x := e$ at the state (γ, p) ends at a state (γ', p'), then elements of v' are guaranteed to contain addresses w.r.t. γ'. This is formalized in Lemma 1. The judgment of a statement S has the from $S : (pts, v, p) \rightarrow (pts', v', p')$ and assures that if the execution of S from a pre-state of the pre-type ends at a post-state, then this post-state is of the post-type. This is proved in Theorem 1.

Comments on the inference rules above are in order. The condition $\sum_{pts} y \geq p_{ms}$ of the rule (y_1^m) assures that when reaching the program point being assigned a type along any of the computation paths whose probabilities greater than or equal to p_{ms}, y will contain an address. The condition $\forall y \in FV(e_1 \oplus e_2)(\sum_{pts} y = 0)$ of the rule $(\oplus^m)$ assures that all free variables of the expression contain integers and hence guarantees the success of calculating the expression $e_1 \oplus e_2$ at any state of the type (pts, v, p_s). In the rules $(* :=^m)$ and $(:= *^m)$ the expression $\Upsilon(pts, pts_1, \ldots, pts_n)$ denotes the reference post-type calculated by the underlying reference analysis[3]. $\Upsilon(pts, pts_1, \ldots, pts_n)$ is naturally a function in $\{pts, pts_1, \ldots, pts_n\}$ and its precise shape does not contribute to the calculations of the inference rules $(* :=^m)$ and $(:= *^m)$. The rule (if_1^m) treats the case when the probability of the *then* path is greater than or equal to the threshold p_{ms} and that of the *else* path is strictly less than p_{ms}. In this case, it is sensible to consider the analysis results of S_t and to neglect that of S_f. The rule (par^m) has this shape in order to treat any possible integrations between the statement threads. For the rule (whl^m), n is an upper bound for the trip-count of the loop. Therefore the post-type of the rule is an upper bound for post-types corresponding to number of iterations bounded by n. The statistical and probabilistic information concerning correctness probabilities of *if* statements and trip counts of loops can be obtained using edge-profiling techniques. Heuristics can be used in absence of edge-profiling methods.

Remark 1. As it is common with a probabilistic reference analysis [10], we assume that our underlying reference analysis satisfies the following condition. Suppose that pts is the reference type assigned to a program point t of a statement S and $\sum_{pts} y = p$. Then for all computational paths of S with probabilities less than p, the variable y contains no address at the point t.

Lemma 1. *1. Suppose $(\gamma, p) \models_{p_{ms}} (pts, v, p_s)$ and $e : (x, pts, v) \rightarrow v'$. Then $[\![e]\!]\gamma \neq !$, and*

$$x := e : (\gamma, p) \rightarrow (\gamma', p') \Longrightarrow \forall y \in Var. (y \in v' \Longrightarrow \gamma'(y) \in Addrs).$$

2. $(pts, v, p_s) \leq (pts', v', p_s') \Longrightarrow (\forall (\gamma, p). (\gamma, p) \models_{p_{ms}} (pts, v, p_s) \Longrightarrow (\gamma, p) \models_{p_{ms}} (pts', v', p_s'))$.

Proof. It is straightforward to prove the second item. The first item is proved by induction on the structure of type derivations:

- The case of the rule (y_1^m): in this case $\gamma' = \gamma[x \mapsto \gamma(y)], p' = p$, and $v' = v \cup \{x\}$. Since $y \in v$, y is guaranteed to contain an address at the program point before the assignment statement. Therefore $\gamma'(x)$ has an address at the program point after the assignment statement. This justifies adding x to v.
- The case of the rule (y_2^m): in this case $\gamma' = \gamma[x \mapsto \gamma(y)], p' = p$, and $v' = v \setminus \{x\}$. Since $\sum_{pts} y < p_{ms}$ and $p_{ms} \leq p_s \leq p$, by Remark 1 y contains no address at the program point before the assignment statement. Hence $\gamma'(x)$ is not assured to contain

---

[3] The interested reader can check [10] for the details of calculating $\Upsilon(pts, pts_1, \ldots, pts_n)$.

an address at the program point after the assignment statement. This legitimizes removing x from v.

- The case of the rule $(\oplus^m)$: in this case $p' = p, \gamma' = \gamma[x \mapsto [\![e_1 \oplus e_2]\!]\gamma]$, and $v' = v \setminus \{x\}$. The condition $\forall y \in FV(e_1 \oplus e_2) (\sum_{pts} y = 0)$ assures that $\forall y \in FV(e_1 \oplus e_2).(\gamma(y) \in \mathbb{Z})$. Therefore $[\![e_1 \oplus e_2]\!]\gamma \in \mathbb{Z}$. Hence $\gamma'(x) \in \mathbb{Z}$ which legitimizes removing x from v.

2.3 Soundness

For a statement S that has types in our probabilistic memory-safety type system, $S : (pts, v, p_s) \to (pts', v', p'_s)$, Theorem 1 assures the following fact about S. It is memory safe to execute S from a positive state (γ, p) of type (pts, v, p_s) w.r.t. p_{ms}, i.e. $(\gamma, p) \models_{p_{ms}} (pts, v, p_s)$. The memory safety means that the program does not abort due to faulty de-referencing (dangling pointers). A *positive* (Definition 5) state is a state that does not start any executions paths with probability less than p_{ms}. Theorem 1 also proves soundness of memory-safety type system.

Theorem 1. *(Soundness and Probabilistic Memory Safety) Suppose $S : (pts, v, p_s) \to (pts', v', p'_s)$. Then*

1. *If $(\gamma, p) \models_{p_{ms}} (pts, v, p_s)$ and S is positive at (γ, p) then S does not abort at (γ, p) i.e. $S : (\gamma, p) \not\leadsto abort$.*
2. *If $S : (\gamma, p) \leadsto (\gamma', p')$ then $(\gamma, p) \models_{p_{ms}} (pts, v, p_s) \Longrightarrow (\gamma', p') \models_{p_{ms}} (pts', v', p'_s)$.*

Proof. The proof is by structure induction on the type derivation. Main cases are shown as follows:

- The case of $(:=^m)$: this case follows from Lemma 1 and the soundness of reference analysis.
- The case of $(* :=^m)$: because $x \in v$, there exists $z \in Var$ such that $\gamma(x) = z'$. And because $(\gamma, p) \models_{p_{ms}} (pts, v, p_s)$, we have $z' \in A_{pts}(x) . z := e$ does not abort at (γ, p) by induction hypothesis and hence neither does $*x := e$. We also have $z := e : (\gamma, p) \leadsto (\gamma', p')$. By assumption, it is true that $z := e : (pts, v, p_s) \to (pts', v', p')$. Hence by soundness of $(:=^m)$, $(\gamma', p') \models_{p_{ms}} (pts', v', p')$.
- The case of (if_1^m): the condition $\forall y \in FV(b)(\sum_{pts} y = 0)$ guarantees that all free variables of the condition b have integers (not addresses) under the state γ. This is so because $(\gamma, p) \models_{p_{ms}} (pts, v, p_s)$. Therefore the semantics of b with respect to γ is a Boolean value. We have the following inequalities
  - $p_t \times p_{if} \geq p_{ms} > p_s \times (1 - p_{if})$, and
  - $p \geq p_s \geq p_t$.
  These inequalities imply that $p \times p_{if} \geq p_t \times p_{if} \geq p_{ms} > p_s \times (1 - p_{if})$ which implies $p \times p_{if} \geq p_{ms} > p_s \times (1 - p_{if})$. Because S is positive at (γ, p) (Definition 5), $[\![b]\!]\gamma =$ true. Now by induction hypothesis S_t does not abort at (γ, p) because $S_t : (pts, v, p_s) \to (pts_t, v_t, p_t), (\gamma, p) \models_{p_{ms}} (pts, v, p_s)$, and S_t is positive at (γ, p) by Lemma 2. Therefore the *if* statement does not abort at (γ, p) which completes the proof of (1) for this case.
  (2) In this case, we have $(\gamma', p') = (\gamma', p_{if} \times p'')$ where $S_t : (\gamma, p) \leadsto (\gamma', p'')$. We also have $(pts', v', p') = (\Upsilon(pts_t, pts_f), v_t, p_t \times p_{if})$ where $S_t : (pts, v, p_s) \to$

(pts_t, v_t, p_t). By induction hypothesis on S_t we have $(\gamma', p'') \models_{p_{ms}} (pts_t, v_t, p_t)$. Therefore $p'' \geq p_t$ which implies $p'' \times p_{if} \geq p_t \times p_{if} \geq p_{ms}$ because $p_t \times p_{if} \geq p_{ms}$. Hence $(\gamma', p') \models_{p_{ms}} (pts_t, v_t, p_t \times p_{if})$ which implies $(\gamma', p') \models_{p_{ms}} (\Upsilon(pts_t, pts_f), v_t, p_t \times p_{if})$ because $(pts_t, v_t, p_t \times p_{if}) \leq (\Upsilon(pts_t, pts_f), v_t, p_t \times p_{if})$. The last inequality holds because $\Upsilon(pts_t, pts_f)$ is an upper bound for pts_t.

- The case of (if_2^m) is pretty much similar to the case of (if_1^m).
- The case of (if_3^m): the condition $\forall y \in FV(b)(\sum_{pts} y = 0)$ guarantees that $[\![b]\!]\gamma$ is a Boolean value. We have the following inequalities
  - $p_{ms} \leq p_t \times p_{if}$,
  - $p_{ms} \leq p_f \times (1 - p_{if})$,
  - $p_t \leq p_s \leq p$, and
  - $p_f \leq p_s \leq p$.
These inequalities imply that

$$p_{ms} \leq p_t \times p_{if} \leq p \times p_{if} \quad \text{and} \quad p_{ms} \leq p_f \times (1 - p_{if}) \leq p \times (1 - p_{if})$$

which implies
  - $p_{ms} \leq \min\{p_t \times p_{if}, p_f \times (1 - p_{if})\} \leq p \times p_{if}$ and
  - $p_{ms} \leq \min\{p_t \times p_{if}, p_f \times (1 - p_{if})\} \leq p \times (1 - p_{if})$.
Now we consider the case $[\![b]\!]\gamma = \text{false}$. In this case $S_f : (pts, v, p_s) \rightarrow (pts_f, v_f, p_f), (\gamma, p) \models_{p_{ms}} (pts, v, p_s)$, and S_f is positive at (γ, p) by Lemma 2. Hence by induction hypothesis S_f does not abort at (γ, p). Consequently the if statement does not abort at (γ, p) which completes the proof of (1) for this case.

(2) In this case, we have $(\gamma', p') = (\gamma', p_{if} \times p'')$ where $S_f : (\gamma, p) \rightsquigarrow (\gamma', p'')$. We also have $(pts', v', p') = (\Upsilon(pts_t, pts_f), v_t \cap v_f, \min\{p_t \times p_{if}, p_f \times (1 - p_{if})\})$ where $S_f : (pts, v, p_s) \rightarrow (pts_f, v_f, p_f)$. By induction hypothesis on S_f we have $(\gamma', p'') \models_{p_{ms}} (pts_f, v_f, p_f)$. Therefore $p'' \geq p_f$ which implies $p'' \times (1 - p_{if}) \geq p_f \times (1 - p_{if}) \geq \min\{p_t \times p_{if}, p_f \times (1 - p_{if})\} \geq p_{ms}$ because $\min\{p_t \times p_{if}, p_f \times (1 - p_{if})\} \geq p_{ms}$. Hence $(\gamma', p') \models_{p_{ms}} (pts_f, v_f, \min\{p_t \times p_{if}, p_f \times (1 - p_{if})\})$ which implies $(\gamma', p') \models_{p_{ms}} (\Upsilon(pts_t, pts_f), v_t \cap v_f, \min\{p_t \times p_{if}, p_f \times (1 - p_{if})\})$ because $(pts_f, v_f, \min\{p_t \times p_{if}, p_f \times (1 - p_{if})\}) \leq (\Upsilon(pts_t, pts_f), v_t \cap v_f, \min\{p_t \times p_{if}, p_f \times (1 - p_{if})\})$. The last inequality holds because $\Upsilon(pts_t, pts_f)$ is an upper bound for pts_t and $v_f \supseteq (v_t \cap v_f)$.

- The case of (par^m): (1) Suppose that $\theta : \{1, \ldots, n\} \rightarrow \{1, \ldots, n\}$ is a permutation. $(\gamma, p) \models_{p_{ms}} (pts, v, p_s)$ implies $(\gamma, p) \models_{p_{ms}} (\Psi(pts, \ldots, pts_j, \ldots \mid j \neq \theta(1)), v \cap \cap_{j \neq \theta(1)} v_j, \min\{p_s, p_j \mid j \neq \theta(1)\})$. Recall that $\Psi(pts, \ldots, pts_j, \ldots \mid j \neq \theta(1))$ is a lower bound for pts. By Lemma 2, $S_{\theta(1)}$ is positive at (γ, p). Therefore $S_{\theta(1)}$ does not abort at γ by induction hypothesis. Hence either the execution of $S_{\theta(1)}$ terminates at a state (γ_2, p_2) such that $(\gamma_2, p_2) \models_{ms} (pts_{\theta(1)}, v_{\theta(1)}, p_{\theta(1)})$ or enters an infinite loop at (γ, p). Therefore $(\gamma_2, p_2) \models_{ms} (\Psi(pts, \ldots, pts_j, \ldots \mid j \neq \theta(2)), v \cap \cap_{j \neq \theta(2)} v_j, \min\{p_s, p_j \mid j \neq \theta(2)\})$. Therefore, clearly (1) is proved via a simple induction on n.

(2) In this case the existence of a permutation $\theta : \{1, \ldots, n\} \rightarrow \{1, \ldots, n\}$ and $n + 1$ states $(\gamma, p) = (\gamma_1, p_1), \ldots, (\gamma_{n+1}, p_{n+1}) = (\gamma', p'')$ such that for every $1 \leq i \leq n$, $S_{\theta(i)} : (\gamma_i, p_i) \rightsquigarrow (\gamma_{i+1}, p_{i+1})$ is guaranteed. In this case $p' = \frac{p''}{n!}$. The fact that $(\gamma_1, p_1) \models_{p_{ms}} (pts, v, p_s)$ implies the fact that $(\gamma_1, p_1) \models_{p_{ms}} (\Psi(pts, \ldots, pts_j, \ldots \mid$

$j \neq \theta(1))$, $v \cap \cap_{j \neq \theta(1)} v_j$, $\min\{p_s, p_j \mid j \neq \theta(1)\}$). Hence $(\gamma_2, p_2) \models_{p_{ms}}$ $(pts_{\theta(1)}, v_{\theta(1)}, p_{\theta(1)})$ by the induction hypothesis. This implies $(\gamma_2, p_2) \models_{p_{ms}}$ $(\Psi(pts, \ldots, pts_j, \ldots \mid j \neq \theta(2)), v \cap \cap_{j \neq \theta(2)} v_j, \min\{p_s, p_j \mid j \neq \theta(2)\})$. Hence Again $(\gamma_3, p_3) \models_{p_{ms}} (pts_{\theta(2)}, v_{\theta(2)}, p_{\theta(2)})$ by the induction hypothesis. Therefore a simple induction on n shows that $(\gamma', p') = (\gamma_{n+1}, p_{n+1}) \models_{p_{ms}} (pts_{\theta(n)}, v_{\theta(n)}, p_{\theta(n)})$ implying $(\gamma', p'') \models_{p_{ms}} (\Upsilon(pts_1, \ldots, pts_n), \cap_i v_i, \min_i p_i)$. Hence because $p_{ms} \leq \frac{\min_i p_i}{n!}$, we get $(\gamma', p') \models_{p_{ms}} (pts', v', p') = (\Upsilon(pts_1, \ldots, pts_n), \cap_i v_i, \frac{\min_i p_i}{n!})$ as required.

- The case of $(par - for^m)$: (1) The proof of this item is in line with item (1) of the (par^m) case.

    (2) In this case there exists n such that $par\{\{S\}_1, \ldots, \{S\}_n\} : (\gamma, p) \rightsquigarrow (\gamma', p')$. We get $S : (\Psi(pts, pts'), v \cap v', \min\{p_s, p'_s\}) \rightarrow (pts', v', p'_s)$ by induction hypothesis. Then by (par^m) we infer that $par\{\{S\}_1, \ldots, \{S\}_n\} : (pts, v, p_s) \rightarrow (pts', v', p'_s)$. Consequently by the soundness of (par^m), $(\gamma', p') \models_{p_{ms}} (pts', v', p'_s)$.

3 Related and Future Work

Related work includes security vulnerabilities, memory management, debugging and testing, garbage collection, failure masking, analysis of multi-core programs, and type systems in program analysis.

A classical trend to reduce vulnerabilities of heaps to security attacks is to use a randomization approach for both choosing the base address [2] of the heap and buffering allocation requests [4]. However this classical approach is believed not to be very effective on 32-bit operating systems [33]. More recent work [29] hides object layouts from attackers in any duplicate.

To maintain fast allocation and low fragmentation, dynamic techniques for memory management scarify strength. Repeated memory frees and heap corruption due to buffer overflows affect most *malloc* implementations. While some memory managers [6,15,16] prevent heap corruption via separating metadata from the heap, other managers [31] just recognize heap corruption.

Via simulation and multiple rewrites on run time, techniques for debugging and testing [27] discover errors of memory in programs. Drawbacks of these techniques include increasing space costs and restrictive runtime overheads. These burdens can only be tolerated during testing. Other techniques significantly reduce runtime overhead and discover memory leaks via using sampling [23].

The drawback of garbage collection [29], a technique helping avoiding errors caused by dangling pointers, is that to perform reasonably it requires an ample amount of space. In particular, the technique of [29] prevents overwrites via separating metadata from heap. This technique, which is probabilistic rather than absolute like most other related techniques, also neglects multiply and faulty frees.

Failure masking [32] is a terminology describing stopping programs from aborting. Pool allocation, a technique of failure masking, classifies objects into pools according to their types and hence guarantees that objects overwrite only dangling pointers of the same type. The drawback of this technique is the unpredictability of behavior of the produced program. Other techniques, failure-oblivious systems, neglect faulty writes and create values for reading uninitialized memory.

None of techniques mentioned above that treat dangling pointers deal with multi-core programs nor provide proofs for correctness of each individual test. Sound type systems for reference analysis and memory safety of Multi-core programs are presented in [9]. However all techniques mentioned so far are absolute; not probabilistic like the technique presented here. Hence our work has the advantage, over all the related work, of being usable in speculative-optimizations sections of modern compilers.

The analysis of multi-core programs is receiving a growing research interest. The possible interactions between various cores significantly complicate analysis of multi-core programs. Work in this area is typically classified into two main categories: techniques designed specifically for optimization or error-detection of multi-core programs and techniques originally designed for analysis of sequential programs and successfully extended to cover multi-core programs.

The work in the first category above includes dataflow frameworks for bitvector problems [26], concurrent static single assignment forms [35], reaching definitions [7], constant propagation [5], code motion [25], file safety [17], faulty function-calls [18]. None of these techniques studies memory-safety of multi-core programs leaving alone probabilistic memory safety of these programs. The work in the other category above includes synchronization analysis [22], race detection [24], reference analysis [9], and deadlock analysis [1].

The use of type systems in program analysis [11,9,20,10,13,14,12,21] is becoming a mainstream approach for applications that require a proof for each individual program analysis like certified code. General methods for transforming monotone data-flow analyses (forward and backward) into type systems are presented in [28]. Type systems for program optimizations based live stack-heap and pointer analyses are presented in [11]. Constant folding, common subexpression elimination, and dead code elimination for *while* language as type systems are presented in [3].

In the area of denotational semantics, data structures and programs are mathematically represented by mathematical domains (sets) and maps between domains. For future work, we are interested in translating concepts of probabilistic memory-safety analysis to the side of denotational semantics [19,8]. This translation will facilitate theoretical studies about probabilistic memory-safety analysis. Obtained theoretical results can be then translated back to the side of data structures and programs.

References

1. Ahmad, F., Huang, H., Wang, X.-L.: Petri net modeling and deadlock analysis of parallel manufacturing processes with shared-resources. J. Syst. Softw. 83, 675–688 (2010)
2. Antonatos, S., Anagnostakis, K.G.: TAO: Protecting against hitlist worms using transparent address obfuscation. In: Leitold, H., Markatos, E.P. (eds.) CMS 2006. LNCS, vol. 4237, pp. 12–21. Springer, Heidelberg (2006)
3. Benton, N.: Simple relational correctness proofs for static analyses and program transformations. In: Jones, N.D., Leroy, X. (eds.) POPL, pp. 14–25. ACM (2004)
4. Bhatkar, S., Sekar, R., DuVarney, D.C.: Efficient techniques for comprehensive protection from memory error exploits. In: Proceedings of the 14th Conference on USENIX Security Symposium, Berkeley, CA, USA, vol. 14, p. 17. USENIX Association (2005)
5. Callahan, D., Cooper, K.D., Kennedy, K., Torczon, L.: Interprocedural constant propagation. SIGPLAN Not 39, 155–166 (2004)

6. Chang, Y.-H., Kuo, T.-W.: A management strategy for the reliability and performance improvement of mlc-based flash-memory storage systems. IEEE Trans. Computers 60(3), 305–320 (2011)
7. Collard, J.-F., Griebl, M.: A precise fixpoint reaching definition analysis for arrays. In: Carter, L., Ferrante, J. (eds.) LCPC 1999. LNCS, vol. 1863, Springer, Heidelberg (2000)
8. El-Zawawy, M.A.: Semantic spaces in Priestley form. PhD thesis, University of Birmingham, UK (January 2007)
9. El-Zawawy, M.A.: Flow sensitive-insensitive pointer analysis based memory safety for multithreaded programs. In: Murgante, B., Gervasi, O., Iglesias, A., Taniar, D., Apduhan, B.O. (eds.) ICCSA 2011, Part V. LNCS, vol. 6786, pp. 355–369. Springer, Heidelberg (2011)
10. El-Zawawy, M.A.: Probabilistic pointer analysis for multithreaded programs. ScienceAsia 37(4), 344–354 (2011)
11. El-Zawawy, M.A.: Program optimization based pointer analysis and live stack-heap analysis. International Journal of Computer Science Issues 8(2), 98–107 (2011)
12. El-Zawawy, M.A.: Abstraction analysis and certified flow and context sensitive points-to relation for distributed programs. In: Murgante, B., Gervasi, O., Misra, S., Nedjah, N., Rocha, A.M.A.C., Taniar, D., Apduhan, B.O. (eds.) ICCSA 2012, Part IV. LNCS, vol. 7336, pp. 83–99. Springer, Heidelberg (2012)
13. El-Zawawy, M.A.: Dead code elimination based pointer analysis for multithreaded programs. Journal of the Egyptian Mathematical Society 20(1), 28–37 (2012)
14. El-Zawawy, M.A.: Heap slicing using type systems. In: Murgante, B., Gervasi, O., Misra, S., Nedjah, N., Rocha, A.M.A.C., Taniar, D., Apduhan, B.O. (eds.) ICCSA 2012, Part III. LNCS, vol. 7335, pp. 592–606. Springer, Heidelberg (2012)
15. El-Zawawy, M.A.: Recognition of logically related regions based heap abstraction. Journal of the Egyptian Mathematical Society 20(2) (2012)
16. El-Zawawy, M.A.: Frequent statement and de-reference elimination for distributed programs. In: Murgante, B., Misra, S., Carlini, M., Torre, C.M., Quang, N.H., Taniar, D., Apduhan, B.O., Gervasi, O. (eds.) ICCSA 2013. LNCS, vol. 7975, pp. 82–97. Springer, Heidelberg (2013)
17. El-Zawawy, M.A., Daoud, N.M.: M. Daoud. Dynamic verification for file safety of multithreaded programs. IJCSNS International Journal of Computer Science and Network Security 12(5), 14–20 (2012)
18. El-Zawawy, M.A., Daoud, N.M.: New error-recovery techniques for faulty-calls of functions. Computer and Information Science 5(3), 67–75 (2012)
19. El-Zawawy, M.A., Jung, A.: Priestley duality for strong proximity lattices. Electr. Notes Theor. Comput. Sci. 158, 199–217 (2006)
20. El-Zawawy, M.A., Partial, H.A.N.: redundancy elimination for multi-threaded programs. IJCSNS International Journal of Computer Science and Network Security 11(10), 127–133 (2011)
21. El-Zawawy, M.A., Nayel, H.A.: Type systems based data race detector. IJCSNS International Journal of Computer Science and Network Security 5(4), 53–60 (2012)
22. Hall, M.W., Amarasinghe, S.P., Murphy, B.R., Liao, S.-W., Lam, M.S.: Interprocedural parallelization analysis in suif. ACM Trans. Program. Lang. Syst. 27, 662–731 (2005)
23. Hauswirth, M., Chilimbi, T.M.: Low-overhead memory leak detection using adaptive statistical profiling. In: Mukherjee, S., McKinley, K.S. (eds.) ASPLOS, pp. 156–164. ACM (2004)
24. Kim, Y.-C., Jun, Y.-K.: Restructuring parallel programs for on-the-fly race detection. In: Malyshkin, V.E. (ed.) PaCT 1999. LNCS, vol. 1662, pp. 446–451. Springer, Heidelberg (1999)
25. Knoop, J., Rüthing, O., Steffen, B.: Lazy code motion. SIGPLAN Not 39, 460–472 (2004)
26. Knoop, J., Steffen, B., Vollmer, J.: Parallelism for free: efficient and optimal bitvector analyses for parallel programs. ACM Trans. Program. Lang. Syst. 18, 268–299 (1996)

27. Langdon, W.B., Harman, M., Jia, Y.: Efficient multi-objective higher order mutation testing with genetic programming. J. Syst. Softw. 83, 2416–2430 (2010)
28. Riis Nielson, H., Nielson, F.: Flow logic: A multi-paradigmatic approach to static analysis. In: Mogensen, T.Æ., Schmidt, D.A., Sudborough, I.H. (eds.) The Essence of Computation. LNCS, vol. 2566, pp. 223–244. Springer, Heidelberg (2002)
29. Novark, G., Berger, E.D.: Dieharder: securing the heap. In: Al-Shaer, E., Keromytis, A.D., Shmatikov, V. (eds.) ACM Conference on Computer and Communications Security, pp. 573–584. ACM (2010)
30. Novark, G., Berger, E.D., Zorn, B.G.: Exterminator: Automatically correcting memory errors with high probability. Commun. ACM 51, 87–95 (2008)
31. Robertson, W.K., Krügel, C., Mutz, D., Valeur, F.: Run-time detection of heap-based over-flows. In: LISA, pp. 51–60. USENIX (2003)
32. Sardiña, S., Padgham, L.: A bdi agent programming language with failure handling, declarative goals, and planning. Autonomous Agents and Multi-Agent Systems 23(1), 18–70 (2011)
33. Shacham, H., Page, M., Pfaff, B., Goh, E.-J., Modadugu, N., Boneh, D.: On the effectiveness of address-space randomization. In: Proceedings of the 11th ACM Conference on Computer and Communications Security, CCS 2004, pp. 298–307. ACM, New York (2004)
34. Da Silva, J., Steffan, J.G.: A probabilistic pointer analysis for speculative optimizations. In: Shen, J.P., Martonosi, M. (eds.) ASPLOS, pp. 416–425. ACM (2006)
35. Srinivasan, H., Hook, J., Wolfe, M.: Static single assignment for explicitly parallel programs. In: Proceedings of the 20th ACM SIGPLAN-SIGACT Symposium on Principles of Programming Languages, POPL 1993, pp. 260–272. ACM, New York (1993)
36. Ungerer, T., Robič, B., Šilc, J.: A survey of processors with explicit multithreading. ACM Comput. Surv. 35, 29–63 (2003)

Appendix: Probabilistic Operational Semantics

This appendix reviews and augments a probabilistic operational semantics that we presented in [10] for the programming language (Figure 1) that we study. This language is the simple *while* language extended with new basic statements for parallel programming: $par\{\{S_1\}, \ldots, \{S_n\}\}$ (fork-join), $par\text{-}if\{(b_1, S_1), \ldots, (b_n, S_n)\}$ (conditionally spawned threads), and $par\text{-}for\{S\}$ (parallel loops). At the begin of the *par* command, the basic parallel command, a main core initiates the run of various concurrent inner cores. The subsequent statement (to the main core) can only be executed when the run of all inner cores are finished. The semantics of conditionally spawned command is akin to that of fork-join. The run of $par\text{-}if\{(b_1, S_1), \ldots, (b_n, S_n)\}$ includes initiating the conditionally concurrent runs of the n cores; only if b_i is *true*, S_i is executed. The following command (to the *par-if* statement) can only be executed when the runs of all conditional cores are finished. The semantics of parallel loop construct $par\text{-}for\{S\}$ includes running concurrently a statically unknown number of cores that all are S.

Semantically a computational state is a pair (γ, p): γ is a mapping from variables to values (integers plus symbolic addresses) and $p \in [0, 1]$. The intuition is that p is the probability of reaching γ. The following is the formal definition for computational states:

Definition 3. *1. Addrs* $= \{x' \mid x \in Var\}$ *and Val* $= \mathbb{Z} \cup Addrs$.
2. $\gamma \in \Gamma = Var \longrightarrow Val$.
3. A state is either an abort or a pair (γ, p) *such that* $p \in [0, 1]$.

We adopt the usual semantics for arithmetic and Boolean expressions, except that we do not allow arithmetic and Boolean operations on pointers.

$$[\![n]\!]\gamma = n \quad [\![\&x]\!]\gamma = x' \quad [\![x]\!]\gamma = \gamma(x) \quad [\![true]\!]\gamma = true \quad [\![false]\!]\gamma = false$$

$$[\![*x]\!]\gamma = \begin{cases} \gamma(y) & \text{if } \gamma(x) = y', \\ ! & \text{otherwise.} \end{cases} \qquad [\![e_1 \oplus e_2]\!]\gamma = \begin{cases} [\![e_1]\!]\gamma \oplus [\![e_2]\!]\gamma & \text{if } [\![e_1]\!]\gamma, [\![e_2]\!]\gamma \in \mathbb{Z}, \\ ! & \text{otherwise.} \end{cases}$$

$$[\![\neg A]\!]\gamma = \begin{cases} \neg([\![A]\!]\gamma) & \text{if } [\![A]\!]\gamma \in \{true, false\}, \\ ! & \text{otherwise.} \end{cases} \qquad [\![e_1 = e_2]\!]\gamma = \begin{cases} ! & \text{if } [\![e_1]\!]\gamma = ! \text{ or } [\![e_2]\!]\gamma = !, \\ true & \text{if } [\![e_1]\!]\gamma = [\![e_2]\!]\gamma \neq !, \\ false & \text{otherwise.} \end{cases}$$

$$[\![e_1 \leq e_2]\!]\gamma = \begin{cases} ! & \text{if } [\![e_1]\!]\gamma \notin \mathbb{Z} \text{ or } [\![e_2]\!]\gamma \notin \mathbb{Z}, \\ [\![e_1]\!]\gamma \leq [\![e_2]\!]\gamma & \text{otherwise.} \end{cases}$$

$$\text{For } \diamond \in \{\wedge, \vee\}, \quad [\![b_1 \diamond b_2]\!]\gamma = \begin{cases} ! & \text{if } [\![b_1]\!]\gamma = ! \text{ or } [\![b_2]\!]\gamma = !, \\ [\![b_1]\!]\gamma \diamond [\![b_2]\!]\gamma & \text{otherwise.} \end{cases}$$

The following are the inference rules of our probabilistic operational semantics (transition relation).

$$\frac{[\![e]\!]\gamma = !}{x := e : (\gamma, p) \rightsquigarrow abort} \qquad \frac{[\![e]\!]\gamma \neq !}{x := e : (\gamma, p) \rightsquigarrow (\gamma[x \mapsto [\![e]\!]\gamma], p)} \qquad \frac{\gamma(x) = z' \quad z := e : (\gamma, p) \rightsquigarrow state}{*x := e : (\gamma, p) \rightsquigarrow state}$$

$$\frac{\gamma(x) \notin Addrs}{*x := e : (\gamma, p) \rightsquigarrow abort} \qquad \frac{}{x := \&y : (\gamma, p) \rightsquigarrow (\gamma[x \mapsto y'], p)} \qquad \frac{\gamma(y) = z' \quad x := z : (\gamma, p) \rightsquigarrow (\gamma', p)}{x := *y : (\gamma, p) \rightsquigarrow (\gamma', p)}$$

$$\frac{\gamma(y) \notin Addrs}{x := *y : (\gamma, p) \rightsquigarrow abort} \qquad \frac{}{skip : (\gamma, p) \rightsquigarrow (\gamma, p)} \qquad \frac{S_1 : (\gamma, p) \rightsquigarrow abort}{S_1; S_2 : (\gamma, p) \rightsquigarrow abort}$$

$$\frac{S_1 : (\gamma, p) \rightsquigarrow (\gamma'', p'') \quad S_2 : (\gamma'', p'') \rightsquigarrow state}{S_1; S_2 : (\gamma, p) \rightsquigarrow state} \qquad \frac{[\![b]\!]\gamma = !}{if\ b\ then\ S_t\ else\ S_f : (\gamma, p) \rightsquigarrow abort}$$

$$\frac{[\![b]\!]\gamma = true \quad S_t : (\gamma, p) \rightsquigarrow abort}{if\ b\ then\ S_t\ else\ S_f : (\gamma, p) \rightsquigarrow abort} \qquad \frac{[\![b]\!]\gamma = true \quad S_t : (\gamma, p) \rightsquigarrow (\gamma', p')}{if\ b\ then\ S_t\ else\ S_f : (\gamma, p) \rightsquigarrow (\gamma', p_{if} \times p')}$$

$$\frac{[\![b]\!]\gamma = false \quad S_f : (\gamma, p) \rightsquigarrow abort}{if\ b\ then\ S_t\ else\ S_f : (\gamma, p) \rightsquigarrow abort} \qquad \frac{[\![b]\!]\gamma = false \quad S_f : (\gamma, p) \rightsquigarrow (\gamma', p')}{if\ b\ then\ S_t\ else\ S_f : (\gamma, p) \rightsquigarrow (\gamma', (1 - p_{if}) \times p')}$$

$$\frac{[\![b]\!]\gamma = !}{while\ b\ do\ S_t : (\gamma, p) \rightsquigarrow abort} \qquad \frac{[\![b]\!]\gamma = false}{while\ b\ do\ S_t : (\gamma, p) \rightsquigarrow (\gamma, p)} \qquad \frac{[\![b]\!]\gamma = true \quad S : (\gamma, p) \rightsquigarrow abort}{while\ b\ do\ S_t : (\gamma, p) \rightsquigarrow abort}$$

$$\frac{[\![b]\!]\gamma = true \quad S : (\gamma, p) \rightsquigarrow (\gamma'', p'') \quad while\ b\ do\ S_t : (\gamma'', p'') \rightsquigarrow state}{while\ b\ do\ S_t : (\gamma, p) \rightsquigarrow state}$$

- **Fork-join:**

$$\frac{}{par\{\{S_1\}, \ldots, \{S_n\}\} : (\gamma, p) \rightsquigarrow (\gamma', \frac{p'}{n!})} \dagger \qquad \frac{}{par\{\{S_1\}, \ldots, \{S_n\}\} : (\gamma, p) \rightsquigarrow abort} \ddagger$$

† there exist a permutation $\theta : \{1, \ldots, n\} \rightarrow \{1, \ldots, n\}$ and $n + 1$ states $(\gamma, p) = (\gamma_1, p_1), \ldots, (\gamma_{n+1}, p_{n+1}) = (\gamma', p')$ such that for every $1 \leq i \leq n$, $S_{\theta(i)} : (\gamma_i, p_i) \rightsquigarrow (\gamma_{i+1}, p_{i+1})$.

‡ there exist m such that $1 \leq m \leq n$, a one-to-one map $\beta : \{1,\ldots,m\} \to \{1,\ldots,n\}$, and $m+1$ states $(\gamma,p) = (\gamma_1,p_1),\ldots,(\gamma_{m+1},p_{m+1}) = abort$ such that for every $1 \leq i \leq m$, $S_{\beta(i)} : (\gamma_i,p_i) \rightsquigarrow (\gamma_{i+1},p_{i+1})$.

- **Conditionally spawned threads:**

$$\frac{par\{\{if\ b_1\ then\ S_1\ else\ skip\},\ldots,\{if\ b_n\ then\ S_n\ else\ skip\}\} : (\gamma,p) \rightsquigarrow (\gamma',p')}{par\text{-}if\{(b_1,S_1),\ldots,(b_n,S_n)\} : (\gamma,p) \rightsquigarrow (\gamma',p')}$$

$$\frac{par\{\{if\ b_1\ then\ S_1\ else\ skip\},\ldots,\{if\ b_n\ then\ S_n\ else\ skip\}\} : (\gamma,p) \rightsquigarrow abort}{par\text{-}if\{(b_1,S_1),\ldots,(b_n,S_n)\} : (\gamma,p) \rightsquigarrow abort}$$

- **Parallel loops:**

$$\frac{\exists n.\ par\{\{S\}_1,\ldots,\{S\}_n\} : (\gamma,p) \rightsquigarrow (\gamma',p')}{par\text{-}for\{S\} : (\gamma,p) \rightsquigarrow (\gamma',p')} \qquad \frac{\exists n.\ par\{\{S\}_1,\ldots,\{S\}_n\} : (\gamma,p) \rightsquigarrow abort}{par\text{-}for\{S\} : (\gamma,p) \rightsquigarrow abort}$$

The following definitions introduce terminologies that are used above to discuss and prove soundness of our proposed type system for probabilistic memory safety.

Definition 4. *For a statement S, a judgement of the form $S : (\gamma,p) \to (\gamma',p')$ is described as a computation (or an execution) path. The quantity p' is the probability of this execution path.*

Definition 5. *Suppose that $S : (pts,v,p_s) \to (pts',v',p'_s)$. Then S is positive at a state (γ,p) of type (pts,v,p_s) if along any execution path of S that starts at (γ,p) whenever an if statement, whose condition is true with probability p_{if}, is encountered at a state (γ'',p'') whose type is (pts'',v'',p''_s) in the proof tree of $S : (pts,v,p_s) \to (pts',v',p'_s)$, i.e.*

$$(\gamma,p) \rightsquigarrow \ldots (\gamma'',p'') \overset{if\ b\ then\ldots}{\rightsquigarrow} \ldots$$

the following are true:

- *if $p_{if} \times p'' \geq p_{ms} > (1-p_{if}) \times p''_s$, then $[\![b]\!]\gamma'' = true$.*
- *if $p_{if} \times p''_s < p_{ms} \leq (1-p_{if}) \times p''$, then $[\![b]\!]\gamma'' = false$.*

A simple structure induction proves the following lemma which is used in the soundness proof above:

Lemma 2. *Suppose that $S : (pts,v,p_s) \to (pts',v',p'_s)$, $(\gamma,p) \models_{ms} (pts,v,p_s)$, and S is positive at (γ,p). Suppose also that along an execution path of S that starts at (γ,p), a sub-statement S' of S is encountered at a state (γ'',p''), i.e.*

$$(\gamma,p) \rightsquigarrow \ldots (\gamma'',p'') \overset{S'}{\rightsquigarrow} \ldots$$

If $S' : (pts_1,v_1,p_{1s}) \to (pts'_2,v'_2,p'_{2s})$ in the proof tree of $S : (pts,v,p_s) \to (pts',v',p'_s)$ and $(\gamma'',p'') \models_{ms} (pts_1,v_1,p_{1s})$ then S' is positive at (γ'',p'').

A Proposal for Native Java Language Support for Handling Asynchronous Events

Carlos Rafael Gimenes das Neves[1], Eduardo Martins Guerra[2],
and Clovis Torres Fernandes[1]

[1] Instituto Tecnológico de Aeronáutica – ITA, Computer Science Department,
São Paulo, Brazil
[2] Instituto Nacional de Pesquisas Espaciais – INPE,
Laboratory of Computing and Applied Mathematics, São Paulo, Brazil
{carlosrafael.prog,guerraem}@gmail.com, clovistf@uol.com.br

Abstract. During early stages of computer software development, depending on the methodology employed, developers usually create the application's initial skeleton based on previously gathered requirements and on generated diagrams. When counting purely on what is provided by languages such as Java, developers tend to come up only with synchronous method calls, making use of coding tricks to achieve asynchronous behavior, which usually disrupts the original system model by adding a series of undesired side-effects such as unnecessary class coupling and error-prone constructions. This work proposes an extension to the Java language to allow for both executing asynchronous methods and handling asynchronous events occurring during normal execution, as a straightforward, class coupling-free and native alternative. With this extension it is expected that developers can natively use asynchronous communication from the beginning of the development cycle without having to make structural modifications to the original system.

Keywords: Asynchronous Event Handling, Language Structures, Compilers, Event Oriented Programming, Exception Handling.

1 Introduction

Developing computer software is a task often performed with the aid of software methodologies and techniques that aim at providing developers with some level of ease and confidence. Along with that, code generation tools have recently become a more and more common choice for developers willing an extra level of confidence and convenience. In this scenario, asynchronous execution and communication is not a worry, as the tool would be in charge of it should it be necessary. Nonetheless, when only languages features can be used, considering the Java language, asynchronous execution and communication usually push developers towards solutions involving creation of new classes or interfaces not present in the original model, together with unplanned implementations and undesired class coupling, adding unnecessary complexity to both development and testing processes.

B. Murgante et al. (Eds.): ICCSA 2013, Part V, LNCS 7975, pp. 531–546, 2013.

There is a series of commercial frameworks, for example Akka framework [1] and JMS API [2], deal with these complexities and provide developers with means to achieve asynchronous execution and communication without too much effort. This topic is also the target of many academic works, such as JR Language, which proposes an extension to the Java language, making it a more suitable language for asynchronous communication and execution [3].

In spite of all the benefits brought by these works, some failed to deliver a transparent solution to developers [1-3], forcing them to change the original software model in some way, either implementing extra code or changing the original class hierarchy, in order to actually obtain the asynchronous behavior. On the other hand, others took care only of communication issues, even with first-class constructs, but left asynchronism aside [4]. Most of them also required the sender of an asynchronous message to know the receiver, creating a possibly undesired coupling between sender and receiver [1-3].

Our aim was to fill the gap between these works, proposing means for developers to naturally use asynchronous execution and communication, through first-class constructs and specially without forcing them to implement extra code, to couple parts of the software and to change the original software model of the software.

That behavior was achieved adding new language constructs to the Java language as well as creating a custom Java compiler capable of interpreting them. These new language constructs enable developers to easily add specific asynchronous behavior to the system without side-effects, as they act just as native structures do. Moreover, the fact that neither extra code nor changes to the class hierarchy are required makes the resulting code more compliant to the original model, improving both code readability and faithfulness to the documentation where the system derives from.

2   Characterizing the Research Problem

A close inspection to a typical Java program shows that the native structures available for communication fall basically into two groups, namely, method calls and thrown exceptions [5], [6]. Both are first-class constructs that work with the call stack and present a synchronous behavior. Considering method calls and thrown exceptions as messages, the execution actually relies on first-class constructs that send synchronous messages between methods in the call stack. A method call differs from a thrown exception, as the throwing of an exception makes the sender method terminate [5].

Existing tools and works regarding message-passing techniques present different approaches. Some provide asynchronous behavior, but force changes to the structure of existing classes, or require implementation of extra code [1-3]. Some come as first-class constructs, but without the asynchronous behavior [4]. Coupling of the sender and the receiver is also present in some works [1-3]. Ability to access the original call stack is provided only by some of them [3], [4], [11], just as the ability to access the original context [4], [5], [11].

Hence, the main problem addressed in this work is how to allow code to execute asynchronously through first-class constructs, giving it the ability to both asynchronously notify other methods in the call stack about the occurrence of general events and access its original context, without the need for interrupting the execution of the sender method, for coupling classes or for implementing extra interfaces.

3   Proposed Solution

One possible solution to the issue previously brought up would be creating extensions to the Java language, in order to provide first-class constructs for developers to handle asynchronous events, without implementing extra interfaces and without worrying about usual matters on asynchronous event handling, such as registration control and capture of the method's context, namely, its closure [7].

Unlike the closure capture performed by Java, up to version 1.7, which includes only immutable local variables, those marked as final [5], [6], the captured closure must be mutable. Moreover, it must be accessible from different threads at the same time, as the asynchronous method execution will be accomplished by creating several concurrent threads.

Since in Java, a method's closure is composed of both local variables and parameters, capturing the closure and making it accessible from another thread means giving this other thread access to the memory area where such information is stored. According to the Java Virtual Machine (JVM) specification [6], the memory is arranged similarly to the diagram in Fig. 1, with all local variables stored inside frames.

New frames are created every time a method is called, and placed at the top of the stack. Frames are destroyed and thus, lost, when its corresponding method terminates [6]. No method, from any thread, can access frames directly, either before or after frame's destruction.

The only way to access frames during runtime is through tools such as the Java Virtual Machine Tool Interface or the Java Debug Interface, both part of the Java Platform Debugger Architecture [8]. Nevertheless, there are three important reasons to avoid using these tools in the solution. First, there is no guarantee that these tools will be part of all platforms where Java is supported [8]. Also, the thread must be in a suspended state to have its frames available for inspection. Last, using those tools involves writing some native code, removing the portability of the solution.

For the solution to be portable it should only make use of native Java commands and structures, therefore, allowing an ordinary Java Virtual Machine to execute it.

Since a traditional Java compiler is capable of capturing only local variables marked as final, in order to capture all local variables and keep them mutable, the key would be to move them into a local class, similar to what has been done in [9]. This local class has one single local instance marked as final, capable of being captured by the traditional Java compiler. As parameters cannot be moved, they are just copied into the new local class.

Next, all references to local variables and parameters are replaced with references to the fields of this newly created class, in a similar fashion to what is shown in Fig. 2. As a consequence of those changes, inline variable initializers must be moved to their own line. Variables and parameters that are marked as final can be left just as they are, since they are immutable and traditional Java compilers can already capture them.

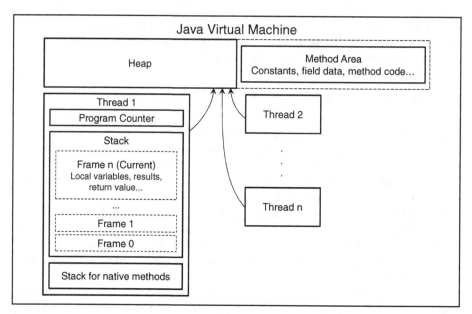

Fig. 1. Simplified internal memory distribution of the Java Virtual Machine

Original Method	Modified Method
| ```java
String m1(int p) {
  final String f =
   "...";
  String s =
   p.toString();
  s += f;
  return s;
}
``` | ```java
String m1(int p) {
  class Closure {
   int p;
   String s;
   Closure(int p) {
    this.p = p;
   }
  }
  final Closure c = new Closure(p);
  final String f = "...";
  c.s = c.p.toString();
  c.s += f;
  return c.s;
}
``` |

Fig. 2. Code modification to allow the method's closure to be properly captured

Although that technique does enable the closure to be captured, it is just a part of the solution being proposed. In order to actually solve the problem, this technique must be performed automatically, while still allowing methods in the call stack to communicate asynchronously through first-class constructs, which involves adding new language constructs to the Java language. There are many tools available to help in the creation of extended Java compilers, capable of generating standard Java

Bytecode [6], executable in any standard Java Virtual Machine. Among all the available tools, JastAddJ [10] was chosen due to its relative simplicity and powerful reference attribute grammars.

Using JastAddJ, seven new language constructs were added to the Java language. They form structures that are translated into ordinary Java code referencing existing precompiled helper classes that are responsible for simulating the expected behavior.

One language construct, *within*, is used to specify regions in the code where to monitor the occurrence of events, similar to what *try* statements do in plain Java. It is used together with two other language constructs, *when* and *xwhen*, which specify the code that is to be asynchronously executed when a particular event happens inside the region delimited by *within*. The difference between *when* and *xwhen* is that the latter serializes simultaneous executions while the former allows as many concurrent executions as necessary. Both language constructs are analogous to the *catch* clause of the Java language. The fourth language construct added was *event*, used for asynchronous event notification, similar to what *throw* statements do with exceptions in plain Java. With these four language constructs, developers are able to produce general event notifications, to specify the region where events are monitored and to asynchronously handle these events, using first-class constructs without implementing extra interfaces or dealing with any kind of registration control.

Another language construct, *arun*, was added to provide developers with means to run ordinary asynchronous code unconditionally, while still allowing events occurred during this asynchronous execution to be properly handled by previous *when* and *xwhen* blocks. The last two constructs, *handlerStack* and *currentHandlerStack*, were added to allow event notifications to be properly handled in scenarios where code must be run asynchronously but *arun* cannot be used, such as when new threads must be explicitly created by the developer.

3.1   *when, within* and *event* Constructs

The *when*, *within* and *event* language constructs constitute the foundation of the proposed solution. A *within* statement behaves similarly to what an ordinary *try* statement does [5]. It is used to determine the region where to monitor the occurrence of events. Any methods called in the *within* block are also monitored, as well as the methods called inside those methods and so on. One or more *when* clauses must precede a *within* statement to specify the behavior for each event that the developer wishes to monitor the occurrences.

A *when* clause behaves like an ordinary *catch* clause [5], specifying what code to execute upon the occurrence of an event of a particular class. Unlike the *catch* clause that can be used only with the class *Throwable* or with its subclasses [5], there are no limitations to what classes can be used with a *when* clause, giving developers greater flexibility and freeing them from changing the current class hierarchy in the system, just to be able to use that class as an event. Code inside *when* blocks always executes asynchronously. Also, if an event of class A is expected, but an event of class B is triggered, and B extends A, the *when* block expecting the event of type A is triggered.

Finally, the *event* statement is used like an ordinary *throw* statement, notifying the occurrence of an event. The main difference is that *event* statements do not interrupt or abort current method's execution, as they execute asynchronously. One single event can trigger multiple *when* blocks, because unlike usual Java exceptions, that are propagated only until the first handler is found [5], events were defined to propagate to all suitable, currently active *when* handlers.

Actual exceptions thrown in a *within* block are not automatically handled, and should be handled manually by the developer. On the other hand, in order to address the issues of handling exceptions in asynchronous environments [11], it has been stipulated that if an exception is thrown during the asynchronous execution of a *when* block, and the developer fails to catch that exception inside the block, the exception is transformed into an event, which is subsequently notified, so that it could be handled by other *when* blocks expecting that exception or one of the exception's super-classes.

While this behavior addresses the problem, it cannot be stated that the problem is actually solved. The proposed behavior is just an initial proposition for future works and discussions around exception handling in asynchronous environments.

3.2   *xwhen* Construct

The *xwhen* clause can be used in place of a *when* clause, to indicate that no more than one simultaneous execution of that block can exist. When an event triggers a *when* block and that block is already executing, another execution of that block starts. On the other hand, when an event triggers an *xwhen* block and that block is already executing, the notification is queued, and the block executes again after having finished its current execution.

3.3   *arun* Construct

The *arun* statement is used to asynchronously execute a block of code. What differs using *arun* and running another thread is that events occurring inside the *arun* block are propagated to the handlers present in the stack, while a new ordinary thread will have an empty stack of handlers.

3.4   handlerStack and currentHandlerStack Constructs

Since all event notifications are propagated to the current stack, should developers need to manually create a new thread, in situations where *arun* cannot be used, that new thread will have a new blank stack of its own, and any events occurred there will not be propagated to the stack of the method creating that thread. The new primitive data type *handlerStack* and the statement *currentHandlerStack* have been created for this kind of situation. The *currentHandlerStack* statement allows developers to obtain a copy of the current stack of handlers and to use that copy as the target of an event notification or as the initial stack used for an execution of an *arun* block. The new data type *handlerStack* is the data type of the copies of the current stack.

4   Implementation Details

As the solution introduces seven new constructs to the Java language, it is necessary to demonstrate what happens behind those constructs, specially the aspects regarding their inner workings, such as how they are converted into normal Java code and how they interact with the precompiled helper classes. Although not all these helper classes are explained in this work, for brevity, the behavior of the main class referenced by the transformed code, *Manager* class, is explained in some detail.

4.1   *when, within* and *event* Constructs

After having defined the behavior of *when, within* and *event* constructs, the following code snippet helps clarify their basic usage.

```java
class A {
  void m1() {
    when (Event e) {
      m3(); //Code to handle events of class Event
    } when (Event2 e2) {
      m4(); //Code to handle events of class Event2
    } within {
      m2();
    }
  }
  void m2() {
    event new Event();
    m5();
  }
  void m3() { event new Event2(); }
  void m4() {...}
  void m5() {...}
}
```

[Sample code demonstrating the usage of *within, when* and *event* constructs]

As it can be seen in the diagram illustrated in Fig. 3, the eventual notification of the event of type *Event* in *m2( )* triggers one of the *when* blocks inside *m1( )*; in fact, the first *when* block inside *m1( )*, which includes only a call to the method *m3( )*. Therefore, that block is executed asynchronously on a new thread, even though *m1( )* is not the current method on the call stack of the thread where the notification occurs; the current method is *m2( )*. By the time *m3( )* executes and eventually notifies about the occurrence of event of type *Event2*, there is no guarantee whether *m1( )* will still be in its corresponding call stack, which does not matter, as the *when* block for *Event2* is triggered and method *m4( )* is executed even if *m1( )* has already terminated.

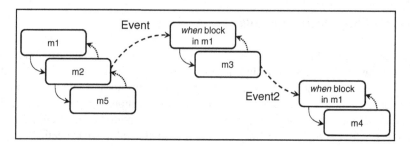

Fig. 3. Block diagram showing the execution of the sample code using *within*, *when* and *event*

Events occurring in one *when* block of a given method are seen as if occurring in the *within* block below it. Consequently, these events can trigger any *when* blocks above that *within* block, as well as other *when* blocks that were registered in the stack at the time the *within* block started executing. This behavior can be better understood by looking at the following code.

```
class A {
  void m1() {
    when (Event e)  { m4(); }
    when (Event2 e)  { m6(); }
    within { m2(); }
  }
  void m2() {
    when (Event e)  { m5(); }
    when (Event2 e)  { m7(); }
    within { m3(); }
  }
  void m3()  { event new Event(); }
  void m4()  {...}
  void m5()  { event new Event2(); }
  void m6()  {...}
  void m7()  {...}
}
```

[Sample code demonstrating the behavior of *within* and *when* in more detail]

Assuming that the execution starts with a call to method *m1( )*, method *m2( )* is the next one to be called, followed by a call to method *m3( )*, which eventually notifies about the occurrence of the event of type *Event*. That notification triggers the *when* blocks in *m1( )* and in *m2( )*, and the methods *m4( )* and *m5( )* are called. Once *m5( )* is called, the event of type *Event2* ultimately occurs, triggering the second *when* block in *m1( )* and in *m2( )*, finally causing methods *m6( )* and *m7( )* to be called.

Again, all this behavior is thanks to the precompiled helper classes referenced by the transformed code, presented in Fig. 4. The transformation does not actually change the source files, it just changes the abstract syntax tree generated by JastAddJ before the code is actually compiled [10].

Original Code	Transformed Code
```void m1(int p) { int a = 0; when (Event e) { a += p; } within { a = 10; m2(); } } void m2() { event new Event(); }```	```void m1(int p) { class Closure { int p, a; Closure(int p) { this.p = p; } } final Closure c = new Closure(p); c.a = 0; Manager.startAppending(); try { Manager.append(true, Event.class, new Handler<Event>() { public void handle(Event e) { c.a += c.p; }}); Manager.endAppending(); c.a = 10; m2(); } finally { Manager.rewind(); } } void m2() { Manager.notify(new Event()); }```

Fig. 4. Transformation applied to the code before compiling *when*, *within* and *event* constructs

4.2   *xwhen* **Construct**

As already explained, the behavior of *xwhen* is different in comparison with the behavior of *when* clauses. In spite of that, there are not too many differences in the transformed code, except for the first parameter of the method *append( )*, shown in Fig. 4, which is changed from *true* to *false*, telling the helper class *Manager* to change the execution behavior during runtime.

4.3   *arun* **Construct**

The next code snippet helps demonstrating the runtime behavior of *arun* blocks, and Fig. 5 shows what *arun* blocks look like after the code transformation.

```
class A {
  void m1() {
    when (Event e) { m5(); }
    within { m2(); }
  }
  void m2() {
    arun { m3(); }
    m5();
  }
  void m3() { event new Event(); }
  void m4() { ... }
  void m5() { ... }
}
```

[Sample code demonstrating the behavior of *arun* blocks]

Assuming that the execution starts with a call to method *m1( )*, method *m2( )* is the next one to be called. In there, the *arun* block is executed asynchronously, and the method *m3( )* is called from another thread, which notifies about the occurrence of the event of type *Event*. That notification triggers the *when* block in *m1( )* and the method *m5( )* is called.

Original Code	Transformed Code
```void m1(int p) { arun { int a = p + 1; m2(a); } } void m2(int x) {...}```	```void m1(int p) { class Closure { int p; Closure(int p) { this.p = p; } } final Closure c = new Closure(p); Manager.fork(new Runnable() { public void run() { int a = c.p + 1; m2(a); }}); } void m2(int x) {...}```

Fig. 5. Transformation applied to the code before compiling the *arun* construct

When an actual exception is thrown during the asynchronous execution of an *arun* block, it is caught and transformed into an event that could be handled by a *when* block expecting that exception or one of its super-classes. This is the same behavior as the one explained in 3.1, therefore, that same assumption is valid here; the behavior is just an initial proposition which needs to be further studied.

4.4   handlerStack and currentHandlerStack Constructs

The statement *currentHandlerStack* is translated into a simple call to a method of the *Manager* class which returns a copy of the current stack. The data type *handlerStack* is translated into a reference to a normal interface. The basic usage of *handlerStack* and *currentHandlerStack* constructs is demonstrated in the following code snippet.

```
handlerStack cs = currentHandlerStack;
event new Event() cs;
arun cs { ... }
```

[Sample code demonstrating the usage of *handlerStack* and *currentHandlerStack*]

Once a copy of the stack is made, it can be used together with the *event* statement, forcing the event to be notified in the given stack, or it can be used with *arun*, forcing the given stack as the initial one used by *arun*.

4.5   Precompiled Classes

The code translation itself is not enough to give the desired behavior to the solution, as this requires some code to be executed. All this code necessary to bring life to the solution was encapsulated in precompiled helper classes. Instead of referencing many classes, it was decided to create the *Manager* class, and make it the only class referenced by the transformed code, apart from the interface referenced after converting the *handlerStack* data type.

The *Manager* class acts as a façade (kind of design pattern [12]) to the remaining precompiled helper classes, calling their methods as necessary. As this helper class is in charge of making the solution actually work during runtime, next we present an overview of its methods called in the transformed code shown in Fig. 4 and Fig. 5:

- startAppending( ) – It creates an empty list of handlers and adds this list to the top of an internal stack of lists, which is maintained on a per thread basis using the concept of thread-locals variables [13].
- append( ) – It appends the handler to the list on the top of the current stack, specifying the class of the event that the handler is prepared for.
- endAppending( ) – It creates a copy of the current internal stack, propagating that copy to all handlers in the list at the top of the stack, so that events occurring during the asynchronous execution of one of those handlers can be properly notified to other handlers in the current stack.
- rewind( ) – It removes from the stack the list of handlers at the top of it.
- notify( ) – It notifies all suitable handlers in the internal stack about the occurrence of an event, starting their asynchronous execution.

5   Discussion

Five discrete topics come out by dividing the problem discussed in this work: asynchronous method execution, asynchronous event notification and handling, class coupling, implementation of extra interfaces and ability to access the original call stack and context. All these five topics were addressed by the implementation previously explained.

Asynchronous method execution is achieved through the *arun* first-class construct. It allows not only for methods to be executed asynchronously but also for any kind of structured code to be executed asynchronously. Once the *arun* block is executed it creates another thread, without developer's interaction, to execute its contents and the code following the *arun* block continues as soon as this new thread starts.

With *event* statements the developer can asynchronously notify about the occurrence of events as these statements do not block the execution of the current method. On the other end, with *when* blocks developers can asynchronously handle these events. The execution of one *when* block is not tied to the execution of any other *when* block, therefore, many *when* blocks can be executing at the same time.

Class coupling is not a concern as the sender of an event communicates with all the handlers present in the call stack, independently of their class. Also, the handlers need not to know which class is responsible for generating an event, they should only care about the area of code where the desired event could happen.

Implementation of extra interfaces is also not a concern for the developer, since all constructs provided by the solution are first-class constructs placed throughout the code along with other ordinary native structures, they do not require any changes to be applied to the declaration of the classes.

The access to the original call stack is guaranteed by the automatic registration technique, shown in Fig. 4, which creates lists of registered handlers on a per thread basis. These lists are propagated to the new threads created by *arun* and *when* blocks, in such a way that the events triggered within those threads can be propagated to the original handlers.

Last, the access to the original context is accomplished through the code transformation shown in Fig. 2 and Fig. 4, which encapsulates all local variables and parameters inside a local class, making them accessible from within the handlers.

The scope of this work was to raise the problem and propose an initial solution to it. Although the proposed solution does provide developers with constructs with the expected behavior, this work neither attempts to exhaust all possible solutions nor assumes the proposed solutions formally solves the raised problem. A formal proof of solution can be made in future works, as well as future works can deeper analyze the performance achieved by the precompiled helper classes, for example, eventually proposing an improved algorithm for handling thread creation and thread reuse.

6   Related Works and Tools

Asynchronous communication and asynchronous method execution is a fascinating broad topic. As such, it is issued by several academic works and commercial tools, some of which are discussed next.

6.1   Akka Framework

Akka [1] is a commercial Java framework focused on parallel and distributed computing, helping in the creation of highly scalable and fault tolerant applications, offering many features that extend beyond the scope covered by this work. The main interest in Akka is its message passing functionality. In a brief description, it is based on the concept of message passing between actors. Actors are the components that actually execute code in the system, similar but not equal to what classes are to object oriented programming, as it is possible to assign a different class to the role of one specific actor at different times [1].

Each actor has one mailbox where the messages are kept until they are processed and they do not need to reside on the same process in order to communicate. Moreover, their distribution can be hierarchical in such a way an actor can create other actors, delegate them tasks and supervise them [1].

Considering the sending of a message as an event and the handling of the message as the handling of the event, Akka provides the developer with asynchronous event handling and with many other features not covered by this work, such as asynchronous execution and communication in physically distributed environments.

Albeit there is no need for the classes to know each other, they must know the actor they want to send a message to, which can be seen as a type of coupling.

Also, the framework requires classes to extend one of the *Actor* classes provided in order to play the role of an actor, changing the original structure of the system. Last, because it is an external framework, upon which software can be developed, Akka does not behave as a first-class construct.

6.2   Spring Framework and the Java Message Service

Spring [14] is a commercial modular framework for developing software applications, which integrates with integrates with the Java Message Service API, JMS API [2], in order to provide an asynchronous and flexible messaging system. The Java Message Sevice works based on two distinct models, Point-To-Point and Publish/Subscribe [2].

In the Point-To-Point model there are message queues acting as mailboxes, where messages are sent to and retrieved from. In the Publish/Subscribe model, everyone interested in one topic subscribe to that topic, so they can receive all messages that are published under that topic. Since both message queues and topics implement one special interface, called Destination [2], the difference between the models becomes transparent for those accessing only the Destination interface.

Considering the sending of a message as an event and the handling of the message as the handling of the event, the integration on Spring and the Java Message Service provides the developer with asynchronous event handling. Actually, this integration offers a highly customizable platform for application development in many aspects that surpass the scope of this work.

In spite of this fact, for the asynchronous event handling to work properly, the sender of the message must know the message's destination, coupling sender and receiver. Also, neither the sender nor the receiver make use of first-class constructs and any classes willing to receive messages must implement one extra interface in order to be able to do so.

6.3   JR Language

The JR language [3] is actually an extension to the Java language, adapted from the SR language [15], [16], aiming at providing better Java support to distributed and concurrent computing. As a part of that effort, the extension provides several features such as remote object creation, remote method invocation, and asynchronous message passing, among others. Its asynchronous message-passing model is based on new constructs introduced by the language, specially *send* and *forward*.

The general format of the commands is as follows:

```
send target.message(arguments) handler handlerObject;
forward target.message(arguments) handler handlerObject;
```

[General format of the *send* and *forward* commands in the JR language]

Both *send* and *forward* commands send the given message to the target. But the *forward* command additionally delegates to the given method the obligation to return a value to its caller. The *handler* part of the command is used to indicate the object responsible for handling any exceptions that may occur during the asynchronous execution of the given method. In order to do so, that object must implement one handler method for each possible exception.

Thereby, the JR language works with two different models of event handling, one for the exception handling and another one for the explicit messages. Its exception handling mechanism behaves synchronously and is similar to the native Java mechanism, with the difference that it relies on methods to handle the exceptions, not on *try-catch* clauses.

In contrast, its asynchronous message-passing model provides a straightforward mechanism that uses first-class constructs and does not require the implementation of any extra interfaces. However, it still requires the sender of the message to know the receiver, coupling both classes.

6.4   Robust Exception Handling in an Asynchronous Environment / ProActive

Robust exception handling in an asynchronous environment proposes transparent means for handling exceptions occurred in asynchronous method calls [11], using regular Java constructs together with the tool ProActive [17]. The main idea is to actually throw the exception generated by the asynchronous method, if any, upon the first use of the object returned by the asynchronous method, similarly to what Java's Future objects [18] do, but in a transparent fashion. If the object returned is not used within the *try* block, the exception is thrown at the end of the *try* block [11].

Similarly to the JR language, this work has two different models of event handling, one for the explicit messages, which is similar to the JR language model, and another for the exception handling, which, in turn, is analogous to the native Java one. As the explicit asynchronous message-passing model is analogous to the proposed by the JR language, it does provide means to communicate asynchronously using first-class constructs and without requiring the implementation of any extra interfaces, at the same time it still requires the sender to know the receiver, coupling both classes.

6.5   Exception Handling with Resumption: Design and Implementation in Java

Exception handling with resumption shows an alternative for the native Java exception handling model [4]. In this proposed model the method where the exception occurs is not necessarily terminated. After the occurrence of the exception, the handler has a chance to try to solve the problem and propose a solution. If the solution is accepted by the originating method, then its execution continues. This behavior is achieved through the use of two new language constructs, *resume*, used within a catch block to try to propose a solution, and *accept*, used with *throw* clauses, to specify acceptable solutions for that particular exception.

Although this work proposes a way to prevent terminating abruptly the execution of a method using first-class constructs and without any extra interfaces, its execution does get interrupted momentarily, as there is no kind of asynchronous behavior. Also, its entire notification model is to be used with exceptional conditions only, not with general events.

7   Conclusion

Asynchronous execution and event notification are truly fascinating topics that can be looked upon from different perspectives. This work proposes a slightly different form of viewing and dealing with those topics, focusing on asynchronous communication between methods in the call stack as well as their asynchronous execution.

While the proposed solution does provide developers with clean, native means for methods to communicate asynchronously, without requiring them to implement extra interfaces, create unnecessary class coupling and control the registration of active event handlers manually, the final code that is actually executed adds several new classes to the program and makes extensive use of handler registration control, but as an automatic process. Even so, the final code neither changes the original classes' hierarchy nor creates unnecessary coupling between them.

Although the presented solution offers enough features to address the aforementioned problem, there is still plenty of room for future improvements and future works such as creating a *multi-when* version of the *multi-catch* present in the Java language [5] or even providing the developer with means for manually creating threads with custom-initialized stacks. Another possible improvement would be providing developers with means to help with synchronization issues and automatic mutual exclusion control of shared resources used inside *when* and *arun* blocks.

The authors would like to thank FAPESP for the support provided, without which it would not be possible for the authors to attend to the congress.

References

1. Typesafe Inc.: Akka Documentation 2.1.2 (2013),
   http://doc.akka.io/docs/akka/2.1.2/
2. Sun Microsystems: Java Message Service Specification 1.1. (2002),
   http://download.oracle.com/otndocs/
   jcp/7195-jms-1.1-fr-spec-oth-JSpec/
3. Keen, A.W., Ge, T., Maris, J.T., Olsson, R.A.: JR: Flexible distributed programming in an extended Java. In: ACM Transactions on Programming Languages and Systems (TOPLAS), pp. 578–608. ACM, New York (2004)
4. Gruler, A., Heinlein, C.: Exception handling with resumption: design and implementation in Java. In: Proceedings of the 2005 International Conference on Programming Languages and Compilers (PLC 2005), Las Vegas, pp. 165–171 (2005)
5. Gosling, J., Joy, B., Steel, G., Bracha, G., Buckley, A.: The Java Language Specification, Java SE 7 Edition (2011), http://docs.oracle.com/javase/specs/
6. Lindholm, T., Yellin, F., Bracha, G., Buckley, A.: The Java Virtual Machine Specification, Java SE 7 Edition (2011), http://docs.oracle.com/javase/specs/
7. Krishnamurthi, S.: Programming Languages: Application and Interpretation (2007),
   http://www.cs.brown.edu/~sk/Publications/Books/ProgLangs/
8. Oracle: Java Platform Debugger Architecture (JPDA) (2012),
   http://docs.oracle.com/javase/7/docs/technotes/guides/jpda/
9. Heinlein, C.: Local Virtual Functions. In: Proceedings of NODe 2005, GSEM 2005, Erfurt, pp. 129–144 (2005)
10. Ekman, T., Hedin, G.: The JastAdd Extensible Java Compiler. In: Proceedings of the 22nd Annual ACM SIGPLAN Conference on Object-Oriented Programming, Systems, Languages, and Applications, Montreal, pp. 1–18 (2007)
11. Caromel, D., Chazarain, G.: Robust exception handling in an asynchronous environment. In: ECOOP Workshop on Exception Handling in Object-Oriented Systems: Developing Systems that Handle Exceptions, number 05050 in Technical Reports - Laboratoire. Sophia Antipolis (2005)
12. Gamma, E., Helm, R., Johnson, R., Vlissides, J.: Design patterns: elements of reusable object-oriented software. Addison-Wesley Longman Publishing, Boston (1995)
13. Oracle: ThreadLocal Class Documentation (2012),
   http://docs.oracle.com/javase/7/docs/
   api/java/lang/ThreadLocal.html
14. Spring Source: Spring Framework 3.2.2 (2012),
   http://static.springsource.org/spring/docs/3.2.x/
   spring-framework-reference/pdf/spring-framework-reference.pdf
   (2012)
15. Olsson, R.A., Andrews, G.R., Coffin, M.H., Townsend, G.M.: SR: A Language for Parallel and Distributed Programming, The University of Arizona, Tucson (1992)
16. Andrews, G.R., Olsson, R.A.: The SR programming language: concurrency in practice. Benjamin-Cummings Publishing, Redwood City (1993)
17. ActiveEon: ProActive Parallel Suite (2011), http://proactive.activeeon.com
18. Oracle: Future Interface Documentation (2012),
   http://docs.oracle.com/javase/7/docs/api/java/
   util/concurrent/Future.html

An Improved Approach to the Recovery of Traceability Links between Requirement Documents and Source Codes Based on Latent Semantic Indexing

Jianwei Shao[1], Wei Wu[2], and Peng Geng[3]

[1,3] Zhejiang Topthinking Information Technology Co., Ltd, Hangzhou, China
[2] Zhejiang Provincial Key Laboratory of Network Technology and Information Security,
Hangzhou, China
`jianwei.shao@zjtzsw.com`, `ww@topcheer.cn`

Abstract. The traceability links among software artifacts plays a very important role for the maintenance of consistency in the evolution of software product lines. This paper introduces an improved approach to the recovery of traceability information between requirement documents and source codes based on Latent Semantic Indexing (LSI) and the special features of object-oriented source codes. In order to obtain accurate traceability links, it employs the hierarchical information caused by the inheritance relationship among classes and recovers the traceability links using class clusters. Moreover, it assigns different weights for the terms in source codes according to their degrees of correlation to the documents. The case demonstrates the improved approach, and the experimental results show that it can increase the extraction precision by 3%~6%, compared with that based on the traditional LSI.

Keywords: traceability links, Latent Semantic indexing, consistency maintenance, information retrieval.

1 Introduction

Establishing the traceability links between requirement documents and source codes has been a challenging task during the whole life cycle of software product line [1]. If a domain expert wants to maintain consistency of software product line for specific systems that he is not familiar with, an essential task is to extract the traceability links between requirement documents and source codes. However, the requirement document is written in natural language with complex grammatical structures. Its semantics are consequently difficult to be accurately analyzed. Meanwhile, the traceability links between requirement documents and source codes are rarely represented explicitly, because requirement documents and source codes which are both used for formalizing the real world are represented at different abstract levels. Therefore, extracting traceability links between requirement documents and source codes is an arduous challenge [2].

At present, there are many approaches to extracting traceability links between requirement documents and source codes. One way is to utilize some modeling

B. Murgante et al. (Eds.): ICCSA 2013, Part V, LNCS 7975, pp. 547–557, 2013.

integrated development tools, which support the extraction and maintenance of traceability links, such as Rational Requisitepro [3] and DOORS [4]. Another way is to employ Information Retrieval (IR) models such as Probability Model (PM), Boolean Model (BM) and Vector Space Model (VSM) [5-6].

However, the above approaches have some limitations. The mentioned tools usually could not be friendly used. In addition, they incur huge cost and are error-prone when analysts deal with a large scale application with long life cycle [7][8]. Although PM is easy to be implemented and has definite physical meanings, it cannot deal with synonymies. Moreover, etyma must be generated when documents are preprocessed before applying PM. As an improved one based on BM, VSN is actually a general representation approach for text documents. Its term-document matrix can be used by any retrieval models as well as powerful tool for IR. However, in the process of using VSM, the high dimension of term-document matrix is a fatal defect that leads to low efficiency and big noises.

Considering the above shortage, this paper presents an approach to extracting traceability links between requirement documents and source codes based on Latent Semantic Indexing (LSI), which is an improved one from VSM. It presents two improved strategies based on traditional LSI to raise the precision of extraction. The suggested approach is based on the followings assumptions: 1) class names, attribute names and method names in source codes are meaningful identifiers; 2) there exist many specific comments in source codes.

The rest of this paper is organized as follows. After showing the overall process of the extracting traceability links between requirement documents and source codes in Section 2, Section 3 presents two improved strategies based on LSI, while Section 4 compares experimental results of LSI-based approach and the improved LSI-based approach. After Section 5 shows related works, last section concludes this paper and outlines the future work.

2   Approach Overview

LSI is an IR model for information retrieval based on VSM. The core concept of LSI is that there exist some implicit relationships among the terms of documents. The semantic structure suggests an abstract semantic form which is composed of semantic category and semantic relationship in natural languages [9]. Therefore, LSI doesn't rely on syntax matching of terms in the documents to compute similarity. Instead it extracts traceability information by analyzing latent semantic structures of terms in the documents. Its advantage over VSM is that the term-document matrix provided by LSI has lower dimensions. In addition, it can filter out lots of noises.

In our context, it is assumed that there are some latent semantic structures in requirement documents and source codes. Fig.1 exhibits the overall process of extracting traceability links by LSI, which is further divided into three main phases, namely phase of preprocessing, phase of mining traceability links and phase of refining traceability links.

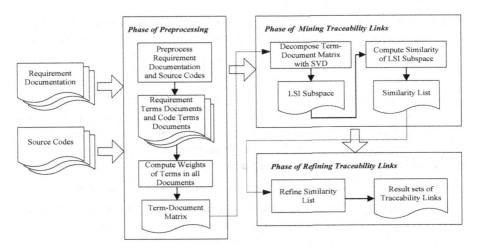

Fig. 1. Process of extracting traceability information based on latent semantic indexing

2.1   Preprocessing

The core idea of LSI is that the similarity of the term vector space represents the semantic similarity. Hence, the requirement documents and their related source codes should be first converted to a set of term vectors.

During the preprocessing phase, all identifiers in source codes, such as class names, method names, attribute names and comments are extracted from source codes. The identifiers are then broken into a list of terms. In addition, the stop words that do not help the retrieval of term semantic are removed, while all uppercase characters are converted to lowercase ones. In this way, all the meaningful code terms are finally identified and preserved in the code-term documents.

On the other hand, the requirement document written in various formats, such as Word and PDF, need to be converted to that in a unified document format. It is then partitioned into requirement-term documents according to paragraphs. Moreover, all stop words are removed and the uppercase letters are converted into the lowercase letters in requirement-term documents as what is done for source codes.

A requirement-term document containing m different terms can be represented as:

$$D_i = \left(d_{i,1}, d_{i,2}, \dots, d_{i,m}\right) = \sum_{j=1}^{m} d_{i,j}\, \bar{t}_j \tag{1}$$

Here, $d_{i,j}$ is the weight of term t_j in the requirement-term document D_i. Similar, the code-term document q can be defined as:

$$q = (q_1, q_2, \dots, q_m) = \sum_{j=1}^{m} q_j\, \bar{t}_j \tag{2}$$

where q_j is the weight of term t_j in the code-term document q. The terms which have bigger weights are regarded to be more representative of the term document.

We employ tf–idf, or term frequency–inverse document frequency, as the weight to reflect how important a term is to a term document, as Equation (3) indicates. The tf-idf value increases proportionally to the number of times a term appears in a certain term document, but is offset by the frequency of the term in all term documents [10].

$$w_i = tf_i \times idf_i = \frac{freq(t_i,d)}{\sum_j^m freq(t_j,d)} \times \log_2 \left(\frac{n}{df_i}\right) \qquad (3)$$

where tf_i represents the frequency that the term t_i occurs in all term documents, m represents the total number of terms in the term document d, $freq(t_i, d)$ represents the total number of term t_i occurs in the term document d, and df_i represents the document frequency, or the ratio between the number of term documents containing term t_j and the total number of term documents. The inverse document frequency idf_i is defined as $\log_2 \left(\frac{n}{df_i}\right)$, where n represents the total number of all term documents.

After computing the weights of terms in all term documents, we utilize Text-Matrix Generator (TMG) [11] to generate a $m \times n$ term-document matrix X as Table 1 shows.

Table 1. The term-document matrix

Query	Document$_1$	Document$_2$	...	Document$_n$
q_1	$d_{1,1}$	$d_{1,2}$	...	$d_{1,n}$
q_2	$d_{2,1}$	$d_{2,2}$	...	$d_{2,n}$
...	...	...	...	...
q_m	$d_{m,1}$	$d_{m,2}$	...	$d_{m,n}$

2.2   Mining Traceability Links

In order to reduce the dimensions of the term-document matrix X, we decompose X into a product of three matrices by the method of Singular Value Decomposition (SVD) [10] as Equation (4) shows.

$$X = Q_0 S_0 D_0^T \qquad (4)$$

where Q_0 is an $m \times r$ left-singular matrix, D_0 is an $r \times n$ right-singular matrix, and S_0 is an $r \times r$ matrix containing the singular values of X. Here r is the rank of X.

The k $(k \leq \min(m,n))$ maximum singular values are selected from S_0 to formalize a new matrix S_k. Meanwhile, the most right columns in the matrix Q_0 and matrix D_0 are removed to formalize the matrix Q_k and matrix D_k with $k -$ $columns$. A matrix X_k can then be formalized to a product of Q_k, S_k and D_k:

$$X_k = Q_k S_k D_k^T \qquad (5)$$

in which matrix X_k is an $r - rank$ approximate matrix of term-document matrix with greatly reduced dimensions A similarity list can be then generated as Equation (6) indicates. Here, the values in similarity list represent precise extent of traceability information between requirement documents and source codes.

$$Similarity(q, D_i) = \sum_{k,j=1}^{m} q_j d_{i,k} (\vec{t_j} \cdot \vec{t_k}) \qquad (6)$$

2.3   Refining Traceability Links

The similarity list generated using Equation (6) needs to be further refined by predefined similarity thresholds due to its large size. The bigger the value in the similarity list is, the more the semantics of requirement documents is close to the set of source codes. In other words, small similarity values imply traceability links of low probabilities. Our approach introduces following two similarity refining methods to filter out some small similarity values.

(1) Result quota method. Set a constant quota value C. Only the top C values in the similarity list are regarded to represent the possible traceability links between requirement documents and source codes.

(2) Similarity threshold method. Set a similarity threshold T. Only the values in the similarity list higher or equal than T are regarded to represent the possible traceability links between requirement documents and source codes.

Through the refining of similarity list using above methods, a series of more precise traceability links between requirement documents and source codes can be obtained.

3   The Improved LSI-Based Approach

Approaches simply based on LSI, or LSI-based approaches, do not take the exclusive features of requirement documents and source codes into account. Hence, there are at least two defects for LSI-based approaches.

(1) Object-oriented programs have the hierarchical structure due to the inheritance relationship among classes. The classes in the same hierarchy structure usually contain methods with similar functionalities. Therefore, if a traceability link exists between a requirement-term document and a parent class inherited by multiple subclasses in source codes, it is more possible that traceability links exist between this requirement-term document and these subclasses. However, LSI-based approaches do not consider the hierarchy structure of source codes.

(2) The weights of all terms in term documents are computed by the same tf-idf formula as Equation (3) shows. The different values of terms therefore cannot be fully reflected. For example, the traceability information extracted by class names is obviously more valuable compared with that extracted by the method names.

In the following two subsections, we present two methods to improve the LSI-based approach.

3.1   Refine Similarity List Based on Class Clustering

The inheritance relationship among the classes in source codes is an important feature for the object-oriented programming. Hence, if one class is the parent of another class, both are assigned to the same class cluster. The classes in the same cluster are assumed to have traceability links of similar probabilities with requirement documents. The pseudo-codes of algorithm for refining similarity list based on class clustering are shown in Table 2.

Table 2. Algorithm of refining the similarity list based on class clustering

Input: code class cluster *cluster*, original similarity list *List*	
Output: refined similarity list *List'* based on class clustering	

```
1    for similarity(q, d_i) in  List  do
2         if (similarity(q, d_i) > t_h) then
3              construct_traceability(d_i,q);
4              similarity(q, d_i) → List';
5         else if (similarity(q, d_i) < t_l) then
6              ignore_traceability(d_i,q);
7         else
8              while (q_j ∈ Cluster[q])
9                   S'+= Similarity(d_i,q_j)/|Cluster[q]|;
10             end while
11             if  (S' > t_e)  then
12                  similarity(q, d_i) = S';
13                  similarity(q, d_i) → List';
14                  construct_traceability(d_i,q);
15             end if
16        end
17        end if
18   end if
19   end for
```

In the table2, $|Cluster[q]|$ is the total number of classes in the cluster, and S' is the average similarity of the classes which belong to the cluster. All of similarity values in the *List* need to be compared with similarity thresholds t_l and t_h. The similarity values which are greater or equal than t_h are picked from *List* and saved in *List'*, while the values smaller than t_l are deleted. For the similarity values between $[t_l, t_h]$, set similarity(q, d_i) as S' and save it in *List'* if the given anchor threshold t_e is smaller than S'. Otherwise, delete similarity(q, d_i). Typically, threshold t_l represents the similarity threshold for high recall rate, or the lowest acceptable similarity, whereas threshold t_h represents the similarity threshold for high precision rate. Threshold t_e is a compromise choice for the recall and the precision.

In order to refine the similarity list based on class clustering, a step about code class clustering needs to be added in the phase of preprocessing. Moreover, the phase of refining traceability links is enhanced by refining similarity list based on class clustering.

3.2   Term Classification

There are many identifiers in source codes. They however play different roles during similarity computation. We classify all the terms into three categories:

(1) Terms in the class name. Terms in the class name describe the functionality of the class, and usually appear in the requirement-term document with high probability than other terms do. Hence, the similarity values of terms in the class name are given a similarity weight sw (sw $\geq$ 1), in order to enhance their importance.

(2) Comments in source codes. Comments in source codes, such as class comments, method comments and attribute comments, usually appear in the requirement-term document with high probability. Hence, after computing weights of terms in the comments according to Equation (3), the weights of these terms are reassigned a weight factor wf (wf $\geq$ 1), in order to enhance their importance.

(3) Generally terms. All terms except these appear in class names and comments do not need to be given any special treatments.

In order to differentiate terms of source codes according to their probabilities of appearing in requirement-term documents, a step about term classification and weight modification needs to be added in the phase of preprocessing.

4   Case Studies

The proposed approach has been successfully applied in the Labor Market Monitoring Software Product Line (LMMSPL) engineering. The applications based on LMMSPL collect and analysis regional employment and unemployment information for the local governments. Currently, these applications are deployed in more than 100 cities located in two provinces in China. Due to the frequent changes of requirements in different cities, to accurately locate the source codes for specific requirements is becoming a prerequisite for the maintenance of LMMSPL in the large scale.

There are altogether 70 pages of requirement documents for the applications based on LMMSPL, whose implementations contain 468 java files. We constructed a code-term document of 565 terms, each representing one class, and a requirement-term document of 258 terms, each representing a paragraph of requirement documents. A 295×565 term-document matrix is generated in the preprocessing phase. Then a similarity list was generated in the traceability mining phase by setting the dimension k of LSI 200. We picked up only top 500 similarity values as the sample in the similarity list. There are 156 correct values representing traceability links in the sample for the LSI-based approach, compared with 180 correct values for the improved LSI-based approach. During the traceability filtering phase, the values in the sample are compared with similarity thresholds. The final experimental results of the traditional and the improved LSI-based approach are shown in Tables 3 and 4, where the recall and precision are defined as (7) and (8) respectively.

$$\text{Recall} = \frac{\text{Number of Correct Tracebility Links Retrieved}}{\text{Number of Correct Tracebility Links Retrieved+Number of Missed Tracebility Links}} \quad (7)$$

$$\text{Precision} = \frac{\text{Number of Correct Tracebility Links Retrieved}}{\text{Number of Correct Tracebility Links Retrieved+Number of Incorrect Tracebility Links Retrieved}} \quad (8)$$

Table 3. Experimental results of the traditional LSI-based approach

Similarity Threshold	Number of Correct Traceability Links Retrieved	Number of Incorrect Traceability Links Retrieved	Number of Missed Traceability Links	Recall	Precision
0.90	50	11	106	32.05%	81.97%
0.85	61	45	95	39.10%	57.54%
0.80	77	79	79	49.36%	49.36%
0.75	91	121	65	58.33%	42.92%
0.70	105	178	51	67.31%	37.10%
0.65	122	235	34	78.21%	34.17%
0.60	146	291	8	94.81%	33.41%
0.55	155	317	1	99.36%	32.84%

Table 4. Experimental results of the improved LSI-based approach

Similarity Threshold	Number of Correct Traceability Links Retrieved	Number of Incorrect Traceability Links Retrieved	Number of Missed Traceability Links	Recall	Precision
0.90	55	9	125	30.56%	85.94%
0.85	67	40	113	37.22%	62.62%
0.80	86	69	94	47.78%	55.48%
0.75	103	106	77	57.22%	49.28%
0.70	119	157	61	66.11%	43.12%
0.65	137	214	43	76.11%	39.03%
0.60	164	266	16	91.11%	38.14%
0.55	178	292	2	98.89%	37.87%

The comparison of recall and precision between the traditional LSI-based approach and the improved LSI-based approach is shown Fig.2.

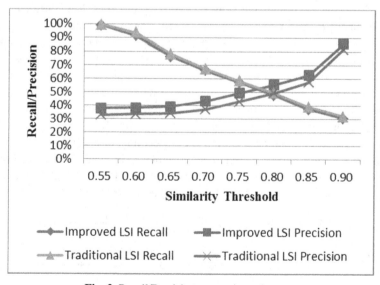

Fig. 2. Recall/Precision comparison chart

As Fig.2 illustrated, the recall of the improved approach is lower than the that of the traditional one 1% to 3%, because the traditional approach doesn't take special features of source codes into account and generates only 156 correct similarity values in the sample, compared with 180 correct values generated by the improved approach. On the other hand, the precision of the improved approach is higher than that of the traditional one 3% to 6%. In addition, the improved approach refines similarity list by the similarity thresholds. In other words, the similarity values higher than 90% are regarded as correct ones, and the similarity values lower than 55% are regarded as incorrect ones. Since there are too many similarity values between [55%, 90%] which are difficult to determine their correctness, it is very important to choose an optimal anchor threshold to guarantee the quality of traceability information. For the case, we can choose 0.75 as the optimal anchor threshold, under which the recall is 57.22% and precision is 49.28% for the improved approach.

The improved approach increases the precision, as Fig. 2 shows, with only a little of loss of recall because of following two reasons: 1) By analyzing the hierarchical structural caused by inheritance relationship among classes, it discovers more possible traceability links which traditional LSI-based approach fails to discover; 2) It weights terms in class names and comments in source codes according to their importance, which enhances the precision of traceability links.

5   Related Works

In order to enhance the quality of traceability information recovered among software artifacts, many IR based approaches have been presented. The following gives some representatives as the related works.

Zou et al. propose in [12] an improved strategy, called Query Term Coverage, which increases the relevance ranking of traceability between artifacts that have more than one unique word in common. Abadi et al give an improved PM, which is driven by a probabilistic approach and hypothesis testing techniques [13]. Baccheli et al. experiment with lightweight methods, involving capturing program elements with regexes. They show that their method of VSM significantly outperforms traditional VSM approach [14]. In [15], Gueheneuc et al. reacts the problem of feature location in source codes as a decision-making problem in the presence of uncertainty. The solution to the problem is then formulated as a combination of a scenario-based probabilistic ranking of events and LSI. Lucia et al. describe the usage of smoothing filters to improve the performance of existing traceability recovery techniques, based on VSM, which removes the common information among artifacts of the same type [16]. Capobianco et al. introduce an improved approach to acting on the artifact indexing considering only the nouns contained in the artifact content to define the semantics of an artifact, which reduces the number of false positives retrieved by IR-based traceability recovery methods [17]. In order to improve the efficiency and precision of finding crosscutting concerns in legacy systems, Yu et al. propose an improved approach to identify the candidates for crosscutting concerns based on execution patterns and fan-in analysis [18].

The above works enhance the quality of extracting traceability information by utilizing and improving PM and VSM. However, the basic defects of the PM and VSM still exist during the extraction of traceability information, such as low efficiency and big noises.

6   Conclusion

This paper presents an improved approach to the recovery of traceability links between requirement documents and source codes based on Latent Semantic Indexing. It employs the special features of object-oriented source codes to raise the recovery precision. Specifically, it recovers the traceability links based on class clustering caused by the inheritance relationship, and assigns different weights for the terms in source codes according to their degrees of correlation to the documents. The experimental results show that improved approach can increase the precision of traceability extraction by 3%~6%, compared with the traditional LSI-based approach.

When traceability information is extracted between requirement documents and source codes in the large system, there still exists the problem of huge volume of computation. Hence, we intend to optimize the performance of the suggested approach when dealing with large system. In addition, we will conduct experiments on some open source software in the future to further verify its effectiveness.

Acknowledgments. The work is supported by the open project foundation of Zhejiang Provincial Key Laboratory of Network Technology and Information Security. The authors would also like to thank anonymous reviewers who made valuable suggestions to improve the quality of the paper.

References

1. Dit, B., Revelle, M., Gethers, M., Poshyvanyk, D.: Feature location in source code: a taxonomy and survey. Journal of Software Maintenance and Evolution: Research and Practice 25(1), 53–95 (2013)
2. Marcus, A., Maletic, J.I., Sergeyev, A.: Recovery of Traceability Links between Software Documentation and Source Code. Int'l Journal of Software Engineering and Knowledge Engineering 15(4), 811–836 (2005)
3. IBM, http://www-306.ibm.com/software/awdtools/reqpro/
4. IBM, http://www.teleogic.com/index.cfm
5. Tai, X.Y.: Introduction to Information Retrieval Technology. Science and Technology (2006)
6. Antioniol, G., Canfora, G., Casazza, G., Lucia, A.D., Merlo, E.: Recovering Traceability Links between Code and Documentation. IEEE Trans. J. Soft. Eng. 28(10), 970–983 (2002)
7. Hayes, J.H., Dekhtyar, A., Osborne, J.: Improving Requirement Tracing via Information Retrieval. In: 11th IEEE Int'l Conf. Requirements Engineering, pp. 138–147 (2003)
8. Cleland-Huang, J., Settimi, R., Duan, C., Zou, X.C.: Utilizing Supporting Evidence to Improve Dynamic Requirements Traceability. In: 13th IEEE Int'l Conf. Requirements Engineering, pp. 135–144 (2005)

9. Marcus, A., Maletic, J.I.: Recovering Documentation to Source Code Traceability Links Using Latent Semantic Indexing. In: 25th Int'l Conf. Software Engineering, pp. 125–135 (2003)
10. Yin, L., Juan, L., Shu, L.M.: Research on Dynamic Requirement Traceability Method and Traces Precision. Journal of Software 20(2), 177–192 (2009)
11. Zeimpekis, D., Gallopoulos, E.: Design of a Matlab Toolbox for Term-Document Matrix Generation. In: Workshop on Clustering High Dimensional Data and Its Applications, pp. 38–48 (2005)
12. Zou, X., Settimi, R., Cleland-Huang, J.: Improving Automated Requirements Trace Retrieval: a Study of Term-based Enhancement Methods. Empirical Software Engineering 15(2) (2010)
13. Abadi, A., Oliveto, R., Tortora, G.: Assessing IR-based Traceability Recovery Tools through Controlled Experiment. Empirical Software Engineering 14(1), 57–92 (2009)
14. Bacchelli, A., Lanza, M., Robbes, R.: Linking e-mails and Source Code Artifacts. In: 32nd ACM/IEEE Int'l Conf. Software Engineering., vol. 1, pp. 375–384 (2010)
15. Gueheneuc, Y.G., Marcus, A., Antoniol, G., Rajlich, V.: Feature Location Using Probabilistic Ranking of Methods based on Execution Scenarios and Information Retrieval. IEEE Trans. J. Software Engineering 33(6), 420–432 (2007)
16. Lucia, A.D., Penta, M.D., Oliveto, R., Panichella, A., Panichella, S.: Improving IR-based Traceability Recovery Using Smoothing Filters. In: 19th Int'l Conf. Program Comprehension, pp. 21–30 (2011)
17. Capobianco, G., Lucia, A.D., Oliveto, R., Panichella, A., Panichella, S.: On the Role of the Nouns in IR-based Traceability Recovery. In: 17th Int'l Conf. Program Comprehension, pp. 148–157 (2009)
18. Yu, D., Yan, D.: Crosscutting concerns identifying based on execution patterns and fan-in analysis. Journal of Huazhong University of Science and Technology (Natural Science Edition) 40(1), 45–48 (2012)

Local Logo Recognition System
for Mobile Devices

Phong Hoang Nguyen, Tien Ba Dinh, and Thang Ba Dinh

Faculty of Information Technology, University of Science, Ho Chi Minh City, Vietnam
nhphong@apcs.vn, {dbtien,dbthang}@fit.hcmus.edu.vn
http://www.fit.hcmus.edu.vn/vn/

Abstract. In this paper, we propose a novel logo recognition system which can process a very large number of logos locally on mobile devices. The system is not only robust against challenging conditions such as different image scale, rotation, and noisy input, but time efficient and low memory consuming as well. The total computation cost is minimized by using a cascade approach, in which the fast algorithm is kept in the first layer to filter most of the testing cases, while the more expensive but robust one is put on the second layer to investigate only the *"confusing"* logos. In this paper, we also propose a "background subtraction" method, which considerably improves the second layer in terms of speed, accuracy, and database size. The system has been tested on a dataset of 3000 logos with promising results. The average running-time is just about 1.7 seconds on an average single core mobile device, which is very potential for many mobile applications.

Keywords: Image matching, logo recognition, local feature, cascade.

1   Introduction

Nowadays, with the rapid development of technology, a smartphone contains not only a good built-in camera, but also a powerful processor. It brings up many interesting applications using visual information on mobile devices such as face recognition, virtual reality, and business card reader. Logo recognition is one of them. Logos can act as a valuable means in website summarization, enterprise identification, entertainment advertising, vehicle recognition, road sign reading, and content-based image retrieval [1–7]. One of the challenges is that logos have arbitrary shapes depending on their design, hence no reasonable assumptions about the shape could be made. Another challenging problem is that different product logos of the same company may be very similar, which makes it hard to distinguish one from the others (an example is shown in figure 1). Moreover, if the number of logos is large, the system requires a lot of memory and matching time which is not applicable for a mobile device which has limited memory and computational power. In this paper, we propose a novel logo recognition system, which is robust against challenging conditions such as different image scale, rotation, and noisy input, while having a very efficient running time.

B. Murgante et al. (Eds.): ICCSA 2013, Part V, LNCS 7975, pp. 558–573, 2013.

Fig. 1. Ten look-alike Google-product logos

The whole system can run locally on a mobile device, where speed, memory and persistent storage capacity are limited. It can also differentiate very similar logos from the same company (as in the case of figure 1). The new approach has combined different keypoint detectors, descriptor calculators, a public majority voting process and a background subtraction method to determine the best matching. A cascade filtering approach is also proposed to improve the efficiency. In our implementation, we use a two-layer cascade, which can be extended easily later.

For recognition purpose, we first extract a set of keypoints from the captured scene, *i.e.* the query image, and calculate their descriptors. Then, a similarity measure is carried out by comparing each query descriptor with the database of training descriptors extracted from the sample logo images. Next, target logos most similar to the query image are retrieved. If those logos give a hint that the result belongs to a group (of *"look-alike"* logos), the query image continues to go through the second layer of testing. At that time, the query image is compared to all group members to determine the best one. The proposed system has been tested on a dataset of 3000 logos with very promising results. The observed execution time on an average single-core smartphone is about 1.7 seconds for each recognition.

This paper is organized as follows. Section 2 presents the related researches, together with a quick analysis on the efficiency of different local feature descriptors. Section 3 describes the system overview. Section 4 focuses on the public majority voting process. In section 5, a new "background subtraction" method is proposed to improve the second testing layer. The experimental results and conclusion are presented in the last two sections.

2   Related Research

In computer vision, especially in pattern recognition, image feature is a very important concept. There are two basic types of features: global and local ones. Global features (such as color moment [8]) work on the entire image and describe it as a whole. Local features (such as corner, blob), on the other hand, are computed at multiple points in the image and hence more robust to occlusion and clutter. Local features tend to be preferred in many recognition systems.

In 1999, David G. Lowe proposed the Scale Invariant Feature Transform (SIFT) [9] detector which later became one of the most well-known algorithms widely used by both research community and industry. SIFT computes a histogram of local oriented gradients around an interest point and stores the bins in a 128-dimensional vector. SIFT performs reliable matching under different geometric transformations (image scale, rotation, viewpoint) and photometric transformations (image blur, noise and illumination). It can also identify objects among clutter and under partial occlusion with high accuracy. However, the high dimensionality of the descriptor is a drawback of SIFT. Being a 128-dimensional vector, it is relatively slow to compute and match.

Speeded-up Robust Feature (SURF) [10] is another robust local feature detector which was first presented by Herbert Bay et al. in 2006. It is inspired by the SIFT and has been shown to have similar performance to SIFT, while at the same time being much faster [11]. SURF is based on sums of 2D Haar wavelet responses [12] and makes an efficient use of integral images [13]. Thus, its computation cost is dramatically reduced while still preserving the discriminative power of SIFT. Although SURF is faster than SIFT, the descriptor is still a 64-dimensional vector of floating-point values. When millions of descriptors need to be stored, especially on mobile devices, it becomes inapplicable.

The descriptor's high dimensionality is a common drawback of SIFT and SURF. One way to tackle this issue is to apply dimensionality reduction methods, such as Principal Component Analysis (PCA) [14] or Linear Discriminant Embedding (LDE) [15], to the original SIFT/SURF descriptors. They are very powerful techniques that can be used to reduce the descriptor size without degrading its recognition performance. Another way to shorten a descriptor vector is to quantize its floating point values into integers using very few bits per number (as shown in [16], [17] and [18]). Quantization is a simple operation that yields not only a memory gain but also faster matching. A third and even more radical reduction is to transform SIFT/SURF descriptors into binary strings whose similarity can be measured by the Hamming distance. On modern CPU that supports bit-wise XOR and bit count operations, Hamming distance can be done extremely fast.

A very good example of binary descriptors is the Fast Retina Keypoint (FREAK) [19]. It has been proposed recently with the goal to make descriptors faster to compute, more compact while remaining robust to scale/orientation variations and noisy input. FREAK is a novel keypoint descriptor which was inspired by the human visual system (or more precisely the retina). In FREAK, a cascade of binary strings is computed by comparing image intensities over a retinal sampling pattern. The authors show that FREAK, in general, is faster to compute with lower memory load and also more robust than SIFT, SURF and BRISK [20] (another famous binary descriptor).

There have been a lot of studies done on the performance evaluation of different local descriptors using a wide range of algorithms (examples can be found at [14], [21] and [22]). After careful consideration, we decided to use the combination of SURF keypoint detector and FREAK descriptor for our system. They

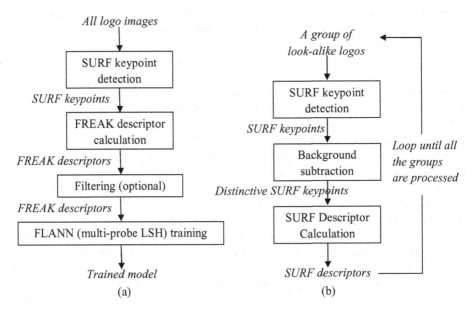

All logo images

SURF keypoint
detection

SURF keypoints

FREAK descriptor
calculation

FREAK descriptors

Filtering (optional)

FREAK descriptors

FLANN (multi-probe LSH) training

Trained model

(a)

*A group of
look-alike logos*

SURF keypoint
detection

SURF keypoints

Background
subtraction

Distinctive SURF keypoints

SURF Descriptor
Calculation

SURF descriptors

*Loop until all
the groups
are processed*

(b)

Fig. 2. The proposed database creation process: (a) Database for the first layer, (b) Database for the second layer

seem to be the most suitable choice, in terms of speed, robustness and memory consumption, for a mobile phone application. To speed up the matching step, we suggest using a Fast Approximate Nearest Neighbor algorithm [23]. Locality Sensitive Hashing (LSH) [24] is one of the nearest-neighbor search techniques that has been shown to be effective for fast matching of binary features [25], so we will utilize it.

3   System Overview

As previously mentioned, we divide the whole recognition process into two stages. Therefore, two different databases are needed at different times. Figure 2 depicts their creation process. As shown in figure 2(a), SURF keypoint detector and FREAK descriptor calculator are chosen as the main tools to build up the first database. The input images in this process should be checked for good quality. If there is any logo that has very few number of keypoints or unreasonably small size, we simply remove it. In order to improve the later matching speed, we suggest using FLANN (Fast Library for Approximate Nearest Neighbors [26]). It is a library that contains a collection of algorithms optimized for fast nearest neighbor search in large datasets and for high dimensional features. We use it in combination with multi-probe LSH [24]. The output is a trained model that will later be used for recognition.

Suppose in the first database, there are different product logos of the same company, such as Google (as shown in figure 1). Those logos look very similar.

They share a big common part, *i.e.* the word "Google", and only differ from one another by a small portion, except **Google Wallet**. As a result, no matter what kind of detector is used, most of the detected keypoints are ambiguous. For example, if the query image is **Google Answers**, it is too easy that the matching step will mis-recognize it as **Google Labs** or **Google Catalogs**. Therefore we need the second layer of testing. All those similar logos should be grouped together and marked with the label "Google-alike group". Whenever a member in this group is detected in the first layer, a re-consideration of the whole group is necessary in the second layer. This method is a kind of cascade filtering approach: first, we consider individual logo within the database of all logos, if the detected result belongs to a group, we then move to that group and do an expensive verification to ensure which group member is the best matching.

In this paper, we also propose a "background subtraction" method that considerably improves the second layer. Not only it enhances the accuracy, it also reduces the memory load, the matching time and the database size (one of the key things when running a large number of logos locally on mobile devices is the database size).

Figure 2(b) illustrates how the second database can be built. Note that the background subtraction step is done here. In particular, it removes the ambiguous keypoints of each logo image in a group. Thus, what remain after the step are the distinctive parts of them. Next, we calculate SURF descriptors of the remaining keypoints and save them to disk. We repeat the creation process for every group in the logo dataset.

Figure 3 shows the full process of our logo recognition system on mobile devices. After extracting a region of interest (ROI) from the captured scene, we detect the SURF keypoints and calculate FREAK descriptors. The trained model which is loaded beforehand is used to carry out a k-nearest-neighbor matching. Each feature point in the query image is compared to the database of feature points (from training images) to find candidate matches. After that, we apply a majority voting process to determine n highest voted logos. It is important to note that the most voted logo is not always the best one, especially when there are several similar logos in the database. Therefore, if the n logos give a hint that the result belongs to a group, we will continue with the second layer. The detected group will be loaded into memory. Then, we calculate SURF descriptors of the SURF keypoints of the query image that we already have before. Then, a normal brute-force matching with 2-nearest-neighbors is applied. The result of that will go through a filter so that false and bad matches are removed. Finally, the logo which has most votes is chosen to be the output result.

4   Public Majority Voting Process

4.1   Voting Process

After extracting feature points from the query image, we let them vote for their best correspondences. Each query keypoint can select up to 5 training keypoints in the database (note that all the keypoints must be transformed into descriptors

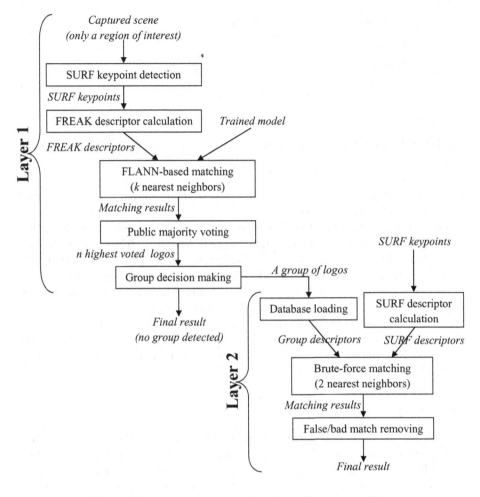

Fig. 3. The proposed system flowchart of logo recognition

before matching). Then, we calculate the distances of those 5 matches and store them in an ascending list. If the ratio of the 1st item to the 2nd item in the list is small, we keep the first closest match and ignore the other fours. Otherwise, we consider the ratio of the 1st item to the 3rd item. Again, if that ratio is small, we keep the first two closest matches and ignore the other threes. Otherwise, we continue to calculate the ratio of the 1st item to the 4th item, and so on.

When we sort the 5 distances in ascending order, the 1st item, *i.e.* the smallest distance, clearly belongs to the closest match, and the 2nd item belongs to the second closest one. Therefore, the ratio between them falls within the range from zero to one. The smaller the ratio is, the more outstanding the first/closest match is. If the ratio is 0.8, it means that the second match is 80% as good as the first one, which seems to be an ambiguity. In that case, we cannot ignore

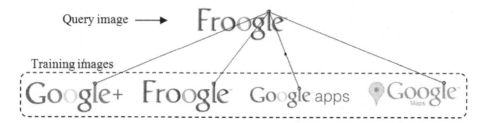

Fig. 4. One query keypoint sometimes has a need to vote for more than one training keypoint

the second match; we have to "respect" it as much as the first one. The same argument also holds for the other matches.

Because there are several similar logos in the database, one query keypoint may consider many training keypoints as its potential matches (figure 4 illustrates that). Hence, rather than restricting each query keypoint to selecting only one best match, we allow it to select $k = 5$ best matches and suggest a threshold $\theta = 0.8$ to filter out bad choices. In this way, each query keypoint can vote for maximum k candidates, depending on how similar they are. When a training keypoint is voted, it means that the corresponding sample/training image which contains that keypoint is voted. Therefore, after the voting process, each sample image will have a number of votes. If there is no group detected, the most voted logo is the winner (see figure 3).

4.2   Group Decision Making

As shown in figure 3, the output of the voting process is the n highest voted logos. Here we choose $n = 5$, because the most voted logo is not guaranteed to be the best one. Next, we filter the n logos using the previous method (with $threshold = 0.45$) to keep only the top logos that have similar number of votes. At this point, if those logos suggest that the result belongs to a certain group, the group descriptors will be loaded into memory and we continue with the second testing layer. Otherwise, we accept the most voted logo and stop.

Let m be the number of logos that remain after the filtering step ($m \leq n$). There are several ways to examine the m logos to determine whether the result belongs to a group. One simple way is: if the most voted logo belongs to a group, the final result also belongs to that group (note that in this case, we are not sure whether the most voted logo is the final result or not, until the second testing layer is done). Another criterion could be: if there is more than one logo (among the m logos) belonging to the same group and one of them is the most voted logo, the result probably belongs to that group. A third (more subtle) way could be: if there are at least 3 logos falling into the same group (although none of them is the most voted logo), the result still probably belongs to that group. Note that in the second testing layer, all members of the group are taken into account, not just those that appear in the list of m logos.

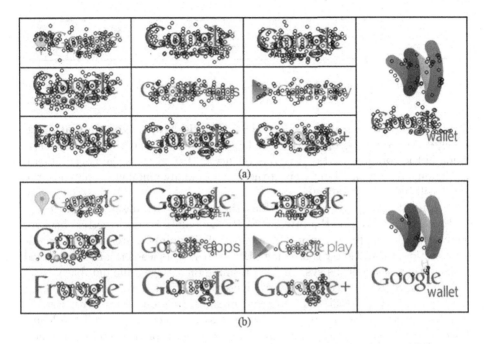

Fig. 5. Keypoints of different Google-product logos: (a) SURF keypoints, (b) Remaining SURF keypoints after applying FREAK

5   Background Subtraction Method

5.1   Necessity of the Second Testing Layer

Before introducing the background subtraction method, it is important to understand why we need the second testing layer.

Although FREAK is robust, it easily fails when the analogy level between logos is high. In the first-database creation process (figure 2(a)), we extract keypoints from the sample logo images by using SURF and compute FREAK descriptors of those keypoints. When computing descriptors, as illustrated in FREAK paper, a few keypoints are removed, keeping only the most stable ones. That is why the number of keypoints (after applying FREAK) is often less than the number of ones that SURF originally detects. Although this reduction improves the speed, it also introduces one big disadvantage, as illustrated in figure 5.

Part (a) of figure 5 shows the logos' SURF keypoints; while part (b) shows their remaining (SURF) keypoints after applying FREAK to calculate descriptors. The figure shows that using FREAK is not good in the "Google" case. As we can see from part (b) of the figure, it is almost impossible to distinguish among **Froogle, Google, Google+** and **Google Play**. None of their keypoints reside in the distinctive parts, so the system does not "see" much difference between them (note that what the system can "see" from a logo is its keypoints, not

Fig. 6. An "ideal" result: all SURF keypoints around the common parts are removed

the logo itself). In case of other logos (*i.e.* **Google Maps**, **Google Catalogs**, **Google Answers**, **Google Apps** and **Google Labs**), most of the keypoints reside in the common parts. The consequence is that most of the keypoints are ambiguous. In the matching step, if there are so many ambiguous matches, the result will be seriously affected.

Looking at figure 5(a), one may think that using SURF descriptor calculator (instead of FREAK) for the original SURF keypoints is much better, because there would be no reduction in the number of keypoints. Later, in the experiment section, we will show that using SURF descriptors leads to another critical problem that restricts the system from running on a large scale database.

Although SURF descriptors cannot be used in the first layer, they can be used in the second layer. Recall that the second testing layer only deals with a group of *look-alike* logos, hence, in that stage, we do not need to load all the logos' descriptors into the memory. The use of SURF descriptors is therefore feasible.

5.2   Background Subtraction Method

In the second-database creation process (figure 2(b)), before calculating the SURF descriptors, we apply the background subtraction method to filter out the ambiguous SURF keypoints. The idea of the filtering method is illustrated in figure 6.

Figure 6 shows the logos' SURF keypoints (like figure 5(a)), but the difference is that all the keypoints in the common parts are removed. Note that the figure does not show the result of the background subtraction method, it only illustrates what we expect for a filtering method. If *all* the ambiguous keypoints can be removed (like that), the accuracy will be considerably improved. Fewer keypoints (after filtering) also means faster matching process, lower memory consumption and smaller database size. However, the result shown in figure 6 is an "ideal" one that we can hardly achieve. The "actual" result of the background subtraction method is shown in figure 7 and 8, which seems very good, but not perfect. The rest of the section will describe how the method works.

When we match a query image A (the captured scene) with a training image B (the sample image), it means that we are trying to "*find B in A*". In other words, we try to find which portion of A that resembles B. Now, suppose we have two different product logos that look alike. Each logo has two parts: the common part and the distinctive/private part. By matching the first logo with the second one, we get their common parts. Removing those parts from each logo yields the distinctive parts. Similarly, suppose we have 10 different product logos (as in figure 1). In other to get the distinctive part of each logo, we sequentially match

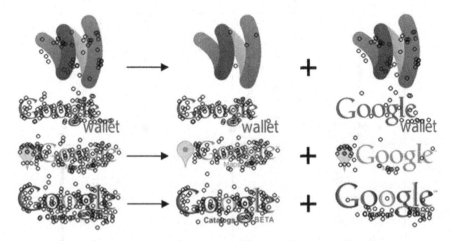

Fig. 7. SURF keypoints of the three *in-grouped* logos are divided into background and foreground ones (this was done by using the background subtraction method)

Fig. 8. Results of the background subtraction method

it with the other 9 logos. Each matching reduces the number of keypoints a bit. After 9 matchings, the remaining keypoints clearly belong to the distinctive part of the logo.

Before giving a detailed explanation on how to find "the common part", we first introduce the three important terms: *good match*, *bad match* and *false match*.

In the matching step between the two images, each keypoint in the first image tries to find the closest keypoint in the second image. For simplicity, let $\mathbf{x}$ be a query keypoint, and $\mathbf{y}$ be a training keypoint that is closest to $\mathbf{x}$. The relation between $\mathbf{x}$ and $\mathbf{y}$ is called a *match*. Basically, there are three kinds of matches: good match, bad match and false match. Let $\mathbf{z}$ be a training keypoint that is second-closest to $\mathbf{x}$. If the ratio of the distance $(\mathbf{x}, \mathbf{y})$ to the distance $(\mathbf{x}, \mathbf{z})$ is small, we consider $(\mathbf{x}, \mathbf{y})$ as a *good match*. On the contrary, if that ratio is big (*e.g.*, bigger than 0.8), $(\mathbf{x}, \mathbf{y})$ is called a *bad match*. Finally, if we match the **Google Apps** logo with the **Google** logo, any query keypoint that resides

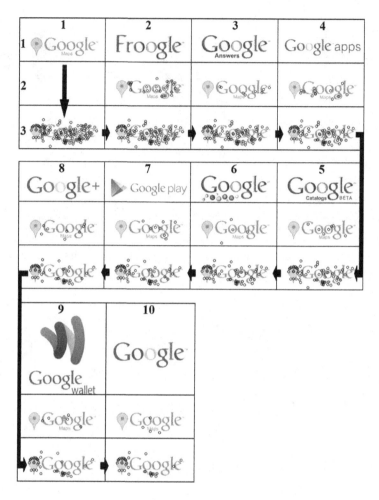

Fig. 9. Simulation of the filtering process for "Google Maps" logo

around the word "Apps" causes a *false match*. In other words, a *false match* is caused when a query keypoint cannot find any "true" corresponding point in the sample image.

With the above definition, SURF keypoints (in figure 5(a)) that cause false matches are exactly the ones we want to keep for each logo image in a *similar* group. The other keypoints, *i.e.* the ones causing good or bad matches, should be removed. However, using the background subtraction method, only the keypoints that cause good matches can be detected and removed. Recall that the method is only based on a simple ratio criterion, hence it cannot differentiate the keypoints causing bad match from the ones causing false match.

Figure 9 shows a full simulation of the filtering process for **Google Maps** logo. Starting from row 1 column 1 and following the black arrows, you can see how the filter works. Row 1, column 1 is the gray-scale image of **Google**

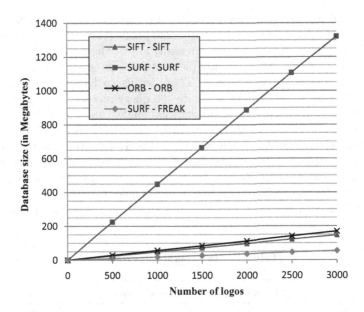

Fig. 10. The size of the first database (by using different types of keypoint detectors and descriptor calculators)

Maps. Row 3, column 1 shows its original SURF keypoints. When matching those keypoints with **Froogle** (row 1, column 2), we get the keypoints that cause good matches (row 2, column 2). After removing them from the original SURF keypoints, we obtain the result shown in row 3, column 2. That is also the end of the first loop. Next, we let the keypoints in row 3, column 2 match with **Google Answers** (row 1, column 3) and continue exactly the same way until all 9 logos are covered.

6   Experimental Results

One of the reasons why FREAK descriptors are preferred (in the first testing layer) is that they generate small-size databases. For a "local" application on mobile devices, the database must be copied to the memory card (or elsewhere in the device) before loading into the memory. Therefore, its size cannot be too big. As shown in figure 10, the combination of SURF and FREAK is the best choice. Their database size is 3 times smaller than SIFT-SIFT and 24 times smaller than SURF-SURF. This also indicates that SURF-SURF is not feasible for running large-scale databases on mobile devices.

Table 1 shows the experimental results. First, we build up the database from 3000 sample logo images (some of them are shown in figure 11, 12). The dataset was downloaded from **Brands of the World** [27]. This is a very famous website which offers the world's largest collection of freely downloadable vector logos.

Each logo is of the size 200×200 pixels. In this experiment, we vary the Region of Interest (ROI) from 240×160 to 240×240 pixels (the application's interface is designed in such a way that users can easily adjust the ROI). The experiment is conducted as follows. We randomly pick out 300 *non-grouped* logos (*i.e.*, each logo does not belong to any group). Then, we use a handheld mobile phone camera to capture those logos to generate 300 query images. The camera device operates at varying viewpoints, scale, rotation angles, and under different lighting environments. Therefore, the query image may differ substantially from the database target images. As shown in table 1, the average running time for each query image is 1579.7 milliseconds. The device we used in this experiment is a Sony Ericsson smartphone, Xperia Arc S, single-core 1.4 GHz. No query image went through the second layer (hence the second layer takes 0 milliseconds). The "*other cost*" field in the table includes majority voting process, group decision

Table 1. The experimental results

	First Layer			Second Layer				
	Keypoint-detection cost (ms)	Descriptor-calculation cost (ms)	Matching cost (ms)	Descriptor-calculation cost (ms)	Matching cost (ms)	Other cost (ms)	Total cost (ms)	Accuracy (%)
Non-grouped (300 tests)	576.5	26.1	941.5	0	0	35.6	1579.7	97.67
Grouped (30 tests)	544.1	6.4	800.4	738	580.2	243.1	2912.2	90
Overall (330 tests)	573.5	24.3	928.7	67.1	52.7	54.5	1700.8	96.97

Fig. 11. Some of the 3000 logos used in the experiment (*non-grouped* selected)

Fig. 12. Some of the 3000 logos used in the experiment (*grouped* selected)

making process, data loading and so on. The observed accuracy is 97.67% (*i.e.*, 7 query images were mis-recognized).

Next we repeat the same process for 30 *grouped* logos (*i.e.*, each logo belongs to a certain group). The average speed is 2912.2 milliseconds and the accuracy is 90%. The overall result for 330 tests is 1700.8 milliseconds (of speed) and 96.97% (of accuracy).

7   Conclusions

In this paper, we proposed a novel logo recognition system for mobile devices. To enhance the performance in both speed and memory usage, our system runs in a cascade manner: the first layer, which is fast but not the best in accuracy, is used for scanning the whole database, and the second layer, which is more accurate but costly, is used for differentiating confusing logos. Also, we proposed a new filtering method, namely "background subtraction", which is used in the second-database building process. The idea is to reduce the number of ambiguous keypoints of each logo in a "similar" group (to achieve higher accuracy in the second layer). The proposed system has combined different effective algorithms (such as SURF, FREAK, FLANN, LSH...) to give an optimal solution for logo recognition (with large-scale databases) on mobile devices. In the future, we plan to build a full system that supports users in getting a certain product's information (using their smart phones) when going shopping.

Acknowledgments. This research is supported by research funding from Advanced Program in Computer Science, University of Science, Vietnam National University - Ho Chi Minh City.

References

1. Baratis, E., Petrakis, E., Milios, E.: Automatic website summarization by image content: A case study with logo and trademark images. IEEE Transactions on Knowledge and Data Engineering 20(9), 1195–1204 (2008)
2. Zhu, G., Doermann, D.: Automatic document logo detection. In: Ninth International Conference on Document Analysis and Recognition (ICDAR), vol. 2, pp. 864–868 (September 2007)

3. Cai, G., Chen, L., Li, J.: Billboard advertising detection in sport TV. In: Proceedings of the 7th International Symposium on Signal Processing and Its Applications, vol. 1, pp. 537–540 (July 2003)

4. Nieto, P., Cozar, J., Gonzalez-Linares, J., Guil, N.: A TV-logo classification and learning system. In: 15th IEEE International Conference on Image Processing (ICIP), pp. 2548–2551 (October 2008)

5. Wang, J., Liu, Q., Duan, L.-Y., Lu, H., Xu, C.S.: Automatic TV logo detection, tracking and removal in broadcast video. In: Cham, T.-J., Cai, J., Dorai, C., Rajan, D., Chua, T.-S., Chia, L.-T. (eds.) MMM 2007. LNCS, vol. 4352, pp. 63–72. Springer, Heidelberg (2006)

6. Yunqiong, W., Zhifang, L., Fei, X.: A fast coarse-to-fine vehicle logo detection and recognition method. In: Proceedings of IEEE International Conference on Robotics and Biometrics, pp. 691–696 (December 2007)

7. Xia, L., Qi, F., Zhou, Q.: A learning-based logo recognition algorithm using SIFT and efficient correspondence matching. In: International Conference on Information and Automation (ICIA), pp. 1767–1772 (June 2008)

8. Duanmu, X.: Image retrieval using color moment invariant. In: 7th International Conference on Information Technology: New Generations, ITNG (April 2010)

9. Lowe, D.: Object recognition from local scale-invariant features. In: The Proceedings of the Seventh IEEE International Conference on Computer Vision (1999)

10. Bay, H., Tuytelaars, T., Gool, L.V.: SURF: Speeded up robust features. In: Proceedings of the Ninth European Conference on Computer Vision (May 2006)

11. Bauer, J., Sunderhauf, N., Protzel, P.: Comparing several implementations of two recently published feature detectors. In: Intelligent Autonomous Systems (2007)

12. Wikipedia: Haar-like features,
    http://en.wikipedia.org/wiki/Haar-like_features (accessed February 8, 2013)

13. Wikipedia: Summed area table,
    http://en.wikipedia.org/wiki/Summed_area_table (accessed February 8, 2013)

14. Mikolajczyk, K., Schmid, C.: A performance evaluation of local descriptors. IEEE Transactions on Pattern Analysis and Machine Intelligence 27(10), 1615–1630 (2005)

15. Hua, G., Brown, M., Winder, S.A.J.: Discriminant embedding for local image descriptors. In: International Conference on Computer Vision (2007)

16. Tuytelaars, T., Schmid, C.: Vector quantizing feature space with a regular lattice. In: IEEE 11th International Conference on Computer Vision, ICCV (2007)

17. Winder, S.A.J., Hua, G., Brown, M.: Picking the best daisy. In: IEEE Conference on Computer Vision and Pattern Recognition (2009)

18. Calonder, M., Lepetit, V., Fua, P., Konolige, K., Bowman, J., Mihelich, P.: Compact signatures for high-speed interest point description and matching. In: International Conference on Computer Vision (September 2009)

19. Alahi, A., Ortiz, R., Vandergheynst, P.: FREAK: Fast Retina Keypoint. In: IEEE Conference on Computer Vision and Pattern Recognition, CVPR (2012)

20. Leutenegger, S., Chli, M., Siegwart, R.: BRISK: Binary robust invariant scalable keypoints. In: Proceedings of the IEEE International Conference on Computer Vision, ICCV (2011)

21. Ievgen: Feature descriptor comparison report,
    http://computer-vision-talks.com/2011/08/
    feature-descriptor-comparison-report/

22. Ievgen: A battle of three descriptors: Surf, Freak and Brisk,
    http://computer-vision-talks.com/2012/08/
    a-battle-of-three-descriptors-surf-freak-and-brisk/
23. Muja, M., Lowe, D.G.: Fast approximate nearest neighbors with automatic algorithm configuration. In: International Conference on Computer Vision Theory and Application (VISAPP). INSTICC Press (2009)
24. Gionis, A., Indyk, P., Motwani, R.: Similarity search in high dimensions via hashing. In: Proceedings of the 25th International Conference on Very Large Data Bases. VLDB 1999, San Francisco, CA, USA, pp. 518–529 (1999)
25. Rublee, E., Rabaud, V., Konolige, K., Bradski, G.: ORB: An efficient alternative to SIFT or SURF. In: IEEE International Conference on Computer Vision (2011)
26. Muja, M.: Fast library for approximate nearest neighbors (FLANN),
    http://www.cs.ubc.ca/~mariusm/index.php/FLANN/FLANN
27. Website: Brands of the world, http://www.brandsoftheworld.com/

M^X-tree: A Double Hierarchical Metric Index with Overlap Reduction

Shichao Jin, Okhee Kim, and Wenya Feng

School of Software and Microelectronics, Peking University, Beijing, China
{shichaojin.cs,anniekim.pku,pkuwenyafeng}@gmail.com

Abstract. Large multimedia repositories often call for a highly efficient index supported by external memories, in order to fast retrieve the desired information. The M-tree, one of the metric trees, is a well-tested and dynamic index structure for similarity search in metric spaces where various distance measures can be applied. Nevertheless, its performance is undermined dramatically by the number of paths it has to traverse, which consequently increases CPU and I/O costs both. In this paper, an analysis has been performed to demonstrate the gravity of this issue. As a result, we propose a novel index structure called the M^X-tree. It introduces the super node, which is inspired by the X-tree in the spatial search area, and the M^X-tree fully extends the super node to metric spaces. Besides, a new node split method is presented in the M^X-tree to meet the need of the low cost of index construction. This proposed method uses only $O(n^2)$ runtime to split the overfull node without tuning any parameter while the search performance of the whole index is still guaranteed compared to the node split policy with $O(n^3)$ in the M-tree. In addition, an internal index is proposed in the M^X-tree to seamlessly handle the CPU costs in the extended leaf nodes due to the introduction of the super node. Compared to other former improvements of the M-tree, the M^X-tree retains all the merits of the M-tree without any post-processing steps or losing the applicability. To survey the proposed index, we conduct extensive experiments, and experimental evaluations illustrate the efficiency of the M^X-tree with regard to both CPU and I/O costs.

Keywords: Similarity Search, Metric Spaces, Distance Based, Range Query.

1 Introduction

With the rapid growth of multimedia information subsuming text, image, video and voice, content-based retrieval plays an increasingly important role in order to effectively and efficiently retrieve the relevant parts of a database. Of most urgency is the design of a high-performance access method to answer queries like finding similar objects in a large database, as the traditional exact-matching is no longer expressive enough in a multimedia environment. However, the traditional spatial access methods, such as R-tree [12] and its variants [13-14] are not applicable to the multimedia database totally, as they are limited to the vector space, and the distance functions they use are restricted by Minkowski distance. In this case, a metric space is much more

B. Murgante et al. (Eds.): ICCSA 2013, Part V, LNCS 7975, pp. 574–589, 2013.

meaningful to represent the objects indexed, and it offers a general way to handle the search problem. For example, in the object oriented programming, one only needs to define an interface to realize the problem, where two parameters are the class of the object and the type of the distance function respectively.

Considering the reality that data are often in a large volume (or the data can be completely loaded into the main memory, while the memory cannot be totally allocated to current program due to multiple processes), a main memory based index is apparently not able to deal with the requests from users and hence the disk based access methods come into being, where data are organized in pages of fixed size resident on the disk. As a result, a reasonable index should take into account the times of I/O operations. Besides, computational costs of a distance function may have a vital impact on the whole performance of an index when the distance function is complicated (for example, the Edit distance [19] or high dimensional data). Hence, CPU costs should be concerned as well in such a case.

Among many existing indexes based on external memories, the M-tree [1] is a kind of dynamic and balanced metric trees which supports consecutive insertion. Nonetheless, its performance is undermined by the overlap between nodes evidently, which increases the possibility of multi-way traverse. Consequently, there is an obvious growth in both CPU and I/O costs (it may perform worse than linear scan with respect to page accessing as illustrated in our experiments) for the M-tree. Moreover, the node split policy in the M-tree is highly CPU bound with $O(n^3)$ complexity, which increasingly impedes the index construction with the growing ratio of the node size and the object size.

In order to effectively handle the problems in the M-tree, we present in this paper the M^X-tree (extendable metric tree), which shares the basic data structure with the M-tree. Inspired by the X-tree [15] in the spatial search area, we introduce the super node into the M^X-tree to avoid an unsatisfying node split, and fully extend it to the metric space. With the employment of the super node, an effective and efficient node split policy is proposed in the M^X-tree, whose time complexity is $O(n^2)$ without tuning any parameter. To further enhance the index's performance in CPU costs, an internal index in the leaves is presented in the M^X-tree.

The rest of this paper is organized as follows. We first review the related work in Section 2. After we discuss the preliminaries and the existing issues in the M-tree in Section 3, the structure and related algorithms of the M^X-tree will be described in Section 4. Section 5 reports the experimental results and the conclusion is made in Section 6.

2   Related Work

Similarity search has attracted a great deal of attention in both applications and academic circles, especially in database and multimedia areas. Although there are a variety of approaches to achieve the search goals, most of them are specialized for a certain category of objects (for example, dimensional data or strings). In addition, for a very large volume of a dataset, effective similarity search might be achieved by a distributed approach [11], where the M-tree is the cornerstone in essence. As for a small quantity of a dataset, many main memory based indexes using a tree or an array structure are available [20]. Hence, in this paper we mainly focus on those indexes applicable to metric spaces, and meanwhile based on the secondary memory.

The M-tree is a kind of dynamic and balanced indexes supported by external memories. There have been plenty of variants and improvements about it in the past. The Slim-tree [3] is an extension of the M-tree, which defines a different node split policy with $O(n^2 log(n))$ complexity, and a slim-down algorithm applying in the post-processing phase. However, only I/O costs are reported in their paper. The DBM-tree [4] is an unbalanced M-tree specialized for data with dense regions. With the additional pivot-filtering, the PM-tree [5] is one kind of the M-tree variants as well, where several parameters have to be tuned manually to make a trade-off between I/O and CPU costs. Likewise, the DF-tree [6] uses additional pivot-filtering to improve the search performance. The M^+-tree [7] has a larger fanout resulting from further partitioning each node into two parts in the M-tree and it is restricted to vector datasets. The BM^+-tree [8] makes use of a binary hyperplane to partition the original node and a post-processing step has to be made. The M^2-tree [9] focuses on the complex similarity queries instead of the performance improvement. The bulk loading algorithm [10] is a procedure, which constructs the M-tree efficiently provided that the dataset is given beforehand.

3   Preliminaries and Problems Statement

In this section we first present an introduction on similarity search in metric spaces. As the M-tree is the cornerstone of our work, the basic knowledge of it will be explained, and an analysis of existing problems in the M-tree will also be stated.

3.1   Metric Space

Formally, a metric space $\mathcal{M}$ is a pair $\mathcal{M} = (\mathcal{D}, d)$, where $\mathcal{D}$ is the domain of objects (or the objects' keys or indexed features) and d is the distance function $d: \mathcal{D} \times \mathcal{D} \rightarrow \mathbb{R}$ with the following properties for all objects $x, y, z \in \mathcal{D}$:

- $d(x, y) \geq 0$ (non-negativity),
- $d(x, y) = d(y, x)$ (symmetry),
- $x = y \Leftrightarrow d(x, y) = 0$ (identity),
- $d(x, z) \leq d(x, y) + d(y, z)$ (triangle inequality).

The most common type of similarity query is the range query $R(q, r)$ which is our focus in this paper, as the k nearest neighbor query can be implemented based on it in the M-tree. This kind of queries retrieves all objects found within distance r of a query object $q \in \mathcal{D}$. Let $X \in \mathcal{D}$ be a finite set of objects indexed. Mathematically, $R(q, r) = \{x \in X | d(x, q) \leq r\}$.

3.2 The M-tree

The M-tree [1] is a kind of balanced and paged metric trees, whose structure is similar to the B-tree [2]. Likewise, the M-tree grows from bottom to top through the node split, and the node's size corresponds to one disk page constantly. A node in the M-tree is composed of several entries. An entry in a non-leaf node is a tuple $\langle p, r^c, d(p, p^p), ptr \rangle$, where p is a pivot and r^c is the corresponding covering radius around p. p^p is the parent pivot of p and $d(p, p^p)$ is the distance between them. An entry in a leaf node is a tuple $\langle p, d(p, p^p) \rangle$, where p is the stored object and r^c can be neglected as it is always 0 in this case. The M-tree concerns not only I/O costs but also CPU costs through the triangle inequality, since in multimedia applications the distance computation may always be CPU bound, such as the Edit distance [19] of strings (the complexity is $O(nm)$, where n and m are lengths of two strings). As for the process of range query in the M-tree, it always starts from the root with a depth-first way. Supposing there is an entry in a non-leaf node, it has following situations:

- If $|d(q, p^p) - d(p, p^p)| - r^c > r$, the entry and its subtree pointed to by ptr are pruned.
- If $|d(q, p^p) - d(p, p^p)| - r^c \leq r$, the entry cannot be pruned meaning that $d(q, p)$ has to be calculated. After this calculation, the subtree is pruned provided that $d(q, p) - r^c > r$ holds. Otherwise, the subtree should be recursively searched.

For an entry in a leaf, the above pruning conditions can be applied again only if r^c is set to 0 and there is no subtree pointed to by ptr.

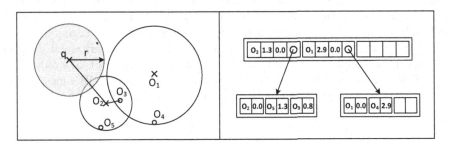

Fig. 1. A 2-D dataset sample and the corresponding M-tree

Since it is intuitive in the two-dimensional space, all the illustrations related to the dataset sample in this paper will be explained under the assumption that objects are 2-D data, and meanwhile the Euclidean distance is employed. Fig. 1 gives an example of a 2-D dataset including 5 objects, and the M-tree for the given dataset, where the structure of entries is illustrated. When a range query $R(q, r)$ begins, $d(q, O_1)$ and $d(q, O_2)$ are first calculated as they are located in the root. However, the right subtree of root is pruned due to $d(q, O_1) - r_1^c > r$, while the left subtree should be searched. The calculation of $d(q, O_3)$ can be avoided as $|d(q, O_2) - d(O_3, O_2)| > r$ holds. On the contrary, calculation of $d(q, O_5)$ is necessary as $|d(q, O_2) - d(O_5, O_2)| < r$ holds.

3.3   Existing Problems in the M-tree

The reason to index the database with a tree-based structure is to attain a sublinear search performance. However, it is barely possible to be so efficient for an index like the M-tree, which has to traverse multi-ways. There are possibly three reasons contributing to this phenomenon:

- As the radius of a range query is always larger than 0, it may overlap more than one region represented by the node in the M-tree.
- The existing overlap between nodes is another factor responsible for the degradation when a query intersects this overlap.
- As the objects stored increasingly occupy more spaces in a node (for example, high dimensional data), the fanout of a node becomes smaller.

Due to the nature of similarity search, the first reason cannot be eliminated since a query radius totally depends on a specific requirement, while we can put emphasis on other two reasons. Let us first examine the degree of the overlap in the M-tree through the point query $R(q, 0)$, which is a special kind of range query with the query radius set to 0. It is an explicit way to measure the overlap, as the size of the query radius will not take effect to degrade the search performance of the M-tree. Note that for an exact match in the B-tree, the amount of accessed nodes always equals to the height of the index (without the buffer between the memory and the disk), which can be the reference standard here. Likewise, in the ideal case of the M-tree (meaning that there is no overlap between nodes), the amount of accessed nodes (or page accessed) in a search process also equals to the height of the index. Thus, we can define the degree of the overlap in the M-tree as the ratio of the amount of visited nodes and the index's height. The examination results are shown in Fig. 2, which is conducted with 10^6 uniformly and normally distributed data in various dimensionalities ranging from 5 to 50 (more specific experimental setting can be referred in the experimental study in this paper). According to Fig. 2, it is obvious to see that the degree of the overlap is so critical that the search performance of the M-tree degrades at least 10^3 times, and its impact grows with the increment of the dimensionality in the vector space.

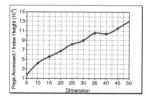 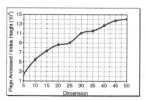

Fig. 2. Degrees of the overlap in the M-tree with uniformly and normally distributed data

Guided by these examination results, our endeavor is to minimize the overlap between nodes in order to be close to the ideal search performance. Please note that we only test I/O costs in this experiment, and CPU costs are not mentioned, because it is expected that there is a strong relationship between the amount of accessed pages and distance calculation times, as one disk page access may cause several distance calculations consequently.

Another drawback of the M-tree is its node split policy. Actually, there exist various split policies in the M-tree, which include the promotion and distribution procedures. Specifically, two objects must be selected from the objects stored in the overfull node as pivots when a node split occurs. After the selection, all the objects in the overfull node are partitioned into two nodes. On the basis of experiments on the M-tree [1], the mM_RAD_2 policy is the best with regard to the efficiency including I/O and CPU costs both, which minimizes the maximum of covering radii of two pivots. The implementation of this policy is straightforward because each pair of objects in the overfull node will be tested, and the complexity of this policy reaches to $O(n^3)$. As a result, it is worth investigating a new node split policy with lower complexity while the search performance should be guaranteed as well.

4   M^X-tree Overview

In this section, we present the structure and related algorithms of the M^X-tree, which is to effectively and efficiently support the similarity query in metric spaces. The M^X-tree has the changeable node size to resolve the overhead resulting from the unnecessary node split. A new node split policy in the M^X-tree is proposed to save the time in the tree building process, especially for large nodes after the introduction of the super node. Besides, we adopt an internal index to further reduce CPU costs in leaves of the M^X-tree.

The structure of the M^X-tree is presented in Fig. 3. Formally, a M^X-tree is a metric tree having the following properties:

- The root is either a leaf or has at least two children.
- Each node must not be empty. For an internal node at least one subtree is needed, and there is at least one object stored for a leaf.
- Each path from the root to a leaf has the same length.
- Each node's size corresponds to a disk page or a multiple of a disk page.
- The objects are all stored in leaves, and internal nodes reserve the space for pivots.
- The internal indexes in leaves act as a second filter.

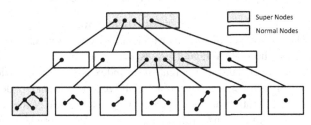

Fig. 3. Structure of the M^X-tree

4.1   Super Node

According to the size of a node, there are two kinds of nodes in the M^X-tree, normal nodes and super nodes respectively. In comparison to a normal node, the size of a super node doubles at least. However, its size still meets the requirement of a multiple of the disk page. From the aspect of position, nodes of the M^X-tree can be classified as internal nodes and leaves, where there are several differences as indicated in Fig. 4. As we can see in the figure, each kind of nodes in the M^X-tree has a header telling the necessary information, such as the level of the current node. In an entry of an internal node, the pivot with the covering radius and the distance are also stored to prune the subtree with the triangle inequality, and the pointer is responsible for finding its subtree. As for an entry in the leaf, it is a simplified version, and there is some information related to the internal index stored in the latter half of the leaf. Details about leaves in the M^X-tree will be discussed in subsection Internal Index.

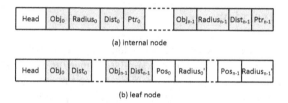

(a) internal node

(b) leaf node

Fig. 4. An internal node and a leaf node of the M^X-tree

Let us first explore the super node in the M^X-tree. To the best of our knowledge, the super node originated in the X-tree, whereas the super node in the M^X-tree differs from it in several ways:

- The super node in the X-tree is only available in vector spaces due to hyper-rectangle partition, while in the M^X-tree it is extended to metric spaces.
- When objects are points, a leaf in the X-tree is impossible to be a super node as the overlap-minimal split in the X-tree guarantees an overlap-free split, while a leaf in the M^X-tree is feasible to be a super node or not.
- When a split occurs in a super node of the X-tree, it will create a new super node with the same size, while either a normal node or a super node will be generated in the M^X-tree depending on the distribution of the objects in the overfull node.

With the introduction of the super node, the inefficient overlap between nodes is possibly avoided. The advantages of the super node are obvious. On the one hand, as the node size increases, the node's fanout grows strengthening the search performance horizontally. On the other hand, the height of an index is shortened, which enhances the performance vertically as the overhead of traversing internal nodes decreases. Besides, the storage utilization of nodes is improved. The reason of the higher storage utilization lies in the gap existing in a node. For example, a normal node with 4KB page size is able to store two 200 dimensional objects with double precision, only two-thirds of the total space utilized. If a 8KB super node is created, 5 objects are stored instead of 4 ones. The higher storage utilization is especially useful in response to a range query with a large radius.

In general, the insertion algorithm of the M^X-tree is similar to the M-tree's. The node split part, however, is different. To be specific, when a node reaches to its capacity, further insertion will cause a split in the M-tree. Different from it, a decision has to be made in the M^X-tree about whether it is beneficial to split or not. When a split is uneconomic, the node will be prevented from split, and instead a super node will be created. Otherwise, a split is performed in the node. When a split occurs in a super node, the same routine will be executed. However, the only difference is that the super node may shrink after a split depending on the partition of objects.

As a secondary memory based index, the underlying free storage management may make a great difference to its performance, especially in I/O costs. In the M-tree, a linked list is maintained to manage free pages with only one page accessed once an allocation process starts (the first page always stores the number of the next free page). In the M^X-tree, however, a super node contains several pages. In this case, allocation with a linked list is not effective as it is unable to guarantee a consecutive allocation of free pages, which may result in seeking disk tracks many times for reading a super node only. For the purpose of minimizing the seek time as far as possible, we employ the bitmap to manage the free pages in the M^X-tree. It is reasonable to make the whole bitmap resident in the main memory as long as the index is opened, since one bit in the bitmap manages one page. As a result, consecutive allocations can be achieved through a sliding window in the bitmap instead of any disk accessing.

4.2   Node Split Policy

As indicated in the paper of the M-tree [1], the preferable node split policy is mM_RAD_2 with the hyperplane partition. The disadvantage of this policy, however, is evident because the process of finding two suitable pivots is significantly CPU-intensive with $O(n^3)$ complexity. As dividing a graph into two parts with the minimum spanning tree (MST) algorithm has been successfully used in the data mining area, the Slim-tree [3] proposes a Kruskal's MST [16] based node split policy in order to reduce the complexity. As a usual routine, objects in an overfull node are viewed as vertexes in a complete graph with $n(n - 1)/2$ edges. A MST is built from the graph, and two sub-trees are generated by cutting the longest edge in the MST. Compared to the method used in data mining, an additional step in the Slim-tree is that in each subtree a pivot is selected with the minimal covering radius. Only with these steps, however, a possible problem is that a highly unbalanced index may be built, where all nodes of the index may emerge from one node in practice. Note that the unbalance here refers to the object distribution instead of the index structure. To address this problem, the Slim-tree suggests using one among longest edges in the MST, which brings into an even distribution in terms of the amount of objects. In summary, the complexity of this algorithm is $O(n^2 log(n) + n^2)$ on the amount of objects in a node.

A difficulty in the node split policy above is tuning the parameters, such as the node utilization that controls the degree of an even object distribution, and the appropriate amount of longest edges. Another problem is the rescuing measure that takes

$O(n^2)$ to find the appropriate longest edge at worst, as the method has to count the amount of vertexes for every possible longest edge. Besides, since the graph is edge-intensive, the Prim's algorithm [17] based on the adjacency matrix is preferable to the MST's construction with only $O(n^2)$ complexity in this case.

Guided by these analyses, we propose a new node split policy in the M^X-tree without tuning any parameter. Meanwhile, its efficiency improves with complexity $O(n^2 + n)$. To be specific, previous steps in this algorithm including MST construction and finding the longest edge are similar with the method used in the data mining and the Slim-tree. The only difference in these steps is the way to generate the MST. As the complete graph is edge-intensive, the Prim's algorithm is used instead of the Kruskal's in the Slim-tree. However, the method making a balanced distribution is totally different, because an attempt is made to enlarge the small subset measured by the covering radius after the partition by cutting the longest edge in the graph. The enlargement is done through searching an appropriate object o in the large subset (marked as S_l). Let us denote the pivot in the small subset as p_s, and correspondingly p_l is the one in S_l. r_s^c and r_l^c are the covering radii for the small and the large subset respectively $(r_s^c < r_l^c)$. Given a set, the function *Farthest Object (FS)* returns the farthest object from the pivot of this set. Hence, we can choose the object o as follows:

- Firstly, try to find an object o in the subset S_l, where $o \neq p_l$, $o \neq FS(S_l)$, and can also minimize $|d(p_s, o) - r_l^c|$.
- Then, check the object o with two conditions listed below:
  - $d(p_s, o) > r_s^c$ (lower bound),
  - $d(p_s, o)/r_l^c < r_l^c/r_s^c$ (upper bound).
- If the object o satisfies all these conditions, move it to the small subset and update r_s^c. Otherwise, do nothing about enlargement of the radius.

The overall target of above steps tries to attain an equal length of covering radius, which provides the same possibility for inserting a new coming object when their locations are unknown beforehand. This avoids the large subset continuously absorbing the new object and the resulting unbalance finally. Note that both the pivot and the farthest object in the large subset cannot be chosen, for the large region is represented by both of them. As for the second step, it limits the enlargement of the original small radius in a reasonable range. Fig. 5 shows an example of our node split policy. In the third part of this figure, the small subset consists of only one object p_s, and thus it manages to enlarge itself through containing O_1 in the large subset. It is worth mentioning that the search performance resulting from this policy is guaranteed compared to mM_RAD_2 with the hyperplane partition in both CPU and I/O costs.

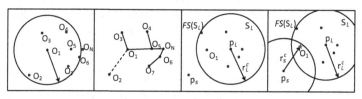

Fig. 5. The node split policy in the M^X-tree

4.3   Internal Index

To a B-tree-like index, there is a natural trade-off between the hierarchical and the linear structure. More specifically, the larger the node of the index is, the flatter the index tends to be, and vice versa. As a result of the employment of the super node in our case, CPU costs in a node may grow linearly with the enlargement of the node size. Although a pruning strategy exploiting the triangle inequality has already been taken in the M-tree, there is still room for further improvement in CPU costs based on the following observations:

- Due to the page alignment, even if a node is already full, there may still exist a gap in it, whose size is less than an entry's size.
- The distribution in the node split process results in two nodes, either of which may have the free spaces.

Therefore, the free spaces in nodes suggest the likely enhancement in CPU costs. As illustrated in Fig. 3, a binary internal index is integrated into the leaf of the M^{X}tree by exploiting the free spaces. In reality, the internal index can be viewed as a variant of the vantage-point tree (vp-tree) [20], which is a kind of binary space partitioning trees that segregates data in a metric space by choosing a vantage point in the space and dividing the data points into two partitions recursively: those that are nearer to the vantage point than a threshold and those that are not. We name our internal index as the modified vp-tree. The left part of Fig. 6 shows a region represented by a leaf in the M^{X}tree containing 5 objects, and the corresponding modified vp-tree is presented in the right part. Suppose O_2 is selected as the root of the modified vp-tree, and $d(O_2, O_1)$ is set as the first threshold. As a result, O_3, O_4 and O_5 go to the right branch of the modified vp-tree. Again, if O_4 is the root of the sub tree, O_5 will be its left child according to $d(O_4, O_5)$. The physical structure of the modified vp-tree is described in Fig. 4 (b). According to the sequence of objects in the leaf of the M^{X}tree, the position and the threshold of a vantage point are stored consecutively in the latter part of the leaf. The position here is recorded as the serial number of the level traversal of a complete binary tree in the main memory, though the modified vp-tree is not always a complete tree. When the free spaces are not enough to store all the objects in the modified vp-tree, we simply build the index with first few objects, and remaining objects are pruned with the triangle inequality only.

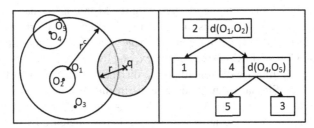

Fig. 6. The region represented by a leaf in the M^{X}tree and the corresponding internal index

Table 1 compares four pruning strategies with a same range query $R(q,r)$ in Fig. 6. How the modified vp-tree works is described in algorithm 1. In essence, it first exploits $d(obj(N), p^p)$ and then $thr(N)$ with the triangle inequality. Note that $d(q, O_2)$ has been calculated in the non-leaf node of the M^X-tree.

Algorithm 1: RangeSearch(*N*)
1. N is a vantage point in the modified vp-tree;
2. p^p is the pivot in the leaf of the M^X-tree;
3. $obj(N)$ returns the object of N;
4. $thr(N)$ returns the threshold of N;
5.
6. **if** N is NULL **then**
7.   **return**;
8. **if** $|d(q,p^p) - d(obj(N), p^p)| \leq r$ **then**
9.   calculate $d(q, obj(N))$;
10.   **if** $d(q, obj(N)) < r$ **then**
11.   add $obj(N)$ to the result;
12.   **if** $d(q, obj(N)) \leq r + thr(N)$ **then**
13.   RangeSearch($N.left$);
14.   **if** $d(q, obj(N)) + r \geq thr(N)$ **then**
15.   RangeSearch($N.right$);
16. **else**
17.   RangeSearch($N.left$);
18.   RangeSearch($N.right$);

Table 1. Calculated distances for different pruning strategies

Strategy	Calculated distances
non-pruning	$d(q, O_2)$ $d(q, O_3)$ $d(q, O_4)$ $d(q, O_5)$
triangle inequality only	$d(q, O_3)$ $d(q, O_4)$ $d(q, O_5)$
vp-tree only	$d(q, O_2)$ $d(q, O_4)$ $d(q, O_3)$
modified vp-tree	$d(q, O_4)$ $d(q, O_3)$

For the purpose of high efficiency, the pivot should be placed in the modified vp-tree as low as possible so that more objects may be filtered. Another point is about how to set a better threshold for a vantage point. According to the suggestion of the original vp-tree, the median object in terms of the distance is a good choice. Last, one may argue that if the space occupied by $d(obj(N), p^p)$ is replaced by the modified vp-tree, there may be a better effect. Actually, this may be true if the leaf is large compared to the object's size, while our strategy ensures that the overall pruning ability will be at least equal to the original one (with triangle inequality only) all the time.

5   Experimental Study

In this section, we provide experimental results of the performance of the M^X-tree. We run experiments comparing the M^X-tree with the M-tree and other variants whenever possible. The implementations of them are all based on the original M-tree and the GiST C++ package [18] with a constant page size 4KB. In our experiments, we mainly focus on CPU and I/O costs of an index, and by convention they are measured with the number of times with regard to the distance calculation and the disk page access respectively. Similar to many other experiments of a metric index, both costs are defined in the corresponding function. For example, once the reading function in the storage management of an index is called, I/O costs increase by one, even if the page is currently in the main memory. Thus, the buffer between the main memory and the hard disk does not take effect at all, and results are totally hardware-independent.

We use both the synthetic and the real datasets as listed in Table 2 to conduct experiments. Duplicates have been eliminated in all datasets. To gain the flexibility, however, most of our experiments are based on synthetic datasets. For vector spaces, the distance measure is the Euclidean distance. As for strings, the Edit distance is employed. All the results about the performance on query processing are averaged over 10^3 queries, and query sets are randomly extracted from the corresponding dataset used to construct an index. Without an additional statement, the size of a synthetic dataset is 10^6 with 20-D data, and for real data the whole set is used. Range queries with $\frac{1}{100}$ of the maximum distance in a dataset are applied by default.

Table 2. Details of datasets in experiments

Dateset	Size	Type	Description
Uniform Data[1]	1.5×10^6	vector, double	Uniformly distributed data in [0,1) with dimensionality ranging from [5,50]
Normal Data[1]	1.5×10^6	vector, double	Clustered data with Gaussian distribution, dimensionality ranging from [5,50]. Coordinates of centers are uniformly distributed in [0, 1). The variance of the Gaussians around the points is 1 and points are assigned to the clusters with uniform probability.
Images (SIFT)[2]	10^6	vector, int	Vectorization of images with scale-invariant feature transformation, dimensionality 128
DBLP Authors[3]	1,162,047	string	Extracted from a XML in DBLP, non-lexicographical ordering
Dictionary[1]	69,069	string	An English dictionary, disordered to avoid dependences on lexicographical ordering

5.1 Comparing the Split Policies

We first compare the node split policy in the M^X-tree with mM_RAD_2 in the M-tree and the one in the Slim-tree. Interestingly, if we set the maximum size of the super node in the M^X-tree as one disk page only and do not use its internal index (we name it as the M^X-tree (split)), the M^X-tree (split) will differ from the M-tree only in the node split policy, which is employed here for a comparison purpose.

A promising feature of the node split policy in the M^X-tree is its high efficiency to split an overfull node. However, testing the node split process only is almost invisible, while the index construction time is a good indicator if we can neglect the effect of disk access. Let us denote the M^X-tree (split) with the node split policy in the Slim-tree as the M^X-tree (slim). Experimental results in terms of the time measurement are shown in Fig. 7. We must point out that in this case, observations are hardware and software dependent (2.5GHz CPU with L2 512KB and L3 3MB, executable program with release version). In order to eliminate the influence of the disk latency, datasets and indexes are stored in a virtual disk made up of the main memory actually, where

---

[1] Metric spaces library, http://www.sisap.org/Metric_Space_Library.html
[2] Datasets for approximate nearest neighbor search,
http://corpus-texmex.irisa.fr/
[3] DBLP website, http://dblp.uni-trier.de/xml/?C=D;O=A

10^5 uniform data are tested in the line chart, and 10^5 DBLP authors and the whole dictionary are used in the column chart of Fig. 7. In vector spaces, the M^X-tree (split) runs significantly faster than the M-tree (up to a factor of 12) and constantly outperforms the M^X-tree (slim). The reason to explain the decline of speedup with growing dimensionality is the decreasing ratio of the page and the object size. As the Edit distance is much more CPU-intensive which may reduce other factors in the index construction (for example, finding the appropriate node to insert an object), the column chart shows a more favorable results about the M^X-tree (split), where it is around 56 times faster than the M-tree, and about 4 times faster than the M^X-tree (slim) in the dictionary test.

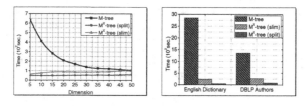

Fig. 7. Time of index construction with different node split policies

Normally, the first priority of a node split policy is to gain a better search performance of the resulting index, which is our focus currently. As we have discussed that the node split policy in the Slim-tree has to tune parameters, the reasonable minimal node utilization is 0.1 through our empirical study in the uniform case, and $\left\lceil \frac{n}{2} \right\rceil$ edges (n is the amount of edges) are tested according to the length (the same as the experiment does in the split policy test). In the normal distributed case, however, we do not find the appropriate minimal node utilization as the search performance of the M^X-tree (slim) is seriously weak due to its large size (up to 6 times of the M^X-tree (split)). If we set a large value for the minimal utilization, the real node utilization may be very small on the contrary, because an uneven node partition by cutting the longest edge will be accepted provided that all the edges tested do not result in an even split.

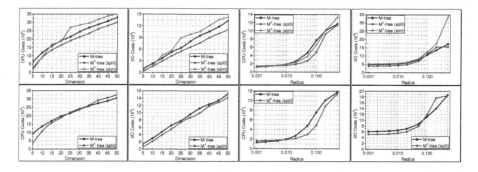

Fig. 8. Range queries with different split policies

Fig. 8 shows CPU and I/O costs with the growing dimensionality (left part) and the increasing radius of a query (right part, the x-axes are given in a log scale of the volume of a range query) respectively. They are conducted with both uniform (top part) and normal (bottom part) synthetic data. It is clear to see that indexes resulting from three node split policies perform similarly in general, and the M^X-tree (split) is slightly better in most cases. Increments with the higher dimensionality for both costs are reasonable as the fanout of an index shrinks and thus the height of the index grows.

5.2   Experiments with Synthetic Datasets

From now we examine the performance of the unrestricted M^X-tree (with the super node and the internal index), and the linear scan is added as a sanity test. We first study the impact of dimensionality. Fig. 9 depicts CPU and I/O costs with both uniform (left part) and normal (right part) data for the growing dimensionality. As for CPU costs, the M^X-tree is strongly robust to the growing dimensionality, while the M-tree shows a logarithmic trend (the M^X-tree outperforms the M-tree at least 58%). In I/O costs, the M-tree is always worse than the linear scan. In contrast, the M^X-tree performs better with a smaller slope than the linear scan (up to 3.6 times faster than the M-tree). A direct reason to explain the weak performance in I/O costs for the M-tree is the query set, where we employ a strict criterion that the answering objects of a query always fall into the index. Assuming an extreme case with a query far from the dataset, the M-tree may compare it with the root only while the linear scan has to go through the entire dataset. This also applies to the M^X-tree and following observations. Reconsidering the left part of Fig. 8 where same queries are performed, the superiority resulting from the super node and the internal index is evident compared to the M^X-tree (split).

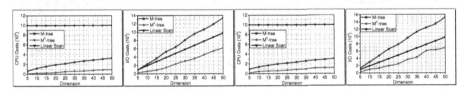

Fig. 9. CPU and I/O costs with the growing dimensionality

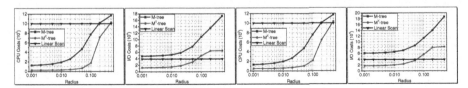

Fig. 10. CPU and I/O costs with the growing radius

Next, we consider how the radius of a range query influences the search performance. Fig. 10 shows CPU and I/O costs with both uniform (left part) and normal (right part) synthetic data for the growing radius. According to Fig. 10, the M^X-tree consistently outperforms the M-tree at least 43% as measured by I/O costs, and the improvement is up to about 75%. As for CPU costs, the substantial advancement of the M^X-tree is observed in the relatively small volume of the range query (up to around 85% compared to the M-tree). To explain these speedups, one of the reasons is the storage utilization, which is a key factor to tackle the query with a large radius, apart from the endeavor to reduce the overlap between nodes and the effort to do a second filter.

Last we study the scalability of the M^X-tree with varying dataset sizes ranging from 0.6×10^6 to 1.5×10^6. Results are shown in Fig. 11, where the left part corresponds to the uniform case and the right part is the normal case. Compared to a slight upward trend in the M-tree, the M^X-tree presents a strong robustness to the growing dataset with respect to CPU costs (at least around 70% faster than the M-tree) and a flatter slope than the linear scan in I/O costs (at least 66% faster than the M-tree).

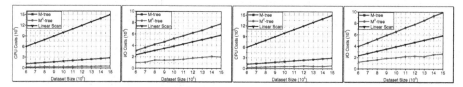

Fig. 11. CPU and I/O costs with the growing dataset size

5.3   Experiments with Real Datasets

We here survey the performance of the M^X-tree with two real datasets (images and DBLP authors) with the varying radius as the datasets are in fixed amount. In the case of testing images, we find that the M^X-tree keeps around 2.8 times faster than the M-tree in CPU costs and meanwhile about 4 times faster in I/O costs when the radius is less than 2% of the maximum distance. For the string dataset, we compare the M^X-tree with the M-tree only, as they are on the same test bed based on the implementation of the fixed entry size. In CPU costs, the maximum improvement of the M^X-tree is up to 56% while it stays at around 10% on average. As for I/O costs, the speedup of the M^X-tree is observed with a large radius (up to 2.3 times). Note that the longest string is 78 including the spaces in the DBLP authors.

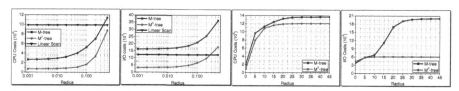

Fig. 12. CPU and I/O costs with images (SIFT) and DBLP authors

6   Conclusions

We have presented the M^X-tree, which is an efficient and scalable metric index. The M^X-tree allows changeable node size to avoid the unnecessary split. In addition, a new node split policy without tuning parameters is proposed in the M^X-tree, whose runtime complexity is $O(n^2)$ and meanwhile the resulting search performance is guaranteed. Besides, an internal index is employed in the leaves to further decrease CPU costs in the M^X-tree. We have implemented the M^X-tree based on the original source code of the M-tree, and an experimental implementation of it is available to the public domain[4].

References

1. Ciaccia, P., Patella, M., Zezula, P.: M-tree: An Efficient Access Method for Similarity Search in Metric Spaces. In: VLDB, pp. 426–435 (1997)
2. Comer, D.E.: The Ubiquitous B-tree. ACM Computing Surveys 11(2), 121–137 (1979)
3. Traina Jr., C., Traina, A.J.M., Seeger, B., Faloutsos, C.: Slim-Trees: High Performance Metric Trees Minimizing Overlap between Nodes. In: Zaniolo, C., Grust, T., Scholl, M.H., Lockemann, P.C. (eds.) EDBT 2000. LNCS, vol. 1777, pp. 51–65. Springer, Heidelberg (2000)
4. Vieira, M.R., Traina Jr., C., Chino, F.J.T., Traina, A.J.M.: DBM-tree: A Dynamic Metric Access Method Sensitive to Local Density Data. In: SBBD, pp. 163–177 (2004)
5. Skopal, T.: Pivoting M-tree: A Metric Access Method for Efficient Similarity Search. In: DATESO, pp. 27–37 (2004)
6. Traina Jr., C., Traina, A.J.M., Filho, R.F.S., Faloutsos, C.: How to Improve the Pruning Ability of Dynamic Metric Access Methods. In: CIKM, pp. 219–226 (2002)
7. Zhou, X., Wang, G., Yu, J.X., Yu, G.: M$^+$-tree: A New Dynamical Multidimensional Index for Metric Spaces. In: ADC, pp. 161–168 (2003)
8. Zhou, X., Wang, G., Zhou, X., Yu, G.: BM$^+$-Tree: A Hyperplane-based Index Method for High-dimensional Metric Spaces. In: Zhou, L.-Z., Ooi, B.-C., Meng, X. (eds.) DASFAA 2005. LNCS, vol. 3453, pp. 398–409. Springer, Heidelberg (2005)
9. Ciaccia, P., Patella, M.: The M^2-tree: Processing Complex Multi-feature Queries with Just One Index. In: DELOS Workshop (2000)
10. Ciaccia, P., Patella, M.: Bulk Loading the M-tree. In: ADC, pp. 15–26 (1998)
11. Jagadish, H.V., Ooi, B.C., Tan, K.-L., Vu, Q.H., Zhang, R.: Speeding Up Search in Peer-to-Peer Networks with a Multi-way Tree Structure. In: SIGMOD, pp. 1–12 (2006)
12. Guttman, A.: R-trees: A Dynamic Index Structure for Special Searching. In: SIGMOD, pp. 47–57 (1984)
13. Sellis, T., Roussopoulos, N., Faloutsos, C.: The R$^+$-tree: A Dynamic Index for Multi-dimensional Data. In: VLDB (1987)
14. Kriegel, H.-P., Schneider, R., Seeger, B., Beckmann, N.: The R*-tree: A Efficient and Robust Access Method for Points and Rectangles. Sigmod Record 19(2), 322–331 (1990)
15. Berchtold, S., Keim, D.A., Kriegel, H.-P.: The X-tree: An Index Structure for High-dimensional Data. In: VLDB, pp. 28–39 (1996)
16. Kruskal, J.B.: On the Shortest Spanning Subtree of a Graph and the Traveling Salesman Problem. The American Mathematical Society 7(1), 48–50 (1956)
17. Prim, R.C.: Shortest Connection Networks and Some Generalizations. Bell System Technical Journal 36(6), 1389–1401 (1957)
18. Hellerstein, J.M., Naughton, J.F., Pfeffer, A.: Generalized Search Trees for Databases Systems. In: VLDB, pp. 562–573 (1995)
19. Levenshtein, V.I.: Binary Codes Capable of Correcting Deletions, Insertions and Reversals. Soviet Physics Doklady 10 (1966)
20. Yianilos, P.N.: Data Structures and Algorithms for Nearest Neighbour Search in General Metric Spaces. In: SODA, pp. 311–321 (1993)

---

[4] M^X-tree Implementation by Authors,
   http://sourceforge.net/projects/mxtree/

Integrating Behavior Driven Development and Programming by Contract

Larry Schoeneman and Jiang B. Liu

Computer Science & Information Systems Department
Bradley University
Peoria, Illinois, U.S.A.
jiangbo@bradley.edu

Abstract. This paper developed a Contracted Behavior Driven Development (CBDD) method that extends and combines the ideas behind Test/Behavior Driven Development (TDD/BDD) and Programming by Contract (PBC) to improve the overall stability and quality of a system. A tool is developed to derive unit tests automatically by analyzing human written specifications for preconditions and post-conditions when coupled with data definitions. These results will be used to generate code to be run by a unit testing framework before deployment, either as part of a continuous integration environment or by individual developers. The tool will also generate wireframe classes implementing pre and post-conditions within the code and using runtime contract analysis to generate information when an exception occurs, thereby helping to automate verification of bug fixes.

Keywords: Software Contracted Behavior Driven Development, Software Development Tools.

Introduction

In many professional development settings, Test Driven Development (TDD), and its evolutionary descendent, Behavior Driven Development (BDD) are increasingly used to improve code quality and reducing error rates. While valuable, they are considerably less rigorous than formal methodologies of program correctness via formal logic. A less rigorous methodology utilizing the ideas of formal methods but without requiring an actual full proof is also in use called "Design by Contract" (DBC) or "Programming by Contract." (PBC). Both Test/Behavior Driven development and Programming by Contract address program correctness, but in different ways. This paper proposed to create.

A combined approach that takes best practices from both methods and ideally will produces a better resulting product with fewer bugs. Combining the two practices will create a resultant practice that meets the following criteria:

- Produces code with fewer bugs than either approach
- Allows expression of features by customers in natural fashion

B. Murgante et al. (Eds.): ICCSA 2013, Part V, LNCS 7975, pp. 590–606, 2013.

- Enables creation of executable acceptance criteria to validate requirements
- Provides a path to generate code from user stories and acceptance criteria
- Encourages user centered design practices
- Works well with existing tools
- Allows building of automated tools to reduce repetitive activities
- Allows limited definition of behavioral contracts at the analysis level

The combined approach, called Contracted Behavior Driven Development (CBDD) enhances TDD/BDD's focus on pre-execution testing with PDC's run time capabilities and code level focus on defining correct behavior. It will examine using both TDD/BDD and PBC to enhance the quality of development throughout the software development life cycle, while aiding in the development and stabilization of sound architecture. After detailing this combined approach, a tool has be designed and implemented to demonstrate automating this approach and integrating it with pre-existing tools.

There is a fairly substantial amount of research which shows that using Test Driven Development/Behavior Driven Development strategies produce code with fewer issues than standard waterfall quality assurance testing or manual unit testing [1-3]. Essentially the largest returns are due to guiding development based on user scenarios. By writing tests first, and using that to guide coding, as opposed to the other way around, considerably less code is written, which in turn yields fewer issues. Similarly, the existence of unit tests acts an invariant, protecting developers from changes introduced via refactoring [5]. In fact, the idea of refactoring without unit testing simply devolves to random rewriting of code [6-7]. Programming by Contract improves code quality by forcing developers to create well defined entry and exit conditions for their methods [8]. This improvement is based solidly in the theory of program correctness [9] and while not equivalent to a formal proof, provides strong support in that direction. PDC does yield positive results via automatic generation as shown by [10].

Programming by Contract and Behavior Driven Development both provide substantial gains in reducing bugs in software and improving correctness. Most developers tend to choose one approach or the other, or most commonly, neither. It is the contention that these two approaches have considerable commonality as well as complimentary techniques. Instead of using them in isolation, using the two approaches together can produce a more desirable result.

While substantial progress has been made using automated tools to integrate the two approaches [11], these solutions attempt to automate the generation of unit testing from existing code and contracts. This approach is certainly valid but it defeats the advantages of test driven development's test first approach and removes easy integration with business acceptance criteria in the language of the problem domain. This research defines a domain driven approach that includes contracts as part of the resulting code and attempts to guide their creation via business acceptance criteria expressed in nearly common language. The net result should be executable behavior specifications, combined with contract-stabilized code to allow both pre-emptive bug elimination and run-time bug analysis.

Contracted Behavior Driven Development (CBDD) is an approach to developing software which emphasizes creating executable specifications from business requirements in order to drive code development, and using programming contracts to ensure that at a technical level, code executes correctly and can be automatically verified when appropriate tools are available. In CBDD, we define a behavior as a discreet unit of functionality corresponding to a user story and its acceptance criteria. This definition is consistent with behavior driven development. In general, a behavior is expressed in BDD as a user story with the following format:

Story:
Feature: [Feature name]
As a [user]
I want to [userAction]
So that [intended result]
Scenario:[Scenario Name]
   Given [Scenario expectations]
   When [an action is taken or event occurs]
   Then [Expected result

CBDD extends this basic story concept with several additional ideas:

1. Entity definition – Defining an entity for establishing contractual obligations.
Given the definition, the developer can now take this story and quickly generate scaffolding code with contracts embedded as specified in the story, and build executable behavior tests as well. A C#-like language for entity definition would allow a developer quickly generates skeletal code with nBehave, C# and code contracts. Additionally, we've created a skeletal behavior that can be filled in and executed in a test first fashion.

2. Analysis
CBDD encourages definition of entities by business analysts/non-technical people. Developers can use these to define simple, contractually enhanced classes as well as a set of executable behaviors. These executable behaviors can be used as part of a continuous integration process to ensure the continued stability of the system as it evolves and as more stories are integrated.

The combination of Behavior Driven Development and Design by Contract enables validation of pre and post conditions both at development time and at execution time, ideally capturing a larger portion of the domain under consideration. Incorporating the concepts of Programming by Contract at the story level requires the ability to declare but not define simple operations. This facilitates creating assertions against the defined entities is used to link entities, contracts and story narratives to produce tests, code and contracts. The relationship between pieces of a user story is shown in Figure 1.

Fig. 1. Relationship between user story components

An important concept to keep in mind is that CBDD is meant as part of a larger methodology and is not a software development life cycle process itself. It deals with design, coding and the relationship between code and user stories. It does not address quality assurance testing (although it does hopefully improve the inputs), deployment, or requirements gathering, just to name a few. It is meant to serve as technique within the larger frame of a process framework such as scrum, DSDM, etc.

Contracted Behavior Driven Development

Programming by contract is a development methodology which specifies that classes and interfaces should be extended with preconditions, postconditions and invariants, collectively called a contract in order to aid in ensuring correct execution of programs.

The idea roughly corresponds Hoare logic [9], of the form {P}Q{R} where {P} is the set of preconditions which are true before execution of Q and {R} is the set of postconditions which are true after the execution of Q, assuming Q terminates. Using preconditions and postconditions allows validation of entry and exit requirements and corresponds well with the Hoare Logic based method for proving program correctness. Typically, during development, the pre-conditions and post-conditions will be active and throw errors when they are violated. They can be turned off if needed in production, or turned on as needed without recompiling. When these pre and postconditions fail, assertions can be thrown to log the problem behavior and notify the user.

Programming by Contract is more granular than test driven development as its focus is at the class and method level. Typically, pre and postconditions do not extend across method boundaries although class level invariants can be enforced. Some of the groundwork for using PBC in programming languages stems from the Eiffel language, created by Bertrand Meyer. This allowed expression of Preconditions, Post-conditions and invariants as first class citizens within code [8]. Recently, features supporting Programming by Contract have been added to the .Net platform, after first being evaluated by Microsoft Research and producing the Code Contracts library [11]. While more restrictive than Eiffel's approach, they are more widely available due to the popularity of the .Net platform. When used with Microsoft's runtime contract checker, this will cause an exception to occur if the contract is violated. Rather than generate code, the default is to generate nothing and the tool causes interception code to be added to run the contracts. This helps reduce any impact to execution speed when the contracts are not needed. If code needs to throw an exception in a given failure case, this can be expressed as a precondition as well, using a generically typed requires method:

Contract.Requires<OutOfRangeException>(...Boolean..., exception string);

This would enable easy custom trapping if needed.

To aid in implementing CBDD a prototyping tool called Kime has been built. Kime provides the following features:

1. A tool to translate user stories to executable behaviors runnable by existing tools
2. A tool to generate stub classes and their contracts as specified in entity declarations.

Implementing Kime

In order to implement Kime, the following assumptions are made:

- Target generation language is C#
- Target BDD framework is nBehave
- Templates will use the latest implementation of Microsoft's Razor engine
- Parsers and lexical analyzers are be generated by Antlr with C# extensions
- Unit tests will be implemented using the MSTest framework
- Heterogeneous Abstract syntax trees will be generated by Antlr

The Kime Project will be organized as a standard .Net solution containing the following projects:

- Kime – The main console application. This takes as input a set of files containing entity definitions and gherkin scenarios and outputs C# containing C# entity definitions annotated with pre and post-conditions as well as BDD scenarios with assertions.

- Kime /CData
    - o Antlr generated parser for the entity definition language
- Kime /Gherkin
    - o Antlr generated parser for the gherkin language
- Kime /Integration Engine – Classes and utilities to blend the gherkin results with the entity definitions. This is the heart of Kime
- Kime /Templates – The primary razor templates used with Kime. These generate entity definition classes and BDD Scenarios in C# for nBehave
- Entity Definitions – Heterogeneous AST classes used to build an in memory representation of the entity definitions. These classes are used by both Integration Engine and the Entity Definition Parser
- Gherkin Definitions – Heterogeneous AST classes used to build an in memory representation of the gherkin scenarios. These classes are used by both Integration Engine and the Gherkin Scenario Parser
- Kime Test – A set of unit tests validating the translation from text to parsed entities for both Entity Definitions and Gherkin Scenarios
- TemplateManager – A set of utility interfaces to use the razor engine

Note that the generated code from Antlr has been refactored to make it easier to understand.

A number of test scripts were built to unit test the parsers as well as the integrator.

Expected Input

Kime will expect input in the form of feature files. A feature file consists of the following parts:

- Feature name – A high level name for the entire feature
- User Story – A story with the standard format of
    As a [user]
    I want to [action]
    So that [intended goal]
- A set of one or more Scenarios of the form:
    Given [initial conditions]
    When [action occurs]
    Then [expected result]
- A set of one or more definitions with assertions of the form:

Given the definitions, Kime will attempt to do the following:

1. Generate classes for each definition with methods, members and contracts
2. Generate behavioral specifications using nBehave
3. Using the various definitions and their points in the stories, generate unit tests which correspond to the contracts defined

Example:

> Feature: Register a student for a course
> As a student
> I want to register for a course
> So that I can complete my degree
>
> Scenario:
> Given a student who must be enrolled in the university
> When I register for course 1234
> Then I should receive confirmation of registration

A corresponding definition file appears below:

```
project KimeTest;
definitions KimeTestDefinitions
{
      definition Student
      {
            members
            {
                 StudentId;
            }
            operations
            {
               Register(int x, int y)
               {
                      contracts
                      {
                           require
                           {
                                 ["for course"]
                                 RequireValid { StudentId!=null };
                           }
                           ensure
                           {
                                 [Confirmation]
                                 RecieveConfirmation { StudentId!=null };
                           }
                      }
               }
            }
      }
}
```

This will generate the following class definitions:

```
using DataTypes;
using System.Collections.Generic;

namespace KimeTest
{
        public class Student
        {
                #region Member Declarations

                public object StudentId { get; set; }

                #endregion

                #region Operation Declarations

                public void Register(int x, int y)
                {
                        // Preconditions
                        Contract.Requires(RequireValid());

                        // Postconditions
                        Contract.Ensures(RecieveConfirmation());
                }

                #endregion

                #region Rule Declarations

                internal bool RequireValid()
                {
                        return StudentId!=null;
                }

                internal bool RecieveConfirmation()
                {
                        return StudentId!=null;
                }

                #endregion

        }
}
```

The following behavioral specification scaffolding to be used with nBehave is also generated:

```
using NBehave.Narrator.Framework;
using NBehave.Spec.NUnit;
using GherkinData;

/* Feature Description
Register a student for a course
As a student
I want to register for a course
So that I can complete my degree

*/

    [ActionSteps]
    public class
    {
        [Given(@"a student who must be enrolled in the university ")]
        public void Given()
        {
            Student __Student = new Student();
            Assert.IsTrue(__Student.BeEnrolledInTheUniversity());

        }

        [When(@"I register for course 1234")]
        public void When()
        {
            Student __Student = new Student();
            __Student.Register();
            Assert.IsTrue(__Student.RequireValid());

        }

        [Then(@"I should receive confirmation of registration")]
        public void Then()
        {
            Student __Student = new Student();
            Assert.IsTrue(__Student.RecieveConfirmation());

        }

    }
```

The generated code is just a stub.   Much remains to be implemented but the basic scaffold is in place. Assertions are available in both the classes defining entities and in the unit test themselves.

Kime consists of the following key components (Figure 2):

- An entity definition lexer and parser
- A Gherkin (Cucumber, 2012) lexer and parser
- A set of internal data structures for managing entity definitions
- A set of internal data structures for managing Gherkin parsed forms
- An interface to the Razor engine for templating
- A set of templates
- A set of unit tests

Each of these components will be examined in detail.

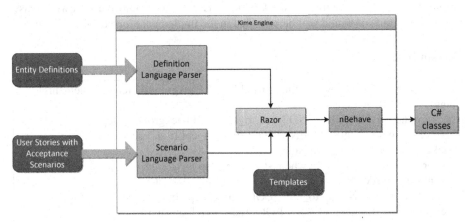

Fig. 2. Kime High Level Architectural Overview

Parsing Entity Definitions
In order to efficiently manage entity definitions a language with data definition features derived from C# was created. This has been named "CData." It provides a simple syntax for both typed and un-typed definition of entities.   To simplify implementation of a simple translator, Antlr was used to generate C# code. The decision was made to proceed with a more programmer friendly definition language primarily due to ease of implementation and clean expression.

Parsing Gherkin
Gherkin is the name of the common, English based language used in Cucumber and its variants. Again, Antlr was used to generate C# code for the parser and lexical analyzer (the generated code can be seen in the appendix).

The EBNF used as the basis for the Antlr grammar is shown in the next page.

Data Structures for Entity Definitions
During parsing, the entity definitions are deserialized into the following class model, the details of which will be provided in the implementation section (Figure 3).

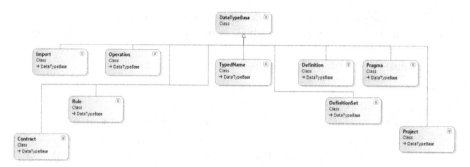

Fig. 3. Entity Definition Class Diagram

Data Structures for Gherkin Parsed Forms

During parsing, the scenarios in the Gherkin file are parsed into the following class model, the details of which will be provided in the implementation section (Figure 4).

Gherkin Grammar

```
feature = tags? FEATURE ':' line_to_eol EOL* background? feature_elements+
tags = (TAG ID)+ EOL
background = BACKGROUND lines_to_keyword background_steps
feature_elements = tags? SCENARIO ':' line_to_eol EOL* scenario_steps
background_steps =   given_step
scenario_steps = given_step when_step then_step
given_step = GIVEN line_to_eol EOL* (and_step | but_step )*
when_step = WHEN line_to_eol   EOL* (and_step | but_step )*
then_step = THEN line_to_eol   EOL* (and_step |but_step )*
and_step = AND line_to_eol EOL*
but_step = BUT line_to_eol EOL*
lines_to_keyword =
~(GIVEN|WHEN|THEN|AND|BUT|FEATURE|SCENARIO| VBAR|TAG)
line_to_eol     = .* EOL


TAG = '@'
COMMENT = '#' ~('\n'|'\r')* '\r'? '\n'
WS = (' '|'\t')
VBAR = '|'
QUOTE = '"'
ID   = ('a'..'z'|'A'..'Z'|'_') ('a'..'z'|'A'..'Z'|'0'..'9'|'_')*
INT = [0-9]+
FLOAT = [0-9]+ '.' [0-9]* EXPONENT? | '.' [0-9]+ EXPONENT?|[0-9]+
EXPONENT
STRING   = '"' ( ESC_SEQ | ~('\\'|'"') )* '"'
CHAR = '\'' ( ESC_SEQ | ~('\''|'\\') ) '\''
EXPONENT : ('e'|'E') ('+'|'-')? [0-9])+
EOL : '\r'? '\n'
```

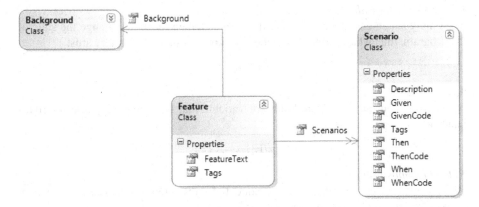

Fig. 4. Gherkin Data Structures

Razor Engine Interface

The Razor engine is the standard implementation provided by Microsoft. Essentially to use it in a non-web scenario, you simply need to wrap it in an interface and inject the proper libraries as well as provide scripts.

A simple provider pattern is used to allow swapping of the Razor Engine if some other templating engine is desired.

Razor Templates for Code Generation

Templates have been created for both the Entity Definition code generation task and the Executable scenario code generation task. Essentially these tasks run through the Entity definition and Gherkin structures respectively and generate appropriate code.

Unit Tests

Unit tests are provided to ensure the parsing of the two grammars produces correct data structures that are then fed into the template engine. Details of the unit tests will be provided in the implementation section.

CBDD Analysis

In the definition of CBDD we defined several criteria where we were hoping for positive improvement. As an evaluation, CBDD will be evaluated against each of these original points to provide an overall review of the solution.

4.1   Larger Validated Code Coverage Area Than BDD or PBC

It can be shown relatively simply that CBDD will cover at least as much of the code with pre and post condition validations as either BDD or PBC alone, using Hoare logic rules.

Hoare Logic Rules

Given a simple language, L with atomic operations corresponding to Hoare logic rules of:

Empty Statement Rule

The empty statement rule asserts that the **skip** statement does not change the state of the program, thus whatever holds true before **skip** also holds true afterwards.

$$\{P\} \text{ skip } \{P\}$$

Assignment rule

After assignment any predicate holds for variable that was previously true for right-hand side of assignment:

$$\{P[x/E]\} \; x{:=}E \; \{P\}$$

Rule of composition

Hoare's rule of composition applies to sequentially-executed programs S and T, where S executes prior to T and is written $S;T$.

$$\frac{\{P\}S\{Q\}, \{Q\}T\{R\}}{\{P\}S;T \; \{R\}}$$

Conditional rule

$$\frac{\{B{\wedge}P\}S\{Q\}, \{{\sim}B{\wedge}P\}T\{Q\}}{\{P\} \text{ if B then S else T endif } \{Q\}}$$

Consequence rule

$$\frac{P'{=}{>}P, \{P\}S\{Q\}, Q{=}{>}Q'}{\{P'\} \; S \; \{Q'\}}$$

While rule

$$\frac{\{P{\wedge}B\} \; S \; \{P\}}{\{P\} \text{ while B do S done } \{{\sim}B \wedge P\}}$$

Here P is the loop invariant.

Unit Test

Given a sequence of code, C_u we define a unit test evaluation function as:

$$T = (C_u, C_t, P, R)$$

$$\text{assert}(r) = \begin{cases} true & \leftrightarrow r \text{ is } true \\ abort & \leftrightarrow r \text{ is } false \end{cases}$$

Where: $P = \{p | \text{Preconditions of } C_u \text{ and assert}(p) \in \{true, false\} \}$
$R = \{r | \text{Postconditions of } C_u \text{ and assert}(r) \in \{true, false\} \}$
C_u is a sequence of code statements devoid of side effects
C_t is a sequence of code statements. $\exists s \in C_t$, such that $s = C_u$
$\quad \forall p \in P, assert(p)$
$\quad C_t$
$\quad \forall r \in R, assert(r)$

Given a unit Test, $T = (C_u, C_t, P, R)$, a unit test can be said to have succeeded if Prior to execution of C_u within C_t,

$$\text{assert}(p) \neq \text{abort } \forall p \in P$$

After execution of C_u

$$\text{assert}(p) \neq \text{abort } \forall r \in R \text{ and execution of } C_t \text{ terminates}$$

Contract Annotated Code
Programming by Contract corresponds to Hoare Logic Rules

Equivalency of validated code between unit tests and Hoare logics
Validated code is the sequence of statements S, validated before execution by precondition P and after execution by postcondition R. If there are multiple preconditions or postconditions they can be combined via the AND operation.

Using the general form of a Hoare Logic Rule:

$$\{P\} \ S \ \{R\}$$

$\{P\}$ = preconditions
$\{R\}$ = postconditions

We define $\{P_h\} = f(P_t)$ where $f(P) = \bigwedge p \ \forall p \in P$ where P_t is the set of preconditions for the unit test T

And $\{R_h\} = f(R_t)$ where $f(R) = \bigwedge r \ \forall r \in R$ where R_t is the set of preconditions for the unit test T

And $S = C_u$

We can then say $\{P_h\} \ C_u \ \{R_h\} \equiv (C_u, C_t, P, R)$

Therefore the combination of BDD and Programming by Contract, CBDD validates code at least to the equivalent level provided by either BDD or Programming by Contract.

4.2   Increased Coverage in Multiple Execution Environments

Since BDD scenarios are executed during the build process, they tend to catch issues in code prior to deployment [2]. Usually this will catch errors that violate our preconditions and postconditions for specific cases.

Programming by Contract assertion sets differ, in that they catch issues during the actual execution of the system.

Given this, it is apparent that CBDD will catch at least the number of issues as BDD or PDC separately and it will most likely be the case that it will catch more issues overall due to the presence of both build time and runtime coverage.

4.3   Code with Fewer Bugs Than Either Approach

It has been shown that both Test Driven Development and Programming by Contract reduce overall bug rates [2]. Since the approaches do not cover the same areas in the same manner, a simple union of the two solutions should produce a larger domain coverage area (or at least one of the same sizes) then either approach individually. Additionally, by blending the two approaches, commonalties can be realized and better testing performed. This is a result of now working at both the story level and the code level. Programmatic guards can be expressed as can domain based feature testing.

Another advantage provided is from tools by other sources. For example, Microsoft's Pex [11] can use contracts in code to improve its ability to automatically generate cogent unit tests.

Expression of Features by Customers in Natural Fashion
Human readable language is used to express the user stories and acceptance criteria. There are multiple formats available and no constraints are imposed unless imposed by an outside tool. The entity and contract declarations propose a more formal language, but even this is simple and natural in terms of language expression. One could imagine extending a tool to allow for even more general language when expressing entities and types.

Creation of Executable Acceptance Criteria to Validate Requirements
As one of the primary steps of Contracted Behavioral Driven Development is taken directly from BDD, it is obvious that executable behaviors can be expressed. The conversion to a form acceptable to a tool such as nBehave is fairly straightforward and can be automated.

Path to Code from User Stories and Acceptance Criteria
CBDD provides a clean route to code. A traceable connection exists between stories, definitions, acceptance criteria, executable behaviors and final code (Figure 5).

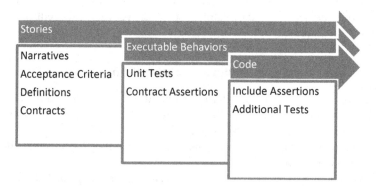

Fig. 5. Traceability from Stories to Code

User Centered Design Practices
In CBDD we begin every development task with the user in mind. If it is not included in a user story and has no benefit to the user, it has no reason to be developed. User stories are written entirely from the user's point of view and the use of a ubiquitous language focusing entirely on the problem domain is highly encouraged.

Integration with Existing Tools
The story framework proposed by CBDD integrates well and is acceptable to nBehave, SpecFlow, jBehave and other common BDD frameworks. Extra information such as contract details will be treated as comments by most of these tools and ignored.

Building of Automated Tools to Reduce Repetitive Activities
It is relatively straightforward to build a tool for CBDD to generate scaffolding code. One is included as part of this Paper. Language flexibility can pose some problems, but it is certainly possible to build high quality code generators to implement the CBDD scaffolding

Limited Definition of Behavioral Contracts at Analysis Level
CBDD lets business analysts define simple behavioral contracts in a fashion that is non-technical. These can then be used to shore up acceptance criteria as well as enhance the final code.

Conclusions

Combining Test Driven/Behavior Driven Development and Programming by Contract to implement a user driven design can produce code of excellent quality and substantially fewer bugs. By following this approach, bugs can be caught during development and after deployment, and considerable analysis can be done using automated post mortem tools. Additionally, the combined approach can be used with tools to reduce coding drudgery and can integrate well with more powerful tools, particularly ones cognizant of code level contracts like Pex. The approach detailed here is already starting to informally and organically appear in the programming world, as the strength of the combined approach starts to become evident as an approach to building better software. When we combine their strengths and hopefully reduce their weaknesses the end result should be stronger as well.

The ideas presented in this paper, as well as the Kime project are only an initial step. The Kime project itself could be enhanced in the following ways:

- Hardened – Currently it is only in the prototype stage
- Better support for analysis of specifications – Analysis is currently performed by text matching aided by hints. Heuristics and other aids could be used to produce code with a stronger relation to the story
- Optimization of resulting specification code – As it stands today, Kime does not examine its own results for redundant object creation or opportunities to improve relating of code to the specification.
- Templates for output in different languages – It would be a straightforward exercise to extend Kime to produce Java or Ruby code for example.

The ideas behind CBDD would benefit from formal analysis of their effectiveness. As Kime is a prototype and CBDD has not been introduced to a large population, no real data is available on the effect of combining the two approaches or the likelihood of development teams embracing the approach. It is my opinion that developers would most likely gravitate to increased usage of code contracts supplemented by existing use of TDD, but again, this is just an opinion.

References

1. Janzen, D., Saidian, H.: A Leveled Examination of Test-Driven Development Acceptance. In: 29th International Conference on Software Engineering, pp. 1–4 (2007)
2. Janzen, D., Saiedian, H.: Does Test-Driven Development Really Improve Software Design Quality? IEEE Software, 77–84 (2008)
3. Kaufman, R., Janzen, D.: Implications of Test Driven Development: A pilot study. In: OOPSLA 2003 Companion of the 18th Annual ACM SIGPLAN Conference, pp. 298–299. ACM, New York (2003)
4. North, D.: (2006), http://dannorth.net/introducing-bdd/, http://dannorth.net (retrieved)
5. Fowler, M.: Refactoring: Improving the design of existing code. Addison Wesley, New York (1999)
6. Beck, K.: Test Driven Development: By Example. Addison Wesley, Boston (2002)
7. Jansen, D.: Test-Driven Development concepts, Taxonomy and Future Direction. Computer, 43-50 (2005)
8. Meyer, B.: Object Oriented Software Construction, 2nd edn. Prentice Hall, Boston (2000)
9. Gries, D.: The Science of Programming. Springer (1981)
10. Lietner, A., Ciupa, I., Oriol, M., Meyer, B., Fivo, A.: Contract driven development = test driven development - writing test cases. In: ESEC-FSE 2007, pp. 425–434. ACM, New York (2007)
11. Microsoft Research (n.d.). Code Contracts. Retrieved from Microsoft Research: http://research.microsoft.com/en-us/projects/contracts/

Personalized Semantic Search Using ODP:
A Study Case in Academic Domain

Trong Hai Duong[1], Mohammed Nazim Uddin[2], and Cuong Duc Nguyen[1]

[1] School of Computer Science and Engineering,
International University - VNU - HCM, Vietnam
ndcuong@hcmiu.edu.vn, haiduongtrong@gmail.com
[2] Samsung Research Institute Bangladesh,
Samsug Electronics, Bangladesh
tonazim@yahoo.com

Abstract. Personalized search utilizes the user context in a form of profile to increase the information retrieval accuracy with user's interests. Recently, semantic search has greatly attracted researchers' attention over the traditional keyword-based search because of having capabilities to figure out the meaning of search query, understanding users' information needs accurately using semantic web technology. In this paper, a ODP (open directory project)-based approach for personalized semantic information search is proposed. A reference ontology is generated by utilizing ODP to model user's interests and semantic search space. User profile is initially derived by matching between user's details and the reference ontology to model his/her interests and preference. Semantic search space is constructed by fuzzily classifying documents into the reference ontology. We also present an evaluation of our proposed method which indicates a considerable improvement of search accuracy with the field of application human judgment.

Keywords: Information retrieval, Ontology, Knowledge base, Semantic search, User profile.

1 Introduction

Though web search has become an essential part of our daily life, it still has considerable challenges to find the necessary information to fulfill the user satisfaction. Personalization is a popular technique applied to information retrieval incorporating with individual interests. Personalization is carried out by modeling each individual details in the form of user profile. The main challenges of personalization are how to gather user's details and model the individual details to improve the search effectiveness. Several approaches are conveyed by the number of researchers to model the user details adapt with the user query and search resources. Major drawback of these approaches are lack of semantics while model the user details more specifically user's interests and preferences. Semantic user model could improve search accuracy by understanding user intent and the concept meaning of terms as they appear in the search space to generate more relevant results. Semantic web is an extension of the current web with meaningful definition of information, provide better undebatable format to computer and human to

B. Murgante et al. (Eds.): ICCSA 2013, Part V, LNCS 7975, pp. 607–619, 2013.

incorporate with it. Ontologies are the backbone of Semantic Web which provides a common vocabularies and formal specification of a particular domain. Ontological approaches have been applied to many retrieval systems to model user background knowledge and query formulation to improve the precision of information extraction in web search application.

In this paper, we proposed a framework for personalized information search with semantic web techniques. We consider the following issues to provide semantic information search. First, how to collect user information and model it with ontological approach to construct a user's initial profile? How to construct semantic search space organizing documents. How to formulate query terms based on user profile to match with semantic search space? To solve aforementioned issues, an ODP (open directory project)-based approach for personalized semantic information search is proposed. A reference ontology is generated by utilizing ODP to model user's interests and semantic search space. To consider the first issue, we proposed a new and efficient techniques to discover the user interest by finding the related information about user using query including the user name and email address. The related information is modeled by the reference ontology called user profile. Most of the user query consists of one or more keywords. However, these simple keywords do not provide any semantic (meaningful content) about information search causing returning too many irrelevant results with lot of ambiguity. Therefore, we try to extend the query keyword in more meaningful manner based on the user profile. In addition, we construct a semantic search space by modeling the scientific research in the field of computer science by analyzing real world data of CiteSeer digital library. Semantic search space is a knowledge base constructed with ontological approach to describe academic information named Academic knowledge Base(AKB). Semantic search space is constructed by fuzzily classifying documents into the reference ontology. We also present an evaluation of our proposed method which indicates a considerable improvement of search accuracy with human judgment.

The rest of the paper is organized as follows. We discuss the related works in Section 2. Section 3 describes methodology including the semantic user modeling and building semantic search space using a reference ontology derived from ODP. Section 4 describes the experimental evaluations and we conclude the paper in Section 5.

2 Related Work

Personalization is a successful technique in the domain of news recommendation [3] to web search [4]. [3] described the algorithm and techniques to recommend Google news based on users search queries. Personalization describes the method of modeling user's details in the form of user profile. A user profile and category hierarchy is created and mapped for improving retrieval effectiveness in personalized web search [4] based on user's search histories. Creation of user profile is collecting the necessary information about individual users. Some of the most popular techniques are outlined in Susan [5] for building user profiles. User profiles can be create as an ontology describing the details information in a hierarchical manner. Possible techniques for building automatic ontology with the help of existing knowledge base discussed in [9]. Ontological user profile [1], [10],[12] can also be constructed by the reference of pre-exiting concept hierarchies of related domain. Major drawback with these approaches is collecting user information to construct the initial user profile. User's information are collected from the user's past browsing history such as analysis of log file which contain lot of irrelevant information to reflect the user interest. Similar approach also presented in [2] to build the initial user profile and learning but log file contains large amount noisy data which increases the time complexity to pre-process and clean to identify user interest. A different and effective approach to collect the user details described in [1] and applied to construct the user initial profile. Similar approach also carried out in [11] for mining researcher's profile from the web. Our approach is similar with [11] in the sense that use world wide web to discover the user interest but we used different techniques to infer the user preferences and interests. In [11] only researcher's name is used to search the information which returned many ambiguous information and apply different modeling techniques to solve it and FOAF scheme is applied to construct the user profile.

Nowdays, semantic web personalization [6],[7], [8] drawn a lot of attention for web search and recommendation system. Ontologies are proven techniques to model users and content in conceptual way. Ontology is a popular way to handle cold start problem in recommedation system and information overload in web repositories. [6] incorporate users rating and item ontology to generate meaningful recommendation in movie domain. A knowledge base is created to describe the item in the form of ontology. In [7], Ontology based contextualize and personalization have been combined to improve the performance of personalized information retrieval. The idea of semantic search aimed at improving traditional web search by augmenting the search results with the concepts in the ontology is presented in [8]. The most of the existing semantic search based on personalization is to filter and rank the traditional search results return by query keyword with the help of semantic user profile [4],[5],[12],[7] where, most of facilities provided by the semantic web techniques not utilized when searching the information. In our framework, we model user details and user query in structural way to utilize the maximum benefit of sematic web techniques and matched it with semantic search space to search the information.

3   Methodology

3.1   Ontological Reference with ODP

ODP[1] is an open content directory of web pages maintained by a community of volunteer editors. The web pages are classified into a hierarchical topics (directories). *Fig.* 1 shows a part of ODP hierarchy of the *Artificial Intelligence* concept. In this approach, a reference ontology is constructed by reusing the ODP hierarchy of topics associated with documents. The reference ontology is useful to initial derive the user modeling by aligning concepts generated from user's selected documents (e.g. publications and personal pages) to the ontological reference. This reference ontology is also used to index documents in a semantic manner (semantic search space) by classifying documents into the reference ontology.

---

[1] http://www.dmoz.org/

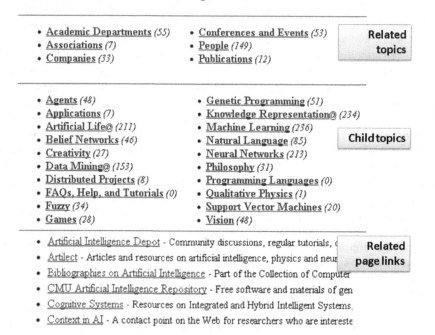

Top: **Computers**: **Artificial Intelligence** *(1,211)*

Related topics

- Academic Departments *(55)*
- Associations *(7)*
- Companies *(33)*
- Conferences and Events *(53)*
- People *(149)*
- Publications *(12)*

Child topics

- Agents *(48)*
- Applications *(7)*
- Artificial Life@ *(211)*
- Belief Networks *(46)*
- Creativity *(27)*
- Data Mining@ *(153)*
- Distributed Projects *(8)*
- FAQs, Help, and Tutorials *(0)*
- Fuzzy *(34)*
- Games *(28)*
- Genetic Programming *(51)*
- Knowledge Representation@ *(234)*
- Machine Learning *(236)*
- Natural Language *(85)*
- Neural Networks *(213)*
- Philosophy *(31)*
- Programming Languages *(0)*
- Qualitative Physics *(7)*
- Support Vector Machines *(20)*
- Vision *(48)*

Related page links

- Artificial Intelligence Depot - Community discussions, regular tutorials,
- Artilect - Articles and resources on artificial intelligence, physics and neu
- Bibliographies on Artificial Intelligence - Part of the Collection of Computer
- CMU Artificial Intelligence Repository - Free software and materials of gen
- Cognitive Systems - Resources on Integrated and Hybrid Intelligent Systems,
- Context in AI - A contact point on the Web for researchers who are intereste

Fig. 1. ODP hierarchy of the topic *Artificial Intelligence*

The implementation of the reference ontology accomplishes using ODP in the following way. The web pages are investigated as documents to represent the reference ontology. Each concept in the reference ontology is associated with related documents. The keywords are extracted from the related documents of a concept, which consider as index terms to create a feature vector representing the concept. The feature vector is made by considering the importance of the terms in the documents.

Document Feature Vector Extraction. To extract the feature vector of a document, the document is segmented into sequences of tokens by a tokeniser which recognizes punctuation, cases, blank characters, digits, etc. The strings underlying tokens then are morphologically analyzed in order to reduce them to normalised basic forms, e.g. tokens $\{has, Hands, Free, Kits\}$ becomes root form $\{have, hand, free, kit\}$. Stop words, which are irrelevant to the content of the document, such as a pronoun, conjunction, preposition, article, and auxiliary verb, are removed. Removing stop words ensures more efficient processing by downsizing and the remaining words are relevant to the content.

The tf–idf weight (term frequency–inverse document frequency) is a weight often used in information retrieval and text mining. This weight is a statistical measure used to evaluate how important a word is to a document in a collection or corpus. Here, we use the traditional vector space model(tf-idf) to define the feature vector of documents. Let $T^d = (t_1, t_2, \ldots, t_n)$ be the collection of all of key-words (or terms) of the document d.

Term frequency $tf(d,t)$ is defined as the number of occurrences of term t in document d. A set of term frequency pairs, $P^d = \{(t,f)|t \in T^d, f > threshold\}$, is called the pattern of the document d. Let $P^i = \{(t_1, f_1), (t_2, f_2), \ldots, (t_m, f_m)\}$ be the pattern of a document i belonging to a set of document ds, $i = 1..n$. A set of term frequency pairs, $P^c = \sum_{i=1..n} P^i$, is called the pattern of the set of documents ds. Let $\overrightarrow{S}$ be the feature vector of the ds and let T be the collection of corresponding terms to the pattern; we then have:

$$\overrightarrow{S^{ds}} = (w_1, w_2, \ldots, w_k) \tag{1}$$

$$T^{ds} = (t_1, t_2, \ldots, t_k) \tag{2}$$

where

$$w_i = \frac{f_i}{\sum_{j=1}^m f_j} * \log \frac{|D|}{|d : t_i \in d|} \tag{3}$$

- $|D|$: the total number of documents in the corpus
- $|d : t_i \in d|$: the number of documents where the term t_i appears (that is). If the term is not in the corpus, this will lead to a division-by-zero. It is therefore common to use $1 + |d : t_i \in d|$.

Ontological Reference Building. The hierarchy of concepts are extracted from ODP base. The process of the reference ontology generation is depicted as $Fig. 2$. There are two following main steps to construct the feature vector for the concept. First step, using the vector space model tf/idf to construct the generalization of the documents. Second step, for each leaf concept, the feature vector is calculated as the feature vector of set of documents associated with the concept. For each none-leaf concept, the feature vector is calculated by taking into consideration contributions from the documents directly associated with the concept and its direct sub-concepts. The detail of these two steps is presented as follows:

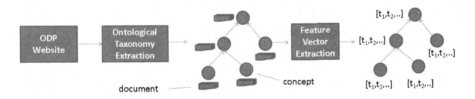

Fig. 2. Reference ontology generation

There is leaf concept and non-leaf concept in the reference ontology. None-leaf concepts are those that have sub-concepts.

– For each leaf concept, the feature vector is calculated as the feature vector of a set of associated documents:

$$\overrightarrow{S^c} = \overrightarrow{S^{ds}} \tag{4}$$

– For each none-leaf concept, the feature vector is calculated by taking into consideration contributions from the documents directly associated with it (D^c), its direct sub concepts ($D^{c'}$, for any c' is a direct sub concept of c):

$$\overrightarrow{S^c} = \alpha\overrightarrow{S^{ds}} + \beta\overrightarrow{S^{ds'}}$$

(5)

where $\overrightarrow{S_c}$ and $\overrightarrow{S_{c'}}$ correspond to the feature vectors of the sets of documents D^c and $D^{c'}$, respectively, $0 \leq \alpha, \beta \leq 1$ and $\alpha + \beta = 1$.

3.2   User Semantic Modeling

User modeling is an important component in personalized search to decrease the search ambiguity and present search results that are more likely to be interesting to a particular user. User's preferences and interest are represented in an ontological model which formed a user profile for an individual user. This approach has been successfully applied and tested by researchers to reduce the cold start problem in recommendation system and personalize search results.

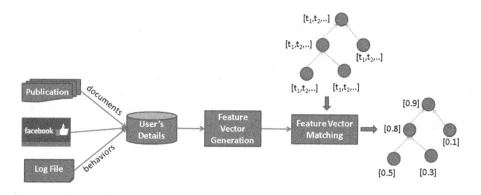

Fig. 3. User Profile Generation Process

User profiles may include general information about users such as name, age, educational background, etc, and specific information such as the interests and preferences. The information may be collected explicitly, through direct user intervention, or implicitly, observing the user's behaviors. The collected details of users can be modeled with ontological model to semantically represent their preferences. A new ontology can be developed or an existing hierarchical taxonomy can be reused as a reference ontology, to represent user preferences. Here, an initial ontological profile is generated by aligning feature vectors deriving from user's information to the reference ontology. The user's information can be derived from user's publications, home pages, and behaviors analyzed through social web and log file [1].

3.3 Semantic Search Space Using Reference Ontology

The reference ontology can be used to model semantic search space covering domains in ODP (see *Fig.* 4). However, to simply demonstrate the proposed approach, we only focus on academic domain as a study case for building semantic space.

Fig. 4. Domain covering by ODP

The information is modeled using ontology with concepts and relationships among them. The ontology for academic knowledge base (AKB) is depicted in *Fig.* 5. Ovals with solid line represents the top classes, dotted ovals represent sub-classes, arcs stand for describing relationship between classes and arrows for subsumed relations. Sub-classes are related to the parent classes with different degree relevancy (weights) which indicate importance with several aspects respectively. For example, a *Journal* is usually more important than a *Proceeding*, thus the concept *Journal* belongs the concept *Publication* with higher degree than the concept *Proceeding*'s. Similarly, attributes belong to classes with different weights, e.g., the attribute *first_Author* belongs to *Publication* with higher degree than the attribute *co_Author*. Detail descriptions about the AKB are described in [18].

Here, we consider a partial ontology of AKB with the root concept *Field*. Actually, the partial ontology is the reference ontology. Each concept in the reference ontology is considered as a topic belonging to the *Field*. This semantic space is constructed by organizing academic documents (e.g. articles and proceedings) in the reference ontology (see *Fig.* 6). Each academic document is represented by a feature vector. The similarity between a document's feature vector and a reference concept's feature vector is

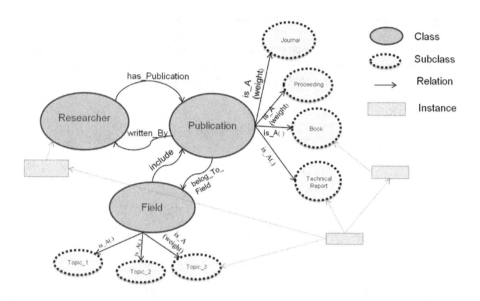

Fig. 5. Academic Semantic Space

considered as a relevant weight of the *belong_To_Field*relation from the document to the reference concept (a topic in the *Field*). The similarity is calculated as follows:

$$similarity = cos(\overrightarrow{d}, \overrightarrow{c}) = \frac{d * c}{\|d\|\|c\|} = \frac{\sum_{(T_i^c, T_i^{c'}) \in K}(S_i^c * S_i^{c'})}{\sqrt{\sum_{i=1}^n (S_i^c)'} * \sqrt{\sum_{i=1}^n (S_i^{c'})'}} \qquad (6)$$

where $K = \{(T_i^c, T_i^{c'}) | sim(T_i^c, T_i^{c'}) = 1\}$

3.4 Personalized Semantic Search

Query processing is a method for improving the initial query formulation using the information that is related to the query intent. The process accomplishes by explicit feedback, where the users explicitly provide information on relevant documents to a query for reformulation, and implicit feedback, in which the information for query expansion is implicitly derived by the system. A traditional query consists of one or more keywords for searching the information on the Web. These keywords does not provide any semantic for searching the information in the traditional search engines like Google or Yahoo!. Hidden concepts behind these keywords can be explored implicitly with the help information from external sources such as thesaurus or using some background knowledge to identify the user interest. Here, a query is submitted to the system by query interface, is expended using ontological profile for generating semantic context (see *Fig.* 7). The context is passed to semantic space (AKB). In the semantic search

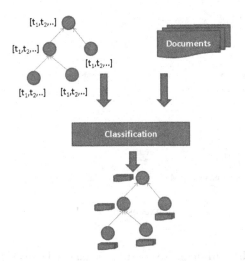

Fig. 6. Semantic Document Annotation

space, the *Field* concept contains all the publications with respective sub-classes in different topics. Context is matched with the topics of the *Field* for retrieval of related publications and search results are ranked based on scores by ranking algorithm [18]. Retrieval of publications for a given query is analogous with searching information on the Web.

4   Experimental Evaluations

The main goal of this experiment is to measure the retrieval accuracy of academic information in computer science using the proposed approach.

4.1   Dataset

Reference Ontology Building. In order to construct ontological user profiles and academic knowledge base using reference ontology for personalized search services, the ODP concept hierarchy has been investigated. The RDF representation of ODP is downloaded for the website (http://www.dmoz.org/). Only top "Computer" concept of ODP is considered as a root concept for the experimental purpose and used a branching factor of ten with a depth of six levels of root concept. The model proposed in this research is targeted to a specific domain of scientific research in computer science field that the top computer concept of ODP has been chosen for the experiment. However, ODP concepts include many sub concepts and documents which are not related to this research domain. Only suitable concepts, sub concepts and documents are included after filtering the RDF version of ODP. Finally, experimental data sets contained 650 concepts in the hierarchy and 15,326 documents that were indexed under various concepts. The indexed documents were pre-processed as described in Section 3 and built reference ontology accordingly.

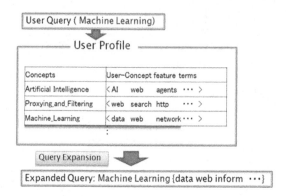

Fig. 7. Query Expansion Using User Profile

Academic Knowledge Building. To construct academic knowledge base, a real world data has been collected from a scientific literature digital library CiteSeerX that focuses primarily on the literature in computer and information science. Metadata are downloaded according to information provided in http://citeseerx.ist.psu.edu/oai2 using java program. The metadata in XML format provide information about scientific publication e.g., abstract, citation relationships, author-coauthor(s) names, author affiliation etc. The metadata includes about 800,000 publications related to computer and information science of total size approximately 2 GB. For the experiment it is need to collect those publications that provide the abstract information to represent the respective publication. Moreover, the dataset has cleansed the dataset by discarding publications that do not provide the abstract. Titles, keywords and abstracts of publications are utilized to make feature vectors that supply useful information to represent the publication.

Ontological Profile Building. In this research, a hybrid approach has investigated for collecting user information. Users whose profile is to be constructed need to register the system with minimum necessary information such as name and email address. Additionally, user can supply more information if agree or not feel burden to explore his/her interest. Details information about user can be searched from the Web using their name concatenate with email address. For example, if a user register the system with name, "Trong Hai Duong" and email address, haiduongtrong@gmail.com then his details can be searched from the Web by submitting the query as "Trong Hai Duong + haiduongtrong@gmail.com" to collect the details information for initially building the profile.

4.2 Evaluation Method

Two widely used statistical methods in information retrieval; Recall and Precision are used to evaluate the accuracy of user profile. The Precision measures the probability that the system mapped the user details to the reference ontology to build profile will be relevant to users whereas recall measures the probability that the classifier will select entire set of relevant documents. Precision and Recall can be defined as:

$$Precision = \frac{the\ number\ of\ relevant\ documents\ mapped}{the\ number\ of\ documents\ mapped} \qquad (7)$$

$$Recall = \frac{the\ number\ of\ documents\ mapped}{the\ total\ number\ of\ relevant\ documents} \qquad (8)$$

Finally, the F-measure is calculated by combining precision and recall as follows.

$$F = 2 * \frac{Precision * Recall}{Precision + Recall} \qquad (9)$$

4.3 Experimental Results

The goal of the experiment of user profile accuracy is to demonstrate that constructed ontological user profile represents user interests and preferences accurately. To construct the ontological user profile, fifty users' details are collected from Google scholar and social network site Facebook by querying their name and e-mail address. A super document is created by combing the collections from these two sites for each user. After pre-processing the documents each user details are mapped with reference ontology to construct the ontological user profile which represents the interests and preferences of a particular user. Mapping between reference ontology and user's details are accomplished by measuring similarities among corresponding feature vectors of respective concepts and document. Ontological user profile contains the weighted concept hierarchy to represents the user interests. Concepts' weights are calculated by summation of its feature vectors along with similarity scores measured by mapping process. Five graduate students were assigned for manual judgments of profiles were relevant or not to the users with necessary information of whose profile have been constructed. According to their opinion precision and recalled were measured. Comparison of precision and recall are presented in $Fig.$ 8.

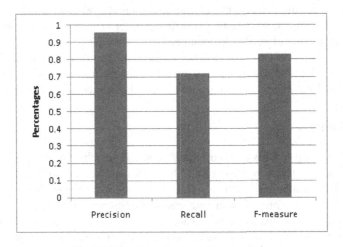

Fig. 8. Evaluation of user profile generation

To evaluate the search results we have investigated the method of pooled relevance judgments with the human judgment. Initially, for a given query top 50 results are collected from Google. Then same query is extended based on semantic profile constructed for 10 users and send to Google for collecting results.

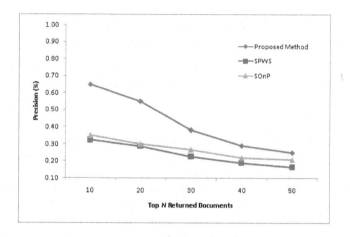

Fig. 9. Search publications with profile

All the results were given to some researchers including research faculty members, post doctoral researchers, doctoral and master students to assess the search results. To help the researchers in evaluation process we have provided the necessary information about users whose profile was constructed. According to the assessment personalized search results returned based on semantic profile out performed over non personalized results.

For evaluation of retrieval accuracy Precision of top n documents are measured. Top fifty documents are divided into five categories using interval of ten and Precisions are measured for top 10, 20, 30, 40 and 50 results. In order to compare the performance of searching and ranking of academic document with other methods the related approach such as SPWS [20] and SOnP [19] have been analyzed. The other approaches SPWS and SOnP are also related to search and ranking information using ontological user profile. $Fig.$ 9 depicts the comparison results with Precision percentage of top n search results.

5   Conclusions

The research findings of this paper offered a semantic search framework for searching and ranking academic information using Ontology and validated by series of experiments. The framework modeled the academic information using Ontology by considering various relations to provide semantic search. Additionally, presented a Ontological profile building method for personalized academic information search such as publications. Empirical results show that the framework outperformed in searching and ranking personalized academic information. The presented framework notably improved the searching accuracy and efficiency compare with other existing methods such as SPWS and SOnP. Comparison results signified the feasibility of model carried out in this paper in terms of personalized approach.

References

1. Duong, T.H., Uddin, M.N., Li, D., Jo, G.S.: A Collaborative Ontology-Based User Profiles System. In: Nguyen, N.T., Kowalczyk, R., Chen, S.-M. (eds.) ICCCI 2009. LNCS, vol. 5796, pp. 540–552. Springer, Heidelberg (2009)
2. Mohammed, N.U., Duong, T.H., Jo, G.S.: Contextual Information Search based on Ontological User Profile. In: Pan, J.-S., Chen, S.-M., Nguyen, N.T. (eds.) ICCCI 2010, Part II. LNCS, vol. 6422, pp. 490–500. Springer, Heidelberg (2010)
3. Abhinandan, D., Mayur, D., Ashutosh, G.: Google News Personalization: Schable Online Collaborative Filtering. In: WWW, pp. 8–12 (2007)
4. Liu, F., Yu, C., Meng, W.: Personalized Web Search For Improving Retrieval Effectiveness. IEEE Computer Society (2004)
5. Gauch, S., Speretta, M., Chandramouli, A., Micarelli, A.: User Profiles for Personalized Information Access. In: Brusilovsky, P., Kobsa, A., Nejdl, W. (eds.) Adaptive Web 2007. LNCS, vol. 4321, pp. 54–89. Springer, Heidelberg (2007)
6. Sarabjot, S.A., Patricia, K., Mary, S.: Generating Semantically Enriched User Profiles for web Personalization. ACMTransaction on Internet Technology 7(4), Article 22 (2007)
7. Myloans, P.H., Vallet, D., Castells, P., Fernandez, M., Avrithis, Y.: Personalized information retrieval based on context and ontological knowledge. The Knowledge Engineering Review 23, 73–100 (2008)
8. Guha, R.: Rob McCool, Eric Miller, semantic search. In: WWW, pp. 20–24 (May 2003)
9. Hyunjang, K., Myunggowon, H., Pankoo, k.: Design of Automatic Ontology Building System about the Specific Domain Knowledge. In: ICACT, pp. 20–22 (2006)
10. Trajkova, J., Gauch, S.: Improving Ontology-Based User Profiles, RIAO (2004)
11. Tang, J., Yao, L., Zhang, D., Zhang, J.: A combination Approach to Web User Profiling. ACM Transaction on Knowledge Discovery from data V(N), 1–38 (2010)
12. Sieg, A., Mobasher, B., Burke, R.: Learning Ontology-Based User Profiles: A Semantic. Approach to Personalized Web Search, IEEE Intelligent Informatics Bulletin 8(1) (2007)
13. Deng, H., King, I., Michael, R.L.: Formal Models for Expert Finding on DBLP Bibliography Data. In: Eight IEEE Internation Conference on Data mining, pp. 155–4786 (2008)
14. Bao, S., Duan, H., Zhou, Q., Xiong, M., Cao, Y., Yu, Y.: A Probabilistic Model for Fine-Grained Expert Search. In: HLT, pp. 914–922. ACL (2008)
15. Bogers, T., Kox, K., Bosch, A.: Using Citation Analysis for Finding Experts inWorkgroups. In: DIR, Maastricht, the Netherlands, pp. 14–15 (April 2008)
16. Tang, J., Zhang, J., Yao, L., Li, J., Zhang, L., Zhong, S.M.: Extraction and Mining of Academic Social Networks. In: KDD, August,Las Vegas,Nevada,USA (2008)
17. Newman, M.E.J.: Coauthorship networks and patterns of scientific collaboration. PNAS 101, 5200–5205 (2004)
18. Mohammed, N.U., Duong, T.H., Jo, G.S.: Semantic Search based on Ontological User Profile. International Journal of Software Engineering and Knowledge Engineering (to be appeared, 2013)
19. Mohammed, N.U., Duong, T.H., Jo, G.S.: Contextual Information Search based on Ontological User Profile. In: Pan, J.-S., Chen, S.-M., Nguyen, N.T. (eds.) ICCCI 2010, Part II. LNCS, vol. 6422, pp. 490–500. Springer, Heidelberg (2010)
20. Sieg, A., Mobasher, B., Burke, R.: Learning Ontology-Based User Profiles: A Semantic Approach to Personalized Web Search. IEEE Intelligent Informatics Bulletin 8(1) (2007)
21. Teresa, T., Harrison, M., Stephen, D.: The Electronic Journalas the Heart of an Online Scholarly Community. Library Trends 43(4) (Spring 1995)

An Effective Keyword Search Method for Graph-Structured Data Using Extended Answer Structure

Chang-Sup Park

Dongduk Women's University, 23-1 Wolgok-dong, Seongbuk-gu, Seoul, Korea
cspark@dongduk.ac.kr

Abstract. This paper proposes an effective approach to ranked keyword search over graph-structured data which is getting much attraction in various applications. To provide more effective search results than the previous approaches, we suggest an extended answer structure which has no constraint on the number of keyword nodes and is based on a new relevance measure. For efficient keyword search, we also use an inverted list index which pre-computes connectivity and relevance information on the nodes in the graph. We present a query processing algorithm based on the pre-constructed inverted lists, which aggregates entries relevant to each node and finds top-k answer trees relevant to the given query. We also enhance the basic search method by storing additional information on the relevance of the related entries in the lists, in order to estimate the relevance score of each node more closely and to find top-k answers more efficiently. We show by experiments that the proposed keyword search method can provide effective top-k search results over large amount of graph-structured data with good execution performance.

Keywords: Graph-structured Data, Keyword Search, Top-k Query Processing.

1 Introduction

Recently, graph-structured data are widely used in various applications such as XML, semantic web, ontologies, social network services, and bio-informatics. Keyword-based search over graph-structured databases has been attracting much attention since it allows users to represent their information need using only a set of keyword terms without understanding and using a query language and underlying database schema [1-7]. Keyword-based query processing has also been studied extensively in the literature of relational databases. Many approaches materialize relational data as a directed graph where tuples are treated as nodes and foreign-key relationships among tuples are represented as edges [8-18].

The previous keyword search methods for graph-structured data usually return a set of connected structures, either sub-trees or sub-graphs, from the database, which represent how the data containing query keywords are interconnected in the database. Given a query, since there can be a significant number of answer structures in a large volume of graph data, search methods usually adopt a scoring function to evaluate and rank the answer structures and return top-k ones most relevant to the query.

B. Murgante et al. (Eds.): ICCSA 2013, Part V, LNCS 7975, pp. 620–635, 2013.
© Springer-Verlag Berlin Heidelberg 2013

To satisfy users' information need by finding more effective and relevant answers to a given query than the previous approaches, we suggest an extended answer structure which has no constraint on the number of keyword nodes chosen for each query keyword and is based on a new relevance measure for nodes in the graph. Then we propose an inverted list index to represent connectivity and relevance information on the nodes, as well as a query processing algorithm exploiting the pre-constructed index to find top-k answer trees. Aiming at improving the efficiency of the proposed method, we also present an enhanced inverted list which stores additional information on the relevance of related entries and an improved search algorithm which estimates the relevance score of each node more closely and can find top-k answers more efficiently.

The rest of the paper is organized as follows. Section 2 presents related work and motivation of our study. Section 3 defines a new answer structure for keyword queries and a relevance measure for it. In Section 4, we propose an inverted list index and describe a top-k query processing algorithm using the index. In Section 5, we present an extension of the inverted list and an enhanced search algorithm to process keyword queries more efficiently. We provide experimental results on the effectiveness and efficiency of the proposed methods in Section 6 and draw a conclusion in Section 7.

2   Related Work and Motivation

In the previous approaches to keyword-based search on a graph-structured database, tree structure is popularly used to describe an answer to a given query [3, 4, 5, 8, 9, 13, 14]. As a sub-tree of the database graph, an answer tree should have nodes directly containing the keywords in the query and its leaves should come from those keyword nodes. To rank the sub-trees satisfying the above conditions, weight functions were proposed in the literature based on two different semantics [19]. The Steiner tree-based semantics defines the weight of an answer tree as the total weight of the edges in the tree. Under this semantics, finding an answer tree with the smallest weight is the well-known optimal group Steiner tree problem which is NP-complete [20]. The previous approaches based on this semantics have limitations on the search result and performance against the large amount of graph data [4, 8, 9, 14].

As an alternative to the Steiner tree semantics, some approaches adopted easier semantics, namely distinct root semantics, to find answer trees rooted at distinct nodes [3, 5, 13]. For each node in the graph, only a single sub-tree is considered a possible answer to the query, which is rooted at the node and has the minimal weight. The weight of a sub-tree is defined as the sum of the shortest distances from the root to the keyword nodes chosen for each query keyword. Under this semantics, given a graph having n nodes, there can be at most n answer trees and thus we can deal with very large graph databases more efficiently than using the Steiner tree semantics. A bidirectional search algorithm proposed in BANKS-II [13] performs backward explorations of the graph starting from nodes containing query keywords, as well as forward explorations from the potential roots of answer sub-trees toward keyword nodes. It uses a heuristic activation strategy to prioritize nodes to expand during the bidirectional search. However, it does not take advantage of connectivity information in the graph hence it may lead to poor performance on certain graphs. In BLINKS approach [5], indexing schemes and query processing algorithms were proposed to speed up the bidirectional exploration of the graph with a good performance

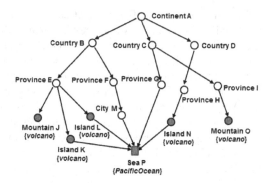

Fig. 1. An example of keyword search on graph data

guarantee. A single-level index, consisting of sorted keyword-node lists and a node-keyword map, pre-computes and indexes all the shortest paths and distances from nodes to keywords in the graph. By exploiting the index, the query processing algorithm performs graph search efficiently and finds top-k answers in a time and space efficient manner. To reduce the index space for a large graph, they also proposed graph partitioning strategies and a bi-level indexing scheme.

The previous approaches mentioned above have a common constraint in the answer structure employed: for each and every keyword in the query, one and only one node directly containing it should be included as a keyword node. For example, assume that a query $Q = \{volcano, PacificOcean\}$ is given on the graph-structured data shown in Fig. 1. The set of nodes directly containing the keywords in Q is $\{J, K, L, N, O, P\}$. Denoting an answer sub-tree by <root node, a set of selected keyword nodes>, possible answer sub-trees containing a keyword node for each keyword are <$E, \{J, P\}$>, <$E, \{K, P\}$>, <$E, \{L, P\}$>, and <$H, \{N, P\}$>. Note that in the previous approaches based on the distinct root semantics, at most one sub-tree is selected among those having the same root and returned as an answer to the query. Assuming that keyword nodes J, K, L, N, and O have the same relevance to keyword $volcano$ and all the edges in the graph have the same weight, the answer trees rooted at E, <$E, \{J, P\}$>, <$E, \{K, P\}$>, and <$E, \{L, P\}$>, will have the same weight and rank as the answer tree rooted at H, <$H, \{N, P\}$>, under the distinct root semantics. Moreover, the answer trees rooted at B, <$B, \{J, P\}$>, <$B, \{K, P\}$>, and <$B, \{L, P\}$>, have larger weights (and thus lower ranks) than the answer tree rooted at C, <$C, \{O, P\}$> since the shortest distance from B to P is longer than the shortest distance from C to P. However, we observe that nodes E and B have more paths to the nodes containing keyword $volcano$ than nodes H and C, respectively, hence they should be considered more relevant to the given query.

To improve the quality of query results, we propose a new answer structure which has no constraint on the number of keyword nodes chosen for each query keyword. In an answer tree, multiple keyword nodes can be chosen for a certain keyword while no keyword node can be included for some keywords. Specifically, given a pre-defined constant p and a node n in the graph, we find top-p pairs of a query keyword k and a node s containing k, to which n is most relevant. The relevance of n to a (k, s) pair is computed by the relevance of s to k and the shortest distance from n to s. Then the answer tree rooted at n is a sub-tree constructed by merging the shortest paths from n to each keyword node s contained in the p pairs of a keyword and node chosen for n. For example, assuming that $p = 4$, possible answers to the query {$volcano$, $PacificOcean$} over the graph in Fig. 1 include the sub-trees that are rooted at E or B and have nodes J, K, L, and P together, i.e. $<E, \{J, K, L, P\}>$ and $<B, \{J, K, L, P\}>$. According to the proposed relevance measure, they have higher scores and ranks than other answer trees such as $<H, \{N, P\}>$ and $<C, \{O, P\}>$. Thus our answer structure can produce more effective top-k search results compared to the previous approaches.

3   Problem Definition

Let $G = (V, E)$ be a directed graph representing a graph-structured database and K be a set of keyword terms extracted from the nodes in V. We define relevance of a node in V to a keyword term in K based on the *tf-idf* weighting scheme [21] which is popularly used in information retrieval. We consider that even if a node n does not contain a keyword k, it can be relevant to k if it has a path to a node s directly containing k, called a keyword node for k. We first define the relevance of n with respect to a pair of keyword k and a keyword node s as follows.

Definition 1. (Relevance of a node n to a keyword k contained in a node s) Given a keyword $k \in K$ and a node $s \in V(G)$, let $tf(k, s)$ be the number of occurrences of k in s and $df(k)$ be the number of nodes in $V(G)$ which contain k. The relevance of s to k is defined by

$$rel(s, k) = \sqrt{tf(k, s)} \cdot \left(1 + \log\left(\frac{N}{df(k) + 1}\right)\right)$$

where N is the number of nodes in $V(G)$. For nodes n and s in $V(G)$, the relevance of n to s is defined by

$$rel(n, s) = \begin{cases} 1, & n = s \\ \frac{1}{|SP(n, s) + 1|}, & n \neq s \text{ and } \exists(\text{a path from } n \text{ to } s) \\ 0, & otherwise \end{cases}$$

where $|SP(n, s)|$ is the length of the shortest path from n to s. The relevance of n with respect to k in s is defined by

$$rel(n, s, k) = rel(n, s) \cdot rel(s, k). \qquad \square$$

According to the above definition, when s does not contain k or there is no path from n to s, $rel(n, s, k)$ becomes 0. Also note that if n and s represent the same node, $rel(n, s, k)$ equals to the relevance of n to the keyword k, i.e., $rel(n, k)$. A node n is considered *relevant* to a keyword k if and only if there exists a keyword node s such that $rel(n, s, k) > 0$.

Given a keyword query $Q = \{k_1, k_2, ..., k_l\}$ and a positive integer p, an answer structure and its relevance to the query are defined based on Definition 1 as follows.

Definition 2. (Answer tree to a query Q and its relevance to Q)) Given a keyword query Q, a node $n \in V(G)$, and a constant p greater than or equal to $|Q|$, let $Top_p(n, Q)$ be the set of p pairs of a node $s \in V(G)$ and a keyword $k \in Q$ such that the relevance of n with respect to (s, k) is in the p highest among all the pairs of a node in $V(G)$ and a keyword in Q. That is,

$Top_p(n, Q) = \{(s, k) | s \in V(G), k \in Q, rel(n, s, k)$ is in the p highest of $rel(n, s_i, k_j)'s$ for all $(s_i, k_j) \in V(G) \times Q\}$

where ties in relevances are broken at random. Let $V(n, Q)$ be the set of keyword nodes selected for $Top_p(n, Q)$, i.e., $V(n, Q) = \{v | (v, k) \in Top_p(n, Q)\}$. An answer tree to the query Q rooted at a node n, denoted by $T(n, Q)$, is a sub-tree of G which contains all the nodes in $V(n, Q)$ and consists of the shortest paths from n to each node in $V(n, Q)$.

The relevance of $T(n, Q)$ to the query Q, denoted by $rel(n, Q)$, is the sum of the relevances of n with respect to the (node, keyword) pairs in $Top_p(n, Q)$, i.e.,

$$rel(n, Q) = \sum_{(s,k) \in Top_p(n,Q)} rel(n, s, k)$$

□

Note that our approach is based on the distinct root semantics and thus there is at most one answer tree $T(n, Q)$ rooted at a node n. It has multiple keyword nodes for some keywords in Q to which the root n is most relevant in terms of $rel(n, s, k)$.

4   Basic Search Method

In this section, we propose a keyword search method including an indexing scheme and query processing algorithm to find k best answers to a given query based on the relevance measure defined in the previous section.

4.1   Inverted List Index

To enable efficient exploration of the graph-structured data, we use an inverted list-style index on the nodes which pre-compute and store information on the relevant nodes for each keyword term. Based on the proposed relevance measure, we find all the relevant nodes, as well as keyword nodes, for each keyword in the graph and then build an inverted list per keyword which is formally defined as follows.

Definition 3. (Inverted list $L(k)$ for a keyword k) For a keyword term k in the graph, let $S(k)$ be the set of nodes in $V(G)$ which contain k. The inverted list for k, denoted by $L(k)$, is a list of triples $(n, s, rel(n, s, k))^1$ obtained from all the pairs of nodes $n \in V(G)$

---

[1] Practically, the node IDs of n and s are stored in the entry.

and $s \in S(k)$ such that $rel(n, s, k) > 0$. The list entries are sorted in a non-increasing order of their relevance values. Formally,

$$L(k) = \langle (n_1, s_1, r_1), (n_2, s_2, r_2), ..., (n_m, s_m, r_m) \rangle, \text{where } n_i \in V(G), s_i \in S(k),$$
$$r_i = rel(n_i, s_i, k) \ (1 \leq i \leq m), \text{and } r_i \geq r_{i+1} > 0 \ (1 \leq i \leq m-1)$$

$\square$

We call a list entry (n, s, r) an entry of node n. As defined above, $L(k)$ stores entries of the nodes that are directly or indirectly relevant to k in a decreasing order of relevance values. Therefore, we can find the nodes most relevant to k by reading the entries in $L(k)$ sequentially. Note that the proposed inverted list is different from the conventional ones used for ranked search over documents or multi-dimensional data [21, 22, 23] by the fact that it can have entries of the nodes that do not contain the keyword of the list in them, in addition to the entries of keyword nodes for k. The proposed list is also distinguished from the keyword-node list suggested in BLINKS [5] since it can have multiple entries of the same node n, one for each keyword node reachable from n, while the latter has only a single entry for each node n which refers to a keyword node in the shortest distance from n.

4.2   Query Processing

Our query processing model is based on the threshold algorithm [22, 23], which is popularly used for top-k query processing on multi-dimensional data, such as similarity search on multimedia objects [24, 25, 26]. Given a query $Q = \{k_1, k_2, ..., k_l\}$, let $L(Q)$ be the set of inverted lists for the query keywords, i.e. $L(Q) = \{L(k_i) \mid k_i \in Q\}$. We perform sequential scans on the inverted lists in $L(Q)$ in parallel by reading their entries in a round-robin manner. During the scan, the query processor maintains the relevance value of an entry at the current scan position in each list $L(k_i)$, denoted by $curScore_i$. The largest one among those is called $maxCurScore$, i.e., $maxCurScore = max_{1 \leq i \leq l} curScore_i$. Note that since the entries in each list are stored in a non-increasing order of their relevance values, $maxCurScore$ can serve as an upper bound of the relevance values of the entries that have not yet been read from the lists in $L(Q)$.

While reading the lists, the query processor also maintains a priority queue per a candidate root node n of an answer tree, called a relevance queue q_n. It stores at most p entries of n retrieved from the lists which have the highest relevances. From a triple (n, s, r), only the pair of s and r is stored in the queue. The list of relevance values in q_n that are greater than or equal to $maxCurScore$ is denoted by R_n, i.e. $R_n = [r \mid (s, r) \in q_n, r \geq maxCurScore]$.

Since $maxCurScore$ is an upper bound of the relevances of the entries currently unseen from the lists in $L(Q)$, we ensure that the relevance values in R_n belong to the p highest ones of all the entries of n in the lists. Thus, the sum of the values in R_n can be a lower bound of $rel(n, Q)$, the relevance of an answer tree rooted at n. Based on the observation, we define the worst score of n as follows:

$$worstScore(n) = \sum_{r \in R_n} r. \tag{1}$$

In addition, assuming that the unknown values in the final top-p relevances of n are the same as $maxCurScore$, an upper bound of $rel(n, Q)$, called the best score of n, can be defined as follows:

$$bestScore(n) = worstScore(n) + maxCurScore \cdot (p - |R_n|). \quad (2)$$

Note that since $maxCurScore$ monotonically decreases as entries are retrieved from the lists, $worstScore(n)$ monotonically increases whereas $bestScore(n)$ monotonically decreases during the list scan. When the relevance queue of n has p entries and all the relevances in them are no less than $maxCurScore$ (i.e., $|R_n| = p$), $bestScore(n)$ and $worstScore(n)$ equal $rel(n, Q)$.

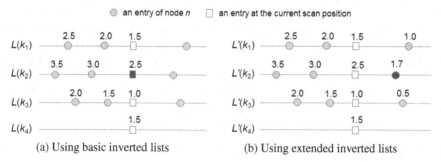

(a) Using basic inverted lists   (b) Using extended inverted lists

Fig. 2. An example of computing the worst and best scores of a node

Example 1. Fig. 2-(a) shows an example of computing the worst and best scores of a node n given a query $Q = \{k_1, k_2, k_3, k_4\}$ and $p = 6$. In the figure, lines represent inverted lists for the query keywords, scanned from left to right in a round-robin manner. In the lists, the entries of n are indicated by closed dots with their relevance values and the entries at the current scan positions are denoted by rectangles. $curScore_i$'s are 1.5, 2.5, 1.0, and 1.5, respectively, hence $maxCurScore = 2.5$. Currently, the relevance queue of n has 6 entries of n retrieved from the lists, i.e. $q_n = [(s_1, 3.5), (s_2, 3.0), (s_3, 2.5), (s_4, 2.0), (s_5, 2.0), (s_6, 1.5)]$, and the list of relevance values in q_n that are greater than or equal to $maxCurScore$ is $R_n = [3.5, 3.0, 2.5]$. Consequently, based on Eq. (1) and Eq. (2), we have $worstScore(n) = 3.5 + 3.0 + 2.5 = 9.0$ and $bestScore(n) = 9.0 + 2.5 \cdot 3 = 16.5$. □

As scanning the inverted lists, we find a set of nodes that can be roots of top-k answer trees using two priority queues.

- A top-k queue T stores at most k nodes having the highest worst scores among those that have been read from the lists. The nodes in T are sorted by their worst scores in a descending order. The minimum (i.e., rank-k) worst score value from the current top-k nodes is called $min\text{-}k$, i.e.,

$$min\text{-}k = \begin{cases} \min_{n \in T}\{worstScore(n)\}, & if \ |T| = k \\ 0, & otherwise \end{cases}$$

- A candidate queue C maintains candidate nodes which have a worst score smaller than $min\text{-}k$ but could still make it into the top-k queue T. A node whose best score is smaller than $min\text{-}k$ cannot belong to the final top-k nodes and thus is removed from C. The nodes in C are sorted in a descending order of their best scores to facilitate looking up a node with the maximum best score.

Whenever the worst score and best score of a node change during the list scan, we check if the node can be entered into the top-k queue T or it should be maintained in the candidate queue C. Query processing can terminate safely with the correct top-k nodes in T when the maximum best score in C as well as the best score of any node n_u currently unseen from the lists is no higher than min-k, i.e., when

$$|T| = k \text{ and } max\Big\{ \max_{m \in C}\{bestScore(m)\}, bestScore(n_u)\Big\} \leq min\text{-}k,$$
$$\text{where } bestScore(n_u) = maxCurScore \cdot p \qquad (3)$$

Then, using each node in T and the set of keyword nodes stored in its relevance queue, we can derive top-k answer trees from the data graph as defined in Definition 2.

Algorithm 1. Basic Search

```
1    For a given query Q = {k₁, k₂, ..., kₗ}, let L(Q)={L(kᵢ) |kᵢ∈ Q} and curScoreᵢ = 0 (1≤i≤l)
2    Initialize a top-k queue T and a candidate queue C empty.
3    repeat {
4        Select a list Lᵢ from L(Q) in a round-robin manner.
5        Read an entry e=(n, s, r) at the current scan position in Lᵢ.
6        curScoreᵢ := r and maxCurScore := max{curScoreᵢ} (1≤i≤l)
7        if (n had been evicted from C or top-p relevances of n had been found) continue
8        Insert (r, s) into the relevance queue of n, i.e., qₙ.
9        Compute worstScore(n) and bestScore(n) based on maxCurScore.
10       if (e is the first entry of n found in L(Q)) {
11           if (worstScore(n) > min-k)
12               Insert n into T (re-calculate min-k).
13           else if (bestScore(n) > min-k) Insert n into C.
14       }
15       else if (n is in T and worstScore(n) increases from the previous value) {
16           Remove and re-insert n (re-calculate min-k).
17       }
18       else if (n is in C) {
19           if (worstScore(n) > min-k)
20               Move n from C into T (re-calculate min-k).
21           else if (bestScore(n) ≤ min-k)
22               Remove n from C.
23           else if (bestScore(n) decreases from the previous value)
24               Remove and re-insert n.
25       }
26       if (a node m was ejected from T in Line 12 or 20 and bestScore(m) > min-k)
27           Insert m into C.
28       if ((C = ∅ or bestScore of the top node in C ≤ min-k) and maxCurScore · p ≤ min-k)
29           break
30       Update T and C periodically after every pre-defined number of entries is read.
31   } until (no entry remains in the lists in L(Q))
32   Build top-k answer trees using the nodes in T and the entries in their relevance queues.
33   return top-k answer trees.
```

Fig. 3. Query processing algorithm

Fig. 3 shows a sketch of the query processing algorithm described above. At each step of reading an entry of a node n from inverted lists in a round-robin manner, the following tasks are performed repeatedly. First, if either the node n had been evicted from the candidate queue C or $rel(n, Q)$ had been already determined, the current entry is ignored (in Line 7). In Line 8~9, the relevance queue q_n of n is updated using the current entry, and $worstScore(n)$ and $bestScore(n)$ are computed based on q_n and $maxCurScore$. If the current entry is the first entry of n found from the lists, n can be inserted into the top-k queue T or candidate queue C depending on its worst and best scores and the current min-k value in T (in Line 10~14). When n is already in T, T should be reorganized based on the new $worstScore(n)$ (in Line 15~17). Or, when n already exists in C, n is moved into T, remains in C, or is eliminated from C depending on its new worst and best scores and min-k value (in Line 18~25). As mentioned earlier, if Eq. (3) is satisfied by the result of the above tasks, query processing stops immediately and top-k answer trees can be derived from the graph using the nodes in T and the entries stored in their relevance queues (in Line 28~32).

In our method, the worst and best scores of the nodes stored in the top-k queue and candidate queue change as the list entries are read since they depend on $maxCurScore$. However, a naïve approach to re-calculating the worst and best scores of all the nodes in two queues and re-organizing the queues in every step of the list scan would incur very large overhead. Therefore, we perform periodic updates and cleaning of the queues after every pre-defined number of entries is read from the lists (in Line 30). We omit detailed algorithm of the queue updates due to the limit of space.

5   Enhanced Search Method

In the basic method described in Section 4, the worst and best scores of each node are estimated assuming that all the unknown relevance values in the entries unseen from the lists are equal to $maxCurScore$, i.e. the largest relevance value of the entries at the current scan positions. This strategy, however, is too conservative since the actual relevances of the entries unseen from a list $L(k_i)$ might be much smaller than the relevance of the entry at the current scan position in the list, i.e. $curScore_i$. We consider that when we read an entry of n from a list, if the relevance of the entry of n appearing *next* in the same list is available, we can predict $worstScore(n)$ and $bestScore(n)$ more closely to the correct relevance score of n, i.e. $rel(n, Q)$, by exploiting it instead of $curScore_i$. Based on the consideration, we propose an extended structure of inverted list which has in each entry of a node, additional information on the relevance of the next entry of the same node, formally defined as follows.

Definition 4. (Extended inverted list $L'(k)$ for a keyword k) For a keyword term k, let $S(k)$ be the set of nodes in $V(G)$ which contain k. For a node n in $V(G)$ and a keyword term k, let $L(n, k)$ be the ordered list of triples $(n, s, rel(n, s, k))$ which are obtained from all the nodes s in $S(k)$ such that $rel(n, s, k) > 0$ and are sorted in a non-increasing order of $rel(n, s, k)$. Formally,

$$L(n, k) = \langle (n, s_1, r_1), (n, s_2, r_2), \ldots, (n, s_m, r_m) \rangle,$$

where $s_i \in S(k), r_i = rel(n, s_i, k)$ $(1 \leq i \leq m)$, and $r_i \geq r_{i+1} > 0$ $(1 \leq i \leq m - 1)$.

Then we consider a list $L'(n, k)$ derived from $L(n, k)$ as follows:

$$L'(n, k) = \langle (n, s_1, r_1, r_1'), (n, s_2, r_2, r_2'), \ldots, (n, s_m, r_m, r_m') \rangle,$$
$$\text{where } r_i' = \begin{cases} r_{i+1}, & 1 \leq i \leq m - 1 \\ 0, & i = m \end{cases}$$

The extended inverted list for k, denoted by $L'(k)$, is a list of quadruples (n, s, r, r') which are merged from the lists $L'(n_i, k)$ for all nodes $n_i \in V(G)$ and sorted in a non-increasing order of the relevance value r. □

Now, we suggest an enhanced query processing algorithm based on the extended inverted lists, which can compute a narrower range of (*worstScore*, *bestScore*) for each node in the lists and thus can find the top-k nodes relevant to a given query earlier. The overall query processing strategy is similar to the basic algorithm described in Section 4.2. Assuming that $L'(Q) = \{L'(k_i) \mid k_i \in Q\}$ for a given query Q, we scan the lists in $L'(Q)$ in parallel by reading entries in a round-robin manner. Like the basic search algorithm, the query processor maintains a relevance value *curScore*$_i$ at the current scan position of each list $L'(k_i)$, as well as the top-k queue and candidate queue of the nodes with the highest worst and best scores. For each node n in the queues, the enhanced method maintains a relevance queue as well as a *next relevance value* of n in each list $L'(k_i)$, denoted by *nextScore*$_{n,i}$, which is obtained from r' in an entry (n, s, r, r') of n read from the list $L'(k_i)$ most recently. It provides the relevance of the entry of n which will be found next when the scan on the list $L'(k_i)$ continues.

When no entry of n has been retrieved from a list $L'(k_i)$ yet and *nextScore*$_{n,i}$ is unknown, the maximum relevance of the entries of n in the list is estimated by the relevance of the entry at the current scan position, i.e., *curScore*$_i$. Therefore, an upper bound of the relevances in the entries of n unseen from the lists in $L'(Q)$ can be obtained from *nextScore*$_{n,i}$ and *curScore*$_i$ for all $i \in [1..l]$ as follows:

$$maxNextScore_n = max \left\{ \begin{array}{l} \{nextScore_{n,i} \mid nextScore_{n,i} > 0, 1 \leq i \leq l\} \cup \\ \{curScore_i \mid nextScore_{n,i} = \infty, 1 \leq i \leq l\} \end{array} \right\} \quad (4)$$

Now, a lower bound and upper bound for the relevances of a node n with respect to Q can be computed based on *maxNextScore*$_n$ instead of *maxCurScore*. We first identify from the relevance queue of n a list R'_n of relevance values that are no less than the current *maxNextScore*$_n$, i.e. $R'_n = [r \mid (s,r) \in q_n, r \geq maxNextScore_n]$. Since *maxNextScore*$_n$ is an upper bound for the relevances of n unseen from the lists, *worstScore*(n) and *bestScore*(n) are defined as follows:

$$worstScore(n) = \sum_{r \in R'_n} r \quad (5)$$

$$bestScore(n) = worstScore(n) + maxNextScore_n \cdot (p - |R'_n|) \quad (6)$$

Since the entries in each list are sorted in a descending order of relevance, *curScore*$_i$ and *nextScore*$_{n,i}$ monotonically decrease during the list scan. Note that when an entry of n is found from $L'(k_i)$ for the first time, its next relevance value is no greater than the previous $curScore_i$. Thus, $maxNextScore_n$ in Eq. (4) monotonically decreases as we proceed with the list scan. Therefore, for each node n, $worstScore(n)$ monotonically increases while $bestScore(n)$ monotonically decreases during the scan.

Example 2. Fig. 2-(b) shows an example of computing the worst and best scores of a node n when evaluating a query using extended inverted lists. Assuming that the graph data are the same as Example 1, at the current scan positions denoted by rectangles, the relevance queue q_n and $curScore_i$'s have the same entries and values as those in Example 1. From the extended inverted lists, however, the relevances of the entries of n which will appear next after the current scan positions in the lists are available, i.e., $nextScore_n = [1.0, 1.7, 0.5, \infty]$. Note that $nextScore_{n,4} = \infty$ since no entry of n has been found from $L'(k_4)$ yet. Therefore, according to Eq. (4), $maxNextScore_n$ becomes 1.7, which is the largest value among $nextScore_{n,i}$ for $i \in [1..3]$ and $curScore_4$, the current relevance value 1.5 in $L'(k_4)$. Based on it, we have $R'_n = [3.5, 3.0, 2.5, 2.0, 2.0]$, and according to Eq. (5) and (6), $worstScore(n)$ and $bestScore(n)$ are 13.0 and 14.7, respectively. Note that this range of the relevance score of n is much narrower than the result [9.0, 16.5] obtained from the basic inverted list in Example 1.   ◻

Fig. 4 shows a sketch of the enhanced query processing algorithm we have described. The overall structure is the same as the basic search algorithm presented in Fig. 3. It should be noted that $nextScore_{n,i}$ is introduced for each node n to maintain the next relevance of n from list $L'(k_i)$ (in Line 9~10) and $maxNextScore_n$ is computed and used to estimate the worst and best scores of the current node n (in Line 11~12). Processing the current node n in the top-k queue or candidate queue, checking the termination condition, and updating the queues periodically, are the same as the basic algorithm. Note that periodic updates of the queues also exploit $maxNextScore_n$ for each node n in the queues instead of $maxCurScore$.

Algorithm 2. Enhanced Search
1   For a given query $Q=\{k_1, k_2, ..., k_l\}$, let $L'(Q)=\{L'(k_i) \mid k_i \in Q\}$ and $curScore_i = 0$ $(1 \leq i \leq l)$
2   Initialize a top-k queue T and a candidate queue C empty.
3   **repeat** {
4   Select a list L_i from $L'(Q)$ in a round-robin manner.
5   Read an entry $e=(n, s, r, r')$ at the current scan position in L_i.
6        $curScore_i := r$ and $maxCurScore := max\{curScore_i\}$ $(1 \leq i \leq l)$
7   **if** (n had been evicted from C or top-p relevances of n had been found) **continue**
8   Insert (r, s) into the relevance queue of n, i.e., q_n.
9   **if** (e is the first entry of n found in the lists in $L'(Q)$) $nextScore_{n,i} := \infty$ $(1 \leq i \leq l)$
10       $nextScore_{n,i} := r'$
11       $maxNextScore_n := max\{\{nextScore_{n,i} \mid nextScore_{n,i} > 0, 1 \leq i \leq l\} \cup \{curScore_i \mid nextScore_{n,i} = \infty, 1 \leq i \leq l\}\}$
12   Compute $worstScore(n)$ and $bestScore(n)$ based on $maxNextScore_n$.
13~33: *Refer to Line 10~30 of* **Algorithm 1** *in Fig. 3.*
34   } **until** (no entry remains in the lists in $L'(Q)$)
35   Build top-k answer trees using the nodes in T and the entries in their relevance queues.
36   **return** top-k answer trees.

Fig. 4. Enhanced query processing algorithm

6   Performance Evaluation

In this section we evaluate effectiveness and efficiency of the proposed keyword search methods by experiments using real datasets. We implemented the proposed methods, Basic Method (BM) and Enhanced Method (EM), in Java. For the performance comparison, we also experimented with BLINKS [5] which adopts distinct root semantics and uses a kind of inverted list index.

For experimentation, we used Mondial[2] and IMDB[3] databases which store geographic data and movie-related data. In IMDB databases, we selected data on 147K movies released in 2006~2010 and derived a directed graph which contains about 831K nodes, 2.8M edges, and 303K keyword terms. We used JGraphT[4] library to construct and manipulate the graph. We also used Lucene[5] library to extract keywords from the nodes in the graph and to compute relevances of nodes to keywords. The top-k queue and candidate queue in the proposed algorithms were implemented using a Fibonacci heap. The top-k queue stores node IDs using the worst score value as a key to facilitate selecting of the min-k node, while the candidate queue uses the best score of a node as a key to find the node with the highest best score in the queue efficiently. We conducted experiments on a server machine with two 2.0GHz Quad Core CPUs and 8GB memory.

Table 1. Test queries on Mondial dataset

query ID	keyword set	query ID	keyword set
Q1	{Michigan, Wisconsin, Toronto}	Q7	{military, communist, Asia, Africa}
Q2	{Atlantic, cape, bay}	Q8	{volcano, island, Pacific, Ocean}
Q3	{monarchy, democracy, Europe}	Q9	{Spain, Morocco, Malta, Gibraltar}
Q4	{salt, lake, Asia}	Q10	{volcano, volcanic, mountain, island}
Q5	{republic, catholic, Europe}	Q11	{Himalaya, China, Nepal, India}
Q6	{APEC, Asia, America}	Q12	{Reykjavic, Ireland, Norwegian, sea}

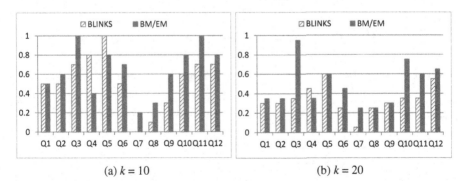

(a) $k = 10$ (b) $k = 20$

Fig. 5. Precisions of the top-k answers obtained by the search methods

---

[2] http://www.dbis.informatik.uni-goettingen.de/Mondial/
[3] http://www.imdb.com/
[4] http://www.jgrapht.org/
[5] http://lucene.apache.org/java/docs/index.html

In the first experiment, we evaluated precision of the search results obtained by our approach and BLINKS. The precision of top-k answers to a query is measured by $P@k = \frac{|Res[1..k] \cap Rel|}{k}$, where $Res[1..k]$ is a set of top-k answer trees returned as a search result and Rel is a set of all the answer trees in the graph which are considered relevant to the given query [21]. In the experiment of our approach, parameter p is set to double the number of keywords in the query.

Fig. 5 shows the precisions of top-10 and top-20 search results for the test queries over Mondial dataset shown in Table 1. We can observe that the precision of the result of our method is higher than or equal to that of BLINKS for the most queries, specifically, for 10 queries in top-10 search and for 11 queries in top-20 search. The average precisions of our method and BLINKS for the given queries are respectively 0.53 and 0.64 in top-10 searches and 0.34 and 0.49 in top-20 searches. Note that for the queries with AND semantics such as $Q4$, $Q5$, and $Q8$, BLINKS shows higher precisions than our method while for the queries having OR semantics such as $Q3$, $Q6$, $Q7$, and $Q11$, our method achieves better performance than BLINKS. This is due to the fact that the previous methods including BLINKS only search for sub-trees containing keyword nodes for each and every keyword in the query while the proposed method finds sub-trees having a different number of keyword nodes for each query keyword which are most relevant to the root node. This enables the proposed method to find more relevant results than the previous methods for some queries such as $Q10$.

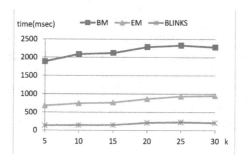

Fig. 6. Execution time of the search methods with varying k

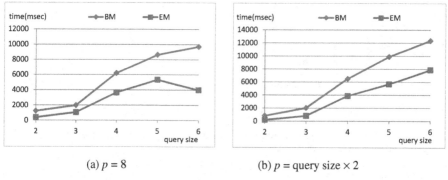

(a) $p = 8$ (b) $p = $ query size $\times$ 2

Fig. 7. Execution time of the proposed methods with varying p

In the second experiment, we evaluated performance of the proposed methods by measuring execution time in processing top-k queries. We generated and processed 20 test queries having 3 or 4 keywords. The parameter p is set to double the number of query keywords and periodic updates of the priority queues T and C are conducted after reading every 15,000 entries from inverted lists. We experimented with the same queries to find top-k answers varying k from 5 to 30. In Fig. 6, the results over IMDB dataset show that the average execution time of EM is about 4.7 times longer than that of BLINKS. This is mainly due to the overhead in finding answers in the generalized structure proposed in the paper. When comparing our basic method and enhanced method, the average execution time of EM based on the extended inverted lists reduces to about 38% of that of BM. Note that as k increases 6 times from 5 to 30, execution time of BM and EM increases only about 21% and 41%, respectively, and the performance gap between EM and BM gets wider about 10%.

Finally, we have evaluated performance of the proposed methods with respect to query size, i.e. the number of keywords in the query. We executed 20 test queries to find top-10 answers over IMDB dataset, varying the queries' size from 2 to 6. The parameter p is fixed to 8 in Fig. 7-(a) and set to double the query size in Fig. 7-(b), and update period of priority queues is 15,000. Fig. 7-(a) shows that as the query size grows from 3 to 6, the execution time of BM increases about 4.9 times while that of EM enlarges about 3.8 times. We can observe that the performance gap between EM and BM increases about 6.3 times and the execution time of EM is about 59% shorter than that of BM when the query size is 6. Thus EM is more efficient than BM for the queries of large size.

7   Conclusion

In this paper, we propose a new keyword search method for graph-structured databases. To find more effective top-k answers relevant to a given query, we suggest a generalized answer tree structure which has no constraint on the number of keyword nodes chosen for each keyword and selects a set of keyword nodes based on the relevance of the root to a query keyword contained in a specific node in the graph. For efficient top-k query processing based on the new answer structure, we propose an inverted list index which stores relevance and connectivity information on the nodes relevant to each keyword term. Then we provide a query processing algorithm exploiting the proposed inverted lists to find top-k answer trees most relevant to the given query. Furthermore, we also present Enhanced Method using an extended inverted list containing additional next relevance information in the list entry. It can estimate the relevance score of each node more closely to the correct value and find top-k answers more efficiently than Basic Method.

The experiments with real datasets and various test queries show that the precision of our approach is higher than that of BLINKS for most of the queries, especially for those with OR semantics. Thus, the proposed answer structure can satisfy users' information need better than the conventional ones. The performance of the proposed search algorithms is shown to be degraded compared to BLINKS mainly due to the overhead incurred by adopting the extended answer structure. However, the execution performance of Enhanced Method based on the extended inverted list is much better than Basic Method, and it is scalable with respect to the number of answers to be found for top-k queries.

Acknowledgments. This work was supported by the Dongduk Women's University grant.

References

1. Amer-Yahia, S., Shanmugasundaram, J.: XML full-text search: Challenges and opportunities. In: 31st Int. Conf. on Very Large Data Bases, pp. 1368–1368 (2005)
2. Chen, Y., Wang, W., Liu, Z., Lin, X.: Keyword search on structured and semi-structured data. In: 2009 ACM SIGMOD Int. Conf. on Management of Data, pp. 1005–1010 (2009)
3. Dalvi, B.B., Kshirsagar, M., Sudarshan, S.: Keyword search on external memory data graphs. The Proceedings of the VLDB Endowment 1(1), 1189–1204 (2008)
4. Golenberg, K., Kimelfeld, B., Sagiv, Y.: Keyword proximity search in complex data graphs. In: 2008 ACM SIGMOD Int. Conf. on Management of Data, pp. 927–940 (2008)
5. He, H., Wang, H., Yang, J., Yu, P.S.: BLINKS: ranked keyword searches on graphs. In: 2007 ACM SIGMOD Int. Conf. on Management of Data, pp. 305–316 (2007)
6. Kim, H., Park, C.-S., Lee, Y.J.: Improving Keyword Match for Semantic Search. IEICE Trans. Inf. & Syst E94-D(2), 375–378 (2011)
7. Li, G., Ooi, B.C., Feng, J., Wang, J., Zhou, L.: EASE: an effective 3-in-1 keyword search method for unstructured, semi-structured and structured data. In: 2008 ACM SIGMOD Int. Conf. on Management of Data, pp. 903–914 (2008)
8. Bhalotia, G., Hulgeri, A., Nakhe, C., Chakrabarti, S., Sudarshan, S.: Keyword searching and browsing in databases using BANKS. In: 18th Int. Conf. on Data Engineering, pp. 431–440 (2002)
9. Ding, B., Yu, J.X., Wang, S., Qin, L., Zhang, X., Lin, X.: Finding top-k min-cost connected trees in databases. In: 23rd Int. Conf. on Data Engineering, pp. 836–845 (2007)
10. Hristidis, V., Gravano, L., Papakonstantinou, Y.: Effcient IR-Style keyword search over relational databases. In: 29th Int. Conf. on Very Large Data Bases, pp. 850–861 (2003)
11. Hristidis, V., Hwang, H., Papakonstantinou, Y.: Authority-based keyword search in databases. ACM Trans. Database Syst. 33(1), 1–40 (2008)
12. Hristidis, V., Papakonstantinou, Y.: DISCOVER: Keyword search in relational databases. In: 28th Int. Conf. on Very Large Data Bases, pp. 670–681 (2002)
13. Kacholia, V., Pandit, S., Chakrabarti, S., Sudarshan, S., Desai, R., Karambelkar, H.: Bidirectional expansion for keyword search on graph databases. In: 31st Int. Conf. on Very Large Data Bases, pp. 505–516 (2005)
14. Kimelfeld, B., Sagiv, Y.: Finding and approximating top-k answers in keyword proximity search. In: 25th ACM SIGACT-SIGMOD-SIGART Symp. on Principles of Database Systems, pp. 173–182 (2006)
15. Liu, F., Yu, C.T., Meng, W., Chowdhury, A.: Effective keyword search in relational databases. In: 2006 ACM SIGMOD Int. Conf. on Management of Data, pp. 563–574 (2006)
16. Luo, Y., Lin, X., Wang, W., Zhou, X.: Spark: top-k keyword query in relational databases. In: 2007 ACM SIGMOD Int. Conf. on Management of Data, pp. 115–126 (2007)

17. Qin, L., Yu, J.X., Chang, L.: Keyword search in databases: The power of RDBMS. In: 2009 ACM SIGMOD Int. Conf. on Management of Data, pp. 681–694 (2009)
18. Yu, J.X., Chang, L., Tao, Y.: Querying communities in relational databases. In: 25th Int. Conf. on Data Engineering, pp. 724–735 (2009)
19. Yu, J.X., Qin, L., Chang, L.: Keyword search in relational databases: a survey. IEEE Data Engineering Bulletin 33(1), 67–78 (2010)
20. Hwang, F.K., Richards, D.S.: The Steiner tree problem. Networks 22(1), 55–89 (1992)
21. Buttcher, S., Clarke, C., Cormack, G.: Information Retrieval: Implementing and Evaluating Search Engine. MIT Press (2010)
22. Fagin, R., Lotem, A., Naor, M.: Optimal aggregation algorithms for middleware. Journal of Computer and System Sciences 66(4), 614–656 (2003)
23. Güntzer, U., Balke, W.-T., Kießling, W.: Towards efficient multi-feature queries in heterogeneous environments. In: 2001 Int. Symp. on Inf, pp. 622–628 (2001)
24. Bruno, N., Gravano, L., Marian, A.: Evaluating top-k queries over web-accessible databases. In: 18th Int. Conf. on Data Engineering, pp. 369–380 (2002)
25. Theobald, M., Weikum, G., Schenkel, R.: Top-k Query Evaluation with Probabilistic Guarantees. In: 30th Int. Conf. on Very Large Data Bases, pp. 648–659 (2004)
26. Best, H., Majumdar, D., Schenkel, R., Theobald, M., Weikum, G.: IO-Top-k: Index-access Optimized Top-k Query Processing. In: 32nd Int. Conf. on Very Large Data Bases, pp. 475–486 (2006)

Semantic Representation of Public Web Service Descriptions

Maricela Bravo[1], Jorge Pascual[1], and José Rodríguez[2]

[1] Systems Department, Autonomous Metropolitan University
Azcapotzalco, DF, CP 02200 - Mexico
mcbc@correo.azc.uam.mx,
gaidar26@gmail.com
[2] Computing Department, CINVESTAV-IPN
Gustavo A. Madero, DF, CP 07300 - Mexico
rodriguez@cs.cinvestav.mx

Abstract. Among the main benefits of service-oriented architectures is the reutilization of software components that may solve specific tasks for complex problems, requiring the composition of multiple Web services. Currently Internet is largely populated with Web services offered by different providers and published in various Web repositories. However, public available Web services still suffer from problems that have been widely discussed, such as the lack of functional semantics. This lack of semantics makes very difficult the automatic discovery and invocation of public Web services, even when the system integrator can obtain a copy of the WSDL file. This paper describes an ontological model representation of public Web service descriptions. The objective of this work is to automatically parse public Web service descriptions and represent extracted data into ontological representations. Experimental results show that the overall process towards the automation of public Web services discovery based on ontology population is feasible and can be completely automated.

Keywords: Public Web Service Descriptions, Ontology Representation, Semantic Web Services.

1 Introduction

In the last decade, many software vendors have deployed and offered software as services using interface description languages, such as the Web Service Description Language (WSDL). In order to make their services available online, providers publish their services descriptions in public repositories, which may be conformant with a specific standard such as the Universal Description Discovery and Integration (UDDI) or the Electronic Business using eXtensible Markup Language (ebXML). When software integrators (service consumers) search for Web services that meet their criteria in public repositories, and try to select and invoke existing Web services, they may face some of the following problems:

B. Murgante et al. (Eds.): ICCSA 2013, Part V, LNCS 7975, pp. 636–651, 2013.
© Springer-Verlag Berlin Heidelberg 2013

- *Lack of well-documented Web service descriptions.* This is a common problem that many public Web service clients or requestors face. In the study reported by Rodríguez et al. [1], authors identify common mistakes in WSDL documents: inappropriate or lacking comments, use of ambiguous names for the main elements, redundant port-types, low cohesive operations in the same port-type, enclosed data model, redundant data models, etc. According to Rodríguez study [1], less than 50% of the studied WSDL files have some documentation. Additionally, the naming of services, operations, messages and parameters does not follow any convention, and the specification of WSDL does not obligate service authors to provide additional functional semantics information. These reasons cause enormous difficulties during search, selection and invocation of services.

- *Lack of semantically enhanced Web services repositories.* There are many public Web service repositories, but they do not offer sufficient semantic information about the service functionality, making very difficult the automated exploitation of deployed Web services. Majority of public repositories offer key-based search mechanisms, and some sort of classifications, but none of them offer semantic-based search, matchmaking and discovery mechanisms between existing services considering a provided service interface (template) information.

- *Heterogeneous Web service description languages.* Existing service description languages share similar objectives and provide common elements for the description of Web services, such as: service name, service address, operations, input and output messages. However, their level of expressivity differs significantly from a simple syntactic-based service description such as WSDL to a semantic-based service description. This variety of service description languages produces that existing Web services are described by heterogeneous service description languages. This heterogeneity makes very difficult the automatic service discovery, service selection, service substitution and service inter-operation. Reuse of existing Web services is harder and the creation of Web services compositions from multiple vendors is not feasible.

Different solutions have been proposed to solve these limitations. The semantic Web has influenced many works by providing logic-based mechanisms to describe, annotate and discover Web services. Within this context, McIlarith, Cao Son & Zeng [2] proposed one of the first initiatives to mark-up Web services based on the ontology language DARPA Agent Markup Language (DAML), which started the important research area of "Semantic Web Services". The term Semantic Web Services is related to the set of technologies and models based on the implementation and exploitation of ontologies as a mechanism to semantically enhance service descriptions, for instance: the ontology language for the semantic description of Web services (OWL-S), the Web Service Modeling Ontology (WSMO), and the Semantic Annotation for WSDL (SAWSDL). However, these semantic approaches require human experts intervention to construct ontologies and annotate Web service descriptions before their deployment.

From the perspective of Web services providers, if they want to take advantage of these semantic-based technologies, they will have to re-design their solutions with the following considerations: in case of annotating semantically their Web services using SAWSDL, they need to construct or select an ontological representation relative to the domain of the services offered; in the case of using OWL-S, service providers need to learn this model and use the tools available to create the corresponding ontological descriptions of their services; and in case of using WSMO, the learning curve is steep because it requires more effort to understand and use the complex framework Web Service Execution Environment (WSMX) with a new ontology language[1]. From the point of view of a service requestor, the first problem that he will face is to find Web services repositories containing semantic Web service descriptions using these technologies. Furthermore, he will not find a universal repository containing all types of semantic service descriptions. Until now there is no reported approach or tool that automatically translates or connects (without human intervention) any pre-existing Web service description to any of the aforementioned solutions.

Despite the increasing popularity of semantic-based technologies, the numerous researchers devoted to them, the great advances and achievements; there is still an important gap between these semantic-based Web service technologies and the pre-existing Web services, which were deployed using only WSDL (including the common mistakes pointed by Rodríguez et al. [1]). There is no doubt that the semantic Web trend will continue and will consolidate in the following years, and if service providers want to stay competitive, they need to adapt and re-deploy their services using these technologies. Meanwhile, from the perspective of computer science, many semantic-based solutions are necessary and technologies are mature enough to facilitate solution design and deployment.

The solution reported in this paper considers that there are essential common elements that all Web services description languages must provide: a general communication interface that the client uses to create a proxy object to invoke the service remotely.

Table 1. Web service description languages and models

Language or model	Syntactic models		Semantic models		
	WSDL 1.1	**WSDL 2.0**	**OWL-S**	**WSML**	**General Model**
Element	Operation	Operation	Profile	Capability	General operation
Input parameters	Input Message	Input	hasInput	Precondition	General input
Output parameters	Output Message	Output	hasOutput	Postcondition	General output
Preconditions	-	-	hasPrecondition	Assumption	General preconditions
Effects	-	-	hasResult	Effects	General effects

---

[1] Considering that service providers are familiar with software development, but not necessarily with ontologies or semantic Web technologies.

This communication interface must describe information about the functions that the service offers (operations in WSDL, profile in OWL-S or capability in WSMO) as well as the correct description of input and output parameters (see Table 1). Taking into account these common elements between services described in any of these languages and that ontologies represent the cutting edge technological movement, this article describes an ontological representation for public Web services. With this motivation in mind, this paper describes the following contributions:

- A set of service language-specific parsers which automatically gather and extract relevant service elements coded into different service descriptions.
- Ontology-based representation of public Web services, which serves as a service repository to allow the dynamic acquisition of more service instances.

The resulting ontologies of service descriptions allow the definition and use of query rules to support complex service tasks such as search, discovery, selection, substitution and composition. An additional benefit of using an ontology-based representation of public Web services is the possibility of inter-connection and inter-operation with existing semantic models.

The rest of the paper is organized as follows: in Section 2, Web service description languages are briefly described; in Section 3, the solution design is presented; in Section 4, 5 and 6 respective semantic representation of different service descriptions are described; in Section 7, implementation and evaluation is described; in Section 8, related work and finally in Section 9, conclusions.

2   Web Service Description Languages

A Service Description Language (SDL) can be defined as a mean by which a service vendor provides the technical description (document or contract) of the service programmatic interface. Such technical description must include the network address where the executable file of the service resides and accepts requests, the name of the operations (functions or methods) available to be remotely invoked, and parameter names and data types. The proposed SDL for the Web Service implementation is named Web Service Description Language (WSDL), which is currently a well-established W3C standard. WSDL defines an XML grammar for describing networked services as collections of communication endpoints capable of exchanging messages. Another important service model proposed for the semantic representation of Web services is the OWL-S service description language. In this Section, a brief overview of WSDL and OWL-S models is presented.

2.1   WSDL

WSDL 1.0 was originally developed by IBM, Microsoft, and Ariba in 2000. This first version resulted from the combination of two service description languages: NASSL (Network Application Service Specification Language) developed by IBM, and the

SDL developed by Microsoft. In 2001 the formal version of WSDL 1.1 was submitted to the W3C, no major changes were made from the previous version. WSDL 1.1 was widely adopted by service providers community and many tools and frameworks were developed to support service management. The latest version is WSDL 2.0, which incorporated important changes in the description of a service. Both SDLs are briefly described here:

— WSDL 1.1 files start the description with the <definitions> tag, which is divided into an abstract section and a concrete section. The abstract section includes the definition of <portType> tags, which define the set of operations offered by the service; for each operation defined in the <portType>, input and output messages are defined. In this abstract section, the message types are also defined. The concrete section defines the service data and the binding style of the service.
— WSDL 2.0 changes the <definitions> tag with the <description> tag. Similarly, WSDL 2.0 is also divided into an abstract section and a concrete section. The main difference between WSDL 1.1 and WSDL 2.0 are: the *targetNamespace* is a required attribute of the definitions element in WSDL 2.0; message constructs are removed in WSDL 2.0; operator overloading is not supported in WSDL 2.0; PortType is renamed as Interface; Interface inheritance is supported by using the extends attribute; and Port is renamed as Endpoint.

Albeit WSDL 2.0 is currently a W3C Recommendation, little software development kits have been developed, majority of integrated development environments offer support for WSDL 1.1.

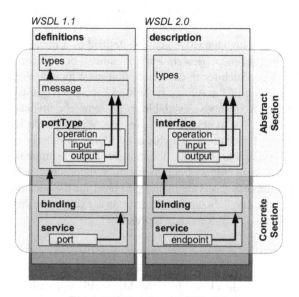

Fig. 1. WSDL 1.1 and WSDL 2.0

2.2   OWL-S

OWL-S is an ontology-based service description language, which supplies Web service providers with a set of constructs for describing the properties and capabilities of their Web services in unambiguous, computer-interpretable form. OWL-S emerged as a promising solution to enhance semantically service descriptions. It was conceived to attend the need for richer semantic specifications of Web services, to enable the automatic service provisioning and use, and to promote the use of reasoning tasks over service descriptions [3]. From a general perspective, an OWL-S Service presents a service Profile, which describes what the service does; is described by a Service Model (or Process Model); and supports a Service Grounding, which specifies how to interact with the service.

3   Solution Design

In this paper Semantic Representation of Public Web Service Descriptions is defined as the automatic task of parsing and extracting significant data from public Web service descriptions aiming at populating a set of ontologies with such descriptions. Specifically the resulting ontologies are designed to support registry functionalities, such as: search, discovery, and selection of Web services. The main differences with semantic service description approaches such as OWL-S and WSML, is that these are languages designed to describe service descriptions, but they were not built as a service repository architecture.

The process of Semantic Representation of Public Web Service Descriptions is depicted in Figure 2. This process involves the following phases:

a) *Retrieving public Web services.* This phase consists of searching and copying service descriptions files from the Web. The objective is to gather information about services available on the Web and maintain a local file directory of retrieved public Web services. To achieve this objective, a common strategy is to program several softbots that seek for services on the entire Web; or visit specific service repositories that manage lists of services.

b) *Parsing Web service descriptions.* This phase consist of reading description files, identify the relevant elements, retrieve and process them. With this regard, two important requirements have to be addressed: heterogeneity of service description languages (SDL) and selection of the relevant data to be retrieved from service files. The former requirement is derived from the existence of various SDLs: WSDL, OWL-S, WSML, and SAWSDL. Both requirements are addressed by identifying the essential common elements that all SDLs must provide: a communication interface that the client uses to create a proxy object to invoke the service remotely. This communication interface must describe information about the functions that the service offers (operation in WSDL, profile in OWL-S or capability in WSML) as well as the descriptions (name and data types) of input and output parameters.

c) *Web service ontology population.* This step consists of identifying the corresponding ontology model in accordance with the specific SDL file implementation and record service instance data. This phase consists of a set of program modules which are responsible for the automatic detection of the specific SDL implementation, parsing and populating the respective ontology model.

d) *Web service ontologies.* This is the outcome of the entire process. Three different ontology models are proposed, each corresponding with a different SDL implementation: WSDL 1.1, WSDL 2.0 and OWL-S. Each ontology design is described in the following sub-sections.

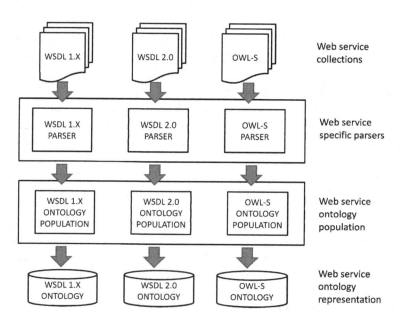

Fig. 2. Semantic representation of public Web services

3.1   Semantic Repository of WSDL 1.0 Services

Taking as a reference the WSDL 1.0 and the WSDL 1.1 specifications, an ontology model was designed and implemented. This ontology model aims at allocating multiple service definitions in order to serve as a repository where users can search and select specific services. Figure 3 shows the upper layer of this ontology model, Figure 4 shows the taxonomy of the *Parameter* class, and Figure 5 shows the semantic relations between the *Operation* class and the *Parameter* class. The core classes and semantic relationships of this ontological model are described here:

1. *Service class.* The central element of this ontology is the Service class, which represents a WSDL service description.
2. *Endpoint class.* Every service instance must define an endpoint address (which specifies a unique network address that the service consumer uses to invoke the methods of the service), therefore a semantic relation (objectProperty in OWL language) named *hasEndpoint* correlates instances of the Service class with instances of the *Endpoint* class.
3. *Binding class.* A service defines bindings, which specify the SOAP binding style and transport. The binding section also defines the operations that the service offers. A semantic relation called *hasBinding* correlates instances of the class Endpoint with instances of the class Binding. In WSDL 1.0 the service binding specifies the message format and protocol details for operations and messages defined by a particular *portType*. There may be various bindings for each *portType*.
4. *Interface class.* The *porType* of a Web service defines the operations that can be invoked, and the input and output messages that are used to execute the operation. However, in the ontological model depicted in Figure 3, instead of a *porType* class, an *Interface* class was created to make this model similar and compatible with the WSDL 2.0 specification. As the process of parsing WSDL 1.0 files and populating its corresponding ontological model is automatically executed, this translation is done with no further implications. A semantic relation called *hasInterface* correlates instances of the Binding class with instances of the Interface class.
5. *Operation class.* This class represents the methods or functions offered by the service interface. The WSDL specification defines an input message and output message for each operation. To create this representation the semantic relation *hasParameterInput* objectProperty correlates instances of the Operation class with instances of the *ParameterInput* class, similarly the semantic relation *hasParameterOutput* correlates instances of the Operation class with instances of the *ParameterOutput* class.
6. *Parameter class.* A detailed diagram of the Parameter class is shown in Figure 4. The Parameter class represents the super class of the *ParameterInput* class and the *ParameterOutput* class, where every parameter defined in the WSDL fits and it is used. This representation is made in order to specify parts of the parameter instance defined by the classification of the Part class.
7. *Part class.* Input and output messages consist of one or more logical parts. The Part class is used to describe the content of the message. The semantic relationships of *hasDataTypeSimpleType*, *hasDataTypeComplexType* and *hasDataTypeSimpleElement* are created to correlate instances of the Part class with an specific data type. With this distinction it is possible to treat the different data type structures that can be described in a WSDL.

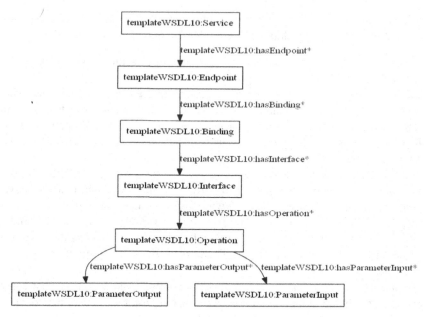

Fig. 3. Ontological model for the representation of services described with WSDL 1.0

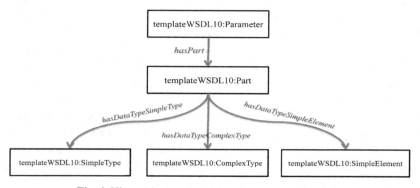

Fig. 4. Hierarchical relationships of the *Parameter* class

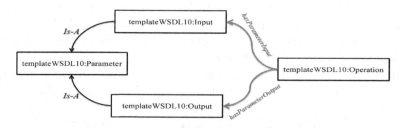

Fig. 5. Semantic relationships of the *Operation* class with the *Parameter* class

3.2   Semantic Repository of WSDL 2.0 Services

Taking as a reference the WSDL 2.0 specification, an ontology model was designed and implemented. This ontology model will allocate multiple service descriptions in order to serve as a repository where users can search and select services described in the WSDL 2.0. Figure 6 shows the upper layer of this ontology model. The main classes of this ontology model are: Service, Endpoint, Binding, Interface, Operation, and Parameter. These classes represent the same ontological model as the WSDL 1.0 model shown in Figure 3.

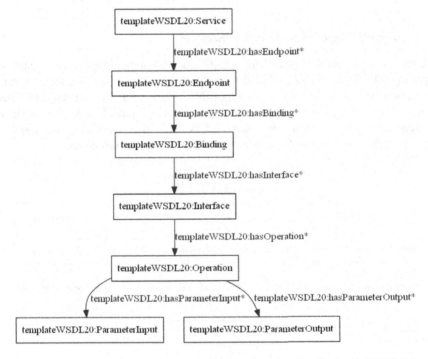

Fig. 6. Ontological model for the representation of services described with WSDL 2.0

A detailed diagram of the *Parameter* class is shown in Figure 7. The Parameter class represents the super class of the *ParameterInput* class and the ParameterOutput class, where every parameter defined in the WSDL fits and it is used. A particular difference between WSDL 1.1 and WSDL 2.0 ontology models is the definition of a *Part* element. WSDL 2.0 does not include a message Part element, however, the Parameter class is correlated with the *ComplexType*, *SimpleType* and *SimpleElement* classes, which are subclasses of the *DataType* class. Providing with this the necessary data types classification present in service descriptions.

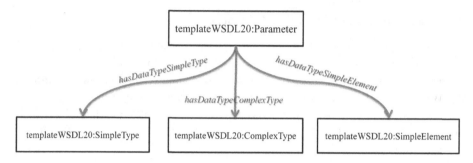

Fig. 7. Ontological representation of the *Parameter* class described with WSDL 2.0

3.3   Semantic Repository of OWL-S Services

Taking as a reference the OWL-S specification, an ontology model was designed and implemented. This ontology model will allocate multiple semantic service descriptions in order to serve as a repository where users can search and select services. Figure 8 shows the upper layer of this ontology model, and Figure 9 shows the semantic relations of the *Parameter* class. The main classes of this ontology model are: *Service, ServiceModel, ServiceProfile,* and *ServiceGrounding.*

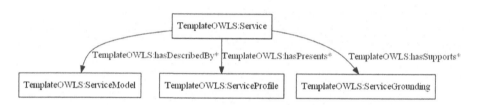

Fig. 8. Ontological model for the representation of services described with OWL-S

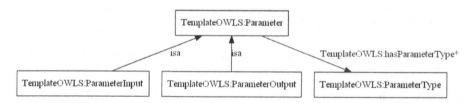

Fig. 9. Ontological representation of the *Parameter* class described with OWL-S

4   Implementation and Evaluation

In order to evaluate the solution design a fully automated population Web system was implemented, which represents an important tool towards the generation of ontological representation models for any given set of public web services. For each type of service description language or model considered in this paper, an object-oriented population module was implemented. Figure 10 shows the class diagram of the WSDL 1.X, which represents all java interfaces that are later implemented in concrete classes to facilitate code extensibility and maintainability.

Fig. 10. Class diagram of the WSDL 1.X population module

In Figure 11, a Semantic Description Generator of Web Services (SDGWS) is depicted. SDGWS is a Web based application which allows the user generate semantic descriptions (in the form of ontologies) of public Web services. The system is executed following a sequence of predefined steps: selection of any set of Web services described with WSDL 1.0, WSDL 1.1, WSDL 2.0, and OWL-S; uploading the set of selected services, parsing the services according to their representation language; populating a new ontology based on the respective T-Box; and finally downloading the new produced ontology.

In order to evaluate the proposed solution three sets of Web services were selected to create their respective ontological representations. Figure 12 shows the system after the execution of the WSDL 1.X parser.

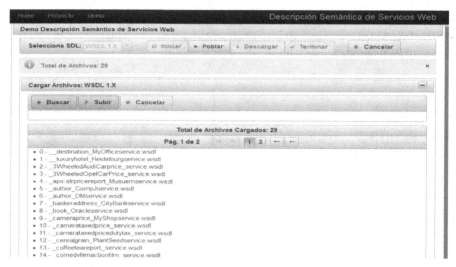

Fig. 11. Ontological representation of the Parameter class described with OWL-S

As a result of the population system, three ontologies were automatically generated. Figure 9 shows the resulting ontology populated with a set of 29 services described in WSDL 1.X.

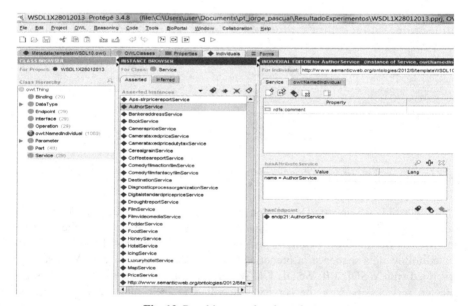

Fig. 12. Resulting populated ontology

Once the three ontologies were generated a new *TransformationOntology* was created, the resulting populated ontologies were imported into this Transformation Ontology and using the following rules a general repository of Web services was created.

Inference rules (1) and (2) are used to populate the ontology class *GeneralService*, which is populated with all Service individuals from the imported ontologies. *p1* and *p2* suffixes represent the different imported ontologies.

$$p1:Service(?op) \rightarrow GeneralService(?op) \tag{1}$$

$$p2:Service(?op) \rightarrow GeneralService(?op) \tag{2}$$

Similarly a *GeneralInterface* class and a *GeneralOperation* Class are populated by executing the following inference rules.

$$p1:Interface(?int) \rightarrow GeneralInterface(?int) \tag{3}$$

$$p2:Interface(?int) \rightarrow GeneralInterface(?int) \tag{4}$$

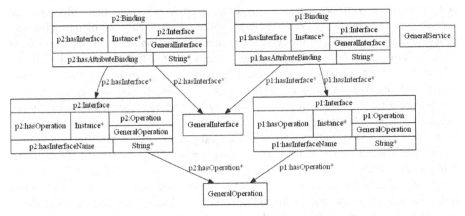

Fig. 13. General Transformation Ontology

Figure 13 shows a partial representation of the *TransformationOntology*, where different Interface definitions are semantically related with the *GeneralOperation* class.

This unified representation does not mean that imported ontologies are really integrated into a unique ontology. The individual service ontologies are maintained separately and the *TransformationOntology* is the means by which inference rules and queries are executed to support complex service tasks such as service discovery.

5   Related Work

In [4] authors argue that there exist a vast number of public available services that cannot be utilized by tools that enable service users to create mashups without programming knowledge. They propose a solution based on the **combination** of existing **description languages** and **learning ontology mechanisms** in order to enable the development of semantic web services compliant with architectural style of RESTful web services. Their ontology learning mechanism is used to extract and cluster service parameters, producing a parameter-based domain ontology which is evaluated against a traditional keyword-based service search mechanism.

In [5] authors propose an ontology-based framework for the discovery of semantic Web services (SWS) using a QoS approach. Their framework describes an ontology manager component which handles the provider and the requester domain or general ontologies. This component merges these ontologies with general ontologies and creates a new generalized ontology, which is used for ranking the resulting list of SWS. Even though authors address automatic discovery of SWS by means of ontologies, their ontologies are domain-oriented. In contrast, in this paper ontologies are used for modeling service programmatic interfaces aiming at supporting automatic search and discovery of public available Web services.

In [6] authors address the problem of annotating Web services from the Deep Web. Deep Web refers to Web pages that are not accessible to search engines. In particular, authors consider Web forms interface pages as Deep Web services that reflect the real content types of the Deep Web. Their proposed solution consists of the automatic generation of a domain ontology for semantic annotation of Web services. Such domain ontology is built based on Web page attributes (any items of descriptive information about the site). Their research main goal is improving the automatic search and discovery of public Web services. However, their service description sources are different as they are using Web form interface pages instead of a formal service description language.

In 2007 [7] Sabou and Pan presented a study of the major problems with Web service repositories (some of them are no longer available, however the result of the study is still relevant). They concluded that Web service repositories use simple techniques of accessing the content of Web services, browsing across services listings relies on few and low quality metadata, and metadata is not fully exploited for presentation. Authors also proposed various semantic-based solutions to enhance semantically service repositories. Retaking the early ideas of these authors, the solution that is reported in this article is to lay the foundations for the automatic construction of public Web services repositories based on ontologies.

Mokarizadeh et al. [8] reported a very similar approach for the automatic construction of ontologies from public Web services. However, in their work they apply Natural Language Processing techniques to extract domain data from the service descriptions, and the resulting ontology model represents the domain rather than the essential elements of multiple service descriptions.

6   Conclusions

There are many Web services scattered in public and private networks that cannot be easily reused for the integration of complex applications mainly because they show different service descriptions.

This heterogeneity makes the automatic system integration based on pre-existing web services a very difficult task. The system engineer is forced to implement particular adaptations to the implementations of the service descriptions. Also, it is really difficult to use the semantic service models because they do not interoperate easily with pre-existing services. Several manual transformation tasks are required to use heterogeneous services descriptions.

Results show advances on automatic ontology construction from public available Web service descriptions. The automatic population of the ontology with existing WSDL files and OWL-S files is a relevant advance towards the automated reutilization and construction of service-based solutions using pre-existing resources.

The system proposed in this paper represents a step towards the solution to these problems by integrating ontological models through which it is possible to define conceptual intermediaries between different description languages, facilitating the automated execution of service-related complex tasks that have always been a research concern. Complex tasks of web services are: discovery, matchmaking, selection, composition and optimization of Web services, regardless of the language used to describe them and regardless of their public or private availability.

References

1. Rodríguez, M., Crasso, M., Zunino, A., Campo, M.: Improving Web Service descriptions for effective service discovery. Science of Computer Programming 75(11) (2010)
2. McIlarith, S., Cao Son, T., Zeng, H.: Semantic Web Services. IEEE Intelligent Systems (2001)
3. Martin, D., Paolucci, M., McIlraith, S.A., Burstein, M., McDermott, D., McGuinness, D.L., Parsia, B., Payne, T.R., Sabou, M., Solanki, M., Srinivasan, N., Sycara, K.: Bringing Semantics to Web Services: The OWL-S Approach. In: Cardoso, J., Sheth, A.P. (eds.) SWSWPC 2004. LNCS, vol. 3387, pp. 26–42. Springer, Heidelberg (2005)
4. Young-Ju, L., Chang-Su, K.: A Learning Ontology Method for RESTful Semantic Web Services. In: Proceedings of the ICWS (2011)
5. Yousefipour, A., Mohsenzadeh, M., Neiat, A.G., Hemayati, M.S.: An Ontology-based Approach for Ranking Suggested Semantic Web Services (2010)
6. An, Y.J., Geller, J., Wu, Y.-T., Chun, S.A.: Automatic Generation of Ontology from the Deep Web. In: Proceedings of the 18th International Workshop on Database and Expert Systems Applications (2007)
7. Sabou, M., Pan, J.: Towards semantically enhanced Web service repositories. Journal of Web Semantics: Science, Services and Agents on the World Wide Web, 142–150 (2007)
8. Mokarizadeh, S., Küngas, P., Matskin, M.: Ontology acquisition from web service descriptions. In: Proceedings of the 28th Annual ACM Symposium on Applied Computing (SAC 2013), pp. 325–332. ACM, New York (2013)

Author Index